AAPG Treatise of Petroleum Geology

Reprint Series

The American Association of Petroleum Geologists
gratefully acknowledges and appreciates the leadership and support
of the AAPG Foundation in the development of the
Treatise of Petroleum Geology.

STRUCTURAL CONCEPTS AND TECHNIQUES II

BASEMENT-INVOLVED DEFORMATION

COMPILED BY
NORMAN H. FOSTER
AND
EDWARD A. BEAUMONT

TREATISE OF PETROLEUM GEOLOGY
REPRINT SERIES, NO. 10

PUBLISHED BY
THE AMERICAN ASSOCIATION OF PETROLEUM GEOLOGISTS
TULSA, OKLAHOMA 74101, U.S.A.

ISBN: 0-89181-409-4

Library of Congress Cataloging-in-Publication Data

Structural concepts and techniques / compiled by Norman H. Foster and
    Edward A. Beaumont.

      p.   cm. -- (Treatise of petroleum geology reprint series ; no.
9-)
Bibliography: p.
Contents: 1. Basic concepts, folding, and structural techniques --
2. Basement-involved deformation. -- 3. Detached deformation.
ISBN 0-89181-408-6 (v. 1) : $34.00. -- $18.00 (v. 1, v. 2 : pbk.)
ISBN 0-89181-409-4 (v. 2)
   1. Geology, Structural. 2. Folds (Geology) 3. Petroleum --
Geology. I. Foster, Norman H. II. Beaumont, E. A. (Edward A.)
III. American Association of Petroleum Geologists. IV. Series:
Treatise of petroleum geology reprint series ; no. 9, etc.
QE601.S8637   1988   39231
                                        CIP
551.8 -- dc19

American Association of Petroleum Geologists Foundation

Treatise of Petroleum Geology Fund*

The Foundation also gratefully acknowledges the many who have supported this endeavor with additional contributions, which now total $12,415.50.

*Contributions received as of November 10, 1988.

Treatise of Petroleum Geology
Advisory Board

INTRODUCTION

This reprint volume belongs to a series that is part of the *Treatise of Petroleum Geology*. The *Treatise of Petroleum Geology* was conceived during a discussion we had at the 1984 AAPG Annual Meeting in San Antonio, Texas. When our discussion ended, we had decided to write a state-of-the-art textbook in petroleum geology, directed not at the student, but at the practicing petroleum geologist. The project to put together one textbook gradually evolved into a series of three different publications: the Reprint Series, the Atlas of Oil and Gas Fields, and the Handbook of Petroleum Geology; collectively these publications are known as the *Treatise of Petroleum Geology*. With the help of the Treatise of Petroleum Geology Advisory Board, we designed this set of publications to represent the cutting edge in petroleum exploration knowledge and application. The Reprint Series provides previously published landmark literature; the Atlas collects detailed field studies to illustrate the various ways oil and gas are trapped; and the Handbook is a professional explorationist's guide to the latest knowledge in the various areas of petroleum geology and related fields.

The papers in the various volumes of the Reprint Series complement the different chapters of the Handbook. Papers were selected on the basis of their usefulness today in petroleum exploration and development. Many "classic papers" that led to our present state of knowledge have not been included because of space limitations. In some cases, it was difficult to decide in which Reprint volume a particular paper should be published because that paper covers several topics. We suggest, therefore, that interested readers become familiar with all the Reprint volumes if they are looking for a particular paper.

We have divided the topic of structural concepts and techniques into three volumes. The first volume contains papers that discuss Basic Concepts, Folding, and Structural Techniques. The first paper in Basic Concepts is the classic 1979 paper by Harding and Lowell, "Structural styles, their plate-tectonic habitats, and hydrocarbon traps in petroleum provinces." We have used Harding and Lowell's classification of structural styles to group papers in volumes II and III. Basic Concepts also includes papers on stress analysis and pore-pressure effects. Folding includes papers describing folding processes and geometries. Structural Techniques is a collection of practical papers included to help the petroleum geologists solve structural problems encountered in exploration and development. These papers relate ways to visualize and predict the three-dimensional arrangement and location of strata in the subsurface.

Volume II contains papers related to Basement-Involved Deformation. These papers are subdivided into Extensional, Compressional, and Strike-Slip Deformation. Extensional Deformation includes papers discussing crustal rifting and normal faulting. Compressional Deformation includes papers discussing foreland deformation. In most cases, these papers use the Rocky Mountains as an example. Strike-Slip Deformation includes papers discussing strike-slip or wrench fault deformational processes and the consequent effects these processes have on folding, faulting, basin formation, and sedimentation.

Volume III, Detached Deformation, is subdivided into three sections: Extensional Deformation, Compressional Deformation, and Salt Tectonics. Extensional Deformation contains papers that discuss listric normal faulting and growth faulting (a species of listric normal faulting). Compressional Deformation contains papers on the processes and mechanics of low-angle thrust faulting. Salt Tectonics is a collection of papers on salt movement and salt dissolution, and their subsequent effect on the structural geometry of associated strata.

Edward A. Beaumont
Tulsa, Oklahoma

Norman H. Foster
Denver, Colorado

TABLE OF CONTENTS

STRUCTURAL CONCEPTS AND TECHNIQUES II

BASEMENT-INVOLVED DEFORMATION

EXTENSIONAL DEFORMATION

COMPRESSIONAL DEFORMATION

STRIKE-SLIP DEFORMATION

TABLE OF CONTENTS

STRUCTURAL CONCEPTS AND TECHNIQUES I

TABLE OF CONTENTS

STRUCTURAL CONCEPTS AND TECHNIQUES III

DETACHED DEFORMATION

SALT TECTONICS

EXTENSIONAL DEFORMATION

Reprinted by permission of the University of Chicago from P. Mann, M. R. Hempton, D. C. Bradley, and K. Burke, *Journal of Geology*, v. 91, no. 5 (1983), p. 529-554.

DEVELOPMENT OF PULL-APART BASINS[1]

PAUL MANN, MARK R. HEMPTON,[2] DWIGHT C. BRADLEY, AND KEVIN BURKE[3]

Department of Geological Sciences, State University of New York at Albany, Albany, New York 12222

ABSTRACT

A comparative study of well mapped active and ancient pull-apart basins suggests a qualitative model for their continuous development. Pull-aparts evolve from incipient to extremely developed basins through a sequence of closely related states. New and compiled map data from several areas, including the northern Caribbean and Turkey, suggests the following stages in pull-apart development: (1) in rigid, intracontinental strike-slip plate boundary zones, larger pull-aparts nucleate at releasing fault bends along segments of the principal displacement strike-slip fault zone which are oblique to the theoretical interplate slip lines; (2) initial opening across releasing fault bends produces spindle-shaped basins defined and often bisected by oblique-slip faults connecting the discontinuous ends of the strike-slip faults; (3) increased strike-slip offset produces basin shapes which we colloquially call "lazy S"-shape for basins between sinistral faults and "lazy Z"-shape for basins between dextral faults; (4) rhomboidal pull-aparts or "rhomb grabens" result from lengthening of an S or Z-shaped basin with increased strike-slip offset and characteristically contain two or more sub-circular deeps within the basin floor; and (5) prolonged strike-slip over tens of millions of years can produce long narrow troughs floored by oceanic crust created at an orthogonal short spreading ridge; basin width does not increase significantly and remains fixed by the width of the releasing bend. Most pull-aparts have low length to width ratios, and this is a consequence of their short lives in rapidly changing strike-slip zones.

INTRODUCTION

"We suggest that the central part of Death Valley is related to tension along a segment of a strike-slip fault that is slightly oblique to the main trend of the fault zone. If this idea is correct, the two sides of Death Valley have been pulled apart and a graben produced between them." (Burchfiel and Stewart 1966).

Since the introduction of the term in 1966, a pull-apart origin has been proposed for about sixty Quaternary basins along active strike-slip faults (Aydin and Nur 1982, their table 1) as well as several ancient basins in Alaska (Fisher et al. 1979), California (Hall 1981), Atlantic Canada (Bradley 1982) and eastern Europe (Royden et al. 1982). To most geologists, the term "pull-apart" retains a meaning similar to the above interpretation of Death Valley: a depression produced by extension at a discontinuity or "step" along a throughgoing strike-slip fault. "Left-step-

ping" fault discontinuities (i.e., an observer looks left along the fault to see the next, approximately parallel fault strand) localize pull-apart basins for sinistral faults and compressional uplifts or "push-ups" for dextral faults. The two approximately parallel strike-slip faults bounding pull-aparts or push-ups are conveniently referred to as "master faults" (Rodgers 1980).

Individual workers have adopted different terms roughly synonymous with both structures: pull-apart basins are also known as *rhombochasms* (Carey 1958); *tectonic depressions* (Clayton 1966); *wrench grabens* (Belt 1968); *rhomb grabens* (Freund 1971); and *releasing bends* (Crowell 1974): push-up blocks are also called *restraining bends* (Crowell 1974); *pressure ridges;* and *rhomb horsts* (Aydin and Nur 1982). The terms releasing and restraining bend indicate that the master fault zones are continuous through the fault bend rather than overlapping or *en echelon*.

In addition to pull-aparts, more complicated strike-slip basins have been recognized at: (1) the intersections of bifurcating faults (*fault-wedge basins*—Crowell 1976; Mann and Burke 1982); (2) parallel to fault traces (*fault angle depressions*—Ballance 1980); (3) between reverse or thrust faults related to strike-slip movement (*ramp valleys*—Willis 1928; Burke et al. 1982); and (4) be-

[1] Manuscript received November 8, 1982; revised May 3, 1983.

[2] Present address: Department of Geology, Carleton College, Northfield, Minnesota 55057.

[3] Also at: Lunar and Planetary Institute, 3303 NASA Road 1, Houston, Texas 77058.

tween transverse secondary folds or normal faults (*fault flank depressions*—Crowell 1976). The reader is referred to Reading (1980) for a complete review of strike-slip sedimentation.

The purpose of this paper is to discuss the development of pull-apart basins. Our discussion is based on our own field studies of pull-aparts in active (northern Caribbean, Turkey) and ancient (Carboniferous of Atlantic Canada) strike-slip zones as well as published descriptions of basins from several other areas. After summarizing some of the major conclusions of previous workers, our aim will be to: (1) describe the plate tectonic setting of active pull-aparts and (2) synthesize field observations from mainly active basins into a qualitative model for their continuous growth from nucleation to extreme development. At the same time, we will compare our qualitative model with the predictions of previously proposed qualitative, theoretical, and experimental models.

Convenience has led some workers to adopt "pull-apart" as a generic term describing all types of strike-slip basins. In our discussion, we retain the genetic use of "pull-apart" for those basins developed between a single discontinuous strike-slip fault trace and adopt "strike-slip basin" as the general term. "Pull-apart" has also been used to refer to rift basins produced by continental breakup (e.g., Klemme 1971), but the usage we follow has precedence. We adopt the term "push-up" as the general term describing compressional uplifts at strike-slip fault discontinuities.

PREVIOUS MODELS OF PULL-APART BASIN DEVELOPMENT

Pull-apart basins along active strike-slip faults have received considerable attention from field geologists, seismologists, and theoreticians since the late 1950's. For geologists, the surficial depression of an active basin is easily recognizable and mapped; seismologists are attracted by the tendency of earthquake swarms to occur at pull-aparts; and theoreticians can easily produce basin models by manipulating the simple fault geometry.

In recent years, the tectonic and economic importance of active and ancient pull-aparts has become recognized (Reading 1980). If divergent relative plate motion produces pull-

aparts, the azimuths of master faults may not be exactly parallel to the direction of plate motion and could introduce errors in models of present-day plate motion (Minster and Jordan 1980). In ancient mountain belts, well mapped pull-aparts indicate a strike-slip tectonic environment as well as the direction, timing, sense, and minimum amount of offset along the master faults. Less information can be obtained from the narrow strike-slip zones themselves or from the uplift history of push-up blocks, where much of the rock record is not preserved. Pull-apart basins are now known to be promising sites of mineral, hydrocarbon, and geothermal resources because of typically high heat flow, volcanism, and immense thickness of clastic reservoir and source rocks, all of which are associated with large amounts of very localized lithospheric extension.

A variety of structural models for pull-apart development have been proposed and reflect the regional or experimental bias of the individual workers (fig. 1). The simplest model was proposed in pioneer field studies of active pull-aparts along the Dead Sea Fault System in the Levant (Quennell 1958) and the Hope Fault Zone in New Zealand (Clayton 1966): a basin nucleates between discontinuous and parallel strike-slip faults and evolves into a "sharp pull-apart" (Crowell 1974) or "rhomb graben" whose width remains fixed and is determined by the initial master fault separation ("S" in fig. 1A). The basin lengthens with increasing master fault offset or overlap ("O" in fig. 1A). Model A has been extensively applied to pull-apart development along the San Andreas Fault System in the Gulf of California-Salton Trough area (see Crowell 1974, for review) and the Dead Sea Fault System (see Garfunkel et al. 1981, for review).

Careful observations of deformed Quaternary terraces in pull-aparts along the Hope Fault Zone in New Zealand allowed Freund (1971) to modify the simple model of rhomboidal basin opening between parallel faults (fig. 1B). He noted the tendency of master faults to be non-parallel with their strikes, differing by several degrees. Moreover, the two master faults were not overlapping but were connected by a short oblique fault segment, which makes a 10–15° angle with the master faults (fig. 1B). Opening across the oblique median fault creates a narrow gap on one side

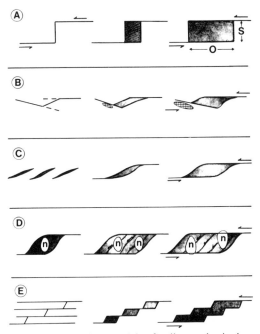

FIG. 1.—Previous models of pull-apart basin development. *A.* Simple model of pull-apart opening between left-stepping and sinistral "master" strike-slip faults; master fault separation (S) or basin width remains constant; master fault overlap (O) or basin length increases with strike-slip offset. *B.* Modification of simple model based on detailed mapping of active pull-aparts by Freund (1971); opening across an oblique median fault and along non-parallel master faults results in an extensional gap on one side of the basin and a compressional overlap or bulge (shown in cross hatching) on the other. *C.* Deformational pattern produced in shear box experiments suggested pull-aparts nucleate on *en echelon* fractures (Koide and Bhattacharji 1977). *D.* Theoretical model of deformation in pull-aparts based on the elastic dislocation theory (Rodgers 1980); "n" designates areas of predicted normal faulting. *E.* Coalescing of adjacent pull-aparts was suggested by Aydin and Nur (1982) as one explanation for widening of pull-aparts with increased offset.

of the basin and an overlap or bulge on the other (fig. 1*B*). This initial master fault pattern is similar to the gentle releasing bend geometry of Crowell (1974).

Pull-apart models based on surface mapping were refined by detailed seismic studies of earthquake swarms localized at California pull-aparts. It has been shown that: (1) opening occurs obliquely across a releasing fault bend geometry; and (2) master faults bounding pull-aparts extend through the seismogenic zone rather than merging into a single fault plane beneath the basin (Weaver and

Hill 1978/79; Segall and Pollard 1980). Field workers had previously assumed a downward tapering of the basin at depth (Clayton 1966; Sharp 1975).

Within the past 10 years, experimentalists and theoreticians have added pull-apart models to those based on geologic and seismic data. On the basis of shear box experiments, Koide and Bhattacharji (1977) suggested pull-apart basins were structurally analogous to *en echelon* extensional fractures produced during the formation of a strike-slip fault in clay model experiments (fig. 1*C*). As the shear fractures joined to form a throughgoing fault, an alternating series of compressional positive areas and extensional negative or pull-apart-like areas developed. A similar model was proposed by Dewey (1978), who suggested that pull-aparts nucleate on large-scale rotated tension gashes or Riedel shears formed during the initial stages of fault offset.

Using a model based on the elastic dislocation theory, Rodgers (1980) simulated fault patterns of pull-apart basins developed between lengthening, parallel master faults (fig. 1*D*). His model suggests pull-apart development is controlled by: (1) the amount of master fault overlap; (2) the amount of master fault separation; and (3) whether or not the faults intersect the surface. An initial basin configuration of no overlap is assumed; the ends of the master faults are connected by a zone of normal faulting ("n" in fig. 1*D*). Increasing fault overlap or basin length results in two distinct zones of normal faulting at the distal ends of the basin. Data from the Cariaco Basin, Venezuelan Borderlands (Schubert 1982), agree well with the predicted theoretical case of Rodgers.

Segall and Pollard (1980) theoretically considered the complete elastic interaction between all faults within a pull-apart rather than assume like Rodgers (1980) that each fault segment moves independently of neighboring faults. They concluded that tensile fractures and normal faults strike about 45° to the ends of the master faults in dextral pull-aparts with no overlap.

It is important to note that the relevance of theoretical models to actual pull-aparts is questionable, because the fault patterns predicted by the models only apply to the initiation of faulting and not to subsequent faulting which may reactivate older faults rather than form new ones (Rodgers 1980). Moreover, the basin shapes and secondary fault patterns shown in figure 1D apply only to the basement rocks of the pull-apart. If the basin is filled faster than the secondary faults can propagate upward through the sediment cover, a tertiary structure pattern in the sediments may develop.

Aydin and Nur (1982) proposed two models for pull-apart development based on a worldwide compilation of the dimensions of 62 active pull-apart basins which ranged widely in size. Several of the basins had been described by previous workers but had not been interpreted as pull-aparts; several new pull-aparts were described for the first time. A plot of log of basin length (fault overlap) against log of basin width (fault separation) for the 62 basins showed a well-defined linear correlation between basin length and width with a ratio of approximately 3. The most common range of ratios determined directly from the data using a relative frequency histogram was between 3 and 4. Two possible mechanisms suggested for the increase in width and uniform basin length/width ratios regardless of basin size are: (1) coalescing of adjacent pull-aparts into a single wider basin with increasing offset (Garfunkel 1981; fig. 1E); and (2) formation of new fault strands parallel to existing ones (Aydin and Nur 1982, their fig. 8).

The approach we present here is similar to Aydin and Nur and compares characteristics of well studied, active basins. Our goal will be to demonstrate that pull-apart basins, like most other structures in the earth's crust, do not suddenly come into existence but evolve through a sequence of closely related states. A single pull-apart, as it is exposed today, represents only one time frame in its development. We identify other time frames in a generalized sequence of pull-apart development by comparing mainly surficial structural characteristics of a number of pull-aparts which are not necessarily of the same age, occurring along the same fault, or formed by the same tectonic process.

TECTONICS OF PULL-APART BASINS

Before embarking on a discussion of how pull-apart basins develop, it is important to ask, "Where and why do pull-apart basins form?" As with other types of basins found in plate boundary zones, the answers involve consideration of relative plate motion across the entire boundary zone, particularly if it is thousands of kilometers long and lies mostly within rigid continental lithosphere. Plate tectonics is less successful in predicting the distribution and origin of pull-aparts on shorter faults in less rigid arc or collisional settings.

Long Strike-Slip Boundary Zones between Continental Plates.—In this setting, most larger, active pull-aparts do not occur singly but rather as part of a series on a segment of the principal displacement fault which is oblique to the theoretical interplate slip vector (fig. 2). Assuming plates are torsionally rigid, non-distorted except at their edges, and structurally homogeneous, major intracontinental strike-slip or transform faults intersect lines of pure strike-slip motion parallel to theoretical small circles which describe the relative motion between the plates (fig. 2). A present-day or "instantaneous" pole of relative plate motion about which the small circles are latitudinal is defined by assuming that the horizontal projection of earthquake slip vectors and the azimuths of well mapped strike-slip faults vary systematically over the entire length of the plate boundary zone. Because most strike-slip boundaries curve, the principal displacement strike-slip zone is parallel to small circles for only part of its length (fig. 2).

Fieldwork along active boundaries has shown that slip-parallel strike-slip fault segments are characterized by relatively narrow fault zones whereas slip-oblique fault segments are wider zones of either plate convergence ("transpression") or divergence ("transtension") (Harland 1971). Convergent or restraining fault bends are push-ups characterized by thrusting and mountains, while divergent or releasing fault bends consist of a staggered arrangement of pull-aparts, which are often submarine or in low valleys (Garfunkel 1981) (fig. 2). Seismically, slip-parallel fault segments tend to rupture in frequent large events ($6.5 < M < 7.5$); restraining bends inhibit plate motion and rupture infrequently during great earthquakes ($M \geq 8$) (Scholz 1977); and pull-apart segments are areas of reduced interplate friction, lengthen steadily, and act to localize volcanism and earthquake swarms ($M < 5$) (Weaver and Hill 1978/79; Segall and Pollard 1980).

Several general characteristics of pull-

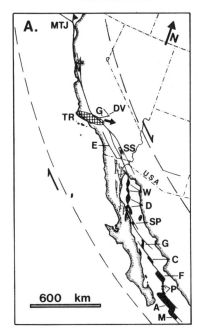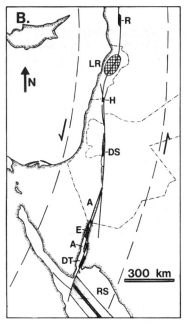

FIG. 2.—Examples of pull-apart basins found within long strike-slip boundaries between rigid continental plates. *A*. Pacific North America Plate Boundary Zone—dashed lines are theoretical interplate slip lines from Minster et al. (1974) and abbreviations are: MTJ Mendocino Triple Junction; TR Transverse Ranges push-up block (shown in cross hatching) along dextral San Andreas Fault Zone; DV Death Valley Basin; G sinistral Garlock Fault Zone forming the northern edge of the triangular Mohave Block (black arrow shows general direction of block displacement); E Lake Elsinore Basin along Elsinore Fault Zone; SS Salton Sea pull-apart area at a right-step between dextral San Andreas and Imperial Fault Zones; W Wagner Basin (separated by a short transform fault) in northern Gulf of California; D Delfin Basin (separated by a short transform); SP San Pedro Martir Basin; G Guaymas Basin (separated by a short transform); C Carmen Basin; F Farallon Basin; P Pescadero Basin Complex (separated by several short transforms); A Alarcon Basin; M Mazatlan Basin. Sources of data include: Crowell (1981); Henyey and Bischoff (1973); Bischoff and Henyey (1974); and Niemitz and Bischoff (1981). Many smaller pull-aparts are omitted from compilation map. *B*. Arabia and Sinai (Levant) Plate Boundary Zone—dashed lines are theoretical interplate slip lines from LePichon and Francheteau (1978) and abbreviations are: R Er Rharb (Gharb) Basin; LR Lebanon Ranges push-up block (shown in cross hatching) along Sinistral Dead Sea Fault Zone: H Hula Basin; DS Dead Sea Basin; A Arava Fault Trough; E Elat Basin in northern Gulf of Aqaba (Elat); A Arnona-Aragonese Basin; DT Dakar-Tiran Basin; RS Red Sea. Sources of data include: Garfunkel (1981) and Ben-Avraham et al. (1979). See Muehlberger (1981) for alternative interpretation of Er Rharb (Gharb) Basin. Many smaller pull-aparts are omitted from compilation map.

aparts appear predictable from their position relative to the pole of interplate motion: (1) *distribution*—a plate boundary zone, like the Dead Sea Fault System, which is closer to its rotation pole than the San Andreas System (fig. 2), is characterized by a smaller radius of curvature and more curvilinear trace, and seems to generate a larger number of oblique segments resulting in more frequent, wider spaced pull-aparts; (2) *length*—the longest pull-aparts (i.e., the basins with greatest fault overlap) seem to occur farthest from the pole of rotation (fig. 2) and may reflect faster relative plate motion farther from the pole; (3)

width—the widest basins (i.e., basins with greatest fault separation) are also the most closely spaced and occur at the greatest changes in strike of the fault with respect to the slip vector (e.g., northern and southern Gulf of California; Gulf of Aqaba—fig. 2); conversely, the narrowest basins are the most widely spaced and occur along segments of the fault which diverges only slightly from their slip vector; (4) *orientation of master faults*—although the overall trend of pull-apart arrays is oblique to the regional slip vector, the strike of master faults appears more or less slip parallel; however, a detailed

534 W. P. MANN ET AL.

study of fault strikes and earthquake first motions in the Gulf of California (fig. 2A) revealed a considerable scatter in the directional data and was unable to determine a single pole of rotation (Sharman et al. 1976); and (5) *coalescing*—the process of pull-apart coalescing (Garfunkel 1981; Aydin and Nur 1982) (fig. 1E) occurs between wider, more closely spaced basins in areas where the slip vector and fault trace are significantly non-parallel (e.g., southern Gulf of California and Gulf of Aqaba—fig. 2); basin coalescing does not appear common between basins along parallel but separate fault zones (Aydin and Nur 1982, their fig. 8) or along faults which closely parallel their slip vectors.

There are several possible mechanisms for the divergence between slip vector and fault trace that leads to pull-apart nucleation between continental plates: (1) changes in relative plate motion across a straight fault in a manner similar to that suggested by Menard and Atwater (1968) for "leaky" oceanic transforms (Holcombe and Sharman in press) (fig. 3A); (2) motion across a fault which is originally oblique to relative plate motion (fig. 3B); or (3) extension across a fault at an oblique segment developed adjacent to a restraining bend (fig. 3C). Model A appears unlikely to occur along faults in cold and thick continental lithosphere; models B and C appear more likely (fig. 2).

The variation in fault overlap or length of pull-aparts (fig. 2) suggests that they are produced by different amounts of horizontal offset and therefore at different times. Because the length of the largest pull-apart is much less than the total offset on the fault (fig. 2), it appears that most active pull-aparts were produced by the last phase of horizontal motion (Garfunkel 1981) (fig. 2). Plate tectonics is less useful in predicting pull-apart geometry in active arcs or zones of continental convergence where plates behave less rigidly and are more susceptible to internal deformation.

Strike-Slip Systems in Active Arcs.—Oblique subduction of oceans at arcs is accommodated in part by intra-arc strike-slip faulting because nearly vertical strike-slip fault planes concentrate horizontal shear more effectively than an inclined Benioff Zone of equal or greater strength (Fitch 1972). Pull-apart basins are common along strike-slip faults in the type locality of this

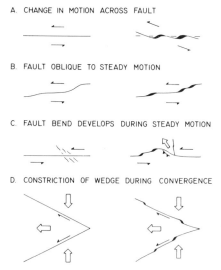

FIG. 3.—Possible tectonic mechanisms for the divergence between the interplate slip vector and strike-slip fault trace. Divergence leads to pull-apart basin nucleation. See text for explanation.

setting—the Sumatra-Andaman Sea margin between the Indian and Chinese plates (Page et al. 1979). Application of rigid plate tectonics to the distribution and origin of these pull-aparts is more difficult than those between two continental plates because (1) arc lithosphere probably behaves non-rigidly and (2) complex relative motions are expected between the oceanic plate and the divided arc (Dewey 1980).

Strike-Slip Systems Bounding Fault Wedges in Convergent Zones.—The inability of buoyant continental lithosphere to be consumed at convergent zones formed either by collision or by development of a strike-slip restraining bend (figs. 2, 3C and 3D) leads to sideways motion of triangular, continental fragments away from the point where the two continents met (McKenzie 1972). Several pull-aparts occur along the North and East Anatolian strike-slip faults in Turkey which accommodate eastward relative motion of the Anatolian fragment away from eastern Turkey and western Iran (Sengor 1981, his fig. 2). Pull-aparts along the Garlock Fault Zone in California (Aydin and Nur 1982) appear to be related to eastwards motion of the triangular Mohave Block away from the Transverse Ranges restraining bend (McKenzie 1972, his fig. 35) (figs. 2 and 3C). The formation of pull-aparts in convergent settings is not well described by rigid plate tectonics and is more likely to be related to constriction of the triangular continental wedge during convergence (Hempton 1982) (figs. 3C and 3D) than to divergent relative motion between stable blocks (fig. 3A and 3B).

In the next several sections, similar structural and sedimentary characteristics of well studied nucleating, S and Z shaped, rhomboidal, and extremely developed pull-aparts are described from the Caribbean, New Zealand, California, Turkey, Sumatra, the Middle East, and the Carpathians.

Similar characteristics of pull-aparts from all settings suggest that their continuous development is similar despite their probable differences in tectonic origin (fig. 3).

PULL-APART BASIN NUCLEATION

Pull-apart basins in the early stages of development can easily be over-looked by field workers because they have undergone a limited amount of opening and are not as geomorphically prominent as the older, more familiar rhomboidal or "rhomb graben" morphology (fig. 1). In this section, we will document common structural and sedimentary characteristics of incipient pull-aparts from two different neotectonic settings: (1) long strike-slip boundary zones between rigid plates—San Andreas Fault Zone in the Salton Trough, California (fig. 2A); Enriquillo-Plantain Garden Fault Zone in Haiti (fig. 4); Hope Fault Zone in New Zealand (see Mercer and Freund 1974, their fig. 1, for regional map) and (2) a back-arc strike-slip system in Sumatra (see Page et al. 1979, for regional map).

Structural Pattern.—Common characteristics of incipient pull-apart basins include: (1) a "releasing bend" initial geometry consisting of master faults with no overlap connected by an oblique median fault with a large strike-slip component, and (2) master faults which are not exactly parallel.

A well defined 25 km long segment of the sinistral Enriquillo-Plantain Garden Fault Zone near Tiburon, Haiti (T in fig. 4) contains four left-stepping "releasing bends" which are interpreted as incipient pull-apart basins (fig. 5). The throughgoing strike-slip fault appears continuous at the bends, although overlapping of the master fault traces at the easternmost bend appears to be occurring (fig. 5).

The amount of fault separation at the bend is less than 0.5 km in all cases; the oblique segment at the bend makes a 30–40° angle with the master faults; and the strike of the master faults differ by several degrees on either side of the bend (fig. 5). The three western fault bends are topographically low areas and are sites of recent alluvial and alluvial fan sedimentation; the eastern bend occupies an elevated ridge and is not sedimented (fig. 5).

The east-west direction of relative motion across the Enriquillo-Plantain Garden Fault Zone (fig. 5) is based on earthquake slip vector and fault strike data from the Oriente-Swan Fault system (fig. 4) (Jordan 1975) and is consistent with divergence and pull-apart nucleation on this west-southwest (261°) trending fault zone (fig. 3B). No pull-aparts occur along the fault where it trends due east (270°) in eastern Haiti and the Dominican Republic. In eastern Haiti at about 72°20'W, the fault trace changes from its east-west azimuth to a more southwesterly "leaky" orientation and is marked by a very prominent fault trough south of Leogane (fig. 4). Large push-up blocks or restraining bends in eastern Jamaica and western Haiti (Mann et al. 1981; this paper fig. 4) may increase the obliquity of the strike-slip fault trace to the interplate slip vector in this area by a fault-bend mechanism (fig. 3C).

The Mesquite Basin at a right-step between the dextral San Andreas and Imperial Fault Zones in the Salton Trough in southern California (SS in fig. 2; fig. 6A) and the Clonard Basin at a left-step in the Enriquillo-Plantain Garden Fault Zone in Haiti (C in fig. 4; fig. 6C) are morphologically similar and are suggested to be a more advanced stage of pull-apart opening than the releasing bends described along the fault in the Tiburon area. Both the Mesquite and Clonard Basins occur as releasing fault bends between master faults with no overlap; the oblique fault segment connecting the ends of the master faults has ruptured the Quaternary sedimentary fill of both basins (fig. 6). In the Mesquite Basin, the more active oblique fault—the Brawley Fault Zone—appears to form one edge of the structural depression, whereas the more conspicuous oblique fault in the Clonard Basin bisects the Quaternary basin (fig. 6). A significant amount of strike-slip displacement on the Brawley Fault Zone is indicated by strike-

FIG. 4.—Regional setting of active pull-apart basins formed by sinistral strike-slip motion between the North American and Caribbean plates (modified from Case and Holcombe 1980). Most of the grey areas shown are known or inferred pull-aparts at left-steps in either the Septentrional-Oriente-Swan-Motagua Fault Zones or the Enriquillo-Plantain Garden Fault Zone. The map pattern of basins north of the Motagua Fault Zone in Central America does not indicate a simple pull-apart origin as suggested by Aydin and Nur (1982). The 1400 km long Cayman Trough, the world's longest active pull-apart, formed by extension and seafloor spreading at the 100 km long Cayman Rise. Much younger and smaller pull-aparts occur along a 400 km long segment of the Enriquillo-Plantain Garden Fault Zone between eastern Jamaica and western Haiti. Abbreviations for Haitian localities are: L fault segment south of Leogane; ML Mirogoane Lakes (detailed map shown in fig. 5); C Clonard Basin (detailed map shown in fig. 6C); and T Tiburon Fault Segment (detailed map shown in fig. 9A). Most of the clastic sediment derived from Central America or push-up blocks like the islands of Jamaica and Hispaniola (Haiti and the Dominican Republic) bypass the Cayman Trough (horizontal ruled areas represent ponded turbidites; dotted area represents proximal submarine fan deposits).

8

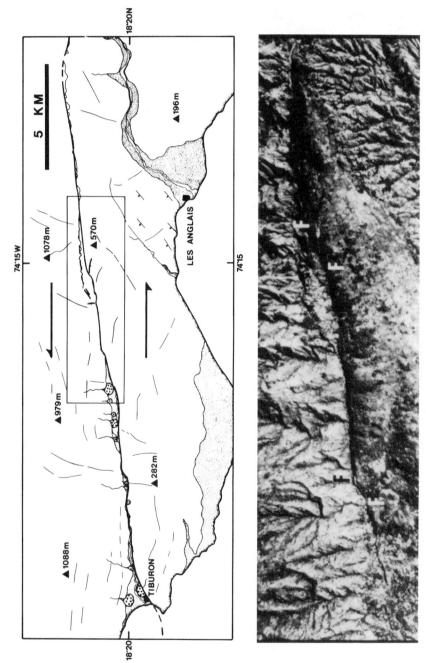

Fig. 5.—Aerial photointerpretation of incipient pull-aparts forming at left-stepping releasing bends along the Enriquillo-Plantain Garden Fault Zone in southwestern Haiti (see fig. 4 for regional setting). Stippled areas represent alluvium; dotted areas are alluvial fans. Inset is aerial photograph showing details of "releasing bend" master fault geometry. Letter F marks approximate termination of dominant oblique fault segments at releasing bends. Overlapping of master faults appears to be occurring at the easternmost bend.

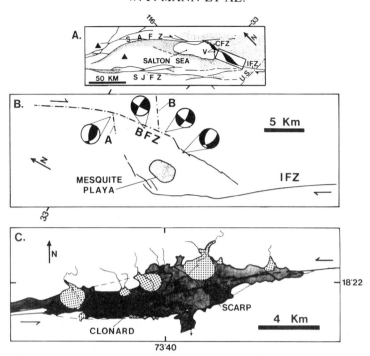

FIG. 6.—Examples of nucleating pull-apart basins along active strike-slip faults. *A*. Regional setting of pull-aparts in Salton Trough, California, from Crowell (1981); abbreviations are: SAFZ San Andreas Fault Zone; SJFZ San Jacinto Fault Zone; CFZ Calipatria Fault Zone; V recent volcanic dome; dots indicate areas of Quaternary basin fill; black triangles are peaks about 3 km in elevation in the western part of the Transverse Ranges restraining fault bend; box indicates Salton Trough map area shown in figure 6*B*. *B*. Fault map and earthquake focal mechanism solutions of Mesquite Basin between Brawley (BFZ) and Imperial (IFZ) Fault Zones from Johnson and Hadley (1976). Scarps are shown by solid lines; buried faults inferred from seismicity are dashed. *C*. Fault and sediment map of Clonard Basin, Haiti, based on photointerpretation of 1:40,000 aerial photographs and field observations. Fault scarps are indicated by solid lines; inferred faults are dashed; grey represents basin alluvium; dotted areas are alluvial fans; letter D indicates downthrown side of oblique fault scarp in basin alluvium.

slip first motions of earthquakes (Johnson and Hadley 1976), pattern of surficial fault breaks (Sharp 1976) and the linearity of both the buried (inferred seismically) and exposed fault trace (fig. 6*B*). In the Clonard Basin, strike-slip displacement on the oblique fault is indicated by a remarkably linear, 1.5 to 2.0 high scarp in the alluvial basin floor (fig. 6*C*).

The most significant difference between the Mesquite and Clonard Basins is their mode of occurrence: the Mesquite Basin is one of at least two and possibly even three (Crowell 1981) staggered pull-aparts within a 30 km right-step between the Imperial and San Andreas Fault Zones (fig. 6*A*) whereas the Clonard Basin occurs singly at a very gradual 2 km left-step in the Enriquillo-Plaintain Garden Fault Zone (fig. 4, 6*C*).

From the previous discussion of the tecton-

ics of pull-apart basins along rigid strike-slip boundaries, the multiple occurrence of pull-aparts in the Salton Trough results from a sharper releasing fault bend that is significantly nonparallel to the regional interplate slip vector (fig. 2); conversely the narrower Clonard Basin results from a gentler initial fault bend which is not significantly nonparallel from the slip vector (fig. 4). In both areas, the presence of nearby push-up blocks or restraining bends may contribute to the obliquity of the fault trace by the mechanism shown in figure 3*C*. Pull-apart basin coalescing (fig. 1*E*) will result with continued offset in the case of the Salton Trough but not in the case of the Clonard Basin.

Several other pull-aparts described from other areas exhibit common structural characteristics with the Mesquite and Clonard Ba-

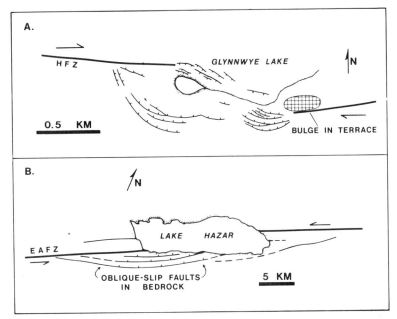

FIG. 7.—A. Map of active Glynnwye Lake Basin, New Zealand, from Clayton (1966) and Freund (1971). Heavy lines are traces of Hope Fault Zone (HFZ). B. Map of active Lake Hazar Basin, Turkey. Heavy lines are traces of East Anatolian Fault Zone (EAFZ).

sins and therefore are interpreted as very young pull-apart basins. The Glynnwye Lake Basin (fig. 7A) occurs between two non-parallel strands of the dextral Hope Fault Zone in New Zealand which do not overlap. Freund suggested that the master faults were connected by a short oblique fault segment which was obscured there by the course of Glynnwye Stream (fig. 7A) but was noted in a similar pull-apart along strike. He also observed a small topographic bulge in a planar terrace at the eastern end of the basin (fig. 7A). He suggested that the basin had opened along a median fault segment which makes about a 15° angle to the ends of the master faults. Because the master faults are not parallel, an extensional gap opens on the west side of the releasing fault bend, and a compressional overlap or tectonic bulge appears on the east (fig. 1B and 7A). Bulges in river terraces were mapped by Freund at the eastern end of four out of five of the Hope Fault pull-aparts. Although semi-circular normal fault scarps are prominent, most occur outside of the zone of master fault separation and are related to gravitational slumping of the sides of the basin into the tectonic depression.

The Siabu Basin along the Sumatra Fault Zone within the Sumatra arc exhibits several of the structural characteristics of a young pull-apart (Tija 1977, his fig. 6). The angle between the non-overlapping master faults (10°), the fault separation (5 km), and the overall spindle shape are very similar to the Clonard Basin (fig. 5). No oblique median fault scarp has been reported from the basin itself, but it may be concealed by the linear course of the Angkola River in the center of the basin.

Sedimentary Pattern—Because of the limited amount of opening, the structural depression of nucleating pull-aparts is generally not as prominent as more evolved rhomboidal pull-aparts: most are above sea level, not covered by a standing body of water, and, in some cases, exhibit low fault scarps which localize only gradual facies changes across the depression.

For example, the Mesquite Basin (fig. 6B) is 40 m below sea level but lies only about 10 m lower than the basin margins. The present depocenter in the Salton Trough (fig. 6A) is not centered on the pull-apart area but rather on the Salton Sea, whose surface is now at an elevation of about 70 m below sea level. Most of the faults defining the basin are buried within about 6 km of Plio-Pleistocene alluvial and lacustrine sediments (Crowell 1981). Active fault scarps defining the depression (fig. 6B) are insignificant with maximum vertical displacements of 15 mm (Sharp 1976). Similarly, the Glynnwye Lake Basin (fig. 7A) occupies a depression that is only about 30 m lower than the surrounding area. Peripheral slumping on down-to-the basin normal faults has considerably widened the low area between the separated master faults.

In some nucleating basins, such as Clonard (fig. 6C) and Siabu (Tija 1977), the structural depression is well defined by scarps with relief ranging from 125 to 200 m. In Clonard, alluvial fans of various sizes are found along the steep basin edges; the floor of the basin is remarkably flat, and deposition occurs in meandering river channels (fig. 6C). The upthrown southern side of the median oblique fault scarp appears to act as a structural barrier to the present meandering river system (fig. 6C).

Comparison to Model Predictions.—The above observations of nucleating pull-aparts suggest they form at releasing bends along throughgoing faults and are unlikely to be analogous to tension gashes or Riedel shears seen in outcrop and generated in shear box experiments (fig. 1C). An oblique fault scarp connecting and at a low angle to the discontinuous ends of the master faults is observed in several active basins and is approximately consistent with secondary oblique faults produced in theoretical models of basement deformation between faults with no overlap (fig. 1D) (Rodgers 1980; Segall and Pollard 1980). It should be noted, however, that theoretical fault patterns calculated for basement deformation (e.g., fig. 1D) cannot be assumed to be identical to scarps observed in the sedimentary cover of the basin. Rodgers (1980) considers the case of master faults which are not overlapped and are overlain by a uniformly thick sedimentary layer (his fig. 6). The theoretical basin fault pattern is: (1) strike-slip faults at a lower angle to the master faults (10°) than the predicted basement deformation pattern (40–50°); and (2) there is no normal faulting within the basin as in the case of the basement deformation case. All of these predictions are consistent with the structural patterns described above.

YOUNG PULL-APARTS: S AND Z-SHAPED BASINS

Following nucleation of pull-aparts at releasing bends along throughgoing strike-slip faults, continued offset produces basin shapes that we colloquially call "lazy S" shape for pull-aparts between sinistral faults and "lazy Z" shape for those between dextral faults (fig. 8). These shapes may be modified by local deformation associated with lengthening non-parallel master faults and peripheral basin slumping. The lazy S and Z shape of pull-apart basins represents a transitional stage between incipient spindle-shaped basins between master faults with no overlap and rhomboidal basins or "rhomb grabens" between overlapping master faults. S and Z-shaped pull-aparts are particularly prominent when the master fault separation is greater than about 10 km and the strikes of the master faults are non-parallel.

Structural Pattern—Unlike incipient pull-aparts, the alluvial cover of many S and Z-shaped pull-aparts is not ruptured by an oblique-slip fault connecting the ends of the master faults (fig. 8). Oblique faults generally trend at an angle of 35 to 50° to the master faults and appear to be dominantly normal faults which define the sides of the basin. The S or Z shape of the basin tends to be exaggerated by extensions of the alluvial cover of the basin along the traces of the master faults (fig. 8).

Fault scarps within the Quaternary basin sediments of the Hanmer Basin along the Hope Fault Zone in New Zealand (Freund 1971) more or less define the edges of the basin (fig. 8A). Oblique scarps are present along the eastern edge of the basin but either terminate or are buried before extending far into the center of the basin. A bulge in the Quaternary terrace at the eastern end of the basin exposes a 750 m thick section of river gravels and silts and was interpreted by Freund as an overlap area resulting from basin opening between non-parallel master faults. Terraces at the western end of the basin are tilted eastwards (fig. 8A).

The Niksar Basin along the North Anatolian Fault Zone in Turkey (Seymen 1975) exhibits a very pronounced Z shape and is not bisected by an oblique fault scarp (fig. 8B). Although the bedrock highlands surrounding the basin are pervasively faulted (Seyman

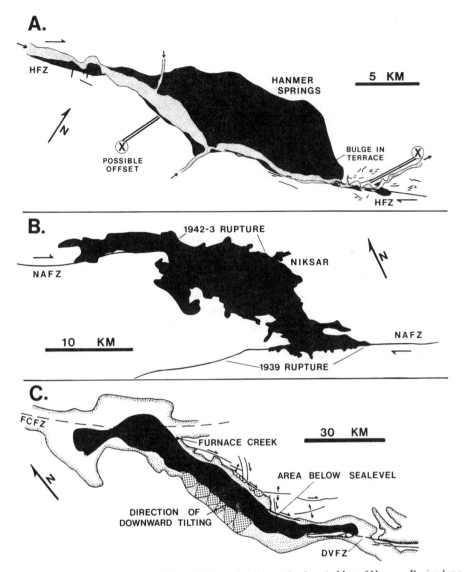

FIG. 8.—Examples of active "lazy" S and Z-shaped pull-apart basins. *A.* Map of Hanmer Basin along the Hope Fault Zone in New Zealand from Freund (1971). Lines marked "x" indicate abrupt contrast in morphology of highlands which appears to have been dextrally offset by about 19 km, the total offset of the Hope Fault Zone. Stippled areas represent braided river deposits; grey area represents locally deformed basin sediments. *B.* Map of active Niksar Basin along the North Anatolian Fault Zone (NAFZ) in Turkey based on Seymen (1975). Heavy lines mark surface ruptures during 1939 and 1941–42 earthquakes; grey area represents Quaternary depression. *C.* Map of active Death Valley, California, from Burchfiel and Stewart (1966) and Hill and Troxel (1966). Abbreviations are: FCFZ Furnace Creek Fault Zone and DVFZ Death Valley Fault Zone. Grey area represents area of basin floor which is below sea level. Eastward tilting of basin floor is indicated by detailed alluvial fan studies by Hooke (1967).

1975, fig. 61), the only faults within the basin sediments are ground breaks associated with the 1939 and 1942–43 earthquakes (fig. 8*B*). The 20 km overlap of the master fault ground breaks suggests that either: (1) the Niksar Basin is anomalous and does not fit the pattern of a young pull-apart between master faults with little or no overlap; or (2) that the master faults propagate faster than the basin can widen.

The Death Valley Basin in California, where the term pull-apart was introduced

(Burchfiel and Stewart 1966), also exhibits a pronounced Z shape (fig. 8C). Unlike the Hanmer or Niksar Basins, there is good evidence for dextral strike-slip motion along oblique faults through the center and along the eastern edge of the basin (Hill and Troxel 1966). Evidence for a component of strike-slip on a complex system of faults forming the eastern margin of the basin include: (1) slickenside striae, (2) drag folds where faults transect basin sediments, and (3) orientation of secondary folds, normal faults, and dikes. Dextral offset of a volcanic cone within the basin sediments at the southern end of the basin suggest the presence of a dextral fault that bisects the sedimentary basin (see Hill and Troxel 1966, their fig. 1). Recent strike-slip displacement along the eastern edge of the valley is indicated by dextral skewing of alluvial fan deposits (Hooke 1972). Eastward tilting of the basin floor is indicated by segmentation of fans on both sides of the valley (fig. 8C).

Sedimentary Pattern.—Increased extension across S and Z-shaped basins along normal or oblique-slip faults at 35–50° to the master faults results in a prominent topographic and sedimentary basin which more or less coincides with the basement pull-apart (fig. 8). The pull-aparts are typically deep depressions surrounded by upfaulted highlands of basement rock: for example, Death Valley (fig. 8C) has a minimum elevation of 85 m below sea level and is about 3 km lower than peaks in the surrounding mountains; the Hanmer Basin (fig. 8A) is more than 1 km lower than the surrounding highlands.

Because of the large amount of relief, sedimentation typically consists of alluvial fan deposits localized along fault scarps forming the sides of the basin; strike-slip motion along these faults may result in progressive displacement of fan bodies from their source. Tilting of the basin floor results in fan segmentation as well as extensive low gradient fans along one valley edge and smaller, steeper gradient fans along the other (fig. 8C). Proximal fan deposits grade into distal sand and mud facies which do not seem to be affected by faulting or localized depocenters (fig. 8). Some facies changes and local unconformities may result from syndepositional bulges near the edges of the basin (fig. 8A).

Comparisons to Model Predictions.—The S or Z-shaped pull-apart structure is a fair approximation of Rodgers' (1980) initial theoretical basement fault pattern developed between master faults with no overlap (fig. 1D). The single area of normal faulting between areas of oblique strike-slip faulting trending 40–50° to the master faults (fig. 1D) is consistent with the structural map patterns shown in figure 8. With the exception of Death Valley (fig. 8C), no oblique strike-slip faults bisecting the sedimentary basin are observed. Again, it should be noted that the theoretical basement fault pattern shown in figure 1D is not necessarily directly expressed in the overlying basin sediments.

The maximum predicted depth of pull-aparts with dimensions very similar to the Hanmer basin (fig. 8A) is about 15% of the offset on the master faults (Rodgers 1980). The measured 1800 m depth (thickness of sediment plus average height of surrounding mountains) of the Hanmer Basin suggests about 12 km of offset on the Hope Fault Zone produced the basin. This offset estimate is less than the 19 km total offset inferred from an abrupt morphological change in the topography of the highlands adjacent to the Hanmer Basin (fig. 8A) and two other offset markers in other areas (Freund 1971).

Because incipient and S and Z-shaped pull-aparts are not rectangular, their basin length to basin width ratios are difficult to measure and compare to the constant length to width ratio of 3 which was suggested for all pull-apart basins by Aydin and Nur (1982).

For example, Aydin and Nur (1982) measured a basin length of 300 m and a basin width of 90 m for the Glynnwye Lake Basin (fig. 7A); Freund (1971) measured a basin length of 2000 m and a width of 600 m; and Clayton (1966) measured a length of 1000 m and width of 600 m. Some variation is attributable to the large amount of gravitational slumping into the depression (fig. 7A).

PULL-APART MATURITY: RHOMBOIDAL BASINS

The rhomboidal shape of many pull-aparts has led to the widespread use of the term "rhomb graben" (e.g., Freund 1971; Garfunkel 1981) and to the observation that all pull-apart basins have a constant length to width ratio of about 3 (Aydin and Nur 1982). We suggest that the "rhomb graben" or rectangular form of pull-aparts results from lengthening of an S or Z-shaped basin with increased master fault offset. Length to width

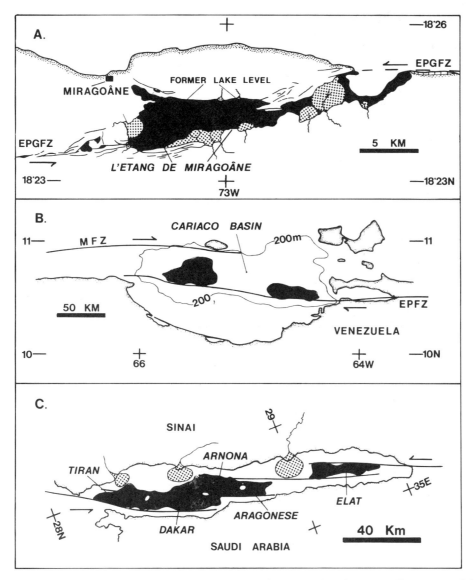

FIG. 9.—Examples of active rhomboidal pull-apart basins. A. Fault and recent sediment map of the Mirogoane Lakes (*L'Etang de Mirogoane*) Basin at left-step of sinistral Enriquillo-Plantain Garden Fault Zone (EPGFZ) in Haiti (see fig. 4 for location). Black areas represent fresh-water lakes; grey area represents basin alluvium; grey and stippled area represents former extent of lakes; stippled areas represent alluvial fans. B. Fault and bathymetric map of Cariaco Basin at a right-step between the dextral Moron (MFZ) and El Pilar (EPFZ) Fault Zones, Venezuelan Borderlands. Map is taken from Schubert (1982). Grey area represents water depths from 1000 to 1400 m; black represents depths greater than 1400 m. C. Fault, bathymetric and recent sediment map of pull-aparts at left-steps of the sinistral Dead Sea Fault Zone in the Gulf of Aqaba (Elat) (see fig. 2B for location). Map is taken from Ben-Avraham et al. (1979). In the Elat Basin, black represents water depths greater than 900 meters; in the Arnona-Aragonese and Dakar-Tiran Basins, black represents water depths from 1300 to 1800 m. Stippled areas represent fan delta complexes.

ratios (i.e., the ratio of fault overlap to fault separation) therefore tends to increase; basin width remains fixed by the initial fault separation or width of the releasing bend (fig. 1A). Increased extension in rhomboidal pull-aparts causes them to be prominent structural depressions which are often below sea level.

Structural Pattern.—Common structural characteristics of rhomboidal pull-aparts include: (1) overlapping master faults that usually form prominent scarps along the edges of the basin, and (2) deeps at distal ends of the basin separated by a shallow sill: the deeps are roughly circular and may be fault-bounded; multiple deeps are often diagonally arranged across the floor of the basin (fig. 9).

The Mirogoane Lakes Basin at a 5 km left-step in the sinistral Enriquillo-Plantain Garden Fault Zone in Haiti (see fig. 4 for location) exhibits an approximately rhomboidal shape (fig. 9A). The northern and southeastern margins are well-defined by several fault scarps parallel to the east-west master faults. The floor of the basin contains four depressions: three are fresh water lakes (the Mirogoane Lakes and a smaller unnamed pond); the fourth is a marine embayment of the Caribbean Sea (fig. 9A). The depressions are diagonally distributed across the floor of the basin. No fault scarps are apparent in the basin floor alluvium, and basin floor relief is less than 5 m. A large alluvial fan forms the eastern sill and may have restricted marine access to the Mirogoane Lakes in the western part of the valley.

The Cariaco Basin in the Venezuelan Borderlands (fig. 9B) is a much larger rhomboidal pull-apart that occurs at a 35 km separation and 125 km overlap between the Moron and El Pilar Fault Zones (Schubert 1982). The northern and southern margins of the basin are sharply defined by the master fault scarps that dip steeply into the basin with relief up to 1 km. The floor of the basin contains two sub-circular deeps which reach depths slightly over 1400 m and are separated by a sill with a depth less than 900 m (fig. 9B). The center of the western deep is 15 km north of the center of the eastern deep and suggests the approximately diagonal distribution of deeps described in the Mirogoane Lakes Basin (fig. 9A). Because acoustic reflection profiles have been run in a north-south direction, only east-west fault scarps have been recognized on the

floor of the basin. None of these are traceable for more than 50 km. The northern and southern edges of both deeps appears to be defined by east-west scarps although the scarps have little surface expression. The reader is referred to Schubert (1982) for more detailed description of the basin.

The Gulf of Aqaba (Elat) in the northern Red Sea is the widest and deepest segment of the sinistral Dead Sea Fault System (fig. 2B) and consists of three closely spaced rhomboidal pull-aparts (Ben-Avraham et al. 1979) (fig. 9C). The eastern and western edges of the basins are sharply defined by the very straight and inwardly dipping master fault scarps. The three basins are separated by low sills; the southern two basins overlap or coalesce (fig. 1E) and contain deeps at their distal ends (fig. 9C). The northern basin is flat-bottomed. Maximum water depths in the deeps range from 1300 to 1850 m. The Arnona and Aragonese Deeps in the central basin occur diagonally across the basin (fig. 9C) but contrast with the diagonal pattern of the Mirogoane Lakes Basin by connecting the opposite corners of the basin (fig. 9A). The floor of the deeps is flat-bottomed, but the intervening sill areas consist of folded and warped basinal sediments. The reader is referred to Ben-Avraham et al. (1979) for more detailed description of the basin.

The Miocene Vienna Basin in the western Carpathians has been suggested as an ancient example of a rhomboidal pull-apart basin (Royden et al. 1982, their figure 4A) and exhibits several common characteristics with the active rhomboidal pull-aparts described above. An isopach map to the base of the Miocene shows that the overlapping master faults are almost vertical and define the straight southeastern and northwestern margins of the basin. Sediment thicknesses change abruptly from less than 1 km to 3–4 km across the master faults (Royden et al. 1982, their figure 4A). The floor of the basin is marked by two sub-circular depocenters with sediment thicknesses greater than 5 km. The depocenters are separated by a sill area of much thinner sediment and, like the Mirogoane Lakes Basin (fig. 9A) and the Cariaco Basin (fig. 9B), are distributed diagonally across the basin. As in the Cariaco Basin, faults within the basin floor which are parallel to the master faults appear to define the edges of the depocenters. Another set of faults trending 30–50° is present within the basins and roughly defines the northeastern and southwestern edges of the basin.

Sedimentary Pattern.—Subsidence between overlapping master faults of rhomboidal pull-aparts generally greatly exceeds sedimentation and can result in fault-bounded deep marine (fig. 2A, 9B, 9C), marine and lacustrine (fig. 9A), or lacustrine (fig. 7B) sedimentary environments. Fault control of sedimentation is most prominent along the master faults where steep gradients are associated with complex zones of oblique-slip faults. In contrast, faults defining the ends of rhomboidal pull-aparts often have no surface expression and little effect on basin facies (fig. 9). Greater vertical displacement on one of the master faults causes tilting of the basin block and asymmetric facies development with smaller, steeper gradient fans and fan deltas on the downthrown side of the basin block, and larger, lower gradient fans on the upthrown side.

For example, master fault scarps bounding the Gulf of Aqaba (Elat) pull-aparts (fig. 9C) separate maximum water depths of about 2 km from 1.5 km high highlands over a distance of 15–20 km. Oblique faults forming the ends of the basin form only gentle slopes. Oblique-slip movements appear greater along the eastern coast with basin slopes as steep as 25–30° (Ben-Avraham et al. 1979). The western slope attains average angles of 16° in its steeper parts, but the shallower parts usually have slopes of 7–11°. No fan deltas are present along the steeper eastern margin, in contrast to the western margin where extensive fan delta and submarine cone complexes are found (fig. 9C). Similarly, master fault scarps in the Lake Hazar Basin are much more prominent than oblique scarps forming the ends of the basin and control steeper gradient fan deltas (fig. 7C; Hempton et al. 1983). The Lake Hazar Basin is also asymmetric with tilting towards the southern master fault zone. Smaller, steeper gradient fans are found along the southern margin. Similar relationships between faulting and sedimentation have been documented for Devonian sedimentation in the rhomboidal Hornelen Basin in Norway (Steel and Gloppen 1980), although a pull-apart origin is not entirely clear.

Basin floor deeps, which may or may not be bounded by fault scarps, exhibit much less relief than the master fault scarps (fig. 9). Subsidence within the deeps is fast enough to keep pace with the high sedimentation rates within the floor of the basin. Sediment thicknesses in the floor of the basin vary significantly across the deeps and reflect syndepositional faulting. Sediment thicknesses are 2–3 km without reaching basement in the Gulf of Aqaba (Elat) and at least 1 km in the Cariaco Basin. The sedimentation rate on the sill area of the Cariaco Basin is 0.5 m/1000 years.

Comparison to Model Predictions.—The existence of deeps or depocenters within the distal ends of active (fig. 9) and ancient rhomboidal pull-apart basins does not support the simple pull-apart model of uniform extension between overlapped master faults (fig. 1A). Schubert (1982) pointed out the similarity of the observed structural pattern in the Cariaco Basin (fig. 9B) to that predicted for rhomboidal basins by the theoretical model of Rodgers (1980) (fig. 1D). The theoretical fault pattern for rhomboidal basins consists of two sub-circular zones of normal faulting which occur roughly on a diagonal connecting opposite corners of the basin. The individual normal faults in the zones strike at a high angle to the master faults. Strike-slip faults dominate in the center and ends of the basin and occur at about 45° to the master faults (fig. 1D).

In addition to the Cariaco Basin, Rodger's model predicts well the distribution of deeps on the floor of the Mirogoane Lake Basin (fig. 9A), the Arnona-Aragonese and Dakar-Tiran Basins in the southern Gulf of Aqaba (Elat) (fig. 9B, 9C), and the Vienna Basin (Royden et al. 1982, their fig. 4A). The Arnona and Aragonese deeps occupy the distal corners of the basin, which are opposite to those predicted by the model (fig. 9C). The close correspondence between predicted and theoretical fault patterns suggests that the elastic dislocation theory may be capable of accurately representing large-scale deformation.

An alternative model to explain the existence of the depocenters involves the existence of two or more smaller pull-aparts in the basin floor of the larger pull-apart (fig. 1E; Aydin and Nur 1982). The smaller pull-aparts occur diagonally across the basin and are each overlain by a discrete deep. Continued offset on the master faults results in coalescing of the smaller basins into a single wider basin (fig. 1E). The width of the overall pull-apart would be limited by initial width of the releasing bend.

In this interpretation, the Mirogoane Lakes Basin (fig. 9A) would consist of four individual pull-aparts beneath each one of the basin floor deeps rather than two coalescing pull-aparts with a deep at each end (fig. 1D); similarly, the Cariaco Basin and the Vienna Basin would consist of two individual pull-aparts connected by a strike-slip fault roughly in the center of the basin. The pattern is difficult to apply to the Dakar-Tiran and Arnona-Aragonese Basins in the Gulf of Aqaba (Elat) because the deeps occur either at the opposite corners expected or are not diagonally arranged (fig. 9C).

Distinguishing between basin floor subsidence related to combinations of normal and strike-slip faulting (e.g., fig. 1D) and multiple smaller pull-aparts (e.g., fig. 1E) remains difficult without more detailed structural and stratigraphic data. In either case, it can no longer be assumed that the center of a rhomboidal pull-apart is the site of greatest extension and subsidence (e.g., Crowell 1974, his fig. 8).

EXTREME PULL-APART DEVELOPMENT: THE CAYMAN TROUGH

With continued strike-slip offset, rhomboidal pull-apart basins may lengthen indefinitely into narrow oceanic basins whose basin length, or master fault overlap, can exceed basin width, or fault separation, by a factor of 10. The Cayman Trough, dominating the North America-Caribbean Plate Boundary Zone (fig. 4), is a unique example of an active pull-apart which has evolved until its master fault overlap of approximately 1400 km greatly exceeds its fault separation of 100–150 km. Although the exact age of nucleation is uncertain, the thickness of sediments and sedimentation rates of Cayman Trough sediments suggests a Late Eocene or Early Oligocene Age (Erickson et al. 1972); pelagic facies were deposited in the trough during the Oligocene (Wadge and Burke in press). The unsedimented eastern floor of the Cayman Trough (fig. 4) provides a unique glimpse into the early structural development of the basin.

Structural Pattern.—The Cayman Trough is presently lengthening by seafloor spreading along a short spreading ridge, the Mid-Cayman Rise, which connects the discontinuous ends of the sinistral master faults, the Oriente and Swan Fault Zones (Holcombe et al. 1973) (fig. 4). The active traces of the master faults form prominent linear deeps; fracture zones that extend beyond the master faults on the northwest and southeast sides of the trough are inactive and partially obscured by sediment fill. The active trace of the Enriquillo-Plantain Garden Fault Zone appears to merge with the southeastern fracture zone near 79°W. A short transform fault zone trending about 45° to the master faults appears to offset the spreading ridge by about 20 km (fig. 4). The fault zone is diffuse and has a width of about 35 km (Macdonald and Holcomb 1978).

The topography of the Cayman Trough in longitudinal profile is virtually indistinguishable from that of normal slow spreading ridges. The only significant difference is the greater depth to basement: the mean depth of the trough is 5 km. The geomorphology of the basement of the trough, which is covered by sediment only in its eastern third (fig. 4), consists of north-south trending bathymetric highs and basement faults which are symmetrically distributed about the Mid-Cayman Rise (Holcombe and Sharman in press). At about 77°W at the eastern end of the trough, basement trends change abruptly from a north-south orientation to a northeast-southwest orientation (fig. 4).

Magnetic anomalies in the central part of the trough (Macdonald and Holcomb 1977), and the rate of subsidence of the oceanic floor of the trough (Holcombe et al. 1973), suggest a full spreading rate of about 2 cm/yr. Identifiable anomalies are absent beyond the central third of the trough (fig. 4). In outer areas, isolated magnetic highs associated with dome-shaped bathymetric features suggest either unusually highly magnetized crust produced at a spreading center or off-axis volcanism (Macdonald and Holcombe 1978). The existence of a short transform offset of the Mid-Cayman Rise may account for distortion of generally north-south magnetic anomalies as well as variations in north-south basement trends.

Sedimentary Pattern.—Most of the clastic sediment derived from land masses to the north (Cuba) or south (Jamaica, Central America in Honduras) is trapped by linear deeps associated with either the transform or fracture zone margins of the basin (fig. 4). A large fan complex derived from Central America has prograded over the eastern end of the trough and has buried most of the basement floor in the west (fig. 4).

Comparison to Model Predictions.—The general orthogonal relationship of north-south basement trends produced by extension at the Mid-Cayman Rise to the trend of the master faults suggests the simple model of uniform pull-apart extension between roughly parallel master faults (fig. 1A). Northeast-southwest basement trends in the eastern end of the trough (fig. 4) are interpreted as oblique faults formed during early basin opening between master faults with little or no overlap (fig. 1B, 1D). Assuming a constant east-west direction of sinistral slip during early opening, the southern master fault north of Jamaica would be under compression in the manner shown in figure 1B; late Oligocene uplift and gradual southward tilting of Jamaica (Robinson 1971) may record compression related to the non-parallelism of the master faults in this area. With increased fault offset and overlapping of master faults, a rhomboidal basin was produced by stretching and orthogonal short ridge spreading. Symmetric variations in north-south spreading fabric may record the presence of an oblique transform offsetting the short spreading ridge (Macdonald and Holcombe 1978). Similar short ridge offsets are found in much younger basins in the Gulf of California (fig. 2A). Symmetric variations in the width of the trough may result from sudden changes in the direction of relative plate motion (Holcombe and Sharman in press) (fig. 4).

The large length to width ratio of the Cayman Trough (~14) does not suggest significant long-term widening of the trough by coalescing mechanisms (Aydin and Nur 1982) (fig. 1E), although a much smaller pull-apart along the parallel Enriquillo-Plantain Garden Fault Zone (Holcombe and Sharman in press) appears to be coalescing with the southern margin of the Cayman Trough near 79°W (fig. 4).

THERMAL ASPECTS OF PULL-APART BASINS

Localized crustal stretching and lithospheric extension at pull-aparts produces high heat flow and volcanism. Gradually declining thermal subsidence of the lithosphere around larger pull-aparts results in extensive overlying sedimentary basins similar to those formed above rifts. Thick sediments in most pull-aparts tends to: (1) reduce predicted heat flow values; (2) prevent magmas from reaching the surface; and (3) cause magnetic anomalies produced at orthogonal short ridge segments to be indistinct.

Pattern of Magmatism.—Because of high rates of sedimentation, volcanoes are found only in a small percentage of active pull-apart basins. The composition of volcanic rocks reflects their tectonic environment: alkali basalts and tholeiites are formed along continental strike-slip boundaries (fig. 2) while calc-alkaline magmas are typical of strike-slip zones behind arcs and in areas of continental collision.

There seems to be no systematic pattern for the distribution of volcanoes in pull-aparts. In some cases, volcanoes do not occur within the pull-apart basin or directly above a buried pull-apart. For example, late Pleistocene to recent rhyolite domes in the Coso Range, California, occur both within and up to 8 km outside of an inferred pull-apart structure (Weaver and Hill 1978/79). Five Quaternary rhyolite domes in the Salton Trough (fig. 6A) do not directly overlie buried pull-aparts but rather are spaced 2–3 km apart and appear to lie along a 7 km long northeast trending lineament (Robinson et al. 1976). In the rhomboidal Erzincan Basin along the North Anatolian Fault Zone (Aydin and Nur 1982, their fig. 2D) about a dozen andesitic volcanic cones and hot springs are found along the traces of both master faults. Late Miocene submarine fissure eruptions in the uplifted Low Layton pull-apart area in Jamaica (Mann and Burke 1980) are arranged roughly *en echelon* and vary in orientation from 45° to parallel with the inferred overlapping master faults (Wadge 1982).

The locus of magmatism with Z-shaped and rhomboidal basins in the Gulf of California (fig. 2A) and the Andaman Sea (Curray et al. 1978) appears to be along orthogonal spreading centers situated roughly in the center of the basin. The short ridges consist of a narrow bathymetric trough often separated (as in the case of the Mid-Cayman Rise) by a short transform segment. Deep sea drilling results in the Guaymas Basin (G in Fig. 2A; Einsele et al. 1980) indicate that basaltic magma forming the new oceanic basement of the pull-apart is intruded as sills rather than extruded as pillow basalt because of several hundred meters of sediment overlying the spreading center. The highest heat flow values measured along the central trough in the Guaymas Basin are about half the value predicted for typical spreading centers (Lawver and Williams 1979). The oceanic floor of the basin may have been formed at multiple sites of intrusion, because sediment thicknesses are uniform and do not thin towards the present central trough (Bischoff and Henyey 1974). In nearby basins like the Farallon (F in fig. 2A), the highest heat flow values are found 15–20 km from either side of the axis of the central trough. Magnetic anomalies are not found in pull-aparts where the sedimentation rate is high because the magma does not reach the surface as highly magnetized pillow basalts.

Comparison to Model Predictions.—Large sediment thicknesses in most active pull-aparts create a variable pattern of volcanism. Most magma is probably intruded as sills but may occasionally reach the surface along the master faults (e.g., Erzincan Basin) or other steep faults cutting the basin sediments.

The existence of short ridge segments in the Gulf of California and the Andaman Sea indicates that extension of some S or Z-shaped and rhomboidal basins results in two or more smaller pull-aparts linked by short transforms (Aydin and Nur 1982) (fig. 1E). This pattern is suggested to develop at releasing bends which make a high angle to the theoretical interplate slip vector (fig. 3B). The variation in heat flow and the uniformity of sediment thickness in some of the Gulf of California basins suggest that these spreading centers are not steady-state features but migrate over time. There seems to be no clear relationship between the amount of master fault overlap or offset and the onset of igneous activity. The large discrepancy between predicted and observed heat flow in pull-aparts can be attributed to (1) the insulating effect of thick overlying sediments; (2) lateral conductive heat loss to the cold sides of the basin (this would be particularly important for smaller basins); and (3) heat transfer mechanisms involving hydrothermal systems (Lawver and Williams 1979).

Pattern of Thermal Subsidence in the Magdalen Basin.—McKenzie (1978) has shown that the subsidence history of many rift basins can be modeled in terms of an initial phase of stretching followed by an episode of gradually declining thermal subsidence. The stretching phase is characterized by rapid, coarse clastic sedimentation and volcanism within a fault bounded basin and is followed by a slower phase during which the basin expands beyond its original bounding faults. The stretching model can be applied to some older, larger pull-aparts to explain their two-phase subsidence history.

Isopachs and geologic evidence indicate that the 250,000 km^2 Carboniferous Magdalen Basin opened as a pull-apart in the Gulf of St. Lawrence between dextral faults in Newfoundland and New Brunswick (Bradley 1982) (fig. 10). The existence and extent of the rhomboidal pull-apart is inferred from drill holes, geophysical studies, and limited exposures on islands in the Gulf of St. Lawrence (fig. 10). These studies have indicated that about 7 km of clastic sediments, evaporites, and volcanics were deposited during a stretching phase that lasted about 20 Ma. Later Carboniferous clastic deposition was not fault controlled and resulted in progressive onlap of basement as far west as central New Brunswick (fig. 10). Reconstructed profiles across the Magdalen Basin suggest that subsidence during this second phase was greater over the buried fault-bounded basin (fig. 10, inset). Continued strike-slip displacement south and east of the Magdalen Basin during this phase resulted in more complex sedimentary patterns there.

The two stage sedimentary history of the Magdalen Basin conforms well with predictions based on the McKenzie stretching model: stretching of the lithosphere within the rhomboidal pull-apart was accompanied by crustal thinning and increased heat flow; following the rift phase, recovery of isotherms caused slower subsidence of an enlarged sedimentary basin that buried the pull-apart.

The application of the stretching model to rhomboidal pull-aparts with two or more recognizable depocenters should consider local variations in stretching and subsidence related either to localized areas of normal faulting (fig. 1D) or to smaller pull-aparts (fig. 1E).

DEFORMATION OF PULL-APART BASINS

The most recent phase of strike-slip displacement produces a new generation of pull-

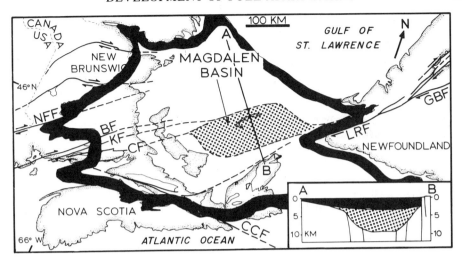

FIG. 10.—Fault and sediment map of Carboniferous Magdalen Basin at a right-step between the dextral Long Range Fault (LRF) in Newfoundland and the Belleisle (BF), Kennebacasis (KF) and Caledonia (CF) Faults in New Brunswick. Dotted area represents sediments deposited in early Carboniferous rhomboidal pull-apart; grey area represents limit of medial to late Carboniferous clastic rocks which progressively onlapped basement rocks. Inset shows "steer's head" profile of Magdalen Basin with effects of late Paleozoic salt diapirism removed. Other abbreviations are: NNF Norumbega-Fredericton Fault; CCF Cobequid-Chedabucto Fault; GBF Green Bay Fault.

apart basins by the types of mechanisms shown in figure 3; the previous generation of pull-aparts often becomes completely obliterated. Although remnants of older strike-slip basins are commonly reported in active and ancient strike-slip zones, a pull-apart origin is clear only for a few ancient basins, like the Vienna Basin, which is virtually undeformed and still slightly active (Royden et al. 1982).

Aydin and Nur (1982) have shown that 64 active pull-aparts have length to width ratios of approximately 3. Although some of their measurements may include basins which are not pull-aparts (see fig. 4), the general observation is convincing. We suggest that the constant 3:1 dimension ratio of active basins may reflect the tendency for older, longer basins with higher ratios to be destroyed in rapidly changing strike-slip environments. Only in very rare cases, such as the Cayman Trough, is the direction of relative motion constant enough to produce long basins with length to width ratios greater than 3.

For example, along the Dead Sea Fault System (fig. 2B), the cumulative strike-slip offset for the last 40 m.y. is 105 km. The active pull-aparts presently found along the fault (fig. 2B) were created by the last 40 km of slip, which probably occurred in the Plio-Plistocene (Garfunkel 1981). Pull-aparts

created during the previous 65 km of displacement are poorly known because they have either been buried by subsequent sedimentation or deformed.

Several different modes of deformation may affect pull-apart basins. Perhaps the simplest is strike-slip offset of the basin in a direction subparallel to the master faults. For example, the Carboniferous Moncton Basin in New Brunswick appears to have formed in the latest Devonian as a rhomboidal pull-apart and was subsequently offset by strike-slip faults into three slices (Bradley and Rowley 1983). More complex deformation involving lateral offset, shortening, and large rotations appears to occur at push-up areas like the Transverse Ranges in California (fig. 2A). A pull-apart origin for Miocene volcanics and sediments on the northern (Hall 1981) and southern (Crowell 1976) flanks of the Transverse Ranges has been proposed but the basin and master fault geometries are not entirely clear. A pull-apart origin appears more certain for the Miocene Low Layton volcanics in Jamaica (Wadge 1981; Mann and Burke 1980) because both the volcanics and master faults were not severely deformed during the uplift of the eastern Jamaica push-up block (Mann et al. 1981). Compressive deformation of pull-aparts may occur quickly and over greater lengths of the plate boundary zone during sudden changes in the overall direction of plate motion (Dewey 1975) (fig. 3A). Holcombe and Sharman (in press) have suggested that symmetrical variations in width of the Cayman Trough (fig. 4) were produced by variations in the direction of relative plate motion: periods of divergent plate motion or transtension widened the trough, while periods of convergent plate motion or transpression narrowed the trough.

Localized, rapid sedimentation and volcanism do not by themselves constitute conclusive evidence for an ancient pull-apart basin setting. Both may occur in different types of strike-slip basin. Convincing evidence for an ancient pull-apart basin consists of a reasonably shaped basin (e.g., S or Z-shaped, rhomboidal) between properly stepping faults which can be shown to be strike-slip and active at the time of basin subsidences. Because of overprinting of later strike-slip deformation, ancient pull-apart basin shapes and master fault patterns are only recognizable after careful reconstruction of sedimentary facies. Because of uplift and erosion of parts of many strike-slip basins, a pull-apart origin may not always be possible to demonstrate.

SUMMARY

Pull-apart basins, like most other structures in the Earth's crust, do not suddenly come into existence but evolve through a sequence of closely related states. Structural, sedimentary, and thermal characteristics of well mapped active and ancient basins can be generalized into a *continuum* model of pull-apart development (fig. 11). Pull-apart basins are valuable indicators of relative plate motions in active and ancient strike-slip environments and are now recognized as being important sites of mineral, hydrocarbon, and geothermal resources.

Active pull-aparts are found at strike-slip fault discontinuities in all types of strike-slip settings: (1) long, strike-slip boundary zones between rigid, continental plates (fig. 2); (2) strike-slip zones found in response to oblique subduction at arcs; and (3) strike-slip systems bounding "escaping" continental fragments in convergent zones (fig. 3C, 3D). In rigid intracontinental settings, larger active pull-aparts do not occur singly but rather as part of a series on a segment of the principal dis-

placement fault zone which is oblique to theoretical interplate slip lines (fig. 2). Wider pull-aparts form at fault segments that (1) are significantly non-parallel (>10°) to the theoretical interplate slip lines, (2) tend to be closely spaced, and (3) are more prone to coalescing with neighboring basins (figs. 11A, 2). Conversely, narrower basins occur along fault segments that diverge only slightly (<10°) from their theoretical interplate slip lines; they tend to be widely spaced and are not prone to coalescing (figs. 11A, 4). Divergence between the strike-slip fault and the direction of interplate slip results from the original oblique orientation of the strike-slip and/or more complex tectonic mechanisms (fig. 3). Plate tectonics is less successful in predicting the distribution and origin of pull-aparts on shorter strike-slip faults in less rigid active arc or collisional settings.

Pull-apart basins nucleate at releasing bend fault segments along active strike-slip faults (fig. 5); initial opening across the bend produces spindle-shaped basins defined and often bisected by oblique-slip faults connecting the ends of the master faults (figs. 11B, 6). Several pull-aparts may nucleate simultaneously across sharper releasing bends which are significantly nonparallel to the theoretical interplate slip lines (fig. 6A). Nonparallelism of master strike-slip faults produces overlaps and gaps of the basin edges and can result in local basin edge deformation (fig. 7A). Most active nucleating basins do not form prominent topographic depressions because of the limited amount of opening. Their surficial structural pattern is approximately consistent with theoretical models of pull-apart faulting (Rodgers 1980; Segall and Pollard 1980).

Increased offset on master strike-slip faults produces active pull-apart basin shapes which we colloquially call "lazy S" shape for basins between sinistral faults (fig. 11C) and "lazy Z" shape for basins between dextral faults (fig. 8). Increased opening produces deep topographic depressions between oblique-slip faults defining the edges of the basin. Extension of the basement beneath S and Z-shaped sedimentary basins may occur across two or more smaller pull-aparts at sharper releasing bends with larger fault separation (fig. 11C). In some cases, smaller basement pull-aparts may be the sites of roughly orthogonal sea floor spreading (e.g.,

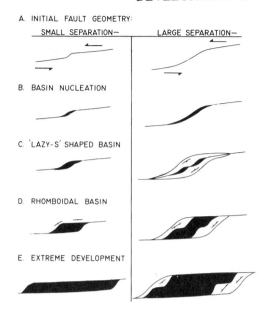

A. INITIAL FAULT GEOMETRY:

SMALL SEPARATION—   LARGE SEPARATION—

B. BASIN NUCLEATION

C. 'LAZY-S' SHAPED BASIN

D. RHOMBOIDAL BASIN

E. EXTREME DEVELOPMENT

F. THERMAL SUBSIDENCE / DEFORMATION

FIG. 11.—*Continuum* model of pull-apart basin development generalized from structural, sedimentary, and thermal characterics of well mapped active and ancient pull-aparts described in this paper. *A*. Pull-aparts nucleate at releasing fault bends along faults which are oblique to the direction of relative plate motion (indicated by arrows); larger bends or areas of greater strike-slip fault separation will develop into wider basins. *B*. Spindle-shaped pull-aparts nucleate on releasing bends. *C*. Continued offset produces "lazy S" sedimentary basin geometry; extension across bends with large fault separation may be relayed by two or more buried pull-aparts (shown in grey). *D*. Continued offset of lazy S-shaped basin produces rhomboidal or "rhomb graben" basin geometry; buried pull-aparts at bends with large fault separation may coalesce to form a basin whose width remains fixed by the original width of the releasing bend; volcanoes (shown by black dots) may be erupted into the basin. *E*. With steady offset, rhomboidal basins can develop into long narrow troughs which at some point develop a short spreading ridge and are floored by oceanic crust; distal ends preserve early opening structures. *F*. Gradually declining thermal subsidence following lithospheric extension at larger pull-aparts results in extensive overlying sedimentary basins similar to those formed above rifts. Deformation in rapidly changing strike-slip environments generally terminates pull-apart development after the basin length triples the width of the initial releasing bend.

Guaymas Basin (G) in fig. 2*A*). The observed shapes and some sedimentary thicknesses of S and Z-shaped pull-aparts are in approximate agreement with theoretical models of pull-apart faulting (e.g., fig. 1*D*).

Rhomboidal pull-aparts or "rhomb grabens" result from lengthening of an S or Z-shaped basin with increased strike-slip offset (fig. 11*D*). Increased extension in rhomboidal pull-aparts causes them to be prominent fault-bounded depressions which are frequently below sea level. The floor of active rhomboidal pull-aparts is often occupied by roughly circular depressions or deeps (fig. 9) which suggest that extension between the overlapped master faults is not uniform. The pattern of two deeps in the distal ends of rhomboidal basins is predicted by the theoretical pull-apart fault model of Rodgers (1980), in which more rapid subsidence occurs within two sub-circular zones of normal faulting (fig. 1*D*). An alternative model to explain the presence of two or more deeps requires the existence of a diagonal array of smaller pull-aparts which relay extension across the floor of the basin (fig. 1*E*). This structural pattern is predicted to occur in rhomboidal basins with larger fault separation (fig. 11*D*). Because of thick sediment cover, it is difficult to distinguish between the two structural patterns in active basins. Although volcanoes are occasionally found in rhomboidal pull-aparts, there seems to be no clear relationship between the onset of volcanism and the amount of strike-slip fault overlap.

With continued strike-slip offset and a constant direction of relative plate motion, rhomboidal pull-aparts infrequently develop into narrow oceanic basins, like the Cayman Trough (fig. 4). Extension occurs across a central orthogonal spreading ridge. Most rhomboidal pull-aparts have average length/width ratios of less than 4 (Aydin and Nur 1982) and reflect the tendency of older pull-aparts to be deformed in rapidly changing strike-slip environments.

Our general model of pull-apart basin development (fig. 11) incorporates some but not all aspects of previous models (fig. 1). Pull-aparts do not appear to nucleate as *en echelon* tension gashes or Riedel shear structures which are observed in outcrop and in shear box experiments (fig. 1*C*). Instead, pull-aparts nucleate on oblique fault segments which constitute releasing fault bends (Freund 1971; Crowell 1974) (fig. 1*B*). Pull-aparts do not steadily widen as they lengthen (Aydin and Nur 1982); their widths remain fixed by the width of the releasing bend fault segment (fig. 11*A*). At sharper releasing bends with greater strike-slip fault separation (fig. 11*A*), smaller pull-aparts may coalesce to form a wider basin (fig. 11*D*), but the maximum width of that basin remains fixed by the initial fault separation.

ACKNOWLEDGMENTS.—We are grateful for helpful reviews by D. Rodgers and H. Reading. Mann thanks F. Maurrasse, C. Jean-Poix and the Haitian Department of Mining and Energy Resources for field support in Haiti. This work was supported by the NASA Geodynamics Program (NAG5155 and NAS527227) and a GSA Grant to Bradley. Part of the research was done while Burke was at the Lunar and Planetary Institute (LPI contribution 505).

REFERENCES CITED

AYDIN A., and NUR, A., 1982, Evolution of pull-apart basins and their scale independence: Tectonics, v. 1, p. 91–105.

BALLANCE, P. F., 1980, Models of sediment distribution in non-marine and shallow marine environments in oblique-slip zones, *in* BALLANCE, P. F., and READING, H. G., eds., Sedimentation in oblique-slip mobile zones: Int. Assoc. Sed. Spec. Pub. 4, p. 229–236.

BELT, E. S., 1968, Post-Acadian rifts and related facies, eastern Canada, *in* ZEN, E-AN; WHITE, W. S.; HADLEY, J. B.; and THOMPSON, J. B., eds., Studies in Appalachian geology, Northern and Maritime: New York, Interscience, p. 95–113.

BEN-AVRAHAM, Z.; ALMAGOR, G.; and GARFUNKEL, Z., 1979, Sediments and structure of the Gulf of Elat: Sediment. Geol., v. 23, p. 239–267.

BISCHOFF, J. L., and HENYEY, T. L., 1974, Tectonic elements of the central part of the Gulf of California: Geol. Soc. America Bull., v. 85, p. 1893–1904.

BRADLEY, D. C., 1982, Subsidence in Late Paleozoic basins in the Northern Appalachians: Tectonics, v. 1, p. 107–123.

———, and ROWLEY, D. B., 1983, Was Carboniferous strike-slip in the northern Appalachians dextral or sinistral?: Geol. Soc. America (Abs. with Prog.), v. 15, p. 143.

BURCHFIEL, B. C., and STEWART, J. H., 1966, "Pull-apart" origin of the central segment of Death Valley, California: Geol. Soc. America Bull., v. 77, p. 439–442.

BURKE, K.; GRIPPI, J.; and SENGOR, A. M. C., 1980, Neogene structures in Jamaica and the tectonic style of the Northern Caribbean Plate Boundary Zone: Jour. Geology, v. 88, p. 375–386.

———; MANN, P.; and KIDD, W., 1982, What is a ramp valley?: Abstract volume, 11th Int. Congr. Sediment.: McMaster Univ., Hamilton, Ontario, p. 40.

CAREY, S. W., 1958, A tectonic approach to continental drift; *in* CAREY, S. W., ed., Continental Drift: A Symposium: Univ. Tasmania, p. 177–355.

CASE, J. E., and HOLCOMBE, T. L., 1980, Geologic-tectonic map of the Caribbean region: U.S. Geol. Survey Misc. Invest. Ser., map I-1100.

CLAYTON, L., 1966, Tectonic depressions along the Hope Fault, a transcurrent fault in North Canter-

bury, New Zealand: N.Z. Jour. Geol. Geophys., v. 9, p. 95–104.

CROWELL, J. C., 1974, Origin of late Cenozoic basins in southern California, *in* DOTT, R. H., and SHAVER, R. H., eds., Modern and ancient geosynclinal sedimentation: SEPM Spec. Pub. 19, p. 292–303.

——— 1976, Implications of crustal stretching and shortening of coastal Ventura Basin, California: Pacif. Sect. Am. Ass. Petrol. Geol. Misc. Pub. 24, p. 365–382.

——— 1981, Juncture of the San Andreas transform system and Gulf of California rift: Ocean. Acta., Proc. 26th Int. Geol. Congr., Paris, p. 137–141.

CURRAY J. R.; MOORE, D. G.; LAWVER, L. A.; EMMEL, F. J.; RAITT, R. W.; HENRY, M.; and KIECKHEFFER, 1979, Tectonics of the Andaman Sea and Burma, *in* WATKINS, J. S.; MONTADERT, L; and DICKERSON, P. W., eds, Geological and geophysical investigations of continental margins: Am. Ass. Petrol. Geol. Mem. 29, p. 189–198.

DEWEY, J. F., 1975, Finite plate motions: some implications for the evolution of rock masses at plate margins: Am. Jour. Sci., v. 275-A, p. 260–284.

——— 1978, Origin of long transform-short ridge systems: Geol. Soc. America (Abs. with Prog.), v. 10, p. 388.

——— 1980, Episodicity, sequence, and style at convergent plate boundaries, *in* STRANGWAY, D. W., ed., The continental crust and its mineral deposits: Geol. Assoc. Can. Spec. Paper 20, p. 553–573.

EINSELE, G.; GIESKES, J. M.; and others, 1980, Intrusion of basaltic sills into highly porous sediments and resulting hydrothermal activity: Nature, v. 283, p. 441–445.

ERICKSON, A. J.; HELSLEY, C. E.; and SIMMONS, G., 1972, Heat flow and continuous seismic profiles in the Cayman Trough and Yucatan Basin, Caribbean Sea: Geol. Soc. America Bull., v. 84, p. 2133–2138.

FISHER, M. A.; PATTON, W. W., JR.; THOR, D. R.; HOLMES, M. L.; SCOTT, E. W.; NELSON, C. H.; and WILSON, C. L., 1979, The Norton Basin of Alaska: Oil and Gas Jour., v. 77, p. 96–98.

FITCH, T. J., 1972, Plate convergence, transcurrent

faults, and internal deformation adjacent to southeast Asia and the western Pacific: Jour. Geophys. Res., v. 77, p. 4432–4460.

FREUND, R., 1971, The Hope Fault: a strike-slip fault in New Zealand: N. Z. Geol. Surv. Bull. (n.s.), v. 86, p. 1–49.

GARFUNKEL, Z., 1981, Internal structure of the Dead Sea leaky transform (rift) in relation to plate kinematics: Tectonophysics, v. 80, p. 81–108.

———; ZAK, I.; FREUND, R., 1981, Active faulting in the Dead Sea Rift: Tectonophysics, v. 80, p. 1–26.

HALL, C. A., JR., 1981, San Luis Obispo Transform Fault and middle Miocene rotation of the western Transverse Ranges, California: Jour. Geophys. Res., v. 86, p. 1015–1031.

HARLAND, W. B., 1971, Tectonic transpression in Caledonian Spitsbergen: Geol. Mag., v. 108, p. 27–42.

HEMPTON, M. R., 1982, The North Anatolian Fault and complexities of continental escape: Jour. Struct. Geol., v. 4, p. 502–504.

———; DUNNE, L. A.; and DEWEY, J. F., 1983, Sedimentation in a modern strike-slip basin, southeastern Turkey: Jour. Geology, v. 91, p. 401–412.

HENYEY, T. L., and BISCHOFF, J. L., 1973, Tectonic elements of the northern part of the Gulf of California: Geol. Soc. America Bull., v. 84, p. 315–330.

HILL, M. L., and TROXEL, B. W., 1966, Tectonics of Death Valley region, California: Geol. Soc. America Bull., v. 77, p. 435–438.

HOLCOMBE, T. L.; VOGT, P. R.; MATTHEWS, J. E.; and MURCHISON, R. R., 1973, Evidence for sea-floor spreading in the Cayman Trough: Earth Planet. Sci. Letters, v. 20, p. 357–371.

———, and SHARMAN, G. F., 1983, Post-Miocene North American-Caribbean motion: evidence from the Cayman Trough: Geology, in press.

HOOKE, R. LEB., 1972, Geomorphic evidence for late-Wisconsin and Holocene tectonic deformation, Death Valley, California: Geol. Soc. America Bull., v. 83, p. 2073–2098.

JOHNSON, C. E., and HADLEY, D. M., 1976, Tectonic implications of the Brawley earthquake swarm, Imperial Valley, California, January, 1975: Bull. Seismol. Soc. America, v. 66, p. 1133–1144.

JORDAN, T. H., 1975, The present-day motions of the Caribbean plate: Jour. Geophys. Res., v. 80, p. 4433–4499.

KLEMME, H. G., 1971, To find a giant, find the right basin: Oil and Gas Jour., v. 69, no. 10, p. 103–110.

KOIDE, H., and BHATTACHARJI, S., 1977, Geometric patterns of active strike-slip faults and their significance as indicators for areas of energy release, in SAXENA, S. K., ed., Energetics of Geological Processes: New York, Springer Verlag, p. 46–66.

LAWVER, L. A., and WILLIAMS, D. L., 1979, Heat flow in the central Gulf of California: Jour. Geophys. Res., v. 84, p. 3465–3478.

LEPICHON, X., and FRANCHETEAU, J., 1978, A plate-tectonic analysis of the Red Sea-Gulf of Aden area: Tectonophysics, v. 46, p. 369–406.

MACDONALD, K. C., and HOLCOMBE, T. L., 1978, Inversion of magnetic anomalies and sea-floor spreading in the Cayman Trough: Earth Planet. Sci. Letters, v. 40, p. 407–414.

MANN, P., and BURKE, K., 1980, Neogene wrench faulting in the Wagwater Belt, Jamaica: Trans. Caribbean Geol. Conf., 7th, Santo Domingo, Dominican Republic, p. 95–97.

———, and ——— 1982, Basin formation at intersections of conjugate strike-slip faults: examples from southern Haiti: Geol. Soc. America (Abs. with Prog.), v. 14, p. 555.

———; ———; DRAPER, G.; and DIXON, T., 1981, Tectonics and sedimentation at a Neogene strike-slip restraining bend, Jamaica: EOS, v. 62, p. 1051.

McKENZIE, D., 1972, Active tectonics of the Mediterranean region: Geophys. Jour. Royal Astron. Soc., v. 30, p. 109–185.

——— 1978, Some remarks on the development of sedimentary basins: Earth Planet. Sci. Letters, v. 40, p. 25–32.

MENARD, H. W., and ATWATER, T. W., 1968, Changes in direction of sea floor spreading: Nature, v. 219, p. 463–467.

MERCER, A. M., and FREUND, R., 1974, Transcurrent faults, beam theory, and the Marlborough Fault System, New Zealand: Geophys. Jour. Royal Astron. Soc., v. 38, p. 553–562.

MINSTER, J. B.; JORDAN, T. H.; MOLNAR, P.; and HAINES, E., 1974, Numerical modelling of instantaneous plate tectonics, Geophys. Jour. Royal Astron. Soc., v. 36, p. 541–576.

———, and ——— 1980, Present-day plate motions: a summary, in ALLEGRE, C. J., ed., Source Mechanism and Earthquake Prediction: Paris, Centre National de la Recherche Scientifique, p. 109–124.

MUEHLBERGER, W. R., 1981, The splintering of the Dead Sea Fault Zone in Turkey: Bull. Inst. Earth Sci. of Hacettepe Univ., Ankara, Turkey, v. 8, p. 125–130.

NIEMITZ, J. W., and BISCHOFF, J. L., 1981, Tectonic elements of the southern part of the Gulf of California: Geol. Soc. America Bull., v. 92, Part II, p. 360–407.

PAGE, B. G. N.; BENNETT, J. D.; CAMERON, N. R.; BRIDGE, D. McC.; JEFFREY, D. H.; KEATS, W.; and THAIB, P., 1979, A review of the main structural and magmatic features of northern Sumatra: Jour. Geol. Soc. London, v. 136, p. 569–579.

QUENNELL, A. M., 1958, The structural and geomorphic evolution of the Dead Sea Rift: Quat. Jour. Geol. Soc. London, v. 114, p. 2–24.

READING, H. G., 1980, Characteristics and recognition of strike-slip fault systems, in BALLANCE, P. F., and READING, H. G., eds., Sedimentation in oblique-slip mobile zones: Int. Assoc. Sed. Spec. Pub. 4, p. 7–26.

ROBINSON, E., 1971, Observations on the geology of the Jamaican bauxite: Jour. Geol. Soc. Jamaica Bauxite Spec. Issue, p. 3–9.

25

ROBINSON, P. T., and ELDERS, W. A., 1976, Quaternary volcanism in the Salton Sea geothermal field, Imperial Valley, California: Geol. Soc. America Bull., v. 87, p. 347–360.

RODGERS, D. A., 1980, Analysis of pull-apart basin development produced by *en echelon* strike-slip faults, *in* BALLANCE, P. F., and READING, H. G., eds. Sedimentation in oblique-slip mobile zones: Int. Assoc. Sed. Spec. Pub. 4, p. 27–41.

ROYDEN, L. H.; HORVATH, F.; and BURCHFIEL, B. C., 1982, Transform faulting, extension, and subduction in the Carpathian Pannonian region: Geol. Soc. America Bull., v. 93, p. 717–725.

SCHOLZ, C. H., 1977, Transform fault systems of California and New Zealand: similarities in their tectonic and seismic styles: Jour. Geol. Soc. London, v. 133, p. 215–229.

SCHUBERT, C., 1982, Origin of the Cariaco Basin, southern Caribbean Sea: Marine Geol., v. 47, p. 345–360.

SEAGALL, P., and POLLARD, D. D., 1980, Mechanics of discontinuous faults: Jour. Geophys. Res., v. 85, p. 4337–4350.

SENGÖR, A. M. C., and YILMAZ, Y., 1981, Tethyan evolution of Turkey: a plate tectonic approach: Tectonophysics, v. 75, p. 181–241.

SEYMEN, I., 1975, Kelkit Valdisi kesiminde Kuzey Anadolu Fay zonunun tektonik özelliği: Unpub. Ph.D. dissertation, Istanbul Technical University, Istanbul.

SHARMAN, G. F.; REICHLE, M. S.; and BRUNE, J. N., 1976, Detailed study of relative plate motion in the Gulf of California: Geology, v. 4, p. 206–210.

SHARP, R. V., 1975, *En echelon* fault patterns of the San Jacinto Fault Zone, *in* CROWELL, J. C., ed., San Andreas Fault in southern California: California Div. Mines Geol. Spec. Rep. 118, p. 147–152.

—— 1976, Surface faulting in Imperial Valley during the earthquake swarm of January-February, 1975: Bull. Seismol. Soc. America, v. 66, p. 1145–1154.

STEEL, R., and GLOPPEN, T. G., 1980, Late Caledonian (Devonian) basin formation, western Norway: signs of strike-slip tectonics during infilling, *in* BALLANCE, P. F., and READING, H. G., eds., Sedimentation in oblique-slip mobile zones: Int. Assoc. Sed. Spec. Pub. 4, p. 79–103.

TIJA, H. D., 1977, Tectonic depressions along the transcurrent Sumatra Fault Zone: Geologi Indonesia, v. 4, p. 13–27.

WADGE, G., 1982, A Miocene submarine volcano at Low Layton, Jamaica: Geol. Mag., v. 119, p. 193–199.

——, and BURKE, K., 1983, Neogene Caribbean plate rotation and associated Central American tectonic evolution: Tectonics, in press.

WEAVER, C. S., and HILL, D. P., 1978/1979, Earthquake swarms and local crustal spreading along major strike-slip faults in California: Pure Appl. Geophys., v. 117, p. 51–64.

WILLIS, B., 1928, Dead Sea problem: rift valley or ramp valley?: Geol. Soc. America Bull., v. 39, p. 490–542.

Development of extension and mixed-mode sedimentary basins

A. Gibbs

SUMMARY: Analysis of onshore and offshore basins using a variety of techniques including deep seismic reflection profiles has shown that basins develop above linked fault systems. These fault systems comprise both steep and gently dipping faults which can have either dip- or strike-slip displacements. Whilst linked fault systems are common to all basins they are developed to different degrees and hence produce a variety of structural expressions within, and between the basins. This allows basins to be classified either in terms of their dominant structural elements or seen as a mechanical and geological continuum.

As the basins grow, changes in crustal thickness in the upper brittle crust can take place, on staircase arrays of low-angle listric faults, on steeper planar (domino) faults or commonly on a combination of both linked by crustally conservative strike-slip systems.

The development of the controlling basement elements affects the distribution and pattern of internal deformation in the evolving sedimentary fill resulting in a characteristic basin architecture. This resulting architecture has profound implications for models of the stratigraphy and sedimentology of basins, and provides an insight into the occurrence of certain facies and stratigraphic changes.

A general model for the evolution of such basin architectures is presented in this paper including both the basement structural elements and the overlying carapace of sediments with both stratigraphic and structural components.

Sedimentary basins and their component fault systems have been described by a number of workers using observational data such as the deep seismic profiles acquired in the UK by the BIRPS group and in the United States by the COCORP group, from theoretical modelling and from field and industry data (see for example Allmendinger *et al.* 1983; Brewer & Smythe 1984; Chadwick *et al.* 1984; Cooke *et al.* 1981; Finckh *et al.* 1984; Graciansky *et al.* 1985; Harding 1984; Jackson & McKenzie 1983; McKenzie 1978). Such work suggests that we can distinguish different tectono-structural categories of basins which operate by extending on linked fault systems. Although apparently simple end members exist, few, if any, basins can be described completely in terms of a single structural style. All, to a greater or lesser extent, must share the same architectural elements. Changes in plate stresses with time and rotational strains may in some cases mean that a basin progresses from one end member to another and complex mixed-mode systems can occur. A first-order classification of basins into extensional on listric faults, extensional on domino faults (pure shear), strike-slip (simple shear), and mixed-mode (general strain) systems is useful. This classification serves to focus attention on the dominant kinematics of the basin architecture and also on the way in which the sedimentary fill may have responded to, and been involved in, basin development. In the UK,

on and offshore there is increasing evidence from commercial seismic data that most basins are not simply extensional but are of mixed-mode types (transpressional and transtensional basins, cf. Sanderson & Marchini 1984; Harland 1971).

Basin architecture

As the basement structure evolves with time the sedimentary fill becomes involved in the crustal deformation process as a 'carapace' overlying the basement architecture. The term 'carapace' is used to include both the sedimentary fill with its stratigraphic elements such as unconformities, growth faults etc. and its developing tectonic structure. Hence a basin will consist of some controlling basement structural pattern, a sediment and deformed sediment shell, the 'carapace', and an overlying undeformed or relatively little deformed 'upper carapace'. The latter corresponding to sedimentation during thermal subsidence of the basin. Together these elements make up the architecture of the basin. In North Sea terms, this division can be seen between basement structures in the upper Palaeozoic and older 'basement', carapace development through the Jurassic and Early Cretaceous, and subsidence phase from mid-Cretaceous to Tertiary times (Sclater & Christie 1980; Gibbs 1984a, figs 8 & 9). This carapace deformation interacts

From COWARD, M.P., DEWEY, J.F., & HANCOCK, P.L. (eds), 1987, *Continental Extensional Tectonics*, Geological Society Special Publication No. 28, pp. 19-33.

19

with the sedimentary processes resulting in the control of the development of sedimentary facies by the tectonic processes which in turn may inherit their style from an earlier basement structure.

A simplified North Sea model for the internal collapse of a sediment wedge as the hanging wall continues to move off on the extension fault is shown in Fig. 1. The sediment may collapse on a series of faults detaching in a low-strength layer such as shale or salt (see Gibbs 1984a) and the roll-overs produced in this way will control the distribution of continuing marginal and axial deposition. This idea can be taken a stage further where a stacked set of detachments develops in a half-graben with zones of shale flowage allowing disharmonic faulting. As the basin margin grows and subsides the sediment overlying these internally faulted wedges may show a variety of gravity driven structures and further complicate margins such as those described by Todd & Mitchum (1977) and Graciansky et al. (1985, fig. 16).

Basins dominated by ramp and flat 'staircase' fault arrays

The work of the BIRPS group published on the MOIST reflection profile (Brewer & Smythe 1984) demonstrates extension on older, reactivated thrust faults producing simple half-graben basins. Such extensional basins develop by dip-slip on staircase-like detachment faults made up of flats and ramps. This is similar to models of some of the Basin and Range type of structure such as the Sevier Desert basin (Allmendinger et al. 1983) and the Diamond Valley basins (Effimoff & Pinezich 1981). Gibbs discussed the structural evolution of basins of this type in terms of sequential development of hanging wall synclines and footwall collapse to produce a complex graben cross-section (Gibbs 1984b fig. 9). Where lower ramps collapse, core complex geometries can be generated and extensional chaos zones (Wernicke & Burchfiel 1982) and extensional duplexes will be common.

The characteristic of such basins is that they are controlled by low-angle extensional faults which remain active throughout basin development. The sole detachment of the staircase must eventually penetrate the brittle upper part of the crust, the 'elastic lid' and detach in a sub-horizontal ductile shear zone. Such fault systems are, as pointed out by Gibbs (1983, 1984b), analogous in the development of their fault systems and hanging wall structures with contractional thrusting.

Mixed-mode examples of these basins can develop where oblique-slip occurs on the detach-

(a)

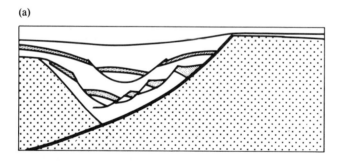

(b)

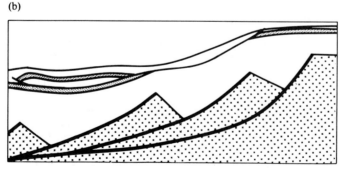

FIG. 1. Structures developed above basement extension faults. The geometries shown are seismic sections, but can be developed at all scales. (a) Distributed extension within sedimentary wedge coupled to basement faults; the faults in the wedge detach on low-strength and over-pressured shales. Tops of basement blocks are eroded. (b) Gravity slide on starved basin margin above extended basement. Note repetition of section.

ment fault with wrench, reverse faults and folding components of deformation in the hanging wall plate (Fig. 2). These will map out systematically in relation to a simple ellipse where its long axis on the map defines the direction of oblique stretch across the basin-margin fault. This geometry can be used to explain many of the apparent anomalies in seismically derived maps and sections of graben margins where the structure does not seem explicable in terms of either a simple dip-slip or strike-slip model. Figures 5 & 6 of Pegrum & Ljones (1984) show part of the eastern margin of the S Viking graben. Their model for this area of extension, then contraction, will work mechanically only where their sections are reinterpreted with gently dipping oblique-slip faults which would allow first extension, and then contraction of the basin. Similar structural patterns may account for some of the features described in the North Sea which cannot be explained by a simple extension model (e.g. Badley *et al.* 1984; Gabrielsen 1984; Price & Rattey 1984).

The general geometry of this class of basin is illustrated in the interpretation of part of the WINCH line shown in Fig. 3. The figure shows a low-angle detachment with the hanging wall

plate moving down a deep-crustal fault detaching at Moho level. The offshore Solway basin, developed above this fault, has relatively steep faults along its northern margin which can be shown in the field to have a strike-slip component. Observed rotations across faults dipping and throwing down towards the basin axis are not entirely compatible with a dip-slip listric extensional model. Steps or transfer faults (Gibbs 1984b) along this margin controlled fan growth in the Carboniferous (Fig. 4) and in the field faulting of the fanglomerates can be shown to be synsedimentary. This same set of Carboniferous transfer faults was reactivated during Permo/Triassic basin growth and was active again in the mid Jurassic and, offshore, in the Tertiary. Further to the E, in the Canonbie and Bewcastle areas, thrusts and thrust-related folds can be identified both in the field and on Survey maps. Thrusts active during Carboniferous times are the result of transpressive shortening in progressive strike-slip. Taken as a whole the evidence indicates that the Carboniferous basin grew dominantly by extension and strike-slip on older faults in a linked system similar to that shown in Fig. 2. A significant component of oblique-slip during deposition of the basin fill is necessary to explain the observed

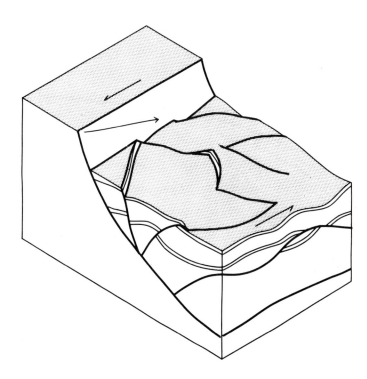

FIG. 2. Pattern of reverse and normal faults developed in the hanging wall block of a sinuous oblique-slip extension fault.

A. Gibbs

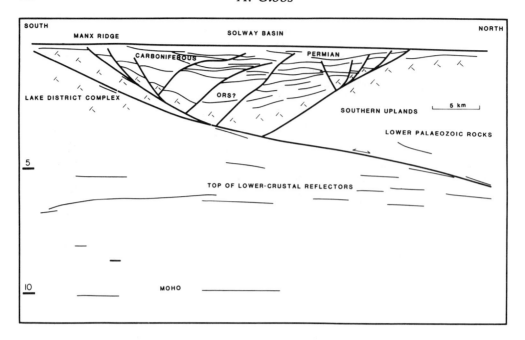

FIG. 3. Schematic interpretation of part of the WINCH deep seismic profile across the outer Solway Firth. Interpretation shows major detachment fault dipping N upon which the basin rides and the steeper North Solway fault system cutting down to join the detachment.

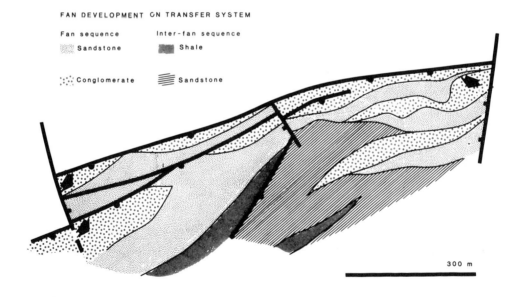

FIG. 4. Simplified map of the fan sequence developed on transfer faults on part of the North Solway fault to the S of Dalbeattie. Arrows show direction of sediment transport from the footwall.

thrusts, folds and obliquely trending normal faults. The basin filled and deformed as the hanging wall moved obliquely down and across the footwall fault. The depositional systems and sediments produced, therefore, became involved in the deformation. Hence gently dipping fault arrays allow for both dip-slip and oblique-slip across the sole detachment. Where the detachment fault is stepped, along-strike combinations of frontal and oblique ramps (Butler 1983) are possible as with thrust faulting, and give rise to apparently complex patterns of thrusts and folds in otherwise extensional basins.

Fault blocks on the margins of extensional basins may develop sequentially (e.g. Beach 1984; Gibbs 1984b). In these cases the locus of sedimentary deposition will be tied closely to the developing footwall fault. As the footwall and hanging wall collapse sequentially there will be a progressive rotation of planar stratigraphic markers across the faults. Changes in age of the sediment fill allow the footwall collapse to be dated and correlated with a corresponding rotation of the sediment wedges as described by Beach (1984).

Continuing growth of the sole fault leads to the development of growth faulting in the sediment fill. This is illustrated in Fig. 5 where depocentres may be stacked in the offset pattern indicated. When a ramp is present in the extension fault the ramp may collapse. Figure 5 shows how arrays of sedimentary wedges are developed, back-rotated, and then moved considerable distances away from later depocentres. The 'asterisk' and 'diamond' ornaments in the figure mark the positions of the axial depocentres in the first two stages of sediment fill. This model is particularly powerful in explaining the structural control of sedimentation axial to and down dip from major extension fault margins. In several areas of the UK continental shelf, complex sand-distribution patterns in the Jurassic were probably controlled by this process of carapace development. Several examples of this stacking of axial and marginal facies linked to a staircase-controlling fault occur onshore in the UK. Some particularly impressive occurrences of this process can be mapped in the Carboniferous of the Midland Valley of Scotland and in the New Red sandstones of the Solway basin (Gibbs & Page in prep.).

The structural similarity of staircase-basin fault systems and contractional thrust faults may suggest that the former are simply inherited from the latter (Coward 1983; Hossack 1984). It is not yet clear whether such systems can evolve in areas where the staircase elements are not already keyed into the pre-extension basement.

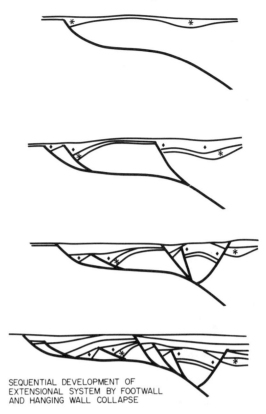

SEQUENTIAL DEVELOPMENT OF EXTENSIONAL SYSTEM BY FOOTWALL AND HANGING WALL COLLAPSE

FIG. 5. Development of faults and sedimentary wedges on a stepped extensional fault. Asterisk and diamond ornaments mark the sequential positions of the axial depositional systems deposited in the first two stages of development.

Indeed those areas where this form of basin seems best represented are just those areas with an earlier history of thrust faulting. An added consequence of the mechanistic similarity of such systems is that staircase basins are easily inverted by driving the system back up the flats and ramps, reversing the geometric development of the hanging wall as illustrated in Fig. 6 (see also Jackson 1980; Davies 1983).

Where faults are used as extensional systems the bed-lengths in the roll-over are increased (see Gibbs 1984a) and sedimentation in the half-graben on the roll-over will, in cross-section, have a longer bed-length section than the pre-extension marker. If the same faults are later recompressed, as for example in the Wessex Basin (Stoneley 1982), this extra bed-length must undergo considerable shortening before the pre-extension marker moves back to its 'null position'. Where the extensional system emerges at

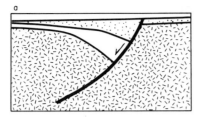

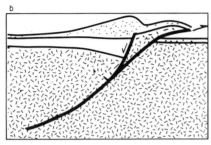

INVERSION OF ROLL-OVER

FIG. 6. Schematic sections showing collapse of roll-over with the development of back-thrusts and a shallower compressional ramp breaking through under the extensional block. The model is developed from field work and seismic data across Purbeck-type fold and fault zones in southern England and the English Channel.

the palaeosurface the ramp will be too steep for the hanging wall to translate over the step and fold simply. Back-thrusting, folding and possible detachment along post-extension sedimentary fill are a geometrical necessity. Purbeck-type monoclines may be the result of pushing a sedimentary growth sequence back up an extensional surface without necessarily pushing the pre-extension marker into net contraction. Figure 6 is a schematic model across a fault such as that shown by Stoneley (1982, fig. 3) with net extension preserved at lower stratigraphic levels (Sherwood sandstone) but with southerly directed back-thrusting at intermediate (Kimmeridgian) and northerly overthrusting at higher levels (Cretaceous). The formation of a new thrust ramp may be critical in developing the prospective structures in this area. These jacked up and back-thrust 'monocline' structures are common at varying scales in many of the UK basins and seem, as in the Wessex Basin example, to be associated either with inversion or oblique-slip on listric extension faults.

Where the step in the extensional system is relatively gently dipping (possibly 20–30° or less) simple roll-off roll-on structures (Fig. 7) may

result without significant development of structural complexity at the emergent ramp. Basins formed by the hanging wall moving off, and then back on to the footwall are the extensional and transtensional equivalents of Ori & Friend's (1984) piggy-back basins seen in thrust systems. These simpler roll-ons occurred in the East Midlands Carboniferous basins during regional transtension and basin formation in the Carboniferous—a modification and mechanistic explanation of the old 'tilt-block' concept. The controlling ramps must, however, have been relatively gently dipping and this may in part be due to the extensive development of overpressured shales throughout much of the Carboniferous sequence. Depths to controlling detachments and ramp positions can be estimated by the construction of regional balanced sections making allowances for oblique-slip where possible. Where the ramp is steeper new faults and folding will nucleate around the old ramp and Purbeck-type monoclines with complex associations of minor folds and faults will result. The Don monocline is probably an example of such a fold produced by shortening obliquely across an older ramp. The dominant features of basins controlled by staircase faults are that they may develop on sites of older basins or re-extended thrust-fold belts (e.g. Chadwick *et al.* 1984). Surprisingly complex patterns of stratigraphic build-ups result from sequential footwall and hanging wall collapse. Additionally the basins can be re-shortened using the same fault array with new faults being largely confined to modifications of the earlier extensional ramps.

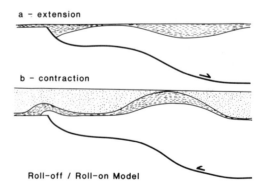

a – extension

b – contraction

Roll-off / Roll-on Model

FIG. 7. Cartoon of hanging wall basins developed during extension (a) and contraction (b) on a stepped fault system during regional subsidence. The depositional highs and lows are inverted.

Sag basins and major hanging wall syncline basins

Basins which develop without appreciable extension of their floors and which in cross-section have simple saucer shapes may be termed 'sag basins'. In the UK continental shelf, onshore and offshore these are probably common but have not yet been described in the literature. A sag basin can only develop if material is removed from lower in the lithosphere by some process which does not stretch the upper leaf of the crust on which the basin sits. At a regional scale this may occur if the thinning of the elastic lid and the ductile lower plate are linked by a low-angle detachment fault (Fig. 8). This diagram, which is derived from regional analysis of North Sea data, shows three types of extensional basin linked by a common detachment. Wernicke (1985) published a similar model derived from analysis of Basin and Range structures.

A gently dipping detachment, as shown in Fig. 9, may result in the faulted basins in the upper crust being offset from the later thermal basin above the thinned lower lithosphere. The inversion of the basement and simple overlap geometry of the basin fill are diagnostic of this type of basin. Faulting will be limited to compactional features in the fill and there may be minor faulting within the basin system. These basin types may be large and remain as yet unidentified, as the crustal geometries and controlling detachments can only be mapped by the use of regional deep seismic grids.

Where the crust is thinned at an intermediate level by the detachment fault, simple hanging wall basins are formed. These occur as synclines above either mid-crustal ramps or above pulled-out wedges of material similar in geometry to under-thrust wedges. The hanging wall syncline basins so formed exhibit a large variation in size from major sub-basins such as the Beryl Embayment on the western margin of the Viking graben system (Fig. 9) to some of the local sub-basins and 'gulfs' in the onshore English Carboniferous.

Figure 9 is a simplified line drawing from a Western Geophysical non-exclusive seismic survey. The base Cretaceous, base Permian and Jurassic seismic markers are identified on the basis of character and only some of the basement seismic markers are shown. In the central part of the diagram the fault planes are picked out along markers on the seismic record, elsewhere they are interpretations. Over the area covered by this survey many of the seismic lines show gently dipping markers in the basement upon which the later extensional faults apparently detached. Half-graben sedimentary basins of Devonian age are present to the S and W of the line illustrated and these seem to have geometries similar to those suggested by Hossack (1984) for the W coast of Norway and to those illustrated from the MOIST profile by Brewer & Smythe (1984) to the NW of Scotland. A similar interpretation of extension and relaxation of Caledonian thrusts to allow the development of Devonian basins is proposed for this area. Some of the gently dipping basement markers can be interpreted as Caledonian thrusts or shear zones.

Following the Devonian extension, Permian basins developed on the sites of the older basins and subsequently the Mesozoic basins appeared to utilize many of the same basement structures. The main Jurassic basin in this area is the S Viking graben which occurs just to the E of the basement high shown at the eastern end of Fig. 10. The margin of the Jurassic graben was probably controlled by a major basement structure of the sort shown by Beach *et al.* (this volume) controlling the development of the N Viking graben. This system therefore forms a double ramp with half-grabens detaching on shallow structures at the western end, a major hanging wall basin and then the main half-graben on a deeper detachment at the western end of the section. The line drawing and seismic profile shown by Allmendinger *et al.*

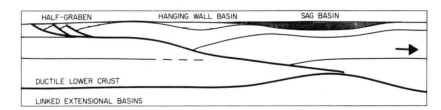

Fig. 8. Linked extensional basins showing possible relationship between a half-graben, hanging wall and sag basins on a crustal detachment. Horizontal length of section variable, ≈ 2–300 km.

A. Gibbs

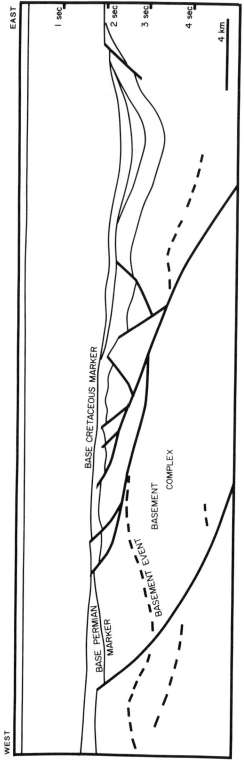

FIG. 9. Interpreted section across part of the Beryl embayment (Viking graben) based on Western Geophysical non-exclusive seismic line FGS 24.

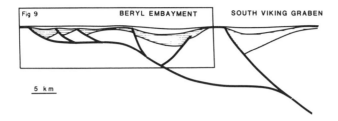

FIG. 10. Model for development of ramp basins on different levels of detachment. Compare with Fig. 9 (shown in area of box).

(1983) for the Sevier desert of Nevada could be interpreted in a similar way. Erosion and further footwall collapse of such a stepped detachment may be responsible for producing some core complex geometries on the upper, shallower detachment (cf. Spencer 1984; Wernicke & Burchfiel 1982; Wernicke 1985).

Basins developed by planar 'domino' fault arrays

Extensive data from currently active seismic faults (Jackson *et al.* 1981; Jackson & McKenzie 1983) suggest that the faults which penetrate the crust in some extensional basins are planar, that is 'domino' faults. These detach at around 10 to 15 km on a ductile zone which in terms of lithospheric structural geometry is a 'fault'.

Several authors (e.g. Wernicke & Burchfiel 1982; Bally 1981) have pointed out that a domino fault array must resolve the 'space problem' not only on its sole detachment, but at the end of the array, and Wernicke & Burchfiel suggested that this may be accomplished by a combined 'listric' and domino array. There is a clear and simple stratigraphic test for this. If the domino array has evolved by unfolding a listric fault system the fill above each successive fault block should young towards the footwall. In a true domino array as described by Jackson and his coworkers, all elements must rotate at the same time. The same stratigraphic sequences will be seen on each fault block. Moreover, the increasing rotations apparent between adjacent blocks as the footwall is approached, which are a necessary characteristic of a listric array, will not be seen in a domino array.

Extension on domino faults, while leading to rapid thinning of the crust, will also flatten the faults progressively and the system will tend to lock up (e.g. Gibbs 1984b). A new set of dominoes may be expected to develop at some stage and these will re-fault the earlier set. Repetition of this process will rapidly result in geometries similar to those described by Proffett (1977) in the Yerington district of the Basin and Range Province. The important implication of this process for the analysis of offshore basins is that only the latest faults will be apparent on conventional seismic data, and that these will form a domino array on already thinned crust.

On reflection seismic data the early faults will appear as sub-horizontal basement events broken by steeper, later faults. As in the Yerington district the early rift-stage sediments must develop very high dips at quite low extensions. Although these would not be imaged directly on reflection data it should be possible to test this where an adequate number of wells reach 'basement'. It is also unlikely that significant internal complexity of the carapace will develop. The sediment cover will simply rotate on top of the basement dominoes. Observation of relative rotation between adjacent basement blocks; stratigraphic younging of fill across the basement; lack of steep (80°) dips in early sediments; and lack of internal carapace faulting in the sediment fill can be employed to suggest that a sequential development of domino faults is not operative in a particular basin. These features are relatively easily tested by conventional geological and geophysical data.

Domino fault arrays which do not end in listric elements must be bounded by zones of more complex strike-slip deformation. Once formed, domino faults may slip laterally in later deformation but will probably not easily restack to rethicken the crust. Cyclic reuse of domino faults other than as strike-slip systems may not be possible. Jackson *et al.* (1982b) reported dips of about 40° for the Aegean fault arrays and these would probably be too steep to act as ramps in compression. Changes in the regional stress field will therefore result in the generation of new faults.

A. Gibbs

Segmentation of extensional basins —transfer or compartmental faults

Along-strike, in both dip-slip ramp and flat staircase faults and domino extensional arrays, transfer faults (Bally 1981; Gibbs 1984a) segment the structure. They act as lateral and sidewall ramps and in staircase arrays will detach not only at the base of the elastic lid but at intermediate detachment levels. In this way they again resemble lateral ramps in thrust systems and may also be confined to either the footwall or hanging wall plate. Figure 11 shows a stepped detachment with a transfer fault joining two detachment levels. Transfer faults have the important attribute of being the longest-lived single fault elements in an extensional array while slip is localized on the sole fault of the extensional array. This process leads to footwall collapse and 'piggy-back' deformation of the overlying plate with the earlier extensional faults becoming inactive. In the case of domino arrays the second extensional fault array will develop bounded by the same transfer fault zone. The transfer system must therefore remain active as long as there is any extension within the compartment.

This longevity of the transfer elements means that they will be important in controlling sediment movement in the basin throughout the basin's history. Brooks *et al.* (1984) argued that in the northern North Sea the transfer elements effectively controlled Palaeocene sand build-ups and the fluid migration routes of hydrocarbons generated late in the burial history of the basin. A more overt effect early in basin development is seen when the transfers act as input channels for sand build-ups derived directly from the tilted fault blocks on the basin margins, as may be the case in the Brae Field of the S Viking graben.

The 'flower' type geometry in cross-section of the transfers as they develop up into the sediment fill is also important. The cross-sectional geometry and evolution of transfer-fault flowers will be as described for strike-slip flower structures (e.g. Harding & Lowell 1979). As the faults propagate upwards into unconsolidated sediment they will flatten into low-strength layers within the carapace. *En échelon* patterns of folding and both reverse and normal faults will occur riding on these concave-downwards faults. Figure 12, a section across the Vale of Pickering, is an example of a steep transfer zone branching upwards into flat-lying fault 'petals' in the carapace of Mesozoic and upper Palaeozoic sediments above a fairly simple basement tear fault. The basement fault allows displacement transfer between the strike-slip zones offshore to the E (Gibbs 1986) and the tilted block and basin geometry of northern England.

Basin formed by a 'mixed-mode' combination of strike-slip and dip-slip arrays—'leaf tectonics'

Many basins seem to be developed not only on gently dipping listric staircase or domino arrays but in association with steeply dipping faults. In the simplest cases these are strike-slip pull-apart basins (e.g. Reading 1980). Deep seismic profiles across such faults do not normally directly image the fault plane. In some cases these faults separate lower crust of different character. They are therefore very deep rooted, possibly at or near Moho level. Faults such as the Great Glen, the Highland Boundary and Southern Uplands Faults of Scotland are examples of such structures. In other cases the deep structure appears unaffected by the strike-slip zone and the fault may ride on an intermediate detachment.

A system of fault-bounded leaves which link to the deep-rooted strike-slip elements are necessary to allow the crust to thin under a pull-apart basin. The deep-rooted faults do not serve to thin or thicken during deformation. Figure 13 shows a composite basin bounded by a deep-rooted strike-slip fault and a gently dipping

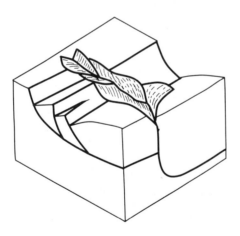

FIG. 11. Isometric model of transfer fault with strike-slip 'flower zone' (shaded) developed in the overlying sediments by differential movement of the blocks in the extensional compartments. The transfer fault steps the level of the extensional detachment and separates compartments with different structural styles.

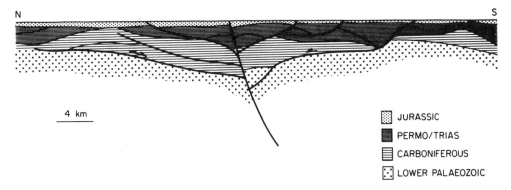

FIG. 12. Simplified geological cross-section across the Vale of Pickering along easting '80'. Section drawn from plunge projections and a detachment geometry derived from balanced-section techniques, applied to series of sections through the basin. Note the development of fault leaves within the sedimentary basin linking on to, and driven by, a deep-crustal strike-slip fault zone along the basin axis which inverted the basin by changing slip on the detachment faults.

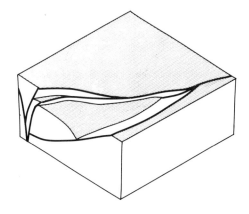

FIG. 13. Isometric sketch of model of linked extensional leaf and strike- or oblique-slip fault system. Compare front panel with Fig. 12.

listric fault forming the floor fault of the basin. This would define the lowest leaf in the fully evolved system. These leaves have a sole fault active during sedimentary growth and then develop a roof fault which forms the sole fault for the next stratigraphic sequence. Later leaves will show less rotation than those formed early in the development of the strike-slip basin. Gibbs (1984b) called these systems partial duplexes because of the stratigraphic break between the development of the sole and roof to the duplex. The results of this process can be observed throughout the Midland Valley of Scotland. Figure 14 is a composite section parallel to the E coast of Arran crossing the Highland Boundary Fault. The Carboniferous and New Red Sandstone sections are developed on a growing fault-controlled unconformity and form an upper leaf to a duplex at Old Red Sandstone levels (Page & Gibbs in prep.). Further eastwards (Fig. 15) complete duplexes are preserved in the Carboniferous section and the Oil Shale and Cementstone sections form an intermediate leaf lying on top of the Old Red Sandstone, overlain in turn by the fault leaf carrying the coalfield. Field work in Scotland on new road sections and in open cast workings provides ample evidence for the gently dipping detachments in low-strength shales that this model requires.

In these mixed-mode basins the whole of the carapace continues to be involved in faulting driven by the deep-crustal faults as the basin develops. This contrasts with dip-slip extensional basins where deformation localizes on the footwall margin. The distinction between this type of basin and a dominantly dip-slip basin is that in the former case although the crust thins on the ramps and flats, the crustal-penetrating faults are essentially strike-slip elements. A simple division of basin-forming faults into extensional and transfer types is not appropriate.

Both the basement and the sedimentary carapace deform by the same structural process of steep ramp faults and compartmental leaves. The dominance of this structural style over

A. Gibbs

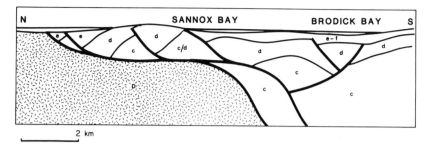

FIG. 14. Simplified composite section along the E coast of Arran across the Highland Boundary fault zone. The section was constructed by projecting out-crop patterns down-plunge on to the plane of section and constraining these plunge predictions with offshore seismic interpretations. (D = Dalradian; d = ORS; c = Carboniferous: e-f = Permo/Trias).

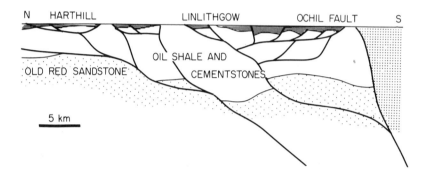

FIG. 15. Simplified geological cross-section of the Central Coal Basin of the Midland Valley of Scotland showing intermediate and upper fault leaves developed as a 'carapace' to the simpler basement strike-slip shears.

deep-rooted domino or listric basin-forming mechanisms is probably controlled by the predominance of layered upper and lower crust beneath the UK continental shelf inherited from Caledonian and Hercynian deformations and by the stratigraphic layering within the carapace.

Composite systems where the crustal-penetrating faults are dominantly strike-slip, but with dips less than 90° and linked to staircase or domino fault-bounded leaves seem to have been important in the development of many UK basins. Variants of this model have been recently proposed for the development of a linked fault system through the Carboniferous province of England into the Gas Area of the Southern North Sea for a dominantly strike-slip basin (Gibbs 1986) and by Beach *et al.* (this volume) for the dominantly extensional Viking graben. Figure 17 is a generalized section across such a basin where the basin has stretched obliquely. Carapace faulting allows the sediment-fill to stretch to accommodate changes in displacement of the basement faults as the basin grows.

In some cases the upper part of the basement may also be involved in these carapace structures as, for example, possibly occurs in the area shown in Fig. 9. Figure 16 illustrates a map of such a mixed extensional and strike-slip basin with an overlying carapace. In this cartoon the

COMPOSITE BASIN OF STRIKE-SLIP AND DOMINO FAULTS
NB Fault throws shown for carapace

FIG. 16. Schematic map of basin such as that illustrated in Fig. 17. The line of section runs from top left to bottom right. Fault throws on this cartoon are as they would be seen at a structural level in the sediment fill. At basement level the faults would all throw the same way as shown in Fig. 17.

section (Fig. 17a) runs from NW to SE and the basement faults define oblique-slip dominoes. The extension seen on the faults at basement level, basin subsidence, and crustal thinning need not balance on a single cross-section of such a basin. Throws on the map indicate the possible geometry of the carapace and it may not simply correspond to basement geometry.

Linked fault systems of this nature appear to provide a more general stuctural framework for upper-crustal deformation. Not only can older fault systems be reutilized but the crust can extend both across and along the basin axis. It also allows a clear linkage between the crustal-penetrating vertical, or dipping domino faults (evinced by neotectonic data) and other types of arrays which are amply supported by geological observations and interpretations of reflection seismic data. Furthermore, this structural pattern precludes neither domino nor staircase dip-slip basins with simple transfer elements, but gives an added dimension of mixing these components with large strike-slip zones.

Conclusions

Basins form on a variety of linked fault arrays which must ultimately join ductile shear zones in the mid and lower lithosphere. The variety of shallow structural geometries and their effect on the sedimentation and deformation of the sedimentary fill can be used to begin to understand the main structural elements and the patterns of linked fault arrays at depth. Deep

seismic reflection profiles provide a means of testing these models. Reflection seismic profiles may show only certain of the geologically necessary elements and important steeper dipping faults with small vertical offsets can be missed unless the basin development is examined regionally. Where the basins have developed in this way, on what are essentially two or more interfering staircase systems, the distinction between transfer and dip-slip elements is one of degree.

The analysis of basin dynamics necessitates merging and reconciling apparently conflicting data from theoretical and observational branches of tectonics, stratigraphy, and geophysics. Ultimately our models must be predictive and be capable of being tested. The implications of such basin models are important in that they begin to show how stratigraphy can evolve in response to structure and how the sediments may distribute with time. This approach provides a new suite of structural and stratigraphic models which have major economic implications to the distribution of, and exploration for, hydrocarbons.

ACKNOWLEDGMENTS: I wish to thank Western Geophysical for permission to publish the line drawing of non-exclusive survey Line FGS-24 (Fig. 9) and British Institutes Seismic Profiling Syndicate (BIRPS) for permission to publish an interpretation of part of the WINCH data (Fig. 3). I would also like to thank those who have contributed to the development of these ideas through their discussions, particularly John Nicholson and Alastair Beach.

(a)

(b)

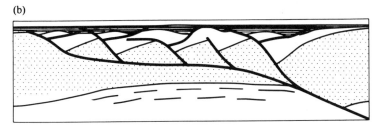

FIG. 17. Schematic cartoon of basin developed on (a) extensional domino faults and (b) extensional ramp system with considerable strike-slip occuring during basin filling. Carapace faults detach in low-strength sediments.

References

ALLMENDINGER, R.W., SHARP, J.W., VON TISH, D., SERPA, L., BROWN, L., KAUFMAN, S., OLIVER, J., & SMITH, R.B. 1983. Cenozoic and Mesozoic structure of the eastern Basin and Range province, Utah, from COCORP seismic-reflection data. *Geology*, **11**, 532–6.

BADLEY, M.E., EGEBERG, T. & NIPEN, O. 1984. Development of rift basins illustrated by the structural evolution of the Oseberg feature, Block 30/6, offshore Norway. *J. geol. Soc. London,* **141**, 639–51.

BALANCE, P.F. & READING, H.G. (eds) 1980. Sedimentation in oblique slip mobile zones. *Spec. Publ. Int. Assoc. Sed. 4.* 337pp.

BALLY, A. W. 1981. Atlantic Type Margins. *In:* BALLY *et al.* (eds) *Geology of Passive Continental Margins, AAPG Course Note Series,* **19**, 1–48.

—— , BERNOULLI, D., DAVIS, G.A. & MONTADENT, A. 1981. Listric Normal Faults. *Oceanologica Acta, 4. Proc. 26th Int. Geol. Cong., Geology of Continental Margins Symp.* Paris. Colloque C3 Geology of Continental Margins. pp. 87–101.

BEACH, A. 1984. Structural evolution of the Wytch Ground Graben, *J. geol. Soc. London,* **141**, 621–8.

BEACH, A. *et al.*

BOYER, S.M. & ELLIOT, D.W. 1982. Thrust systems. *Bull. Am. Assoc. Pet. Geol.* **66**, 1196–230.

BREWER, J.A. & SMYTHE, D.K. 1984. MOIST and the continuity of crustal geometry along the Caledonian–Appalachian orogen. *J. geol. Soc. London,* **141**, 105–20.

BROOKS, J., CORNFORD, C., GIBBS, A.D. & NICHOLSON, J. 1984. Geologic controls on occurrence and composition of Tertiary Heavy Oils, Northern, North sea. *Bull. Am. Assoc. Pet. Geol.* **68**, 793 (abstract).

BUTLER, R.W.H. 1982. The terminology of structures in thrust belts. *J. struct. Geol.* **4**, 239–45.

—— 1983. Balanced cross-sections and their implications for the deep structure of the northwest Alps. *J. struct. Geol.* **5**, 125–37.

CHADWICK, R.A., KENOLTY, N. & WHITTAKER, A. 1984. Crustal structure beneath southern England from deep seismic reflection profiles. *J. geol. Soc. London,* **140**, 893–912.

COOK, F., BROWN, D., KAUFMAN, S., OLIVER, J. & PETERSEN, T. 1981. COCORP seismic profiling of the Appalachian orogene beneath the Coastal Plain of Georgia. *Bull. geol. Soc. Am.* **92**, 738–48.

COWARD, M.P. 1983. Thrust tectonics, thin skinned or thick skinned, and the continuation of thrusts to deep in the crust. *J. struct. Geol.* **5**, 113–23.

DAVIES, V.M. 1983 Interaction of thrusts and basement faults in the French external Alps. *Tectonophysics,* **88**, 325–31.

EFFIMOFF, I. & PINEZICH, A.R. 1981. Tertiary structural development of Celeste Valleys based on seismic data: Basin and Range Province, North-

eastern Nevada. *Phil. Trans. R. Soc. London,* A.**283**, 289–312.

FINCKH, P., ANSORGE, J., ST MUELLER, J. & SPRECHER, CHR. 1984. Deep crustal reflections from a vibroseis survey in Northern Switzerland. *Tectonophysics,* **109**, 1–14.

GABRIELSEN, R.H. 1984. Long-lived fault zones and their influence on the tectonic development of southwestern Barents Sea. *J geol. Soc. London,* **141**, 651–63.

GIBBS, A.D. 1983. Balanced cross-section constructions from seismic sections in areas of extensional tectonics. *J. struct. Geol.* **5**, 152–60.

—— 1984a. Clyde Field Growth Fault secondary detachment above basement faults in the North Sea. *Bull. Am. Assoc. Pet. Geol.* **68**, 1029–39.

—— 1984b. Structural Evolution of extensional basin margins. *J. geol. Soc. London,* **141**, 609–20.

—— 1986. Strike-slip basins and inversion: a possible model for the Southern North Sea Gas Area. *In:* BROOKS, J., GOFF, J.C. & VAN HORN, B. (eds) *Spec. publ. geol. Soc. London,* **23**, 23–36.

GRACIANSKY, P.C., POAG, C.W., CUNNINGHAM, R., LOUBERE, P., MASSON, D.G., MAZZULLO, J.M., MONTADERT, L., MULLER, C., OTSUKA, K., REYNOLDS, L.A., SIGAL, J., SNYDER, S.W., TOWNSEND, H.A., VAOS, S.P. & WAPLES, D. 1985. The Goban Spur transect: Geological evolution of a sediment-starved passive continental margin. *Bull. geol. Soc. Am.* **96**, 58–76.

HARDING, T.P. 1984. Graben hydrocarbon occurrences and structural style. *Bull. Am. Assoc. Pet.* **68**, 333–62.

—— & LOWELL, J.D. 1979. Structural styles, their plate tectonic habitats, and hydrocarbon traps in petroleum provinces. *Bull. Am. Assoc. Pet. Geol.* **63**, 1016–59.

HARLAND, W.B. 1971 Tectonic transpression in Caledonian Spitsbergen. *Geol. Mag.* **108**, 27–42.

HOSSACK, J.R. 1984. The geometry of listric growth faults in the Devonian basins of Sunnfjord, W Norway. *J. geol. Soc. London,* **141**, 629–38.

JACKSON, J.A. 1980. Reactivation of basement faults and crustal shortening in orogenic belts. *Nature, Lond.* **283**, 343–6.

—— & MCKENZIE, D. 1983. The geometrical evolution of normal fault systems. *J. struct. Geol.* **5**, 471–82.

—— , KING G. & VITA-FINZI, C. 1982a. The neotectonics of the Aegean: an alternative view. *Earth planet. Sci. Lett.* **61**, 303–18.

—— , GAGNEPAIN, J., HOUSEMAN, G., KING, G.C.P., PAPADIMITRIOU, P., SOUFLERIS, C. & VIRIEUX, J. 1982b. Seismicity, normal faulting, and the geomorphological development of the Gulf of Corinth (Greece): the Corinth earthquakes of February and March 1981. *Earth planet. Sci. Lett.* **57**, 377–97.

MCKENZIE, D.P. 1978. Some remarks on the development of sedimentary basins. *Earth planet. Sci. lett.* **40**, 25–32.

ORI, G.G. & FRIEND, P.F. 1984. Sedimentary Basins formed and carried piggyback on active thrust sheets. *Geology*, 12, 475–8.

PEGRUM. R.M. & LJONES, T.E. 1984. 15/9 Gamma Gas Field offshore Norway, new trap style for North Sea Basin with regional structural implications. *Bull. Am. Assoc. Pet. Geol.* 68, 874–902

PRICE, I. & RATTEY R.P. 1984. Cretaceous tectonics off mid-Norway: implications for Rockall and Faeroe-Shetland troughs. *J. geol. Soc. London*, 141, 985–92.

PROFFETT, J.M. 1977. Cenozoic geology of the Yerington district, Nevada, and implications for the nature and origin of Basin and Range faulting. *Bull. geol. Soc. Am.* 88, 247–66

READING, H.G. 1980. Characteristics and recognition of strike-slip fault systems. pp. 7–26. *In*: BALLANCE, P.F. & READING, H.G. (eds) *Sedimentation in obliques slip mobile zones. Spec. Publ. 4 Int. Assoc. Sed.*

SANDERSON, J. & MARCHINI, W.R.D. 1984. Transpression. *J. struct. Geol.* 6, 449–58.

SCLATER, J.G. & CHRISTIE, P.A.F. 1980. Continental stretching: an explanation of the post mid-Cretaceous subsidence of the Central North Sea Basin. *J. geophys. Res.* 85, B.

SPENCER, J. 1984. Role of tectonic denudation in warping and uplift of low angle normal faults. *Geology*, 12, 95–8.

STONELEY, R. 1982. The structural development of the Wessex Basin. *J. geol. Soc. London*, 139, 543–54.

TODD, R.G. & MITCHUM, R.M. 1977. Seismic Stratigraphy and Global Changes of Sea level, part 8; Identification of Upper Triassic, Jurassic, and Lower Cretaceous Seismic Sequences in the Gulf of Mexico and offshore West Africa. *In*: PAYTON, C.E. (ed.) *Seismic Stratigraphy—Applications to hydrocarbon exploration, Mem. Am. Assoc. Pet. Geol.* 26, 145–63.

WERNICKE, B. 1985. Uniform sense of normal simple shear of the continental lithosphere. *Can. J. Earth Sci.* 22, 108–25.

—— & BURCHFIEL, B.C. 1982. Modes of extensional tectonics. *J. struct. Geol.* 4, 105–15.

ALAN GIBBS, Midland Valley Exploration, 14 Park Circus, Glasgow G3 6AX, UK.

Reprinted by permission of the Australian Petroleum Exploration Association from *APEA Journal*, v. 25, pt. 1 (1985), p. 344-361.

EXTENSIONAL BASIN — FORMING STRUCTURES IN BASS STRAIT AND THEIR IMPORTANCE FOR HYDROCARBON EXPLORATION

by M.A. Etheridge, J.C. Branson and P.G. Stuart-Smith
Bureau of Mineral Resources, Geology and Geophysics, Canberra, ACT

ABSTRACT

The Bass, Gippsland and Otway Basins of southeastern Australia were initiated by north-northeast to south-southwest lithospheric extension, largely during the Early Cretaceous. The extensional stage was followed by a Late Cretaceous to Pliocene thermal subsidence stage and a late stage of compressional tectonic overprinting.

The extensional stage was dominated by two orthogonal fault sets — shallow to moderately dipping, rotational, normal faults and steeply dipping, transfer (transform) faults. Thermal subsidence involved vertical rather than horizontal movements, and consequently generated a discrete fault geometry, comprising steep, down-to-basin, normal faults with small displacements. The major extensional structures exerted a range of controls on both sedimentation and structuring during the subsidence stage. Likewise, the location and style of late Tertiary compressional structures overprinted on the Gippsland and, to a lesser extent, Bass and Otway Basins are controlled by reactivation of major early normal and transfer faults. In particular, the Kingfish, Mackerel, Halibut, Flounder and Tuna fields in the Gippsland Basin overlie a single Early Cretaceous transfer fault zone that was a basinwide structural boundary during extension. These fields occupy en echelon compressional structures generated by left-lateral wrench reactivation of the transfer zone during late Tertiary northwest-southeast compression. The major extensional structures have had an important influence on all stages of the evolution of these basins. It is contended that a thorough understanding of their extensional framework is an important factor in hydrocarbon exploration of these and other basins.

INTRODUCTION

Lithospheric extension has recently become widely accepted as a key process in the evolution of passive continental margins and some intracontinental sedimentary basins. McKenzie (1978) established the broad thermomechanical concepts of basin formation resulting from lithospheric extension. At around the same time, the importance of rotational normal faulting in accomplishing substantial extension in the upper crust was being demonstrated, both in actively extending continental regions (Wernicke & Burchfiel, 1982) and on passive margins (Bally, 1981; Le Pichon & Sibuet, 1981). Together, these advances have led to the development of a conceptual framework for the structural evolution of extensional sedimentary basins. Recently, Gibbs (1984a) and Harding (1984) have shown how these structural concepts enhance our understanding of petroleum-rich basins and specific hydrocarbon traps within them.

Extensional basins have two discrete structural/depositional stages — *the extensional or rift stage*, followed by a thermally induced *subsidence stage*. The kinematics of these two stages are quite different, and so, therefore, are their structural geometries. A key feature of the extensional stage is that it induces large fault zones that may penetrate a large fraction of the crust and be of basin-wide extent. Such structures provide long-lived zones of weakness in the sub-basinal crust which can influence the subsequent structural evolution of the basin. In particular, they can be reactivated during a later *tectonic overprint stage* to produce a range of structural traps within the subsidence stage sediments. In this paper, structures are described from each of these stages in the Bass, Gippsland and Otway Basins of southeastern Australia. In particular, examples are used from these basins to develop concepts for the structural interpretation of extensional basins elsewhere, and to emphasize the importance of these concepts for evaluation of hydrocarbon prospectivity. The paper concentrates on the extensional and tectonic overprint stages, because they have been the dominant influence on the structural evolution and hydrocarbon trap development in these basins.

In the Bass and Gippsland Basins in particular, analysis of the early rift faulting has been hampered by poor quality of seismic data, due to the depth of the structures and to seismic dispersion caused by the overlying Tertiary coal measures. The fact that all known hydrocarbon shows are located within the subsidence sequence has also focussed attention on that structural interval with consequent neglect of older structures, except at the basin margins. The 1982 BMR Bass Strait seismic survey has greatly improved the definition of the early extensional structures, due to (a) improved survey and processing parameters to enhance the deeper data, (b) recording of data to 6 seconds two-way time (TWT), and (c) the regional nature of the survey, with its basinwide survey lines. This study was based on a regional structural analysis of the Bass Basin, utilising the BMR survey data, all post-1974 and limited pre-1974 company seismic data. This was augmented by analysis of the BMR data and a large quantity of the available company data in specific parts of the Gippsland and Otway Basins where the early structures are well illustrated. The study is part of a complete reinterpretation of the Bass Basin by the BMR, with other aspects of the work being presented elsewhere in this volume (Williamson *et al.*, 1985)

REGIONAL SETTING AND PREVIOUS STRUCTURAL MODELS

The Bass, Gippsland and Otway Basins (Fig.1) consist of a largely Early Cretaceous rift-fill, which is highly faulted and unconformably overlain by a Late Cretaceous to Tertiary sequence which occupies broader, more uniform depressions. The stratigraphy and structure of these basins is described in Douglas & Ferguson (1976). Descriptions of the geology of individual basins are given by: Bass — Robinson (1974), Brown (1976), Nicholas *et al.* (1981); Gippsland — James & Evans (1971), Hocking (1972), Threlfall *et al.* (1976); Otway — Reynolds *et al.* (1966), Glenie (1971), Boeuf & Doust (1975), Denham & Brown (1976).

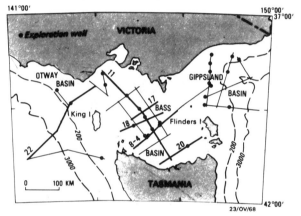

Figure 1 — Track map for the BMR 1982 Bass Strait geophysical survey. Survey lines referred to in the text are heavily drawn and numbered.

The structural and tectonic evolution of the three basins has been extensively discussed, especially in the context of the plate tectonic setting of the southern and eastern Australian margins (Carey, 1970; Griffiths, 1971; Elliot, 1972; Burke & Dewey, 1973; Falvey, 1974; Boeuf & Doust, 1975; Gunn, 1975; Davidson, 1980). There is general agreement that the basins resulted in some way from the rifting of Antarctica and Lord Howe Rise — New Zealand from Australia in the Cretaceous and early Tertiary. However, there is a range of views as to the detailed timing, kinematics and dynamics of individual basin formation in the broader tectonic context. For example, early rifting or extension are invoked by Griffiths (1971), Elliot (1972) and Gunn (1975), among others, although there has been no adequate description of normal faults capable of producing large extensions. In contrast, Falvey (1974) and Middleton (1982) attributed basin formation solely to the thermal effects of mantle upwelling, with the rifting associated with only small horizontal extension (i.e. essentially vertical movements). Davidson (1980) proposed a complex model of wrench faulting and oblique extension, in the only comprehensive attempt to explain the apparently different trends of the three basins and their structures. At the individual basin scale, cross-sections such as that in Figure 2 have been widely published for each of the basins. The section illustrates the bipartite nature of basin development, with a relatively narrow rift sub-basin overlain by a more extensive thermal sag or subsidence basin, giving rise to the classical "steer horn" geometry. However, the large vertical exaggeration of such sections obscures the nature of the faulting, and makes it impossible to determine the gross kinematic history. Similarly, trend maps of the basins have been published, but the structures on such maps have a range of ages. Thus, the geometry of the rift stage faulting is further obscured. However, it has been widely recognised that the rift faults have westerly to northwesterly trends in the three basins, although

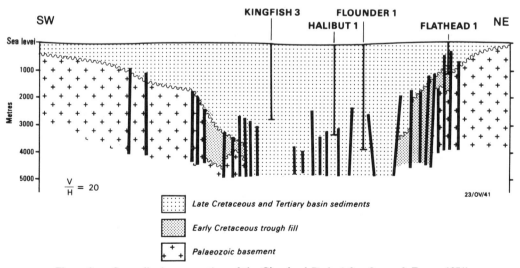

Figure 2 — Generalised cross-section of the Gippsland Basin (after James & Evans, 1971).

discrete trends within this range are commonly ascribed to each basin.

EXTENSIONAL STAGE STRUCTURES

This study has confirmed that the Bass, Gippsland and Otway Basins were initiated by approximately contemporaneous extension or rifting during the Early Cretaceous. Two major sets of faults were developed during this stage — *rotational normal faults* and an orthogonal set of near vertical *transfer faults*. The geometry of these structures indicates that the sub-basinal crust underwent extension of 60 to 80 per cent, prior to a thermal subsidence phase. In the Otway Basin, continued rifting and margin collapse associated with the separation of Antarctica has overprinted the early extensional structures.

Extensional Normal Faults

As pointed out by Wernicke & Burchfiel(1982), normal faults which accomplish substantial extension must be either initially shallow dipping or rotational. Rotational normal faults may start with steep dips, but they will rotate towards shallower dips as extension proceeds. Shallow dipping normal faults bounding major Early Cretaceous half-grabens are seen throughout Bass and Otway Basins, and on the margins of the Gippsland Basin. The main characteristics of the extensional normal faults in these basins are:—

1) They commonly dip less than 45°, and form the boundaries to rotated tilt blocks in which the basement dips up to 35° (Fig. 3).
2) They are approximately planar to the maximum depth of the seismic sections, but may be listric to a deeper level detachment fault.

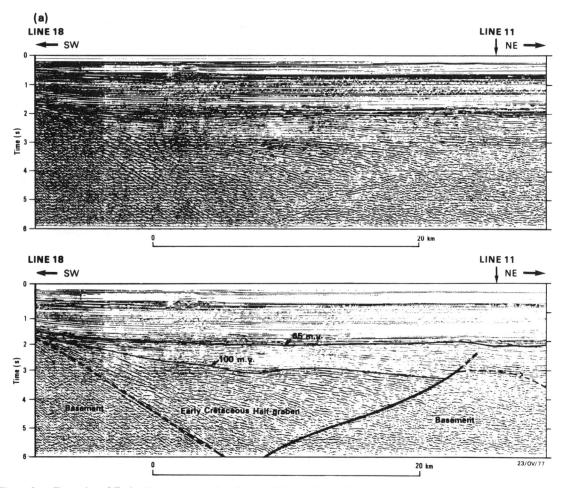

Figure 3 — Examples of Early Cretaceous rotational normal faults with shallow to moderate dips in the three Bass Basins. Note the virtually straight traces of the faults over several km of vertical extent, and the magnitude of the displacements.

(a) Part of BMR line 18 from Bass Basin, illustrating the largest of the Early Cretaceous half-graben and rotational normal faults confidently identified from the BMR survey; at 35 TWT, the vertical and horizontal scales are approximately equal. This is an unmigrated section, since migration destroys information in the lower second or more of the record. (Figure 3 is continued on next two pages.)

(b)

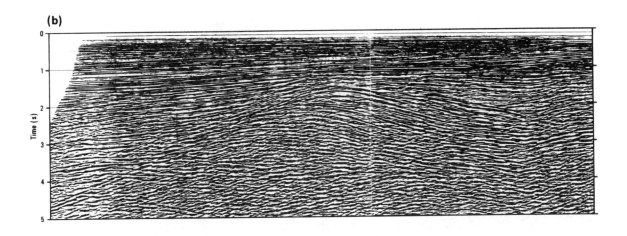

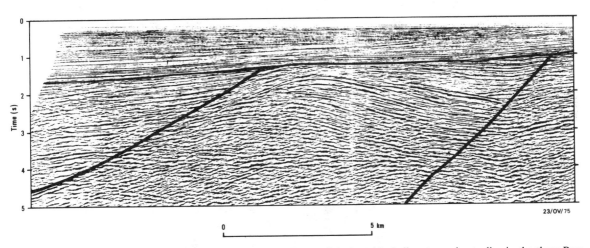

23/OV/75

0 —————————————— 5 km

Figure 3 (cont.) — Examples of Early Cretaceous rotational normal faults with shallow to moderate dips in the three Bass Basins. Note the virtually straight traces of the faults over several km of vertical extent, and the magnitude of the displacements.

(b) Seismic section from the northern margin of the Gippsland Basin. This section is reproduced without specific location information with the permission of Shell Development (Australia) Pty Ltd, and is approximately true scale at 2 s TWT. Two faults are shown on the section; the right hand fault has smaller displacement (approx 3 km), and bounds an Early Cretaceous half-graben about 2 km deep, whereas the left hand fault passes off the base of the section, with Cretaceous sediment on its south side. A minimum displacement of about 10 km is implied by this interpretation, and this must therefore have been a major extensional structure during the early evolution of the Gippsland Basin.

3) Relatively few major normal faults or fault zones are present across the basin, and they therefore tend to have large displacements (several km). On any section, the faults tend to dip the same way across the whole basin, but the dip direction may change from one section to the next.

4) In all three basins, the normal faults strike consistently between 290° and 300°, but individual fault segments have short strike extents.

Seismic sections illustrating shallow-dipping, rotational normal faults in each of the three basins are shown in Figure 3. In these cases, the fault surface is fairly clearly defined, and it bounds a half-graben of substantially tilted Early Cretaceous sediment. In areas where data quality is not as good as in Figure 3, the dip of the pre-to syn-rift fill is a useful guide to normal fault orientation and position. We have used the disposition of packages of dipping reflectors at this level to interpret approximate fault positions in parts of all three basins. More importantly, however, the commonly recognised consistency of Early Cretaceous dip direction across the basin requires that the rotational fault dip also be consistent, but in the opposite direction.

It is also important to realise that the upwards opening angle between a rotational normal fault and adjacent syn-rift fill must be greater than 90°, since the fault can-

(c)

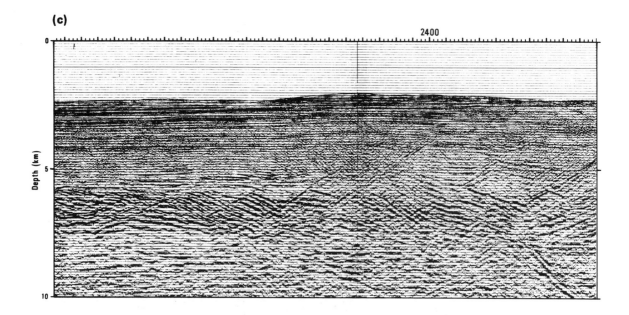

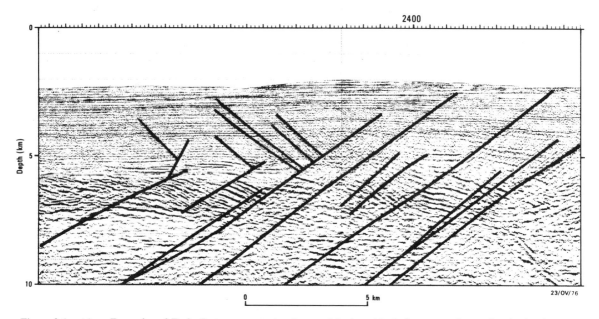

Figure 3 (cont.) — Examples of Early Cretaceous rotational normal faults with shallow to moderate dips in the three Bass Basins. Note the virtually straight traces of the faults over several km of vertical extent, and the magnitude of the displacements.

(c) A depth migrated seismic section from the deeper water part of BMR Line 22 in the Otway Basin. A set of domino-style normal faults have rotated through about 30° during Cretaceous extension of this margin. Extension calculated from this section using the method described for planar faults by Wernicke & Burchfiel (1982) is about 50 per cent.

not have rotated through the vertical. This simple relationship limits faults to moderate to shallow dips in regions where substantially tilted sediment occurs, although care must be taken to ensure that approximately true scale sections are used.

Using these techniques, the gross extensional structural framework was established for the three basins, even

where continuous high quality data are not available. Figure 4 shows adjacent BMR sections across the centre of Bass Basin which were constructed in this way (note 3 times vertical exaggeration). In these cases, volcanics within the Tertiary sequence markedly reduced the quality of deeper data along parts of the sections. However, utilizing strike lines and intervening company data to

348

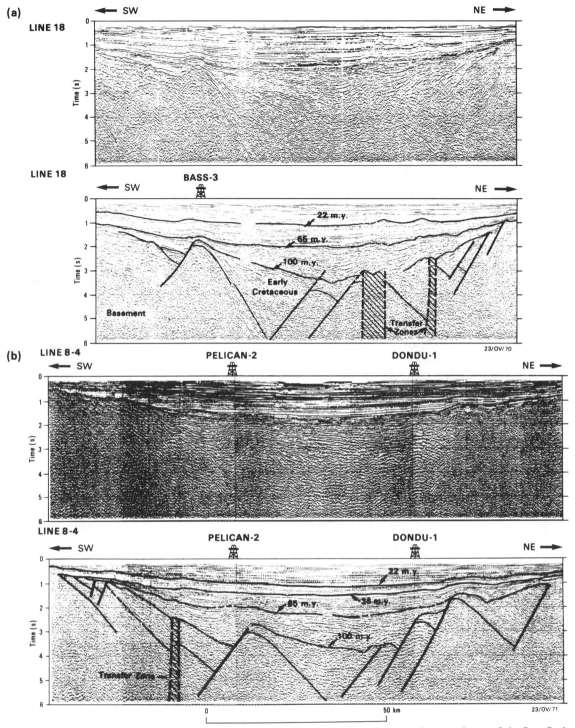

Figure 4 — (a) Compressed seismic section and interpretation of BMR line 18, across the central part of the Bass Basin. Vertical exaggeration is approximately 3 at the Early Cretaceous level. This section crosses a major transfer fault zone, which is characterised by an absence of regular structure below the Late Cretaceous. Note the low amplitude structuring and the concentration of volcanics above the transfer fault zone.

(b) Compressed seismic section and interpretation of BMR line 8-4, immediately southeast of line 18 (see Fig. 1). Vertical exaggeration is 3. Note the difference in location and spacing of major normal and transfer faults between lines 18 and 8-4, even though they are only about 40 km apart.

define dip directions in the Early Cretaceous, we have arrived at a consistent interpretation of the extensional normal fault pattern. Mapping of these normal faults demonstrated their short strike extents, and led to recognition of the second set of extensional structures — transfer faults.

Transfer Faults

The concept of transfer (or continental transform) faults in substantially extended continental regions was first discussed in detail by Bally (1981), and has most recently been applied to the North Sea Basin by Gibbs (1984). Transfer faults are not simple strike-slip faults, but are accommodation structures analogous to oceanic transforms, in that they allow variations in the geometry of extension along the strike of a rift. Major normal faults can terminate against them (and vice versa) without the necessity for distributed strain throughout the rock mass.

Transfer faults were first recognised in this region within the Bass Basin, when attempting to map the major normal faults. The normal faults were found to terminate abruptly, and be apparently displaced across transverse zones that spanned the width of the basin. Figure 5 illustrates the regional character of the transfer faults within the Bass Basin, emphasizing their role as accommodation structures during oblique extension, as discussed by Etheridge et al. (1984). Similar structures have been mapped along the margins of the Gippsland

Basin (Fig. 6), and interpreted from normal fault distributions within the Otway Basin.

Transfer faults are generally steep to vertical, and trend at high angles to the extensional normal faults, The location, spacing and even dip direction of the normal faults may change from one side of a transfer fault to the other (Bally, 1981, Figs. 21 & 22; Gibbs, 1984, Fig. 15). Therefore, seismic sections which cross transfer faults may have complex geometries, which may only be interpretable if the extensional structural concepts developed by Bally (1981) and Gibbs (1984a) are applied.

Since transfer faults generally have an apparent dip slip component and are steeply dipping within a rift-like sequence, they are commonly identified as normal faults. However, as Figure 7 demonstrates, the apparent dip slip component of a single transfer fault may change along strike, making correlation between sections and fault mapping difficult. If the transfer fault illustrated in Figure 7 is correctly correlated on the three sections shown, it would probably be described as a hinge fault, which would not accurately reflect its kinematic history and significance. Seismic sections at high angles to transform faults in Bass and Gippsland Basins are characterised by abrupt displacements of the inferred basement level across a near vertical fault (Fig. 8). Reflectors within the Early Cretaceous sequence are nearly horizontal and terminated abruptly at the fault.

Seismic sections oblique to transfer faults may be more complex, if they also intersect one or more normal faults.

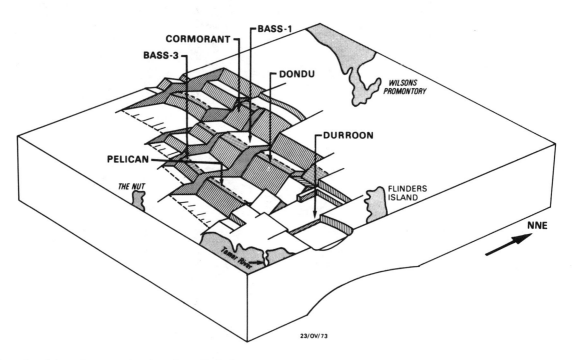

Figure 5 — Schematic perspective view of post-Early Cretaceous structure in the Bass Basin, showing the major fault-bounded half-graben and the consistent right-lateral offset across the transfer faults. Crustal thinning must have decreased substantially towards the southeastern extremity of the basin, as shown by the dramatic change in structural style (after Etheridge et al., 1984)

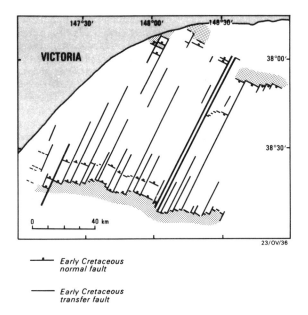

—▲— Early Cretaceous
normal fault

——— Early Cretaceous
transfer fault

Figure 6 — Generalised map of the major Early Cretaceous extensional structures in the Gippsland Basin, showing the short, west-northwest trending normal fault segments bounded by basinwide north-northeast trending transfer faults. No mapping was possible in the basin centre because of the virtual absence of 6 s TWT data, and because Tertiary channeling obscured deeper structure. However, the geometry of the structures is well constrained along the entire southern margin and much of the northern margin.

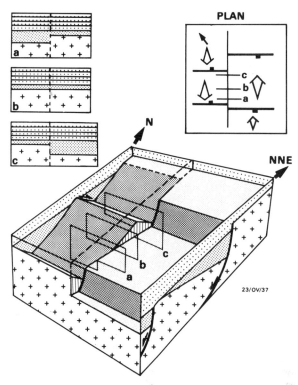

Figure 7 — Block diagram and schematic sections indicating how the seismic character of a transfer fault may change dramatically along its length.

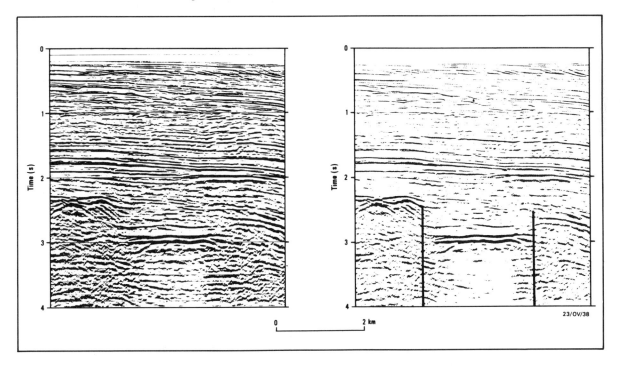

Figure 8 — Seismic section normal to a pair of transfer faults near the southern margin of Gippsland Basin. This section is reproduced without location information with the kind permission of Esso Australia Ltd.

The southern margin of the Gippsland Basin is marked by a major north-northeast dipping, Early Cretaceous normal fault in the west, and a north-northeast dipping shelf in the east (Fig. 6). The dip of the Early Lower Cretaceous section changes from south-southwest in the west to north-northeast in the east, and the eastern segment is therefore inferred to contain south-southwest dipping rotational normal faults in the basin centre. This change in fault dip has also been documented along the northern margin of the basin, and occurs across a major transfer fault zone that must traverse the whole basin (Fig. 6). North-south seismic sections in the vicinity of the transfer zone near the southern margin trend at about 60° to the strike of the normal faults and 30° to the transfer faults, and display unusual geometries within the Early Cretaceous sequences.

Three of these sections are reproduced in Figure 9. The westernmost section (Fig. 9a) shows a growth fault dipping to the right with a thick Early Cretaceous sequence dipping into it with increasing dip down the section. The central section (Fig. 9b) shows similar features in the lower part, but the character of the Early Cretaceous basin margin changes to an onlapped shelf to the left. This change occurs across a sharp vertical boundary, and is also seen on the easternmost section (Fig. 9c), but lower in the section. The vertical boundary coincides with the major transfer fault in this region and the relationship between seismic character and three dimensional structure is explained in Figure 10. Similar complexities are commonly ascribed to offside reflections or unusual diffractions, but may well, as in this case, result from the specific geometry of extensional normal and transfer faults.

Elsewhere in the Gippsland Basin, a seismic section close to the strike of a major rotational normal fault that is offset by a series of closely spaced transfer faults is shown in Figure 11. The section is close to the basin margin, where the Tertiary sequence is thin, and Early Cretaceous reflectors are clearly visible below a marked unconformity (Fig.11). The complex structure within the Early Cretaceous sequence is difficult to explain in terms of the relatively simple tilt-block geometry on perpendicular lines. However, mapping of the normal fault and tilt-block trends on the cross lines shows up offsets on small transfer faults. The seismic section shown in Figure 11 appears complex because it crosses the same normal fault and tilt block three times between the transfer faults (see plan view, Fig. 11).

Transfer faults are also shown on the BMR survey lines across the centre of the Bass Basin (Fig.4). Major transfer faults such as these appear to be vertical zones up to 5 km wide with little or no internal structure, and which terminate against the base of the subsidence sequence. Detailed horizon and fault mapping within the Bass Basin has shown that the widest zones (Fig. 4a; Etheridge *et al.*, 1984, Fig. 5) consist of a number of discrete fault strands (Williamson *et al.*, this volume). Whereas the transfer faults should ideally stop abruptly at the base of the sag or subsidence sequence, larger ones in particular are commonly overlain by small inflections in the

sag sediments (Fig.12). These inflections may result from reactivation of a transfer fault (see below), but may also be due to either, 1) small differences in thermal subsidence across the fault because of initial differences in the extensional geometry, or 2) post-extension steps across transfer faults (Figs 5 & 7).

Finally, transfer faults are deep-seated vertical fracture zones that provide potential paths for magma (and other fluid) migration and are therefore commonly the locus of concentrations of volcanic and high-level intrusive bodies (Fig.12).

SUBSIDENCE STAGE STRUCTURES

During extension, isotherms within the lithosphere are raised, so that, when extension ceases, the stretched portion of the lithosphere cools and subsides (McKenzie, 1978). This subsidence has a different gross geometry from the extension, and, most importantly, from the structural viewpoint, a sharply contrasting displacement field. Displacements are near vertical rather than horizontal, and tend to be significantly smaller during subsidence. The displacement field is smoothly varying, stepping down from the flanks of the basin to the centre. Normal faults which accommodate this movement will therefore tend to be steeply dipping towards the basin centre, (down-to-basin), and relatively closely spaced with small displacements (Fig. 13). It is these faults which dominate many seismic sections, primarily because the depth of exploration interest is commonly within the subsidence sequence, and extensional structures are either not penetrated or poorly displayed deep in the section.

The foregoing assumes that the stretched lithosphere acts as a homogeneous elastic body during subsidence. Since the upper crust contains at least two sets of major faults induced by the extension, this assumption is clearly not warranted, and the extensional structures may impose a range of controls on the details of subsidence and faulting. Various types of control are possible:—

1) Some reactivation of extensional normal faults may take place during subsidence, although their generally shallow dips preclude significant movement without further extension. For example, subsidence stage faults in the Gippsland Basin trend parallel to the underlying extensional structures, rather than to the overall, more east-west basin trend (Fig. 13; Threlfall *et al.*, 1976). However, unlike the extensional faults, they are predominantly down-to-basin (Fig. 13).

2) Differential subsidence may take place across transfer faults, particularly to accomplish the subsidence variation along the basin axis. This will give rise to a set of transverse subsidence faults localized above the transfer faults.

3) Variously eroded tilt-block corners will provide basement and/or topographic relief after extension. During subsidence, compaction over these (commonly discontinuous) basement ridges is a major cause of structuring in extensional basins (Harding & Lowell, 1979). The best examples in this region are along the marginal tilt-blocks in the Bass Basin (Fig.4), although similar structures are also developed over a tilt-block

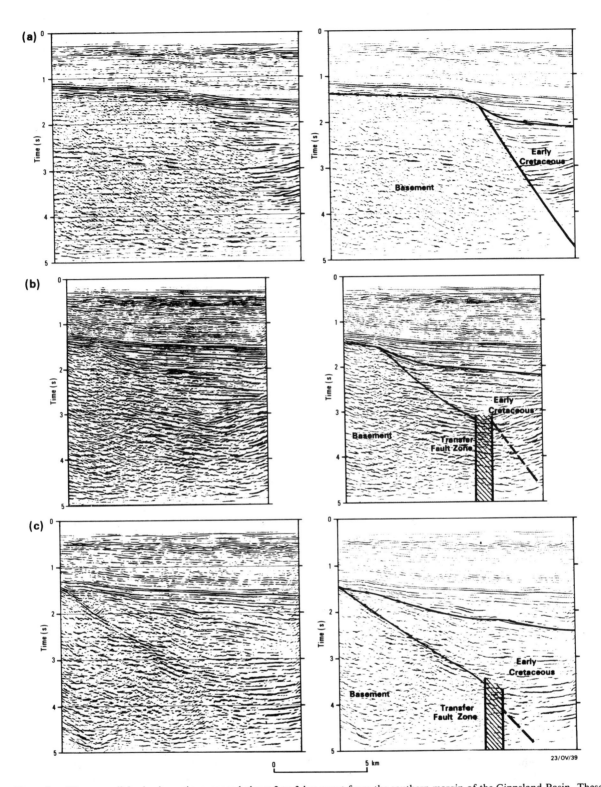

Figure 9 — Three parallel seismic sections, spaced about 2 to 3 km apart from the southern margin of the Gippsland Basin. These sections trend at about 30° to the major transfer fault shown in Figure 6, and illustrate some of the geometrical complexities that arise where a seismic section intersects both normal and transfer faults obliquely. See Figure 10 and text for further explanation. These seismic sections are reproduced without location information with the permission of Phillips Australian Oil Company.

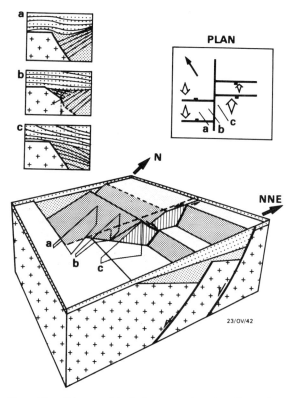

Figure 10 — Schematic block diagram and sections illustrating the three dimensional structure in the region of the seismic sections of Figure 9.

near the basin centre (Etheridge *et al.*, 1984). Transfer faults may play a key role in generating hydrocarbon traps in these drape or compactional structures, by providing along strike closure where tilt-blocks are offset.

TECTONIC OVERPRINT STAGE STRUCTURES

At any time during the evolution of an extensional basin, a change in tectonic setting can modify the regional stress field. The new stress field may superimpose a separate generation of structures on the basin sediments, and may terminate or only interrupt evolution of the basin. Since the larger extensional structures are faults that penetrate a substantial fraction of the crust, they provide discontinuities and zones of weakness that are likely to be reactivated in various ways by the new stress field.

In the Bass Strait region, tectonic overprinting has been minor in the Bass Basin (Etheridge *et al.*, 1984), very significant in the Gippsland Basin, and widespread and complex in the Otway Basin. This paper will concentrate on the Gippsland Basin, where the reactivation of Early Cretaceous normal and transfer faults during mid-Eocene and younger movements has been the primary control on structural development and hydrocarbon entrapment in the Tertiary sequence.

The important tectonic overprint in the Gippsland Basin was recognised very early in its exploration history

(Hocking, 1972; Threlfall *et al.*, 1976), and the structures it produced contain virtually all the hydrocarbons so far discovered in the basin. In most of the earlier literature, this overprint was related to right-lateral wrench movement predominantly on the Rosedale Wrench Fault, an east-west structure along the northern margin of the basin (Threlfall *et al.*, 1976). The anticlines and related reverse faults in the basin centre trend approximately northeast, and were considered to be en echelon to the Rosedale Fault. However, this model did not explain why these structures occur largely on one side of and up to 50 km from the Rosedale Fault, and why a number of structures appear to be aligned along other trends.

Detailed studies of seismicity in southern Victoria (Denham *et al.*, 1981; Gibson *et al.*, 1981), and of fracture geometry and in situ stresses in the Latrobe Valley (Barton, 1981) have shown that the current and late Tertiary stress field in the region was one of near horizontal northwest to north-northwest compression. It will be assumed that this stress field was first imposed in the mid-Eocene, and explore its consequences for reactivation of the (admittedly incomplete) extensional fault array shown in Figure 6.

A horizontal, northwest-southeast maximum principal compressive stress makes an angle of about 70° with the transfer faults, and one of about 15° with a 45° — dipping normal fault trending 295°. Therefore, a significant resolved shear stress would be present on both fault sets, but it would be somewhat greater on the transfer faults. North-northwesterly directed compression would result in higher resolved shear stress on both fault sets. Reactivation of the normal faults in response to this stress field would have led to oblique reverse movement, whereas the transfer faults would have undergone simple wrench movement (Fig. 14). Inclination of the principal compressive stress direction could give rise to a component of dip slip on the transfer faults (Fig. 14). In the Gippsland Basin, we have mapped a large number of transfer faults, but most of them produce only small displacements of the normal faults, and are presumably structures of limited width and/or extent. Reactivation of these has been limited. However, two major transfer fault zones have been mapped, one in the east and one in the west of the basin. The eastern transfer zone separates regions in which the rotational faults have opposite dips (Figs 6&10), and it has been mapped over 10 km or more at each basin margin. In addition, it appears to consist of an array of possibly braided faults up to a kilometre wide. Whereas the western zone is not nearly as well characterized, it appears to mark a substantial offset of the Early Cretaceous rift basin. It is also closely aligned with the transfer fault at the southeastern extremity of the extended zone in Bass Basin, and may represent a significant regional structure.

Wrench reactivation of these transfer faults in this stress field would have been left-lateral (Fig. 15), and would have produced en echelon anticlines and/or reverse faults with a northeasterly trend in the overlying subsidence stage sediments. Figure 16 demonstrates that there is an excellent correlation between the anticlinal structures in the Gippsland basin and these major transfer fault

354

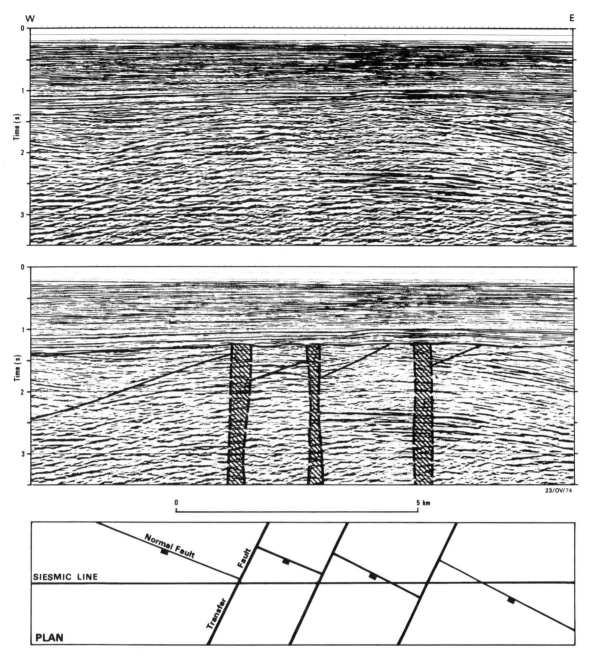

Figure 11 — Seismic section from near the northern margin of the Gippsland Basin. This section is close to being parallel to the strike of a rotational normal fault where it is progressively offset by a series of minor transfer faults (see plan). Reproduction of this seismic section without location information is with the permission of Shell Development (Australia) Pty Ltd.

zones. In particular, the Kingfish, Mackerel, Halibut, Flounder and Tuna fields are aligned above the eastern transfer zone referred to above, and comprise en echelon anticlinal and/or reverse fault structures with the northeasterly to easterly trend expected from left-lateral reactivation of the transfer fault. The alignment of the Bream and Snapper structures indicates that a third reactivated transfer zone may exist across the central region of the

basin. It is also apparent from Figure 16 that there has been some reactivation of the Early Cretaceous rotational normal faults along the northern margin of the basin. Because of the oblique reverse reactivation of these structures, their interpretation in seismic section is complex. Once again, the technique of mapping inclined reflector packages in the Early Cretaceous sediments has been used to locate the extensional faults, giving rise to the struc-

LINE 11

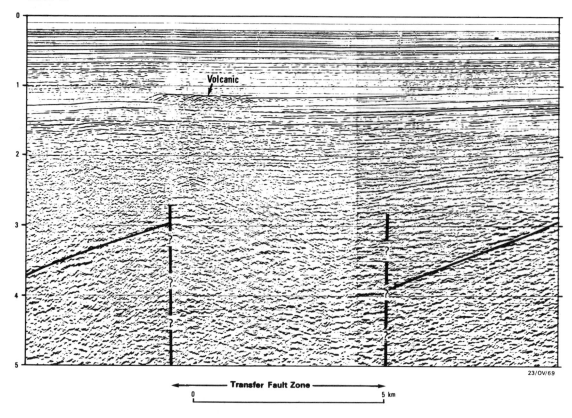

Figure 12 — Part of BMR line 11, along the axis of the Bass Basin, near its southeastern end. The major transfer fault zone that terminates the Early Cretaceous extensional region of the basin (Etheridge *et al.*, 1984, Fig. 5) is marked on the section. Note the small volcanic accumulation above the zone, the termination of dipping reflectors in the Early Cretaceous sequence against the zone, and the virtual absence of reflectors within it. Section is approximately true scale at 3 s TWT.

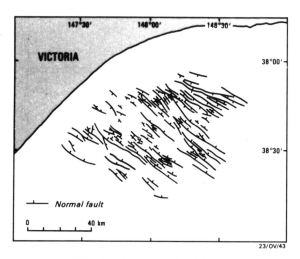

Figure 13 — Map showing traces of mainly subsidence stage faults in Gippsland Basin (after Threlfall *et al.*, 1976). Note the largely down-to-basin displacements, in contrast to the extensional faults shown in Figure 6.

tures shown in the northwest corner of the basin in Figure 16. Examples of reverse reactivation of normal faults from the northern and southern margins of the Gippsland Basin are illustrated in Figure 17.

Current mapping of the late structures along the northern margin of the Gippsland Basin casts doubt on the existence of the Rosedale Wrench Fault as a discrete late structure. Wherever this mapping has been in detail, this proposed feature and its relatives to the north (Threlfall *et al.*, 1976) are not readily distinguishable from reactivated Early Cretaceous rotational normal faults and the short transfer fault segments that connect them (Fig.16). The east-west fault systems figured by Threlfall *et al.* (1976) correspond closely to the traces of known Early Cretaceous normal faults. The overall east-west trend results from progressive left-lateral offset of the normal faults along the contemporaneous transfer faults during the somewhat oblique extension that initiated the basin. Reactivation of the early structures seems to reduce in intensity southwards, presumably reflecting a significant north to south decrease in the post-Eocene differential stress magnitude. A similar tendency is seen in the Bass

356

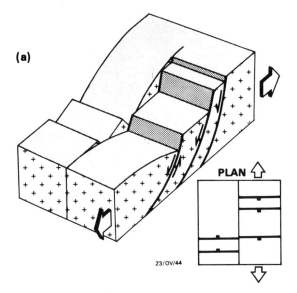

(a)

PLAN

23/OV/44

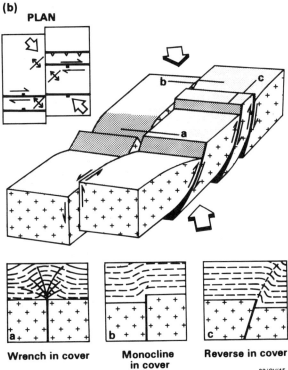

(b)

PLAN

Wrench in cover   Monocline in cover   Reverse in cover

23/OV/45

Figure 14 — Schematic block diagrams illustrating the role of reactivation of extensional normal and transfer faults during Tertiary northwest-southeast compression.

(a) Extensional normal and transfer fault array in the basement to an extensional basin such as the Gippsland Basin.

(b) Reactivation of normal and transfer faults during oblique compression (equivalent to northwest-southwest in Gippsland) to produce an array of structures in the thermal subsidence stage sediments overlying the major structures (see text for further explanation).

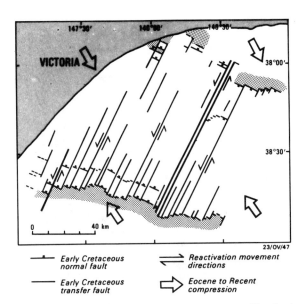

VICTORIA

23/OV/47

⊢—— Early Cretaceous normal fault

—— Early Cretaceous transfer fault

⇌ Reactivation movement directions

⇨ Eocene to Recent compression

Figure 15 — Map of Early Cretaceous structures, Gippsland Basin (after Fig 6), showing how Tertiary northwest-southeast compression gives rise to left-lateral wrench reactivation of transfer faults, and oblique reverse reactivation of normal faults.

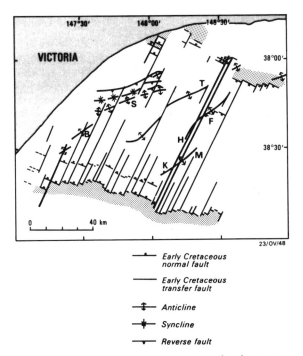

VICTORIA

23/OV/48

⊢—— Early Cretaceous normal fault

—— Early Cretaceous transfer fault

⚓ Anticline

⚓ Syncline

▽— Reverse fault

Figure 16 — Map of Late Tertiary compressional structures (most of them hydrocarbon reservoirs) superimposed on the Early Cretaceous extensional structures of the Gippsland Basin. Note especially the alignment of the major oil-producing structures (Mackerel-M, Kingfish-K, Halibut-H, Flounder-F, and Tuna-T) along the major transfer fault zone that separates regimes in which the dip of the rotational normal faults is opposed. This zone is known to contain an array of vertical faults that extend to depths of at least 12 to 15 km.

357

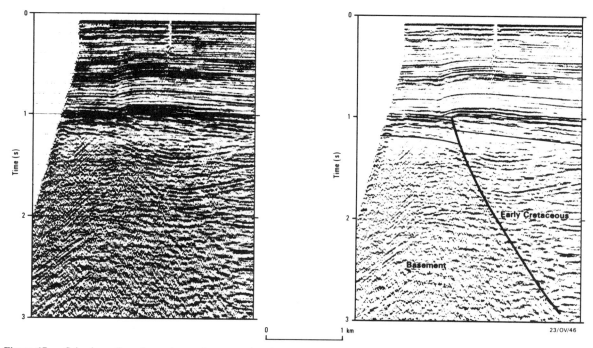

Figure 17 — Seismic sections from the southern margin of the Gippsland Basin showing evidence of Tertiary reverse movement on Early Cretaceous normal faults. This section is reproduced without location information with the permission of Australian Aquitaine Petroleum Pty Ltd.

Basin, where young structures (e.g. Snail, Nerita, Konkon, Cormorant), probably associated with reactivation of older faults, are restricted to the northern part of the basin. Little reactivation of the major normal and transfer fault systems in the central and south-eastern parts of the basin has taken place.

DISCUSSION

Recognition of Extensional Regimes

In all but the most sediment-starved extensional basins, the major normal and transfer faults have been buried beneath several km of sediment deposited during post-stretching thermal subsidence. The depth of these important basin-forming faults (commonly greater than 3 seconds TWT) means that they are either poorly resolved on or largely below the display depth of routine seismic reflection profiles, except near basin margins. The importance of extensional structures in the development of both structural and stratigraphic traps has been widely demonstrated (Harding & Lowell, 1979; Harding, 1983, 1984; Gibbs, 1984a, 1984b; Etheridge et al., 1984; Pegrum & Ljones, 1984). The recognition and accurate reconstruction of the extensional fault geometries is therefore an important factor in the deeper level exploration of extensional basins.

The interpretation of seismic sections is commonly model-dependent, because of poor data quality and/or ambiguity of horizon picking, especially at deeper levels (Gibbs, 1983). It is therefore important to have criteria for recognising extensional regimes, so that the relevant structural models can be applied. Based on experience in the Bass Strait basins, the following criteria are suggested:

(i) Shallow-dipping normal faults.

ii) Significantly tilted basement or syn-rift sediment packages, especially where the dip direction is consistent from fault block to fault block.

(iii) Two orthogonal fault sets, where one set comprises more shallowly dipping normal faults, and the other set comprises steeply dipping faults with variable displacement.

The most important criterion is the presence of shallow-dipping normal faults. As shown in Figure 18, substantial extension can only be accomplished by initially shallow faults or by faults which rapidly rotate towards shallower dips as extension proceeds (Wernicke & Burchfiel, 1982). The corollary is that shallow-dipping normal faults with large (cumulative) displacements require significant upper crustal extension. The first two criteria are commonly related, in that many shallow faults are rotational, and the concommitant rotation of the blocks between faults gives rise to substantially tilted basement and rift-fill sequences. We regard the presence of systematically tilted reflectors, especially beneath the so-called rift unconformity (Falvey, 1974), to be a key indicator of rotational normal faulting, and therefore horizontal extension. Figure 19 illustrates two possible interpretations of an array of dipping seismic reflectors below a rift unconformity. Rotation of the lower sedimentary sequence back to horizontal must be accompanied by sympathetic rotation of the fault. This reconstruction causes the fault in Figure 19a to rotate

358

THE APEA JOURNAL, 1985

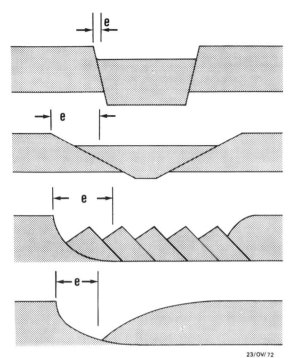

Figure 18 — Relationships between fault geometry and crustal displacement field, demonstrating that shallow-dipping faults, whether rotational or not, require substantial horizontal extension (e), and vice versa.

(a)

(b)

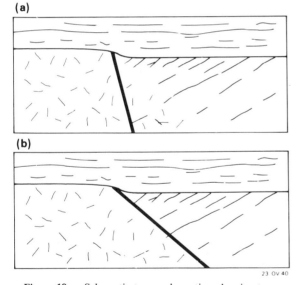

Figure 19 — Schematic true scale section showing two alternative interpretations of a major normal fault separating basement to the left from a dipping syn-rift sequence on the right. Discontinuity and/or overlap of reflectors near a fault commonly allows such ambiguity of fault orientation. In this case, the interpretation in (a) is likely to be incorrect, because rotation of the syn-rift sediments back to horizontal results in rotation of the fault through the vertical. See text for further explanation.

through vertical and reverse its dip direction, a most unlikely scenario that requires early reverse movement in an extensional regime. Reconstruction of Figure 19b, however, gives rise to an initial fault dip of about 70°, which is within the expected range for initiation of a near-surface normal fault.

Reconstructions such as that described above form the basis of the technique of balancing cross-sections. The concept of balanced cross-sections was first widely applied in the analysis of terrains substantially shortened by thrusting (Dahlstrom, 1969). However, as Gibbs (1983,1984b) has recently demonstrated, the concept applies equally to substantially extended terrains. All significantly faulted sections should be tested for balance (e.g. using constant bed length), and should preferably be palinspatically reconstructed (e.g. Gibbs, 1984b; Beach, 1984).

The third criterion emphasises the importance of the transfer fault concept (Gibbs, 1984a; Etheridge et al., 1984). In the authors' view, extensional basins will generally contain transfer faults, in the same way that transform faults are a general feature of oceanic basins. The density of transfer faults will tend to increase with the obliquity of the extension direction to the basin margin trend. In branched rift systems such as in Bass Strait and the North Sea, the importance of transfer faults is greater in the branches (e.g. Bass and Gippsland Basins; Viking and Moray-Firth Graben) than in the trunk rift (Otway Basin; Central Graben). In fact, there is a complete gradation from symmetrical rifts with spreading perpendicular to their margins and low transfer fault density, to transtensional rift basins associated with dominantly transcurrent motion (Harding & Lowell, 1979). The authors maintain that transfer faulting is an almost universal result of crustal extension, because of the difficulty of developing continuous normal faults along the whole rift length. Normal faults are likely to develop independently at various locations along the rift as extension begins, and transfer faults will then be required to accommodate these along-strike variations in normal fault geometry. Even in relatively symmetrical extensional basins such as the Otway Basin, transfer faults may be widespread, mainly as accommodation structures bounding offsets in the normal fault system.

Extensional Structures and Hydrocarbon Exploration

In the Bass Strait basins, the major extensional normal and transfer faults have played a key role in a number of important facets of basin evolution that are relevant to hydrocarbon exploration.

1) On the gross scale, recognition of the extensional regime of the basins is important for modelling and understanding their thermal and subsidence histories. The thermomechanical history of the Bass Basin has been modelled by Karner et al. (1984), and they have demonstrated that the gravity profile and overall geometry of the basin are consistent with approximately 60 percent horizontal extension during the Early Cretaceous. The thermal consequences of this model were used to derive the maturation/subsidence scenario for the play near Bass 1 described by

Etheridge *et al.* (1984). Of particular importance in this context is the timing of hydrocarbon generation relative to the structural evolution in the basins.

2) During extension, the faults and related structures impose the primary control on facies distribution. Tilted half-grabens bounded by normal and transfer faults (Fig. 5) have the potential for both up-dip and along-strike closure within the syn-rift fill. This style of play is very important in the North Sea, and other extensional basins (Harding, 1984) but is generally discounted in the Bass Strait basins because of the apparently uniformly low porosity and permeability of the dominantly volcanolithic Otway and Strzelecki Groups. However, as pointed out by Williamson *et al.* (this volume), there has been so little well penetration of the offshore extent of these Groups that it may be premature to dismiss their prospectivity.

3) Most importantly, extensional structures may exert a significant influence on both the sedimentation and deformation of the overlying thermal subsidence sequence. Initially, the extension is likely to give rise to substantial basement topography of the style of the Basin and Range province. Major horst blocks and tilt block corners provide both local controls on sedimentation and the loci for drape/compactional structures (Etheridge *et al.*, 1984). In the basin centre, where the extensional structures are very deep, these controls may be obscured by poor seismic data quality. As subsidence proceeds, minor reactivation of the larger scale extensional structures exerts subtle, but potentially important influences on facies distribution. In the Bass Basin, a sedimentological analysis of the late Tertiary marine sequence has demonstrated that the main transfer fault zones influenced facies distributions through to the Pliocene (B. Radke, BMR, personal communication, March 1984). In parts of the Otway Basin, Early Cretaceous transfer faults are overlain by narrow channels that extend up through much of the subsidence sequence. Such reactivation may continue to the present day, as in the southwestern portion of the Gippsland Basin, where straight tributaries of the Bass Canyon follow transfer fault traces (Fig. 20). The most dramatic influence of the extensional structures on younger deformation in the Bass Strait basins is on the location and style of the Tertiary compressional structures that contain the bulk of the petroleum reserves discovered to date in the Gippsland Basin (Figs 14 to 16). Most extensional basins will contain large normal and transfer fault zones that penetrate much of the sub-basinal crust. It is likely, therefore, that the Gippsland situation is not unusual, and that early faults will exert a similar control on younger structures in other extensional basins (Pegrum & Ljones, 1984). Finally, reworking of the extensional structures is likely to induce zones of fracturing in the overlying sequence. These fractured zones may affect fluid migration, and, in particular, allow hydrocarbon migration into the overlying structures.

In summary, it has been attempted to demonstrate that the most prospective sequences in the Bass, Gippsland and Otway Basins are underlain by an array of structures

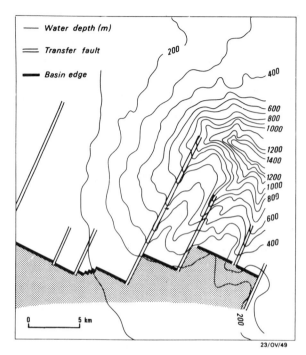

Figure 20 — Bathymetric map of the southeastern part of the Gippsland Basin superimposed on the traces of Early Cretaceous transfer faults. Recent reactivation of the transfer faults has apparently controlled the orientation and position of the tributary channels of the Bass Canyon.

produced during substantial (60 to 100%) crustal extension. Further, these extensional structures have exerted a range of important controls on both sedimentation and deformation of the basins, and on the generation, migration and location of hydrocarbons within them. Clearly, knowledge of the extensional structural framework was not a prerequisite for the discovery and successful exploitation of the major top-Latrobe Group petroleum-producing structures in the Gippsland Basin. However, it is suggested that deeper, more subtle, but still attractive plays may be essentially untested in all three basins (Etheridge *et al.*, 1984) and that an understanding of the extensional structural framework is an important factor in exploration for these less obvious targets.

ACKNOWLEDGMENTS

The Bass Strait project involves a large group of Bureau of Mineral Resources scientists, who have contributed in various ways to the final outcome of this paper, without necessarily agreeing with all of its conclusions. We would like to thank, in particular, Jim Colwell, David Falvey, Anne Felton, Gordon Lister, Keith Lockwood, Evelyn Nicholas, Bruce Radke, Steve Scherl and Paul Williamson. In addition we acknowledge the cooperation of the exploration companies in the Bass, Gippsland and Otway Basins, who have so freely granted access to their seismic data. Published with the permission of the Director, Bureau of Mineral Resources Geology and Geophysics.

REFERENCES

BALLY. A.W., 1981 — Atlantic — Type Margins in Geology of Passive Continental Margin. *American Association of Petroleum Geologists* Education Course Note Series, 19, 1-48

BARTON, C.M., 1981 — Regional stress and structure in relation to brown coal open cuts of the Latrobe Valley, Victoria. *Journal of the Geological Society of Australia*, 28, 333-340

BEACH, A., 1984 — The structural evolution of the Witch Ground Graben. *Journal of the Geological Society of London*, 141, 621-628

BOEUF, M.G. DOUST, H., 1975 — Structure and development of the Southern margins of Australia. *APEA Journal*, 15(1), 33-43

BOYER, S.E., & ELLIOT, D., 1982 — Thrust systems. *American Association of Petroleum Geologists Bulletin*, 66, 1196-1230

BROWN, B.R., 1976 — Bass Basin — some aspects of the petroleum geology. *In:* LESLIE, R.B., EVANS, H.J., & KNIGHT, C.L. (Eds), *Economic Geology of Australia and Papua New Guinea, 3. Petroleum.* Australasian Institute of Mining and Metallurgy, Monograph 7, 67-82

BURKE, K., & DEWEY, J.F., 1973 — Plume-generated triple junction, key indicators in applying plate tectonics to old rocks. *Journal of Geology*, 81, 406-433

CAREY, S.W., 1970 — Australia, New Guinea and Melanesia in the current revolution in concepts of the evolution of the earth. *Search*, 1, 178-189

DAHLSTROM, C.D.A., 1969 — Balanced cross sections. *Canadian Journal of Earth Sciences*, 6, 743-757

DAVIDSON, J.K., 1980 — Rotational displacements in southern Australia and their influences on hydrocarbon occurrence. *Tectonophysics*, 63, 139-153

DENHAM, J., & BROWN. B.R., 1976 — A new look at Otway Basin. *APEA Journal*, 16(1), 91-98

DENHAM, D., WEEKES, J., & KRAYSHEK, C., 1981 — Earthquake evidence for compressive stress in southeast Australian crust. *Journal of the Geological Society of Australia*, 28, 323-332

DOUGLAS, J.G., & FERGUSON, J.A., 1976 — Geology of Victoria. *Geological Society of Australia, Special Publication*, 5

ELLIOT, D., & JOHNSON, M.R.W., 1980 — Structural evolution in the northern part of the Moine Thrust Zone. *Transactions of the Royal Society of Edinburgh*, 71, 69-96

ELLIOT, J.L., 1972 — Continental drift and basin development in south Eastern Australia. *APEA Journal*, 12(1), 46-51

ETHERIDGE, M.A., BRANSON, J.C., FALVEY, D.A., LOCKWOOD, K.L., STUART-SMITH, P.G., & SCHERL, A.S., 1984 — Basin forming structures and their relevance to hydrocarbon exploration in Bass Basin, southeastern Australia. *Bureau of Mineral Resources Journal of Australian Geology and Geophysics*, 9 (in press)

FALVEY, D.A., 1974 — The development of continental margins in plate tectonics theory. *APEA Journal*, 14(1), 95-106

GIBBS, A.D., 1983 — Balanced cross-section construction from seismic sections in areas of extensional tectonics. *Journal of Structural Geology*, 5, 153-160.

GIBBS, A.D., 1984a — Structural evolution of extensional basin margins. *Journal of the Geological Society of London*, 141, 609-620

GIBBS, A.D., 1984b — Clydefield growth fault, secondary detachment above basement faults in North Sea. *American Association of Petroleum Geologists Bulletin*, 68, 1029-1039

GIBSON. G., WESSON, V., & CUTHBERTSON. R., 1981 — Seismicity of Victoria to 1980. *Journal of the Geological Society of Australia*, 28, 341-356

GLENIE, R.C., 1971 — Upper Cretaceous and Tertiary rock-stratigraphic units in the Central Otway Basin. *In:* WOPFNER, H., & DOUGLAS, J.G. (Eds), *The Otway Basin of Southeastern Australia*, Special Publication of the Geological Surveys of South Australia and Victoria, 193-215.

GRIFFITHS, J.R., 1971 — Continental Margin tectonics and the evolution of Southeast Australia. *APEA Journal*, 11(1), 75-79

GUNN, P.J., 1975 — Mesozoic-Cainozoic tectonics and igneous activity — Southeastern Australia. *Journal of the Geological Society of Australia*, 22, 218-222

HARDING, T.P., 1983 — Graben hydrocarbon plays and structural style. *Geologie en Mijnbouw*, 62, 3-23

HARDING, T.P., 1984 — Graben hydrocarbon occurrences and structural style. *American Association of Petroleum Geologists Bulletin*, 68, 333-362

HARDING, T.P., & LOWELL, J.D., 1979 — Structural styles, their plate-tectonic habitats and hydrocarbon traps in petroleum provinces. *American Association of Petroleum Geologists Bulletin*, 63, 1016-1058

HOCKING, J.B., 1972 — Geologic evolution and hydrocarbon habitat, Gippsland Basin. *APEA Journal*, 12(1), 132-137

JAMES, E.A., & EVANS, P.R., 1971 — The stratigraphy of the offshore Gippsland Basin. *APEA Journal*, 11(1), 71-74

KARNER, G., ETHERIDGE, M.A., BRANSON, J.C., & SCHERL, A.S., 1984 — Sedimentary basin modelling: constraints from the Bass Basin. *Geological Society of Australia Abstracts*, 12, 291-292

LE PICHON, X., & SIBUET, J.C., 1981 — Passive margins: a model of formation. *Journal of Geophysical Research*, 86, 3708-3720

McKENZIE, D.P., 1978 — Some remarks on the development of sedimentary basins. *Earth and Planetary Science Letters*, 40, 25-32

MIDDLETON, M.F., 1982 — The subsidence and thermal history of the Bass Basin, Southeastern Australia. *Tectonophysics*, 87, 383-397

NICHOLAS, E., LOCKWOOD, K.L., MARKS, A.R., & JACKSON, K.S., 1981 — Petroleum potential of the Bass Basin. *Bureau of Mineral Resources Journal of Australian Geology and Geophysics*, 6, 199-212

PEGRUM, R.M., & LJONES, T.E., 1984 — 15/9 Gamma Gas Field offshore Norway, new trap type for North Sea Basin with regional structural implications. *American Association of Petroleum Geologists Bulletin*, 68, 874-902

REYNOLDS, M.A., EVANS, P.R., BRYAN, R., & HAWKINS, P.J., 1966 — The stratigraphic nomenclature of Cretaceous rocks in the Otway Basin. *Australian Oil and Gas Journal*, 13, 26-33

ROBINSON, V.A., 1974 — Geological History of the Bass Basin. *APEA Journal*, 14(1), 44-49

THRELFALL, W.F. BROWN, B.R., & GRIFFITH, B.R., 1976 — Gippsland Basin, Offshore. *In:* LESLIE, R.B., EVANS, H.J., & KNIGHT, C. L., (Eds), *Economic Geology of Australia and Papua New Guinea, 3. Petroleum.* Australasian Institute of Mining and Metallurgy, Monograph 7, 41-67

WERNICKE, B., & BURCHFIEL, B.C., 1982 — Modes of extension tectonics. *Journal of Structural Geology*, 4, 105-115

WILLIAMSON, P.E., PIGRAM, G.J., COLWELL, J.B., SCHERL, A.S., LOCKWOOD, K.L., & BRANSON, J.C., 1985 — Pre-Eocene structure and stratigraphy of the Bass Basin. *APEA Journal*, this volume

Journal of Structural Geology, Vol. 4, No. 2, pp. 105 to 115
Printed in Great Britain

0191–8141/82/020105–11 $03.00/0
© 1982 Pergamon Press Ltd.

Modes of extensional tectonics

BRIAN WERNICKE and B. C. BURCHFIEL

Department of Earth and Planetary Sciences, Massachussetts Institute of Technology, Cambridge, MA 02139, U.S.A.

(*Received* 1 *September* 1981, *accepted in revised form* 10 *March* 1982)

Abstract—Although hundreds of papers have been devoted to the geometric and kinematic analysis of compressional tectonic regimes, surprisingly little has been written about the details of large-scale strain in extended areas. We attempt, by means of quantitative theoretical analysis guided by real geological examples, to establish some ground rules for interpreting extensional phenomena. We have found that large, very low-angle normal faults dominate highly extended terranes, and that both listric and planar normal faults are common components of their hanging walls. The very low-angle normal faults may have displacements from a few kilometres up to several tens of kilometres and we regard their hanging walls as extensional allochthons, analogous (but with opposite sense of movement) to thrust-fault allochthons. Differential tilt between imbricate fault blocks suggests listric geometry at depth, whereas uniformly tilted blocks are more likely to be bounded by planar faults. The tilt direction of imbricate normal-fault blocks within large extensional allochthons is commonly away from the transport direction of these sheets, but in many cases tilts are in the same direction as transport, thus limiting the usefulness of the direction of tilting as a transport indicator. The presence of *chaos* structure, a structural style widely recognised in the Basin and Range Province, implies large scale simple shear on very low-angle normal faults and does not necessarily form as a result of listric faulting.

INTRODUCTION

WORK in extensional terranes has shown that they contain a great variety of faults active during deformation. Extensional faults can be grouped into two broad categories: (1) those which produce extension accompanied by rotation of beds, and in a subgroup are additionally accompanied by rotation of the faults and (2) those which produce extension without rotation of faults or beds. In this paper we present an analysis of the geometric and kinematic properties of these fault types and examples of them from extensional terranes, principally the Basin and Range Province of the western United States. The Basin and Range Province is important in the study of the rifting process because it has not thermally subsided and been covered by post-rift sediments, it is well exposed and locally well mapped, it contains a great variety of extensional fault-types, and finally, because vertical movements and erosion have been substantial, a great variety of structural levels are exposed within the extensional complexes. We believe that the analysis of faults and extensional features in this region can be applied to other extensional terranes, particularly to passive continental margins.

Whereas in most extensional terranes the amount of extension is poorly constrained, there are regions within the Basin and Range Province where a minimum amount of extension can be measured without making any assumptions about cross-sectional fault geometry. Such analyses indicate province-wide extension of 64–100% and local areas which have extended well over 100% (e.g. see Hamilton & Myers 1966, Davis & Burchfiel 1973, Hamilton 1978, Wernicke *et al.* 1981). Exten-

sion of this order has been suggested for continental margins, but their fault geometries are less well-known. Our analysis indicates that certain fault types (or combinations of them) can easily produce extension of this magnitude, but these fault types are not the ones commonly assumed to have produced the extension. The conclusion we reach is that widespread imbricate normal-fault blocks (Anderson 1971) and subjacent, large low-angle normal faults provide an explanation for the large magnitudes of extension observed. Listric normal faults function to relieve space problems between families of planar fault blocks, but are not the only contributor to the extension. The geometric and kinematic models we present are testable in the field and by geophysical techniques, especially seismic reflection profiling.

EXTENSIONAL MECHANISMS

We divide extensional structures into two broad classes: rotational and non-rotational (Table 1).

Rotational extension

Rotational extension is defined as a kinematic

Table 1. Types of normal faults

Group	Structures rotated	Fault geometry
Non-rotational	Nothing	Planar
Rotational	Beds	Listric
Rotational	Beds and faults	Planar or listric

mechanism in which extension occurs predominantly by progressive rotation of geological features (hereafter called beds). There are two possible fault geometries in this class: planar and listric.

Planar geometry is depicted in its simplest form in Fig. 1 (all the models presented here are highly simplified and assume no penetrative deformation, pressure solution or bedding-plane slip). As the rock mass is extended, both the faults and the beds rotate (Emmons & Garrey 1910) (Table 1). The simple calculation, the basis of which is presented in Fig. 2 (Thompson 1960), allows us to determine the amount of extension if the attitudes of the faults and beds are known. Figure 3 is a convenient graphical representation of the relationship derived in Fig. 2. Because field examples of rotational extension are invariably more complex than the geometry shown in Fig. 1, it is possible to use Fig. 3 only for approximate determinations. Accurate determination of extension in any given area can be done only by palinspastic restoration of geological maps and cross-sections.

A simple rotational listric fault (Table 1) is shown in Fig. 4. Extension occurs by the separation of two blocks on a curved surface. Because hanging wall strata move down a curved surface convex toward the footwall strata and maintain a constant orientation relative to that surface, listric faults produce differential tilt between hanging wall and footwall, i.e. bedding dips are steeper in the hanging wall than in the footwall. Thus, a series of imbricate listric fault blocks should display successively steeper tilts as one traverses the blocks in the direction of downthrow (Fig. 5), whereas a row of planar fault blocks should all be tilted by the same amount. Another distinction between rotational-planar and listric faults is that planar faults must rotate with bedding, whereas listric faults may remain fixed, rotation occurring only if their footwalls are rotated by structurally-lower faults. Therefore, two groups of rotational faults may be distinguished, one in which both faults and beds rotate, including both listric and planar faults, and a second in which the faults remain fixed while the beds rotate, a situation restricted to listric fault geometry (Table 1).

A geometrically possible situation in which both faults and beds rotate but which is difficult to classify in terms of planar or listric faulting is shown in Fig. 6. The

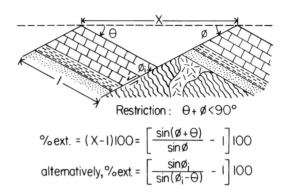

Restriction : $\theta + \emptyset < 90°$

$$\% \text{ext.} = (X-1)100 = \left[\frac{\sin(\emptyset + \theta)}{\sin \emptyset} - 1 \right] 100$$

alternatively, $\% \text{ext.} = \left[\frac{\sin \emptyset_i}{\sin(\emptyset_i - \theta)} - 1 \right] 100$

Fig. 2. Geometrical derivation of the relationship between fault dip ($\emptyset$) and percentage extension of the rock mass, as proposed by Thompson (1960).

extreme displacement on a series of imbricate listric-fault blocks (fault displacement being about the same as the length of the block) may result in a series of uniformly tilted planar-fault blocks. Even though the faults were initially listric, calculation of extension can be done assuming a planar geometry (Figs. 2 and 3), because the final geometry is that of a series of sub-planar blocks.

Consider a situation, similar to that illustrated in Fig. 5, in which the lowest block is rotated on a listric fault but the remainder of the blocks rotate on planar faults above a basal detachment surface (Fig. 7). It is important to note that even though the planar faults may curve into the detachment surface, rotation and thinning of the mass above it is not a result of listric faulting. Curvature of the faults near the base of the blocks, and the 'gap' (Fig. 1) created by rotation of the planar fault blocks above a detachment surface, are space problems which in real geological situations are accommodated by small-scale faulting, pervasive brecciation and, possibly, plastic flow. The key diagnostic feature of fault geometry is differential tilt between blocks, not necessarily the geometry of the fault near a basal detachment surface.

Syntectonic sedimentary deposits which show increas-

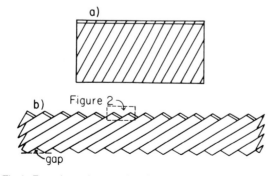

Fig. 1. Extension and attenuation of a rock mass by rotational planar normal faulting.

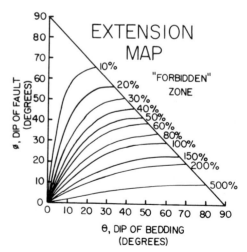

Fig. 3. Graphical representation of relations derived in Fig. 2.

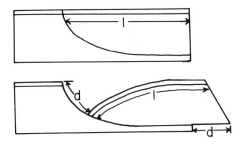

Fig. 4. Listric normal fault with reverse drag (Hamblin 1965).

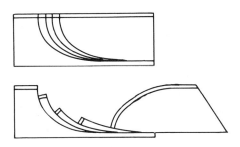

Fig. 5. Imbricate listric normal faults.

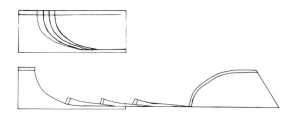

Fig. 6. Thin, imbricate listric fault blocks unflexed by extreme extension, the displacement on each fault being roughly the same as the length of the block.

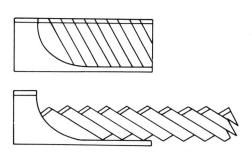

Fig. 7. Listric normal fault bounding a family of planar fault blocks.

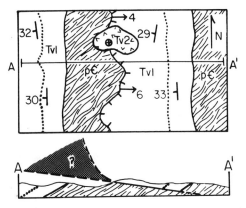

Fig. 8. Hypothetical geological map and section depicting the difference in amount of extension determined palinspastically between planar and listric rotational normal faulting.

ing dip with age, commonly referred to as growth-fault deposits, may develop in any setting involving rotational normal faulting. These deposits always dip toward the fault, except near the fault at the depositional surface where they may dip away from the fault.

A hypothetical situation shown on the geological map and section of Fig. 8 demonstrates the importance of diagnosing fault geometry. A low-angle normal fault dipping about 5° east offsets a section of 41 Ma volcanic rocks (TV1) and is overlain unconformably by a near-horizontal section of volcanics (TV2) dated at 39 Ma. Because the dip of bedding, and hence net rotation, of the two blocks is equal, the fault is best interpreted as planar, and its projection above the cross-section would be a straight line. However, if one were to assume that the fault was the flat portion of a listric fault and assumed it to steepen above the section, the estimate of extension represented by the fault would be considerably less than that predicted by assuming a planar fault geometry (for an application see Le Pichon & Sibouet 1981). To emphasize this point, we have constructed a geometric model of a listric fault (Fig. 9). The model assumes that the angle between bedding and the fault surface remains constant during deformation, and that curvilinear seg-

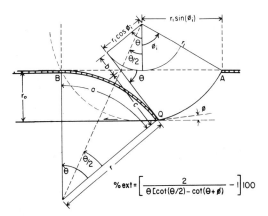

Fig. 9. Listric fault model. See text for explanation.

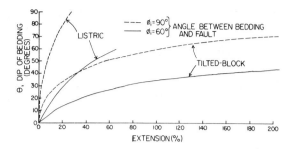

Fig. 10. Comparison of extension for listric and planar rotational normal faults.

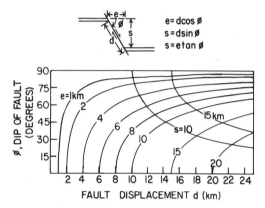

$$e = d\cos\phi$$
$$s = d\sin\phi$$
$$s = e\tan\phi$$

Fig. 11. Non-rotational normal fault, with equations for, and graphical representation of, the relations between displacement (d), extension (e), stratigraphic omission (s) and fault dip (ϕ).

ments of the diagram are arcs of circles. The percentage extension is calculated by comparing the distance between A and B with the length of *a*. By expressing the extension in terms of the dip of the beds next to the fault and the dip of the fault, we may compare the listric model with the planar model derived in Fig. 2. The result, shown in Fig. 10, demonstrates that for a given maximum stratal rotation and fault dip, the listric geometry yields far less extension than the planar geometry.

Non-rotational extension

Non-rotational extension is defined as a kinematic mechanism in which extension takes place without rotation of geological features, for example the high-angle normal fault in Fig. 11 (Table 1). For convenience we have plotted the dip of the fault vs fault displacement for various values of net extension. For non-rotational faults which have a very gentle dip, determination of displacement is difficult in most geological situations. Consider for example an undeformed sequence of strata cut by a very low-angle normal fault. The relationships displayed in Fig. 11 show that displacement and extension on the fault are specified if the dip of the fault and the stratigraphic omission across it are known. Thus, a more convenient representation of the magnitudes of extension possible on these faults is a plot of stratigraphic

omission vs fault dip (Fig. 12). In real geological situations, low-angle normal faults, in common with thrust faults, may show a ramp-décollement geometry if developed in sedimentary sequences of variable competence (Fig. 13) (Dahlstrom 1970). The fault dip appropriate for use in Fig. 12 in this situation would be the fault–bed angle averaged along the direction of transport. Introduction of listric faulting by ramps will cause rotation of strata, but the gross picture of one large sheet moving over another is most easily visualized as non-rotational, since there is no net rotation of the hanging wall. The distinguishing characteristic of low-angle normal faults, as opposed to thrust faults, is the juxtaposition of younger rocks on older with omission rather than repetition of strata. Thus, the rules of interpretation are the inverse of thrust faulting.

Large-displacement, very low-angle normal faults may show a number of movement planes, just as large displacement thrusts do. For example, consider the situation depicted in Fig. 14(a) where a large, low-angle normal fault is initiated in an undeformed sedimentary sequence. After an offset of one stratigraphic unit, movement is initiated on a slightly higher plane (Fig. 14b). The thin sheet between the two faults is accreted to the footwall of the first fault, and movement on the second of one more stratigraphic unit creates the configuration shown in Fig. 14(c), an attenuated stratigraphic section. The total displacement on the fault system is the sum of that across the two faults, and thus, no matter how complex the system of faults, the total stratigraphic omission may be used in Fig. 12 to determine how much extension the stack represents, provided the average fault–bed angle is reasonably well-known.

EXAMPLES

We believe that the modes of extension discussed above can be found in the geological record, and present here some examples of each type.

Perhaps the most completely documented type of extensional fault is the simple listric normal fault, shown here in a reflection profile from a continental margin setting (Fig. 15). Although Fig. 15 is a time section only, and thus cannot be regarded as a true geological section,

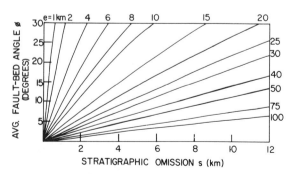

Fig. 12. Plot of stratigraphic omission (s) vs fault dip or average fault–bed angle (ϕ) using non-rotational fault model and equations from Fig. 9.

Fig. 13. Ramp-décollement geometry for large normal faults. Introducing a component of listric faulting forces some rotation of bedding (after Dahlstrom 1970).

the differential tilt between hanging wall and footwall, and the growth fault character are clear. We interpret the hanging wall as a series of uniformly tilted fault blocks bounded by planar faults, which serve to extend it without creating differential tilt between the blocks, that is the configuration is analogous to the geometry shown in Fig. 7. It is important to note that in moving downward from the steeply-dipping to the more gently-dipping portions of a listric fault, differential rotation should gradually decrease until the fault has the characteristics of a very low-angle non-rotational fault.

A series of imbricate normal-fault blocks showing successively steeper tilts was mapped by Anderson (1971) (Fig. 16). The configuration requires initially curviplanar fault blocks, although some have apparently been 'straightened out' by the extension (cf. Fig. 6).

Figure 17 shows a small-scale example of rotational planar faulting from the Rawhide Mountains of west-central Arizona, where measured and theoretical determinations of percent extension using Fig. 2 agree at about 25–30%. The first documented planar rotational normal faulting was recognized by Emmons & Garrey (1910) (Fig. 18). They mapped an impressive sequence of evenly-tilted fault blocks in Tertiary volcanic rocks in the Bullfrog Hills of southern Nevada. They interpreted the structures as having formed like a row of tilted dominoes in which both faults and beds rotated simultaneously and concluded that they could be most easily explained by extension of the crust. Fifty years later the relations between fault dip, stratal dip and extension, were derived by Thompson (1960), and were derived again by Morton & Black.(1975). Despite three separate conceptualizations of this mechanism, it has been almost completely ignored in the literature on the Basin and

Range Province, where beginning with Longwell (1945) virtually all small-scale imbricate normal faults were described as listric (except Thompson 1971), who speculated that some of Anderson's 1971 normal faults may be planar) to the point that low-angle normal fault and listric normal fault came to be used interchangeably by many authors. We believe that true listric normal faults as envisioned by most workers form only a portion of extended terranes, and that much (if not most) extensional strain in the earth's crust is accommodated by both rotational and non-rotational planar normal faults.

Non-rotational, high-angle extensional faults are described abundantly by many geologists (e.g. Stewart 1971), but much less attention has been given to their low-angle counterparts. One of us (Wernicke 1981) has emphasized the potential importance of these faults in accommodating lithospheric extension, and we suspect that structures produced by this type of fault are extremely common in the Basin and Range Province. For example, Dechert (1967) (Fig. 19) mapped a stack of fault slices in Palaeozoic miogeoclinal and Tertiary volcanic strata in the Schell Creek Range of east-central Nevada across which about 5 km of strata are missing. Because the faults are nearly parallel with bedding, the structure is most logically thought of as having been produced by the mechanism shown in Fig. 14. Consideration of Fig. 12 suggests that these sheets record tens of kilometres of extension. Noble (1941) coined the term *chaos* for this type of structure after an example he mapped in the Death Valley region (Amargosa chaos), and attributed its formation to a large, regional thrust sheet. Wright & Troxel (1969) recognised the extensional nature of the Amargosa chaos, and proposed that it developed in a zone of coalescing listric normal faults. Although their interpretation is reasonable for portions of the Amargosa chaos, we would like to emphasize that a comparable structural assemblage may form without rotational faulting simply by peeling sheets off the base of the hanging wall, that is the allochthon, of large displacement, non-rotational normal faults.

Another example of a large-scale extensional allochthon can be seen on a seismic reflection profile from the Sevier Desert area of central Utah (McDonald 1976)

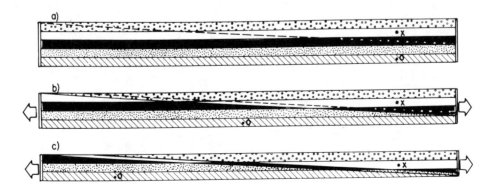

Fig. 14. Formation of *chaos* structure. (a) Unfaulted sedimentary sequence, with reference points marked × and ○. (b) Fault displaced one stratigraphic unit. (c) New fault with an additional offset of one stratigraphic unit. See text for discussion.

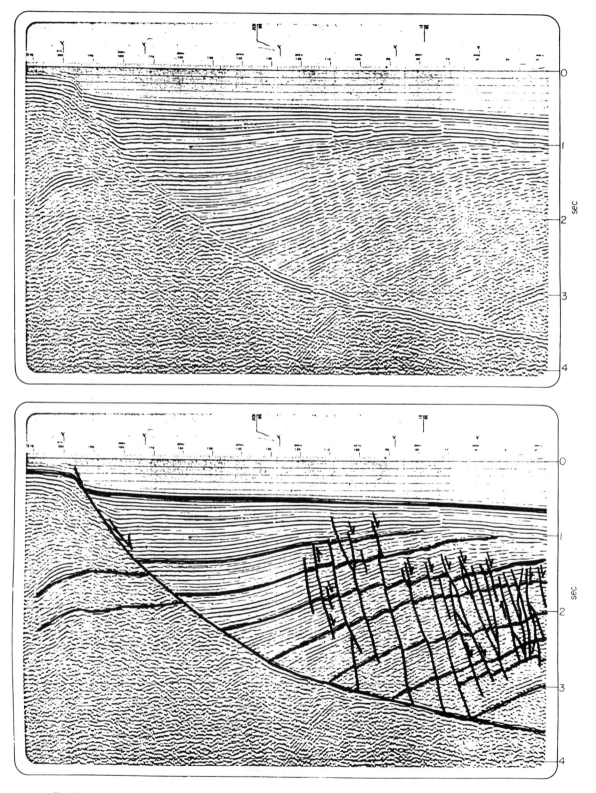

Fig. 15. Listric normal fault revealed by seismic reflection profiling. Reproduced with permission of A. G. Wintershall.

(Fig. 20). The section reproduced convincingly demonstrates that parts of a given hanging wall may remain unrotated while other parts may rotate in opposite directions. We conclude from this that (1) the sense of rotation in a given hanging wall does not indicate its transport direction and (2) the boundaries between tilted and non-tilted parts of extensional allochthons give no information as to the extent of their basal faults beyond those boundaries. In other words, large areas of seemingly intact rock at the surface may be underlain by large, low-angle non-rotational normal faults.

DISCUSSION

The extensional mechanisms described above are not mutually exclusive and can operate contemporaneously within a given extensional system. Listric normal faults, because of their geometry, can separate areas undergoing differential extension and merge at depth with low-angle normal faults. Above a low-angle normal fault further extension can be accommodated by imbricate normal faulting in which both faults and beds rotate (Fig. 21). In areas where large-magnitude extension has occurred, the low-angle normal faults and superjacent imbricate rotational faults may be the main contributors to the overall extension. This simple scheme can be modified and become more complex where differential extension has occurred in the rotated block sequence, in which case listric faults should be present. The low-angle fault or faults at the base of the faulted sequence anastomose leading to the development of *chaos*-type structure. Low-angle faults may also be present at different structural levels, thus dividing the crust into an imbricate stack of allochthonous slices each with both rotational and non-rotational fault blocks in their hanging walls.

One of the important problems is the geometry and character of the low-angle normal fault at the base of a series of faulted blocks. Wernicke (1981) has suggested such faults may involve the entire lithosphere. If this is the situation, the lower crust may be extended simply by divergence of two rigid slabs separated by a gently-dipping shear zone. Brittle shear would occur at shallow levels and grade downward along the low-angle fault to a zone of ductile shear (Fig. 21). The ductile shear would be restricted to the shear zone rather than distributed uniformly throughout the footwall crustal block, as envisioned by Eaton (1979) and Le Pichon & Sibouet (1981). With such a geometry, continental crust could be attenuated to any thickness, and unmetamorphosed sedimentary rocks could be juxtaposed by large-displacement low-angle normal faulting with any part of the crust. In this type of extensional system, rocks which were formerly sheared in a ductile state at deep structual levels along the shear zone early in the history of deformation may be reworked in the brittle regime as shown in Fig. 21.

It is possible, for example, that the prominent reflector at the base of the imbricate normal fault blocks in the Bay of Biscay (reflector 's', Fig. 22) is simply a crustal-scale low-angle normal fault, rather than a boundary between brittle extension and penetrative ductile stretching as suggested by de Charpal et al. (1978) and Le Pichon & Sibouet (1981). Such an interpretation is consistent with geometrically identical examples in the Basin and Range Province where the rocks below the basal detachments behaved as rigid plates during emplacement of the extensional allochthons (e.g. Misch 1960, Davis et al. 1980, Wernicke 1981). Thus, although Le Pichon & Sibouet (1981) demonstrated that the degree of extension by imbricate normal faulting was as large as a factor of two or three, their assumption that the lower crust stretched penetratively by that amount may be incorrect; it is geometrically possible that the crust beneath the basal reflector and mantle lithosphere had not extended at all! This view has rather dramatic implications for geophysical models of passive-margin rifting using subsidence history because it violates their fundamental assumption that the lithosphere stretches like a large elastic band (e.g. McKenzie 1978). It implies that a certain amount of crust which originally lay above reflector 's' is now incorporated in the complex southern margin of the Bay of Biscay and in the western Pyrenees. Furthermore, it implies the lower crust can be very heterogeneous.

Large displacement, low-angle normal faults which serve as boundary faults to both rotational and non-rotational extensional fault mosaics have been termed detachment faults by Davis et al. (1980) and Dokka (1981), and Davis et al. (1981) have followed this usage when describing a number of terranes throughout southern California, Arizona, northern Sonora and Mexico. We believe the overall geometry and kinematics of many extensional terranes have much in common, but because their three-dimensional geometries are

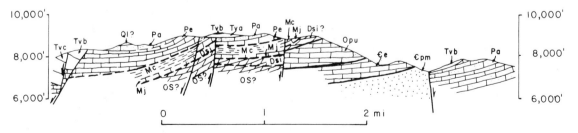

Fig. 19. Cross-section from the Schell Creek Range of east-central Nevada, showing the development of *chaos* structure (after Dechert 1967). Note that the horizontal and vertical scales are in miles and feet, respectively.

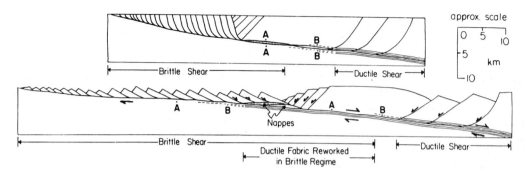

Fig. 21. Large, low-angle normal fault bounding an extensional fault mosaic comprised of listric and planar rotational faults, and an example of how *chaos*-structure might form (two imbricate nappes) beneath an imbricate pile of rotational normal faults. Also shown is a means by which a brittle-fault mosaic may be juxtaposed upon a slightly earlier-formed penetrative ductile fabric (e.g. Snoke 1980). From Wernicke (1981).

poorly known, the use of such terminology should remain informal.

Non-rotational high-angle normal faults are present in extensional terranes, but contribute only a small amount to the overall extension. They may be superimposed on older, large-scale extensional fault systems like those described above (Eberly & Stanley 1978, Zoback *et al.* 1981) and/or be the dominant fault type in areas that have undergone lesser amounts of extension in an inhomogeneously extended terrane.

CONCLUSIONS

The geometry and kinematics of faults in extensional regions can be grouped into two broad categories: (1) those which produce extension accompanied by rotation of layers, and in a subgroup are accompanied by rotation of the faults as well and (2) those which produce extension without rotation of faults or layers (Table 1). Our analysis suggests that large-scale extension is accomplished by large displacements on low-angle faults of the second group and rotated faults and fault blocks (both listric and planar) of the first group. Non-rotated listric normal faults are geometrically important as 'space fillers' but may not be as significant as the other fault types in producing large-scale extension. Several fault

types are related and form contemporaneously: listric normal faults, rotated faults and fault blocks, and large-displacement, non-rotational low-angle normal faults may all form a single fault system with the individual fault types unequally developed from place to place.

While many of the examples presented here are from the Basin and Range Province of the United States, the ideas are probably valid for any terrane which has undergone large magnitude extension. Passive continental margins are probably regions of large magnitude extension, and similar complex fault systems may have developed during their formation.

Our ideas are ultimately testable. Greater detailed mapping coupled with geophysics and drill-hole information should provide the necessary three-dimensional control. From such data we should be able to palinspastically reconstruct the extended terrane, just as we do thrust-fault terranes.

Acknowledgements—This paper benefited from the critical reviews of R. E. Anderson, A. W. Bally, P. L. Hancock and James Helwig. We thank M. S. Beaufait, R. G. Bohannon, G. A. Davis, R. K. Dokka, E. G. Frost, P. L. Guth, W. B. Hamilton, L. Royden and T. J. Shackleford for fruitful discussions. This work was supported by NSF Grant 7713637 awarded to B. C. Burchfiel and NSF Grant EAR7926346 awarded to B. C. Burchfiel and P. Molnar.

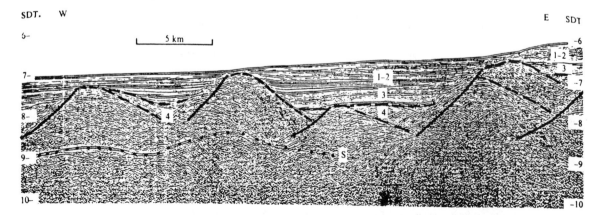

Fig. 22. Seismic reflection profile from the Bay of Biscay (from de Charpal *et al.* 1978).

REFERENCES

Anderson, R. E. 1971. Thin-skin distension in Tertiary rocks of southeastern Nevada. *Bull. geol. Soc. Am.* **82**, 43–58.

Dahlstrom, C. D. A. 1970. Structural geology in the western margin of the Canadian Rocky Mountains. *Bull. Can. Petrol. Geol.* **18**, 332–406.

Davis, G. A. & Burchfiel, B. C. 1973. Garlock fault: an intracontinental transform structure, southern California. *Bull. geol. Soc. Am.* **84**, 1407–1422.

Davis, G. A., Anderson, J. L., Frost, E. G. & Shackleford, T. J. 1980. Regional Miocene detachment faulting and early Tertiary mylonitization, Whipple–Buckskin–Rawhide Mountains, southeastern California and western Arizona. *Mem. geol. Soc. Am.* **153**, 79–130.

Davis, G. H., Gardulski, A. F. & Anderson, T. H. 1981. Structural and structural petrological characteristics of some metamorphic core complex terranes in southern Arizona and northern Sonora. In: *Geology of Northwestern Mexico and Southern Arizona, Field Guides and Papers* (edited by Ortlieb, L. & Roldan, Q.). Univ. Nat. Auton de Mexico, Inst. de Geologia, Hermosillo, Sonora, Mexico, 323–366.

de Charpal, O., Montadert, L., Guennoc, P. & Roberts, D. G. 1978. Rifting, crustal attenuation and subsidence in the Bay of Biscay. *Nature, Lond.* **275**, 706–710.

Dechert, C. P. 1967. Bedrock geology of the northern Schell Creek Range, White Pine County, Nevada. Unpublished Ph.D. thesis, University of Washington, Seattle.

Dokka, R. K. 1981. Thin-skinned extensional tectonics, central Mojave Desert, California. *Geol. Soc. Am. Abs. with Prgms* **13**, 52.

Eaton, G. P. 1979. Regional geophysics, Cenozoic tectonics and geologic resources of the Basin and Range Province and adjoining regions. In: *Basin and Range Symposium* (edited by Newman, G. W. & Goode, H. D.). Rocky Mountain Ass. of Geologists, Denver, 11–40.

Eberly, L. D. & Stanley, T. B., Jr. 1978. Cenozoic stratigraphy and geologic history of southwestern Arizona. *Bull. geol. Soc. Am.* **89**, 921–940.

Emmons, W. H. & Garrey, G. H. 1910. General Geology: In: *Geology and Ore Deposits of the Bullfrog District* (Ransome, F. L. *et al*). *Bull. U.S. geol. Surv.* **407**, 19–89.

Hamblin, W. K. 1965. Origin of "reverse-drag" on the downthrown side of normal faults. *Bull. geol. Soc. Am.* **76**, 1145–1164.

Hamilton, W. 1978. Mesozoic tectonics of the western United States. In: *Pacific Coast Paleogeography Symposium 2*. Pacific Section, Society of Economic Paleontologists and Mineralogists, 33–70.

Hamilton, W. & Myers, W. B. 1966. Cenozoic tectonics of the western United States. *Rev. Geophys.* **5**, 509–549.

LePichon, X. & Sibouet, J. 1981. Passive margins: a model of formation. *J. geophys. Res.* **86**, 3708–3720.

Longwell, C. R. 1945. Low-angle normal faults in the Basin and Range province. *Trans. Am. geophys. Un.* **26**, 107–118.

McDonald, R. E. 1976. Tertiary tectonics and sedimentary rocks along the transition, Basin and Range province to plateau and thrust belt province. In: *Symposium on Geology of the Cordilleran Hingeline* (edited by Hill, J. G.). Rocky Mountain Assoc. Geologists, Denver, 281–318.

McKenzie, D. 1978. Some remarks on the development of sedimentary basins. *Earth Planet. Sci. Lett.* **40**, 25–32.

Misch, P. 1960. Regional structural reconnaissance in central-northeast Nevada and some adjacent areas: observations and interpretations. *Intermountain Ass. Pet. Geol.*, 11th Ann. Field Conf. Guidebook, 17–42.

Morton, W. H. & Black, R. 1975. Crustal attenuation in Afar. In: *Afar Depression of Ethiopia, Inter-Union Commission on Geodynamics* (edited by Pilgar, A. & Rosler, A.). International Symposium on the Afar Region and Related Rift Problems, E. Schweizerbart'sche Verlagsbuchhandlung, Stuttgart, Germany, Proceedings, Scientific Report No. 14, 55–65.

Noble, L. F. 1941. Structural features of the Virgin Spring area, Death Valley, California. *Bull. geol. Soc. Am.* **52**, 941–1000.

Shackelford, T. J. 1980. Tertiary tectonic denudation of a Mesozoic-early Tertiary(?) gneiss complex, Rawhide mountains, western Arizona. *Geology* **8**, 190–194.

Snoke, A. W. 1980. The transition from infrastructure to suprastructure in the northern Ruby Mountains, Nevada. *Mem. geol. Soc. Am.* **153**, 287–334.

Stewart, J. H. 1971. Basin and Range structure: a system of horsts and grabens produced by deep-seated extension. *Bull. geol. Soc. Am.* **82**, 1019–1044.

Thompson, G. A. 1960. Problem of late Cenozoic structure of the Basin Ranges. *Proc. 21st Int. Geol. Congr., Copenhagen* **18**, 62–68.

Thompson, G. A. 1971. Thin-skin distension of Tertiary rocks of southeastern Nevada: discussion. *Bull. geol. Soc. Am.* **62**, 3529–3532.

Wernicke, B. 1981. Low-angle normal faults in the Basin and Range province: nappe tectonics in an extending orogen. *Nature, Lond.* **291**, 645–648.

Wernicke, B., Spencer, J. E., Burchfiel, B. C., Guth, P. L. & Davis, G. A. 1981. Magnitude of crustal extension in the southern Great Basin. *Geol. Soc. Am. Abs. with Prgms* **13**, 578.

Wright, L. A. & Troxel, B. W. 1969. Chaos structure and Basin and Range normal faults: evidence for a genetic relationship. *Spec. Pap. geol. Soc. Am.* **121**, 580–581.

Zoback, M. L., Anderson, R. E. & Thompson, G. A. 1981. Cenozoic evolution of the state of stress and style of tectonism of the Basin and Range province of the western United States. In: *Extensional Tectonics Associated with Convergent Plate Boundaries* (edited by Vine, F. J. & Smith, A. G.). *Phil. Trans. R. Soc.* **A300**, 407–434.

Structural geology

To stretch a continent

from I. R. Vann

THAT large areas of the Earth's crust have been subjected to compression has been known since before the discovery of plate tectonics. But, outside the world of hydrocarbon exploration, extensional tectonics has been a topic of intensive study for less than a decade. A recent conference* illustrated the remarkable progress that has arisen from two very different scientific approaches, the first a deduction straight from the world of physics, the other an induction rooted firmly in the geological tradition of careful field observation.

Seven years ago, D. McKenzie made what in retrospect seems a trivially simple prediction (*Earth. planet. Sci. Lett.* **40**, 25;1978) — if the lithosphere is stretched then it must thin. Since the base of the lithosphere is an isotherm, a thermal anomaly is created that slowly decays causing subsidence of the crust as an inverse exponential function of time. The power of this model is that it connects the stratigraphy and thermal history of the sediments filling a basin with a regional structure. At around the same time, field studies in the 'Basin and Range', the core of the Rocky Mountains of the western United States, began to show that the structure of the area is dominated by flat-lying extensional faults (detachments) with horizontal offsets across them of tens of kilometres.

This observation dispelled an old geological prejudice that sub-horizontal faults are all compressional structures (thrusts). Despite this a number of questions remain, concerning the original geometry of the flat detachments, the relevance of the Basin and Range model to other extensional basins and the relationship between the extension recorded near to the surface in rift basins and deeper structures.

At the conference the problems of unraveling the original geometries of flat detachments was exemplified by the contributions of G. H. Davis (University of Arizona), and E. L. Miller (Stanford University) and her co-workers. Davis suggested that the faults and the 'metamorphic core complexes' that underlie them were originally moderately dipping ductile zones of simple shear formed at 10 km or deeper within the crust. As extension progressed, the material overlying those 'faults' was moved laterally by tens of kilometres. The overall effect of this is to bring the faults nearer the surface and thus reduce the confining pressure and temperature. This causes a change in the deformation from a ductile to an increasingly brittle process. According to Davis, younger cross-cutting extension faults later rotated the dipping zones to horizontal.

*"Continental Extensional Tectonics", University of Durham, UK, 18–20 April 1985.

E. L. Miller's model is markedly different. She suggested that the Snake Range detachments in Nevada are neither faults nor shear zones but the transition, formed at a depth of 12 km, between an upper brittle plate and a lower ductile one, which latter deformed by pure shear expressed as horizontal extension and vertical thinning. Many other speakers accepted in modified form one or other of these apparently incompatible models but it was disconcerting for the outsider to see that this marvellously exposed area remains so enigmatic.

Even if the Basin and Range ultimately yields a single unified model it will remain a unique area of extreme extension within a region of relatively recent crustal thickening in the core of a thrust mountain belt. A possible ancient analogue — the Devonian (380 Myr old) basins of western Norway — was described by M. Serannem, M. Seguret and Ph. Laurent (University of Montpelier II). Here 50 km or more of extension occurs above a shear zone, apparently a reactivated thrust, which dips westwards at 5° to 25°, thereby creating a basin containing sediments up to 25 km thick. Geometrically similar basins, observed on deep seismic profiles (Brewer, J. A. & Smythe, D. K. *J. geol. Soc. Lond.* **141**, 105; 1984), occur above similarly reactivated thrusts on the western margin of the Caledonian mountain belt in north-west Scotland, which appear to flatten as they enter the lower crust at a depth of around 18 km. These fragmentary observations suggest that an extensional province similar to the Basin and Range existed within the Caledonian mountains some 380 Myr ago.

Is it possible to carry the Basin and Range analogy further? Certainly there are great differences between the extension within mountain belts and that which creates great sedimentary basins (rifts) in old stable crust (such as the East African Rift). Perhaps the most obvious of these differences is that almost all these rifts are covered by water or thick blankets of sediment. Hence their structure is not accessible to the geologist's hammer and our understanding depends largely on seismic reflection data that are even more ambiguous than surface observation. A. D. Gibbs (Midland Valley Exploration, Glasgow) presented models of rift fault systems developed in the North Sea that exemplify the belief of a growing number of workers that extension is achieved by movement on a linked series of faults that have flatter and steeper portions producing a 'staircase' geometry in the upper crust. These faults systems are strongly asymmetric giving rise to basins bounded

on one side only by major faults. Along the length of any basin this asymmetry flips from side to side across complex transfer zones.

Unfortunately the seismic data, which has been acquired almost exclusively by oil companies, images only the upper 10 km of the crust. Basin and Range models are applicable, if at all, to rift systems at depths of more than 15 km where it is thought that the crust changes from a generally brittle to a ductile material. Evidence about deformation at these depths below rifts comes from a very few deep reflection experiments conducted mainly on the UK continental shelf. A. Beach (Britoil, Glasgow) presented one such profile across the northern North Sea. His interpretation suggests two major periods of extension in this rift with active faulting in Triassic and early Jurassic times (220 – 200 Myr) above a single low-angle shear zone akin to the Basin and Range model of G. Davis being followed by extension at the end of the Jurassic along steep faults that cut entirely through the crust. The profile demonstrates that the crust below the rift is only half as thick as it is on the flanks and hence supplies one of the few direct tests of the McKenzie stretching model.

Support for Miller's interpretation of the Basin and Range came from a discussion by J. A. Jackson (Cambridge University) of deep deformation processes within active regions of extension. He noted that all major earthquakes in active regions occur on faults dipping at 30 – 60° at depths of 8 – 12 km and are grossly planar from the surface to these depths. Below this earthquakes are rare, but when they do occur (usually as aftershocks to large shallower events) they seem to do so on horizontal fault planes. Jackson concludes that extension in the lower crust normally takes place by ductile diffuse creep with only minor brittle effects.

Despite unanswered fundamental questions, and amidst much conflicting evidence and irreconcilable theory, many participants left the conference feeling that they had attended an important event. Among the conflict it seems there is a convergence that in a very few years will lead to a unifying 'general theory' of extensional tectonics. □

I. R. Vann, is a geologist with BP Petroleum Development, Britannic House, Moor Lane, London, EC2Y 9BU, UK.

Reprinted with permission from *Journal of Structural Geology*, v. 5, p. 471-482, J. A. Jackson and D. P. McKenzie, The geometrical evolution of normal fault systems, copyright 1983, Pergamon Press plc.

Journal of Structural Geology, Vol. 5, No. 5, pp. 471 to 482, 1983
Printed in Great Britain

0191–8141/83 $03.00 + 0.00
Pergamon Press Ltd.

The geometrical evolution of normal fault systems

JAMES JACKSON

and

DAN MCKENZIE

Bullard Laboratories, Madingley Rise, Madingley Road, Cambridge CB3 0EZ, U.K.

(*Received* 31 *January* 1983; *accepted in revised form* 6 *May* 1983)

Abstract—The purpose of this paper is to examine the kinematic behaviour of normal fault systems and see what general conditions govern their geometrical evolution. We pay particular attention to seismological and surface data from regions of present day active normal faulting, as the instantaneous three-dimensional geometry at the time of fault movement is better known in active regions than in areas where the faults are now static.

Most normal faults are concave upward, or listric. This shape can be produced by geometric constraints, either because the faults reactivate curved thrusts, or because they must be curved to accommodate rotations. Another effect which will produce curved faults is the variation of rheology with depth: brittle failure at shallow depths produces less fault rotation than does distributed creep in the lower part of the crust. An important geometric feature of normal faulting is the uplift of the footwall. The amount of such uplift is related not only to the elastic properties of the lithosphere, but also to the throw and dip of the fault. A striking feature of active normal faults is that they occur in groups in which all the faults dip in the same direction. This behaviour arises because the faults cannot intersect: if they do, one must cease to be active. The rotation which such fault systems produce reduces the dip of the faults until a new steeply dipping fault is formed. Once a new fault cuts pre-existing faults the earlier faults become locked, and a new set of faults must propagate rapidly across the whole region involved. Many of these geometric constraints also apply to thrust faulting.

INTRODUCTION

RECENT attention to the details of plate creation on oceanic ridges and to mechanisms of sedimentary basin formation in continental interiors has emphasized the importance of normal faults. It is therefore of interest to understand the evolution of isolated faults and the interaction of such faults with each other when they form an extensional system. The purpose of this paper is to examine the geometry of normal faults and the kinematic consequences of their movement. We have not attempted to address the more difficult question of their dynamics, and will therefore be concerned with displacements and strains, not stresses.

Although some of the ideas we discuss have been known for some time, this paper is not intended to be a comprehensive review of normal faulting, and is rather different in emphasis from most other treatments of this subject (e.g. Bally *et al.* 1981, Wernicke & Burchfiel 1982). We will be particularly concerned with constraints on fault kinematics imposed by observational data from areas of *active* present day faulting, such as the Aegean and central Greece. There are several reasons for concentrating on active regions.

(1) Most seismic faulting takes place during large earthquakes, which are comparatively rare. By studying the faulting associated with the largest earthquakes in a region we can be sure that we are dealing with the major faults responsible for the large-scale motions, and are not being misled by minor features such as the internal deformation of blocks bounded by the major normal faults.

(2) Seismological observations of the earthquake source tell us the fault orientation and slip vector (fault plane solution) at the depth of rupture initiation. This depth is typically 8–12 km in continental areas. By combining these observations with surface data we can know the fault geometry from the surface to depths of about 10 km *at the time of movement*; information that is simply not available in older, static regions where the present-day fault orientation has been affected by substantial uplift or rotation of the vertical.

(3) The seismic activity on a regional scale allows us to see the interaction between the major faults in a fault system at the time of movement. This is also difficult to deduce from static structures.

(4) A particular advantage of the Aegean is that much of it is below sea level. Ironically, although this obscures exposure, it provides a horizontal reference level against which to measure both vertical motions (uplift and subsidence) and tilt. The associated sedimentation also permits the accurate dating of structures and their associated vertical motions. As will be seen, these advantages are substantial.

Many of the ideas developed in this paper were suggested from investigations of normal faulting in central Greece, the Aegean, and western Turkey. Many faults are active in this region and their behaviour is dominated by their interaction. In particular, they can only remain active where they meet or intersect if certain conditions are fulfilled. These conditions, which are discussed in a later section, are geometric in origin and related also to the mechanical properties of rock. They are therefore general and impose kinematic constraints on both fault geometry and evolution in all areas of large-scale normal faulting where mass is conserved (but not necessarily in places like Iceland, where mass addition and dyke injection reach the surface). Because these conditions are kinematic, they apply regardless of the driving forces responsible for the normal faulting and extension. This allows us to compare structures in areas of active faulting with those in older geological terranes, or, specifically, to use features in the Aegean to guide us in the interpretation of structures in the Basin and Range Province and North Sea. All three are areas of large-scale continental extension and normal faulting, though in each case the *cause* of the extension may well have been different. As will be seen, these conditions also impose constraints on the geometry of thrust systems.

In the next (second) section, we discuss the geometry of individual normal faults. Two features of isolated normal faults which are of geological importance are footwall uplift and fault curvature. Both can be quantitatively related to the amount of extension. The third section of our paper is concerned with the interaction between faults. The second and third sections are not intended to be exhaustive reviews, and we deliberately address only what we regard as relevant to our discussion of fault kinematics. In the fourth section, we review the geometrical evolution and interaction of normal fault systems in the Aegean.

Throughout the discussion we assume that the regions between major faults form rigid blocks which cannot deform. Though aftershock locations and antithetic faulting show that this assumption is not generally valid, it does allow us to discuss kinematic problems relatively simply.

INDIVIDUAL NORMAL FAULTS

Listric faults and crustal rheology

The dip of many normal faults appears to flatten with depth. This listric geometry is often ascribed to the existence of a weak layer within the crust or sediment pile. In some cases the existence of weak layers can be demonstrated. Growth faults in the sediment piles on the northern margin of the Gulf of Mexico and in the Niger delta sole-out within layers of evaporite or overpressured shale (see e.g. Garrison & Martin 1973, Buffler *et al.* 1978). The movement on these faults is driven by gravity alone, and no plate motions are involved. On a smaller scale landslips have a similar geometry. In both these cases conservation of mass requires that the downward displacement of the material at the top of the moving body should be balanced by upward movement of the material at its bottom. Also in both cases the material beneath the fault is not deformed by its movement, and hence the displacement surface must intersect the Earth's surface at the base of the moving body as well as at its top. These conditions require growth faults and slip surfaces to be concave upward.

The same conditions do not apply to listric faults when they take up the motion at plate boundaries. There is then no requirement that the faults intersect the Earth's surface at its base, as well as at its top. In central Greece and western Turkey there is evidence that the active normal faults are approximately planar to depths of about 10 km. The most conclusive support for this claim comes from two large earthquake sequences that were studied in great detail: the 1978 series near Thessaloniki (Northern Greece) and the 1981 sequence in the eastern Gulf of Corinth. In both cases the fault plane solutions showed that the fault dips at the focii of the mainshocks were approximately 45°. Various seismological techniques were used to demonstrate that the focal depths of the mainshocks in each sequence were about 8–10 km and that the epicentres (the vertical projections of the focii to the surface) were all about 8–10 km from the surface outcrop of the faults which moved. The projections of the fault dips at the focii thus intersected the surface at the outcrop of the active faults. Although the dip of young normal faults is often very steep at the surface, this is invariably because the top few tens of metres near the Earth's surface fail in tension, not shear, commonly leaving open fissures and cracks. At Thessaloniki and Corinth, where the slip vector could be measured at the surface by matching displaced lineations such as wheel-tracks across the fissures, the dip of the slip vector at the surface agreed well with that calculated at the focal depth from seismological observations. For details of the observational data at Thessaloniki and Corinth, the reader is referred to Mercier *et al.* (1979a), Soufleris & Stewart (1981), Soufleris *et al.* (1982) and Jackson *et al.* (1982a). Thus, in these two earthquake sequences the active faults appear to have a roughly constant dip of 40–50° between the surface and a depth of about 10 km. Most, if not all, large normal faulting earthquakes in Greece and western Turkey appear to nucleate at depths of 8–12 km, and the great majority have dips of 40–50° at their focii (Jackson 1980). Although it is likely that these faults are also planar from the focal depth to the surface, only at Thessaloniki and Corinth were the necessary surface slip vector measurements and seismic observations made which allow this inference to be confirmed.

The planar nature and 40–50° dip of these active normal faults have a number of important consequences. Firstly, the dip is much steeper than the low angle dips of 0–10° that are commonly seen in the Basin and Range Province and which have received so much recent attention (e.g. Wernicke 1981, Wernicke & Burchfiel 1982, Frost & Martin 1982). In the Yerington district of western Nevada, Proffett (1977) pointed out that many of the very low angle normal faults, which are responsible for so much crustal extension, have been rotated flat by movement on later faults, and that at the time they were active had much steeper dips. He also estimated that at the time they were active, the faults in his area became horizontal only at substantial depths (16 km or more). This observation agrees well with the nature of the active normal faults in Greece: if they do flatten with depth

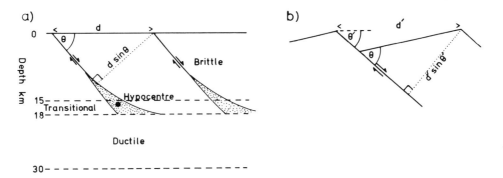

Fig. 1. (a) Sketch of the mechanical properties of the continental crust. The dotted line is perpendicular to the faults. (b) Sketch of the geometry after movement on the normal faults, used to obtain equation (1). See text for details.

they must do so below the earthquake nucleation depth of about 10 km, and any weak zone or layer within the crust must also be below this depth.

It has long been thought that the variation of crustal rheology with depth could be responsible for the listric geometry of major normal faults on the continents. We will now examine this notion quantitatively and attempt to relate it to the amount of crustal extension. The upper region of the crust fails by brittle failure, and generates seismic waves when it does so. It presumably does not have a sharp lower boundary, but merges with the underlying layer where the deformation still principally occurs on faults, but by creep rather than seismically (Sibson 1977). With increasing depth these faults cease to be slip surfaces and become slip zones (Fig. 1). Finally, if the crust is sufficiently thick, or at a high enough temperature, the lowest crustal layer creeps homogeneously and aseismically, without any discrete faults being formed. The depths at which these transitions occur must depend on the temperature gradient. Chen & Molnar (1983) used the maximum depths of continental earthquakes and Caristan's (1982) measurements on diabase to argue that the transition from seismic to aseismic behaviour occurs at a temperature of 350°C. This temperature occurs at a depth of between 10 and 15 km in many stretched basins, and it appears that most large continental earthquakes nucleate near this depth: that is, near the base of the brittle layer. Extrapolation of Caristan's experiments carried out at 1000°C suggests that the transition zone between brittle behaviour and distributed creep is less than 5 km thick in the crust, although more experiments at lower temperatures are needed before this estimate can be used with confidence. Three similar zones, of brittle, transitional and ductile behaviour, exist within the mantle, but, because the melting temperature of peridotite is greater, the temperatures of the transition regions are higher. We can now use this simplified model of crustal rheology to show that faults which are initially planar will, under most circumstances, become concave upward when the region is stretched. The geometry of the motion in the brittle seismic zone is straightforward if the faults are initially planar (Fig. 1) and rotate like a stack of books or toppling dominoes, as Ransome et al. (1910) and Morton & Black (1975) have proposed. It is worth noting that

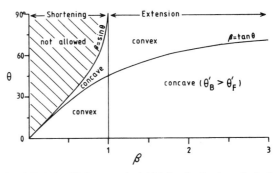

Fig. 2. Relationship between the initial dip, θ, of a planar fault, the amount of extension, β, and its dip after movement in the brittle (θ'_B) and ductile (θ'_F) layers. If a fault with a particular value of θ and β plots below (above) the curve it will be concave (convex) upward after deformation.

fault motion without accompanying rotation fails to produce the steep regional dips common to most extensional terranes. From Fig. 1,

$$d \sin \theta = d' \sin \theta'$$

or

$$\beta = \frac{d'}{d} = \frac{\sin \theta}{\sin \theta'},$$

where β is the amount of extension. Hence, the final dip θ'_B of the faults is related to the initial dip θ by

$$\sin \theta'_B = \frac{1}{\beta} \sin \theta. \qquad (1)$$

If the extension of the ductile layer is homogeneous it is also straightforward to determine the final inclination, θ'_F, of a line drawn in the fluid at an angle θ (see Appendix 1) when the lower layer is extended by the same amount

$$\tan \theta'_F = \frac{1}{\beta^2} \tan \theta. \qquad (2)$$

Clearly, if $\theta'_B > \theta'_F$ the faults will become concave upward, whereas if $\theta'_B < \theta'_F$ the opposite geometry will be produced. It is easy to show that $\theta'_B = \theta'_F$ when

$$\beta = \tan \theta. \qquad (3)$$

Figure 2 shows the location of the line given by (3), where $\theta'_B = \theta'_F$, on a β/θ plot. When the initial dip is less

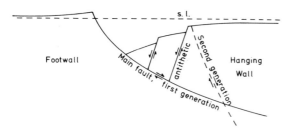

Fig. 3. Sketch of a normal fault system to show uplift of the footwall and to define the terminology used. Although the main fault is shown as a gentle curve, the arguments presented in the text suggest that, in areas where major crustal extension occurs as a result of normal faulting, a significant change in dip is likely at the brittle-ductile transition, near the base of the largest antithetic faults.

than 45° the dip of the faults in the brittle seismic zone is always steeper than the line in the ductile zone. This argument is only intended to be illustrative: it assumes that the fault is initially planar with the same dip in both brittle and ductile layers. The fault could, of course, initially be flatter in the ductile layer beneath the earthquake focal depth; in which case the arguments above will simply accentuate this initial concave geometry. Although the change in dip in the ductile layer is calculated for a line, this line may be a fault provided the slip vector always remains in the fault plane. This simple requirement is necesary to prevent the formation of voids at depth and is a fundamental and powerful constraint on fault geometry and interaction. It is discussed further in the section on interaction between faults. In this case it requires the slip vector in the ductile layer to differ from that at the surface and is one of the major causes of antithetic faulting and internal deformation in the hanging-wall block (see Fig. 3). The arguments above suggest a dramatic change in dip at the base of the brittle layer, and it is here (at a depth of 8–10 km) that movement on a large antithetic fault nucleated in the 1981 Gulf of Corinth earthquake sequence (see Jackson *et al.* 1982a). Small aftershocks, representing minor movement on small faults, typically occur throughout the hanging-wall block (Soufleris *et al.* 1982), presumably representing internal deformation caused by the change in slip vector on the main fault with depth. In practice there is likely to be a smooth change in dip between the brittle seismic region and the ductile region, producing a fault which is concave upwards. These arguments show that most faults will become listric in extensional areas, even if they start planar. The shape of the shallow part of the faults will, however, not be affected by extension, though the dip will be.

When $\beta < 1$, then the region is undergoing shortening. Under these conditions the dip of the fault will always steepen with increasing shortening, though the amount of steepening will not generally be the same in the brittle and ductile layers. Figure 2 shows that planar reverse faults which originally dipped at angles of 45° or more become concave upward. Normal faults that are buried to great depth, such as those in the stretched basement of passive continental margins, may have their original shallow planar geometry buried below the

brittle–ductile transition. If they are then reactivated as thrusts (reverse faults) during subsequent compression, as Jackson (1980) suggested, their initially steep dip will cause them to evolve a concave upwards shape.

In contrast to this process, which causes a fault that is initially planar to become curved, there are various other conditions which require the faults to be concave upward initially. The most obvious is one which Wernicke & Burchfiel (1982) have discussed. Because each fault block in Fig. 1 must rotate anticlockwise through an angle $\theta_B' - \theta$ between the initial and final state, the boundary between the stretched and unstretched regions cannot be a plane fault. If it were, one side of the fault would rotate, the other would not, and a void would form. Hence the fault bounding the undeformed region must be curved, to allow one side to rotate relative to the other. The same argument applies if the amount of extension varies with position in a basin. It is also possible that the initial shape of a major extensional normal fault may be listric, though the earthquake studies discussed earlier show that such a geometry is less common than we expected.

Several authors (e.g. Kanizay 1962, Hose & Daneš 1973, Price 1977) have suggested that the curvature of normal faults is related to the stress field, and hence has a dynamic origin. They argue that the fault geometry is similar to that of slip lines within a plastic layer under vertical compression. Unfortunately, the slip lines within such a layer are not discrete surfaces on which slip is concentrated and across which there is a velocity discontinuity. In this description the velocity is continuous throughout the deforming region and cannot concentrate on faults. Hence, there is no close analogy between slip lines and faults.

In many cases it is clear that listric normal faults are produced by the reactivation of thrusts as normal faults. A good example of such behaviour is the Flathead Valley Fault in the Rocky Mountains, which Bally *et al.* (1966) showed was produced by the reversal of movement on one of the major thrusts in the area. More recently, Smythe *et al.* (1982) have proposed that the normal faults which bound the basins north of Scotland coincide at depth with pre-existing thrusts whose age is Caledonian or older. In the East Shetland Basin many large listric faults, whose geometry controls the major oil accumulations, also follow the Caledonian trend (Bowen 1975) and dip to the east. Therefore, these faults may also be reactivated Caledonian thrusts like those to the southwest. The reverse of this process has been discussed by Jackson (1980), Stoneley (1982) and Cohen (1982), who argue that listric normal faults can be reactivated as thrusts. There is no theoretical limit on the number of times the direction of motion on a fault can reverse.

The fault geometry also controls the extent to which the blocks between the faults must deform during the motion. When the faults are planar, as in Morton & Black's (1975) model and Fig. 1, their movement can be taken up without internal deformation. The rotation of the base of each faulted block is accommodated by creep

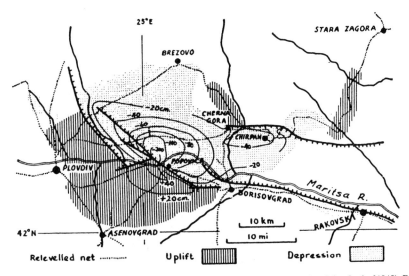

Fig. 4. Faulting and vertical movement produced by the April 1928 Bulgarian earthquakes (after Jankof 1945). Reproduced, with permission, from fig. 3.12 in *Elementary Seismology* by Charles F. Richter (W. H. Freeman & Company) © 1958.

within the crust. When, however, the faults are curved, Bally (1982) has pointed out that their movement causes the Earth's surface to become concave unless the blocks undergo internal deformation. Figure 3 shows the type of deformation required. The subsidiary faults which are required by the geometry of the motion are called antithetic faults. Movement on antithetic faults is frequently associated with earthquakes on the main faults (Myers & Hamilton 1964, Jackson *et al.* 1982a) (Fig. 4). As Proffett (1977) pointed out, continued motion on the main fault may rotate the antithetic faults past the vertical so that they appear as high-angle reverse faults.

Uplift and subsidence in normal faulting

One important feature of normal faulting which has, in our view, been insufficiently emphasized, is the uplift of the footwall. This uplift has probably helped give rise to the widespread belief that rifting is accompanied by doming. Such movement occurs for two reasons. When the fault slips during an earthquake the stress across the fault is reduced. The change in stress on the footwall is the same as that on the hanging wall, but because of their different shape the hanging wall moves down considerably further than the footwall moves up. This elastic rebound is illustrated in Fig. 4, which shows the vertical motion associated with two earthquakes in Bulgaria in 1928. This motion was determined by re-levelling soon after the earthquake and was described by Jankhof (1945) and summarised by Richter (1958). As in the 1981 Corinth earthquakes discussed by Jackson *et al.* (1982a), there were two fault breaks: a southern fault, which is probably the main fault, and a northern fault which is probably an antithetic fault. Although there is minor uplift in the footwall of the antithetic fault, there is net subsidence of the whole hanging-wall block as it slides down the main normal fault dipping north (see Fig. 3). This gives rise to the strong asymmetry in the uplift and

subsidence shown in Fig. 4. The relative displacements illustrated in Fig. 4 were presumably elastic, and will be reversed (i.e. the footwall near the fault will subside relative to the footwall farther from the fault) as continued extension restores the stress to its value before the event.

There is, however, a different mechanism by which long-term uplift can be produced, which was discussed by Vening Meinesz (1950) (see Heiskanen & Vening Meinesz 1958). We use their model to estimate the ratio of the uplift to the downdrop in Appendix 2. As the hanging wall slides down the fault it unloads the footwall which is then uplifted by isostasy. The hanging wall correspondingly sinks. If the fault is completely buried beneath uniform sediment, the upward movement of the footwall is similar in magnitude to the downward movement of the hanging wall. In this case, because of the absence of a horizontal reference level, the displacements are difficult to measure. The only case where the uplift is easily measured is when the footwall is exposed above the sea surface and the hanging wall is beneath the water, where the sedimentation rate is usually sufficiently fast to fill the depression caused by faulting. The elevation of the footwall is then about one tenth of the subsidence of the hanging wall (Appendix 2) and can be determined from the topography or stratigraphy. Though the general form of the vertical motion in this case is similar to the elastic rebound, the detailed geometry is different. In particular, the movements are permanent and are not reversed by the accumulation of elastic strain. Figure 5 shows an example of permanent uplift produced by movement on a fault in the North Sea (Linsley *et al.* 1980). The uplift which forms the trap of the Beatrice oil field was produced by the unloading of a normal fault. Because the displacement on the fault has a maximum, and decreases both to the east and the west, the uplift also dies away along strike. In the southeastern direction, the structure is closed by the downthrown

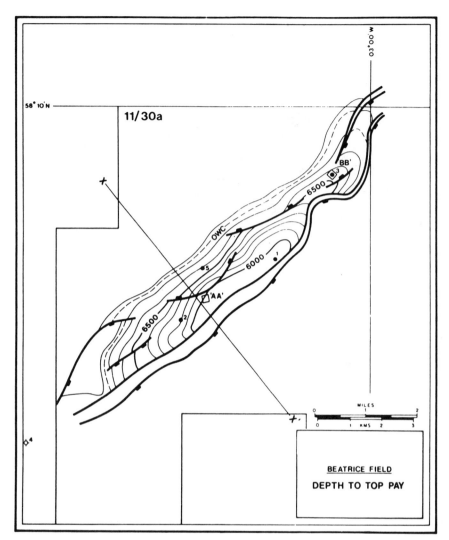

Fig. 5. Depth (in feet) to the upper surface of the reservoir of the Beatrice Field, North Sea. The heavy lines are normal faults, with rectangles on the downthrown side. Reproduced from Linsley *et al.* (1980), with permission of the author and the American Association of Petroleum Geologists.

shales, and in the northwestern direction by the decay of the uplift with increasing distance from the fault. Hence, the uplift associated with the normal fault produces closure in all four directions to form a structural trap. Progressive sediment onlap in the footwall may also give rise to potential stratigraphic traps. Exactly the same arguments can be used to estimate the uplift of the hanging wall which results from thrusting (Appendix 2).

An important feature of normal fault systems is their tendency to form new generations of high-angle faults when the dip of the original faults has been reduced by the rotation associated with their movement. New high-angle faults are required because the contribution which gravity can make to overcoming the friction on the fault plane is reduced as the dip of the fault decreases. The geometry of this process has been discussed by Morton & Black (1975) and Proffett (1977) but is rarely apparent on seismic reflection profiles. Only one generation of

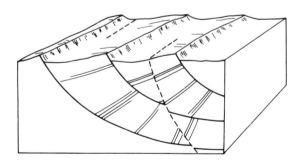

Fig. 6. Block diagram to show the interaction between the first generation of normal faults, shown by heavy lines, and the second generation, shown dashed. Movement on the second-generation fault seen on the front face of the block will lock that of the two faults it intersects. Its propagation into the plane of the page can produce another fault en échelon with the first, shown by the dashed line on the surface. Although the first-generation faults are shown with a listric geometry, they could be planar as long as the faulting extends further to the left of the block.

faults can be seen on the profiles published by Effimoff & Pinezich (1981) and de Charpal *et al.* (1978), perhaps because the stacking used during data reduction is designed to emphasize sub-horizontal strata, not those tilted at steep angles. Unlike these authors, Proffett (1977) used cores from closely spaced drill holes to interpret the structures in the Yerington area. A common, but not universal, feature of younger, steeper faults is that they cut the hanging wall (Fig. 6). This geometry has the important consequence of uplifting the sediments which were deposited close to the earlier fault and which have therefore been tilted through the greatest angles. There are two obvious explanations of this tendency. Antithetic faults of the first generation can be rotated until they dip in the same direction as the main faults, when they become planes of weakness which the second generation can exploit by reversing their displacements. Another explanation is that the second-generation faults exploit part of the fault plane of the first generation. This geometry could allow both generations to continue to move for a limited time. Angelier & Colletta (1983) suggest that vertical tension gashes, present as internal deformation within hanging-wall blocks, may grow to become later steep faults with significant displacement. Although commonly seen at the surface, these cracks can only be present as tensional fissures to the relatively shallow depths at which void formation is permissible (<500 m; see fourth section of this paper). Thus, although they may be exploited as planes of weakness near the surface, they cannot account for steep second-generation faulting at depth.

INTERACTION BETWEEN SEVERAL FAULTS

The principal constraint imposed by the existence of several faults in an area, all of which are active, is that they must not intersect. If they do cross, movement on one fault, A, will offset the other, B, whose slip vector will then no longer lie in the plane of the fault. If B were then to move a void would form. The stress drops involved in earthquakes are typically in the range 10–100 bars (see e.g. Kanamori & Anderson 1975). Only within about 500 m of the surface are these stresses sufficiently large to overcome the lithostatic pressure and form voids. As we have noted already, this inability to form voids at depth is a powerful constraint on the evolution of fault geometry. The only circumstance in which active faults can cross and both remain active is when the slip vector of one fault lies in the plane of both faults. An example of such a system, illustrated by Bally (1982), is a vertical strike–slip fault offsetting a listric normal fault. If the fault system in Fig. 3 is crossed by a vertical strike–slip fault at right angles to the strike of the normal fault and in the plane of the paper, the slip vectors on both the main and the antithetic faults lie in the plane of the strike–slip fault. Hence, their motion will not offset the strike–slip fault and all faults can continue to move. Notice that the same is not true of a strike–slip fault

whose dip is not vertical. The general requirement that most active faults do not cross, and that when they do so, one ceases to be active, controls the geometry of active fault systems. An obvious consequence of this, emphasized by many authors including Bally (1982), Proffett (1977) and Stewart (1980) is that normal faults dip in the same direction over large regions. By doing so they can avoid crossing. Similarly the antithetic faults must not cross the main fault, and so most grabens are asymmetric at depth. Exactly the same rules apply to systems of thrust faults. They also must not cross each other if they are to remain active. Like normal faults, their motion is not affected if they are crossed by a vertical strike–slip fault, provided the slip vectors on the thrusts lie in the plane of the strike–slip fault.

One important consequence of these rules governing the simultaneous activity of several faults is that the initiation of a steep second generation fault can completely change the geometry of the fault movements. If the first-generation faults depicted in Fig. 6 are sufficiently closely spaced, one steep second-generation fault can lock more than one earlier fault. The first steep new fault will presumably start where the rotation of the earlier ones has been greatest, and it will lock all the earlier faults which it intersects. Hence, the new fault must take up the motion of these older faults. It will therefore rapidly propagate along its strike, in the same way as a tear or dislocation does, by concentrating the stress at its tip. If the propagating tip encounters a hard region, another new fault will be formed en échelon with the first (Fig. 6). En échelon propagation of this type, combined with footwall uplift, can produce the type of structure shown in Fig. 5. Another mechanism which can produce the same effect is the propagation of two different sets of second-generation faults into an area, each of which originated in a place remote from the other. Movement on the new faults will rapidly lock the older faults over a wide area. Once this has happened a series of second-generation faults which do not cross each other will form to take up the motion. Such behaviour will only occur if, at least somewhere in the region, the steep new faults cross more than one old fault. Otherwise the faults will move independently.

This type of interaction between faults may be the cause of the regular variation in the direction of fault dip reported from the Basin and Range Province by Stewart (1980). He remarked that the steep second-generation faults bounding the present basins and ranges dip in the same direction over large regions, and are often separated from regions with faults dipping in the opposite direction by discontinuities parallel to the slip vector. As discussed at the beginning of this section, this is the only geometry which allows all faults to move together. Zoback *et al.* (1981) pointed out that the steep second-generation faults cut through the earlier faults at the same time over large regions, which is also in agreement with the idea that they lock the first-generation faults and propagate rapidly. However, propagation along strike will be interrupted by the strike–slip faults, hence each large region in which the faults have the same dip

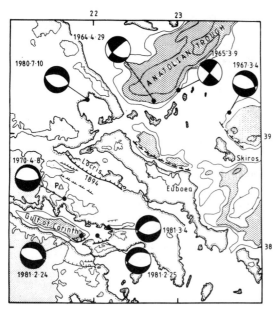

Fig. 7. Map of central Greece showing fault plane solutions (from Jackson *et al.* 1982b), bathymetry (from Morelli *et al.* 1975), and topography (from ONC chart G3). Bathymetric contours are at 400, 600 and 1000 m and shaded with horizontal lines beneath 600 m. Topography higher than 615 m is stippled. Epicentres are those reported by NOAA and its successors, except for those in the Gulf of Corinth, which are from Jackson *et al.* (1982). Fault plane solutions are lower-hemisphere projections, with compressional quadrants shown in black. Northward-dipping normal faults that are known to have moved at the surface during earthquakes are marked by thick lines with teeth on the downthrown side. Those north of Euboea and Skiros are conjectural, based on observations of uplift and seismicity (see text). Older faults, active in the Pliocene, are marked by thinner lines in Locris and south of Corinth. The fault marked north of Mt. Parnassus (P) is based on topography and is of unknown age. The faulting marked on this map is not complete and should not be used in seismic risk assessment: it is designed to illustrate the presence of at least four sub-parallel asymmetric grabens bounded by major normal faults dipping northeast. In particular, many large southward-dipping antithetic faults are present in northwest Euboea and on the northern side of the Gulf of Corinth; only the antithetic fault which moved in the earthquake of 1981.3.4 is marked on this map.

constraint on the fault kinematics are geometric that the evolution of fault systems in the two areas can be compared.

THE AEGEAN SEA AND CENTRAL GREECE

These is no region of the Earth's surface which is well enough known to test all the proposals in the two preceding sections. Though many sedimentary basins have been investigated using reflection seismology, in most cases little is known about the detailed history of the underlying faults. Two areas which are presently being deformed by normal faults are the Aegean Sea in Greece and the Basin and Range Province in the western U.S.A. Though the geology of the second is better known, the first is much more seismically active. In particular, studies of the 1978 Thessaloniki and 1981 Gulf of Corinth earthquakes and their aftershocks have demonstrated the present relationship between surface faulting and that at depths of about 10 km (see second section). Also, enough is known of the post-Miocene history of the area to illustrate some of the processes involved. Another major advantage of the Aegean is that large parts are covered with sea water, which forms a convenient horizontal reference surface (Jackson *et al.* 1982a). We will therefore attempt to relate some of the features in the Aegean to the ideas discussed above. Too little is yet known about the instantaneous motions in three dimensions and their evolution in time for such a comparison to provide a detailed test of our suggestions. Nonetheless, the kinematic ideas outlined in the preceding sections provide a useful framework within which to discuss the observations.

The area of central Greece shown in Fig. 7 has been shaken by a number of recent large earthquakes, three of which had surface breaks. Of these the most carefully studied was the event in 1981 at the eastern end of the Gulf of Corinth (Jackson *et al.* 1982a). Vertical movement of the coast was clearly visible, and corresponded well to the motion expected from elastic rebound. Furthermore the topography of the epicentral region is clearly controlled by faulting, with the footwall probably uplifted by between 0.5 and 1 km. The argument presented in Appendix 2 suggests that the total fault displacement may be between 5 and 10 km. As Jackson *et al.* (1982b) have pointed out, many islands in the Aegean are the uplifted footwalls of major normal faults, and their elevation may also provide a simple guide to the fault displacements. Another earthquake shook the western Gulf of Corinth on 1861.12.26 (see Richter 1958). This event was associated with a surface break on the south shore of the Gulf, similar to that of the 1981 sequence, with the north side moving down. The existence of major faults and uplifted Plio-Quaternary sediments on the southern side of the Gulf suggests that the active faults at the eastern end, which have dips of 40–50° in the brittle layer, are steep second-generation faults. Jackson *et al.* (1982a) believed that the 1981.3.4 event occurred on an antithetic fault, and the 1970.4.8 shock

direction should behave independently. The observations of Stewart (1980) and Zoback *et al.* (1981) would only be expected if the faults of each system are all active at the same time. The clear expression of many of the faults in the topography also suggests that many are now active (Wallace 1977, 1978), although the level of seismicity is much weaker than in the Aegean. The Basin and Range Province has probably been stretched by about a factor of two (Hamilton & Myers 1966, Proffett 1977, Priestley & Brune 1978), though the amount of extension at any particular location is variable. In this respect it is similar to the Aegean, where measurements of crustal thickness suggest similar amounts of extension in the areas which have been most strongly stretched (Makris 1976, Makris & Vees 1977, McKenzie 1978). Hence the degree of deformation in the two areas is also likely to have been similar. It is important to note that none of the arguments in this or the preceding section is affected by the nature of the driving forces responsible for the crustal extension, which may well be different in the Aegean and Basin and Range. It is because the may well have done so too, since one of the nodal planes of its fault plane solution dips steeply to the south.

Another large event took place in Locris on 1894.4.27 and was also accompanied by a surface break (Richter 1958). The footwall of this active fault contains Plio-Quaternary sediments which have been strongly tilted (Philip 1976) and large normal faults that did not move during the earthquake; again suggesting the active fault is a second-generation fault. The existence of uplifted marine terraces on northeast Euboea (Lemeille 1977), and of closed basins to the northeast of the island, together with the presence of the island itself, suggests that another major fault exists offshore, also dipping northeast. The bathymetry and seismicity of the escarpment northeast of Skiros (Fig. 7) makes it likely that this too is an active normal fault. Similar features on land suggest that another fault separates Mount Parnasos from the basins immediately north of it. The spacing between each of these faults is between 30 and 40 km; very similar to those of the Basin and Range. All of them dip to the northeast, have total lengths of about 100 km (though perhaps made up of individual segments 15–20 km long), and are likely to have throws of at least 5 km. The scale of the fault system is very large. The northwestern boundaries of all the faults lie on the prolongation of the Anatolian Trough, which is at right angles to the strike of the normal faults and which is known to have a large strike–slip component of motion from fault plane solutions (Jackson *et al.* 1982b). However, no western extension of this trough is known on land which could form this northwestern boundary. All these normal faults are sub-parallel to the major thrusts of the Hellenides (Auboin 1965). It is therefore possible that they share the same faults at depth and that the present motions are simply produced by reversing the sense of movement on the deeper thrusts, as apparently occurred in the basins north of Scotland (Smythe *et al.* 1982). However, in the absence of deep seismic reflection data this remains merely plausible in central Greece.

The geometry of the active normal faults suggests that no tectonic basement can exist anywhere in the area, and that the subsurface movements must resemble those shown in Fig. 6. Major earthquakes at Corinth in 1981 and Thessaloniki in 1978 all had depths and locations known with sufficient accuracy to show (second section of this paper) that the faults are approximately planar to a depth of about 8–10 km and dip at about 45°. This observation, the existence of tilted Plio-Quaternary rocks in the footwall of the 1894 Locris fault (Philip 1976, Mercier *et al.* 1979b), and the occurrence of normal faults in the uplifted footwalls of the active Corinth and Locris faults, as well as that northeast of Euboea, suggests that the active faults are steep second-generation faults. Their movement must have reduced or stopped the motion on the older systems in their footwalls. If the geometry of the earlier faults resembled those shown in Fig. 6, the second generation, once initiated, would have rapidly propagated throughout the area. This propagation need not, however, cross any strike–slip faults parallel to the extension direction.

The sequence of events outlined above can explain a surviving enigma in the Pliocene–Quaternary history of the Aegean. Careful structural and paleogeographical work on Neogene and Quaternary sediments in and around the Aegean led several people (e.g. Mercier *et al.* 1979b, Mercier 1981) to suggest that the normal faulting which began in central Aegean in the middle or late Miocene was interrupted by a short period of regional compression, lasting only about one million years, in the earliest Quaternary. As Jackson *et al.* (1982b) point out, some of the observations leading to this suggestion, particularly the widespread uplift seen in the islands and coastal areas of the central Aegean at the base of the Quaternary (Keraudren & Mercier 1977, Keraudren 1979), are more likely to be an artefact of renewed normal faulting. The question remains: why was the footwall uplift seen in the islands and coastal regions apparently synchronous over the whole central Aegean region? The rapid propagation of steep second-generation faults over the area in the early Quaternary appears to provide the answer. If this is so, then the earlier, now flatter, faults are presumably responsible for most of the extension in the brittle layer, whereas the steeper, new, and now active faults are responsible for the present day topography and bathymetry, as steep faults produce more footwall uplift than flatter faults (see Appendix 2).

It is not yet clear how common fault geometries like those shown in Figs 6 and 7 are. An essential feature is that many faults must be simultaneously active, otherwise the kinematic constraints do not apply. The regional tilt patterns (Stewart 1980) and the distributed seismicity in the Basin and Range suggest that many of the range-bounding faults are now active, and that this area is deforming in the same way as the Aegean. In western Nevada, Proffett (1977) describes a sequence of events in which most of the crustal extension was achieved in the middle Miocene by faults that are now rotated to a low angle of dip and are inactive. This was followed in the late Miocene by the creation of the present-day Basin and Range topography by high-angle faulting which cuts the older faults. However, in the Basin and Range, large earthquakes with surface breaks are much less frequent than they are in the Aegean, and it is not straightforward to demonstrate simultaneous activity by any other method. One other region where a similar system of normal faults may now be active is western Turkey. The known active graben systems in the area are discussed by McKenzie (1978), but unfortunately the earthquake locations in the area are not yet accurate enough to demonstrate the geometry of the subsurface motions. It is, however, already clear that these must be more complicated than those in central Greece.

CONCLUSIONS

Continental deformation involves simultaneous movement on a variety of different faults, some planar and some curved. For the reasons discussed in the second section of this paper it is probable that faults in extensional or compressional environments become curved at depth as the deformation progresses, even if they were planar to start with. Though both types can become concave upward or downward, normal faults with dips of 45° should become concave upward. The constraint that no voids should form during the motion of the faults imposes severe constraints on the geometry of fault systems and on their evolution. Only vertical strike–slip faults are likely to cut systems of normal faults, and all such systems must contain faults dipping in the same direction. When faults of a new system, with a steeper dip, cross more than one fault of an earlier system, the new system must propagate rapidly throughout the deforming area because it prevents the first generation of faults from moving. Motion of either the new or old generation of faults will uplift the footwalls of faults, and such uplift has generated the present islands in the Aegean. The same process produced the islands on the flanks of the North Sea grabens during Cretaceous and Jurassic rift development (Wood 1982). In the uplifted footwalls of second-generation faults strongly rotated sediments deposited during the first-generation faulting may occur.

One of the most important features of the normal fault system in Greece is the horizontal scale of the faults. Most of central Greece is probably underlain by at least one normal fault. Under such conditions the labels autochthon and allochthon become meaningless. These labels remained in use after plate tectonics became accepted because many geologists believed that much continental deformation consisted of thin sheets of deforming material travelling over essentially intact lithosphere. If normal faults frequently underlie each other, are active at the same time, and chop up the whole lithosphere, the distinction between the two terms is not useful. In areas such as central Greece the terms should be redefined or abandoned.

Acknowledgements—This work would not have been possible without the detailed studies of major earthquakes which have been carried out by Geof King, Christos Soufleris, Manuel Berberian, Graham Yielding and a number of other colleagues. We realized the need for a detailed explanation of these ideas during discussions with Frank Richter and later on a field trip in western Turkey with Celal Şengör, to whom we are particularly grateful for his generous hospitality both during the trip and afterwards at Bursa, where this paper was written. We thank Graham Yielding, Peter Molnar and Jon Brewer for helpful comments on the manuscript. This work was supported by a grant from the Natural Environmental Research Council, and is contribution number 383 of Cambridge University Department of Earth Sciences.

REFERENCES

Angelier, J. & Colletta, B. 1983. Tensional fractures and extensional tectonics. *Nature, Lond.* **301**, 49–51.
Aubouin, J. 1965. *Geosynclines*. Elsevier, Amsterdam.
Bally, A. W. 1982. Musings over sedimentary basin evolution. *Phil. Trans. R. Soc.* **A305**, 325–328.
Bally, A. W., Gordy, P. L. & Stewart, G. A. 1966. Structure, seismic data and orogenic evolution of the southern Canadian Rocky Mountains. *Bull. Can. Petrol. Geol.* **14**, 337–381.

Bally, A. W., Bernoulli, D., Davis, G. A. & Montadert, L. 1981. Listric normal faults. *Oceanol. Acta*, Proc. 26th Int. Geol. Cong. Geology of Continental Margins Symposium, Paris, 87–101.
Bowen, J. M. 1975. The Brent Oil Field. In: *Petroleum and the Continental Shelf of Northwest Europe* (edited by Woodland, A. W.). Applied Science Publishers, 353–362.
Buffler, R. T., Shaub, F. J., Watkins, J. S. & Worzel, J. L. 1978. Anatomy of the Mexican Ridges, southwestern Gulf of Mexico. *Mem. Am. Ass. Petrol. Geol.* **29**, 319–327.
Caristan, Y. 1982. The transition from high-temperature creep to fracture in Maryland diabase. *J. geophys. Res.* **87**, 6781–6790.
de Charpal, O., Guennoc, P., Montadert, L. & Roberts, D. G. 1978. Rifting, crustal attenuation and subsidence in the Bay of Biscay. *Nature, Lond.* **275**, 706–711.
Chen, W-P. & Molnar, P. 1983. The depth distribution of intracontinental and intraplate earthquakes and its implications for the thermal and mechanical properties of the lithosphere. *J. geophys. Res.* **88**, 4183–4214.
Cohen, C. 1982. Model for a passive to active continental margin transition: implications for hydrocarbon exploration. *Bull. Am. Ass. Petrol. Geol.* **66**, 708–718.
Effimoff, I. & Pinezich, A. R. 1981. Tertiary structural development of selected valleys based on seismic data: Basin and Range Province, northeastern Nevada. *Phil. Trans. R. Soc.* **A300**, 435–442.
Frost, E. G. & Martin, D. L. (Eds.) 1982. Mesozoic–Cenozoic evolution of the Colorado River region, California, Arizona and Utah. Cordilleran Publishers, San Diego, California.
Garrison, L. E. & Martin, R. G. 1973. Geologic structures in the Gulf of Mexico basin. *Prof. Pap. U. S. geol. Surv.* **773**, 1–85.
Hamilton, W. & Myers, W. B. 1966. Cenozoic tectonics of the western United States. *Rev. Geophys.* **4**, 509–549.
Heiskanen, W. A. & Vening Meinesz, F. A. 1958. *The Earth and its Gravity Field*, McGraw-Hill New York.
Hose, R. K. & Daneš, Z. F. 1973. Development of the late Mesozoic to early Cretaceous structures of the Great Basin. In: *Gravity Tectonics* (edited by De Jong, K. A. & Scholten, R.). John Wiley & Sons, New York, 429–441.
Jackson, J. A. 1980. Reactivation of basement faults and crustal shortening in orogenic belts. *Nature, Lond.* **283**, 343–346.
Jackson, J. A., Gagnepain, J., Houseman, G., King, G. C. P., Papadimitriou, P., Soufleris, C. & Virieux, J. 1982a. Seismicity, normal faulting and the geomorphological development of the Gulf of Corinth (Greece): the Corinth earthquakes of February and March 1981. *Earth Planet. Sci. Lett.* **57**, 377–397.
Jackson, J. A., King, G. & Vita-Finzi, C. 1982b. The neotectonics of the Aegean: an alternative view. *Earth Planet. Sci. Lett.* **61**, 303–318.
Jankhof, K. 1945. Changes in ground level produced by the earthquakes of April 14 to 18 1928 in southern Bulgaria. In: *Tremblements de Terre en Bulgarie*, Nos. 29–31, Institut meteorologique central de Bulgarie, Sofia, 131–136 (in Bulgarian).
Kanamori, H. & Anderson, D. L. 1975. Theoretical basis of some empirical relations in seismology. *Bull. seism. Soc. Am.* **65**, 1073–1095.
Kanizay, S. P. 1962. Mohr's theory of strength and Prandtl's compressed cell in relation to vertical tectonics. *Prof. Pap. U.S. geol. Surv.* **B414**, B1–16.
Keraudren, B. 1979. Le Plio-Pleistocene marin et oligihalin en Grèce: stratigraphie et paleogeographie. *Revue Géogr. phys. Géol. dyn.* **21**, 17–28.
Keraudren, B. & Mercier, J-L. 1977. Palaogeographie Plio-Pleistocene et néotectonique de l'arc Egéen. Recherches Francais sur le Quaternaire, INQUA 1977. Supplement au Bulletin AFEQ, **50**, 135–140.
Lemeille, F. 1977. Etudes néotectoniques en Grèce Centrale nord-occidentale et dans les Sporades du Nord. Thèse de 3ème cycle, University de Paris.
Linsley, P. N., Potter, H. C., McNab, G. & Racher, D. 1980. The Beatrice Field, Inner Moray Firth, U.K. North Sea. In: *Giant Oil and Gas Fields of the Decade* 1968–1978 (edited by Halbouty, M. T.). American Association Petroleum Geologists, 117–129.
Makris, J. 1976. A dynamic model of the Hellenic Arc deduced from geophysical data. *Tectonophysics* **36**, 339–346.
Makris, J. & Vees, R. 1977. Crustal structure of the central Aegean Sea and the islands of Evia and Crete, Greece, obtained by refraction seismic experiments. *J. geophys. Res.* **42**, 329–341.
McKenzie, D. P. 1978. Active tectonics of the Alpine–Himalayan belt: the Aegean Sea and surrounding regions. *Geophys. J. R. astr. Soc.* **55**, 217–254.
Mercier, J-L., Mouyaris, N., Simeakis, C., Roundoyannis, T. & Angelidhis, C. 1979a. Intraplate deformation: a quantitative study

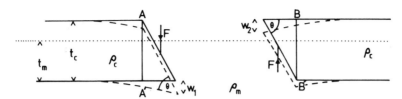

Fig. 8. Continental crust, cut by a planar normal fault, is shown separated at a fault. The constants are defined in Appendix 2. The dotted line shows the level that the fluid of density ρ_m, in which the crustal blocks are floating, would reach.

ot the faults activated by the 1978 Thessaloniki earthquakes. *Nature, Lond.* **278**, 45–48.

Mercier, J.-L., Delibassis, N., Gauthier, A., Jarrige, J., Lemeille, F., Philip, H., Sebrier, M. & Sorel, D. 1979b. La néotectonique de l'arc Egéen. *Revue Géogr. phys. Géol. dyn.* **21**, 67–92.

Mercier, J.-L. 1981. Extensional-compressional tectonics associated with the Aegean arc: comparison with the Andean cordillera of south Peru–north Bolivia. *Phil. Trans. R. Soc.* **A300**, 337–355.

Morelli, C., Gantar, C. & Pisani, M. 1975. Geophysical studies in the Aegean Sea and eastern Mediterranean. *Boll. Geof. Teor. e appl.* **17**, 66.

Morton, W. H. & Black, R. 1975. Crustal attenuation in Afar. In: *Afar Depression of Ethiopia* (edited by Pilger, A. & Rösler, A.). Inter-Union Commission on Geodynamics, Sci. Rep. No. 14. E. Schweizerbart'sche Verlagsbuchhandlung, Stuttgart, 55–65.

Myers, W. B. & Hamilton, W. 1964. Deformation accompanying the Hebgen Lake earthquake of August 17, 1959. *Prof. Pap. U.S. geol. Surv.* **435**, 55–98.

Philip, H. 1976. Un épisode de déformation en compression à la base du Quaternaire en Grèce centrale (Locride et Eubée nord-occidentale) *Bull. Soc. géol. Fr.* **18**, 287–292.

Price, N. J. 1977. Aspects of gravity tectonics and the development of listric faults. *J. geol. Soc. Lond.* **133**, 311–327.

Priestley, K. & Brune, J. 1978. Surface waves and the structure of the Great Basin of Nevada and western Utah. *J. geophys. Res.* **83**, 2265–2272.

Proffett, J. M. 1977. Cenozoic geology of the Yerington district, Nevada, and implications for the nature of Basin and range faulting. *Bull. geol. Soc. Am.* **88**, 247–266.

Ramsay, J. G. 1967. *Folding and Fracturing of Rocks.* McGraw-Hill, New York.

Ransome, F. L., Emmons, W. H. & Garrey, G. H. 1910. Geology and ore deposits of the Bullfrog district, Nevada. *Bull. U.S. geol. Surv.* **407**.

Richter, C. F. 1958. *Elementary Seismology*, W. H. Freeman, San Francisco.

Sibson, R. H. 1977. Fault rocks and fault mechanisms. *J. geol. Soc. Lond.* **133**, 191–213.

Smythe, D. K., Dobinson, A., McQuillan, R., Brewer, J. A., Matthews, D. H., Blundell, D. J. & Kelk, B. 1982. Deep structure of the Scottish Caledonides revealed by the MOIST reflection profile. *Nature, Lond.* **299**, 338–340.

Soufleris, C. & Stewart, G. S. 1981. A source study of the Thessaloniki (northern Greece) earthquake sequence. *Geophys. J. R. astr. Soc.* **67**, 343–358.

Soufleris, C., Jackson, J. A., King, G., Spencer, C. & Scholz, C. 1982. The 1978 earthquake sequence near Thessaloniki (northern Greece). *Geophys. J. R. astr. Soc.* **68**, 429–458.

Stewart, J. H. 1980. Regional tilt patterns of late Cenozoic basin-range fault blocks, western United States. *Bull. geol. Soc. Am.* **91**, 460–464.

Stoneley, R. 1982. On the structural development of the Wessex basin. *J. geol. Soc. Lond.* **139**, 543–554.

Vening Meinesz, F. A. 1950. Les grabens africains, resultat de compression ou de tension dans le croute terrestre? *Inst. R. colonial Belge Bull.* **21**, 539–552.

Wallace, R. E. 1977. Profiles and ages of young fault scarps, north-central Nevada. *Bull. geol. Soc. Am.* **88**, 1267–1281.

Wallace, R. E. 1978. Geometry and rates of change of fault-generated range fronts, north-central Nevada. *J. Res. U.S. geol. Surv.* **6**, 637–650.

Wernicke, B. 1981. Low-angle normal faults in the Basin and Range Province: nappe tectonics in an extending orogen. *Nature, Lond.* **291**, 645–648.

Wernicke, B. & Burchfiel, B. C. 1982. Modes of extensional tectonics. *J. Struct. Geol.* **4**, 105–115.

Wood, R. 1982. Subsidence of the North Sea. Unpublished Ph.D. thesis, University of Cambridge.

Zandt, G. & Owens, T. J. 1980. Crustal flexure associated with normal faulting and implications for seismicity along the Wasatch Front, Utah. *Bull. seism. Soc. Am.* **70**, 1501–1520.

Zoback, M.-L., Anderson, R. E. & Thompson, G. A. 1981. Cainozoic evolution of the state of stress and style of tectonism of the Basin and Range province of the western United States. *Phil. Trans. R. Soc.* **A300**, 407–434.

APPENDIX 1

Dip variation due to pure shear

To determine the change in dip of a line drawn in a homogeneous fluid region extended by an amount β we need the 2×2 matrix $\mathbf{F}$ which converts an initial vector u' after deformation

$$u' = \mathbf{F}u. \tag{A1}$$

$\mathbf{F}$ is easily found by considering two special cases. When $u = (1, 0)$, a horizontal unit vector, the definition of β, the amount of extension, requires $u' = (\beta, 0)$. Similarly conservation of mass requires a vertical unit vector $(0, 1)$ to become $(0, 1/\beta)$. These two cases gives four equations for the four elements of $\mathbf{F}$:

$$\mathbf{F} = \begin{pmatrix} \beta & 0 \\ 0 & \dfrac{1}{\beta} \end{pmatrix}. \tag{A2}$$

An initial unit vector at an angle θ to the horizontal $(\cos \theta, \sin \theta)$ is converted into a vector, which is no longer a unit vector, at an angle θ'_F to the horizontal. Combining (A1) and (A2) shows that

$$\tan \theta'_F = \frac{1}{\beta^2} \tan \theta. \tag{A3}$$

Equation (A3) is similar to equation (3.34) in Ramsay (1967), who describes the change of dip in terms of the axes of a strain ellipse, rather than in terms of β. The expression in (A3) does not allow for any contribution to β from motion on the line which deforms by pure shear.

APPENDIX 2

Vertical movement during faulting

If an isolated normal fault of dip θ cuts through a plate with an elastic thickness t_c, movement will produce uplift of the footwall and subsidence of the hanging wall. The easiest method of determining these movements is to consider the equilibrium of each side of the fault separately, and to imagine that each floats in a fluid whose density ρ_m is that of the mantle. We also suppose that the fault cuts through crust of density ρ_c and thickness t_c overlain by sediment of density ρ_s and of no strength. If we neglect the tilting of the fault, the total downwards force F_1 acting on the vertical section AA' through the end of the plate is (Fig. 8a):

$$F_1 = F + \frac{(\rho_c - \rho_s)gt_c^2}{2 \tan \theta} - \left\{ \frac{(\rho_m - \rho_s)g}{2 \tan \theta} \right.$$
$$\left. \times (t_c^2 - (t_c - t_m - w_1)^2) \right\} \tag{A4}$$

where F is the downwards force exerted by the hanging wall on the footwall, g is the acceleration due to gravity, w_1 is the displacement of the footwall and is measured downwards and

$$t_m = \frac{\rho_c - \rho_s}{\rho_m - \rho_s} t_c \qquad (A5)$$

is the level which the mantle fluid would reach, measured from the base of the crust. If we assume $w_1 \ll t_m, t_c$

$$F_1 \simeq F - \frac{(\rho_m - \rho_s)g}{2 \tan \theta} (t_m + 2w_1)(t_c - t_m). \qquad (A6)$$

Similarly the downward force F_2 on the vertical section BB' of the hanging wall, whose displacement is w_2, is (Fig. 8b):

$$F_2 = -F + \frac{g(\rho_c - \rho_s)t_c^2}{2 \tan \theta} - g(\rho_m - \rho_s) \frac{(t_m + w_2)^2}{2 \tan \theta}$$

$$\simeq -F + \frac{g(\rho_m - \rho_s)}{2 \tan \theta} t_m(t_c - t_m - 2w_2) \qquad (A7)$$

where $w_2 \ll t_m, t_c$ is the downward displacement of the hanging wall. As Heiskanen & Vening Meinesz (1958) show, the displacement w must satisfy

$$\frac{d^4w}{dx^4} + \frac{4}{l^4} w = 0 \qquad (A8)$$

where

$$l = \left(\frac{Et_e^3}{3g(\rho_m - \rho_s)(1 - \sigma^2)} \right)^{1/4} \qquad (A9)$$

and has the dimensions of length, E is Young's modulus and σ Poisson's ratio. The flexural forces on the ends of each plate are related to the displacements through

$$F_1 = \frac{g(\rho_m - \rho_s)l}{2} w_1 \qquad (A10)$$

and

$$F_2 = \frac{g(\rho_m - \rho_s)l}{2} w_2. \qquad (A11)$$

If the ratio of footwall motion to hanging wall motion is f, then

$$w_1 = fw_2 \qquad (A12)$$

substitution of (A10), (A11) and (A12) into (A6) and (A7), followed by addition, leads to an equation for f whose solution is

$$f = -\left(\frac{\rho_c - \rho_s + h}{\rho_m - \rho_c + h} \right) \qquad (A13)$$

where

$$h = \frac{l \tan \theta(\rho_m - \rho_s)}{2t_c}. \qquad (A14)$$

If the footwall is covered with water, rather than sediment, the ratio is approximately f_w where

$$f_w = \frac{\rho_c - \rho_s}{\rho_c - \rho_w} f \qquad (A15)$$

where ρ_w is the density of water. If the footwall is subaerial, the ratio approximates to f_a where

$$f_a = \frac{\rho_c - \rho_s}{\rho_c} f. \qquad (A16)$$

Substitution of $\rho_m = 3.3$ g cm^{-3}, $\rho_c = 2.8$ g cm^{-3}, $\rho_s = 2.5$ g cm^{-3}, and $\rho_w = 1.03$ g cm^{-3} gives

$$f_a = -0.06, \qquad f_w = -0.10$$

if $l = 0$ and

$$f_a = -0.08, \qquad f_w = -0.13$$

if $l = t_c$ and $\theta = 45°$. The negative sign of f indicates that the signs of w_1 and w_2 are different.

As (A13) shows, the ratio of the uplift to the subsidence is independent of the dip of the fault if the elastic thickness, and hence l, is zero. If the elastic thickness is not zero, h and f will increase with increasing θ. Hence the steeper second-generation faults will produce greater uplift of their footwalls than will the first-generation faults. Heiskanen & Vening Meinesz (1958) carried out a similar calculation but ignored the terms which depend on w_1 and w_2 in (A6) and (A7), respectively. The expressions in this Appendix are not exact, and in particular the effect of asymmetric loading is only approximated in (A15) and (A16). The purpose of these calculations is to estimate the ratio of footwall uplift to hanging-wall subsidence; which is about 10%. For a more elaborate discussion, involving more complex viscoelastic rheologies, the reader is referred to Zandt & Owens (1980).

Though the calculation above was motivated by observations of vertical movement during normal faulting, the sign of the fault displacement does not affect the result, which therefore applies equally to thrust faulting. As in the case of normal faulting, the expressions relating to asymmetric loading, (A15) and (A16), will only be approximate.

Reprinted by permission of the Geological Society of America
from G. S. Lister, M. A. Etheridge, and P. A. Symonds, *Geology*, v. 14 (1986), p. 246-250.

Detachment faulting and the evolution
of passive continental margins

G. S. Lister

M. A. Etheridge

P. A. Symonds

Bureau of Mineral Resources, Geology and Geophysics, Canberra City, A.C.T. 2601, Australia

ABSTRACT

Major detachment faults play a key role in the lithospheric extension process in the Basin and Range province and may also be important in other continental extension terranes. Such detachment faulting leads to an inherent asymmetry of extensional structure and of uplift/subsidence patterns. Detachment models developed for the formation of metamorphic core complexes can also be applied to the formation of passive continental margins. We therefore suggest the existence of *upper-plate* and *lower-plate* passive margins. These give rise to a complementary asymmetry of opposing margins after continental breakup. Transfer faults offset marginal features and allow margins to switch from upper-plate to lower-plate characteristics along strike.

INTRODUCTION

There is widespread acceptance of the lithospheric stretching model proposed by McKenzie (1978) to explain the crustal thinning, rifting, and subsidence that predate and accompany continental breakup and lead to the development of passive continental margins. Evidence for such stretching comes from (a) crustal thinning and (b) normal fault geometries that require large extensions (Bally, 1981; Le Pichon and Sibuet, 1981). Geophysical modeling of this phenomenon has been based entirely on symmetrical, pure-shear extension models (McKenzie, 1978; Sclater and Christie, 1980; Le Pichon and Sibuet, 1981). Such models, with variations induced by depth-dependent strain (Keen et al., 1982) or depth-dependent rheology (Vierbuchen et al., 1982), allow prediction of the crustal thickness, subsidence histories, and gravity profiles of extended terranes.

Symmetric extension models, however, do not predict the wide variation in gross continental margin architecture, crustal thinning, and continental uplift reported, for example, by Kinsman (1975) or Falvey and Mutter (1981). Features such as marginal plateaus, outer highs, detached continental ribbons, and submerged continental fragments remain largely unexplained, although sophisticated multilayer modeling (Keen et al., 1982; Vierbuchen et al., 1982) addresses some of these questions (e.g., the origin of outer highs; Schuepbach and Vail, 1980). More important, there is a notable absence of symmetrical rift structures in reflection seismic profiles (as pointed out by Bally, 1981, 1982), and opposing margins do not generally exhibit identical structures. We conclude, therefore, that symmetrical extension models have limited applicability. Structural asymmetry may be a general feature of passive margin development.

Structural asymmetry on a range of scales is a feature of many of the models recently proposed for continental extension in the Basin and Range province of the western United States. These models are based on detachment faults and/or shallow-dipping crustal shear zones (Wernicke,

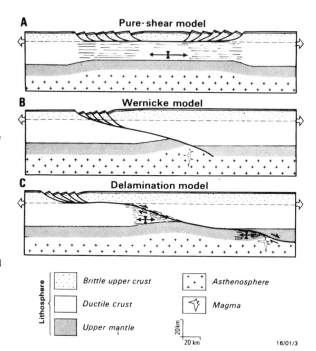

Figure 1. Three models for continental extension.

1981, 1985; Wernicke and Burchfiel, 1982; Davis, 1983). In this paper we explore the consequences of continental separation as the result of the operation of such shallow-dipping detachment faults, and we describe the architecture of passive margins that would result from these inherently asymmetric models for continental extension.

DETACHMENT MODELS FOR CONTINENTAL EXTENSION

Several authors have recognized similarity of passive margin structures to those recognized in the Basin and Range province. However, there is an important element of Basin and Range–style tectonics that has not been recognized on passive margins: detachment faults associated with the formation of metamorphic core complexes and/or (mylonitic) detachment terranes (Crittenden et al., 1980). Metamorphic core complexes, or mylonitic detachment terranes, consist of a largely brittle upper plate overlying ductilely deformed igneous and metamorphic rocks. The upper plate is truncated at its base by low-angle faults of large areal ex-

246

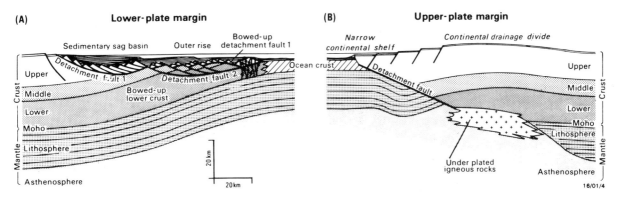

Figure 2. Detachment-fault model of passive continental margins with lower-plate or upper-plate characteristics. Lower-plate margin (left) has complex structure; tilt blocks are remnants from upper plate, above bowed-up detachment faults. Multiple detachment has led to two generations of tilt blocks in diagram shown. Upper-plate margin (right) is relatively unstructured. Uplift of adjacent continent is caused by underplating of igneous rocks. Opposing passive margin pairs exhibit marked but complementary asymmetry.

tent which appear to be normal-slip *detachment* faults on which substantial relative displacements have occurred (e.g., Reynolds and Spencer, 1985). The upper plate has been extended 100%–400% (see Davis et al., 1980; Miller et al., 1983) as the result of movements on listric normal faults and/or dominolike rotations of fault blocks bounded by initially high-angle normal faults (see Wernicke and Burchfiel, 1982; Jackson and McKenzie, 1983). As it is dragged to the surface, the lower plate is subject to intense ductile deformation in intracrustal zones of noncoaxial laminar flow (Lister and Davis, 1983). The lower-plate rocks record a history of rapid uplift while, enigmatically, sedimentation may continue on the upper plate.

There is still some argument about the exact role detachment faults play in the continental extension process. Symmetric pure-shear models assume that the detachment fault represents the brittle-ductile transition (e.g., Miller et al., 1983). The brittle upper crust, typified by rotated tilt blocks, is shown extended over a more uniformly stretched ductile lower crust (Fig. 1A). There is increasing support, however, for models based on shallow-dipping movement zones (Wernicke, 1981, 1985; Davis, 1983; Davis et al., 1986). Wernicke suggested that detachment faults represent low-angle normal faults that cut through the entire lithosphere (Fig. 1B). An alternative separation geometry (Fig. 1C) would involve delamination of the lithosphere, the detachment zone running horizontally below the brittle-ductile transition, steepening, and then again running horizontally at the crust-mantle boundary. We are aware that detachment faults may be merely upper crustal manifestations of major ductile shear zones at depth (Davis et al., 1986) and that the concept of a single lithospheric dislocation may be a gross oversimplification. However, limited space prevents us from developing these concepts further without considerable elaboration. Therefore, in this paper we confine discussion to detachment faults that penetrate the entire lithosphere.

DETACHMENT MODELS FOR THE EVOLUTION OF PASSIVE MARGINS

Asymmetric detachment models imply that continental extension will result in highly asymmetric structure on all scales as the middle to lower crust is dragged out from underneath the fracturing and extending upper crust. The asymmetry of the extended terrane becomes increasingly obvious as continental extension continues. Ongoing extension will eventually lead to continental breakup and to the formation of an ocean basin. The resultant passive margins will, in consequence, also display a

marked but complementary asymmetry. On the crustal or lithospheric scale, the asymmetry of the margin is determined by whether the underlying master detachment fault originally dipped toward the ocean or away from it. In consequence, we predict that there will be two broad classes of passive margins. *Upper-plate margins* comprise rocks originally above the detachment fault. *Lower-plate margins* comprise the deeper crystalline rocks of the lower plate, commonly overlain by highly faulted remnants of the upper plate (Fig. 2). Upper-plate and lower-plate margins will differ primarily in their rift-stage structure and in their uplift/subsidence characteristics. Secondary controls on their character arise from variation in the location of the ocean-continent boundary, complex detachment-fault geometries, and displacement of master detachment faults on transfer faults.

The basement of a lower-plate margin is generally highly structured and has the rotational normal faults, tilt blocks, and half-graben typical of the so-called rift phase of passive margin development (Fig. 2, left). The structure of an upper-plate margin is relatively simple by comparison (Fig. 2, right). Normal faulting on upper-plate margins is generally only weakly rotational.

During detachment faulting the lower plate must warp upward as the load formerly exerted by the upper plate is removed, and a broad arch or culmination will develop. The final form of this culmination is the result of several factors, including the initial geometry of the detachment-fault system (Spencer, 1984) and the effects of multiple generations of detachment faults (Fig. 2). Isostatic stresses acting on the relatively undisrupted upper-plate margin lead to different effects. The detachment system attains its deepest structural levels on the continent side of an upper-plate passive margin where the detachment system passes into the upper mantle. Extension causes horizontal translation of relatively cool and dense lithosphere toward the developing ocean basin, thus exposing the base of the crust to warmer, and hence relatively less dense, rising asthenosphere. The result will be uplift of the continental land surface adjacent to the upper-plate margin. Seaward of the uplifted area, the gravitational response will be reversed because here movement on the detachment results in substitution of mantle for lower crustal material. The upper-plate margin is therefore subjected to an isostatically derived torque. The uplifted area will pass seaward into a sequence of normal faults that drop the land surface abruptly toward sea level. This marginal flexure of the upper-plate margin may be accentuated by sediment loading on the leading edge of the upper plate.

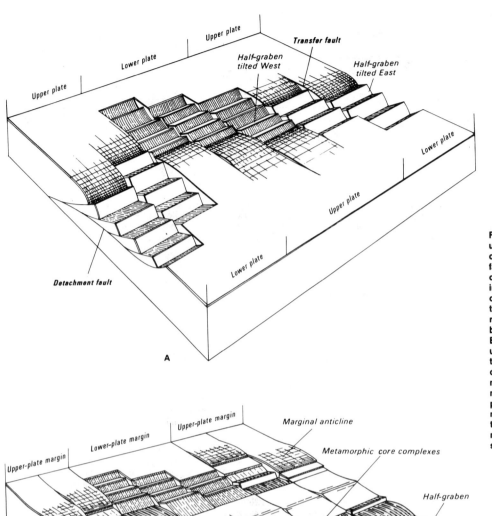

Upper plate

Lower plate

Upper plate

Upper plate

Transfer fault

*Half-graben
tilted West*

*Half-graben
tilted East*

Lower plate

Upper plate

Lower plate

Detachment fault

A

**Figure 3. Changes from
upper plate to lower plate
occur across transfer
faults. A: Half-graben
complex. When underly-
ing detachment faults
change dip across
transfer faults, sense of
rotation of overlying tilt
blocks also changes.
B: If extension continues
until ocean basin forms,
transfer faults mark
changes from *upper-plate*
margins to *lower-plate*
margins. Architecture of
passive margin is deter-
mined by where final con-
tinental separation began
relative to detachment
system.**

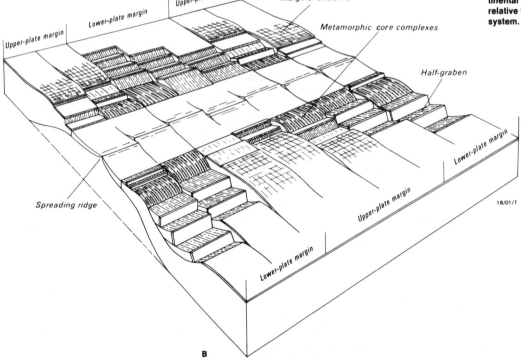

Upper-plate margin

Lower-plate margin

Upper-plate margin

Marginal anticline

Metamorphic core complexes

Half-graben

Lower-plate margin

Spreading ridge

Upper-plate margin

Lower-plate margin

16/01/7

B

In the basic model, depicted in Figure 2, continental breakup is predicted within or close to the culmination in the bowed-up lower plate because that is where the crust is thinnest. This explains one of the more enigmatic features of passive margin development. As noted by Winterer and Bosellini (1981) and Falvey and Mutter (1981), the locus of final continental breakup frequently does not coincide with the rift basins but occurs in the oceanward basement blocks. The rift basins represent the extended part of the upper plate, whereas the oceanward basement blocks represent deeper levels of the crust exposed in the lower-plate culmination where the crust is thinnest. Breakup near this central culmination separates the margins into dominantly upper-plate and lower-plate types. Breakup significantly to either side of this location will produce margin pairs with upper-plate and lower-plate structures on a single margin (Fig. 3).

If breakup occurs seaward of the lower-plate culmination (as shown in Fig. 2), the lower-plate margin will be typified by rift basins defined by half-graben, inboard of an external basement high. The lower-plate culmination could well be the reason for the existence of the *outer highs* that are commonly recognized in passive margins (see Schuepbach and Vail, 1980; Symonds et al., 1984). The outer high is directly analogous to a metamorphic core complex. If erosion of this culmination occurs during the uplift phase, the upper-plate remnants will be lost, and the metamorphic basement will be exposed. If the basement culmination is not eroded, strongly rotated tilt blocks, remnants of the upper plate, will overlie the metamorphic basement.

The uplift-subsidence pattern predicted for lower-plate margins can explain the uplift and denudation of outer highs while rapid deposition occurs in adjacent rift-basin troughs. This is difficult to explain on the basis of conventional models (Falvey and Mutter, 1981). Rift basins defined by half-graben complexes form major sediment traps toward the continent side of the culmination defined by the bowed-up lower plate (Fig. 2, left). The outer culmination will slowly subside after the extension phase as the thermal anomaly caused by extension relaxes. Subsidence will be greatest where the thermal anomaly was strongest, and sediment thicknesses above the culmination will obscure the original structure.

ROLE OF TRANSFER FAULTS

Major normal faults in extensional terranes commonly terminate at orthogonal strike-slip faults or shear zones, which perform a function similar to that of oceanic transform faults (Bally, 1981). These faults have become known as transfer faults (Gibbs, 1984). As shown in Figure 3A, the transfer faults divide the extending terrane into segments. Detachment faults in individual segments may terminate at major transfer faults, in which case the detachment fault may be stepped, or even reverse dip. Other (minor) transfer faults may be confined to the upper plate, affecting only the tilt-block geometry.

Where the detachment fault changes dip direction across one or more transfer faults, the passive margin will change along its length from an upper-plate to a lower-plate margin (Fig. 3B) and will exhibit rectilinear variations in its architecture. Occasionally such variation can be discerned in the bathymetry—e.g., in the Exmouth Plateau area of the northwest Australian margin. We emphasize that transfer faults are a general feature of extended terranes; therefore, they should be expected to occur commonly in passive continental margins. Transfer faults result because both high- and low-angle normal faults nucleate at different places along strike, and mismatches between different fault blocks must be accommodated. Transfer faults are also important in accommodating oblique extension.

Significant variation in margin architecture along strike is therefore a natural consequence of the detachment-fault model, since changes

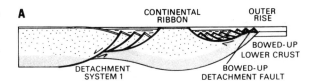

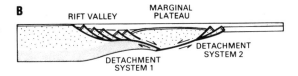

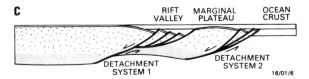

Figure 4. More than one detachment system may be involved in continental extension. Paired detachment systems lead to formation of marginal plateaus, internal rift valleys, or isolated ribbons of continental crust.

along strike at the rift stage (Fig. 3A) are subsequently reflected geometry of the sag phase of the passive margin (Fig. 3B). Transfer faults may pass laterally into oceanic transform faults, but this is not necessarily the case.

ORIGIN OF MARGINAL PLATEAUS AND RIBBON CONTINENTS

Continental separation accomplished by more than one detachment system allows explanation of other aspects of continental margin architecture. Davis and Hardy (1981) described "crustal megaboudins" bounded by (oppositely dipping) shallow-dipping movement zones. If extension continued, a relatively unthinned ribbon continent would result (Fig. 4A), bounded by either ocean crust or an inland rift system.

Another phenomenon results when one shallow-dipping movement zone is cut by another movement zone with the opposite dip (Fig. 4B). One movement zone deactivates, and operation of the transecting movement zone results in removal of the lower part of the crust beneath relatively undeformed parts of the upper plate. Instead of a high-standing region, a low plateau develops (Fig. 4B). Alternatively, two movement zones may develop with the same dip direction, separated by a relatively undeformed region. This event also gives rise to a low plateau (Fig. 4C). If the strike length of these relatively undeformed, high-standing regions is limited by transfer faults, relatively equidimensional highs like many marginal plateaus may develop. Where their strike length is greater, elongate highs (ribbon continents) are formed. Other potential complexities in low-angle extensional fault systems have been described by Gibbs (1984); these also have implications for continental margin architecture when applied on the scale of major detachment faults.

COMPLEMENTARY ASYMMETRY OF OPPOSING PASSIVE MARGINS

The major prediction of the detachment-fault model is that opposing margins should exhibit complementary asymmetry. The upper-plate margin is relatively devoid of structure (Fig. 2, right) and may be uplifted if simple underplating models apply. In contrast, the upper parts of a lower-plate margin are highly structured. The basement to the postextension sag basin on a lower-plate margin (Fig. 2, left) should consist of highly faulted and extended upper-plate remnants overlying the detachment fault and hence will be characterized by highly tilted fault blocks adjacent to half-graben filled with synrift sediments. These rocks are overlain by the gently dipping strata deposited during the postextension, subsidence, or sag phase of margin development. This complementary asymmetry of opposing margins will be easiest to recognize if the margins have straightforward upper-plate or lower-plate characteristics and if relatively precise prebreakup reconstruction can be achieved.

RECOGNITION OF DETACHMENT FAULTS IN PASSIVE MARGINS

We suggest that detachment faulting is an integral part of the continental extension process; therefore, detachment faults are to be expected beneath passive continental margins. Detachment faults, and mylonites in deeper shear zones, may be significant seismic reflectors either because of velocity contrast between upper and lower plates or because of the seismic anisotropy of the mylonites (Fountain et al., 1984). Vaguely defined subhorizontal reflectors have been recognized below highly faulted sequences from several margins (see Fig. 4 of Montadert et al., 1979; Le Pichon and Sibuet, 1981).

Lower-plate passive margins are characterized by rift basins lying inboard of external basement highs. This outer rise represents the culmination formed by the bowed-up lower crust and is therefore the location where the detachment fault is closest to the surface. The dip of the normal faults separating tilt blocks will decrease and may approach horizontal or even reverse dip in the most extended and bowed-up regions above the culmination. Such extreme rotations on the normal faults will give rise to steep dips in the synrift sediments; therefore, these structures will be difficult to resolve on conventionally processed seismic data. An important test of the detachment-fault model would be to mount a drilling program into basement rocks in outer highs and to demonstrate the presence of deeper level sediments and/or metamorphic rocks.

REFERENCES CITED

Bally, A.W., 1981, Atlantic-type margins, *in* Geology of passive continental margins: American Association of Petroleum Geologists Education Course Note Series no. 19, p. 1–48.
—— 1982, Musings over sedimentary basin evolution: Royal Society of London Philosophical Transactions, v. A305, p. 325–328.
Crittenden, M.D., Coney, P.J., and Davis, G.H., eds., 1980, Cordilleran metamorphic core complexes: Geological Society of America Memoir 153, 490 p.
Davis, G.A., Anderson, J.L., Frost, E.G., and Shackelford, T.J., 1980, Mylonitization and detachment faulting in the Whipple-Buckskin-Rawhide Mountains terrane, southeastern California, and western Arizona, *in* Crittenden, M.D., et al., eds., Cordilleran metamorphic core complexes: Geological Society of America Memoir 153, p. 79–130.
Davis, G.A., Lister, G.S., and Reynolds, S.J., 1986, Structural evolution of the Whipple and South Mountains shear zones, southwestern United States: Geology, v. 14, p. 7–10.
Davis, G.H., 1983, Shear zone model for the origin of metamorphic core complexes: Geology, v. 11, p. 342–347.
Davis, G.H., and Hardy, J.J., 1981, The Eagle Pass detachment, southeastern Arizona: Product of mid-Miocene listric(?) normal faulting in the southern Basin and Range: Geological Society of America Bulletin, v. 92, p. 749–762.
Falvey, D.A., and Mutter, J.C., 1981, Regional plate tectonics and evolution of Australia's passive continental margins: Australia Bureau of Mineral Resources, Geology and Geophysics Bulletin, v. 6, p. 1–29.

Fountain, D.M., Hurich, C.A., and Smithson, S.C., 1984, Seismic reflectivity of mylonite zones in the crust: Geology, v. 12, p. 195–198.
Gibbs, A.D., 1984, Structural evolution of extensional basin margins: Geological Society of London Journal, v. 141, p. 609–620.
Jackson, J., and McKenzie, D., 1983, The geometrical evolution of normal fault systems: Journal of Structural Geology, v. 5, p. 471–482.
Keen, C.E., Beaumont, C., and Boutilier, R., 1982, A summary of thermomechanical model results for the evolution of continental margins based on three rifting processes, *in* Watkins, J.S., and Drake, C.L., eds., Studies in continental margin geology: American Association of Petroleum Geologists Memoir 34, p. 725–730.
Kinsman, D.J.J., 1975, Rift valley basins and sedimentary history of trailing continental margins, *in* Fischer, A.G., and Judson, S., eds., Petroleum and global tectonics: Princeton, New Jersey, Princeton University Press, p. 83–128.
Le Pichon, X., and Sibuet, J.-C., 1981, Passive margins: A model of formation: Journal of Geophysical Research, v. 86, p. 3708–3720.
Lister, G.S., and Davis, G.A., 1983, Development of mylonitic rocks in an intracrustal laminar flow zone, Whipple Mountains, SE California: Geological Society of America Abstracts with Programs, v. 15, p. 310.
McKenzie, D.P., 1978, Some remarks on the development of sedimentary basins: Earth and Planetary Science Letters, v. 40, p. 25–32.
Miller, E.L., Gans, P.B., and Garing, J., 1983, The Snake range décollement: An exhumed mid-Tertiary brittle-ductile transition: Tectonics, v. 2, p. 239–263.
Montadert, L., Roberts, D.G., De Charpal, O., and Guennoc, P., 1979, Rifting and subsidence of the northern continental margin of the Bay of Biscay, *in* Montadert, L., Roberts, D.G., et al., eds., Initial reports of the Deep Sea Drilling Project: Washington, D.C., U.S. Government Printing Office, v. 48, p. 1025–1060.
Reynolds, S.J., and Spencer, J.E., 1985, Evidence for large-scale transport on the Bullard detachment fault, west-central Arizona: Geology, v. 13, p. 353–356.
Schuepbach, M.A., and Vail, P.R., 1980, Evolution of outer highs on divergent continental margins, *in* Continental tectonics: Washington, D.C., National Academy of Sciences, p. 50–61.
Sclater, J.G., and Christie, P.A.F., 1980, Continental stretching: An explanation of the post–mid-Cretaceous subsidence of the central North Sea basin: Journal of Geophysical Research, v. 85, p. 3711–3739.
Spencer, J.E., 1984, Role of tectonic denudation in warping and uplift of low-angle normal faults: Geology, v. 12, p. 95–98.
Symonds, P.A., Fritsch, J., and Schluter, H.-U., 1984, Continental margin around the western Coral Sea basin: Structural elements, seismic sequences and petroleum geological aspects, *in* Watson, S.T., ed., Transactions, Third Circum-Pacific Energy and Mineral Resources Conference, Hawaii: Tulsa, Oklahoma, American Association of Petroleum Geologists, p. 243–252.
Vierbuchen, R.C., George, R.P., and Vail, P.R., 1982, A thermal-mechanical model of rifting with implications for outer highs on passive continental margins, *in* Watkins, J.S., and Drake, C.L., eds., Studies in continental margin geology: American Association of Petroleum Geologists Memoir 34, p. 765–778.
Wernicke, B., 1981, Low-angle normal faults in the Basin and Range province: Nappe tectonics in an extending orogen: Nature, v. 291, p. 645–648.
—— 1985, Uniform-sense normal simple shear of the continental lithosphere: Canadian Journal of Earth Sciences, v. 22, p. 108–125.
Wernicke, B., and Burchfiel, B.C., 1982, Modes of extensional tectonics: Journal of Structural Geology, v. 4, p. 105–115.
Winterer, E.L., and Bosellini, A., 1981, Subsidence and sedimentation on a Jurassic passive continental margin, Southern Alps, Italy: American Association of Petroleum Geologists Bulletin, v. 65, p. 394–421.

ACKNOWLEDGMENTS

Peter Davies, John Branson, Cliff Ollier, and Steve Reynolds provided useful comments. The Australian Bureau of Mineral Resources drawing office produced the illustrations. This paper is published with the permission of the Director of the Bureau of Mineral Resources as part of the research program on Extension Tectonics in Australia.

Manuscript received June 13, 1985
Revised manuscript received November 15, 1985
Manuscript accepted December 4, 1985

TECTONICS, VOL. 2, NO. 3, PAGES 239-263, JUNE 1983

THE SNAKE RANGE DÉCOLLEMENT:
AN EXHUMED MID-TERTIARY
DUCTILE-BRITTLE TRANSITION

Elizabeth L. Miller, Phillip B. Gans and
John Garing[1]

Department of Geology, Stanford
University, Stanford, CA 94305

Abstract. The Snake Range décollement
(SRD) in east-central Nevada separates
supracrustal rocks extended by normal
faulting from ductilely deformed igneous
and metamorphic rocks. A well-known
stratigraphy unaffected by earlier
faulting permits analysis of both upper
and lower plate strain leading to a
better understanding of how vastly
different rock types and deformational
styles are juxtaposed along low-angle
faults in metamorphic core complexes.
Middle Cambrian to Permian upper plate
rocks are cut by two generations of NE
trending, east directed normal faults.
Both generations were initiated as
high-angle (60°) planar faults that
flattened abruptly into the SRD and
rotated domino style to low angles,
yielding a total rotation of bedding of
about 80-90°. Faulting is Tertiary in
age as 35-m.y.-old volcanic rocks are
involved and resulted in about 450-500%
extension in a N55W-S55E direction. The
SRD developed as a subhorizontal surface
6-7 km deep at the top of the Cambrian
Pioche Shale. Lower plate granitic

rocks and their late Precambrian-
Cambrian metamorphic country rocks were
involved in progressive ductile to
brittle extension at low greenschist
grade, forming a penetrative
subhorizontal foliation and N55-70W
lineation that increases in intensity
eastward and upward toward the SRD.
Stretching and thinning in the lower
plate is coaxial and comparable in
magnitude to upper plate extension, and
is interpreted as synchronous. K-Ar
ages ranging from 20 to 40 m.y. in the
lower plate suggests the N. Snake Range
represents a Tertiary thermal anomaly.
We conclude that the SRD developed as a
ductile-brittle transition zone at 6-7
km depth. Gravity data suggests that
the gently domed SRD is cut by younger
Basin and Range faults, but the geology
of adjacent ranges suggests that the SRD
does not continue more than 60 km in any
given direction. The lack of
stratigraphic omission across the SRD
rules out large amounts of movement on a
surface that originally cut downsection,
and we suggest that extensional
detachment faults such as the SRD can be
developed locally as boundaries between
brittlely extended rocks and underlying
ductile extension and intrusion.

[1] Now at Amoco Production Company,
Denver, Colorado 80202.

Paper number 3T0536.
0278-7407/83/003T-0536$10.00

INTRODUCTION

Gently dipping detachment faults in
metamorphic core complexes of the U.S.
Cordillera juxtapose brittlely deformed

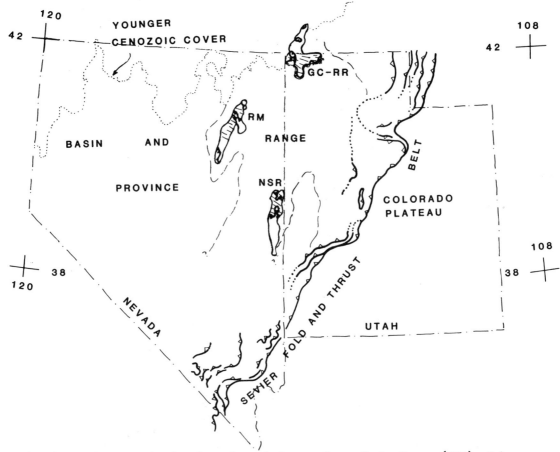

Fig. 1: Index map showing location of the northern Snake Range (NSR), Ruby
Mountains (RM), and Grouse Creek-Raft River (GC-RR) metamorphic core complexes with
respect to the Sevier fold and thrust belt. Lower plates of core complexes show
lineations developed parallel to the direction of extension. Shaded areas
represents highly extended region in NE Nevada and adjacent Utah. Data from Compton
(1980), Snoke (1980), and King (1969).

supracrustal rocks and underlying
metamorphic and igneous rocks inferred
to have originated at much deeper levels
(see review by Coney (1979) and papers
by Crittenden et al. (1980). The age
and tectonic significance of this
striking juxtaposition are topics of
current and lively debate.

Prior to geochronologic evidence for
a Tertiary age, the detachment faults in
these terranes were commonly interpreted
as Mesozoic thrust faults. For example,
Misch (1960) speculated that the Snake
Range décollement in east-central Nevada
(Figure 1) represented a regionally
extensive shearing off fault whose
frontal breakthrough occurred far to the
east as one of the thrusts of the Sevier
fold and thrust belt (Miller, 1966).

Hose and Danes (1973) and Hintze (1978)
also assigned a Mesozoic age to the
Snake Range décollement but recognized
that overlying rocks had been
drastically thinned. They proposed that
east-central Nevada and Utah represented
an uplifted, extended hinterland behind
a gravity-driven thrust belt to the east.

Other workers in the northern Basin
and Range province have applied a gneiss
dome model to core complexes. Armstrong
(1968a) described deep seated Mesozoic
mantled gneiss domes in the Albion
Range, Idaho, that were later affected
by Tertiary uplift and denudation.
Howard (1980) and Snoke (1980) suggested
that the attenuation of rocks beneath
the Ruby Mountains décollement
(Figure 1) was related to the formation

of a Mesozoic abscherungszone between mobile infrastructure and allochthonous brittle suprastructure. Compton et al. (1977), Compton (1980) and Todd (1980) concluded that the Raft River-Grouse Creek core complex (Figure 1) formed by the gravitational denudation of a rising gneiss dome, but documented that both the ductile and brittle deformation was largely Tertiary, not Mesozoic, in age. Similarly, new data from the Ruby Mountains highlight the role of Tertiary ductile deformation and magmatism in the formation of the Ruby Mountains décollement (Snoke et al., 1982).

Within the context of Tertiary extensional tectonics, several models have been proposed for detachment faults in metamorphic core complexes. G. A. Davis et al. (1980) interpreted the Whipple Mountains detachment fault in southern California as the sole of an extensive, far-traveled gravity slide complex. Wernicke (1981) introduced the concept of an 'extensional allochthon' whereby the juxtaposition of supra- and midcrustal rocks is effected along a very low angle zone of simple shear that ultimately involves the entire crust. G. H. Davis (1980) and Hamilton (1982), though differing in the specifics of their models, envision detachment faults as the boundary between an upper crust extended by normal faulting and underlying crustal blocks or lenses that have been pulled apart along ductile shear zones. Rehrig and Reynolds (1980) linked supracrustal extension above detachment faults to underlying, deeper seated penetrative stretching and dilation by intrusions.

In order to evaluate the validity of these models, two basic questions must be answered: (1) How does the age, magnitude, and geometry of the strain in the upper plate of detachment faults compare with that of the lower plate? (2) What is the sense and amount of relative movement between upper and lower plate rocks along detachment faults and how extensive are such faults? Unfortunately, in most metamorphic core complexes, superimposed thermal and structural events have obscured the answers to these questions.

This paper focuses on the age, three-dimensional extent, and tectonic significance of the northern Snake Range décollement (NSRD) in east-central Nevada. The northern Snake Range is particularly well suited for testing models of detachment faulting because (1) a straightforward miogeoclinal stratigraphy in both the upper and lower plates of the NSRD permits an accurate analysis of the strain that has affected these rocks, and (2) the demonstrable lack of predetachment faulting deformation in this region makes structural and stratigraphic relations across the NSRD unambiguous and allows estimation of paleodepths.

REGIONAL SETTING

The Snake Range and surrounding region was the site of relatively continuous continental shelf sedimentation from the Late Precambrian through the early Triassic (Stewart and Poole, 1974; Hose and Blake, 1976). During this timespan, about 10-12 km of strata were deposited above thinned Precambrian crystalline rocks (Figure 2).

Mesozoic thrust faults are well documented to the east in the Sevier orogenic belt (Armstrong, 1968b) and to the west in western Nevada (Speed, 1978) but did not breach the surface in east-central Nevada (see Armstrong, 1972, and discussion by Gans and Miller (1983). The principal evidence for this is that early Tertiary rocks rest disconformably and exclusively on upper Paleozoic strata. Beneath the early Tertiary unconformity, conformable sections that span the entire Upper Precambrian to late Paleozoic interval effectively rule out regional Mesozoic décollement at any of the present levels of exposure (see discussion by Gans and Miller (1983)). Thus, 'stratigraphic depths' within the miogeocline are an accurate estimate of Mesozoic to early Tertiary 'structural depths'.

At shallow levels, Mesozoic shortening of small magnitude is recorded by gentle folds in upper Paleozoic strata, whereas at deeper structural levels, upper Precambrian and locally Lower to Middle Cambrian strata were intruded by plutons and penetratively deformed during regional dynamothermal, greenschist to amphibolite grade metamorphism (Misch, 1960; Misch and Hazzard, 1962; Gans and Miller, 1983). The Snake Range lies within a much broader belt of mid-Tertiary extension and mountain ranges within this belt are

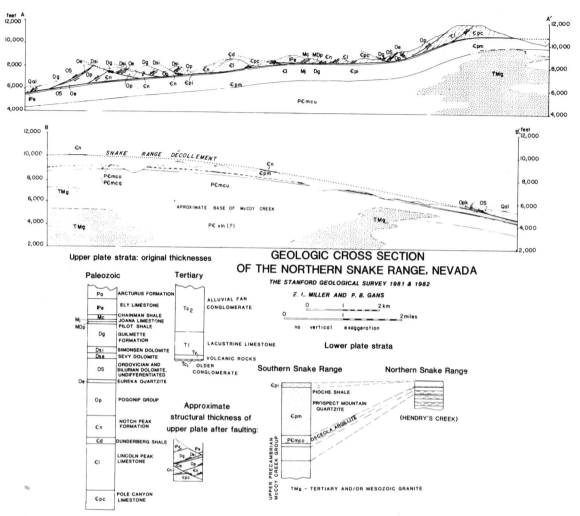

Fig. 2. Geologic cross section and summary of upper and lower plate stratigraphy, northern Snake Range. Symbols as in Plate 1.

characterized by complex arrays of imbricate low- and high-angle normal faults (Gans and Miller, 1983).

The Geology of the Northern Snake Range

Portions of the Snake Range were previously mapped by Drewes (1958), Nelson (1959), Whitebread (1969), Hose and Blake (1976), and Hose (1981). The Snake Range décollement was first described by Hazzard et al. (1953) and has prompted additional interpretations by Misch (1960), Armstrong (1972), Coney (1974), Hose and Whitebread (1981), and Wernicke (1981). Our views differ from these workers and are based on two summers of mapping by Miller, Gans, and the Stanford Geological Survey in the Snake Range, mapping in adjacent ranges

by Gans (1982; Gans and Miller, 1983), and gravity surveys of the region encompassing the Snake Range by Garing.

In the northern Snake Range, vast expanses of penetratively stretched Upper Precambrian and Lower Cambrian metasedimentary rocks and granitic plutons of unknown age are exposed beneath the northern Snake Range decollement (NSRD). Lithologic contacts and foliation in the lower plate are structurally concordant to the gently arched NSRD which generally follows the top of the Lower Cambrian Pioche Shale (Figure 2 and Plate 1). In contrast, Middle Cambrian to Permian and Tertiary strata in the upper plate are broken and tilted by imbricate normal faults that do not cut the decollement.

Upper Plate Faulting

Geometric relations. The geometry of upper plate faults is best documented in the southwestern part of the northern Snake Range, where the upper plate is largely preserved. Exposures of the NSRD to the north, east, and in a window along the Negro Creek drainage constrain its subsurface geometry and provide critical views of the relations between upper plate faults and the décollement (Figure 2 and Plate 1).

Despite the extremely complex map pattern, a systematic structural style is evident in the upper plate. Structural sections that 'young' to the west are repeated eastward on east dipping faults (Figure 2 and Plate 1). Older, west dipping faults within these structural sections typically omit units. The two generations of faults are even more apparent in cross section (Figure 2). The younger faults are spaced approximately 1 km apart, dip 10 to 20° eastward, and merge with but do not offset the NSRD. The older faults are more closely spaced, dip 10° to 30° westward, and are truncated by either the NSRD or the younger faults. In three dimensions, the upper plate faults resemble faults in the Egan Range described by Gans (1982) and Gans and Miller (1983). They define shovel-shaped scoops that are 1.5-5 km across but are relatively planar in the direction of movement.

Hanging wall strata are displaced eastward relative to footwall strata on both generations of faults. The younger faults are clearly down-to-the-east normal faults, whereas the older faults presently have apparent reverse offset. The younger faults typically juxtapose upper Paleozoic formations on lower Paleozoic formations. Their actual offsets (0.2-2.0 km) are generally much less than their stratigraphic offsets (up to 4 km) because they displace sequences that were previously attenuated by the older faults.

Bedding attitudes in the upper plate are variable. Most strata strike N10E to N45E and dip northwest, but the amount of tilting ranges from 0 to 90° or even overturned (Figure 2). Bedding tilts are generally low in incompetent units and near faults. Tilts become progressively steeper away from fault planes, and the steepest westward dips occur in the more massive limestone units between widely spaced faults. Only these steepest dips reveal the true amount of westward rotation and the original bedding-to-fault angles; all lesser dips are demonstrably the result of normal drag on upper plate faults.

Faulting is more complex in the Miller Basin area (Plate 1), where domains of conjugate, down-to-the west faults and southeastward tilting occur as well as domains of down-to-the-east faults. High-angle strike-slip fault zones separate domains of opposite tilting. South of Miller Basin, faults become more widely spaced and the total amount of extension appears to diminish rapidly (Plate 1 and Figure 3).

Kinematic interpretations. The fact that movement on relatively planar, high-angle faults in the upper plate was accommodated by a subhorizontal detachment plane at depth requires that stratal rotation must have accompanied upper plate faulting. Figure 4 illustrates a sequence of faulting and tilting events that would result in the present bedding and fault attitudes. These simplified sequential cross sections do not attempt to show the effect of normal drag. The first generation of faults originated as east dipping, high-angle (50° to 60°) normal faults that subsequently rotated 'domino style' (Thompson, 1960; Morton and Black, 1975) to low angles. Second-generation faults also originated as high-angle faults and were superimposed on previously faulted and tilted strata. As the second-generation faults rotated to low angles, segments of the first-generation faults were rotated through horizontal into westward dips, thus causing the apparent reverse offsets. Note that the toes of the first generation faults were rotated away from the NSRD, whereas higher segments were rotated downward and are presently truncated by the decollement. Nonetheless, the first-generation faults must have interacted with the same basal detachment because they affect the entire range of stratigraphic units in the upper plate. Later doming of the NSRD probably added an additional 5° to 10° of westward tilt to the Negro Creek area.

If both generations of faults originated at the same angle with respect to horizontal, then the average angular difference between them (40°)

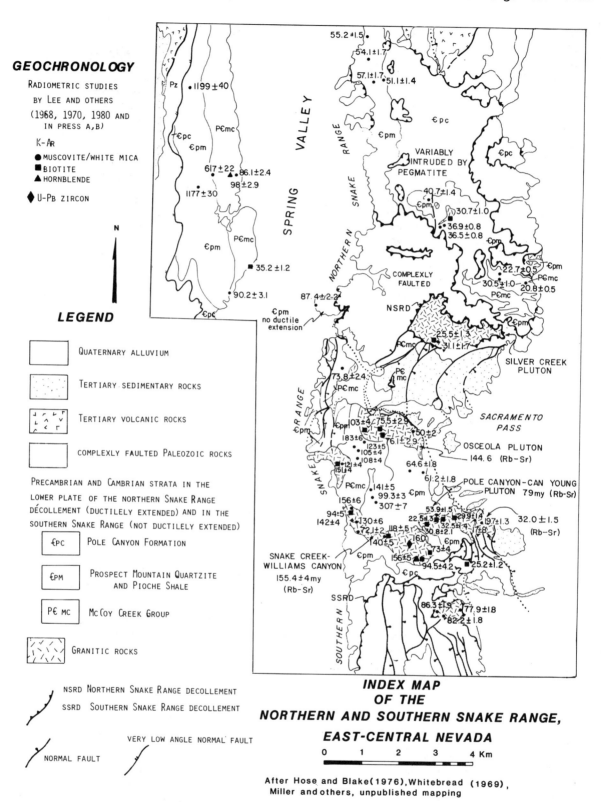

GEOCHRONOLOGY

Radiometric studies
by Lee and others
(1968, 1970, 1980 and
in press a,b)

K-Ar

● Muscovite/white mica
■ Biotite
▲ Hornblende

◆ U-Pb zircon

N

LEGEND

☐ Quaternary alluvium

▫ Tertiary sedimentary rocks

☐ Tertiary volcanic rocks

☐ Complexly faulted Paleozoic rocks

Precambrian and Cambrian strata in the
lower plate of the northern Snake Range
décollement (ductilely extended) and in the
southern Snake Range (not ductilely extended)

€pc   Pole Canyon Formation

€pm   Prospect Mountain Quartzite
       and Pioche Shale

P€mc   McCoy Creek Group

☐ Granitic rocks

nsrd Northern Snake Range decollement
ssrd Southern Snake Range decollement

                    very low angle normal fault

normal fault

INDEX MAP
OF THE
NORTHERN AND SOUTHERN SNAKE RANGE,
EAST-CENTRAL NEVADA

0   1   2   3   4 Km

After Hose and Blake(1976),Whitebread (1969),
Miller and others, unpublished mapping

Fig. 3. Index map of the northern and southern Snake Range, Nevada, showing
location of published radiometric dates.

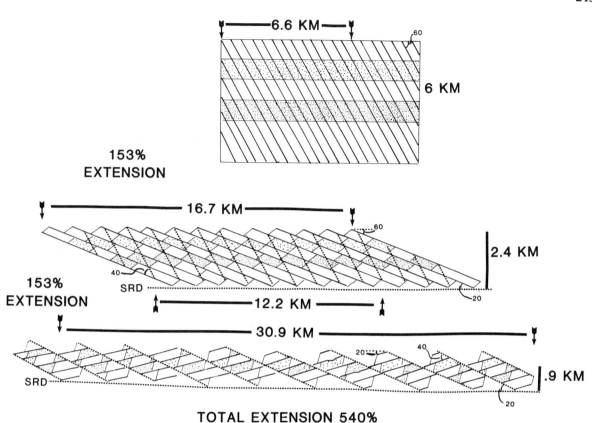

153%
EXTENSION

153%
EXTENSION

TOTAL EXTENSION 540%

Fig. 4. Simplifed geometric model showing how two generations of upper plate faults interact with the decollement (SRD) to produce the bedding and fault attitudes observed in the Snake Range.

is precisely the amount of rotation that occurred on the older faults alone. Once these faults had rotated to dips as low as 20°, the resolved shear stress on the fault planes apparently no longer exceeded the frictional resistance to movement, and a new generation of high-angle normal faults was developed.

Large bedding-to-fault angles at all stratigraphic levels in the upper plate indicates that both generations of faults originally intersected the NSRD at high angles (50° to 60°). Space problems at the toes of fault blocks were apparently relieved by (1) brecciation of the more massive lithologies, (2) warping and folding of less competent units, and (3) low-angle splays at the toes of major fault blocks (see discussion by Gans and Miller 1983). The third process may have been particularly important during movement on the younger, more widely spaced faults as segments of older, rotated faults were reactivated.

Direction of extension in the upper plate. The direction of extension of the upper plate on the west flank of the Snake Range was estimated to be N55W-S55E. Poles to bedding planes form a diffuse great circle whose pole is oriented about N35E and subhorizontal (Figure 5) parallel to the strike of normal faults. These bedding attitudes are compatible with tilting and/or drag folding along southeast directed dip-slip movement faults. Similarly, the orientation of the sides of 'scoops' or 'shovels' and the trends of strike-slip faults between conjugate domains of faulting suggest NW or SE directed movement on upper plate faults.
Amount of extension in the upper plate. We have estimated the amount of extension in the upper plate for the Negro Creek area by three independent methods:

1. Sequentially restoring the faults along our line of cross section (Figure 6) yields approximately 125%

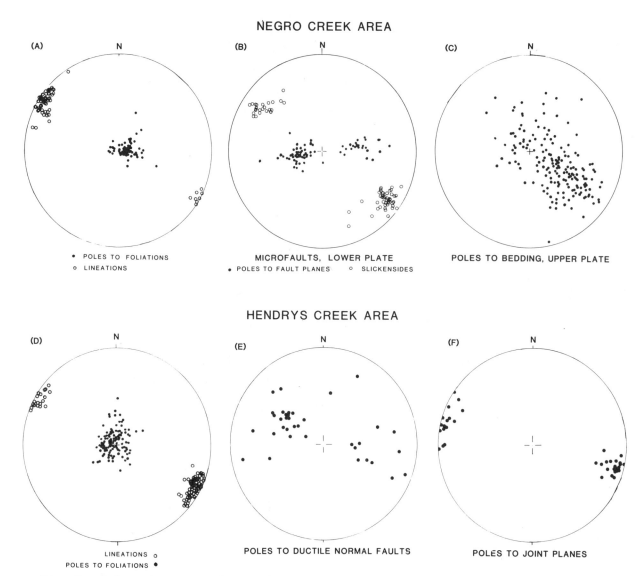

Fig. 5. Selected structural data from the northern Snake Range (lower hemisphere equal area projection). (a.) Attitude of foliation and lineation in the lower plate of the NSRD, Negro Creek area. (b.) Late-stage ductile-to-brittle microfaults in the Prospect Mountain Quartzite, Negro Creek area. (c.) Poles to bedding in the upper plate, Negro Creek area. (d.) Attitude of lower plate foliation and lineations in the middle reaches of Hendry's Creek. (e.) Poles to closely spaced or penetrative conjugate ductile normal faults (extensional cleavage) in McCoy Creek Group schist units along Hendry's Creek. Here, the east dipping set is best developed. (f.) Poles to joint planes, youngest of mesoscopic structures, lower plate rocks in Hendry's Creek. For location of Negro and Hendry's Creek, see Plate 1.

extension by second generation faults and 155% extension by first generation faults for a total of 480% extension (i.e., ((2.55 x 2.25) - 1.0) x 100 = 480%). This method has obvious problems because of the immense amount of

small-scale faulting, brecciation, and folding that has changed the shape of the larger fault slices.

    2. Our best estimate of the average structural thickness of the upper plate after normal faulting but prior to

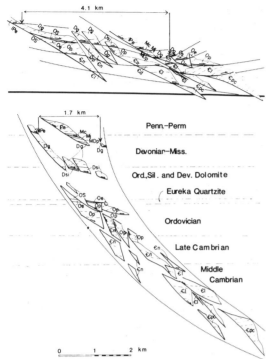

Fig. 6. "Cut and paste" restoration of Snake Range upper plate faults along cross section A-A'. Unit symbols serve as in Figure 2 and Plate 1. Heavy solid lines are second-generation faults, dashed lines are early generation faults. Erosion and change of shape due to brecciation, folding and small-scale faulting is largely responsible for the imperfect match across fault planes.

erosion is approximately 1.1 km compared to an original stratigraphic thickness of 6-7 km (Figure 2), similar to values calculated by Hose (1981). Assuming constant-volume plane strain, this change in thickness is equivalent to about 450% extension.

3. Calculation of the percent extension by using bedding-to-fault angles and the amount of tilting (Figure 6)(Thompson, 1960) yields 153% extension by both generations of faults for a total of 540% extension. This estimate is probably somewhat excessive because it does not account for deviations from two simple generations of faults or for the internal deformation in fault blocks.

Considering the large uncertainties in all of these estimates, we are impressed by how closely they agree. We conclude that 450-500% is a reasonable,

perhaps even conservative, estimate of the amount of extension in the upper plate of the northern Snake Range.

Age of upper plate extension. The best constraints on the age of upper plate faulting come from the Tertiary rock sequence exposed in the Sacramento Pass area (Figures 2 and 3, and Plate 1). This sequence has been described by Hose and Blake (1976), Hose and Whitebread (1981) and most recently by S. Grier (1983; manuscript in progress, 1983). Here, 35-m.y.-old volcanic rocks rest depositionally on only upper Paleozoic strata. Overlying lacustrine and alluvial fan deposits contain debris derived from late Precambrian to Permian miogeoclinal strata, Mesozoic(?) plutons, and Tertiary rocks, suggesting that they were deposited synchronously with nearby faulting and uplift. The sequence does not, however, contain clasts derived from the lower plate of the NSRD. The Tertiary rocks in Sacramento Pass are cut by imbricate normal faults that merge with the NSRD (Hose and Whitebread, 1981)(Figure 3 and Plate 1) and geometrically resemble the second generation of faults in the Negro Creek area. The younger age limit for these faults is unknown. This data is consistent with relations in adjacent ranges that bracket much of the normal faulting in east-central Nevada as Oligocene (Gans, 1982; Gans and Miller, 1983). We differ with Hose and Whitebread (1981) in that we see no evidence for significant faulting and uplift prior to the deposition of the basal Tertiary units.

Lower Plate Deformation

Introduction. In marked contrast to the upper plate, rocks beneath the NSRD are relatively flat-lying and are not cut by faults (Figure 2 and Plate 1). Instead, they reveal a complex history of magmatism, metamorphism, and ductile deformation that ended with a transition into a brittle regime. Argon 40/Ar 39 and U-Pb dating in the northern Snake Range are presently underway; at this time it is not clear what proportion of the high grade metamorphism and plutons are Mesozoic or Tertiary. Most K-Ar mineral dates from lower plate rocks are between 20 and 40 m.y. old (Lee et al., 1968, 1970, 1980, and in press a,b), (Figure 3) but these dates may in part reflect Tertiary reheating of older metamorphic rocks and plutons.

Lower plate metasedimentary rocks are amphibolite grade, and the metamorphism appears to increase both with depth and to the north. Pelitic units in the lowermost part of the Cambrian Pioche Shale along the west flank of the range contain the assemblage staurolite-garnet-muscovite-biotite-quartz. At a deeper stratigraphic level on the east flank of the range, the Osceola Argillite contains kyanite-muscovite-biotite-quartz and locally garnet (Rowles, 1982). It is noteworthy that on the basis of their stratigraphic position, these assemblages appear to have formed at a significantly shallower depth (6-8 km) than the experimentally derived minimum depths (e.g., Holdaway, 1971).

Both in outcrop and in thin section, this high grade metamorphic fabric is clearly cut by or transposed into parallelism with a penetrative subhorizontal foliation which developed at lower metamorphic grade (Rowles, 1982). The older, higher grade metamorphism could be Mesozoic in age, like mid-Jurassic fabrics described in late Precambrian rocks in adjacent ranges (Misch, 1960; Gans and Miller, 1983). However, the younger K-Ar dates from the Snake Range suggest it may well be as young as Tertiary. The strain associated with the younger deformation is so intense that the orientation of older fabrics and magnitude of older strain can no longer be ascertained in the map area. Furthermore, it is unclear what the relationship of plutons are to this metamorphism, as they too have been intensely deformed by this younger event.

Plutons in the lower plate are principally granitic in composition. The largest of these, the Silver Creek pluton (Plate 1), grades from a 2-mica granite margin to a biotite granodiorite interior and, in its eastern exposures includes minor hornblende diorite. Swarms of muscovite-garnet bearing pegmatite dikes intrude both the plutons and metasedimentary rocks and locally comprise up to 80% of the lower plate. Although pegmatite dikes are common immediately below the decollement, they are not present in exposures of the upper plate.

Progressive (Ductile to Brittle) Extension. Superimposed on all lower plate metamorphic and igneous rocks is a younger penetrative subhorizontal foliation and lineation that increases in intensity both eastward and upward toward the decollement. This deformation was accompanied by retrograde low greenschist grade metamorphism. In the western-most exposures of lower plate(?) Prospect Mountain Quartzite (Figure 3 and Plate 1), foliation and lineation are weakly developed, whereas on the east flank of the northern Snake Range, stratigraphic sequences originally about 3 km thick have been ductilely thinned to less than 0.5 km (Figure 2 and Plate 1). Highly attenuated, mylonitic sections of Pioche Shale and the very basal part of the Pole Canyon Limestone are present nearly everywhere beneath the decollement (Figure 2), but are often too thin to be shown on our geologic map (Plate 1). Although lower plate strain diminishes away from the decollement, it does not die out within the present levels of exposure. Since the original thickness and the amount of strain of deeper McCoy Creek Group rocks is unknown, depth to Precambrian crystalline basement is speculative (Figure 2).

In detail, lower plate strain is quite complex and heterogeneous. At the low greenschist grade conditions under which deformation occurred, quartz behaved ductilely while feldspars and micas behaved brittlely. Micas in schist units are either kinked (if the older metamorphic layering is at an angle to the younger subhorizontal fabric), or have been mechanically rotated into parallelism with the new fabric. Metamorphic porphyroblasts such as garnet, staurolite and kyanite are kinked, pulled apart, or rotated, but trains of inclusions in these minerals still preserve an older fabric (Rowles, 1982). Granitic rocks and pegmatite dikes are variably deformed depending on their quartz to feldspar ratio, grain size and/or dike attitude. The plutons, because of their high feldspar content, are in general less deformed than metasedimentary country rocks and have essentially behaved as giant 'augen' during deformation. The only rocks that are not involved in lower plate

STYLES OF PROGRESSIVE
CO-AXIAL DEFORMATION

CRUSTAL COMPRESSION

CRUSTAL EXTENSION

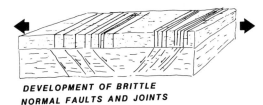

**DEVELOPMENT OF BRITTLE
NORMAL FAULTS AND JOINTS**

*D3: SMALL SCALE KINK FOLDING
OF OLDER FOLIATION SURFACES
ABOUT STEEP AXIAL PLANES*

*THRUST FAULTING:
OLDER OVER YOUNGER,
HIGHER OVER LOWER GRADE*

crinkle lineations

S1   S2

DUCTILE TO BRITTLE

*D2 :UPRIGHT TO OVERTURNED
TIGHT TO OPEN CO-AXIAL FOLDS ,
REFOLD S1 SURFACES*

**DEVELOPMENT OF LAYERING-PARALLEL FAULTS
AND DUCTILE NORMAL FAULTS**

S1

So

*D1 ISOCLINAL TO TIGHT
RECUMBENT FOLDING ,
TRANSPOSITION OF ORIGINAL LAYERING INTO
PARALLELISM W/ AXIAL PLANE CLEAVAGE*

**HORIZONTAL STRETCHING OF UNITS,
FEW FOLDS DEVELOPED IF LAYERING IS
SUBHORIZONTAL, ABUNDANT FOLDS IF STEEP.
PROLATE STRAIN ELLIPSOID**

Fig. 7. Comparison of ductile to brittle progressive deformational fabrics
produced during compression and extension.

deformation are rare, fine-grained
quartz porphyry dikes that cut the
subhorizontal fabric yet in one locality
are truncated by the NSRD. Similar
quartz porphyry dikes are locally
present in the upper plate and are
involved in the faulting.

Mesoscopic structures associated
with the subhorizontal fabric in the
lower plate record ductile to brittle
progressive extension (Figures 5 and
7). Stretched pebbles and mineral
grains indicate extension in a NW-SE
direction and flattening in a vertical
direction. The direction of stretching

in the lower plate rocks of the NSRD
varies from N55W on the west side of the
range to N70W on the east side. Aspect
ratios of stretched pebbles in McCoy
Creek Group strata on the east flank of
the range are commonly of the order of
8-10: 1:0.1. Occasional mesoscopic and,
more rarely, map-scale folds are present
in the lower plate. Folds in the
Hampton Creek area (Plate 1) have been
described by Rowles (1982). Here, fold
axes are everywhere subparallel to the
direction of extension, and are
overturned both to the north and south.
Most of these folds probably formed

during extension and reflect (1) local changes in the magnitude and configuration of the intermediate and minor strain axes, and/or 2) superposition of the extensional strain ellipsoid on originally nonhorizontal layering (Rowles, 1982).

Late stage brittle structures such as microscopic and mesoscopic ductile normal faults, brittle normal faults and joints in the lower plate indicate a NW-SE direction of extension, parallel to the direction of mineral grain and pebble elongation (Figure 5). East dipping mesoscopic and microscopic ductile normal faults are sometimes preferentially developed over their west dipping conjugate sets, particularly along the east flank of the range.

The ductile to brittle progressive deformation recorded in lower plate rocks of the Snake Range is similar to extensional fabrics described by G. H. Davis (1980) in many of the Arizona metamorphic core complexes, and unmistakably different from that typically developed in rocks during progressive compressional deformation (Figure 7).

Amount of extension in the lower plate. We have estimated the amount of extension in the lower plate by the change in thickness of the Lower Cambrian Prospect Mountain Quartzite. Undeformed Prospect Mountain Quartzite sections in adjacent ranges are typically 1200 m thick. On the east flank of the northern Snake Range, complete sections of this unit are generally only 100-200 m thick. On the west side of the range, the base of the Prospect Mountain Quartzite is not exposed. However, the reduction in average bedding thickness (from 45 cm to 15 cm) suggests that it has been thinned to approximately one third its original thickness. An average of 330% extension across the range was derived by restoring the Prospect Mountain Quartzite to its original thickness while preserving its cross-sectional area.

Age of lower plate extension. The age of lower plate extension is not yet bracketed radiometrically, but we strongly suspect that it is Tertiary and synchronous with upper plate normal faulting. Indirect arguments that

support this age assignment include (1) the similarity in magnitude and precise parallelism of lower and upper plate strain axes, (2) anomalously young Tertiary K-Ar mica ages from all rock types in the lower plate as compared to non-stretched but equivalent structural levels in adjacent ranges (Figure 3), and (3) well-documented mid-Tertiary ages for identical fabrics in the nearby Raft River Range and Ruby Mountains (Compton et al., 1977; Snoke et al., 1982).

The northern Snake Range décollement. The northern Snake Range décollement is an extremely sharp break between brittlely and ductilely deformed rocks. In most places, the same formation (Pole Canyon Limestone) is present immediately above and below this fault. Assuming that lower plate stretching was synchronous with upper plate faulting, then the NSRD represents an exhumed mid-Tertiary, ductile-brittle transition zone. Irregular zones of ductile deformation and recrystallization in the lower portions of upper plate fault slices (Pole Canyon Limestone) suggest that, initially, the transition was fairly diffuse. As extension continued, the transition was rapidly localized within a narrow (probably less than 100 m) interval. Local shattering and brecciation of lower plate mylonitic rocks suggest that the NSRD ultimately evolved into a strictly brittle fault.

REGIONAL EXTENT OF THE SNAKE RANGE 'DUCTILE-BRITTLE TRANSITION ZONE'

Gravity Models of the Snake Range and Adjacent Valleys

The extent of the Snake Range décollement beyond the northern Snake Range can be inferred from subsurface data in the adjacent valleys and structural relations in adjacent ranges. We have compiled a Bouguer gravity anomaly map of the Snake Range and surrounding areas (Figure 8) in order to evaluate the geometry of the adjacent basins and to judge whether the associated 'Basin and Range' faults cut the NSRD. Bouguer anomaly values in the map region range from -250 to -150 mGaL; conspicuous northerly trending gravity lows correspond to valleys underlain by low density Tertiary and Quaternary

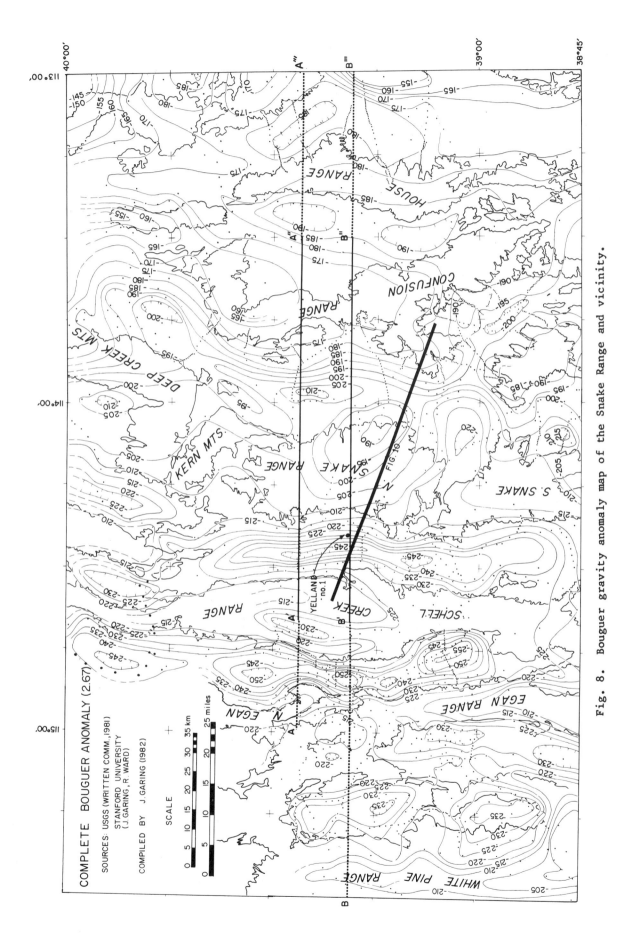

Fig. 8. Bouguer gravity anomaly map of the Snake Range and vicinity.

COMPLETE BOUGUER ANOMALY (2.67)

SOURCES: USGS (WRITTEN COMM., 1981)
STANFORD UNIVERSITY
(J. GARING, R. WARD)

COMPILED BY J. GARING (1982)

SCALE

103

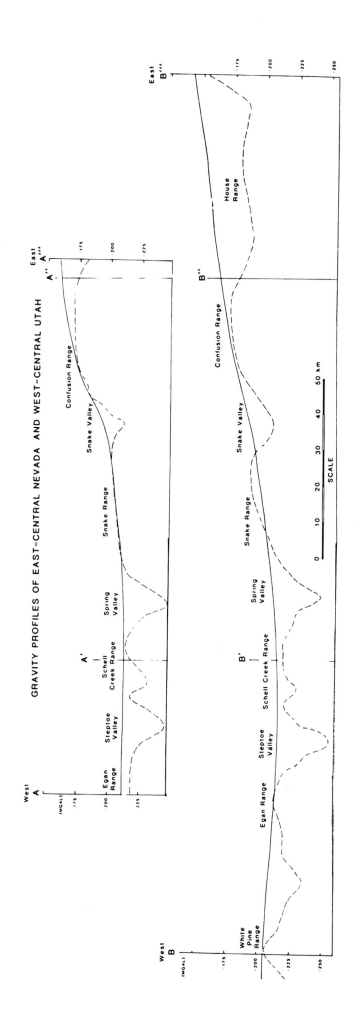

GRAVITY PROFILES OF EAST-CENTRAL NEVADA AND WEST-CENTRAL UTAH

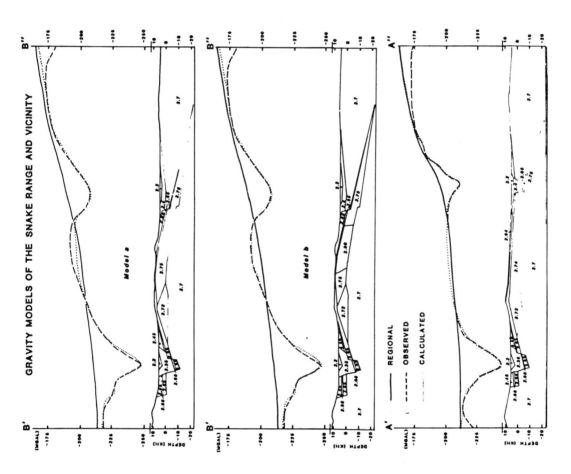

Fig. 9. Gravity profiles and models across the Snake Range and adjacent valleys, see Figure 8 for locations.

TABLE 1. Summary of Sample Densities for the Snake Range Region

Sedimentary and Metamorphic Rocks	Range of Densities	Number of Samples	Average
Quaternary Valley fill	2.1-2.4	well logs*	...
Tertiary Lake Deposits	2.375-2.59	3	2.49
Upper Plate Paleozoic Carbonate	2.57-2.76	9	2.68
Lower Plate Marble	2.59-2.85	8	2.66
Cambrian Prospect Mountain Quartzite	2.615-2.66	7	2.64
PC McCoy Creek Group			
Quartzite	2.64-2.655	2	2.65
Quartzose Schist	2.65-2.68	4	2.67
Schist	2.65-3.25	15	2.78
Igneous Rocks			
Hornblende diorite	2.895-2.925	4	2.91
Biotite granite	2.62-2.665	8	2.65
Muscovite-bearing granite pegmatite	2.57-2.64	2	2.61
Tertiary/Quaternary volcanic rocks	2.3-2.4	well logs*	...
Precambrian crystalline basement	...	...	2.7

Wells used: 1-20 Federal SE NW Sec 20   T12N   R67E; #1 Yelland NW SE Sec
              23 T17N   R67E.

*   Source of information: Dome Petroleum (written communication, 1982).

The relatively undeformed Paleozoic section beneath the Confusion Range
was assigned an average density of 2.70 g/cm^3, while faulted and
brecciated upper Paleozoic rocks in the Snake Range were assigned a
density of 2.68 g/cm^3.

Mylonitic marble beneath the Snake Range decollement (Cambrian Pole
Canyon Limestone) is modeled with a density of 2.64 g/cm^3, the sample
normative value instead of the average density of 2.66 g/cm^3.

The density of the McCoy Creek Group is modeled as varying laterally as
the schist to quartzite ratio increases eastward. Also, highest grade
schists on the east flank of the Snake Range are more dense than any
sample collected from equivalent units in the adjacent Schell Creek
Range to the west.

A density of 2.70 g/cm^3 is used for inferred Precambrian crystalline
rocks at depth which is the average for the United States calculated by
Wollard (1962).

alluvial and lacustrine sediments.
Within the intervening mountain ranges,
the effect of the valley fill is reduced
sufficiently for more regional gravity
anomaly values to be measured. Relative
gravity anomaly highs within the
mountain ranges increase from west to
east, suggesting a strong regional
gradient in the Bouguer anomaly
(Figure 9, B-B"). This regional trend
is thought to be due largely to
variations in thickness of the
lithosphere beneath this region (Eaton
et al., 1978; Thompson and Zoback,

1979). The regional gravity profiles in Figure 9 were qualitatively estimated and were not determined by our modeling which focuses instead on features in the upper 10 km of crust. Density assignments for the gravity models are based on measurements of rock samples collected in the ranges and density log data from wells in Spring Valley (Table 1). Our density models for the Snake Range follow our geological cross sections and exceptions to density assignments for particular lithologies are listed in Table 1.

The gravity low observed over Spring Valley is best fit by a model involving three high-angle normal faults that displace the pre-Tertiary basement surface (Figure 9). The most significant of these faults occurs along the flank of the Schell Creek Range and has a down-to-the-east displacement of approximately 6-8 km (Figures 9 and 10). Gravity data in Snake Valley to the east of the Snake Range suggests the presence of a down to the east range front fault with about 1.5 km or more of offset (Figure 9). According to our data, the thickness of basin fill, and thus the displacement along this fault, increases toward the north. If the NSRD extends beneath these valleys, it seems most likely that it is cut and offset by the high-angle Basin and Range faults.

Using the values listed and discussed in Table 1, the calculated gravity anomaly (model a) along line B'-B" does not fit the observed Snake Range anomaly well and implies that the model needs more mass beneath the east side of the range. An alternate density model for section B'-B" is shown (model b) with mass added in order to generate a calculated anomaly that matches the observed anomaly. The added mass could be due to more mafic intrusive rocks at depth, either within the metasedimentary sequence or in the underlying crystalline basement. Alternatively, the added mass could be explained by a greater percentage of particularly dense garnet-staurolite schist in the McCoy Creek Group (d max = 3.25 gM/cm^3, Table 1).

Section A'-A", on the other hand, requires a reduction in the mass of the density model for a match between the observed and calculated gravity anomalies. The excess mass could be removed by lowering the density

assignment for the McCoy Creek Group rocks, suggesting a decrease in the schist:quartzite ratio northward, or could be due to the presence of lower density granitic plutons at depth (d = 2.65 g/cm^3).

Westward continuation of the NSRD

Yelland Well #1 in Spring Valley (Figure 8) penetrated 1650 m of Quaternary and Tertiary deposits and 150 m of Paleozoic rocks before reaching quartzite at a depth of 1800 m. (Dome Petroleum, private communication, 1982). This depth is compatible with the projected depth of the NSRD based on its attitude along the west flank of the Snake Range, and compatible with our gravity data that suggests no major faults cut the decollement along this side of the valley (Figure 10), but we do not know if this quartzite is penetratively stretched. Further west, an impressively thick 30°-45° west dipping section of Precambrian McCoy Creek Group and Prospect Mountain Quartzite is exposed along the east flank of the Schell Creek Range (Misch and Hazzard, 1962; Young, 1960; Drewes, 1967; Hose and Blake, 1976) (Figure 10). These rocks are not ductilely extended like their counterparts in the lower plate of the northern Snake Range, and are in places conformably overlain by Pioche Shale and Pole Canyon Limestone. In the Conners Pass Area, Drewes (1967) mapped low-angle younger on older faults within the middle Cambrian portion of the section (the 'Schell Creek Thrust') and between Cambrian and Ordovician strata. Further north, Hose and Blake (1976) mapped a low-angle fault within the Pole Canyon Limestone along part of the range. If either of these faults represent the westward continuation of the NSRD, it has cut upsection and has lost its Snake Range character in that it no longer juxtaposes brittlely and ductilely stretched rocks. However, these faults may not be 'basal detachments' like the NSRD but rotated 'upper plate' type faults that have locally dragged bedding into parallelism with the fault planes. To the north of our line of section, McCoy Creek Group strata are clearly involved in down-to-the-southeast normal faulting and have steep (up to 50°) westward tilts, suggesting that they overlie a deeper detachment

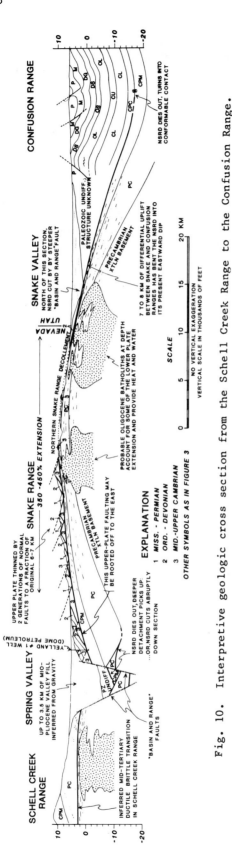

Fig. 10. Interpretive geologic cross section from the Schell Creek Range to the Confusion Range.

fault. We suggest that the NSRD dies out somewhere beneath Spring Valley and that a deeper mid-Tertiary ductile-brittle transition must underlie the Schell Creek Range (Figure 10).

Eastward Continuation of the NSRD

Paleozoic strata in the Confusion Range to the east of the Snake Range are involved in a broad, open synform (Hose, 1977) (Figure 10). Based on known stratigraphic thicknesses of miogeoclinal rocks, the depth to the top of the Cambrian Prospect Mountain Quartzite in the Confusion Range is approximately 7 km (Figure 10). The 7-9 km of differential uplift between the Confusion Range and the northern Snake Range originally pointed out by Hose (1981) might be accomplished either by a fault or by a large-scale upward bend or warp in the decollement surface. We do not know for sure whether the NSRD and the lower plate strain continue beneath the Confusion Range, but we interpret the lack of 'upper plate type' deformation in the Confusion Range as indicative that the decollement also dies out to the east. Further east in the House Range, conformable Lower to Middle Cambrian sections (Hintze, 1980) demand that if it extends that far it must cut downsection.

Northward and Southward Continuation of the NSRD

North of the map area (Figure 3), the décollement cuts stratigraphically up section, and marbles as young as Middle(?) Cambrian are present in a lower plate position (Nelson, 1959, Hose and Blake, 1976) (Figure 3). At the northernmost reaches of the Snake Range it dips gently (5° or less) northward beneath faulted Paleozoic strata that can be traced continuously into the Kern Mountains (Nelson, 1959, Hose and Blake, 1976). If the NSRD continues this far north, it may truncate the Cretaceous Kern Mountain pluton (Best et al., 1974; Lee et al., 1983) at depth, as upper plate strata of the NSRD are apparently intruded by the pluton.

Along the southern flank of the northern Snake Range, the NSRD plunges abruptly southward beneath the Sacramento Pass area (Figure 3, Plate 1). Here, late Precambrian and Cambrian strata are present in both upper and lower plate positions. The upper plate

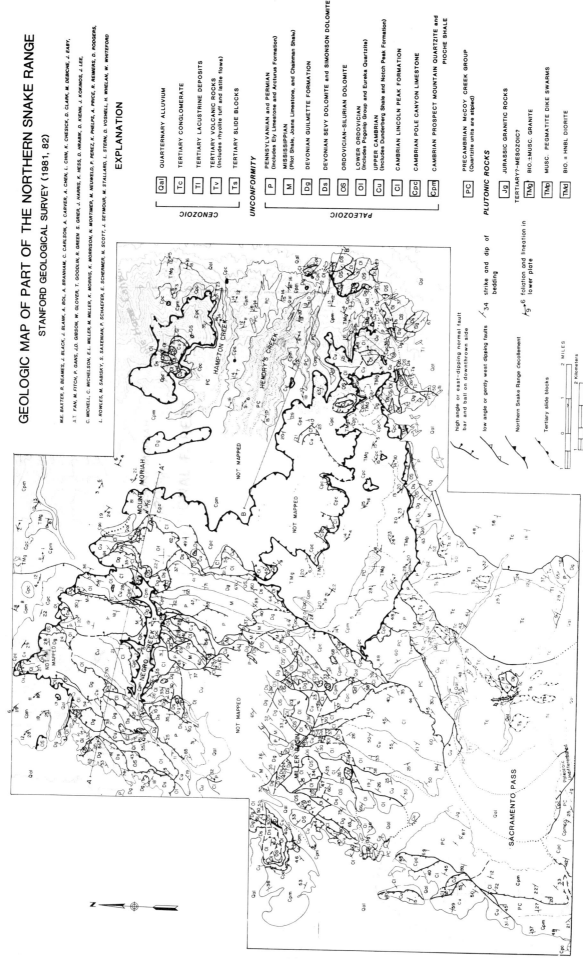

GEOLOGIC MAP OF PART OF THE NORTHERN SNAKE RANGE

STANFORD GEOLOGICAL SURVEY (1981, 82)

M.E. BAXTER, R. BEAMES, J. BLACK, J. BLANK, A. BOL, A. BRANHAM, C. CARLSON, A. CARVER, A. CHEN, L. CHIN, K. CHESICK, D. CLARK, M. DEBICHE, J. EABY, A.T. FAN, M. FITCH, P. GANS, J.D. GIBSON, W. GLOVER, T. GOODLIN, R. GREEN, S. GRIER, J. HARRIS, K. HESS, D. HRABIK, D. KIEHN, J. KOKINOS, J. LEE, C. MICHELL, C. MICHELSON, E.L. MILLER, M. MILLER, K. MORRIS, K. MORRISON, N. MORTIMER, M. NEUWELD, P. PEREZ, K. PHELPS, A. PRICE, R. REIMERS, D. RODGERS, L. ROWLES, M. SABISKY, S. SAXEMAN, P. SCHAEFER, E. SCHERMER, N. SCOTT, J. SEYMOUR, M. STALLARD, L. STERN, D. VOSHELL, H. WHELAN, W. WHITEFORD

EXPLANATION

CENOZOIC

Qal	QUATERNARY ALLUVIUM
Tc	TERTIARY CONGLOMERATE
Tl	TERTIARY LACUSTRINE DEPOSITS
Tv	TERTIARY VOLCANIC ROCKS (includes rhyolite tuff and latite flows)
Ts	TERTIARY SLIDE BLOCKS

UNCONFORMITY

PALEOZOIC

P	PENNSYLVANIAN and PERMIAN (Includes Ely Limestone and Arcturus Formation)
M	MISSISSIPPIAN (Pilot Shale, Joana Limestone, and Chainman Shale)
Dg	DEVONIAN GUILMETTE FORMATION
Ds	DEVONIAN SEVY DOLOMITE and SIMONSON DOLOMITE
OS	ORDOVICIAN–SILURIAN DOLOMITE
Ol	LOWER ORDOVICIAN (Includes Pogonip Group and Eureka Quartzite)
Cu	UPPER CAMBRIAN (Includes Dunderberg Shale and Notch Peak Formation)
Cl	CAMBRIAN LINCOLN PEAK FORMATION
Cpc	CAMBRIAN POLE CANYON LIMESTONE
Cpm	CAMBRIAN PROSPECT MOUNTAIN QUARTZITE and PIOCHE SHALE

PC	PRECAMBRIAN McCOY CREEK GROUP (Quartzite units are stipled)

PLUTONIC ROCKS

Jg	JURASSIC GRANITIC ROCKS

TERTIARY?–MESOZOIC?

TMg	BIO.±MUSC. GRANITE
TMp	MUSC. PEGMATITE DIKE SWARMS
TMd	BIO. ± HNBL DIORITE

high angle or east-dipping normal fault bar and ball on downthrown side

low angle or gently west dipping faults

Northern Snake Range decollement

Tertiary slide blocks

/34   strike and dip of bedding

foliation and lineation in lower plate

2 MILES

2 Kilometers

Plate 1. Simplified geologic map of the northern Snake Range.

109

THIS PAGE INTENTIONALLY BLANK

units are not ductilely deformed and can be traced southward to the southern Snake Range where they are in the lower plate of another 'décollement' mapped by Whitebread (1969). The southern Snake Range 'décollement' does not separate ductile and brittle deformation and cannot be structurally equivalent to the NSRD. Thus if the NSRD continues south from Sacramento Pass, it must cut to deeper structural levels, as both McCoy Creek Group strata and Prospect Mountain Quartzite in the northern part of the southern Snake Range are involved in upper plate type faulting.

Synthesis

Structural and stratigraphic relations in the northern Snake Range help constrain the kinematics of crustal extension that led to development of the Snake Range decollement. These relations provide important documentation of how supracrustal extension by high-angle normal faults is accommodated at depth. Rocks above the NSRD were extended approximately 450% in a NW-SE direction by two generations of high-angle normal faults that rotated to low angles as they moved. Relations in the Sacramento Pass area and in adjacent ranges (Gans and Miller, 1983; Grier, 1983) indicate that this faulting is largely mid-Tertiary in age. Rocks beneath the NSRD were penetratively stretched approximately 350% parallel to the direction of extension in the upper plate. The importance of dilation by plutons in the lower plate is as yet unknown. The coaxial nature of upper and lower plate deformation, together with K-Ar dates that indicate the northern Snake Range was anomalously hot during the mid-Tertiary, suggest that lower plate ductile extension was synchronous with upper plate normal faulting. We conclude that the NSRD represents a mid-Tertiary ductile-to-brittle transition zone.

The consistent stratigraphic position of the Snake Range décollement at the Pioche Shale-Pole Canyon Limestone boundary (Figures 2,10) indicates that it originated at 6-7 km depth and was originally subhorizontal. The precise amount and direction of movement of the upper plate with respect to the lower plate is enigmatic as there are no offset markers. Although the NSRD, like other core complex detachment faults, presently juxtaposes rocks that represent radically different structural levels and deformational styles, this juxtaposition can be explained by two generations of normal faults that thin the upper plate and by the collapsing of isograds by ductile thinning in the lower plate. We emphasize that there need not be a great amount of offset on the decollement as the amount of extension above and below is approximately the same! Thus extension by normal faulting in the upper plate may have been largely accommodated in situ by penetrative stretching and magmatism in the lower plate, similar to models proposed by Rehrig and Reynolds (1980), and Eaton (1982).

On the other hand, several arguments can be made for some movement on the NSRD:

1. The metamorphic grade of the youngest units in the lower plate may locally be appreciably higher than that of the oldest units in the upper plate.

2. The amount of extension in the upper plate appears to be somewhat higher than the amount of extension in the lower plate.

3. The overall assymmetry of lower plate deformation (Figure 2) and the preferential development of down-to-the-east normal faults is compatible with eastward movement of the upper plate with respect to the lower plate. In this view, a component of the extension in the upper plate along the west flank of the range and under Spring Valley may be rooted off to the east where the maximum lower plate strain occurs (Figure 10).

4. The observed strain gradient towards the decollement indicates that lower plate deformation may have involved a component of simple shear. If so, upper plate rocks would be increasingly allochthonous with respect to progressively deeper horizons in the lower plate. Alternatively, the gradient in penetrative stretching may simply reflect increasing dilation by plutons with increasing depth.

Although some movement may have occurred on the NSRD, there is strong evidence that this surface did not cut downsection in the direction of movement of the upper plate, a geometry required by Wernicke's (1981, 1982) model of 'low-angle, rooted normal faults'. Across the entire width of the range, the oldest unit at the toes of upper plate fault slices (Middle Cambrian Pole Canyon Limestone) is in stratigraphic continuity with the youngest unit in the lower plate (Lower Cambrian Pioche Shale). The lack of stratigraphic (i.e., structural) omission across the decollement effectively rules out large amounts of movement on a surface that originally cut downsection.

It is interesting to note that a similar lack of stratigraphic omission is evident in other metamorphic core complexes and may in fact characterize detachment faults in general. In the Ruby Mountains, the youngest intruded and deformed strata in the lower plate and the oldest "upper plate" strata are both Devonian in age (Snoke et al., 1982, Snoke, 1980) and in the Raft River Range these are both Permian in age (Compton et al., 1977). Howard et al. (1982) describe normal fault slices in Arizona-California that involve 10 'km of crustal section, which is also the estimated depth of origin of lower plate rocks of the adjacent Whipple Mountains detachment fault (Davis et al., 1980, 1982).

Relations in surrounding areas suggest that the NSRD need not extend very far beyond the limits of the northern Snake Range. As a single detachment fault, it probably is not more than 60 km across in any given direction. We conclude that extensional detachment faults like the NSRD develop as subhorizontal boundaries at midcrustal levels between overlying rocks extended by normal faulting and underlying rocks that are penetratively stretched and intruded by plutons, similar to models proposed by Eaton (1982) and Rehrig and Reynolds (1980). These boundaries may develop as local dislocations and hence do not need to 'surface' or 'root' deep in the mantle.

Acknowledgments. We appreciate the encouragement and financial support given to us by Shell Oil and Noranda Mining companies and the financial support of the Stanford University Earth Sciences McGee Fund in the beginning stages of our work in East Central Nevada. Our work is presntly funded by NSF grant (EAR-82-06399) awarded to E. L. Miller and G. A. Mahood, which we gratefully acknowledge. Gans in addition acknowledges support of his studies by Tennecco Oil Company. Many thanks go to the members and Teaching Assistants of the 1981 and 1982 Stanford Geological Survey for their enthusiasm and hard work in the Snake Range. We are extremely grateful to Sohio Oil Company for their recent funding of the Stanford Field Camp. Rich Ward energetically helped collect the gravity data for this paper. Our understanding of the geology of east central Nevada has benefited from discussions with R. R. Compton, J. Garing, J. Hakkinen, R. K. Hose, G. . Mahood, A. Snoke, B. Robinson, G. Thompson, and especially the members of the Stanford Geological Survey. This manuscript has benefited from critical reviews by B. C. Burchfiel, W. Hildreth, J. Lee, and G. Thompson. Thanks to Susan Wilgus for typing and assembling the author-produced copy.

REFERENCES

Armstrong, R. L., Mantled gneiss domes in the Albion Range, southern Idaho, Geol. Soc. Am. Bull., 79, 1295-1314, 1968a.
Armstrong, R. L., Sevier orogenic belt in Nevada and Utah, Geol. Soc. Am. Bull., 79, 429-458, 1968b.
Armstrong, R. L., Low-angle (denudational) faults, hinterland of the Sevier Orogenic Belt, eastern Nevada and western Utah, Geol. Soc. Am. Bull., 83, Earth Sci. Bull., 1729-1754, 1972.
Best, M. G., R. L. Armstrong, W. C. Graustein, G. F. Embree, and R. C. Ahlborn, Two-mica granites of the Kern Mountains pluton, eastern White Pine County, Nevada: Remobilized basement of the Cordilleran miogeosyncline?, Geol. Soc. Am. Bull., 85, 1277-1286, 1974.
Compton, R. R., Fabrics and strains in quartzites of a metamorphic core complex, Raft River Mountains, Utah, Cordilleran Metamorphic Core Complexes, edited by M. D. Crittenden, M.D., Jr., et al., Geol. Soc. Am. Mem., 153, 385-398, 1980.
Compton, R.R., V. R. Todd, R. E. Zartman, and C. W. Naesser, Oligocene and Miocene metamorphism, folding, and low angle faulting in northwestern Utah, Geol. Soc. Am. Bull., 88, 1237-1251, 1977.

Coney, P. J., Structural analysis of the Snake Range decollement, east-central Nevada, Geol. Soc. Am. Bull., 85, 973-978, 1974.

Coney, P. J., Tertiary evolution of Cordilleran metamorphic core complexes, in Cenozoic Paleogeography of the Western United States, edited by J. W. Armentrout, et al., pp. 15-28, Society of Economic Paleontologists and Mineralogists Pacific Section, Tulsa, Okla., 1979.

Crittenden, M. D. Jr., P. J. Coney, and G. H. Davis (Eds.), Cordilleran Metamorphic Core Complexes, Mem. Geol. Soc. Am., 153, 490 pp., 1980.

Davis, G. A., J. L. Anderson, E. G. Frost, and T. J. Shackelford, Mylonitization and detachment faulting in the Whipple-Buckskin-Rawhide Mountains terrane, southeastern California and western Arizona, Cordilleran Metamorphic Core Complexes, edited by M. D. Crittenden, Jr., P. J. Coney, and G. H. David, Mem. Geol. Soc. Am., 153, 79-130, 1980.

Davis, G. A., J. L. Anderson, D. L. Martin, D. Krummenacher, E. G. Frost, and R. L. Armstrong, Geologic and geochronologic relations in the lower plate of the Whipple detachment fault, Whipple Mountains, Southeastern California: A progress report, in Mesozoic-Cenozoic Tectonic Evolution of the Colorado River Region, California, Arizona, and Nevada, edited by E. G. Frost and D. L. Martin. Geological Society of America, Anderson-Hamilton Symposium Volume, pp. 408-432, Boulder, Colo., 1982.

Davis, G.H., Structural characteristics of metamorphic core complexes, southern Arizona, edited by M. D. Crittenden, Jr., P. J. Coney, and G. H. Davis, G.H., Cordilleran Metamorphic Core Complexes, Mem. Geol. Soc. Am., 153, 35-78, 1980.

Drewes, H., Structural geology of the southern Snake Range, Nevada, Geol. Soc. Am. Bull., 69, 221-240, 1958.

Drewes, H., Geology of the Connors Pass quadrangle, Schell Creek Range, east-central Nevada, U.S. Geol. Surv. Prof. Paper, 557, 93 pp., 1967.

Eaton, G. P., The Basin and Range Province: Origin and tectonic significance, Ann. Rev. Earth Planet. Sci., 10, 409-440, 1982.

Eaton, G. P., R. R. Wahl, H. J. Prostaka, D. R. Mabey, M. D. Kleinkopf, Regional gravity and tectonic patterns: their relations to late Cenozoic epeirogeny and lateral spreading in the western Cordillera: Geol. Soc. Am. Mem., 152, edited by R. B. Smith and G. P. Eaton, 51-92, 1978.

Gans, P. B., Mid-Tertiary magmatism and extensional faulting in the Hunter District, White Pine County, Nevada: M.S. thesis, Stanford Univ., Stanford, Calif., 1982.

Gans, P.B., and E. L. Miller, Style of Mid-Tertiary extension in east-central Nevada, in Guidebook Part 1, Geol. Soc. Am. Rocky Mountain and Cordilleran Sections Meeting, Utah Geol. and Mining Survey Spec. Studies, 59, 107-160, 1983.

Grier, S., Tertiary stratigraphy and geologic history of the Sacramento Pass Area, Nevada, Geol. Soc. Am., Rocky Mountain and Cordilleran sections meeting, Utah Geol. and Mineral Survey Spec. Studies, 59, 139-144, 1983.

Hamilton, W., Structural Evolution of the Big Maria Mountains, Northeastern Riverside County, Southeastern California, in Mesozoic-Cenozoic Tectonic Evolution of the Colorado River Region, California, Arizona, and Nevada, Anderson-Hamilton Symposium Volume, edited by E. G. Frost and D. L. Martin, pp. 1-28, Geological Society of America, Boulder, Colo., 1982.

Hazzard, J. C., P. Misch, J. H. Wieses, and W. C. Bishop, Large-scale thrusting in northern Snake Range, White Pine County, north-eastern Nevada (abstract), Geol. Soc. Am. Bull., 64, 1506-1508, 1953.

Hintze, L. F., Sevier orogenic attenuation faulting in the Fish Springs and House Ranges, western Utah, Brigham Young Univ. Geol. Stud., 25, pt. 1, 11-24, 1978.

Hintze, L. F., Geologic map of Utah, scale 1:500,000, Utah Geol. and Miner. Surv., Salt Lake City, 1980.

Holdaway, M. J., Stability of andalusite

and the aluminum silicate phase diagram, Am. J. Sci., 271, 97-131, 1971.

Hose, R. K., Structural geology of the Confusion Range, west-central Utah, U.S. Geol. Surv. Prof. Pap., 971, 9 p., 1977.

Hose, R. K., Geologic map of the Mount Moriah further planning (Rare II) area, eastern Nevada, Map MF-1244A, U.S. Geol. Surv., Reston, Va, 1981.

Hose, R. K., and M. C. Blake, Jr., Geology and mineral resources of White Pine County, Nevada, I, Geology, Nev. Bur. Mines Geol. Bull., 85, 105 pp., 1976.

Hose, R.K., and Z. E. Danes, Development of late Mesozoic to early Cenozoic structures of the eastern Great Basin, in Gravity and Tectonics, edited by K. A. DeJong, and R. Scholten, John Wiley, New York, pp. 429-441, 1973.

Hose, R. K., and D. H. Whitebread, Structural evolution of the central Snake Range, Eastern Nevada during the mid-to-late Tertiary, Geol. Soc. Am. Abstr. Programs, 113, 62, 1981.

Howard, K. A., Metamorphic infrastructure in the northern Ruby Mountains, Nevada, Cordilleran Metamorphic Core Complexes, edited by M. D. Crittenden, Jr., P. J. Coney, and G. H. Davis, Mem. Geol. Soc. Am., 153, 335-347, 1980.

Howard, K. A., J. W. Goodge, and B. E. John, Detached crystalline rocks of the Mohave, Buck and Bill Williams Mountains, Western Arizona, in Mesozoic-Cenozoic Tectonic Evolution of the Colorado River Region, California, Arizona, and Nevada, Anderson-Hamilton Symposium Volume, edited by E. G. Frost and D. L. Martin, pp. 377-392, Geological Society of America, Boulder, Colo., 1982.

Kerr, R. A., New gravity anomalies mapped from old data, Science, 215, 5, 1220-1222, 1982.

King, P. B., Tectonic map of North America, scale 1:5,000,000, U.S. Geol. Surv., Reston, Va., 1969.

Lee, D. E., R. F. Marvin, T. W. Stern, and R. E. Mays, and R. E. Van Loenen, Accessory zircon from granitoid rocks of the Mount Wheeler mine area, Nevada, U.S. Geol. Surv. Prof. Pap. 600 D, D197-D203, 1968.

Lee, D. E., R. F. Marvin, T. W. Stearn,

and Z. E. Peterman, Modification of K-Ar ages by Tertiary thrusting in the Snake Range, White Pine County, Nevada, U.S. Geol. Surv. Prof. Pap., 700 D, D93-D102, 1970.

Lee, D. E., R. F. Marvin, and H. H. Mehnert, A radiometric age study of Mesozoic-Cenozoic metamorphism in eastern White Pine County, Nevada and nearby Utah, U.S. Geol. Surv. Prof. Pap., 1158C, C17-C28, 1980.

Lee, D. E., R. W. Kistler, and A. C. Robinson, The strontium isotope composition of granitoid rocks of the southern Snake Range, Nevada, Shorter Contributions to Isotope Research, U.S. Geol. Surv. Prof. Pap., in press, 1983 or 1984a.

Lee, D. E., J. S. Stacey, and L. Fischer, Muscovite-phenocrystic two-mica granites of northeastern Nevada are late Cretaceous in age, Shorter Contributions to Isotope Research, U.S. Geol. Surv. Prof. Pap., in press, 1983 or 1984b.

Miller, G. M., Structure and stratigraphy of southern part of Wah Wah Mountains, Southwest Utah, Bull. Am. Assoc. Pet. Geol., 50, 858-900, 1966.

Misch, P., Regional structural reconnaissance in central-northeast Nevada and some adjacent areas: Observations and interpretations, Interm. Assoc. Pet. Geol., Annu. Field Conf. Guideb., 11th, 17-42, 1960.

Misch, P., and J. C. Hazzard, Stratigraphy and metamorphism of Late Precambrian rocks of central northeast Nevada and adjacent Utah, Bull. Am. Assoc. Pet. Geol., 46, 289-343, 1962.

Morton, W. H., and R. Black, Crustal attenuation in Afar, Afar Depression of Ethiopia, International Symposium on the Afar Region and Related Rift Problems, edited by A. Pilger and A. Resler, Inter Union Comm. Geodyn., Sci. Rep., 14, 55-65, 1975.

Nelson, R. B., The stratigraphy and structure of the northernmost part of the northern Snake Range and Kern Mountains in eastern Nevada and the southern Deep Creek Range in western Utah, Ph.D. thesis, Univ. of Wash., Seattle, 1959.

Rehrig, W. A., and S. J. Reynolds, Geologic and geochronologic reconnaissance of a northwest-trending zone of metamorphic complexes in southern Arizona, Cordilleran Metamorphic Core Complexes, edited by M. D. Crittenden, Jr., P. J. Coney, and G. H. Davis, Mem. Geol. Soc. Am., 153, 157, 1980.

Rowles, L., Deformational history of the Hampton Creek Canyon Area, N. Snake Range, Nevada, M.S. thesis, Stanford Univ., Stanford, Calif., 1982.

Snoke, A. W., Transition from infrastructure to suprastructure in thenorthern Ruby Mountains, Nevada, Cordilleran Metamorphic Core Complexes, edited by M. D. Crittenden, Jr., P. J. Coney, and G. H. Davis, Mem. Geol Soc. Am., 153, 287-334, 1980.

Snoke, A. W., S. L. Durgin, and A. P. Lush, Structural style variations in the northern Ruby Mountains-East Humbolt Range Nevada, Geol. Soc. Am. Abstr. Programs, 14 (4), 235, 1982.

Speed, R. C., Paleogeographic and plate tectonic evolution of the early Mesozoic marine province of the western Great Basin, Mesozoic Paleogeography of the Western United States, edited by D. G. Howell and K. A. McDougall, pp. 253-270, Society of Economic Paleontologists and Mineralogists Pacific Section, Tulsa, Okla., 1978.

Stewart, J. H., and F. G. Poole, Lower Paleozoic and uppermost Precambrian of the Cordilleran Miogeocline, Great Basin, Western United States, edited by W. R. Dickinson, Tectonics and Sedimentation, Spec. Publ. Soc. Econ. Paleontol. Mineral., 22, 28-57, 1974.

Thompson, G. A., Problems of late Cenozoic structure of the Basin Ranges, Int. Geol. Congr. Rep. Sess. Norden, 21st, 62-68, 1960.

Thompson, G. A. and M. L. Zoback, Regional Geophysics of the Colorado Plateau, Tectonophysics, 61, 149-181, 1979.

Todd, V. R., Structure and petrology of a Tertiary gneiss complex in northwestern Utah, Cordilleran Metamorphic Core Complexes, edited by M. D. Crittenden, Jr., et al., Mem. Geol. Soc. Am., 153, 287-333, 287-333, 1980.

United States Geological Survey, Simple Bouguer gravity anomaly data from the United Defense Mapping Agency, St. Louis, Missouri. Machine terrain corrections and curvature corrections made by the U.S. Geol. Survey, Menlo Park office, written comm., 1981.

Wernicke, B., Low angle normal faults in the Basin and Range Province: Nappe tectonics in an extending orogen, Nature, 291, 645, 1981.

Wernicke, B., Cenozoic dilation of the Cordilleran orogen and its relation to plate tectonics, Eos Trans. AGU, 63 (45), 914, 1982.

Whitebread, D. H., Geologic map of the Wheeler Peak and Garrison quadrangles, Nevada and Utah, Map I-578, U.S. Geol. Surv., Reston, Va., 1969.

Woolard, G. P., The relation of gravity anomalies to surface elevation, crustal structure and geology, Univ. Wisc. Res. Report 62-9, Madison, 329 p., 1962.

Young, J. C., Structure and stratigraphy in the north central Schell Creek Range, eastern Nevada, Ph.D thesis, Princeton Univ., Princeton, N. J., 1960

(Received January 21, 1983; revised March 11, 1983; accepted March 22, 1983.)

Reprinted by permission from *Nature*, v. 291, p. 645-648.
Copyright © 1981 Macmillan Magazines Ltd.

Low-angle normal faults in the Basin and Range Province: nappe tectonics in an extending orogen

Brian Wernicke

Department of Earth and Planetary Sciences, Massachusetts Institute of Technology, Cambridge, Massachusetts 02139, USA

Cenozoic normal fault mosaics bounded beneath by a basal fault in the Basin and Range Province (BRP) have traditionally been described either in terms of large-scale surficial gravity sliding or by some form of *in situ* lower plate accommodation. I suggest here that these areas may be an extensional analogue to thin-skinned compressional orogens, a process which may even dominate active BRP tectonics.

These terranes consist of a mosaic of rock sheets bounded by high- and low-angle normal faults[1,2] which resemble (and have been misinterpreted as) compressional nappe piles. In many areas, the normal faults 'bottom out' into a single, subhorizontal fault zone below which either minor or no tectonism has occurred[3-6], or a roughly synchronous, more ductile style of deformation is present[7,8]. The minimum areal extent of the basal faults is typically measured in thousands of square kilometres.

For example, in east-central Nevada, normal fault systems developed in Palaeozoic miogeoclinal strata, Mesozoic and Tertiary intrusives, and Tertiary volcanics end abruptly at a basal fault (Snake Range decollement[3], Fig. 1) accompanied by a thin (0–100 m) zone of tectonite marble. Offsets on the basal and higher faults are difficult to contrain, but Nelson reports a 20-km minimum horizontal offset for one fault alone in the Deep Creek Range[9]. My palinspastic reconstruction of the Northern Egan and Cherry Creek ranges (Fig. 2) indicates a 16.5-km offset for one sheet in the mosaic, and the per cent increase in original width of the area is ~300%. As the basal and higher faults in Fig. 1 are commonly subparallel to bedding and often cut out thousands of metres of section, total offsets are probably at least in the range of several tens of kilometres. Most authors except one[10] who have mapped in the area outlined in Fig. 1 agree that the transport direction of the allochthonous sheets was eastwards. The faults also tend to cut consistently downsection to the east[11,12], (Fig. 2). This terrane apparently has every attribute of a thin-skinned compressional orogen (basal decollement, far-travelled allochthons, consistent direction of section transgression and transport) except (1) the low-angle faults habitually omit rather than repeat section, (2) associated high-angle faults are predominantly normal instead of reverse, and (3) folding is of subdued importance[13]. Though early workers attempted to correlate these faults with Mesozoic thrusting in the Sevier belt to the east[3], there is considerable evidence to support a Tertiary, extension-related origin for most of the deformation[13-15].

Models for these areas must account for three observations: (1) the basal and higher faults predominantly represent extension; (2) the rocks below the basal faults are tectonically inert during transport of the allochthonous sheets and high-angle fault blocks; and (3) the areal extent and transport distances of the allochthons are large. Current models fall into two categories. The first is a localized gravity-slide model in which domal upwarps shed their cover in a megalandslide fashion[15-18] (Fig. 3*a*). The second invokes penetrative ductile stretching and/or igneous intrusion to drive extension in the immediately overlying brittle fault mosaic[2,7,8,19,20] (Fig. 3*b*). Landslide models are weakened by the scarcity of geometrically required areas of shortening correlative in age, size and direction of transport with the extensional terranes. Thus, the model agrees with observations (2) and (3), but not with observation (1). *In situ* ductile stretching or intrusion models lose appeal when it is noted that most metamorphic fabrics and igneous bodies (if any) beneath the basal faults are of inappropriate age or geometry to be rigorously linked to normal faulting[5,6,21,22]. Thus, the model is in harmony with observations (1) and (3), but not with observation (2).

The key to the problem may lie in our understanding of compressional orogens, in which the horizontal movement of thin sheets of rock above an underthrusting slab may occur many

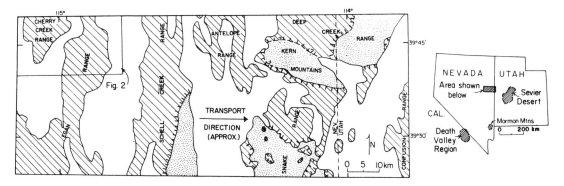

Fig. 1 Location map and tectonic interpretation of east-central Nevada, showing autochthon (stippled), basal fault zone (line with tick marks, queried where uncertain) and normal fault mosaic (hatched). Basal fault in the Schell Creek Range is locally tilted steeply westwards, implying continued deformation after it ceased moving, possibly on a lower basal fault. Tilting and faulting in the Northern Egan Range may be related to this second event rather than to movement on the basal fault shown here.

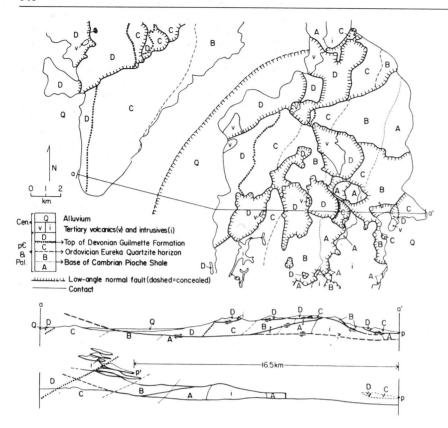

Fig. 2 Highly simplified map, cross-section and palinspastic reconstruction of the area indexed in Fig. 1. Reconstruction does not remove tilting, which was probably synchronous with faulting. Armstrong[15] interpreted the sheets as having formed exclusively by high-angle normal faulting, then subsequently rotated to the horizontal. Geology by Fritz[11].

tens (even hundreds) of kilometres from the nearest possible site of coeval lower plate shortening. Applying this concept to extensional tectonics to explain relationships observed in the BRP, I suggest that the basal faults 'root' into the crust (and perhaps the mantle) at a low angle, thereby serving as a narrow zone of decoupling between two thin sheets of rock (Fig. 3c). Towards the thick end of the upper plate, extensional faulting may be negligible or absent, but towards the thin end it loses its ability to remain coherent and becomes a thin-skinned extensional fault terrane. Lower plate extension may therefore be far removed horizontally (perhaps a distance many times the areal width of upper plate normal faulting) and at much greater depth

than the thin-skinned extensional belt. This hypothesis allows for a rigid autochthon and eliminates the need for immense terranes of coeval shortening, thus satisfying all three observations.

As shown in Fig. 3c, rock sheets (nappes[23]) may be carried along the basal fault and accreted to the autochthon, kinematically similar to nappe emplacement along the soles of overthrust masses in compressional belts. Note that the nappes in Fig. 3c have converged on one another, a relationship that superficially resembles shortening. However, the amount of movement between them is actually a measure of crustal divergence. If the faults had developed in a pristine sedimentary

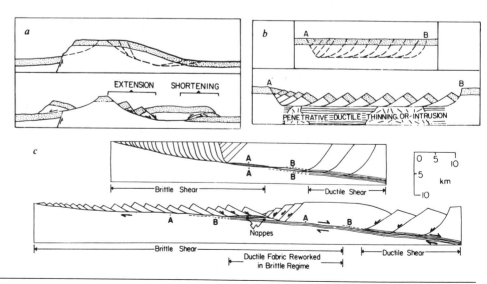

Fig. 3 First-order kinematic schemes to generate low-angle normal fault mosaics above abasal fault. *a*, Megalandslide model; *b*, *in situ* ductile stretching or intrusion model; *c*, rooted, low-angle normal fault model.

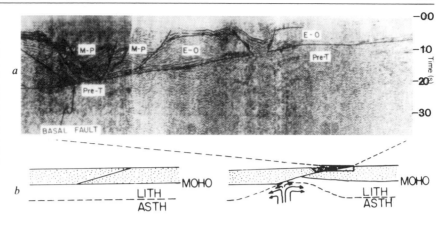

Fig. 4 *a*, North-east trending, 40-km long two-way time section through the Sevier Desert area[29]. Pre-T, pre-Tertiary; E–O, Eocene–Oligocene; M–P, Miocene–Pliocene. *b*, Lower crust–mantle extrapolation of faults such as that in *a*. Note that the geometry of the Moho is indistinguishable from how it would look if the crust had penetratively stretched. (Seismic line reproduced with permission of the Rocky Mountain Association of Geologists.)

sequence, the fault between the nappes would place younger rocks on top of older rocks, a direct consequence of the fault rooting in the same direction as transport. Had the same juxtaposition occurred along a fault which rooted opposite its transport direction, movement between the nappes would be an expression of true crustal convergence. Large, rooted low-angle normal faults may form zones of ductile shear along deeper segments. If the displacement is large enough, these zones may shallow and be tectonically reworked in brittle conditions[21]. This scheme of nappe emplacement, combined with superjacent 'domino-style'[24] and listric normal faulting, are major mechanisms of extension above the basal fault. Although some aspects of this model are not new[25,26], its application to low-angle normal fault terranes in the BRP has been neglected in favour of gravity and *in situ* models.

Distinction between low-angle normal fault terranes and active BRP tectonics has been strongly emphasized by some workers[27], and others have related the apparent difference in style (low-angle compared with high-angle) to plate tectonics[28]. However, the neotectonics of the BRP are apparently sufficiently ill-characterized at depth to entertain the hypothesis that rooted, low-angle normal faults are an important mode of Plio-Quaternary extension, perhaps the fundamental control of its famous topography. For example, the Amargosa Chaos[4], a striking example of low-angle normal faulting[1], involves mid- to late-Pliocene volcanics, and hence was active well within what some consider to be a province-wide period of high-angle faulting. Further evidence suggesting that the present mechanism may control active BRP tectonics comes from seismic reflection profiles of the Sevier Desert area[29]. Figure 4*a* and three other profiles reveal a minimum area of ~5,000 km² underlain by a gently west-dipping low-angle normal fault, above which Miocene–Pliocene basin fill and Quaternary basalt flows are offset by normal faults[29-31].

The basal fault in Fig. 4*a* has been interpreted as a reactivated thrust fault[29], a common interpretation of low-angle normal faults situated in older thrust terranes[25,32,33]. However, recent studies of some exhumed thin-skinned extensional areas suggest that pre-existing anisotropies are not a prerequisite for low-angle normal faults[5], and thrust faults, if present, need not control their geometry. In the Mormon Mountains of southern Nevada (Fig. 1), a Miocene low-angle normal fault event is superposed on a system of Mesozoic thrusts, a situation geometrically identical to that in the Sevier Desert[34]. The normal faults there truncate the thrusts at high angle, and clearly involve Precambrian crystalline rocks of the thrust autochthon. Based on this study, the normal faults in the Sevier Desert may be unrelated to Mesozoic thrusts prominent in the surface geology of the area.

Figures 3*c* and 4*a* suggest that conservative estimates of province-wide extension (10–30%) based on its proportionality to regional tilts of Tertiary strata may have no meaning[35], because large, horizontal translations of rock masses may occur

without appreciable rotation of strata (as well as opposite tilts within the same allochthon). The present model therefore compliments larger estimates (50–100%) deduced by strike-slip fault reconstruction (work in preparation) and crustal thickness arguments[20].

Another critical aspect of the model is that it presents an alternative to penetrative ductile thinning or intrusion to accommodate extension at crustal (and mantle) levels not exposed in the BRP. Whereas igneous intrusion may increase the width of the crust, it cannot thin it; at best, it can retain its original thickness, if the intrusions are vertical dykes. Alternatively, it can actually thicken, if the intrusions are sill-like, a property of the large early- to mid-Tertiary batholiths of southern Arizona[22]. Because intrusion cannot therefore thin deeper crustal levels, it is difficult to envisage it as a major mechanism of BRP extension. Although taffy-like stretching of crustal dimensions does not encounter these geometrical difficulties, there is no compelling evidence to suggest that it is a major process at depth. An attractive alternative, consistent with surface processes, is that large, rooted normal faults extend at a low angle deep into the lithosphere, and that extension at the surface is due to discrete shear between large coherent sheets (Fig. 4*b*). Pull-apart may then ultimately be accommodated by asthenospheric convection rather than any form of stretching.

I thank colleagues at MIT for discussions. This work was supported by NSF grant EAR 7713637 awarded to B. C. Burchfiel.

Received 18 December 1980; accepted 17 March 1981.

1. Wright, L. A. & Troxel, B. W. In *Gravity and Tectonics* (eds de Jong, K. A. & Scholton) 397–407 (Wiley, New York, 1973).
2. Anderson, R. E. *Bull. geol. Soc. Am.* **82**, 43–58 (1971).
3. Misch, P. *Intermountain Ass. petrol. Geol. 11th a. Field Conf. Guidebook*, 17–42 (1960).
4. Noble, L. F. *Bull. geol. Soc. Am.* **52**, 941–1000 (1941).
5. Davis, G. A., Anderson, J. L., Frost, E. G. & Shackelford, T. J. *Geol. Soc. Am. Mem.* **153**, 79–129 (1980).
6. Compton, R. R., Todd, V. R., Zartman, R. E. & Naeser, C. W. *Bull. geol. Soc. Am.* **88**, 1237–1250 (1977).
7. Rehrig, W. A. & Reynolds, S. J. *Geol. Soc. Am. Mem.* **153**, 131–158 (1980).
8. Davis, G. H. *Geol. Soc. Am. Mem.* **153**, 35–78 (1980).
9. Nelson, R. B. *Bull. Am. Ass. petrol. Geol.* **53**, 307–339 (1969).
10. Fritz, W. H. *Nevada Bureau of Mines Map No. 35* (1968).
11. Bick, K. F. *Bull. Utah Geol. miner. Soc.* **77** (1966).
12. Dechert, C. P. thesis, Univ. Washington (1967).
13. Young, J. C. *Intermountain Ass. petrol. Geol. 11th a. Field Conf. Guidebook*, 158–172 (1960).
14. Lee, D. E., Marvin, R. F., Stern, T. W. & Peterman, Z. E. *U. S. geol. Surv. Prof. Pap.* **700-D**, D92–D102 (1970).
15. Armstrong, R. L. *Bull. geol. Soc. Am.* **83**, 1729–1754 (1972).
16. Drewes, H., *U. S. geol. Surv. Prof. Pap.* **557** (1967).
17. Moores, E. M., Scott, R. B. & Lumsden, W. W. *Bull. geol. Soc. Am.* **79**, 1703–1726 (1968).
18. Seager, W. R. *Bull. geol. Soc. Am.* **81**, 1517–1538 (1971).
19. Proffett, J. M. Jr. *Bull. geol. Soc. Am.* **88**, 247–266 (1977).
20. Hamilton, W. B. *Pacif. Coast Paleogeogr. Symp.* **2**, 33–70 (1978).
21. Snoke, A. W. *Geol. Soc. Am. Mem.* **153**, 287–334 (1980).
22. Kieth, S. B. *et al. Geol. Soc. Am. Mem.* **153**, 217–268 (1980).
23. Gary, M., McAfee, R. Jr. & Wolf, C. L. (eds) *Glossary of Geology* (Americal Geological Institute, Falls Church, 1972).
24. Ransome, F. L., Emmons, W. H. & Garrey, G. H. *Bull. U. S. geol. Surv.* **407** (1910).
25. Bally, A. W., Gordy, P. L. & Stewart, G. A. *Bull. Can. petrol. Geol.* **14**, 337–381 (1966).
26. Hunt, C. B. & Mabey, D. R. *U. S. geol. Surv. Prof. Pap.* **494-A** (1966).
27. Reynolds, S. J. & Rehrig, W. A. *Geol. Soc. Am. Mem.* **153**, 159–176 (1980).

Nature Vol. 291   25 June 1981

28. Thompson, G. A. & Zoback, M. L. *Tectonophysics* **61,** 149–181 (1979).
29. McDonald, R. E. *Symp. on Cordilleran Hingeline,* 281–317 (Rocky Mountain Assocation of Petroleum Geologists, 1976).
30. Hoover, J. D. *Brigham Young Univ. Geol. Stud.* **21,** 3–72 (1974).
31. Clark, E. *Brigham Young Univ. Geol. Stud.* **24,** 87–114 (1976).
32. Dahlstrom, C. D. A. *Bull. Can. petrol. Geol.* **18,** 332–406 (1970).
33. Royse, F. Jr, Warner, M. A. & Reese, D. C. *Deep Drilling Frontiers in Central Rocky Mountains,* 41–54 (Rocky Mountain Association of Petroleum Geologists, 1975).
34. Wernicke, B. P. *Geol. Soc. Am. Abstr. Prog.* **13,** 113–114 (1981).
35. Stewart, J. H. *Bull. geol. Soc. Am.* **91,** 460–464 (1980).

COMPRESSIONAL DEFORMATION

Reprinted by permission of University of Wyoming from *University of Wyoming Contributions to Geology*, v. 18, no. 2 (1980), p. 83-100.

Foreland deformation: compression as a cause

D. L. BLACKSTONE, JR.   *Department of Geology, The University of Wyoming, Laramie, Wyoming 82071*

ABSTRACT

Three examples of foreland deformation in which compressional stress played the dominant role are discussed. The examples include the Elk Mountain anticline and the Arlington thrust and related folding of southeastern Wyoming, and East Pryor Mountain, south central Montana. All three examples exhibit faulted margins, overturned strata associated with the faulting, and crustal shortening. Preservation of original bed length and original volume are honored in all cross sections.

The writer concludes that the most rational explanation for deformation described is response of a rigid basement by rupture, lateral movement and shortening of the crust, and displacement of the overlying sedimentary rocks without change in volume or thickness. Compressional stress was the activating force in all three cases.

INTRODUCTION

Publication of Memoir 151 of the Geological Society of America (Matthews, 1978) has focused attention again on the origin of large uplifts in the foreland section of the Rocky Mountains. The characteristic structures are uplifts on a variety of scales in which the Precambrian basement is exposed in the core of the fold. In many cases the uplift is bounded by faults which reach the surface. Similar structural features with no basement exposed are probably controlled by displacement of the basement at depth.

The uplifts are for the most part asymmetrical, with the variation in asymmetry occurring in strips that cross the general north-northwest orientation of the major mountain ranges. The sedimentary cover has responded to the displacement of the Precambrian basement upon faults of magnitude varying in both length and displacement. Relative movement of the crust during the Laramide orogeny can be established by using the displacement of the youngest Cretaceous marine rocks which were deposited at sea level approximately 70 m.y. ago. Subsequent crustal movement has elevated and deformed this datum so that in general the datum is above sea level, indicating a general positive movement of the continental crust since the close of Cretaceous. Adjacent to the Elk Mountain anticline the present elevation of the top of the last marine Cretaceous unit, the Lewis Shale, ranges in elevation from 1800 to 2400 m (6000 to 8000 ft).

Interpretation of the cause of the differential elevation varies primarily as to whether vertically directed forces were dominant or whether horizontal forces had a major role in the deformation.

Stearns (1978) has adopted the terminology "forced fold" to describe the type of deformation which characterizes the "Wyoming foreland" province.[1] He reviewed the general problem, but avoided some geologic data contrary to his hypothesis. The term "forced fold" is intuitively descriptive of the activity (process), but aborts the classic definition of force as used in mechanics.

SPECIFIC EXAMPLES OF FOLDS RELATED TO HORIZONTAL MOVEMENT

Elk Mountain Anticline, Carbon County, Wyoming

Introduction

McClurg and Matthews (1978) proposed that the Precambrian-cored, faulted, asymmetric Elk Mountain anticline in southeastern Wyoming was generated by a stress field oriented in a vertical sense, and that horizontal movement is precluded. The essence of these investigators' argument is graphically shown in their Figure 4 (p. 162), which is reproduced as Figure 4. The upper surface and faulted margins of the Elk Mountain anticline are shown as completely encased by the overlying sediments with no interuption.

The problem is to find a source for the added bed length necessary to cover the flanks of the blocks after faulting and elevation. The writers' analysis of the bed length problem is shown in Figure 5. If the fault bounding the east side of Elk Mountain is vertical, it is necessary to find about 2530 m (8400 ft) of added bed length to cover the faulted face as shown in McClurg and Matthews (1978) Figure 4. The investigators offer no explanation as to the source of the added bed length.

The writer will present data for an alternate explanation.

Location

Elk Mountain anticline is located in Ts. 19-20 N., Rs. 80-81 W., Carbon County, Wyoming. The anticline, as shown on Figure 1, is the northernmost in a series of comparable features constituting a north-trending prong of the northern Medicine Bow Mountains (Houston and others, 1968). The allied features from north to south are: Sheephead Mountain, Bear Butte, Coad Mountain, and Pennock Mountain. Hanna basin

---

[1] The term "Wyoming Province" has been used in two senses. Prucha and others (1965) used the term to describe the area in which the structural style consists of Precambrian cored uplifts. Engel (1963) named the Superior-Wyoming Province as that area in which the exposed Precambrian basement rocks were 2.5 b.y. old or older. Houston (1971) shortened the term Superior-Wyoming Province to Wyoming Province, and considered it to apply to the area in Wyoming where the Precambrian rocks are 2.5 b.y. old or older.

Contributions to Geology, University of Wyoming, v. 18, no. 2, p. 83-100, 15 figs., 2 tables, April, 1980.   83

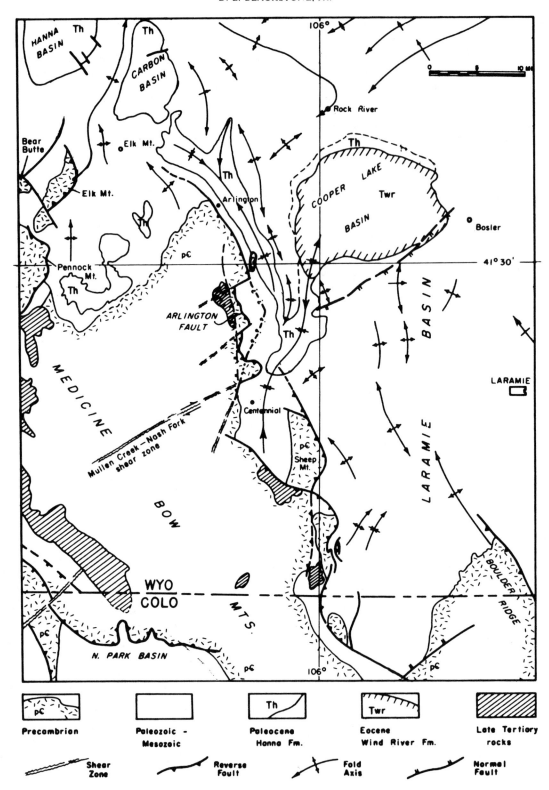

Figure 1. Tectonic index map of Medicine Bow Mountains and Laramie Basin area.

Contributions to Geology, University of Wyoming, v. 18, no. 2, p. 83-100, 15 figs., 2 tables, April, 1980.

lies to the north, Carbon basin to the northeast, and Pass Creek basin directly to the east. Associated with the Elk Mountain anticline is a number of large complex folds in Ts. 19-21 N., Rs. 80-83 W, which plunge to the north or northwest into the Hanna-Carbon basin. Basic stratigraphy is presented in Table 1.

TABLE 1. THICKNESS IN METERS (FEET IN PARENTHESES) OF STRATIGRAPHIC UNITS IN REGION OF NORTHERN MEDICINE BOW MOUNTAINS.

			Meters	Feet
Cenozoic	Miocene	North Park Formation	0- 150	(0- 500)
	Oligocene	White River Fromation	0- 120	(0- 400)
	Eocene	Wind River Formation	360	(1200)
	Paleocene	Hanna Formation	150-2400	(500-8000)
		major unconformity		
Mesozoic	Cretaceous	Medicine Bow Formation	500-1200	(3000-4000)
		Lewis Shale	750- 900	(2500-3000)
		Mesaverde Group	900-1050	(3000-3500)
		Steele Shale with "Shannon Sandstone"	660- 750	(2200-2500)
		Niobrara Formation	225	(750)
		Frontier Formation-Wall Creek Sandstone at top	150	(500)
		Mowry Shale	39	(130)
		Thermopolis Shale-Muddy Sandstone at top	30	(100)
		Cloverly Formation	36	(120)
		Morrison Formation	90	(300)
	Jurassic	Sundance Formation	54	(180)
		Jelm Formation	33	(110)
	Triassic	Chugwater Formation	195	(650)
Paleozoic	Permo-Triassic	Goose Egg Formation	126	(420)
	Pennsylvanian	Tensleep Sandstone (Casper Fm.)	108- 112	(360- 375)
		Amsden Formation	45	(150)
	Mississippian	Madison Limestone	30	(0- 100)
Precambrian		metasediments and crystalline basement		

Geologic Investigations

The primary geologic mapping at a scale of 1:62,-500 was done by Beckwith (1941). Basic mapping north of the area was done by Dobbin and others (1929). Subsequent, more detailed mapping at 1:24,000 has been done adjacent to Elk Mountain by Banks (1970), Barton (1974), Barnes and Houston (1969), Chadeayne (1966), Gries (1964), Hyden and McAndrew (1967), and Sever (1975). Houston and others (1968) compiled a geologic map of the entire northern Medicine Bow Mountain area.

Wells drilled north, northeast, east, and south of the Elk Mountain anticline provide critical information relative to fold geometry, and are listed on Table 2. Detailed geologic mapping at 1:24,000 shows no significant difference from that presented by Beckwith (1941). Figure 2 is a generalized geologic map of the area.

McClurg and Matthews (1978) presented a very simplified map of the Elk Mountain anticline taken from previous data, but added two faults across the area of the outcrop of Precambrian rocks. The writer has not

been able to verify these faults from cross sections or from high-elevation, colored photography. The investigators did not present detailed cross sections of the anticline to justify the vertical faults which they considered to bound the uplift. Subsurface data from drilled wells are not noted in their interpretation.

The present investigation consists of constructing eleven serial interpretive cross sections (Figs. 6-9) at a scale of 1:24,000 (scale reduced for publication) across Elk Mountain anticline and associated folds. Previous geologic mapping, including unpublished proprietary data in addition to all available well data, have been utilized. The locations of the cross sections are shown upon Figure 2, a simplified geologic map derived from Beckwith (1941) and other sources. Figure 3 is a photograph of the northeast face of the mountain, and illustrates its asymmetry to the east.

Discussion of Cross Sections

Cross sections A-A' through D-D' (Fig. 6) trend northwest to southeast across the strike of a series of large, north-plunging folds associated with Elk Mountain anticline, and which are expressed in the Upper Cretaceous Medicine Bow Formation, Lewis Shale, Mesa Verde Group, Steele Shale, and Niobrara Formation. The folds respectively, from west to east are Rattlesnake Creek syncline, Halleck Ridge anticline, Ft. Halleck syncline, Bloody Lake anticline, and Simpson Ridge anticline (Fig. 2). Exploratory test wells for oil and gas are shown on the cross sections. All wells but two have been logged mechanically, and formational boundaries chosen are consistent from well to well.

The dominant folds are the very steep-limbed Halleck Ridge and Simpson Ridge anticlines. The former is asymmetrical to the northwest and is ruptured by a high angle, east-dipping reverse fault. The existence of a fault is mandated by data from the Tesoro Petroleum Co., Champlin 1-5 well in sec. 5, T. 20 N., R. 81 W. (well no. 3 on section B-B'), for which both directional and dip meter surveys are available. Simpson Ridge anticline is an essentially upright, symmetrical anticline with a very low angle of plunge of the axis to the northeast.

Because the deepest penetration by drilling is only to the Cloverly Formation, the position of the Precambrian basement in section A-A' can only be inferred. This is done on the basis of assuming folds to be concentric in form and by using the known thicknesses of strata involved. Precambrian basement was reached in the Texas Co. #1 Mill Creek Unit, sec. 14, T. 20 N., R. 81 W. (well no. 5, on section B-B'). Chugwater Formation was penetrated in the Tesoro Petroleum Co. Champlin 1-5, sec. 5, T. 20 N., R. 81 W. Two wells reached the Pennsylvanian Tensleep Formation; these are (1) Skelly Oil Co. UPRR-Harris, sec. 24, T. 20 N., R. 81 W., (well no. 6) and (2) Occidental Petroleum Co. #1 Govt. Weber, sec. 24, T. 20 N., R. 81 W. (well no. 8 on section C-C').

Cross section E-E' (Fig. 6) was constructed to define the faulted north margin of the block.

Contributions to Geology, University of Wyoming, v. 18, no. 2, p. 83-100, 15 figs., 2 tables, April, 1980. 85

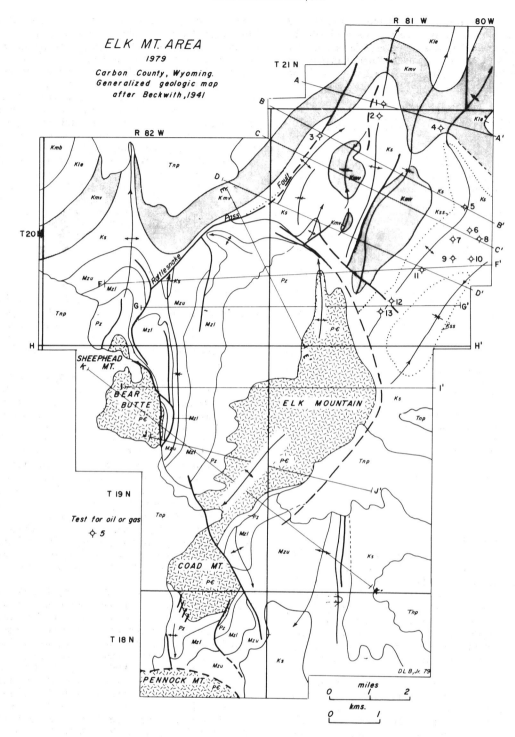

Figure 2. Legend for generalized map of Elk Mountain area, Wyoming. *Tnp*-Miocene(?) North Park Formation; *Kmb*-Cretaceous Medicine Bow Formation; *Kle*-Cretaceous Lewis Shale; *Kmv*-Cretaceous Mesa Verde Group; *Ks*-Cretaceous Steele Shale; *Kss*-"Shannon" Sandstone in Steele Shale; *Mzu*-upper Mesozoic including Niobrara Formation, Frontier Formation, Mowry Shale, Muddy Sandstone, and Thermopolis Shale; *Mzl*-lower Mesozoic including Cretaceous Cloverly Formation and Jurassic Jelm and Chugwater formations; *Pz*-Paleozoic including Pennsylvanian Tensleep (Casper) Sandstone, Pennsylvanian Amsden and Fountain formations, and Mississippian Madison Limestone; and *PC*-Precambrian undifferentiated.

TABLE 2. LIST OF WELLS CODED BY NUMBER TO ACCOMPANY
GENERALIZED GEOLOGIC MAP OF ELK MOUNTAIN
AREA, FIGURE 2.

Well No.	Operator	Well Designation	Sec.	Town-ship	Range
1	True Oil Co,	44-34 Curry	34	21 N	81 W
2	Kansoming	#4 Elk Mtn.	4	20 N	81 W
3	Tesoro Petroleum	1-5 Champlin	5	20 N	81 W
4	Carr and Wrath	#1 Federal	2	20 N	81 W
5	Texas Company	Mill Creek Unit 1	14	20 N	81 W
6	Skelly Oil Co.	UPRR-Harris	24	20 N	81 W
7	J. G. Brown Assoc.	UPRR-West	23	20 N	81 W
8	Occidental Petrol.	#1 Gov't. Weber	24	20 N	81 W
9	J. G. Brown Assoc.	#2 West	23	20 N	81 W
10	J. W. West	#1 Edgar West	24	20 N	81 W
11	Davis Oil Co.	Champlin-Halleck 1	27	20 N	81 W
12	Davis Oil Co.	#1 Halleck	27	20 N	81 W
13	Ohio Oil Co.	#1 UPRR	33	20 N	81 W

Sections *F-F'* through *K-K'* (Figs. 6-8) cross Elk Mountain proper and the minor folds to the west as well as extending several kilometers to the east of the surface trace of the Elk Mountain fault.

Observations Based Upon Cross Sections

In cross sections *A-A'* through *K-K'* four critical observations can be made, and these are particularly well demonstrated in sections *G-G'* and *H-H'* (Fig. 7).

(1) Cretaceous strata as young as the "Shannon" Sandstone of the Upper Cretaceous Steele Shale crop out immediately (150-300 m; 500-1000 ft) east of the trace of the Elk Mountain fault.

(2) Madison, Amsden, Tensleep, Goose Egg, and Chugwater formations overturned as much as 60° to the southwest or west occur west of, and

Figure 3. View to southwest from Hanna exit on Interstate Highway 80. Pennock Mountain at extreme left, Elk Mountain center, and Sheephead Mountain at extreme right.

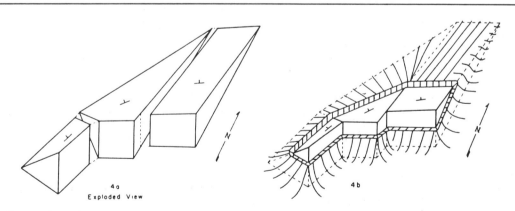

4a
Exploded View

4b

Figure 4. Block diagrams showing configuration of basement blocks (a) that are responsible for fold geometry in overlying sedimentary strata (b). Reproduced from McClurg and Matthews (1978), with permission of Geological Society of America.

Contributions to Geology, University of Wyoming, v. 18, no. 2, p. 83-100, 15 figs., 2 tables, April, 1980.

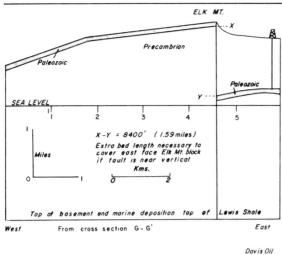

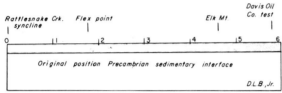

Figure 5. Diagrammatic cross sections illustrating bed length problem at Elk Mountain, Wyoming.

at a higher elevation than, the trace of the Elk Mountain fault.

(3) Two wells located east of the fault trace at a distance of approximately 1200 m (4000 ft) define the position of the Tensleep Formation in the footwall.

(4) The west limb of the Elk Mountain anticline consists of four asymmetrical folds, the axial surfaces of which: (a) dip to the west; (b) have overturned strata on the limbs; and (c) are broken by west-dipping reverse faults.

Interpretation of Data from Cross Sections

The question at issue is the nature of the structural boundary condition along the northwest, northeast, and southeast flanks of Elk Mountain. The existence of overturned strata on one limb of an asymmetric anticline is usually interpreted to indicate tectonic transport such that the overturned strata (reversed dips) face the direction from which the movement developed. All the minor folds on the west flank of the Elk Mountain major fold fit this concept (*i.e.*, tectonic transport was from west to east, beds were omitted [stratal shortening] by reverse faulting, and the northeast asymmetry is consistent throughout).

The overturned strata on the east face of Elk Mountain are a logical part of the geometry of a large asymmetrical fold ruptured by a west dipping fault. The overturned strata are exactly analogous to the situation seen in the lesser folds on the west flank of the major

fold, and are consistent with the geometry of the Coad Mountain and Pennock Mountain folds to the south (Beckwith, 1941; Barton, 1974).

The contact of the Mississippian Madison Limestone on the Precambrian basement is more logically interpreted as a normal, depositional contact than, as suggested by McClurg and Matthews (1978), as a fault which for miles lies exactly at the Madison-Precambrian interface. The Madison-Precambrian contact along the southeast flank of the fold in secs. 24, 25, and 26, T. 19 N., Rs. 81 and 82 W. has been examined by David R. Lageson (Wyoming Geological Survey) in connection with a detailed stratigraphic study of the Madison Limestone. Lageson *(personal communication)* considers the contact to be a normal depositional one, and not a fault.

If the overturned Madison Limestone on the east face of Elk Mountain is in normal depositional contact with the Precambrian basement then it is imperative: (1) that the trace of the Elk Mountain fault lie east of the zone of overturned strata; and (2) that the fault dip to the west.

The attitude of the Elk Mountain fault cannot be definitely defined by three-point, strike and dip solutions based upon elevations taken along the trace of the fault. The trace is concealed by vegetation, pedimented surfaces, and large areas that have undergone mass gravity movement; such concealment is to be expected in the Steele Shale. The writer has shown the Elk Mountain fault as dipping to the west at a lower angle of dip than the overturned strata in the hanging wall, and has terminated the overturned strata down dip against the fault (as shown in section *G-G'*).

Subsurface control of the fold geometry is established by two wells. These are Ohio Oil Company #1 UPPR, sec. 33, T. 20 N., R. 81 W. and the Davis Oil Company #1 Halleck, sec. 27, T. 20 N., R. 81 W. (wells no. 12 and 13, section *G-G'*). Formations correlate excellently between the two wells and demonstrate the existence of reverse faults of small displacement. The position of the top of the Tensleep Sandstone is established at a horizontal distance not to exceed 600 m (2000 ft) east of the surface trace of the Elk Mountain fault.

If the Elk Mountain fault is vertical as shown by McClurg and Matthews (1978), then the entire stratigraphic interval from the top of the Precambrian basement to the "Shannon" Sandstone (1745 to 1806 m [5800 to 6000 ft] of strata) must be exposed at a high angle of dip and pass vertically downward to connect with the rocks drilled in the two wells described above.

This demands that the Steele Shale at a distance approximately 1800 m (6000 ft) east of the Precambrian-Madison Limestone contact be overturned or dip very steeply to the east. The space available between that contact and the two wells (nos. 12 and 13, sec. *G-G'*) precludes placing a vertical section in that location.

Two reverse faults can be documented on the electric logs in wells 12 and 13, sec. *G-G'*. It is possible that other similar faults exist between the wells and the trace of the Elk Mountain thrust fault, but this cannot be demonstrated with available data.

88

Contributions to Geology, University of Wyoming, v. 18, no. 2, p. 83-100, 15 figs., 2 tables, April, 1980.

128

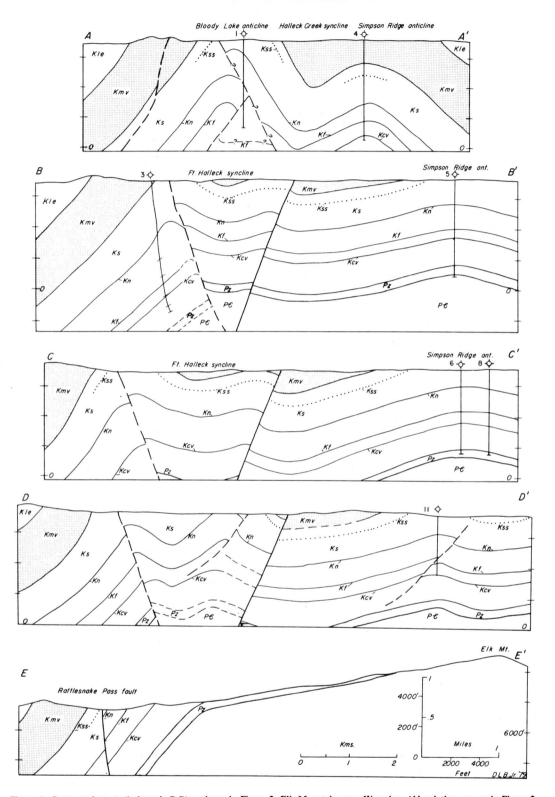

Figure 6. Cross sections *A-A'* **through** *E-E'* **as shown in Figure 2, Elk Mountain area, Wyoming. Abbreviations are as in Figure 2.**

Contributions to Geology, University of Wyoming, v. 18, no. 2, p. 83-100, 15 figs., 2 tables, April, 1980.

89

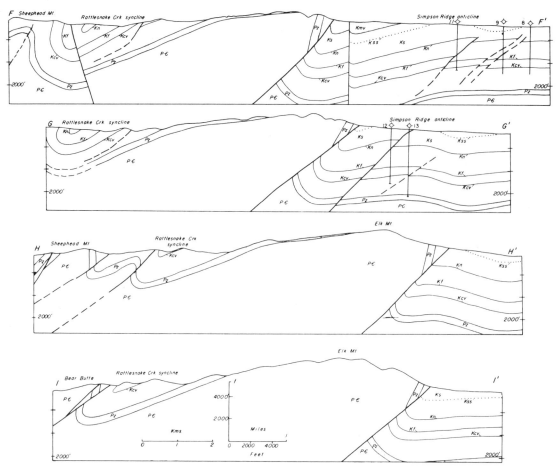

Figure 7. Cross sections *F-F'* through *I-I'* as shown in Figure 2, Elk Mountain area, Wyoming. Abbreviations are as in Figure 2.

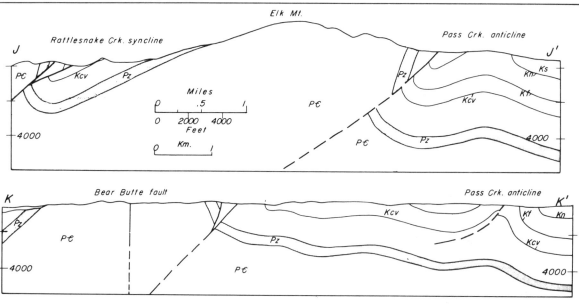

Figure 8. Cross sections *J-J'* and *K-K'* as shown in Figure 2, Elk Mountain area, Wyoming. Abbreviations are as in Figure 2.

90   Contributions to Geology, University of Wyoming, v. 18, no. 2, p. 83-100, 15 figs., 2 tables, April, 1980.

The "Shannon" Sandstone in the footwall of the Elk Mountain fault, on the basis of the reconstruction must be in fault contact with the Chugwater Formation in the hanging wall leaving no space to accommodate (at the surface) the rocks between the Chugwater Formation and the "Shannon" Sandstone (approximately 1080 m [3600 ft] of section).

An added problem in providing space for the entire section in vertical position is shown on Figure 2. The Halleck syncline and anticline strike approximately N45°E, and butt directly into the overturned strata in the hanging wall of the Elk Mountain block (which strikes approximately N40°W, almost at right angles to the footwall beds). The proximity of the Mesa Verde Group (with northeast strike) to the Chugwater Formation (which strikes northwest) leaves no room to accommodate the full section between these two stratigraphic units.

The most logical interpretation of the data which will provide the necessary geometry and space for the sediments is to consider the Elk Mountain fault as dipping to the west. The unresolved problem is the nature of the adjustment of the Precambrian crystalline basement rocks in the core of the fold. That question is not addressed here; only the problem of preserving the volume of the sedimentary rocks in the footwall is considered.

The interpretive cross sections indicate that the crystalline basement is folded, which appears to violate the concepts that the basement must behave as a brittle material and that the overlying sediments must exhibit ductile behavior. The basement no doubt has behaved in a brittle manner, and therefore is highly fractured (as has been noted in many cases in which basement rocks are exposed in the cores of foreland folds).

The adjustment of the brittle rocks by movement on innumerable smaller fractures appears to offer a means of producing an antiform in the basement rocks underlying the sediments. Eventually, such flexing of the body of crystalline rocks into a fold or flex is inadequate to adjust to the stress system, and major rupture occurs and develops into reverse faulting.

McClurg and Matthews (1978), found the matter of the overturned dips to be unresolved by their model, as revealed in the following statement: "It is not clear to us why the strata in both limbs are overturned some distance from the contact" (p. 161).

The Problem of Bed Length Before and After Folding

An additional problem of McClurg's and Matthews' (1978) interpretation is the source of the extra bed length (stratal) around the margins of the Elk Mountain anticline. Their Figure 4 (reproduced here as Fig. 4) shows three elevated segments of Precambrian basement completely surrounded by a continuous envelope of overlying sedimentary strata. The implication is that these strata were continuous over the now-exposed core of the fold after faulting.

Such a model has a very serious deficiency in that it does not explain the added bed length. The differential elevation (structural relief) along the east face of Elk Mountain is in the order of 2530 m (8400 ft). If the Elk Mountain fault is vertical, then the cross section in Figure 4 should be representative of the field relationship. The obvious question is, what is the source of the 2530 m of added bed length needed to cover the east face of the block? If the block was simply lowered to the pre-fracture position the strata match exactly. Once raised as shown in Figure 4 (McClurg and Matthews, 1978), however, from what source was the continuous sheath of sediments derived?

The problem of added bed length cannot be explained by ductility, because there is no evidence of penetrative deformation on that scale. Solution of the problem by involving ductile deformation would require the extension of the original bed length by 26 percent; to retain constant volume under these conditions would demand a drastic thinning in a local area, towit the vertical limb of the fold.

The problem is further compounded by the fact that McClurg and Matthews (1978) did not consider in their model or block diagram the Rattlesnake Pass fault (Beckwith, 1941; Sever, 1975; unpublished mapping; cross section E-E' of this paper). That fault trends N60°E and dips beneath the block along the north edge. It is the north boundary for the Elk Mountain block, and it may have had oblique slip. Additional bed length would be needed to cover the northwest edge of the Elk Mountain block.

Conclusions on Elk Mountain Anticline

The writer believes that the development of the Elk Mountain anticline is best explained by deformation of the Precambrian basement under conditions of horizontally-directed stresses. The basement fractured, or responded to pre-existing fractures (Bekkar, 1973), and yielded along these planes of failure. The result was a number of asymmetrical folds in the overlying sedimentary rocks. The overturned limbs of the folds were sheared off and elevated to their present position. The necessary stratal shortening is accomplished by movement on the fault surface, thereby obviating the need for additional bed length. Such conclusions agree with analyses of similar structures such as Sheephead Mountain (see Barnes and Houston, 1969; Banks, 1970; and Sever, 1975).

Arlington Thrust Fault, Albany and Carbon Counties, Wyoming

Location

The northern Medicine Bow Mountains are a foreland, asymmetrical uplift cored with crystalline and metasedimentary rocks of Precambrian age (Houston and others, 1968). The asymmetry is to the east and the uplift is bounded by a reverse fault system (as shown on Fig. 1) which extends from near Arlington, Wyoming (T. 19 N., R. 78 W.) to the North Fork of the Little Laramie River at a point about 3.2 km (2 mi) northwest

Contributions to Geology, University of Wyoming, v. 18, no. 2, p. 83-100, 15 figs., 2 tables, April, 1980.   91

of Centennial, Wyoming (T. 16 N., R. 78 W.). The fault system has been mapped by Houston and others (1968), Blackstone (1970, 1973, 1976), Hyden and others (1968), and McCallum (1968).

Data are presented below to suggest that the structural features of the northern Medicine Bow Mountains have resulted from horizontally-directed rather than vertically-directed stress fields and that the causative-structural situation is analogous to that previously described for Elk Mountain anticline.

Geologic Investigations

A summation of previous mapping along the Arlington fault is available in Blackstone (1976). The Arlington fault was originally named by Darton and Siebenthal (1909) and mapped as a single, continuous fault bounding the east side of the northern Medicine Bow Mountains. Subsequent mapping demonstrated that the fault is more complex, and affected by several shear zones in the Precambrian rocks of the core of the range.

Geologic Structure

The Arlington thrust fault: (1) trends essentially north-south; (2) is segmented by tear faults; and (3) terminates at the south end against the Mullen Creek-Nash Fork shear zone (Houston and others, 1968). The northern termination of the Arlington thrust fault is concealed by the unconformably overlying Paleocene Hanna Formation at a point about 3.2 km (2 mi) north of Arlington, Wyoming (Blackstone, 1976).

The southern part of the Arlington fault and related folding, including the Mill Creek syncline, is critical to the present discussion. The important exposures lie in the Centennial and Rex Lake, Wyoming 7½ minute quadrangles. Part of the Centennial quadrangle has been mapped by Beckwith (1941), McCallum (1968), and by the writer. The Rex Lake quadrangle was mapped by Blackstone (1970).

The northern Medicine Bow Mountains are divided into two distinct districts insofar as the Precambrian basement rocks are concerned. The division is effected by the Mullen Creek-Nash Fork shear zone, a northeast-trending zone of extreme cataclasis of the basement rocks (Houston and others, 1968; Houston, 1971) with right lateral offset. The location of the shear zone relative to the Arlington fault is shown in Figure 1. The essential point is that all representation of the Arlington thrust fault is restricted to the region north of the shear zone.

An interpretive cross section has been constructed across the toe of the Arlington thrust (Fig. 9), along an east-west line passing through the centers of sections 21, 22, and 24, T. 16 N., R. 78 W. and sec. 19, T. 16 N., R. 77 W. The most easterly part of the Arlington thrust plate crops out in Wards Gulch just west of Bald Mountain, as shown on the Rex Lake map (Blackstone, 1970). The toe of the thrust plate rests upon the vertical to slightly overturned Mesa Verde Formation, which is exposed in the west flank of the large, northeast-plunging asymmetrical Mill Creek syncline. The core of the syncline is occupied by the lower 120-150 m (400-500 ft) of the Cretaceous Medicine Bow Formation, which, in turn, is overlain with 90° unconformity by the upper part of the Paleocene Hanna Formation.

The Arlington fault dips to the west at less than five degrees. A photograph (Fig. 10), taken normal to the strike of the dip, clearly shows the attitude of the fault.

The strata exposed in the footwall are vertical to slightly overturned to the northwest so that the fault plane truncates the sedimentary section at essentially 90°. The fault plane also truncates the earlier Centennial Ridge fault, which crops out along the east face of Centennial Ridge just west of the Centennial, Wyoming townsite.

Observable eastward displacement of the Precambrian basement in the Arlington thrust plate is approximately 3600 m (12,000 ft). Because no post-Cambrian sedimentary section is preserved in the hanging wall toe of the thrust plate, the above value is a measure of the *minimum* eastward movement of the plate. The vertical differential elevation of the Precambrian basement between the crest of the Medicine Bow Mountains and the Laramie basin is at least 4500 m (15,-000 ft).

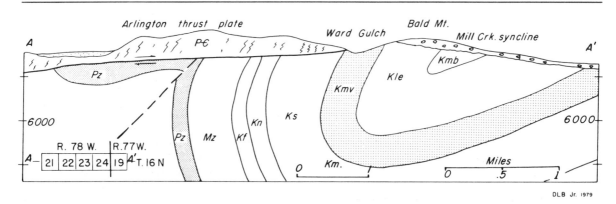

Figure 9. Cross section of Arlington fault near Centennial, Wyoming. Abbreviations are as in Figure 2.

Contributions to Geology, University of Wyoming, v. 18, no. 2, p. 83-100, 15 figs., 2 tables, April, 1980.

Figure 10. View to north from sec. 35, T. 16 N., R. 78 W., Rex Lake 7½ minute quadrangle. Bald Mountain at left and Ward Gulch right center. Arlington fault is near horizontal line across photograph. *PÇ*-Precambrian undifferentiated; *Kmv*-Cretaceous Mesa Verde Group; and *Kle*-Cretaceous Lewis Shale.

Conclusions on Arlington Thrust Fault

The geometry of the Arlington thrust fault (Fig. 9) and the magnitude of the movement clearly preclude explanation of the displacement of the Precambrian basement by purely vertically-directed stress. The tectonic movement took place on an essentially horizontal plane, and the necessary stratal shortening is the result of the movement on that plane. The relationship is analogous in origin and in geometry to the previously discussed Elk Mountain anticline.

East Pryor Mountain, Carbon County, Montana

Location

East Pryor Mountain is located in Ts. 7 and 8 S., Rs., 27 and 28 E. Carbon County, Montana, and extends southward into T. 58 N., R. 95 W. in Big Horn County, Wyoming. The mountain is one of four major segments which comprise the Pryor Mountains. Access to East Pryor Mountain is by way of Wyoming State Highway 137 to the Bighorn Canyon National Recreation Area, and thence northward into Dryhead Basin by dirt road. The area discussed lies within the East Pryor Mountain and Big Mystery Cave 7½ minute topographic quadrangle sheets.

Geologic Investigations

Geologic mapping of the flanks of the Pryor Mountains was done by Knappen and Moulton (1931), and Thom and others (1935). The central part of the uplift was mapped by Blackstone (1940), and the east flank of the uplift and the adjacent Dryhead and Garvin basins were mapped by Stewart (1958). Uranium prospecting in the area led to additional work by Hauptman (1956).

When 1:24,000 scale topographic mapping became available, all previous work by the writer was transferred to the new base and additional checking was done on the ground and with aerial photographs. The East Pryor Mountain geologic quadrangle was published (Blackstone, 1975), and eleven 7½ minute geologic quandrangles were placed on open file with the Montana Bureau of Mines and Geology.

General Geologic Structural Pattern

The Pryor Mountains are a unique foreland structural entity consisting of four roughly equidimensional segments, each of which has the greatest structural elevation at the northeast corner. The segments are bounded on their north sides by fault systems which strike essentially east-west. East Pryor Mountain and Red Pryor Mountain (the southwest segment) are bounded on their east sides by faults. The block margins are folds or faulted folds (Blackstone, 1940, 1975).

The Pryor Mountain composite uplift is separated from the northern Bighorn Mountains by a flat floored syncline, the Dryhead-Garvin Basin, which establishes the Pryor Mountains as a separate uplift (Stewart, 1958). In gross aspect, the uplift extends the general Bighorn Mountain trend to the northwest.

Specific Structural Features

East Pryor Mountain has been elevated higher than other segments of the uplift. The Precambrian crystalline basement is exposed for a distance of 6.4 km (4 mi) in a north-south direction along the east face of the uplift, and is in fault contact with Paleozoic strata in the footwall. The elevation of the Precambrian-Cambrian interface in this area ranges from an elevation

Contributions to Geology, University of Wyoming, v. 18, no. 2, p. 83-100, 15 figs., 2 tables, April, 1980.

93

of 1925 to 2045 m (6400 to 6800 ft). The basement is overlain by Cambrian strata including the Flathead Sandstone, Gros Ventre Shale, and Gallatin Limestone (Shaw, 1954). Figure 12 provides the essentials of the stratigraphy. The essential surface geology for this discussion is shown on Figure 11, which also shows the location of all cross sections (*A-A'* through *J-J'*, Fig. 13).

Strata in the hanging wall of the East Pryor fault dip to the southwest or west at angles of 2° to 8°. Extensive dip slopes are capped by the Madison Limestone, into which deep canyons have been incised. Along the north margin of the block the dips are to the north at angles from 30° to 45°.

The question to be addressed here is whether the form of the folds and faulted folds in the Pryor Mountain uplift are the sole result of vertical uplift, or represent a response to horizontally-directed stresses.

Northeast Block

Figure 11 and cross sections *A-A'* and *B-B'* (Fig. 13) include the extreme southeast corner of the northeast block of the Pryor Mountain uplift. Section *A-A'* roughly parallels Punch Bowl Creek, in which the top of the Ordovician Bighorn Dolomite is exposed. The strata dip slightly to the west (2°-3°) on the west flank of the fold, and then reverse to dips of approximately 20° east into the Dryhead Basin. There is no faulting, and all units are smoothly folded across the hinge of the fold. All folds can be readily explained by slight flexing of the Precambrian basement without rupture.

East Pryor Mountain

The Precambrian basement crops out along the east face of the mountain from the southwest corner of sec. 30, T. 7 S., R. 28 E. at the intersection of the Sage Creek fault zone and the Dryhead fault to a point near the center of sec. 15, T. 8 S., R. 28 E. Section *B-B'* crosses both the Sage Creek fault and a splay from the Dryhead fault. Strata in the hanging wall of the Sage Creek fault dip 5°-7° to the southwest. Strata dip about 45° to the northeast in the segment between the faults, and they are overturned and dip to the west in the footwall of the Dryhead fault.

The critical relationship relative to the nature of the movement on the Dryhead fault is the zone of overturned strata in the footwall that extend for 5.2 km (3¼ mi) along strike in the footwall (see cross sections *B-B'* through *F-F'*, Fig. 13). The oldest rocks exposed in the overturned section in the footwall are part of the Ordovician Bighorn Dolomite, and the youngest rocks are in the Jurassic "Sundance" section.

The fault plane can be observed in canyon walls, but, as is common in such cases, only over a vertical distance of about 300 m (1000 ft, Figs. 14-15). The close proximity (150-600 m [500-2000 ft]) of the Cambrian-Precambrian contact in the hanging wall to the fault trace and the overturned section in the footwall leaves little room for interpolation or alternate explanations.

Truncation of the overturned strata in the footwall requires that the dip of the fault be equal to, or less than, the dip of the strata involved.

The actual attitude of the fault as projected at dip is subject to interpretation, since neither bore holes nor seismic data are available. The fault attitudes presented are based upon the limited surface exposure and the relationship to the overturned strata. The fault movement has been dip slip, and the displacement reflects the requirement of maintaining the original bed length when the hanging wall is restored to a matching position with the footwall.

South of the area of Precambrian exposure, the strike of the Dryhead fault changes abruptly from N30°W to essentially north-south. The writer interprets this sharp change in strike as indicating the existence of two original fracture or failure zones in the Precambrian basement, both of which preceded the Laramide movement. The fault plane in the southern segment (cross sections *H-H'* through *J-J'*) dips steeply to the west, and the separation decreases to the south. At Layout Creek (sec. *I-I'*), erosion has cut deeply enough to expose the Cambrian Gallatin Formation in the hanging wall, in fault contact with the Mississippian Madison limestone in the footwall. There are no overturned strata exposed in the footwall.

Conclusions on East Pryor Mountain

The major conclusions that the writer draws from the surface geology and the cross sections are that the northeast corner of East Pryor Mountain moved northeastward on a reverse fault; in the process, the strata in the footwall were overridden and overturned to the west, and truncated by the fault. Because all segments of the Pryor Mountains have a common pattern, the same mechanism probably acted on all units, with the maximum effect shown in East Pryor Mountain.

An oft-cited model for "drape" or "forced" folds in the Rocky Mountain foreland is that presented by Stearns (1978, Fig. 5, p. 11) for the asymmetric Rattlesnake Mountain anticline west of Cody, Wyoming. The model includes a series of normal faults downthrown to the west with adjustment in the hinge area by reverse faults. However, because no subsurface control exists for the western footwall of the uplift, the model is suspect. It is possible that Rattlesnake Mountain is an atypical case, and that in actuality it is "the exception that proves the rule."

The model for Rattlesnake Mountain, as proposed by Stearns (1978), is not appropriate for application to the Pryor Mountains. Structural conditions at the Pryor Mountains are best explained as response of movement on reverse faults under horizontally-directed stress fields.

GENERAL SUMMARY

The problem within the foreland area of southern Montana, Wyoming, and Colorado is to find a unifying explanation that will explain the bulk of the cases deal-

94   Contributions to Geology, University of Wyoming, v. 18, no. 2, p. 83-100, 15 figs., 2 tables, April, 1980.

134

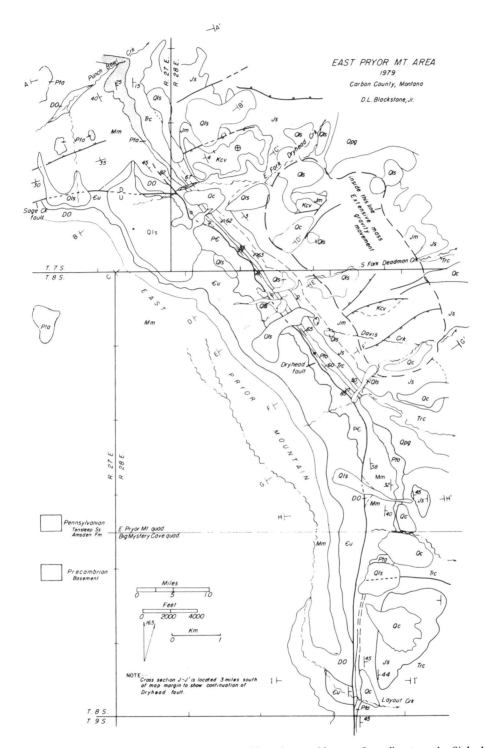

Figure 11. Legend for generalized geologic map of East Pryor Mountain area, Montana. *Qpg*-pediment gravels; *Qls*-landslides; *Qc*-colluvium; *Kcv*-Cretaceous Cloverly Fm.; *Jm*-Jurassic Morrison Fm.; *Js*-Jurassic "Sundance" Fm. (includes the Swift, Rierdon, and Piper fms.); *Trc*-Triassic Chugwater Fm.; *Pta*-Pennsylvanian Tensleep and Amsden fms.; *Mm*-Mississippian Madison Ls.; *DO*-Devonian Jefferson Dol. and Beartooth Butte Fm. and Ordovician Bighorn Dol.; *Єu*-Cambrian Flathead Ss., Gros Ventre Shale, and Gallatin Ls.; and *PЄ*-Precambrian crystalline basement.

Contributions to Geology, University of Wyoming, v. 18, no. 2, p. 83-100, 15 figs., 2 tables, April, 1980.

95

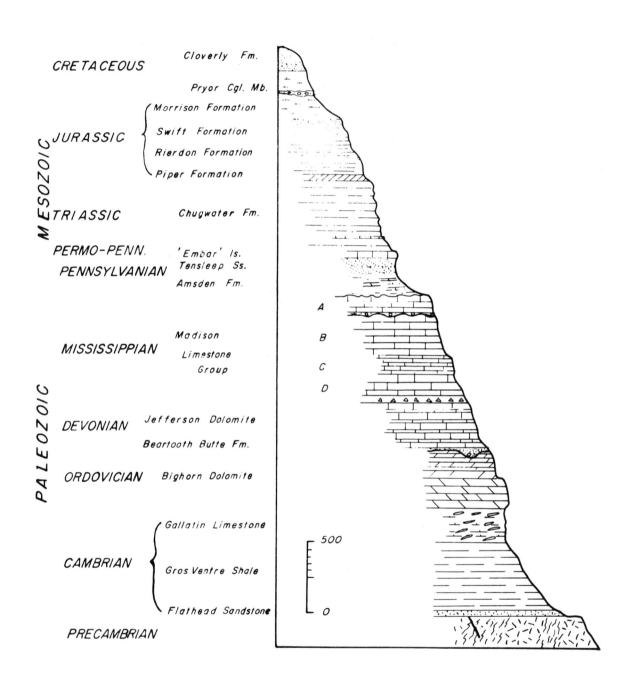

Figure 12. Stratigraphic column in East Pryor Mountain Quadrangle, Carbon County, Montana. Reproduced from Blackstone (1975).

Contributions to Geology, University of Wyoming, v. 18, no. 2, p. 83-100, 15 figs., 2 tables, April, 1980.

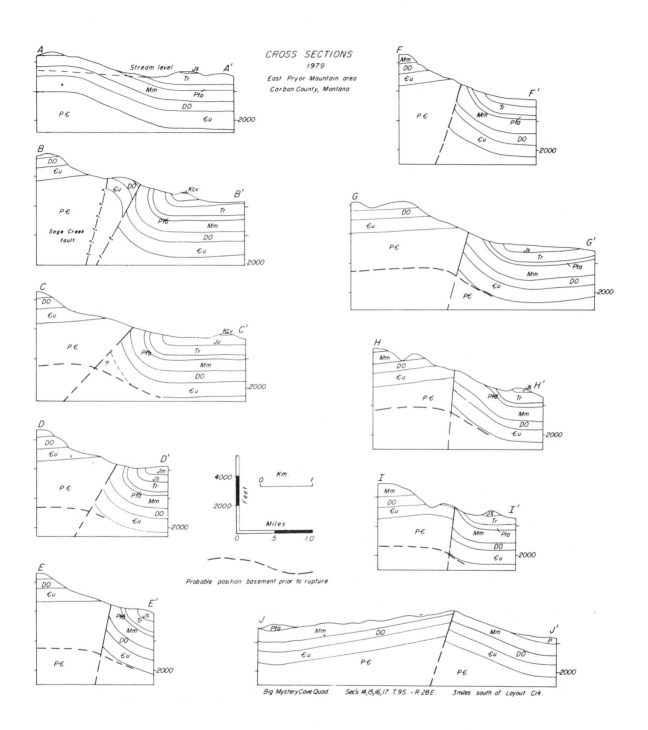

CROSS SECTIONS
1979
East Pryor Mountain area
Carbon County, Montana

Probable position basement prior to rupture

Big Mystery Cove Quad.   Sec's 14,15,16,17 T.9S. - R.28E.   3 miles south of Layout Crk.

Figure 13. Cross sections shown in Figure 11, East Pryor Mountain area, Montana. Abbreviations are as in Figure 11.

Contributions to Geology, University of Wyoming, v. 18, no. 2, p. 83-100, 15 figs., 2 tables, April, 1980.   97

Figure 14. View to northwest from near center of sec. 31, T. 7 S., R. 28 E., Carbon County, Montana. Precambrian (P€) in fault contact with overturned Paleozoic section (Pz). Trace of Dryhead fault is a solid line.

Figure 15. View to north along Dryhead fault from sec. 10, T. 8 S., R. 28 E., Carbon County, Montana. Flathead quartzite (€f) on Precambrian (P€) in normal position. Dryhead fault shown by solid line. Madison Limestone (*Mm*) in fault contact with Precambrian basement.

 Contributions to Geology, University of Wyoming, v. 18, no. 2, p. 83-100, 15 figs., 2 tables, April, 1980.

ing with Laramide deformation. The reversal of asymmetry of folds in extensive northeast-trending strips across the general north-northwest structural grain must be due to some cause other than mere vertical stresses and resultant movement. The region was subjected to stress fields wherein volumetric competition for space was relieved by motion in opposite directions in the strips of asymmetry. Special cases certainly must have existed, as they do in any subcontinental-sized region. The writer believes, however, that general stress fields oriented in more or less horizontal positions provide the most probable causes for the majority of observed Laramide phenomena in the Wyoming Province.

ACKNOWLEDGMENTS

The writer thanks the Department of Geology of The University of Wyoming for use of its facilities. The Geological Survey of Wyoming provided assistance with drafting. Drs. Robert R. Berg, James E. McClurg, and Vincent Matthews III reviewed a draft of the manuscript and made valuable suggestions for its improvement. The Geological Society of America granted permission to reproduce material used in Figure 4. Interpretations and conclusions are the sole responsibility of the author.

REFERENCES CITED

Banks, C. E., 1970, Precambrian gneiss at Sheephead Mountain, Carbon County, Wyoming, and its relationship to Laramide structure [M.S. thesis]: Laramie, Wyoming, The University of Wyoming, 36 p.

Barnes, C. W., and Houston, R. S., 1969, Basement response to Laramide orogeny at Coad Mountain, Wyoming: Contributions to Geology, University of Wyoming, v. 8, p. 37-41.

Barton, R., 1974, Geology of the Kennady Peak-Pennock Mountain area, Carbon County, Wyoming [M.S. thesis]: Laramie, Wyoming, The University of Wyoming, 74 p.

Beckwith, R. W., 1941, Structure of the Elk Mountain district, Carbon County, Wyoming: Geological Society of America Bulletin, v. 52, p. 1445-1486.

Bekkar, H., 1973, Laramide structural elements and relationships to Precambrian basement, southeastern Wyoming [M.S. thesis]: Laramie, Wyoming, The University of Wyoming, 70 p.

Blackstone, D. L., Jr., 1940, Structure of the Pryor Mountains, Montana: Journal of Geology, v. 47, p. 590-617.

_____1970, Structural geology of the Rex Lake Quadrangle, Laramie Basin, Wyoming: Geological Survey of Wyoming Preliminary Report 11, 17 p.

_____1973, Structural geology of the eastern half of the Morgan Quadrangle, the Strouss Hill Quadrangle, and the James Lake Quadrangle, Albany and Carbon Counties, Wyoming: ibid., 13, 45 p.

_____1975, Geology of the East Pryor Mountain Quadrangle, Carbon County, Montana: Montana Bureau of Mines and Geology Special Publication 69, 13 p.

_____1976, Structural geology of the Arlington-Wagonhound Creek area, Carbon County, Wyoming: Geological Survey of Wyoming Preliminary Report 15, 15 p.

Chadeayne, D. K., 1966, Geology of the Pass Creek Ridge, Saint Marys, and Cedar Ridge anticlines, Carbon County, Wyoming [M.S. thesis]: Laramie, Wyoming, The University of Wyoming, 87 p.

Darton, N. H., and Siebenthal, C. E., 1909, Geology and mineral resources of the Laramie Basin, Wyoming—A preliminary report: U.S. Geological Survey Bulletin 364, 81 p.

Dobbin, C. E., Bowen, C. F., and Hoots, H. W., 1929, Geology and coal and oil resources of the Hanna and Carbon basins, Carbon County, Wyoming: ibid., 804, 98 p.

Engel, A. E. J., 1963, Geologic evolution of North America: Science, v. 140, p. 143-152.

Gries, J. C., 1964, The structure and Cenozoic stratigraphy of the Pass Creek basin area, Carbon County, Wyoming [M.S. thesis]: Laramie, Wyoming, The University of Wyoming, 69 p.

Hauptman, C. M., 1956, Uranium in the Pryor Mountain area of southern Montana and northern Wyoming: Uranium, v. 3, p. 14-15 and 18-21.

Houston, R. S., 1971, Regional tectonics of the Precambrian rocks of the Wyoming Province and its relationships to Laramide structure, in Renfro, A. R., ed., Symposium on Wyoming tectonics and its economic significance: Casper, Wyoming Geological Association 23rd Annual Field Conference Guidebook, p. 19-27.

Houston, R. S., and others, 1968, A regional study of rocks of Precambrian age in that part of the Medicine Bow Mountains lying in southeastern Wyoming: Geological Survey of Wyoming Memoir 1, 167 p.

Hyden, H. J., Houston, R. S., and King, J. S., 1968, Geologic map of the Arlington Quadrangle, Carbon County, Wyoming: U.S. Geological Survey Map GQ 643, scale 1:24,000.

Hyden, H. J., and McAndrew, H., 1967, Geologic map of the T L Ranch Quadrangle, Carbon County, Wyoming: ibid., GQ 637, scale 1:24,000.

Knappen, R. S., and Moulton, G. F., 1931, Geology and mineral resources of parts of Carbon, Big Horn, Yellowstone and Stillwater counties, Montana: U.S. Geological Survey Bulletin 822, 70 p.

McCallum, M. E., 1968, The Centennial Ridge gold-platinum district, Albany County, Wyoming: Geological Survey of Wyoming Preliminary Report 17, 13 p.

McClurg, J. E., and Matthews, V., III, 1978, Origin of Elk Mountain anticline, Wyoming: in Matthews, V., III, ed., Laramide folding associated with basement block faulting in the western United States: Boulder, Geological Society of America Memoir 151, p. 157-163.

Matthews, V., III (editor), 1978, Laramide folding associated with basement block faulting in the western United States: Geological Society of America Memoir 151, 370 p.

Prucha, J. J., Graham, J. A., and Nickelsen, R. P., 1965, Basement-controlled deformation in Wyoming Province of Rocky Mountains foreland: American Association Petroleum Geologists Bulletin, v. 49, p. 969-992.

Sever, C. K., 1975, Structural geology of the Sheephead Mountain area, Carbon County, Wyoming [M.S. thesis]: Laramie, Wyoming, The University of Wyoming, 96 p.

Shaw, A. B., 1954, The Cambrian and Ordovician of the Pryor Mountains, Montana and northern Bighorn Mountains, Wyoming, in Richards, P. W., ed., Pryor Mountains-northern Bighorn Basin, Montana: Billings Geological Society 5th Annual Field Conference Guidebook, p. 32-37.

Stearns, D. W., 1978, Faulting and forced folding in the Rocky Mountain foreland, in Matthews, V., III, ed., Laramide folding associated with basement block faulting in western United States: Boulder, Geological Society of America Memoir 151, p. 1-38.

Stewart, J. C., 1958, Geology of the Dryhead-Garvin Basin, Big Horn and Carbon counties, Montana: Montana Bureau of Mines and Geology Special Publication 17, 1 sheet.

Thom, W. T., Jr., Hall, G. M., Wegemann, C. H., and Moulton, G. F., 1935, Geology of Big Horn County and the Crow Indian Reservation with special reference to water, coal, oil and gas resources: U.S. Geological Survey Bulletin 856, 200 p.

SUPPLEMENTAL REFERENCES

Beckwith, R. H., 1938, Structure of the southwest margin of the Laramie Basin, Wyoming: Geological Society of America Bulletin, v. 49, p. 1515-1544.

Berg, R. R., 1977, Deformation of Mesozoic shales at Hamilton Dome, Bighorn Basin, Wyoming: American Association of Petroleum Geologists Bulletin, v. 60, p. 1425-1433.

Blackstone, D. L., Jr., 1975, Late Cretaceous and Cenozoic history of the Laramie Basin region, southeast Wyoming, in Curtis, B. F., ed., Cenozoic history of southern Rocky Mountains: Geologic Society of America Memoir 144, p. 249-279.

Dahlstrom, C. D. A., 1969, Balanced cross sections: Canadian Journal of Earth Sciences, v. 6, p. 743-757.

Hodgson, R. A., 1965, Genetic and geometric relations between structures in basement and overlying sedimentary rocks, with examples from Colorado plateau and Wyoming: American Association of Petroleum Geologists Bulletin, v. 49, p. 935-949.

Neeley, J., 1934, Geology of the north end of the Medicine Bow Mountains, Carbon County, Wyoming: Geological Survey of Wyoming Bulletin 25, 15 p.

Oliver, K. L., 1970, Gravity study of the Hanna Basin [M.S. thesis]: Laramie, Wyoming, The University of Wyoming, 46 p.

Rubey, W. W., 1926, Determination and use of thickness of incompetent beds in oil field mapping and general structural studies: Economic Geology, v. 21, p. 333-351.

Weitz, J. L., and Love, J. D., 1952, Geologic map of Carbon County, Wyoming: U.S. Geological Survey in Cooperation with Wyoming Geological Survey and the University of Wyoming, without identifying number, scale 1 in:2.5 mi.

MANUSCRIPT RECEIVED DECEMBER 28, 1979
REVISED MANUSCRIPT RECEIVED JANUARY 11, 1980
MANUSCRIPT ACCEPTED JANUARY 17, 1980

Contributions to Geology, University of Wyoming, v. 18, no. 2, p. 83-100, 15 figs., 2 tables, April, 1980.

Reprinted by permission of the Geological Society of America
from J. D. Lowell, *Geology*, v. 2 (1974), p. 275-278.

Plate Tectonics and Foreland Basement Deformation

James D. Lowell
Exxon Company, U.S.A.
P. O. Box 120
Denver, Colorado 80201

ABSTRACT

The extent and complexity of Laramide foreland basement deformation in the Wyoming province adjacent to the Idaho-Wyoming thrust-fold belt have led to numerous tectonic hypotheses, including tangential crustal compression (manifest as thrusting), vertical uplift, and strike slip; actually, all three types of deformation have occurred. With the advent of plate tectonics, tangential compression related to subduction has been mentioned most frequently as the ultimate cause of foreland deformation. Tangential compression is presumably operative in all subduction-related orogenic belts, but because most orogenic belts do not have foreland basement deformation comparable with that of the Wyoming province, compression *alone* does not seem sufficient to account for foreland basement structuring. Apparently a more specific condition is necessary. It is here proposed that *in addition* to tangential compression and strike slip, a lithospheric slab subducted well beneath the foreland is a fundamental requisite for deformation there, providing the buoyancy necessary for uplift of basement blocks. Recent work based on K_2O/S_iO_2 ratios of Cenozoic andesites suggests that paleosubduction did occur far east within the Wyoming province (Lipman and others, 1971).

INTRODUCTION

The most famous and extensive area of foreland basement deformation known along any orogenic belt in the world is the Wyoming province (Prucha and others, 1965), where Laramide basement block movement has occurred more than 400 km cratonward from the Idaho-Wyoming thrust-fold belt of partially the same age (Fig. 1). The structural style of the Wyoming province has been given many different interpretations, though all concede that uplift has been an important part of the deformation. Blackstone (1963) summarized published orientations of fault planes bounding basement blocks by saying they vary from normal through vertical and upthrust to low-angle thrust. (In fact, fault attitudes are highly variable. At Rattlesnake Mountain west of Cody, Wyoming, a low-angle thrust passes immediately into a high-angle normal fault at depth [Pierce, 1966]; whereas along the southwest flank of the Wind River Range, a low-angle thrust persists with miles of overlap of Precambrian crystalline basement [Berg, 1962]). In view of the differences in interpretation of structural style, it is not surprising that at least as many hypotheses of types of deforma-

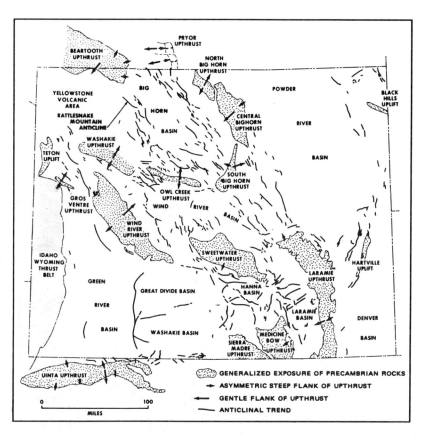

Figure 1. Map of portion of Wyoming province showing position of foreland basement structures (major mountain ranges and anticlines) with respect to Idaho-Wyoming thrust belt. Designation, "upthrust" intended to cover all degrees of thrust overlap from small amount at Rattlesnake Mountain to great horizontal component along southwest flank of Wind River Range.

tion have been proposed. Berg (1962) reviewed hypotheses of foreland basement deformation by block uplift (the bounding fault is a simple upthrust), thrust uplift, and fold-thrust uplift, preferring the last. The ultimate mode of uplift has generally been considered the result of one of three stress systems: (1) tangential crustal compression with important vertical adjustments or wedge uplifts (Chamberlin, 1945; Thom, 1955; Grose, 1972) or with important lateral coupling (Sales, 1968; Thomas, 1971), (2) vertical uplift (Bengston, 1956; Osterwald, 1961; Eardley, 1963; Harms, 1964); and (3) wrenching or strike slip (Stone, 1969).

Almost certainly, no single hypothesis is sufficient to satisfy the structural complexity of the Wyoming province, and elements of all three seem to be needed. Tangential compression seems required for the Wind River Range but not for Rattlesnake Mountain; conversely, vertical uplift probably works for Rattlesnake Mountain but does not explain the thrust overlap of the Wind River Mountains. Moreover, strike slip seems necessary, for example, for the en echelon structures along the north and south sides of the Wind River basin (Keefer, 1970).

Although much attention has been given in the past few years to the relations between plate tectonics and the evolution of orogenic belts, little attempt has been made to relate plate motion and the stresses that cause basement deformation in the foreland immediately adjacent to an orogenic belt. The purpose of this brief paper is to try to set foreland basement deformation in a plate-tectonics perspective.

FORELAND BASEMENT DEFORMATION AND SUBDUCTION

Equally as remarkable as the basement structuring of the Wyoming foreland province adjacent to the Idaho-Wyoming thrust-fold belt, is that along the same orogenic belt, at the latitude of the U.S.–Canadian border, and northward there is no foreland basement deformation whatsoever. Indeed, observable foreland basement deformation is rare along most of the world's orogenic belts. In the context of plate tectonics, many orogenic belts have been attributed to crustal compression produced by the interaction of continental and oceanic lithosphere. The paucity of foreland basement deformation adjacent to orogenic belts worldwide and, specifically, its absence northward adjacent to the same orogenic belt where it is best developed, mean that tangential compression was probably not

the sole cause of foreland basement structuring. Though compression is undoubtedly important (Fig. 2), apparently a more specific condition must be satisfied, and I propose it is subduction directly beneath the foreland (Fig. 3).

On the basis of K_2O/S_iO_2 ratios of Cenozoic andesites, Lipman and others (1971) have postulated that subduction occurred beneath the western United States in early and middle Cenozoic time along two subparallel imbricate zones (Fig. 3). Although the use of potassium content versus depth curves for a given silica content as a basis for determining the depth to proposed paleosubduction zones has recently been questioned (Nielson and Stoiber, 1973), the possible structural implications of this technique applied to the western United States seem too great to ignore, and the assumption is made in this paper that Lipman and others (1971) are correct.

According to Lipman and others (1971), a western subduction zone emerged at the continental margin, while an eastern zone was entirely beneath and decoupled from the continental plate; both zones dipped about 20° eastward.

The eastern zone essentially coincides with the foreland upthrusting of New Mexico, Colorado, Wyoming, and Montana, and it tapers in Montana to possible absence in Alberta. The frequent association of magmatic and volcanic activity with modern subduction zones is a priori evidence of buoyancy forces related to the latter. Moreover, the large Bouguer gravity minima (Woollard and Joesting, 1964) roughly coincident with the proposed eastern paleosubduction zone may even suggest the possible lingering existence of a relatively low-density and, therefore, buoyant plate at depth. Finally, the eastern subducted plate may have been sialic in part (see Fig. 4 caption). The presence of a detached lithospheric slab at depth may have exerted an appreciable buoyant force directed more or less vertically against the superjacent continental lithospheric plate. Tectonic implications are considerable:

1. Buoyancy acting on the continental plate, which includes pre-existing steep to vertical basement faults and fractures, should be manifested as uplifted, differentially rotated and tilted, upthrust-bounded basement blocks (Lowell, 1969).

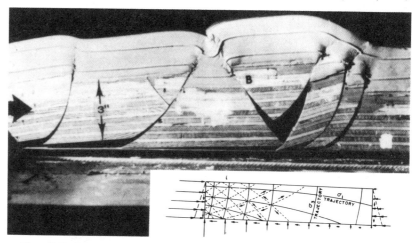

Figure 2. Reverse movement on pre-existing listric faults attained by tangential compression of rigid base as indicated by arrow. In context of this paper, model is likened to horizontal continental lithospheric plate of Figure 3b. Compression commences at one end of plate or rigid base at an early stage of subduction and is transmitted toward interior. Deformation begins in foreland closest to adjacent orogenic belt and progresses cratonward as subducted slab descends and moves eastward. Wind River Range, having greatest amount of thrust overlap (such as that simulated at "B" in model) of any of Wyoming province foreland basement blocks, is also closest to Idaho-Wyoming thrust belt, which presumably was locus of compression. For lithospheric plate, pre-existing curved fault surfaces would not necessarily be a prerequisite to thrusting, because stresses will produce curving thrusts. Inset shows for a homogeneous elastic medium the trajectories of principal stresses (solid lines) and potential reverse fault surfaces (broken lines) compatible with indicated boundary stresses of lateral compression (Hubbert, 1951). Model requires pre-existing surfaces, because no internal deformation could occur in clay if block below were not segmented. Response to compression is mainly upward movement of blocks, and compressional structures are restricted mostly to zone above edge of block. Asymmetry of structures is natural consequence of thrust-fault configuration. (If any pre-existing normal faults were later compressed, asymmetry should result.) Model somewhat reminiscent of wedge-uplift hypothesis of Thom (1955), in which vertical uplift is main response to deep-seated compression. Note that compression accumulated through entire rigid base of model prevents formation of large tensional features in overlying clay, which roughly corresponds to sedimentary rocks above crystalline basement.

2. If a gently dipping lithospheric slab beneath the foreland is a requisite for basement block movement and is at the same time a requirement that is achieved only rarely in the geologic record (for example, subduction occurs far landward around the present Pacific Ocean, but zones with gentle dips—15° to 20°—are not common), then foreland basement deformation would not be a common event. Basement deformation also need not be present continuously in the foreland along the same thrust belt; for example, the lithospheric slab underlying Wyoming may terminate northward by a deep-seated left-lateral arc-arc–type transform fault, surficially expressed by the various central Montana lineaments such as the Lake Basin fault zone. Hence, deformation would not have occurred to the north; presumably a comparable (right-lateral) termination also existed on the south.

3. The age of deformation in the Wyoming province foreland proper should have progressed from west to east as the lithospheric slab descended and moved eastward. This suggested age progression could be checked and would constitute one test of the present hypothesis.

CONCLUSIONS

Foreland basement deformation is envisioned as the response to a complexly varying stress system that includes tangential crustal compression, wrenching or strike slip, and vertical uplift. Perhaps most important, however, is the possible requirement well beneath the foreland of a subducted lithospheric slab that provided much of the impetus for uplift. Horizontally directed maximum principal compressive stress, generated at the earliest stages of subduction or before any significant lithospheric underthrusting (Figs. 2 and 4) and subsequently, equally or more important vertical maximum principal compressive stress produced by the buoyancy of an underthrust lithospheric slab (Figs. 3 and 4), would have both been expressed mainly as dip-slip movement on reverse faults and upthrusts. In the former case, as least principal compressive stress varied from vertical to horizontal, lesser amounts of strike slip could have also occurred, especially along pre-existing structural elements oriented at approximately 30°

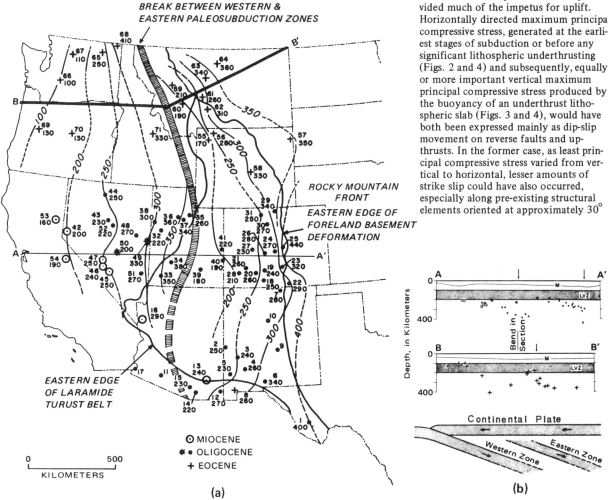

Figure 3. (a) Contoured depths (in kilometers) to inferred middle Cenozoic subduction zones of western United States. Upper number at each point identifies locality (see original reference); lower number gives depth to subduction zone, obtained from K_2O-depth plots for modern arc volcanics. Hachured line indicates discontinuity in contours. Most of samples are from Eocene and Oligocene rocks that are somewhat younger than thrust-belt-foreland deformation of Laramide age that terminated in Eocene time. A lag could be expected, however, between emplacement of an underthrust lithospheric slab and time that slab furnished volcanic material at or near surface by partial melting. Note that although Black Hills (loc. 57) is shown to be east of limit of foreland basement deformation (Rocky Mountain Front), it may be a foreland structure. (b) Sections showing inferred early and middle Cenozoic subduction zones. Horizontal scale same as vertical scale; surface topography not shown. All Oligocene (solid dot) and Miocene (circle with dot) points within 100 km north or south of section A-A are projected into this section. All points (+) for Eocene igneous rocks in northwestern United States are projected into section B-B′. M = Mohorovičić discontinuity, LVZ = approximate location of a low-velocity zone, and arrows are structural province boundaries from (a). Bottom section is schematic interpretation of lithospheric plate geometry suggested by sections A-A′ and B-B′. Presumably, eastern zone is older of two (Lipman and others, 1971).

DEFORMATION:

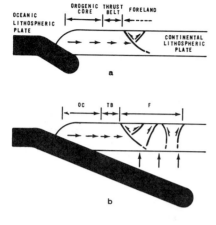

Figure 4. Conceptual diagram of Laramide foreland basement deformation. (a) Initial underthrusting of oceanic plate* causes tangential compression and beginning of foreland deformation in continental plate. Horizontal arrows represent direction and amount of compression; solid lines with half arrows indicate thrusting with direction of displacement. *(Character of eastern underthrust plate [oceanic versus continental] was never distinguished by Lipman and others [1971]. It is here termed "oceanic" by convention; it may have had a significant continental aspect. Eastern subducted slab might have included greatly thinned continental lithosphere [compare Lowell and Genik, 1972] that originated during a much earlier period of rifting [Stewart, 1972]. Greater sialic component would have increased buoyancy potential of the underthrust plate and also helped to explain intracontinental position of Idaho-Wyoming orogenic belt [Burchfiel and Davis, 1972].) (b) With greater length of underthrusting, buoyancy (vertical arrows) of underthrust slab causes uplift and deformation farther continentward in foreland. Actual deformation in orogenic core and adjacent thrust belt (which preceded foreland deformation) is not shown. Subsequent to stage b, continental plate overrides oceanic plate and another subduction zone is formed at lead edge of continental plate; buoyancy from subducted detached oceanic slab may be effective long after it has been decoupled, providing explanation for late Cenozoic uplift that also affected region.

to the direction of any horizontal maximum principal compressive stress (compare Stone, 1969, Pl. 1). Limited amounts of strike slip in a generally east-west direction might be provided by segmentation of the underthrust lithospheric slab by transform faults that would in turn impart lateral movement to the overlying lithosphere somewhat as proposed by Hoppin and Jennings (1971). Such a complex stress system with so many attendant possibilities for structural variety would not only explain the geology but also the proliferation of hypotheses and disagreement between hypotheses of foreland basement deformation.

REFERENCES CITED

Bengtson, C. A., 1956, Structural geology of the Buffalo Fork area, northwestern Wyoming and its relation to the regional tectonic setting: Wyoming Geol. Assoc. Guidebook 11th Ann. Field Conf., p. 158–168.

Berg, R. R., 1962, Mountain flank thrusting in Rocky Mountain foreland, Wyoming and Colorado: Am. Assoc. Petroleum Geologists Bull., v. 46, p. 2019–2032.

Blackstone, D. L., 1963, Development of geologic structure in central Rocky Mountains, in Backbone of the Americas: Am. Assoc. Petroleum Geologists Mem. 2, p. 160–179.

Burchfiel, B. C., and Davis, G. A., 1972, Structural framework and evolution of the southern part of the Cordilleran orogen, western United States: Am. Jour. Sci., v. 272, p. 97–118.

Chamberlin, R. T., 1945, Basement control in Rocky Mountain deformation: Am. Jour. Sci., v. 243-A, p. 98–116.

Eardley, A. J., 1963, Relation of uplifts to thrusts in Rocky Mountains, in Backbone of the Americas: Am. Assoc. Petroleum Geologists Mem. 2, p. 209–219.

Grose, L. T., 1972, Regional geology, tectonics, in Geologic Atlas of the Rocky Mountain Region, United States of America: Rocky Mountain Assoc. Geologists, p. 35–44.

Harms, J. C., 1964, Structural history of the southern Front Range: Mtn. Geologist, v. 1, no. 3, p. 93–101.

Hoppin, R. A., and Jennings, T. V., 1971, Cenozoic tectonic elements Big Horn Mountain region, Wyoming-Montana: Wyoming Geo. Assoc. Guidebook 23d Ann. Field Conf., p. 39–47.

Hubbert, M. K., 1951, Mechanical basis for certain familiar geologic structures: Geol. Soc. America Bull., v. 72, p. 355–372.

Keefer, W. R., 1970, Structural geology of the Wind River basin, Wyoming: U.S. Geol. Survey Prof. Paper 495-D, 35 p.

Lipman, P. W., Prostka, H. J., and Christiansen, R. L., 1971, Evolving subduction zones in the western United States: Science, v. 174, p. 821–825.

Lowell, J. D., 1969, On the origins of upthrusts: Geol. Soc. America, Abs. with Programs for 1969, Pt. 7 (Ann. Mtg.), p. 138.

Lowell, J. D., and Genik, G. J., 1972, Seafloor spreading and structural evolution of southern Red Sea: Am. Assoc. Petroleum Geologists Bull., v. 56, no. 2, p. 247–259.

Nielson, D. R., and Stoiber, R. E., 1973, Relationship of potassium content in andesitic lavas and depth to the seismic zone: Jour. Geophys. Research, v. 78, p. 6887–6892.

Osterwald, Frank W., 1961, Critical review of some tectonic problems in Cordilleran foreland: Am. Assoc. Petroleum Geologists Bull., v. 45, no. 2, p. 219–237.

Pierce, W. G., 1966, Geologic map of the Cody quadrangle, Park County, Wyoming: U.S. Geol. Survey Map GQ-542.

Prucha, J. J., Graham, J. A., and Nickelsen, R. P., 1965, Basement controlled deformation in Wyoming province of Rocky Mountain foreland: Am. Assoc. Petroleum Geologists Bull., v. 49, no. 7, p. 966–992.

Sales, J. K., 1968, Cordilleran foreland deformation: Am. Assoc. Petroleum Geologists Bull., v. 52, p. 2016–2044.

Stewart, J. H., 1972, Initial deposits in the Cordilleran geosyncline: Evidence of a Late Precambrian (>850 m.y.) continental separation: Geol. Soc. America Bull., v. 83, p. 1345–1360.

Stone, D. S., 1969, Wrench faulting and Rocky Mountain tectonics: Mtn. Geologist, v. 6, no. 2, p. 67–79.

Thom, W. T., Jr., 1955, Wedge uplifts and their tectonic significance, in Poldervaart, Arie, ed., Crust of the Earth—A symposium: Geol. Soc. America Spec. Paper 62, p. 369–376.

Thomas, G. E., 1971, Continental plate tectonics: Wyoming Geol. Assoc. Guidebook 23d Ann. Field Conf., p. 103–123.

Woollard, G. P., and Joesting, H. R., 1964, Bouguer gravity anomaly map of the United States: Am. Geophys. Union and U.S. Geol. Survey, scale 1:2,500,000.

ACKNOWLEDGMENTS

Reviewed by R. A. Hoppin.
Exxon Company, U.S.A., granted permission to publish this article.

MANUSCRIPT RECEIVED FEB. 13, 1974
MANUSCRIPT ACCEPTED MARCH 11, 1974

Basement balancing of Rocky Mountain foreland uplifts

Eric A. Erslev

Department of Earth Resources, Colorado State University, Fort Collins, Colorado 80523

ABSTRACT

Restorable cross sections of foreland basement uplifts must contain faults whose curvatures are consistent with the relative slip and tilt between adjacent basement blocks. Cylindrical fault surfaces can explain the uniform dip of strata on the back side of foreland uplifts; local zones of shortening and extension occur in hinge zones above transitions in fault curvature. High-curvature fault splays form fault wedges of basement which ease the transition from thrust and reverse faulting of basement blocks to folding in the sedimentary cover.

INTRODUCTION

The fact that cross sections containing the plane of tectonic slip should be restorable to possible initial geometries provides an important tool for the evaluation and testing of structural hypotheses. Unfortunately, the classical techniques of stratum length and area balancing utilized in fold and thrust belts (Dahlstrom, 1969; Royse et al., 1975; Boyer and Elliott, 1982; Suppe, 1983) are neither commonly nor easily applied to foreland structures involving basement because of the large mechanical contrasts between sedimentary strata and crystal-line rocks. However, tightly constrained foreland structural geometries can be constructed by combining these methods with the requirement that individual basement blocks be restorable to a continuous, interlocking mosaic below planar sedimentary units.

The structure of the Laramide uplifts in the Rocky Mountain foreland has been extensively debated for decades (Fig. 1). Many different models have been proposed, including up-thrusts (Prucha et al., 1965), planar thrusts (Berg, 1962; Gries, 1983), listric thrusts (Black-stone, 1940), and planar normal faults (Pierce, 1966; Stearns, 1971). However, none of the models for the classic Rattlesnake Mountain structure reviewed by Stone (1984) can be restored without a major redistribution of sedimentary cover or basement. Reconstructing the basement so that the sedimentary cover is restored to a planar configuration, an easy task with scissors if one assumes block motion on the faults, typically reveals large overlaps or voids caused by unaccounted rotational motion on the faults.

This paper presents a method for balancing cross sections of foreland uplifts that allows the restoration of the sections to their original geometry with minimal material overlap or omission while realistically portraying the considerable mechanical differences between layered sedimentary rocks and crystalline basement. Regional tilt of basement blocks and local zones of faulting can be explained as logical results of structural balance on curved faults. The resulting models show combinations of coexisting vertical and horizontal motions during compression, with the local dominance of one type of motion determined by fault-plane curvature, crustal depth, and constraints of strain compatibility.

BALANCING TECHNIQUES

This method assumes that basement blocks are translated and rotated on narrow brittle and ductile shear zones without being penetratively deformed. Cross sections are oriented perpendicular to the rotational axes defined by the characteristic anticlinal flexures in cylindrical parts of foreland structures. This allows the modeling of curved faults as circular arcs in cross sections without vertical exaggeration. So long as the structure is cylindrical adjacent to the line of section, oblique slip can be balanced as circular motion perpendicular to the fold axis. Sedimentary strata are assumed to deform by cataclasis in fault zones and flexural slip on bedding within the foreland structure without bedding detachments of regional extent.

Figure 2 shows some simplistic yet balanced models of rigid block motions on variously curved faults. The first three models (Fig. 2: A, B, C) illustrate that the difference in tilt between two blocks (Δ tilt) can be related to fault curvature ($1/R$ where R is the radius of curvature) and slip perpendicular to rotational axis (S_{RA}) by the following relationship:

$$\Delta \text{ tilt} = 180 \, S_{RA}/\pi R. \tag{1}$$

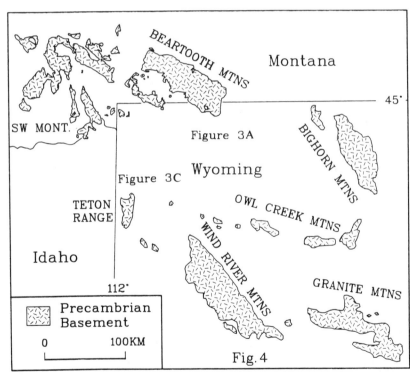

Figure 1. Map of northern Rocky Mountains foreland showing locations of Figures 3 and 4.

GEOLOGY, v. 14, p. 259–262, March 1986

259

This relationship has been used by Wernicke and Burchfiel (1982) to illustrate the diversity of fault curvatures in the Basin and Range. Brown (1984) used similar logic to explain the difference in tilt of strata at Rattlesnake Moun-

tain but failed to balance the cross section locally because of extra basement material in the footwall.

All three fault curvatures illustrated in Figure 2 exist in the Rocky Mountain foreland. The

small difference in dip between strata on either side of the plateau uplifts near Livermore, Colorado (Matthews and Work, 1978), suggests nearly planar faults with very large radii of curvature (Fig. 2A). Upthrusts (concave-

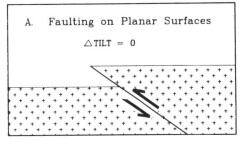

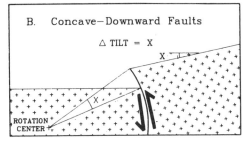

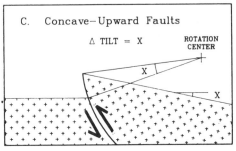

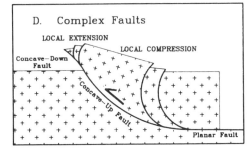

Figure 2. Restorable models of simple basement faults illustrating relationship between fault slip, curvature, and changes in tilt. D shows possible basement configuration resulting from changes in fault curvature. These models were generated by cutting paper rectangles along analytical arcs so that slip on these arcs does not create large overlaps or voids.

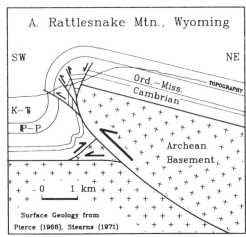

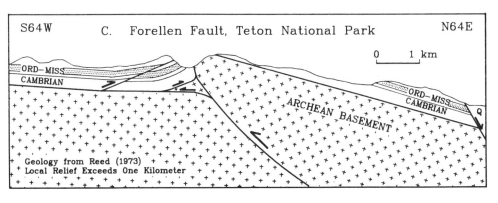

Figure 3. Balanced, unexaggerated cross sections of Forellen and Rattlesnake Mountain structures in northwestern Wyoming. B shows fault dip calculation. Pinning points A and C are located at transitions from folded to planar beds on footwall and hanging wall, respectively. Point B is located where extrapolated bed in hanging wall intersects level of same bed in footwall.

downward thrusts) appear to be common where basement has been thrust over sedimentary strata (e.g., the western Wind River Mountains and southern Owl Creek Mountains, Wyoming). Concave-upward faults (Fig. 2C) below Rattlesnake Mountain and the Wind River Mountains of Wyoming (Figs. 3 and 4) can generate the uniform dips on the back side of the thrust blocks.

However, no curved fault can maintain the same radius of curvature without describing a complete circle in cross section. Figure 2D shows that changes in curvature on faults cause local space problems because of the juxtaposition of two surfaces with different curvature during faulting. At each change of curvature, a strain incompatibility (or space problem) develops, which may be accommodated by the intrabasin thrusts and localized normal faults illustrated in Figure 2D. These minor faults are assumed to be tangent to the master fault, so

they will not offset it. Geometric constraints allow these faults to face in opposite directions within the same hinge zone. For example, the Golden fault dips west and the Milner Mountain fault dips east on the eastern flank of the Front Range in Colorado.

Data needed to balance a foreland cross section by stratum length and basement geometry include one fault orientation and the fault slip in the plane perpendicular to the rotational axis. This information can be derived from surface and subsurface measurements or from the following logic based on structural balance (Fig. 3B).

If a complete, restorable fold occurs over the basement fault, the dip of the basement fault creating the fold should equal the arc tangent of the fold throw over the fold heave (Fig. 3B). The fold throw is the stratigraphic throw in planar faults where both blocks have the same inclination. If the dips of the fault blocks differ,

the construction and equations in Figure 3B can be followed. The total heave equals the original bed length of folded unit between two pinning points in undistorted sections of the strata minus the present horizontal distance between the same points. The fold heave is equal to the total heave minus the shortening caused by tilting. The dip of this fault-equivalent of a fold can be plotted in the center of the fold. Correlative horizons from unfolded strata on either side of the structure can be projected to the fault trace to estimate net slip in the plane of the cross section. In many cases, the existence of bedding-parallel slip and elongation makes the calculated fault inclination a maximum dip.

The rotational center of the fault will lie on a line perpendicular to the fault orientation, and equation 1 can be solved to give the radius of curvature. This center can be used to draw the fault trace as a circular arc which will allow the restoration of the basement blocks. The arc of the fault will generally transfer to another surface with a different curvature whose location will be indicated by small-scale faulting, changes in dip, and local adjustments because of space problems, as shown in Figure 2D.

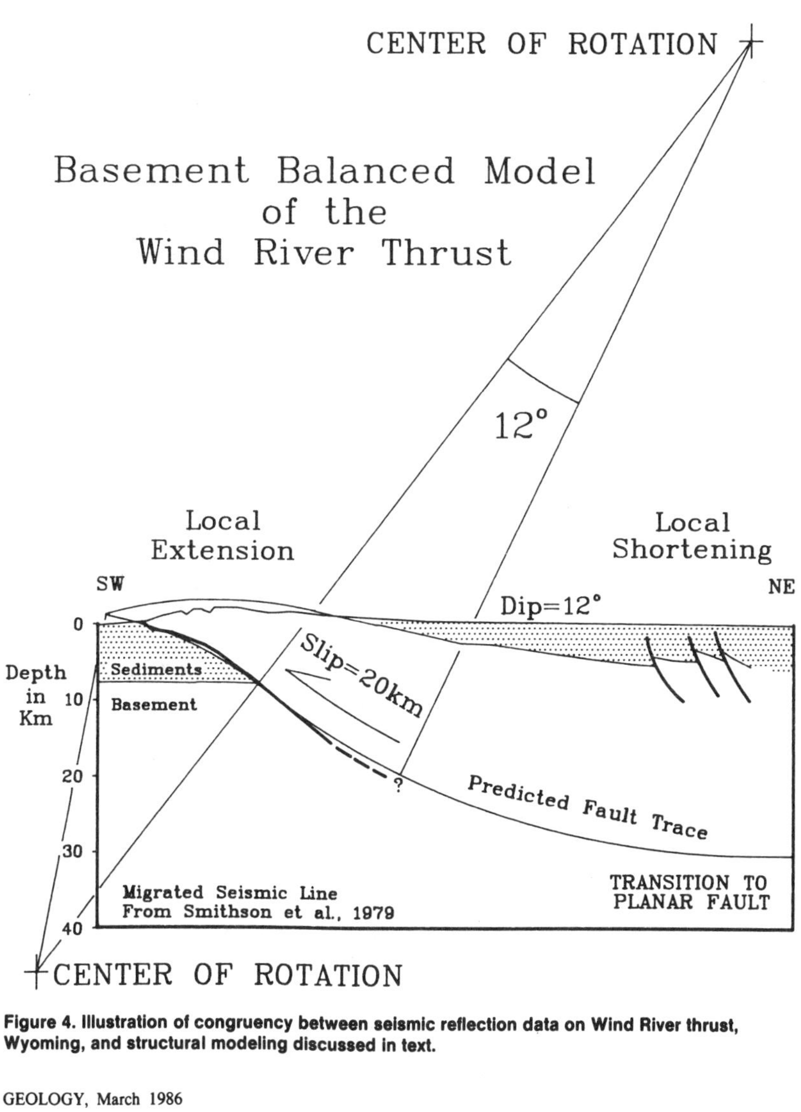

FORELLEN AND RATTLESNAKE MOUNTAIN STRUCTURES

The Forellen fault in the Teton Range, Wyoming, and the Rattlesnake Mountain structure west of Cody, Wyoming, represent different structural levels of similar basement structures. Both uplifts present the following challenges to balancing: (1) fault dips are commonly very steep at the surface and do not match with dips predicted from balancing folded sedimentary strata, and (2) major space problems occur in lower strata because of the transition from angular basement blocks to concentric folds.

The Forellen fault is spectacularly exposed on the walls in Webb Canyon in the northern part of Grand Teton National Park (Reed, 1973) (Fig. 3C). Balancing this cross section requires compatible rotation on the fault separating the footwall from the hanging wall as well as the fault segments defining the coherent basement wedge. Conceptually, the basement wedge must fit on the tip of the hanging wall, and this amalgamation must then fit on the footwall without major gaps or overlaps of material. In reality, the abrupt changes in curvature result in considerable cataclasis during the rotation of the wedge.

The Rattlesnake Mountain structure has been proposed as a type structure for many models of foreland uplifts (see Stone, 1984). Figure 3A gives a balanced interpretation of the structure, using the basement wedge concept to explain the lack of greatly overturned strata in the line of section used by Stearns (1971). On strike, beds with overturned dips of 42° (Pierce,

Figure 4. Illustration of congruency between seismic reflection data on Wind River thrust, Wyoming, and structural modeling discussed in text.

1966) are probably localized where the underlying wedge is poorly developed. The transition between block rotation and concentric folding is accommodated by the subthrust wedge, normal faulting in the basement tip of the hanging wall, fault and fold thickening in the lower strata of the footwall, and local attenuation of strata over the lip of the basement overhang. Fault curvatures are calculated to allow the restoration of a flat basement surface with minimal cataclasis. Space for the fault wedge is made by high-angle, high-curvature imbricate reverse and normal faults in the hanging-wall lip.

WIND RIVER MOUNTAINS

To test the preceding method, selected data from the Wind River COCORP seismic reflection line (Smithson et al., 1979) were used to generate a cross section of the Wind River Mountains (Fig. 4). The migrated depth data (Smithson et al., 1979, Fig. 17, p. 5969) were used, except that strata in the footwall were assumed to be horizontal. The cross section was schematically extended to the west; unmigrated seismic time sections presented by Smithson et al. (1979) were used. The data used in the structural model include a maximum dip of 12° on the east side of the uplift, the surface location of the Wind River thrust and the fault dip of 38° at the Cambrian unconformity in the footwall. A total slip of 20 km was estimated from the fact that Paleozoic rocks occur in the hanging wall north of the cross section (Berg, 1962).

The resulting cross section (Fig. 4) balances by geometry in the basement and contains several relationships that suggest its validity. (1) The predicted fault trace closely parallels the seismic information and curves to the horizontal at a depth of 33 km, nearly identical to the estimate given by the seismic interpretation of Lynn et al. (1983). (2) The position of the transition to planar faulting coincides with the area of local shortening on intrabasin faults, similar to the geometry predicted in Figure 2D on the grounds of strain compatibility. (3) The predicted area of local extension due to a reversal in fault curvature coincides with the "shattered zone" of Couples and Stearns (1978).

This illustration shows that regional structural balance can explain many local features, such as extension in overhangs and intrabasin thrusts, as well as provide a basis for the interpretation of the large-scale geometry of foreland structures.

DISCUSSION

In reality, the assumption of block motion in basement represents one end member of rheological behavior in foreland orogens. In the Rocky Mountains, basement lithologies range from granulite facies gneisses and charnockites

to Proterozoic sedimentary rocks. However, the uniformity of bedding orientations in strata immediately overlying crystalline basement in many areas, including the areas discussed in this paper, the Front Range in Colorado (Matthews and Work, 1978), and southern Beartooth Mountains in northern Wyoming (Pierce, 1966), support the hypothesis that basement block motion predominated in the Rocky Mountains during the Laramide orogeny. Apparent folding of basement surfaces in seismic time sections may be the result of velocity variations (Skeen and Ray, 1983) and small-scale block faulting without penetrative basement deformation, as documented by Wise (1963) in the Wind River Canyon south of Thermopolis, Wyoming.

Basement balancing provides a basic geometric foundation for the interpretation of foreland structures. That a cross section can be restored into an undeformed configuration without major structural mismatches shows that the interpretation is possible but does not prove that it is correct. Thus, structural balancing alone can only invalidate an interpretation, because balanced configurations are seldom unique. The relative magnitudes of the strain incompatibilities in published cross sections of the Rocky Mountain foreland show that principles of structural balance strongly favor thrust and reverse fault interpretations because most folded strata represent extra material telescoped by lateral compression. The cross sections presented in this paper illustrate that rigid block rotations are consistent with horizontal compression during the Laramide orogeny in the Rocky Mountain foreland.

REFERENCES CITED

Berg, R.R., 1962, Mountain flank thrusting in Rocky Mountain foreland, Wyoming and Colorado: American Association of Petroleum Geologists Bulletin, v. 46, p. 2019–2032.

Blackstone, D.L., Jr., 1940, Structure of the Pryor Mountains, Montana: Journal of Geology, v. 48, p. 590–618.

Boyer, S.E., and Elliott, D., 1982, Thrust systems: American Association of Petroleum Geologists Bulletin, v. 66, p. 1196–1230.

Brown, W.G., 1984, A reverse fault interpretation of Rattlesnake Mountain anticline, Wyoming: Mountain Geologist, v. 21, p. 31–36.

Couples, G., and Stearns, D.W., 1978, Analytical solutions applied to structures in the Rocky Mountains foreland on local and regional scales, in Matthews, V., III, ed., Laramide folding associated with basement block faulting in the western United States: Geological Society of America Memoir 151, p. 313–335.

Dahlstrom, C.D.A., 1969, Balanced cross sections: Canadian Journal of Earth Sciences, v. 6, p. 743–757.

Gries, R., 1983, Oil and gas prospecting beneath the Precambrian of foreland thrust plates in the Rocky Mountains: American Association of Petroleum Geologists Bulletin, v. 67, p. 1–26.

Lynn, H.B., Quam, S., and Thompson, G.A., 1983, Depth migration and interpretation of the CO-CORP, Wind River, Wyoming, seismic reflection data: Geology, v. 11, p. 462–469.

Matthews, V., III, and Work, D.F., 1978, Laramide folding associated with basement block faulting along the northeastern flank of the Front Range, Colorado, in Matthews, V., III, ed., Laramide folding associated with basement block faulting in the western United States: Geological Society of America Memoir 151, p. 101–124.

Pierce, W.G., 1966, Geologic map of the Cody quadrangle, Park County, Wyoming: U.S. Geological Survey Geologic Quadrangle Map GQ542, scale 1:62 500.

Prucha, J.J., Graham, J.A., and Nickelson, R.P., 1965, Basement-controlled deformation in Wyoming province of Rocky Mountains: American Association of Petroleum Geologists Bulletin, v. 49, p. 966–992.

Reed, John C., Jr., 1973, Geologic map of the Precambrian rocks of the Teton Range, Wyoming: U.S. Geological Survey Open-File Report, scale 1:62 500.

Royse, F., Jr., Warner, M.A., and Reese, D.L., 1975, Thrust belt structural geometry and related stratigraphic problems, Wyoming–Idaho–northern Utah, in Bolyard, D.W., ed., Symposium on deep drilling frontiers in the central Rocky Mountains: Laramie, Wyoming Geological Society, p. 41–54.

Skeen, R.C., and Ray, R.R., 1983, Seismic models and interpretation of the Casper Arch thrust: Application to Rocky Mountain foreland structure, in Lowell, J.D., ed., Rocky Mountain foreland basins and uplifts: Denver, Colorado, Rocky Mountain Association of Geologists Guidebook, p. 99–124.

Smithson, S.B., Brewer, J., Kaufman, S., and Oliver, J., 1979, Structure of the Laramide Wind River uplift, Wyoming, from COCORP deep reflection data and from gravity data: Journal of Geophysical Research, v. 84, p. 5955–5972.

Stearns, D.W., 1971, Mechanisms of drape folding in the Rocky Mountains: Wyoming Geological Association 23rd Annual Field Conference Guidebook, p. 125–143.

Stone, D.S., 1984, The Rattlesnake Mountain, Wyoming, debate: A review and critique of models: Mountain Geologist, v. 21, p. 37–46.

Suppe, J., 1983, Geometry and kinematics of fault-bend folding: American Journal of Science, v. 283, p. 648–721.

Wernicke, B., and Burchfiel, B.C., 1982, Modes of extensional tectonics: Journal of Structural Geology, v. 4, p. 105–115.

Wise, D.U., 1963, Keystone faulting and gravity sliding driven by basement uplift of Owl Creek Mountains, Wyoming: American Association of Petroleum Geologists Bulletin, v. 47, p. 586–598.

ACKNOWLEDGMENTS

Supported by the Donors of The Petroleum Research Fund, administered by the American Chemical Society. I thank Donald Blackstone, John C. Reed, Jr., Robbie Gries, Scott Smithson, Frank Royse, Elizabeth Robbins, and Vincent Matthews III for their helpful reviews and comments.

Manuscript received July 17, 1985
Revised manuscript received October 22, 1985
Manuscript accepted November 14, 1985

BULLETIN OF THE AMERICAN ASSOCIATION OF PETROLEUM GEOLOGISTS
VOL. 46, NO. 11 (NOVEMBER, 1962), PP. 2019-2032, 9 FIGS.

MOUNTAIN FLANK THRUSTING IN ROCKY MOUNTAIN FORELAND, WYOMING AND COLORADO[1]

ROBERT R. BERG[2]
Denver, Colorado

ABSTRACT

The principal mountain ranges of the Wyoming and Colorado foreland were raised asymmetrically by Laramide uplift which began in latest Cretaceous time as dominantly vertical movement along arcuate trends. Uplift continued into the Paleocene, and the steeper flanks of some ranges developed into large overturned folds by local compressive forces marginal to the main uplifts. Along segments of maximum uplift, overturned folds were broken and thrust far over the basin synclines in latest Paleocene or earliest Eocene time. Throughout the long period of uplift the adjacent basins were downwarped continuously and received sediment from the rising mountains.

The process of uplift by folding and thrusting better explains observed structure of mountain flanks than the older ideas of block uplift along high-angle faults or thrust uplift by regional compression. Fold-thrust structures are best known along the major Wyoming thrust zones bordering, on the south, the Wind River and Granite Mountains and the Washakie and Owl Creek Mountains. Other examples of faulted overturned folds occur in Colorado along the Golden thrust of the Front Range and Willow Creek thrust in western Colorado. In all these areas thrusts are well documented by subsurface control which includes both deep wells and seismic data.

INTRODUCTION

For many years it was believed that the mountain ranges of the Rocky Mountain foreland originated primarily as vertical block uplifts bounded by high-angle faults. This idea was gradually replaced by the concept of thrust uplift by regional compression after many low-angle reverse faults were recognized in surface mapping. However, in recent years additional data from the subsurface lead to the conclusion that the thrust margins of uplifts originated by a process of deformation which began as vertical uplift but resulted in some places in the growth of overturned folds and culminated in thrusting. This fold-thrust idea gives a better understanding of the genetic relationships between mountain flank structures and adjacent basins and makes easier the interpretation of local structure as well as regional tectonic patterns. However, for some geologists the old idea of vertical block uplift persists today, as an exclusive process of foreland mountain building, to the detriment of tectonic understanding. Some recently published papers

have concluded that thrusting is not common or significant in the Wyoming foreland. Therefore, this paper is a defense of thrusting and presents details of known thrust zones which are significant mountain flank structures. It is concluded that folding and thrusting on a grand scale were major factors in mountain flank deformation.

Exploration for oil and gas in the Rocky Mountains, especially in Wyoming, often has extended into the folded and thrust-faulted mountain flanks. In these areas structural interpretation is not simple, and in most places is exceedingly difficult because of the common occurrence of post-Laramide sediments which mask structure in the older rocks. A wider knowledge of mountain flank structure will aid exploration in these complex belts. Furthermore, the study of flank structure undoubtedly will lead toward a better understanding of Laramide history and the influence of uplifts on related sediments of adjacent basins, in which increasing amounts of oil and gas are now being found.

As used in the Rocky Mountain area, "thrust fault" is a low-angle reverse fault, especially one which has formed by compression, and "thrust zone" is a fault zone believed to consist of thrust faults as its major structure. It is shown that the large thrust zones of the Wyoming foreland have fault plane dips which range from an average low dip of about 20° to nearly vertical. Also, parts of these thrust zones appear to have originated by folding; that is, great vertical and horizontal displacements were attained primarily by overturn-

[1] Manuscript received, November 22, 1961.

[2] Embar Oil Company. The writer is grateful to Cosden Petroleum Corporation for the use of geophysical data essential to the interpretation of the Wind River thrust, and to F. A. Romberg of Texas Instruments, Inc., and to E. B. Wasson of Embar Oil Company for assistance in the structural interpretation of these data. Many others have given helpful advice. Among them are G. G. Anderman, C. J. Gudim, J. M. Parker, and T. C. Wilson. The manuscript has benefited from the criticism of D. L. Blackstone, Jr., and G. R. Downs.

ing of mountain flanks with minor amounts of displacement achieved by faulting alone. Therefore, in the following discussion, the use of "thrust" and "thrust zone" admittedly is not invariably precise, but no other names are suitable. However, there is the deliberate attempt to describe relative uplift in terms of total displacement of the Precambrian (sub-Cambrian) surface, and the use of the terms throw and heave is avoided except where specifically applied to faults.

WYOMING FLANK STRUCTURE

MAJOR THRUST ZONES

Zones of thrusting are significant mountain flank structures in the Wyoming foreland. All are the result primarily of deformation of Precambrian crystalline rocks which makes them distinctly different from the thrust slices of sediments in the Overthrust Belt of western Wyoming.

The two major thrust zones of central Wyoming are dominantly west- to northwest-trending complexes of multiple thrust faults which form the south and southwest flanks of principal asymmetric mountain ranges (Fig. 1). The individual thrusts within these zones are arcuate in plan, convex toward the south or southwest, and form an *en échelon* pattern. The longest of these is the Wind River-Seminoe thrust system which bounds the central mountain masses from the Gros Ventre Range in the northwest, past the Wind River Mountains, and along the Granite Mountains in the southeast. Maximum vertical displacement is on the order of 40,000 feet and maximum horizontal displacement is estimated to be about 50,000 feet. A second zone is the Washakie-Owl Creek thrust system which extends along the south flanks of the Washakie and Owl Creek Mountains to the southern end of the Bighorn Mountains where the zone turns southeast and may continue along the west margin of the Casper arch. Maximum displacements along this zone are probably equal to those of the Wind River-Seminoe thrust zone, but here displacements are estimated because more precise data are not available.

These two major thrust zones are better known than others in Wyoming because of available detailed surface maps and a greater amount of subsurface information from seismic surveys, mountain flank test holes, and deep wells in the adjacent basins.

Two similar thrust zones are known in adjacent areas. One is the flanking thrust system of the Uinta Mountains in northeast Utah and northwest Colorado, which consists of the Henrys Fork, Uinta, and Sparks thrusts. The other is the Beartooth thrust zone extending from southwest Montana into northwest Wyoming, which is perhaps continuous with the steep west flank of the Bighorn Basin as far south as Oregon Basin. Both zones include thrust faults, but the nature of these faults at depth is still a matter of argument. Recent diagrams of both the Uinta thrust (Ritzma, 1959) and the Beartooth thrust (Foose et al., 1961) show them as markedly steepening with depth, essentially high-angle faults.

Other zones of less significant thrusting in Wyoming are located along the eastern flanks of the Bighorn and Medicine Bow Mountains, whereas the steep monoclines on the west side of the Rock Springs and the Rawlins-Sierra Madre uplifts are believed to be faulted at depth. However, the monoclinal zones are covered by a great thickness of younger Cretaceous and Tertiary sediments, and the exact nature of deep structure is unknown.

WIND RIVER THRUST

The Wind River Mountains are a northwest-trending asymmetric anticline that has an extensive exposed core of Precambrian crystalline rocks (Fig. 1). At the northwest end of the range significant thrust faults are exposed (Richmond, 1945), but for a distance of about 100 miles the southwest flank is overlapped by Eocene sediments which extend mountainward from the Green River Basin and obscure structure in the older rocks.

For many years it was apparent that great structural relief exists between the southwest flank and the adjacent syncline of the Green River Basin, but the nature of flank structure was unknown. It was suggested that the mountain flank is bounded by a reverse fault (Coffin, 1946; Eardley, 1951), and Love (1950) inferred that the flank is bounded by a thrust fault having 20,000 feet or more of throw. Then more definite evidence became available from seismograph surveys in the Big Sandy area (Berg, 1961b). From this data the Wind River thrust is interpreted to have an average dip of 20° NE., a vertical displacement of 30,000 feet or more, and a horizontal displacement approaching 50,000 feet (Fig. 2). The seismic survey does not extend far enough mountainward to show the relationships in the root zone of the

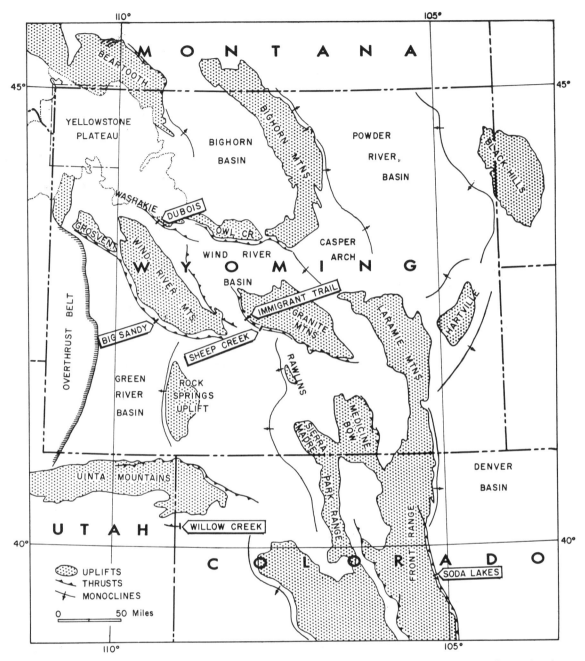

FIG. 1.—Index map of foreland, Wyoming and adjacent areas, showing principal Laramide uplifts, major thrust zones, and location of cross sections.

thrust, but steepening of the fault plane here is expected, based on analogy with other thrust faults at the surface (Berg, 1961b, p. 74).

A prominent feature of the seismograph profile is a strong reflection band that dips northeast and is believed to represent the base of the Precambrian thrust wedge (Berg, 1961b). The thickness, uniformity, and persistence of this band suggest that the thrust is not a simple plane but is a zone

1,000–2,500 feet thick which consists of layered rocks. By comparison with similar thrusts that have been drilled, it is likely that the thrust zone consists of a sheet of sediments, largely Paleozoic in age, deformed by folding and minor faults, but essentially in regular but inverted stratigraphic sequence. Therefore, the Wind River thrust appears to be the faulted limb of a huge, overturned fold. Projection of the possible original Precam-

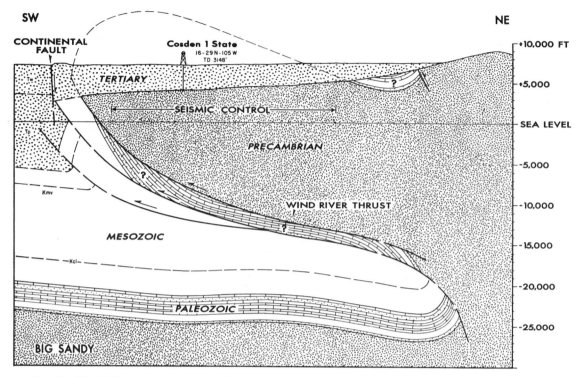

Fig. 2.—Wind River thrust interpreted from seismograph data in Big Sandy area, Sublette County, Wyoming. Kcl is Lower Cretaceous Cloverly Sandstone, Kmv is horizon near top of Upper Cretaceous Mesaverde Formation. Modified from Berg, 1961b, Fig. 3.

brian surface shows that the crest of the fold might have been displaced as much as 45,000 feet above the adjacent syncline.

Other structures associated with the Wind River flank are of interest. The root zone of a second thrust is located toward the northeast. This fault is inferred from gravity data, and relations shown along it are hypothetical. Just southwest of the granite wedge is the Continental fault, a normal fault along which throw has been estimated to be 250–1,000 feet (Nace, 1939; Berman, 1955). The fault dies out downward and represents post-Miocene collapse along the toe of the thrust wedge. The trace of the Continental fault zone along the south flank of the Wind River Range approximately parallels the pre-Wasatch trace of the thrust.

The subthrust sediments of the Green River Basin syncline are remarkably uniform in dip and show no evidence of strong deformation. A gentle, low relief fold is shown in the subthrust section, but this fold may be merely a velocity anomaly resulting from laterally increasing velocity in the subthrust section.

The interpretation of seismic data, without deep well control, does not give an infallible interpretation of structural details, but the seismic control does show a significant thrust fault underlying the mountain flank. From scattered surface and subsurface control, it is probable that the thrust as here depicted is typical of the entire length of the southwest flank of the Wind River Mountains. The relationships along the thrust zone must be inferred from other areas in which similar faults have been drilled.

IMMIGRANT TRAIL THRUST

In other areas of central Wyoming, drilling has revealed overturned sections beneath thick Precambrian thrust wedges. One such area is Immigrant Trail where a Carter Oil Company test well drilled through the Immigrant Trail thrust which forms the southwestern bounding fault of the Granite Mountain uplift. The Carter No. 1 Unit (sec. 32, T. 30 N., R. 93 W.) penetrated 1,200 feet of flat Tertiary beds then 1 900 feet of Precambrian granite, and 880 feet of overturned rocks ranging in age from Cambrian to Jurassic before encountering a normal sequence of Upper Creta- ceous sandstone and shale which dip 13° NE. (Fig. 3). The overturned section itself is partly repeated by minor faults, and the Paleozoic formations are represented by limited thicknesses of only the most competent beds. In the cross section, no at- tempt is made to show the structural complexity of this section. The Immigrant Trail thrust dips 20° NE., an estimate based on all available data; these include a published cross section drawn from seismic data (Wyoming Geological Association, 1951), shallow core holes in the vicinity of the pre- Eocene trace of the thrust, and a second well re- cently drilled through the same thrust zone, The Atlantic Refining Company No. 1 Unit (sec. 9, T. 29 N., R. 93 W.). Total vertical displacement of the Precambrian surface has been about 13,000 feet, and horizontal displacement is estimated to be 16,000 feet. The known extent of the inverted section below the granite is nearly half of the horizontal displacement; therefore, it appears that a significant part of this displacement has been accomplished by overturning. The Precambrian surface has been restored in the structure section to show the outline of an overturned fold.

Osterwald (1961, p. 229) interprets the Immi- grant Trail fault as dominantly high-angle despite the relatively abundant subsurface control to the contrary. His interpretation rests largely on the fact that one well, the Stanolind's Scarlett Ranch Unit No. 1 (sec. 22, T. 31 N., R. 94 W.), drilled 2,600 feet of Tertiary and Paleozoic sediments and 5,400 feet of Precambrian granite without penetrating the fault. This well is located 9 miles northwest of the Carter No. 1 Unit, generally

along strike of the thrust, and only 1½ miles from the pre-Eocene fault trace. The position of the Scarlett Ranch well relative to the Immigrant Trail thrust is indicated by its projected location in Figure 3.

Rather than disproving a thrust, the Scarlett Ranch well shows that the fault is deep at this location and that it is overlain by a greater thick- ness of Precambrian in the superthrust fold. Therefore, both the overlying fold and the sub- thrust syncline plunge northwestward between the two wells. These conclusions are supported by other subsurface data from nearby wells and seismic surveys.

SHEEP CREEK THRUST

The Sheep Creek thrust, another major fault of central Wyoming, bounds the Granite Mountains on the south from the vicinity of Crooks Gap to Muddy Gap. The Immigrant Trail thrust may be the northwestward extension of the Sheep Creek thrust, but the relation between the two is not clearly established. Osterwald (1961, p. 230) pre- sents a cross section in the vicinity of the Sheep Creek oil field which shows the Sheep Creek thrust as a high-angle fault at depth, an interpretation based only on surface exposures and incompatible with available subsurface data. The exposed thrust dips 5°–18° NE. (Reva, 1959), and seismic control indicates that this dip persists at least 1 mile north (Fig. 4). Underlying the granite thrust wedge are steeply overturned Paleozoic and Meso- zoic sediments, bounded below by a second thrust.

Immediately in front of the thrust zone is the strongly asymmetric and faulted Sheep Creek

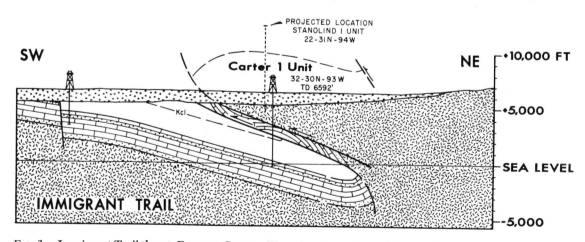

FIG. 3.—Immigrant Trail thrust, Fremont County, Wyoming. Formation symbols and patterns same as in Figure 2.

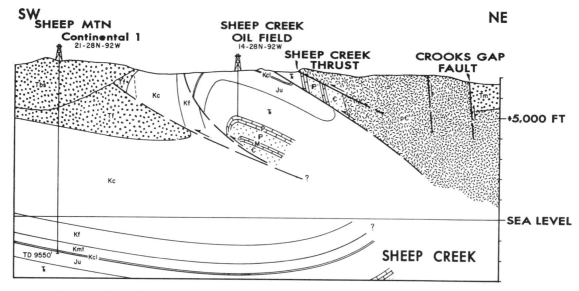

Fig. 4.—Sheep Creek thrust, Fremont County, Wyoming. Surface geology from Reva (1959); structure of Sheep Creek anticline simplified.

anticline. This fold may lack a basement core. Its limited areal extent, sharp asymmetry, and absence of deep folding all indicate that it is truncated below by another relatively shallow thrust. Therefore, the fold lies within a strongly deformed slice in front of and below the main thrust. Not shown on the section (Fig. 4) is another major thrust, located below and south of the Continental's Sheep Mountain well No. 1 (sec. 21, T. 28 N., R. 92 W.). This unnamed fault separates the Crooks Gap-Happy Springs folds from the deep syncline on the south in which the top of the Precambrian surface lies at an elevation of about −15,000 feet.

The root zone of the Sheep Creek thrust is unknown because of limited subsurface control, but enough data are available to show that the thrust is a major low-angle fault. More definitely, the same thrust has been drilled recently at a location 8 miles east of the Sheep Creek field. The Sinclair's Cooper Creek well No. 1 (sec. 19, T. 28 N., R. 90 W.) drilled 2,100 feet of Precambrian granite below which were near-vertical Paleozoic beds, a thrust zone section similar to that shown at Sheep Creek. At a depth of 3,444 feet the well passed through the thrust zone and entered a normal Upper Cretaceous section to total depth of 12,225 feet in the Frontier Sandstone. This well is situated only 1 mile north of the surface trace of the Sheep Creek thrust and confirms an average low dip of about 25° NE. for the fault plane.

EA THRUST

Another drilled thrust is located near Dubois in the northwestern Wind River Basin (Fig. 1). Here a Shell Oil Company test (sec. 9, T. 42 N., R. 105 W.) penetrated the EA thrust, one of the bounding faults of the Washakie Mountain uplift (Fig. 5). This well drilled 7,400 feet of Precambrian schist and found an inverted section 800 feet thick which consisted of 500 feet of deformed and sheared rocks of Mississippian, Pennsylvanian, and Permian age underlain by 300 feet of Triassic redbeds. Below this was about 700 feet of steeply dipping Cretaceous Mowry shale, possibly dragged along the fault, and then a normal section of Lower Cretaceous and Jurassic rocks which dip 15° NE. Here again is an inverted section which exists for a considerable distance beneath a thrust. Folded Precambrian rocks above the EA thrust are well shown by the overlying Paleozoic sediments, and the total vertical elevation of the granite is estimated to be 17,500 feet above the sub-thrust syncline. Horizontal displacement is believed to be about 22,000 feet because seismic surveys indicate that sediments extend at least that far beneath the thrust. The dip of the fault plane is not definitely known, but projection of all subsurface information indicates an average dip of 20° NE.

A subsidiary structure is the additional slice directly in front of the main thrust mass. This frontal zone was drilled nearby and consists of

Cretaceous shales and sandstones, in part folded and overturned. It is similar to the Sheep Creek frontal slice (Fig. 4) but more strongly deformed.

OTHER THRUSTS

Few areas in Wyoming show the nature of mountain flank thrusts as well as those described where wells have actually penetrated a Precambrian thrust wedge. However, similar thrusts of smaller size are known, and one of these was drilled at Sage Creek anticline in the southwest Wind River Basin. The Amerada's Tribal No. 1 (sec. 21, T. 1 N., R. 1 W.) penetrated a Paleozoic section on the steep west flank of the fold. Near the base of the Orodovician the well passsed through a thrust fault and encountered an overturned section of Pennsylvanian and Permian rocks nearly 1,200 feet thick. Then it drilled a second thrust and entered a normal Paleozoic section to bottom in the Tensleep Sandstone at a total depth of 7,390 feet. This is another low-angle fault with dip estimated to be 30° E. and with a typical overturned section in the thrust zone. Here total vertical displacement is only 6,500 feet. The thrust penetrated is part of a fault zone which extends northward along the entire Sage Creek-Steamboat Butte line of folding on the west flank of the Wind River Basin (Fig. 1).

COLORADO FLANK STRUCTURE

GOLDEN THRUST

The Front Range of the Southern Rocky Mountains west of Denver is bounded by a complex reverse fault zone, the Golden thrust. Along much of its extent fault relationships must be inferred, for nowhere is it well exposed so that dip may·be accurately determined. The Golden thrust previously was thought to be a high-angle reverse fault (Zeigler, 1917), but later it was believed to be a low-angle thrust based largely on sparse seismic data (Stommel, 1951). Osterwald (1961, Fig. 3) used a single well for subsurface control to show that the fault dips westward at a rather high angle, about 60°. Recently Harms (1961) has further concluded that "the major structures outlining the flank of the range south of Denver are high-angle reverse faults whose dips steepen with depth." However, when all well control is used, the Golden thrust appears to be less a simple thrust and more a faulted and overturned fold (Fig. 6). The overturned limb is complexly thrust faulted, but the individual thrusts have relatively small throws of approximately 1,000 feet or less, whereas the total vertical displacement of the Precambrian surface is more than 13,000 feet. The thrusts dip 35°–50° W. between the wells. They disappear upward in the thick Cretaceous Pierre Shale (Kp) and are not seen in the near-vertical beds at the surface.

Unfortunately, the Golden fault must be interpreted only from the two deep wells and their relation to outcrops, because other data, such as adequate seismic control and dipmeter surveys, are lacking. It appears, however, that the overturned fold is the primary mountain flank structure and that the faults are a secondary feature of the overturned limb. At the town of Golden, 7 miles north of Soda·Lakes, the thrust has greater throw, for the Fountain Formation (IPf) is faulted against Pierre Shale (Kp) (Van Horn, 1957). Therefore, the fold at Soda Lakes passes northward into a more prominent thrust fault.

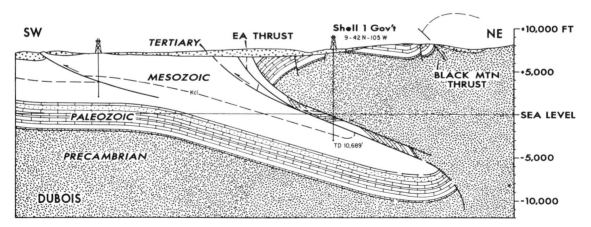

FIG. 5.—EA thrust near Dubois, Fremont County, Wyoming. Kcl is Lower Cretaceous Cloverly Sandstone; surface geology from Love (1939).

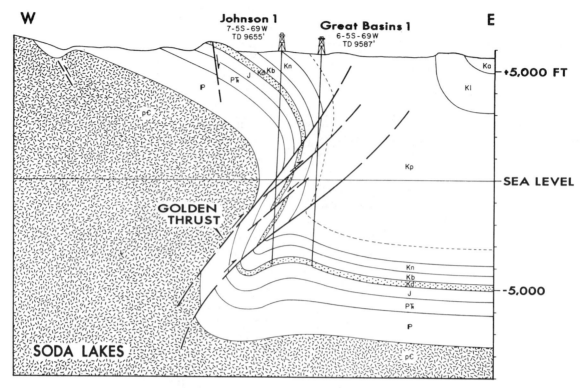

FIG. 6.—Golden thrust in the Soda Lakes area, Jefferson County, Colorado. From Berg (1962).

WILLOW CREEK THRUST

A smaller but better known thrust was drilled in western Colorado on the south flank of the Blue Mountain anticline, a subsidiary fold nearly parallel with the east plunge of the major Uinta Mountain uplift. Here, at Willow Creek, subsurface control shows a major overturned fold with thrust-faulted limb (Fig. 7). The Tennessee's Gov't. well No. 1-A (sec. 3, T. 3 N., R. 103 W.) drilled the overlying sediments and then penetrated 2,000 feet of Precambrian rock, the Uinta Mountain quartzite. Below the first thrust was a 500-foot inverted section of Pennsylvanian, Permian, and Triassic beds, and then a second thrust followed by an 800-foot normal section to total depth of 9,371 feet in the Pennsylvanian Weber Sandstone. Figure 7 is adapted from a cross section by Anderman and DeChadenedes (Anderman, 1961, Fig. 2) which was constructed with the aid of dipmeter and core dip control. The Willow Creek thrust has an average dip of 25° N. and grades upward into steeply dipping Upper Cretaceous Mancos Shale (Kmc). Here again is a low-angle fault zone associated with the overturned limb of a major fold.

ORIGIN OF THRUSTS

BLOCK UPLIFT

The mountain ranges of Wyoming and Montana have long been the subjects of tectonic study, but strong differences of opinion still exist on the interpretation of flank structure. Geologists responsible for early geologic mapping did not recognize thrust faulting. Later when extensive low-angle faults were mapped many geologists persisted in the belief that these were not significant flank structures but only secondary phenomena which occurred in the least competent of the overlying sediments.

The concept developed that the ranges were the result of vertical block uplift during which upward movement of rigid basement blocks took place along high-angle faults. Such mountain ranges were described as block-like in outline, and their forms were believed to be controlled by basement fractures, or conjugate sets of joints. The block uplift hypothesis was widely used to explain regional tectonic patterns, and its foremost proponents were Thom (1923) and R. T. Chamberlin (1945). As block uplift gradually came to be accepted and applied to nearly all of the major uplifts of the foreland, marginal thrusting was relegated to a minor role in deformation. Thus, the thrust structures of the Gros Ventre Range were described as "trap door" uplifts, an asymmetric variation of block uplift, by Horberg, Nelson, and Church (1949). Later, Bengtson (1956) applied the same interpretation and offered a mathematical explanation for thrust faults related to vertical movement. Recently, Osterwald (1961) has presented the extreme view that thrust faults are rare, and he has applied vertical block uplift principles to structural interpretation throughout the entire Cordilleran foreland.

The principles of block uplift are illustrated in Figure 8A. Great vertical displacement is attained by movement along an essentially high-angle reverse fault. The major conclusions of this idea are that the mountain ranges moved upward as rigid basement blocks; that uplift was along vertical fractures developed from pre-existing conjugate joint patterns in the basement; and that thrusts are not major flank structures.

According to the "block-uplifters," all features not compatible with vertical uplift are considered to be of minor importance. Thus Richmond

(1945), in discussing thrust faults at the northwest end of the Wind River Mountains, states that "they do not extend beneath the range for any great distance . . . and that more likely they represent local planes of lateral relief" formed during vertical uplift. Osterwald (1961, p. 234) says that "structures that superficially resemble overthrusts are probably subsidiary breaks resulting from rigid basement blocks crowding aside thick sequences of relatively non-resistant sedimentary rocks." These views seem to be extensions of the older idea expressed by Chamberlin (1945, p. 110) that "in a later episode of the orogeny, further easement by horizontal thrust movements became increasingly prominent." In other words, Chamberlin and other proponents of vertical block uplift did not deny the occurrence of thrusts as many have done, but merely assigned their formation to a late period of mountain uplift.

THRUST UPLIFT

Beginning in the 1930s, large-scale thrust faults were recognized more widely in central Wyoming, and it became generally accepted that low-angle

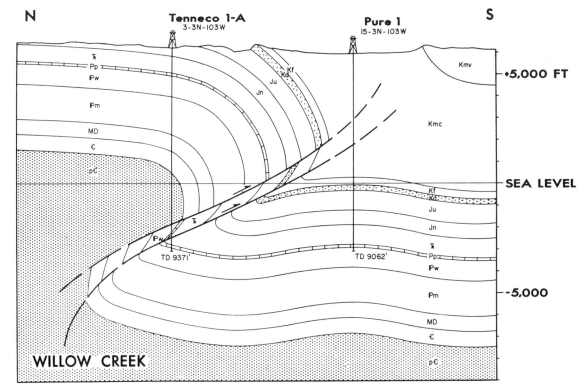

FIG. 7.—Willow Creek thrust south of Blue Mountain, Moffat County, Colorado. After Anderman (1961).

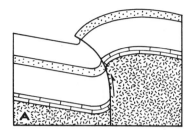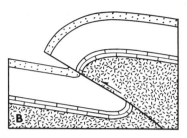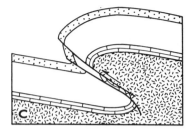

FIG. 8.—Diagrams illustrating hypotheses of mountain flank deformation: A—Block
uplift, B—thrust uplift, C—fold-thrust uplift.

faulting observed or inferred at the surface was actually an indication of significant thrust structure at depth. Thrust faults of great magnitude were believed to occur on the flanks of the arcuate ranges which are asymmetric toward the south and southwest, such as the Owl Creek Mountains (Fanshawe, 1939), the Washakie Range (Love, 1939), and the ranges bordering the Granite Mountain uplift (Lovering, 1929, Carpenter and Cooper, 1951, Heisey, 1951). Extensive thrusting was also mapped in other areas, such as along the east flank of the Medicine Bow Range (Beckwith, 1938), and in the Beartooth Mountains. The general character of the thrust-bounded uplift now is known rather well, although the details of the thrust zones at depth are obscure in most places. For many, the idea of thrust uplift has replaced the theory of block uplift. Thrust uplift is illustrated in Figure 8B, in which great vertical displacement is attained entirely along a low-angle fault. The chief difference between thrust uplift and block uplift is the amount of inferred horizontal displacement; genetically, the difference is one of compressive as opposed to vertical forces. Many geologists believe that compression of the entire foreland was necessary to produce large-scale thrusts, and those who favor thrust uplift retain the idea of basement rigidity (D. L. Blackstone, Jr., personal communication).

FOLD-THRUST UPLIFT

A characteristic feature of all major thrusts seems to be the overturned beds beneath them, both in exposures and in the few drilled sections. The subsurface evidence shows that these beds have too great an areal extent to be merely isolated blocks dragged along fault planes, and therefore they must represent the distorted limbs of folds. This leads to the conclusion that thrusts developed from uplift by folding, which progressed into overturning that produced great horizontal displacement, and was followed by thrusting to give an additional but minor amount of horizontal displacement. This mode of mountain flank deformation has been termed fold-thrust uplift (Berg, 1961a, b) and is illustrated in Figure 8C, where great vertical and horizontal displacements are attained primarily by overturning.

Fold-thrust uplift explains an extensive overturned sequence beneath the fault plane whereas uplift by thrusting alone does not. It suggests moderate to extreme deformation of the Precambrian crystalline rocks as opposed to the rigid basement of both block and thrust uplift.

The idea of related overturning and thrusting is not new in Rocky Mountain structure, but its significance has not been recognized. In fact, folding is commonly found in association with major thrusting, and Beckwith (1941) actually showed overturned fold axes above the Elk Mountain thrust while attributing major horizontal displacement to thrusting alone. Large-scale folding which requires extreme deformation of the Precambrian crystalline rocks seems to suggest plastic deformation at relatively shallow depths. Bucher (1933) pointed out that "plastic"-like deformation of the Precambrian does occur in the Southern Rocky Mountains where the Ute Pass fault of the Colorado Front Range was observed to pass laterally into a recumbent fold in the granite. The Rocky Mountains exhibit only a mild form of "plastic" deformation when compared with other great mountain systems of the world, but the mountain flanks do show a marked degree of "plastic" yielding similar to Bucher's mountain welts.

Although basement rigidity during Laramide deformation has been accepted previously, basement folding is possible without plastic flow. On the northwest flank of the Bighorn Mountains, Wilson (1934) observed slickensided, secondary joints in Precambrian granite along a flexure in the Five Springs Creek area. Because there was no evidence of internal deformation, he concluded that granite folding was accomplished by small movements along a large number of closely spaced joints. This may be the manner of deformation of crystalline rocks along thrust zones.

A possible objection to large-scale overturned folding is the missing volume of sediment on the overturned limb. Some of this sediment was eroded during an early stage of uplift when movement was dominantly vertical. Some of it was pushed ahead of the fold in thick frontal slices and partly eroded. Also, there was probably compaction of younger sediments beneath the heavy Precambrian thrust wedge. However, there seems to be a remaining bulk of sediments not accounted for beneath the thrust zones such as in the Wind River thrust (Fig. 2). This may be a failure in structural interpretation rather than in mechanics of folding.

FOLD-THRUST CHARACTERISTICS

The features of the major foreland thrust zones are particularly useful in local structural studies as well as in regional tectonic interpretation. Besides the thrust faults themselves and great displacements, these characteristics may be important, individually or severally, in diagnosing the significance of local structure in exploration along the flanks of major uplifts.

1. Multiple fault planes are common, but usually two faults are prominent, one between Precambrian and Paleozoic rocks, and another between Paleozoic rocks and Late Cretaceous or early Tertiary beds.

2. The fault zone consists of overturned Paleozoic and early Mesozoic sediments, generally a reduced thickness of only the most competent rocks. The inverted section may be relatively undisturbed, somewhat folded and faulted, or highly deformed and jumbled. Its thickness ranges from about 500 feet to several thousand feet.

3. Dip of the main fault plane is variable, and its amount at any place depends on the position of the exposed or drilled part of the thrust. Dips range from very high to vertical at the point of greatest horizontal displacement and in the root zone, to only a few degrees at a location along the middle of the overturned limb. The observed angle of dip may be a real help in determining local position along a fault plane.

4. Stratigraphic throw is variable when measured in different places along the thrust, and apparent throw is least in the root zone.

5. One or more frontal thrusts are known in some places. These frontal slices involve chiefly Mesozoic sediments, are up to 5,000 feet or more in thickness, and do not include Precambrian rocks. They may be strongly folded and faulted, defying a simple structural interpretation even after drilling.

6. Transverse shear faults of major horizontal displacement up to several miles may offset the thrust zones. These result from breaking and lateral movement in a late stage of fold-thrust development. Examples of lateral shearing occur along the Beartooth thrust near Red Lodge, Montana (Foose et al., 1961).

7. A large anticlinal fold involving basement rocks may suggest the presence of deep-seated thrusts. Conversely, the presence of significant thrust faulting may indicate that major Laramide folding is present in the exposed Precambrian core, a feature usually not mapped but often suggested by structures along mountain fronts. Such folding can be observed in Precambrian sediments along a portion of the Uinta thrust (Hansen, 1957).

8. The overlying major fold may have back-thrusts on its flank, but these are relatively minor when compared with the total displacement of the mountain flank.

9. Significant normal faults are associated with those fold-thrust zones that have greatest horizontal displacement, such as the Continental fault at the toe of the Wind River thrust (Fig. 2). These faults represent post-Miocene collapse of the thrust wedges. Other normal faults occur behind the thrusts and are downthrown toward the mountains, such as the Crooks Gap fault north of the Sheep Creek thrust (Fig. 4) and the Pathfinder fault north of the Seminoe thrust (Carpenter and Cooper, 1951). Significant normal faults are not recognized where horizontal displacements are small (Figs. 6, 7).

10. Subthrust sediments of the deep basin synclines are not strongly deformed. In most cases the subsurface evidence indicates only more or less uniform dip into the root zones, thereby arguing against the possibilities of subthrust anticlinal prospects for drilling.

TIME OF DEFORMATION

The time of deformation in the Wyoming Mountain ranges seems to be largely Laramide, that is, uplift began in late Cretaceous time, Lancian or even earlier, and ended in latest Paleocene or earliest Eocene time. Evidence from the basins suggests continuous downwarping and more significant, a continuously growing uplift. This is indicated by the influx of coarse clastics of increasing age which were derived from the rising mountain flank. The sedimentary evidence leads to the conclusion that uplift of the mountains progressed uniformly with downwarp of the basin (Berg, 1961, a, b). This conclusion stands in opposition to some former interpretations of Laramide history derived from areas of uplift. These have stressed numerous repeated uplifts followed by long periods of widespread erosion. Locally, tectonic events may have been episodic and deformation may have been limited and intense, but these interpretations give an erroneous regional history. In considering the basin-mountain relationship on a larger scale, a continuous sequence of uplift and downwarp is suggested.

As defined here, "Laramide" deformation is restricted to the period in which greatest uplift was achieved; it does not include subsequent minor movements of the middle and late Tertiary. Some smaller thrusts are post-early Eocene, even post-Oligocene in age. For example, the Piney Creek thrust on the east flank of the Bighorn Mountains pushed Paleozoic rocks over the early Eocene Wasatch conglomerates. Such faults are not major flank structures but are the result of final lateral push toward the basin following the main period of uplift. Perhaps the motivation for some of these thrusts was gravity. In the case of the Piney Creek thrust, it does not effect great vertical displacement, but rather it forms a cap on the monoclinal fold which is the main flank structure.

Although fold-thrust structure may be present along the margins of some ranges, it is an extreme condition for the foreland as a whole. Great horizontal displacement is not a universal condition, and thrusting is limited or absent in parts of many uplifts such as the Bighorn Mountains, Laramie Range, Hartville Uplift, Black Hills, and the Colorado Front Range. Therefore, fold-thrust may represent the final phase of mountain flank deformation opposite only the areas of maximum downwarp of the intermontane basins. Maximum uplift probably was attained only along relatively limited parts of the major thrust zones. It is assumed that less intensely deformed structures are present laterally from areas of maximum deformation.

Uplift began during the late Cretaceous with dominant vertical movement, not of blocks but along linear and arcuate trends. This resulted in asymmetry of basin and range and possibly some

high-angle reverse faults at points of great stress (Fig. 9A). Vertical displacement of the Precambrian surface probably attained 10,000–15,000 feet relative to the adjacent basins. During this early period compressional features were weakly developed.

Upwarp continued during the Paleocene. Moderate to high-angle reverse faulting and perhaps gentle folding in the basin sediments accompanied strong arching in the Precambrian rocks of the uplift (Fig. 9B). Vertical relief of 15,000–25,000 feet was attained, and moderate compression of the mountain flank resulted in overturning of major folds.

Uplift culminated in latest Paleocene or earliest Eocene time with dominant compression of the flank which resulted in strong overturning and finally thrusting of the overturned limb (Fig. 9C). Maximum foreland relief ranging from 25,000 to 40,000 feet was achieved. Later Tertiary uplift was largely epeirogenic.

Some uplifts did not progress beyond the first stage while others reached the second stage, and along the major Wyoming thrust zones the mountain flanks reached the final stage of maximum overturning and structural elevation. As shown in Figure 9, erosion of the rising uplifts was continuous, and deposits in the adjacent basins consisted of three facies. Nearest the uplift were conglomeratic sandstones of the flank facies containing pebbles derived first from the sediments, and then from the Precambrian core of the rising mountains. In the basin syncline was a fine-grained shale and siltstone facies representing a non-marine, quiet-water environment of basinal deposition. This facies constitutes the major part of the Hoback Formation in the Green River Basin and Waltman shale member of the Fort Union Formation in the Wind River Basin (Keefer, 1961). Toward the gentle flank of the basin were the interbedded sandstones and coal-bearing shales of the marginal facies that was deposited in a fluvial and fresh-water swamp environment which completely surrounded the non-marine basin. Overturning and thrusting of the mountain flank in the final stage of uplift brought older rocks over the Paleocene sediments that were deposited earlier.

Summary

The structures observed on the mountain flanks of the foreland show that there are thrust zones along the major asymmetric uplifts. Great struc-

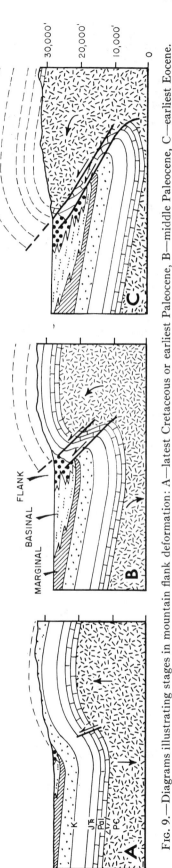

FIG. 9.—Diagrams illustrating stages in mountain flank deformation: A—latest Cretaceous or earliest Paleocene, B—middle Paleocene, C—earliest Eocene.

tural elevation of the uplifts was attained by folding and along significant low-angle faults. There is no subsurface evidence for an exclusive block uplift theory with its rigid basement rocks and dominantly high-angle faults. Neither is compressive thrusting of Precambrian rock wedges an adequate answer, for it does not explain the extensive overturned sequence beneath the thrust planes. Rather, it appears that Laramide basin and mountain flank deformation began with vertical movement along arcuate trends and developed linear upwarps and downwarps with local compressive features. In some places uplift culminated in overturned folding and thrusting. All of this was accompanied by an increasing degree of deformation of the crystalline crust. The entire process was carried on continuously throughout the long duration of Laramide time, from late Cretaceous into the early Tertiary with no significant interruptions either of uplift or of deposition in the adjacent basins.

REFERENCES

Anderman, G. G., 1961, Structure contour map of Colorado, *in* Symposium on Lower and Middle Paleozic rocks of Colorado: Rocky Mtn. Assoc. Geologists, 12th Ann. Field Conf. Guidebook, p. 101–105.

Beckwith, R. H., 1938, Structure of the southwest margin of the Laramie Basin, Wyoming: Geol. Soc. America Bull., v. 49, p. 1515–1544.

———— 1941, Structure of the Elk Mountain District, Carbon County, Wyoming: Geol. Soc. America Bull., v. 52, p. 1445–1486. .

Bengtson, C. A., 1956, Structural geology of the Buffalo Fork area, northwestern Wyoming, and its relation to the regional tectonic setting: Wyo. Geol. Assoc., 11th Ann. Field Conf. Guidebook, p. 158–168.

Berg, R. R., 1961a, Laramide sediments along the Wind River thrust, Wyoming (abs.): Am. Assoc. Petroleum Geologists Bull., v. 45, p. 416.

———— 1961b, Laramide tectonics of the Wind River Mountains, *in* Symposium on Late Cretaceous rocks: Wyo. Geol. Assoc., 16th Ann. Field Conf. Guidebook, p. 70–80.

———— 1962, Subsurface interpretation of Golden fault at Soda Lakes, Jefferson County, Colorado: Am. Assoc. Petroleum Geologists Bull., v. 46, p. 704–707.

———— and Wasson, E. B., 1960, Structure of the Wind River Mountains, Wyoming (abs.): Geol. Soc. America Bull., v. 71, p. 1824.

Berman, J. E., 1955, Geology of the Tabernacle Butte area, Sublette County, Wyoming: Wyo. Geol. Assoc. 10th Ann. Field Conf. Guidebook, p. 108–111.

Bucher, W. H., 1933, The deformation of the earth's crust: p. 161–171, Princteon, N.J., Princeton Univ. Press.

Carpenter, L. C., and Copper, H. T., 1951, Geology of the Ferris-Seminoe Mountain area: Wyo. Geol. Assoc. 6th Ann. Field Conf. Guidebook, p. 77–79.

Chamberlin, R. T., 1940, Diastrophic behavior around the Bighorn Basin: Jour. Geology, v. 48, p. 673–716.

———— 1945, Basement control in Rocky Mountain deformation: Am. Jour. Sci., v. 243-A, p. 98–116.

Coffin, R. C., 1946, Recent trends in geological-geophysical exploration: Am. Assoc. Petroleum Geologists Bull., v. 30, p. 2031-32.

Eardley, A. J., 1951, Structural geology of North
America: fig. 203, p. 348, New York, Harper and
Brothers.

Fanshawe, J. R., 1939, Structural geology of Wind
River Canyon area, Wyoming: Am. Assoc. Petroleum
Geologists Bull., v. 23, p. 1439-92.

Foose, R. M., Wise, D. U., and Garbarini, G. S., 1961,
Structural geology of the Beartooth Mountains,
Montana and Wyoming: Geol. Soc. America Bull.,
v. 72, p. 1143-1172.

Hansen, W. R., 1957 Structural features of the Uinta
Arch: Intermtn. Assoc. Petroleum Geologists 8th
Ann. Field Conf. Guidebook, p. 35-39.

Harms, J. C., 1961, Laramide faults and stress distribution in Front Range, Colorado (abs.): Am. Assoc.
Petroleum Geologists Bull., v. 45, p. 413.

Heisey, E. L., 1951, Geology of the Ferris Mountains-
Muddy Gap area: Wyo. Geol. Assoc. 6th Ann. Field
Conf. Guidebook, p. 71-76.

Horberg, C. L., Nelson, V. E., and Church, H. V., Jr.,
1949, Structural trends of central western Wyoming:
Geol. Soc. America Bull. v. 60, p. 183-216.

Keefer, W. R., 1961, Waltman shale and Shotgun members of Fort Union formation (Paleocene), Wyoming:
Am. Assoc. Petroleum Geologists Bull., v. 45, p.
1310-1323.

Love, J. D., 1939, Geology along the southern margin
of the Absaroka Range, Wyoming: Geol. Soc. America Special Paper 20.

—— 1950, Paleozoic rocks on the southwest flank
of the Wind River Mountains near Pinedale, Wyoming: Wyo. Geol. Assoc. 5th Ann. Field Conf. Guidebook, p. 25-27.

Lovering, T. S., 1929, The Rawlins, Shirley, and

Seminoe iron ore deposits, Carbon County, Wyoming: U. S. Geol. Survey Bull. 811-d.

Nace, R. L., 1939, Geology of the northwest part of the
Red Desert, Sweetwater and Fremont Counties,
Wyoming: Wyo. Geol. Survey Bull. 27.

Osterwald, F. W., 1961, Critical review of some tectonic problems in Cordilleran foreland: Am. Assoc.
Petroleum Geologists Bull., v. 45, p. 219-37.

Reva, J. P., Jr., 1959, Geology of the Sheep Creek-
Middle Cottonwood Creek Area, Fremont County,
Wyoming: Univ. Wyo. M.S. thesis.

Richmond, G. M., 1945, Geology of northwest end of
the Wind River Mountains, Sublette County,
Wyoming: U. S. Geol. Survey Map 31, Oil and
Gas Inv. Ser.

Ritzma, H. R., 1959, Geologic atlas of Utah, Daggett
County: Utah Geol. and Mineralog. Survey Bull. 66,
p. 71-75.

Stommel, H. E., 1951, Seismic investigation in the
Golden-Denver area: Colo. School Mines D.Sc.
thesis no. 729.

Thom, W. T., Jr., 1923, The relation of deep-seated
faults to the surface structural features of central
Montana: Am. Assoc. Petroleum Geologists Bull., v.
7, p. 1-13.

Van Horn, Richard, 1957, Bedrock geology of the
Golden quadrangle, Colorado: U. S. Geol. Survey
Map GQ 103.

Wilson, C. W., Jr., 1934, A study of the jointing in the
Five Springs Creek area, east of Kane, Wyoming:
Jour. Geology, v. 42, p. 498-522.

Wyoming Geological Association, 1951, North-south
cross section, The Carter Oil Company No. 1, Immigrant Trail Unit, Fremont County, Wyoming:
6th Ann. Field Conf. Guidebook, p. 122.

Zeigler, Victor, 1917, Foothills structure in northern
Colorado: Jour. Geology, v. 25, p. 728-733.

Reprinted by permission of the Wyoming Geological Association from W. W. Boberg, ed., *Wyoming Geological Association 34th Annual Field Association Guidebook: Geology of the Bighorn Basin*, 1983, p. 53-62.

Thirty-Fourth Annual Field Conference — 1983
Wyoming Geological Association Guidebook

HORIZONTAL COMPRESSION AND A MECHANICAL INTERPRETATION OF ROCKY MOUNTAIN FORELAND DEFORMATION

J. R. SCHEEVEL[1]

ABSTRACT

A mechanical model of the continental crust of the western U.S. during the Laramide is proposed. It is given the attribute of being an elastic continuum and is endowed with boundary conditions placing it under uniform horizontal compression. The stress-system derived from the model, used in conjunction with data supplied by experimental rock mechanics, yields a model scenario of structural development that closely parallels that inferred from the observed structures of the Rocky Mountain Foreland. Those features predicted by the model are small basement-cored folds (less than 1500 meters - 5,000 feet amplitude) whose location and trend is determined by the loci of basement faults. The small folds are found in basins formed by later basin-bounding fault sytems with as much as 13,000 meters (43,000 feet) of vertical relief.

The model predicts that basement faults should form at 30 to 35 degrees to the horizontal. The rock mechanics data suggest that, under uniform horizontal compression (response of the elastic model), faults will initiate at the upper basement surface and propagate downward (at 30 to 35 degrees dip) with increased shortening.

The model, combined with the rock mechanics data, accurately predicts the regional basement features of the Rocky Mountain Foreland. The model also yields insight into the chronological development of foreland deformation.

INTRODUCTION

The Laramide Rocky Mountain Foreland has long been a subject of interest to geologists. Among the subjects of discussion about the region are those of structural geometry (Thom, 1952, Blackstone, 1963), kinematics (Berg, 1962) and structural geometries as they relate to tectonics of the region (Alpha and Fanshawe, 1954, Hoppin, 1961, Prucha and others, 1965, Stearns, 1971, Samuelson, 1974, Berg, 1981).

A topic which has generated much interest throughout the years is that of the deep geometry of foreland structures, how do they develop and *what is* their ultimate or "first-cause" from a regional perspective (e.g. Sales, 1968). The curiosity in this region is generated by the fact that

structural features in the foreland are notably different from those of the adjacent fold and thrust belt. Specifically, the majority of foreland folds involve the crystalline basement rock. Because of this difference, workers have attempted to determine possible shallow (layered rock) and deep (basement) structural geometries and in turn relate them to the regional framework (Prucha and others, 1965, Berg, 1962, Sales, 1968, Stearns, 1978).

The goal of this paper is to attack the problem of "first-cause" by using 1) general principles of theoretical mechanics, 2) data from experimental rock mechanics, and 3) observations of regional and global tectonics. These principles will be used to justify a regional model. The validity of the model is tested by constraining it with these known mechanical properties, and comparing it with the natural analog.

[1]Chevron U.S.A. Inc., 700 South Colorado Blvd., Denver, Colorado 80201

Basement Rocks and Mechanical Continuity

Because basement rocks are the principal concern in this paper, a discussion of their key mechanical properties is warranted. Basement rocks are upper-crustal crystalline rocks and may have the following properties:

1) localized regions of highly oriented anisotropic fabric elements.
2) previously inherited discontinuities or fracture patterns.
3) abrubt or gradual boundaries with other basement rock types.
4. some degree of time-dependent deformational character.

All of the above are usually present in some degree of abundance in all basement rock provinces. Despite this fact (and contrary to intuition), mechanical continuity over large areas of continental basement appears to be a common condition. This is suggested by present-day stress-orientation continuity across large regions where basement composition and structure vary irregularly in 1) North America (Sbar and Sykes, 1973, Zoback and Zoback, 1980), 2) Central Europe (Greiner, 1975), and 3) Australia (Stephenson and Murray, 1970). Continental-scale, kinematic continuity in the basement is suggested by modern sub-plate motions in China (Molnar and Tapponier, 1975) and the Aegean Sea region (Dewey and Sengor, 1979). (Examples of continental-scale mechanical continuity such as these are important justification for treatment of the large area of foreland basement as an elastic-continuum in the mechanical models below).

Basement Rock and Deformation Mechanism

Basement rock types have been shown in the past to be relatively time independent in their experimental response: elastic with brittle failure (Heard, 1962). However, recent work on experimental deformation of basement rocks has shown time-dependent failure at low temperatures and pressures (Krantz, 1980 and Krantz and others, 1982). In addition, Hansen (1982) has shown that time-dependent behavior of continental basement rocks deformed under simulated deep conditions (40 kilometers - 25 miles depth) can be adequately described with viscous flow equations.

This work could be used erroneously to argue in favor of some homogeneous fluid-like deformation mechanism to explain the long-term behavior of continental basement rocks. However, observation of those experimentally and other naturally deformed crystalline rocks, reveals the primary strain mechanism to be faulting (Heard, 1962, Mitra and Frost, 1981, Borg and Handin, 1966). Even under lower-crustal conditions (40 kilometers - 25 miles depth) strain is accommodated by discrete displacements across narrow zones with lesser strain between the zones

(cataclastic flow) (Hansen, 1982). Inasmuch as the bulk of the deformation can be correlated with the faults, the faults themselves must be considered the time-dependent members of the rock. They serve to connect the relatively rigid, time-independent regions of rock.

Cataclastic flow on many scales has been suggested in the past as a deformation mechanism for continental basement rocks (Stearns, 1969, Gallagher, 1981). On the basis of these experimental and natural examples of deformation mechanisms in continental basement rocks, large-scale deformation of the basement rocks of the foreland can only be accomplished by large-or small-scale configurations of faults.

Sedimentary Rocks and Folding

Sedimentary rock deformation is not treated in depth here, however its importance in foreland folds is far from trivial. Sedimentary rocks are mechanically differentiated from basement rocks in the foreland for two reasons: 1) most sedimentary rocks are considerably weaker, and more time dependent than basement rocks and thereby more

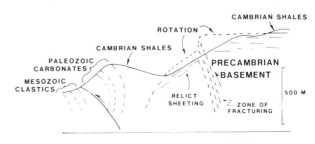

FIVE SPRINGS CREEK, WYOMING

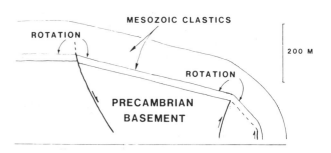

NORTH ENTRANCE CANYON, COLORADO

FIGURE 1: Cross sections of basement cored folds show sharp transitions in basement block rotations. The upper cross section (from Hoppin, 1970) is in the Five Springs Creek area of Wyoming. The lower cross section (from Stearns, 1978) is in North Entrance Canyon of the Colorado National Monument, Colorado. The basement geometries may be indicative of the influence of the layered rocks on the deformation of the basement.

susceptable to failure under the same conditions (Handin and Hager, 1957, Borg and Handin 1966); 2) sedimentary rocks exhibit extreme layered anisotropy in the form of bedding. The second factor plays a large role in the geometry and kinematic development of folds (Donath and Parker, 1964).

It will be argued below that fault systems in the basement control the location, orientation, and magnitude of folds in the Rocky Mountain Foreland. However, the anisotropy of the layered rocks may in turn influence the geometry of the basement surface as it forms the folds. This may be reflected by the sharp transitions in basement block rotations at Five Springs Creek, Wyoming (Hoppin, 1970), on the Uncompahgre Uplift, Colorado (Stearns, 1978) (Figure 1), and in other smoothly folded basement surfaces (Blackstone, 1981, Berg, 1981). This "smoothing" may occur because layered rocks in contact with basement-fault block edges, effectively "spread" the fault displacement to a wider zone of cataclasis.

Mechanics and Synthesis of "First-cause"

Many workers in recent years have identified the mechanical properties of rocks involved in foreland folds and used aspects of mechanics to synthesize an explanation of the "first-cause" of foreland structures (Prucha and others, 1965, Howard, 1966, Stearns, 1971, Couples, 1978, Mathews and Work, 1978). The conclusions of the bulk of

these workers are that the foreland is primarily a region of differential vertical uplift (meaning that vertical [fault] motions dominate over horizontal [fault] motions). Most of the workers that disagree with this final interpretation link the mechanical principles used in the synthesis to the regional interpretation and consequently regard both the principles and the interpretation as invalid. This is improper logic as the concepts of mechanics described above are valid on their own merits and are independent of the final interpretation which they are used to support.

In the text that follows, this author will attempt to develop a case, using these same mechanical concepts (along with observational data) for "first-cause" of Rocky Mountain Foreland deformation. Throughout, it is implicit that the mechanical concepts are used in support of, but are also independent of the regional interpretation.

A REGIONAL FORELAND MODEL

Mechanical Models

Models of elastic continua have been used in the past to represent pre-failure basement deformation in the Rocky Mountain Foreland and to make inferences about the geometry of faults produced by various applied load configurations (Stearns, 1971, Couples, 1978).

Those analyses make use of an elastic-continuum

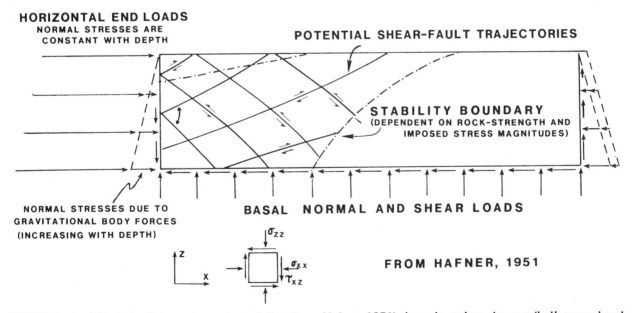

FIGURE 2: Analytical elastic boundary-value solution (from Hafner, 1951) shows boundary stresses (both normal and shear), potential shear-fault trajectories and hypothetical stability boundaries. The presence of a stability boundary is indicative of the fact that stress magnitudes decrease from left to right in the model. The positive x and z directions are shown as well as the stress directions taken as negative (two diagrams at the bottom of the figure). The mathematical solution for the model is given in the Appendix.

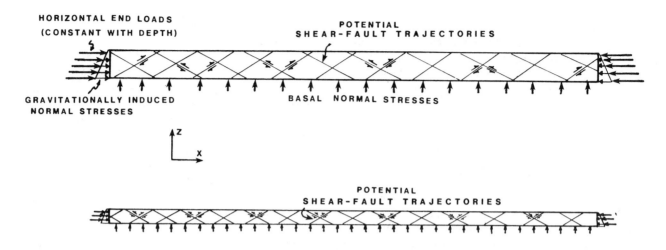

FIGURE 3: Analytical elastic boundary-value solution shows boundary stresses and potential shear-fault trajectories. In these models no stability boundaries are drawn because the stress magnitudes are constant along the length of the model, regardless of scale. The mathematical solution for this model is given in the Appendix.

boundary-value model first introduced to the geologic literature by Hafner (1951). Hafner used the model to explain fault orientations which diverge from those dynamically justified by Anderson (1942). By the use of Hafner's method one can analyze the stress-system present in a continuous, elastic, isotropic two-dimensional model where all strains are in the plane of the model.

It has been argued above that a large region of continental basement may be considered as a continuum. It has also been shown that the bulk of all basement rock in the continental crust (between faults) is relatively time-independent (elastic or near-elastic). Given these criteria, then the foreland can be accurately modeled as an elastic-continuum. Furthermore, if there is little or no strain in the strike direction of the structures, one can justifiably use Hafner's method to model the Rocky Mountain Foreland (e.g. Couples, 1978; Stearns, 1978).

Stress-Transmission in the Foreland

With the advent of the concepts of plate tectonics and subduction/collision-related orogeny, it has been suggested that the deformation in the foreland of the Rocky Mountains was related to a Laramide-aged east-dipping subduction zone (Sales, 1968, Grose, 1972). Justification of a mechanism for transmitting subduction-related stresses long distances into the continent (up to 1200 kilometers -750 miles) proved to be a conceptual barrier to acceptance of this hypothesis. Intuitively, large differential stresses should have been dissipated far to the west of the Laramide Rocky Mountains. An elastic boundary-value solution in

Figure 2 (from Hafner, 1951) can be used in support of this argument as stresses are shown to diminish rapidly with distance from the applied horizontal load.

A much different result can be obtained by changing the boundary conditions (by altering the stress function) in the model. The models in Figure 3 are boundary-value solutions using a simpler stress function (derived in Appendix). In this model there are no shear loads applied at the base as in Figure 2. The result is that all horizontal loads are transmitted unchanged in magnitude and orientation to the opposite end of the model. Any aspect ratio or model dimensions yield the same result (Figure 3). A difference in shear loads at the base is the only major difference between the two models (Figure 3 vs. Figure 2), but it is of critical importance.

To maintain its applicability, the model's boundary conditions should represent those which were imposed on the continental crust of the Rocky Mountain region during the Laramide. The amount of end-loading during the Laramide is an unknown. However, shear loads at the base of the crust, generated at geologic strain-rates are known to be vanishingly small (.1 to 10 MPa) (Carter, 1976 for summary). Consequently, the model in Figure 2 is an invalid representation of the boundary-conditions imposed on the Laramide Foreland. The alternative model in Figure 3, on the other hand, is consistent with the constraints on shear stresses at the base of the crust.

Dickenson and Snyder (1978) present a case for a shallow dipping subducted plate in the Laramide which eventually became horizontal and produced "a regime in which the subducted plate scrapes horizontally beneath the surficial plate." Dickenson and Snyder propose that this interaction, directly beneath the foreland, produced the Laramide deformation. Jordan and others (1983) describe active foreland-style deformation in Argentina (quite similar to that in the Rocky Mountains) where the subducted slab (as defined by the Benioff zone) dips 5 to 10 degrees east. This data suggests that a shallow dipping (but not horizontal) subducted-slab is sufficient to produce Andean Foreland deformation. The *shallow dip* of the subducted slab is required to transmit sufficient horizontal load to the overriding continent, however a *horizontal* slab is *not* necessary (or observed) because the horizontal loads applied at the continental margin can be transmitted easily inboard (as in Figure 3) when there is little or no shear loading at the base of the crust. The above is further justification for applying the model (Figure 3) to the Laramide Foreland of the Rocky Mountains.

According to the model (Figure 3), horizontal stresses

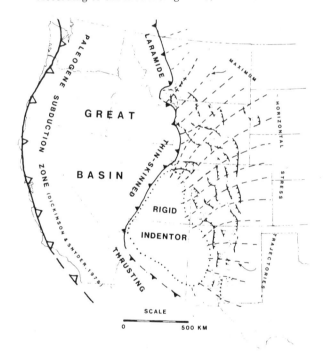

FIGURE 4: Map of the Western U.S. shows major Laramide basement uplifts and bounding faults as well as major tectonic elements of the region. The Colorado Plateau is identified as a rigid indentor which produced Laramide maximum horizontal stress trajectories shown as dashed lines. These maximum stress trajectories are determined by orientations of fault and fold trends, crestal extension faults on folds, and intrusive dike orientations. The position of the subduction zone shown may have been closer to the foreland during Laramide on the basis of restored post-Laramide extension in the Great Basin.

could ideally be transmitted unchanged in magnitude any distance into the continent. The objection could be raised that if the model is accurate, one should observe the same intensity of foreland deformation far to the east of its present location, the Rocky Mountains. In order to overcome this objection, one must consider the Rocky Mountain Foreland in the third dimension. Foreland deformation is anomalous along the length of the Cordillera, and therefore not strictly a two-dimensional mechanical problem as assumed in the model. The small amount of strain along strike in the natural situation may allow the differential stress to be dissipated very gradually to the east. As a result , the maximum horizontal-stress trajectories diverge to the east (Figure 4). In this scenario (Figure 4), the Colorado Plateau is viewed as a "rigid-indentor" or "stress-concentrator" producing a divergent pattern of maximum horizontal stress trajectories (Figure 4) (Kelley, 1955, Sales, 1968; Woodward, 1976; and Hamilton, 1977 suggest similar scenarios). This pattern is modified to the north of the Colorado Plateau by stresses imposed as a result of the impinging fold and thrust belt. Orientations of the maximum horizontal-stress trajectories in Figure 4 were constrained by orientations of Laramide age fault planes, structural trends, cross-crestal extension faults, and intrusive dike orientations, all of which were taken from published state and quadrangle-geologic maps. Despite the apparent three-dimensional strain achieved in nature, the model (Figure 3) is still considered a valid representation of the Laramide Foreland crust. The small strains parallel to strike cause only a very gradual dissipation of stress magnitudes along the length of the model, and do not affect the stress orientations in the plane of the model.

Predictions about Basement Structures

The maximum principal stress orientations in the model are horizontal, suggesting that all faults at the onset of their formation (when the basement is pushed beyond its elastic-limit) will be oriented at 30 to 35 degrees to the horizontal[1].

All rocks can be expected to exhibit an increase of strength and pre-failure strain (ductility) with depth (Handin and Hager, 1957). Therefore, in the case of uniform horizontal shortening (implicit in the model), rocks nearer the surface should tend to fail first. Failure (in the form of faulting) should progress downward with increased crustal shortening (Figures 5, 6). Figure 5 shows curves which plot the maximum ductility (pre-failure strain) of various rock types against depth of burial for experimentally-deformed

[1]Assuming a Coulomb failure criterion and an angle of internal friction of 30 to 20 degrees.

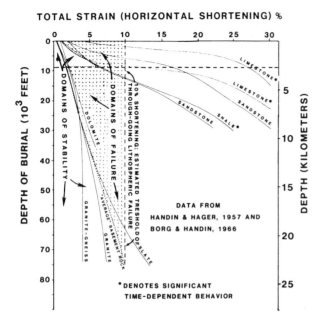

TOTAL STRAIN (HORIZONTAL SHORTENING) %

DATA FROM
HANDIN & HAGER, 1957 AND
BORG & HANDIN, 1966

*DENOTES SIGNIFICANT
TIME-DEPENDENT BEHAVIOR

FIGURE 5: Curves show the relationship between pre-failure strain (ductility) and depth of burial (temperature and pressure effects) for common foreland rock types. The curve connects the points at which strain changes from recoverable (elastic) to non-recoverable (faulting), from left to right of the curve respectively. Basement rock types show much less ductility. Because of this characteristic basement rock types would be the first to fail during uniform horizontal shortening. The domains of failure represent the situation given that only sedimentary rocks were present above approximately 3 kilometers (2 miles) depth (dashed line) at the onset of Laramide deformation.

rocks (Handin and Hager, 1957, Borg and Handin, 1966). By observing the curves relating to each rock type,[2] the pre-failure strain (elastic) and post-failure strain (faulting) can be determined for a given total-strain and depth. By following a curve, one can estimate at what strain-level faulting occurs for a given depth and at what depth faults can be expected to die out for a given total-strain. An important feature of this plot is that it predicts that the first feature to form due to uniform horizontal shortening (with the exception of minor sedimentary rock failure at very shallow depths) should be basement faults. Furthermore, these faults should initiate at the upper basement-surface and diminish in displacement with depth.

Because of their early formation, the basement faults

[2]Prior to the Laramide, only sedimentary rocks were present above approximately 3 kilometers (10,000 feet) depth in the Rocky Mountain Foreland. With the exception of the Uinta Mountains area and portions of southwest Montana, only crystalline basement rocks were present below approximately 3 kilometers (Stearns, 1972).

are expected to determine the location and orientation of the folds that occur in the overlying section (Figure 6). As deformation progresses in the region some of these faults die out with varying degrees of displacement yielding folds of varying amplitudes. This is the case in the Pryor Mountains (Stearns and Stearns, 1978), on the Colorado Plateau (Davis, 1978, Reches 1978, Huntoon, 1981), in the Black Hills (Lisenbee, 1979), and in the northern Front Range (Mathews and Work, 1979).

Through-going Crustal Failure

The absolute value of the long term strength of the lithosphere is a controversial topic (Hanks and Raleigh, 1980 for review). *Relative* values of crustal component strengths are more important to this discussion. Figure 7

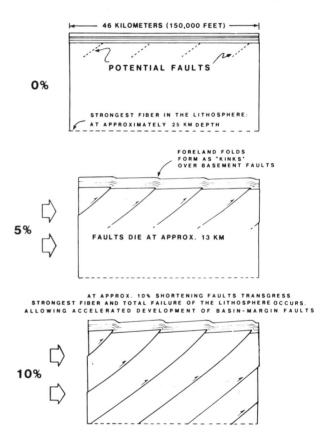

FIGURE 6: Cross sections of segments of continental crust represent predicted fault relationships at various magnitudes of horizontal shortening. At any given strain, basement faults can be seen to lose displacement with depth. This is due to the fact that proportionally more strain is taken up by elastic deformation at greater depth. Ten percent strain is proposed as that which is necessary to cause faults to transgress the strongest fiber of the lithosphere at 25 kilometers (15 miles) depth (bottom cross section).

shows the relative strengths of various crystalline basement rocks as determined experimentally by constant strain-rate and constant load (creep) tests (Borg and Handin, 1966, Geotze and Evans, 1979). The strengths (tested at various temperatures and pressures) are plotted as a function of depth (Figure 7). This plot shows that the strength of the crust increases with depth to approximately 25 kilometers (15 miles). The upper-to lower-crust transition occurs at approximately 25 kilometers (15 miles) depth in the western United States (Prodehl, 1979) and the curves in Figure 7 reflect this compositional change. It has been shown that uniform horizontal compression in the foreland basement will produce faults that initiate shallow and propagate deeper. Accordingly, the crust should maintain its strength-integrity until faults (domains of failure in Figure 5) extend to a depth of 25 kilometers (15 miles). After selected faults cut the strongest fiber (approximately 25 kilometer depth) they should continue to develop more easily, penetrating all the way through the lithosphere. These selected faults are manifest at the surface as large-offset, basin-bounding fault systems. Examples of these include the Wind River/Emigrant Trail, the Beartooth/Oregon Basin, the Owl Creek/Casper Arch, and the Front Range/Golden fault systems.

Predictions of the model and data presented above can be summarized chronologically as follows:

1) A small amount of horizontal elastic shortening occurs (2 percent in Figure 4) before faults initiate at the basement/sedimentary sequence interface. These faults determine the position and orientation of basement-cored foreland fold features.

2) Fault displacements and associated fold amplitudes continue to increase with additional horizontal shortening. During this phase some faults cease movement (displacement may be taken up by more widely spaced faults).

3) At some finite shortening (10 percent in Figure 5), some faults transgress the strongest fiber in the lithosphere (25 km depth in Figure 7) effectively rupturing the crust. These faults develop into large-offset basin-bounding fault systems and serve to generate additional detached fold and fault features in the sediments of the adjacent basin.

The end result of this history is a configuration of small (usually less than 1500 meters - 5,000 feet amplitude) basement cored folds, contained within basins bounded by large fault-systems (up to 13,000 meters - 43,000 feet vertical relief). Subsidiary, detached structures are in turn

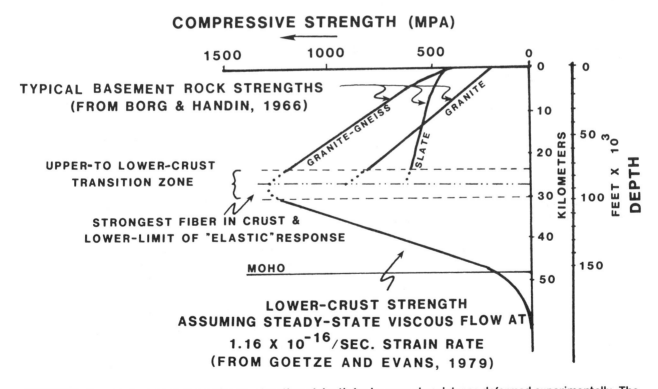

FIGURE 7: Curves represent strengths as a function of depth for basement rock types deformed experimentally. The upper crustal rock types were deformed in constant strain-rate tests (Borg and Handin, 1966) and the lower crustal rocks were deformed in constant stress tests (creep tests) (Goetze and Evans, 1979).

associated with the faulted margins. Observation reveals this to be the general configuration in the Rocky Mountain Foreland (e.g. Hoppin and Jennings, 1971, Samuelson, 1974).

Colorado Plateau

One area of notable exception to this description is the Colorado Plateau. In this region only small structures (1500 meters - 5,000 feet or less amplitude) dominate the deformation. In Figure 4 the Colorado Plateau is represented as a rigid indentor. In the context of the model this would suggest that the Colorado Plateau is more resistant to deformation than the adjacent foreland, and as a result it should show predominantly the "early" stage structural features described by the model (basement cored folds of approximately 1500 meters - 5,000 feet amplitudes or less) as appears to be the case (Kelley, 1955, Davis, 1978).

CONCLUSIONS

The model presented in Figure 3 can be used to help explain regional features of the Rocky Mountain Foreland which developed during the Laramide. The model is validated by 1) showing the continental basement to be essentially an elastic-continuum, 2) assuming very small or no shear loads at the base of the crust, 3) assuming faulting as the primary deformation mechanism in the basement, and 4) justifying some mechanism for horizontal loading at the continental margin (i.e. a shallow dipping subducted slab).

The scenario of structure development inferred from the experimental rock-mechanics data (Figures 5, 7), the assumption of uniform horizontal shortening, and the pre-failure stress-system provided by the model (Figure 3), matches the observed regional features of the foreland well. Considering the accuracy of the fit of the model to the foreland, horizontal compression may be considered as a possible first-cause of Laramide Rocky Mountain Foreland deformation.

ACKNOWLEDGMENTS

I would like to thank Chevron U.S.A. for allowing publication of this manuscript and providing graphics and clerical services towards its completion.

I gratefully acknowledge my numerous colleagues both at Chevron and elsewhere that have helped shape my ideas and ellucidate my naivetes. Thanks are due to Chuck Kluth and Bruce Smith who critically edited the manuscript.

APPENDIX

Solution of an elastic boundary-value problem requires that one establish a governing stress function (known as the Airy stress function) from which the values of all internal and boundary stress magnitudes can be derived.

As shown by Hafner (1951), if the strains are contained in only two dimensions, (x and z) the Airy stress functions $(\phi_{(x,z)})$ must satisfy the following compatibility equation:

$$\frac{\partial^4 \phi}{\partial x^4} + \frac{2\partial^4 \phi}{\partial x^2 \partial z^2} + \frac{\partial^4 \phi}{\partial z^4} = 0 \qquad (1)$$

If equation (1) is satisfied then the stress components can be shown to be:

$$\sigma xx = -\frac{\partial^2 \phi}{\partial z^2} \qquad (2)$$

$$\sigma zz = -\frac{\partial^2 \phi}{\partial x^2} \qquad (3)$$

$$\tau xz = \frac{\partial^2 \phi}{\partial x \partial z} \qquad (4)$$

where σ_{xx} is the normal stress on the x plane, σ_{zz} is the normal stress on the z plane, and t_{xz} is the shear stress on the plane normal to x in the z direction. The x and z axes with positive directions are shown on Figures 2 and 3, and the directions of stresses shown on the box at the bottom of Figure 2 are taken as negative.

For the stress system in Figure 2, Hafner (1951) chose the Airy stress function

$$\Phi = Cf_1(z) x + Ax + Bf_2(z) + D \qquad (5)$$

where A, B, C and D are arbitrary constants and $f_1(z)$ and $f_2(z)$ are functions whose fourth derivatives are zero.

From this stress function the following stress values are derived (by application of equations 2, 3, and 4):

$$\sigma xx = -Cf_1''(z)x - Bf_2''(z) \qquad (6)$$
$$\sigma zz = 0 \qquad (7)$$
$$t_{xz} = Cf_1'(z) \qquad (8)$$

Because no shear stresses are exerted on horizontal planes exposed at the surface of the earth, t_{xz} must be zero when z=0, therefore $f_1'(z)$ can be further limited to zero (equation 8 must equal 0 when z=0).

From these conditions Hafner (1951) sets up three subgroups of possible stress equations.

Two of the subgroups are used here:

for $f_2''(z) = z + D$
$$\sigma xx = -Bz + D$$
$$\sigma zz = 0$$
$$t_{xz} = 0 \qquad (9a)$$
and

$$f_2''(z) = z$$
$$\sigma xx = - Cx$$
$$\sigma zz = 0 \qquad (9b)$$
$$txz = Cz$$

Since any valid stress system can be added to (superposed on) any other valid stress system to produce an additional valid stress system, Hafner (1951) adds the stress system represented by equations 9b to the stress system due to the weight of the overlying rock (σ_{xx} = pgz, σ_{zz} = pgz). The sum yields the stress system shown graphically in Figure 2:

$$\sigma xx = - Cx + pgz$$
$$\sigma zz = pgz \qquad (10)$$
$$txz = Cz$$

Conversely, by adding the stress system represented in 9a to the stress system due to the overburden, one can form the stress system used in Figure 3. (Choosing the constant B as zero):

$$\sigma xx - Bz + D$$
$$\sigma zz = 0 \qquad (11)$$
$$txz = 0$$

Equations 11 show that D determines the constant value of the horizontal normal stress in the model. If B was chosen to be a non-zero value then the horizontal stress could be allowed to increase or decrease with depth (as a linear function). This option was not used in Figure 3.

REFERENCES CITED

Alpha, A.G. and Fanshawe, T.R., 1954, Tectonics of the northern Big Horn Basin area, Billings Geological Society: Fifth annual field conference guidebook, p. 72-79.

Anderson E.M., 1942, The dynamics of faulting: London, Oliver and Boyd, 183 p.

Berg, R.R., 1962, Mountain flank thrusting in the Rocky Mountain Foreland, Wyoming and Colorado: American Association of Petroleum Geologists Bulletin, v. 46, p. 2019-2032.

Berg, R.R., 1981, Review of thrusting in the Wyoming foreland: Contributions to Geology, University of Wyoming, v. 19, p. 93-104.

Blackstone, D.L. Jr., 1963, Development of geologic structure in the Central Rocky Mountains: American Association of Petroleum Geologists Memoir 2, p. 160-179.

Blackstone, D.L. Jr., 1981, Compression as an agent in deformation of the east-central flank of the Big Horn Mountains, Sheridan and Johnson Counties, Wyoming: Contributions to Geology, University of Wyoming, v. 19, p. 105-122.

Borg, I. and Handin, J., 1966, Experimental deformation of crystalline rocks: Tectonophysics, v. 3, p. 249-268.

Carter, N.L., 1976, Steady-state flow of rocks: Reviews of Geophysics and Space Physics, v. 14, p. 301-360.

Couples, G., 1978, Comments on applications of boundary-value analyses of structures of the Rocky Mountains foreland in Matthews V., III., Ed., Laramide folding associated with block faulting in the Western United States: Geological Society of America Memoir 151, p. 337-354.

Dewey, J.F. and Sengor, A.M.C., 1979, Aegean and surrounding regions: complex multiplate and continuum tectonics in a convergent zone: Geological Society of America Bulletin, v. 90, p. 84-92.

Dickenson, W.R., and Snyder, W.S., 1978, Plate tectonics of the Laramide orogeny: Geological Society of America Memoir 151, p. 355-366.

Donath, F.A. and Parker, R.B., 1964, Folds and Folding: Geological Society of America Bulletin, v. 75, p. 45-62.

Gallagher, J.J., 1981, Tectonics of China: Continental scale cataclastic flow: American Geophysical Union Geophysical Monograph 24, p. 259-274.

Goetze, C. and Evans, B., 1979, Stress and temperature in the bending lithosphere as constrained by experimental rock mechanics: Geophysical Journal of the Royal Astronomical Society, v. 59, p. 463-478.

Greiner, G., 1975, In-situ stress measurements in southwest Germany: Tectonophysics, v. 29, p. 265-274.

Grose, L.T., 1972, Tectonics, in Mallory, W.W., Ed., Geologic atlas of the Rocky Mountain region: Denver, Colorado, Rocky Mountain Association of Geologists, p. 35-44.

Hafner, W., 1951, Stress distributions and faulting: Geological Society of America Bulletin, v. 62, p. 373-398.

Hamilton, W., 1977, Mesozoic tectonics of the western United States, in Howell, D.G. and McDougall, K.A., Eds., Mesozoic Paleogeography of the Western United States: Los Angeles, California, Society of Economic Paleontologists and Mineralogists, p. 33-70.

Hanks, T.C. and Raleigh, C.B., 1980, The conference on magnitude of deviatoric stresses in the earth's crust and uppermost mantle: Journal of Geophysical Research, v. 85, p. 6083-6085.

Handin, J., and Hager, R.V., Jr., 1957, Experimental deformation of sedimentary rocks under confining: test at room temperature on dry samples: American Association of Petroleum Geologists Bulletin, v. 41, p. 1-50.

Hansen, F.D., 1982, Semibrittle creep of selected crustal rocks at 1000 MPa (PhD. dissertation): College Station, Texas A & M University, 224 p.

Heard, H.C., 1962, The effect of large changes in strain rate in the experimental deformation of rocks (PhD. dissertation): Los Angeles, University of California, 202 p.

Hoppin, R.A., 1961, Precambrian rocks and their relationship to Laramide structure along the east flank of the Big Horn mountains near Buffalo, Wyoming: Geological Society of America Bulletin, v. 49, p. 993-1003.

Hoppin, R.A., 1970, Structural development of the Five Springs Creek area, Bighorn Mountains, Wyoming: Geological Society of America Bulletin, v. 81, p. 2403-2416.

Hoppin, R.A., and Jennings, T.V., 1971, Cenozoic tectonic elements, Bighorn Mountains region, Wyoming and Montana: Wyoming Geologic Assoc., 23rd Annual Field Conference Guidebook, p. 39-47.

Howard, J.H., 1966, Structural development of the Williams Range thrust, Colorado: Geological Society of America Bulletin, v. 77, p. 1247-1264.

Huntoon, P.W., 1981, Grand Canyon monoclines: vertical uplift or horizontal compression?: Contributions to Geology, University of Wyoming, v. 19, p. 127-134.

Jordan, T.E., Isacks, B.L., Allmendinger, R.W., Brewer, J.A., Ramos, V.A. and Ando, C.J., 1983, Andean tectonics related to geometry of subducted Nasca plate: Geological Society of America Bulletin, v. 94, p. 341-361.

Kelley, V.C., 1955, Monoclines of the Colorado Plateau: Geological Society of America Bulletin, v. 66, p. 789-803.

Krantz, R.L., 1980, The effects of confining pressure and stress difference on static fatigue of granite: Journal of Geophysical Research, v. 85, p. 1854-1866.

Krantz, R.L., Harris, W.J., and Carter, N.L., 1982, Static fatigue of granite at 200°C: Geophysical Research Letters, v. 9, p. 1-4.

Mathews, V., III, and Work, D.F., 1978, Laramide folding associated with basement block faulting along the northeastern flank of the Front Range, Colorado, in Mathews, V. III, ed., Laramide folding associated with basement block faulting in the Western United States: Geological Society of America Memoir 151, p. 101-124.

Mitra, G., and Frost, B.R., 1981, Mechanisms of deformation within Laramide and Precambrian deformation zones in basement rocks of the Wind River Mountains: Contributions to Geology, University of Wyoming, v. 19, p. 161-173.

Molnar, P. and Tapponier, P., 1975, Cenozoic tectonics of Asia: effects of continental collision: Science, v. 189, p. 419-426.

Prodehl, C., 1979, Crustal structure of the Western United States: United States Geological Survey Professional Paper 1034.

Prucha, J.J., Graham, J.A., and Nickelsen, R.P., 1965, Basement-controlled deformation in the Wyoming province of the Rocky Mountains foreland: American Association of Petroleum Geologists Bulletin, v. 49, p. 966-992.

Reches, Z., 1978, Development of monoclines: Part I. Structure of the Palisades Creek branch of the East Kaibab monocline, Grand Canyon, Arizona: in Mathews, V., III, ed. Laramide folding associated with basement block faulting: Geological Society of America Memoir 151, p. 235-271.

Sales, J.K., 1968, Crustal mechanics of Cordilleran foreland deformation: A regional and scale model approach: American Association of Petroleum Geologists Bulletin, v. 52, p. 2016-2044.

Samuelson, A.C., 1974, Introduction to the geology of the Big Horn Basin and adjacent areas, Wyoming and Montana in Barry Voight, ed., Rock Mechanics, the American Northwest, Special Publication, College of Earthland Mineral Sciences, Pennsylvania State University.

Sbar, M.L. and Sykes, L.R. 1973, Contemporary compressive stress and seismicity in Eastern North America: an example of intraplate tectonics: Geological Society of America Bulletin, v. 84, p. 1861-1882.

Stearns, D.W., 1969, Fracture as a mechanism of flow in naturally deformed layered rocks in Baer, A.J. and Norris, D.K., eds., Proceedings, conference on research in tectonics: Canada Geological Survey Paper 68-52, p. 79-95.

Stearns, D.W., 1971, Mechanisms of drape folding in the Wyoming province: Wyoming Geological Association, 23rd Annual Field Conference Guidebook, p. 125-143.

Stearns, D.W., 1978, Faulting and forced folding in the Rocky Mountains foreland in Mathews, V., III, ed., Laramide folding associated with block faulting in the western United States: Geological Society of America Memoir 151, p. 1-37.

Stearns, M.T., and Stearns, D.W., 1978, Geometric analysis of multiple drape folds along the northwest Big Horn Mountains front, Wyoming in Mathews, V., III, ed., Laramide folding associated with basement block faulting in the western United States: Geological Society of America Memoir 151, p. 139-156.

Stephenson, B.R., Murray, K.J., 1970, Application of the strain rosette relief method to measure principal stresses throughout a mine: International Journal of Rock Mechanics and Mineral Science, v. 7, p. 1-22.

Thom, W.T., Jr., 1952, Structural features of the Big Horn Basin rim: Wyoming Geological Association 4th Annual Guidebook, p. 15-18.

Woodward, L.A., 1976, Laramide deformation of the Rocky Mountain foreland: geometry and mechanics in Woodward, L.A., ed., Tectonics and mineral resources of southwestern North America: New Mexico Geological Society Special Publication 6, p. 11-18.

Zoback, M.L. and Zoback, M., 1980, State of stress in the conterminous United States: Journal of Geophysical Research, v. 85, p. 6113-6156.

Reprinted by permission of the Rocky Mountain Association of Geologists from R. R. Gries and R. C. Dyer, eds., *Seismic Exploration of the Rocky Mountain Region*, 1985, p. 165-174.

GEOLOGIC INTERPRETATION OF SEISMIC PROFILES, BIG HORN BASIN, WYOMING, PART I: EAST FLANK

Donald S. Stone
Sherwood Exploration

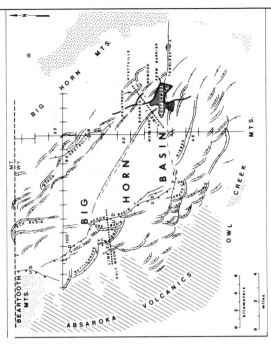

Figure 1: Tectonic sketch map of the Big Horn Basin, showing the location of Line 1, the major thrust-fold structures, and the Cottonwood Creek producing area with its up-dip "permeability barrier".

INTRODUCTION

The seismic profile presented and discussed here (Line 1, Fig. 1) is a dip line on the east flank of the Big Horn Basin, in Washakie and Big Horn Counties, Wyoming. The profile trends northeast from the relatively undeformed deeper basin area, to the west flank of the unproductive Hyattville Anticline, traversing a structural relief of more than 11,000 ft (3353 m). Clearly illustrated on this line (Figs. 2 through 5) is the fundamental, two-dimensional, seismic expression of typical Paleozoic structural oil fields in the Wyoming foreland province. These oil fields are associated with basement-supported "thrust-fold" structures (Stone, 1983d), including the Marshall Field which is part of the Cottonwood Creek group of stratigraphic traps in the Upper Phosphoria Formation. True-scale structural cross sections (Figs. 6 and 7), constructed from a deep-migration of the profile and tied to the well control, illustrate the undistorted two-dimensional structure along the line of the profile.

REGIONAL GEOLOGY

Tectonic Setting

The Big Horn Basin of northwestern Wyoming and southwestern Montana lies within the foreland province of the central Rocky Mountains and contributes more than any other foreland basin to annual oil production totals. It is a topographic and structural basin surrounded by Precambrian-cored mountain ranges on the northwest (Beartooth Mountains), the east (Big Horn Mountains), and the south (Owl Creek Mountains), and by the volcanic Absaroka Mountains on the west (Fig. 1). On the north, the basin axis emerges from beneath the eastern thrust front of the Beartooth Mountains and extends for at least another 50 mi (80 km) northward across the Nye-Bowler Lineament into southern Montana where it is expressed as the Reed Point syncline (Stone, 1983b, Fig. 1).

The Big Horn Basin contains an abundance of significant Paleozoic oil fields distributed mostly around the basin rim. With few exceptions, these fields are structurally controlled. Oil traps are formed in west to northwest-trending, asymmetric, anticlinal thrust-fold closures at the deeper Paleozoic level (Stone, 1983a, c, d), and the oil accumulations generally encompass several Paleozoic formations in a common pool with vertical fluid communication through fractures and faults (Stone, 1967).

The arcuate map patterns produced by these thrust-fold structures (Fig. 1) are primarily an expression of tear fault terminations and suggest the action of a regional left-couple. During the Laramide Orogeny, renewed movement on Precambrian shear zones (e.g., on the Tensleep fault; Allison, 1984) probably included a left component of slip where the older fault planes trend nearly concordant with the theoretical east-west left-wrench direction (Stone, 1969). Northeast-southwest regional Laramide compression was the ultimate cause of thrust-fold development.

Stratigraphy

A stratigraphic column for the eastern Big Horn Basin is shown in Figure 8, along with a sonic log and synthetic seismogram from a well in the Worland Field Area. A larger-scale sample from the seismic profile at the well location is also shown. Lithologic descriptions and formation thicknesses of the Pennsylvanian Tensleep Formation through Eocene Willwood Formation from surface mapping in the Worland-Hyattville area are presented in Rogers et al. (1948).

The Big Horn Basin was part of the stable Wyoming shelf east of the Cordilleran geosyncline during Paleozoic time and most of Mesozoic time, although the basin shape probably was incipient as early as Permian time (Peterson, 1984).

Overlying the Precambrian basement complex (a poor reflector), the competent Paleozoic section is dominated by the massive, cliff-forming carbonates of Ordovician (Bighorn Dolomite), Devonian (Jefferson Formation) and Mississippian (Madison Limestone) age. However, at the bottom of Paleozoic section the Cambrian contains incompetent green shales with thin interbedded limestones (Gallatin Limestone and Gros Ventre Formation). This Cambrian shale section sometimes serves as a thrust detachment zone as illustrated by the interpreted geometry of the Cottonwood thrust shown in Figures 5 and 7. Consistent reflecting horizons are associated with the top of Madison and with the Bighorn Dolomite-Gallatin Limestone contact (Fig. 8).

At the top of the Paleozoic section the massive, cross-bedded, Pennsylvanian Tensleep Sandstone, which is the most important reservoir rock in the Big Horn Basin, is overlain unconformably by the marine, source/reservoir carbonates and black shales of the Permian Phosphoria Formation or its equivalent red bed facies, the Goose Egg Formation. Thickness changes in the Phosphoria and Tensleep formations are more or less compensatory because sandstone buildups in the Tensleep, usually described as dunes or "buried hills", were preserved under basal Permian sediments (Lawson and Smith, 1966; Curry, 1967; Curry, 1984; Moore, 1984).

Regional stratigraphic studies indicate that the facies change in the Phosphoria-Goose Egg formations, which provides the essential eastern, up-dip permeability barrier for the Cottonwood Creek group of Phosphoria Formation oil fields, trends generally northward between Worland and Bonanza fields (Fig. 1). The facies change should therefore cross Line 1 near Marshall (Figs. 2 and 3). However, the Phosphoria Formation is seldom associated with a reflection of reliable or consistent amplitude or character, and the expected Phosphoria Formation facies change cannot be recognized on the profile.

All of the Phosphoria-Tensleep formation reservoirs along the line of profile are capped by Triassic Chugwater Formation red shales and mudstones. These rocks are incompetent in a structural sense and often yield to "contraction" thrusting (Price, 1967) in the crestal area of the thrust-folds and bed attenuation on the steeper limb. Figure 9 is a true-scale log section along Line 1 at Paintrock Anticline illustrating the effect of these contraction thrusts on the attitude of the crestal surface. Identification of small thrust duplications in the shales of the Chugwater Formation is facilitated by the regional continuity of correlation markers on wireline logs through the Chugwater Formation.

A consistent reflecting horizon is produced by acoustic impedence contrast between the Curtis Sandstone and Alcova Limestone at the top of the Chugwater Formation. Other reflectors are correlated with the Gypsum Springs, Sundance and Morrison formations of the overlying Jurassic section. A strong

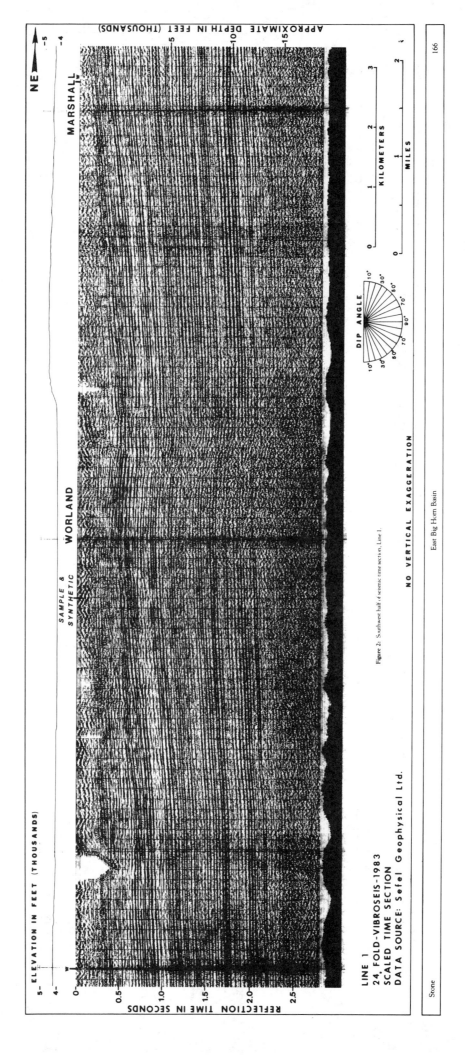

Figure 2: Southwest half of seismic time section, Line 1.

NO VERTICAL EXAGGERATION

East Big Horn Basin

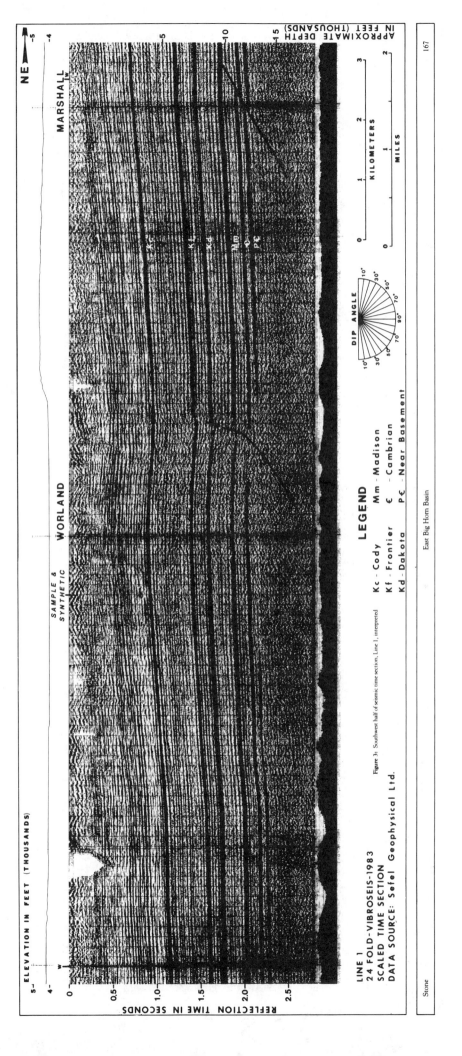

Figure 3: Southwest half of seismic time section, Line 1, interpreted

LEGEND

Kc - Cody   Mm - Madison

Kf - Frontier   € - Cambrian

Kd - Dakota   P€ - Near Basement

LINE 1
24 FOLD-VIBROSEIS-1983
SCALED TIME SECTION
DATA SOURCE: Sefel Geophysical Ltd.

East Big Horn Basin

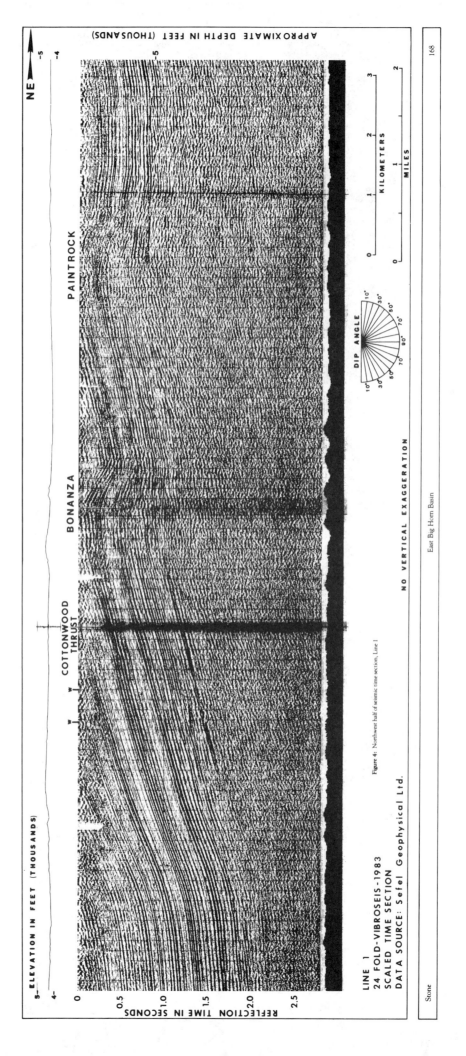

Figure 4: Northwest half of seismic time section, Line 1

LINE 1
24 FOLD-VIBROSEIS-1983
SCALED TIME SECTION
DATA SOURCE: Sefel Geophysical Ltd.

NO VERTICAL EXAGGERATION

East Big Horn Basin

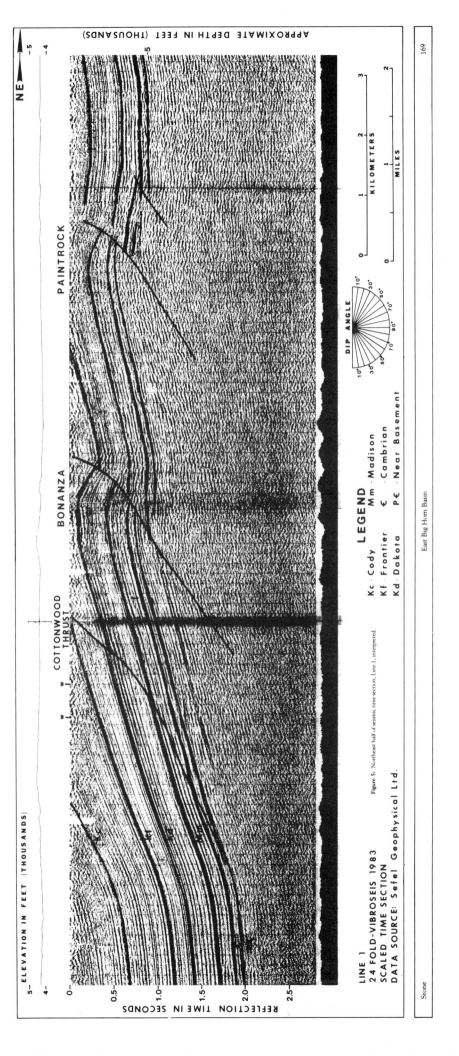

LINE 1
24 FOLD-VIBROSEIS 1983
SCALED TIME SECTION
DATA SOURCE: Sefel Geophysical Ltd.

LEGEND

Kc - Cody   Mm - Madison
Kf - Frontier   € - Cambrian
Kd - Dakota   P€ - Near Basement

Figure 5: Northeast half of seismic time section, Line 1, interpreted.

ELEVATION IN FEET (THOUSANDS)

REFLECTION TIME IN SECONDS

APPROXIMATE DEPTH IN FEET (THOUSANDS)

NE

PAINTROCK

BONANZA

COTTONWOOD THRUST

DIP ANGLE

KILOMETERS

MILES

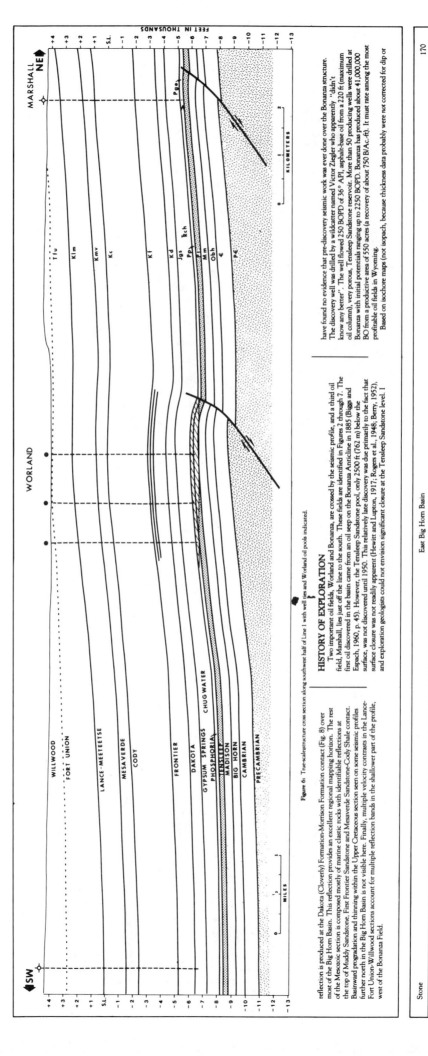

Figure 6: True-scale structure cross section along southwest half of Line 1 with well ties and Worland oil pools indicated.

reflection is produced at the Dakota (Cloverly) Formation-Morrison Formation contact (Fig. 8) over most of the Big Horn Basin. This reflection provides an excellent regional mapping horizon. The rest of the Mesozoic section is composed mostly of marine clastic rocks with identifiable reflections at the top of Muddy Sandstone, First Frontier Sandstone and Mesaverde Sandstone-Cody Shale contact. Basinward progradation and thinning within the Upper Cretaceous section seen on some seismic profiles further north in the Big Horn Basin is not visible here. Finally, multiple velocity contrasts in the Lance-Fort Union-Willwood sections account for multiple reflection bands in the shallower part of the profile, west of the Bonanza Field.

HISTORY OF EXPLORATION

Two important oil fields, Worland and Bonanza, are crossed by the seismic profile, and a third oil field, Marshall, lies just off the line to the south. These fields are identified in Figures 2 through 7. The first oil discovered in the basin came from an oil seep on the Bonanza Anticline in 1885 (Biggs and Espach, 1960, p. 45). However, the Tensleep Sandstone pool, only 2500 ft (762 m) below the surface, was not discovered until 1950. This relatively late discovery was due primarily to the fact that surface closure was not readily apparent (Hewitt and Lupton, 1917; Rogers et al., 1948; Berry, 1952), and exploration geologists could not envision significant closure at the Tensleep Sandstone level. I

have found no evidence that pre-discovery seismic work was ever done over the Bonanza structure. The discovery well was drilled by a wildcatter named Victor Ziegler who apparently "didn't know any better". The well flowed 250 BOPD of 36° API, asphalt-base oil from a 220 ft (maximum oil column), very porous, Tensleep Sandstone reservoir. More than 50 producing wells were drilled at Bonanza with initial potentials ranging up to 2250 BOPD. Bonanza has produced about 41,000,000 BO from a productive area of 550 acres (a recovery of about 750 B/Ac.-ft). It must rate among the most profitable oil fields in Wyoming.

Based on isochore maps (not isopach, because thickness data probably were not corrected for dip or

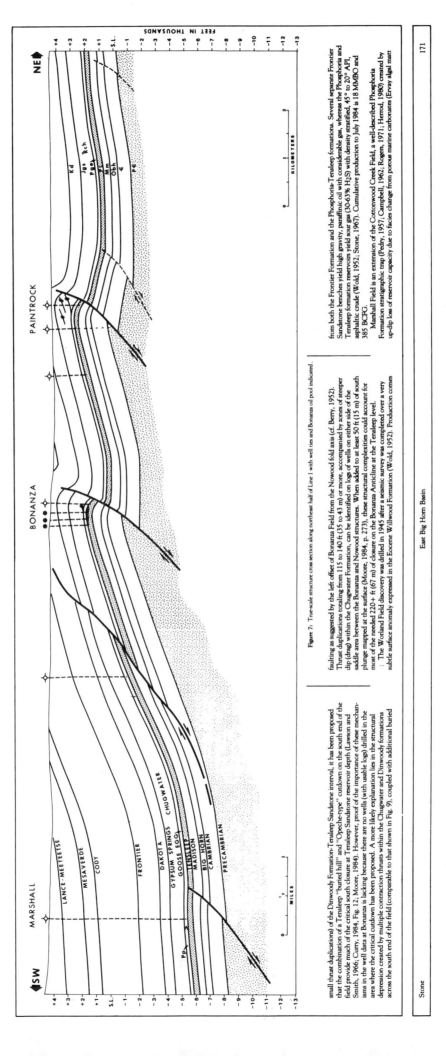

Figure 7: True-scale structure cross section along northeast half of Line 1 with well ties and Bonanza oil pool indicated.

small thrust duplications) of the Dinwoody Formation-Tensleep Sandstone interval, it has been proposed that the combination of a Tensleep "buried hill" and "Opeche-type" curdown on the south end of the field provide much of the critical south closure at Tensleep Sandstone reservoir depth (Lawson and Smith, 1966; Curry, 1984, Fig. 12; Moore, 1984). However, proof of the importance of these mechanisms in the well data at Bonanza is lacking because there are no wells (with usable logs) drilled in the area where the critical curdown lies in the structural depression created by multiple contraction thrusts within the Chugwater and Dinwoody formations across the south end of the field (comparable to that shown in Fig. 9, coupled with additional buried

faulting as suggested by the left offset of Bonanza Field from the Nowood fold axis (cf. Berry, 1952). Thrust duplications totaling from 115 to 140 ft (35 to 43 m) or more, accompanied by zones of steeper dip (drag) within the Chugwater Formation, can be identified on logs of wells on either side of the saddle area between the Bonanza and Nowood structures. When added to at least 50 ft (15 m) of south plunge mapped at the surface (Moore, 1984, p. 273), these structural complexities could account for most of the needed 220+ ft (67 m) of closure on the Bonanza Anticline at the Tensleep level.

The Worland Field discovery was drilled in 1945 after a seismic survey was completed over a very subtle surface anomaly expressed in the Eocene Willwood Formation (Wold, 1952). Production comes

from both the Frontier Formation and the Phosphoria-Tensleep formations. Several separate Frontier Sandstone benches yield high gravity, paraffinic oil with considerable gas, whereas the Phosphoria and Tensleep formation reservoirs yield sour gas (30-63% H$_2$S) with density stratified, 45° to 20° API, asphaltic crude (Wold, 1952; Stone, 1967). Cumulative production to July 1984 is 18 MMBO and 385 BCFG.

Marshall Field is an extension of the Cottonwood Creek Field, a well-described Phosphoria Formation stratigraphic trap (Pedry, 1957; Campbell, 1962; Rogers, 1971; Hernod, 1980) created by up-dip loss of reservoir capacity due to facies change from porous marine carbonates (Ervay algal matt

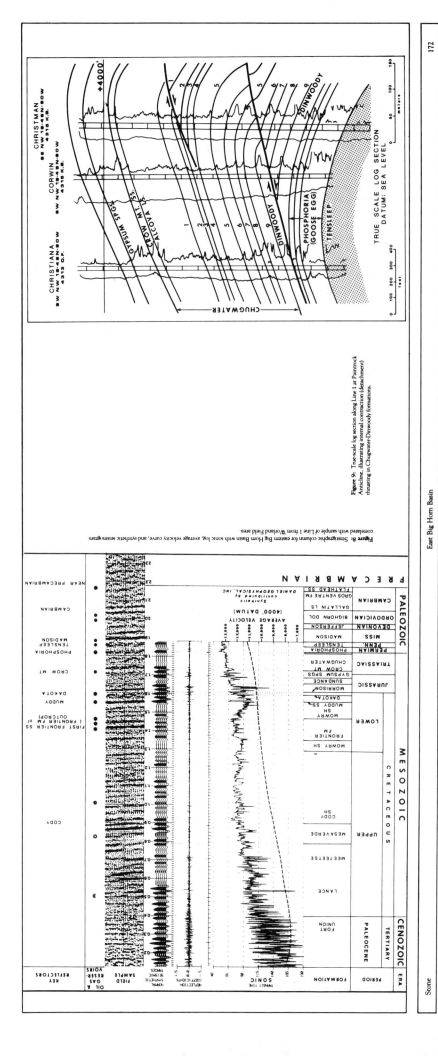

Figure 9: True-scale log section along Line 1 at Paintrock Anticline, illustrating internal contraction (detachment) thrusting in Chugwater-Dinwoody formations.

Figure 8: Stratigraphic column for eastern Big Horn Basin with sonic log, average velocity curve, and synthetic seismogram correlated with sample of Line 1 from Worland Field area.

SHORTENING: 3.5% SHORTENING: 7% SHORTENING: 11.5% SHORTENING: 16%

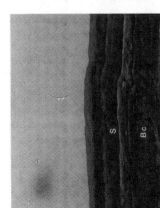

A. MARSHALL B. WORLAND C. PAINTROCK D. BONANZA

Figure 10: Experimental models, simulating the four thrust-fold structures traversed by the seismic profile. The lateral shortening (in percent) required to produce each replication is indicated at the top of each photograph. "Bc" is oil-base modeling clay and represents "brittle" Precambrian basement. "S" represents the sedimentary layers and is a mixture of salt, cornstarch, and water.

facies) to tight dolomite mudstones, red shales, and evaporites. Natural fracturing is also important to production (Herrod, 1980, Allison, 1984). The productive limits of the Marshall Field do not actually intersect the seismic profile but lie about one mi (1.6 km) to the south. The Marshall Field was discovered in 1967 and cumulative production to July 1984 is 506 MBO.

Many test wells have been drilled on the Paintrock Anticline, the next line of thrust-folding east of Bonanza, but the structure remains unproductive. Following several very early tests (1914-1947), Shell (1951 and 1955), Amerada (1963), Husky (1979) and several independents have tested the structure without success. With over 1000 ft (304.8 m) of surface closure (Rogers et al., 1948), the absence of entrapped oil at Paintrock is hard to explain. Hill et al (1961) proposed that flushing of oil from the Tensleep Sandstone reservoir has occurred by vertical migration through crestal faults due to the fact that Paintrock is located in a high pressure environment where potentiometric surfaces for all reservoirs tested lie above the ground surface. However, no evidence for crestal faulting that could allow

Tensleep oil to escape to the surface has been presented. A more likely explanation may be found in the probability that Paleozoic oil would have had to migrate into Paintrock from marine source rocks of Phosphoria Formation downdip and west of Bonanza since the non-source rock Goose Egg Formation red beds are present in these two structures. Bonanza may have captured most of this migrating oil as the first line of structure up-dip from Phosphoria Formation source rocks. The anomalously high gravity (gasoline content 36%) of the shallow Bonanza crude oil and the chemical similarity of the Paleozoic oils from Bonanza and Worland fields, together with the fact that the Worland structure appears to be filled to the spill point, suggests that the oil accumulation at Bonanza was derived through spillover and up-dip, lateral migration from the Worland Paleozoic pool.

INTERPRETATION

Interpretation of the basis seismic time section is shown in Figures 3 and 5. These line segments

tie and do not overlap at Marshall as do the geologic structure sections. After depth migration using a wavefront chart for the area (curved-ray path), the migrated horizons were tied to well control and thereby corrected for lateral velocity changes to serve as a basis for the true-scale structural cross sections presented as Figures 6 and 7.

The seismic data are of good quality and show clearly the two-dimensional geometry of major structural features along the profile. Only the interpreted Cottonwood thrust does not appear to involve Precambrian basement. All of the other structures are basement-supported thrust-folds. These thrust-fold structures are considered to have been forcibly produced by movement on Precambrian thrust wedges where thrusting began initially at the Precambrian surface and propagated into the overlying sediments. The causal thrusts appear to cut the Precambrian surface at an angle of about 30° and steepen upward due largely to rigid body rotation. Thrust displacement also attenuates upward so that the causal thrust plane is seldom seen at the surface. Most of the resultant thrust-induced anticlines are asymmetric with their steeper flanks toward the mountains (to the east on Line 1), a natural result of folding and stretching of beds over the basinward-dipping thrusts.

The basement thrust fault shown on the east side of Figure 3 at Marshall is causal to an incipient thrust-fold, with only 300 to 400 ft (91-122 m) of vertical separation on the fault at the Precambrian level. This thrust appears to die out before reaching the Madison horizon and little or no folding is present. The seismic data across Worland reveal a more advanced stage of thrust-fold development with obvious dip reversal into the causal thrust. Unfortunately the seismic data seem to attenuate near the fault zone at the deeper Paleozoic level, perhaps due partly to geometrical spreading (spherical dispersion), so that structural details around the fault zone are ambiguous. Bonanza and Paintrock anticlines show more advanced thrust-fold development, with estimated vertical separations on the causal thrusts at the Precambrian level of a minimum 1800 ft (549 m) and 1300 ft (396 m) respectively. Geometrical spreading also obscures the seismic response in the axial regions of these anticlines, and footwall "pull-up" is probably present.

EXPERIMENTAL MODELS

Simple, "pressure-box" model experiments with materials of high ductility contrast can simulate the development of foreland thrust-fold structures very closely. In Figure 10, photographs taken of progressive stages in sequential development of an experimental thrust-fold structure were chosen as closest to the replicated configuration of the four thrust-folds appearing on the seismic profile. In each photograph, the amount of horizontal shortening and the analogous structure on the seismic profile are indicated.

In the model experiment, the cross-hatched oil-base modeling clay represents the brittle, Precambrian basement, and a mixture of salt, cornstarch, and water represents the ductile sedimentary layers. A 30° thrust fault plane was precut in the clay block, and the fault plane and basal detachment surface of the hanging wall block were lubricated with petroleum jelly. The faulted "basement block" was then covered with (unfaulted) layers of the colored salt-cornstarch mixture and lateral pressure (compression) was applied to the hanging-wall end of the model. The percentage of lateral shortening was calculated for each half to one centimeter reduction in the original 43 cm length, and several photographs taken after each teach incremental movement. Even though the simulated Paleozoic layers in the model are probably too ductile relative to the clay basement (because they fold rather than fault in the incipient stages of basement thrusting, see Fig. 10A) the cross section views of the experimentally deformed, thrust-fold structures of the models seem quite comparable to the replicated structures on the seismic profile. Clearly no claim can be made for quantitative scaling in the models, but the results are instructive. (The lack of confining pressure seems particularly regrettable.) A more complete discussion of these experiments is the subject of a future paper.

CONCLUSIONS

The fundamental thrust-fold geometry of typical structural oil fields of the foreland province is confirmed by data from a high quality seismic profile across several important Paleozoic oil fields of the eastern Big Horn Basin. Simple model experiments with materials of high ductility contrast simulate the essential characteristics of the basement-supported structures along the profile (Fig. 10). The reproduction of these structures in the model required horizontal shortening of from 3.5% for Marshall Field to 16% for Bonanza Field.

A depth migration of the key horizons on the time section has been tied to the well control, and true-scale structural cross sections have been constructed showing the structural configurations and strong relief between the undeformed deeper basin and the rotated thrust-fold structures of the eastern basin flank.

Unfortunately, the facies change from porous Phosphoria marine carbonates to tight Goose Egg dolomite mudstones and red shales between the Worland and Bonanza field areas is not visible in the seismic data.

SEISMIC PARAMETERS

Client: Frontier Petroleum & StratSeis, Inc.   Contractor: Sefel Geophysical   Year: 1983

Channels: 96	Group Int.: 110 ft	Sample Rate: 2 ms
Source: Vibroseis	Spread: 5940-770-X-770-5940	Recording System: FTI-DFSV
Sweep Freq.: 10-90 Hz	Sweep Length: 24 sec	Scaling: 500 ms AGC
CDP Fold: 24	Final Filter: 14-85 Hz	

Special Processing: Refraction statics (dynamite), spectral balancing (No migration provided)

ACKNOWLEDGEMENTS

The seismic profile was provided by Frontier Petroleum Services and StratSeis, Inc., and I thank particularly Tricia Dark for her cooperation. Daniel Geophysical, Inc., prepared and contributed the synthetic seismograms; I owe special thanks to Dan Shearer. Scott Stockton of Professional Geophysics, Inc., George Lewis, and Charlie Berg also contributed to the project. The drafting and typing were done by my daughter Katherine Johnson.

REFERENCES

Allison, M.L., 1984, Structural controls on a stratigraphic trap: the Cottonwood Creek Field, Big Horn Basin, Wyoming: Wyoming Geological Assoc. Guidebook, 35th Ann. Field Conference, p. 355-367.

Berry, R.G., Jr., 1952, The geology of the Bonanza pool, Big Horn County, Wyoming: Wyoming Geological Assoc. Guidebook, 7th Ann. Field Conference, p. 121-122.

Biggs, P., and Espach, R.H., 1960, Petroleum and natural gas fields in Wyoming: U.S. Bur. Mines Bull. 582, 538 p.

Campbell, C.V., 1962, Depositional environments of Phosphoria Formation (Permian) in southeastern Bighorn Basin, Wyoming: American Association of Petroleum Geologists Bull., v. 46, p. 478-503.

Curry, W.H. III, Paleotopography at the top of the Tensleep Formation, Bighorn Basin, Wyoming: Wyoming Geological Assoc. Guidebook, 35th Ann. Field Conference, p. 199-212.

Herrod, W.H., 1980, Relationship of Cottonwood Creek Field, Washakie County, Wyoming, to carbonate facies of Permian Goose Egg Formation eastern Big Horn Basin: Wyoming Geological Assoc. Guidebook, 31st Ann. Field Conference, p. 67-89.

Hewett, D.F., and Lupton, C.T., 1917, Anticlines in the southern part of the Big Horn Basin, Wyoming: U.S. Geol. Survey Bull. 656, 192 p.

Hill, G.A., Colburn, W.A., and Knight, J.W., 1961, Reducing oil-finding costs by use of hydrodynamic evaluations, in Petroleum exploration, gambling game or business venture: Englewood, N.J., Prentice-Hall, Inc., p. 38-69.

Lawson, D.E., and Smith, J.R., 1966, Pennsylvanian and Permian influence on Tensleep oil accumulation, Big Horn Basin, Wyoming: American Association of Petroleum Geologists Bull., v. 50, p. 2197-2220.

Moore, D.A., 1984, The Tensleep Formation of the southeastern Big Horn Basin, Wyoming: Wyoming Geological Assoc. Guidebook, 35th Ann. Field Conference, p. 273-280.

Pedry, J.J., 1957, Cottonwood Creek Field, Washakie County, Wyoming, carbonate stratigraphic trap: American Association of Petroleum Geologists Bull., v. 41, p. 823-838.

Peterson, J., 1984, Permian stratigraphy, sedimentary facies, and petroleum geology, Wyoming and adjacent area: Wyoming Geological Assoc. Guidebook, 35th Ann. Field Conference, p. 25-64.

Price, R.A., 1984, The Cordilleran foreland thrust and fold belt in the southern Canadian Rocky Mountains, in North American thrust-faulted terranes: American Association of Petroleum Geologists Reprint Series 27, p. 345-366.

Rogers, C.P., Jr., Richards, P.W., Conant, L.C., Vine, J.P., and Notley, D.F., 1948, Geology of the Worland-Hyattville area, Big Horn and Washakie counties, Wyoming: U.S. Geol. Survey Oil & Gas Investigation Map 84.

Rogers, J.P., 1971, Tidal sedimentation and its bearing on reservoir and trap in Permian Phosphoria strata, Cottonwood Creek Field, Big Horn Basin, Wyoming: Mountain Geologist, v. 8, p. 71-80.

Stone, D.S., 1967, Theory of Paleozoic oil and gas accumulation in Big Horn Basin, Wyoming: American Association of Petroleum Geologists Bull., v. 51, p. 2056-2114.

_____, 1969, Wrench faulting and central Rocky Mountain tectonics: Mtn. Geologist, v. 6, p. 67-79; also in Wyoming Geological Assoc. Earth Sci. Bull., v. 2, p. 27-41.

_____, 1983a, The Greybull Sandstone pool (Lower Cretaceous) on the Elk Basin thrust-fold complex, Wyoming and Montana, in Lowell, J.D., ed., Conference on Rocky Mountain foreland basins and uplifts: Rocky Mountain Assoc. of Geologists, p. 345-356.

_____, 1983b, Detachment thrust faulting in the Reed Point Syncline, south-central Montana: Mountain Geologist, v. 20, p. 107-112.

_____, 1983c, The Rattlesnake Mountain, Wyoming, debate: a review and critique of models: Mountain Geologist, v. 21, p. 37-46.

_____, 1983d, Seismic profile: South Elk Basin area, Big Horn Basin, Wyoming, in Bally, A.W., ed., Seismic expression of structural styles, v. 3: American Assoc. of Petroleum Geologists, p. 2.0.24.

Wold, J.S., 1952, Report on Worland Field, Wyoming, in Southern Big Horn Basin: Wyoming Geological Assoc. Guidebook, p. 117-119.

Reprinted by permission of the Rocky Mountain Association
of Geologists from J. D. Lowell, ed., *Rocky Mountain Foreland Basins and Uplifts,* 1983, p. 99-124.

SEISMIC MODELS AND INTERPRETATION OF
THE CASPER ARCH THRUST:
APPLICATION TO ROCKY MOUNTAIN FORELAND STRUCTURE

RILEY C. SKEEN[1]
R. RANDY RAY[2]

ABSTRACT

The Casper arch thrust fault marks the eastern structural boundary of the Wind River Basin of central Wyoming.
New well data combined with regional seismic lines indicate the southwest side of the Casper arch is bounded by a
large-displacement foreland thrust with sedimentary section of the adjacent eastern Wind River Basin continuing underneath thrusted Precambrian rocks.

As part of the interpretation procedure, a series of seismic models were generated using a basic geologic model
of Rocky Mountain foreland thrusts. The models aid in understanding the velocity and ray path effects that distort
seismic reflection data and complicate its interpretation. The dominant effect is the large seismic time pull up created by
the overthrusting of high-velocity Precambrian rocks. The pull-up effect can both create and destroy structure on subthrust horizons on the reflection time section. The velocity distortions can be corrected by converting seismic times
to depth using a sequential layer approach.

Interpretation of the seismic data indicates the horizontal displacement on the Casper arch thrust decreases
from north to south with the fault angle increasing from about 20° in the north to nearly 40° in the south. A sliver of
overturned Paleozoic and Mesozoic rocks is present beneath the Precambrian thrust and the sliver varies in thickness
laterally along the fault. Two major left-lateral shear zones divide the Casper arch suprathrust into three distinct
plates which are characterized by differing styles of deformation.

Many of the interpretation techniques and results from seismic modeling of the Casper arch thrust can be applied to
the interpretation of other Rocky Mountain foreland structures.

INTRODUCTION

A thrust fault, here named the "Casper arch thrust fault",
marks the eastern structural boundary of the Wind River Basin
of central Wyoming. It has been interpreted as the southeastward extension of the Owl Creek Mountains fault (Keefer, 1970),
created by the uplift of the Casper arch. New well data combined with regional seismic lines (Fig. 1) indicate the southwest
side of the Casper arch is bounded by a large-displacement
foreland thrust with sedimentary section of the adjacent eastern
Wind River Basin continuing underneath thrusted Precambrian.

Approximately 300 mi of multifold seismic data were interpreted in this area as part of an exploration project undertaken
by Beren Corporation and Energetics, Inc., in 1981. The displays
and seismic maps presented in this paper are part of that exploration project.

The outline of this paper will essentially follow the same
procedure as was used in interpreting the seismic data. The
interpretation procedure began with the assumption of a basic
geologic model of Rocky Mountain foreland thrusts. Recog-

nizing that the interpretation of complex structure from seismic
data would have many pitfalls, we chose to seismically model
the geologic ideas to better understand velocity and ray path
distortion effects. Then, based on modeling results, we carefully
interpreted the seismic data base and constructed time-structure maps. Finally, all of the seismic maps were converted to
depth.

BASIC GEOLOGIC MODEL

Data from recent exploratory drilling through Precambrian
rocks combined with regional seismic lines provide a refined
geologic model of mountain front uplifts. Low-angle reverse
faulting with miles of horizontal displacement places potentially
productive sedimentary section below thick wedges of thrusted
Precambrian rocks (Fig. 2). Critical information about rock velocities, rock thicknesses, attitude of fault planes, and the existence of overturned rock packages was integrated into the evolving geologic model. The interpretation of seismic reflection
data is based on understanding the separate components and
characteristics of the assumed geologic model of mountain
front thrusts.

Structure in the subthrust sedimentary rocks is anticipated,
but is likely distorted on the seismic time response record by
velocity effects. To obtain the most accurate structural interpretation in this complex mountain front play, extensive revisions of the geologic model and concurrent seismic modeling
are essential. The results of the seismic modeling are compared
to the seismic profiles to interpret the true structural configuration of the subthrust section.

SEISMIC MODELING

The primary purpose for constructing seismic models is to
determine the seismic reflection time response in a given geo-

[1] Consulting Geophysicist, Mountain Eagle Exploration, Inc., Littleton, CO 80120

[2] Consulting Geophysicist, R³ Exploration Corporation, Lakewood, CO 80228.

We express thanks to all of the many individuals who contributed ideas, data,
time, and encouragement in the writing of this paper. Special thanks go to
Robert C. Busby for his individual work devoted to this interpretation.
  Energetics, Inc., is acknowledged for the release of proprietary interpretation
maps, models, and results for publication in this paper. Energetics, Inc., sup-
ported this project through the contribution of in-house drafting personnel,
supplies, etc.
  We thank the following companies for the release of seismic data: Beren
Corporation, Coastal Oil and Gas Corporation, Chevron U.S.A., Inc., Internorth,
Inc., S.T.M. Geophysical Corporation, and Sheehan Exploration Company.
  Compagnie Generale de Geophysique is thanked for the regeneration of one
of the ray trace models.
  We are grateful to Robbie Gries and James D. Lowell for editorial comments
and encouragement in the trenches.

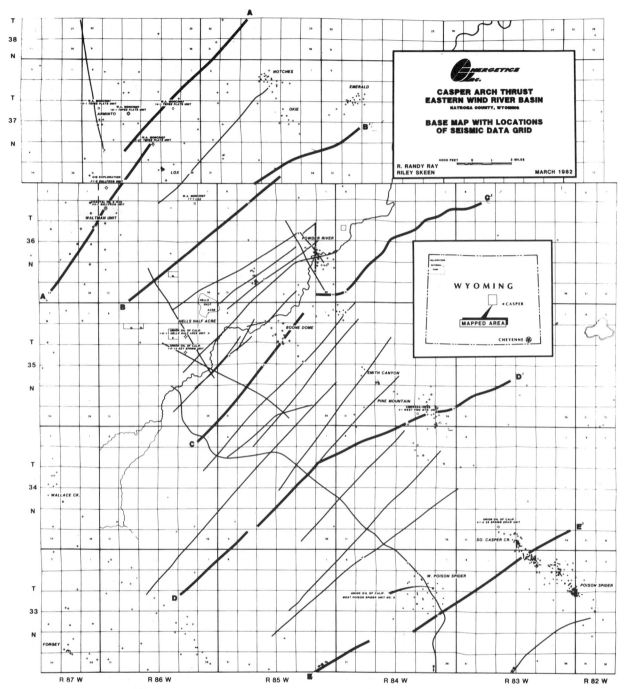

Figure 1.   Index map of study area showing seismic data used in this study and five seismic profiles A-A' through E-E'.

logic depth model. The dominant effect is the large seismic time "pull up" created by the thrust wedge of high velocity Precambrian rocks. The velocity effect can both create apparent time structure or mask real structure beneath the fault. Furthermore, the juxtaposition of different velocity rocks across a dipping fault plane can drastically bend seismic ray paths. This bending results in the mispositioning of seismic reflections relative to their true subsurface position and can some-

times create areas which are uninterpretable or blind to detection.

Velocity and ray path distortions were also studied as part of the interpretational procedure for the Casper arch thrust fault area. Seven geologic models were prepared, beginning with a simple thrust fault geometry and systematically increasing in complexity. Zero offset and normal incidence seismic responses were generated for each model and are displayed in

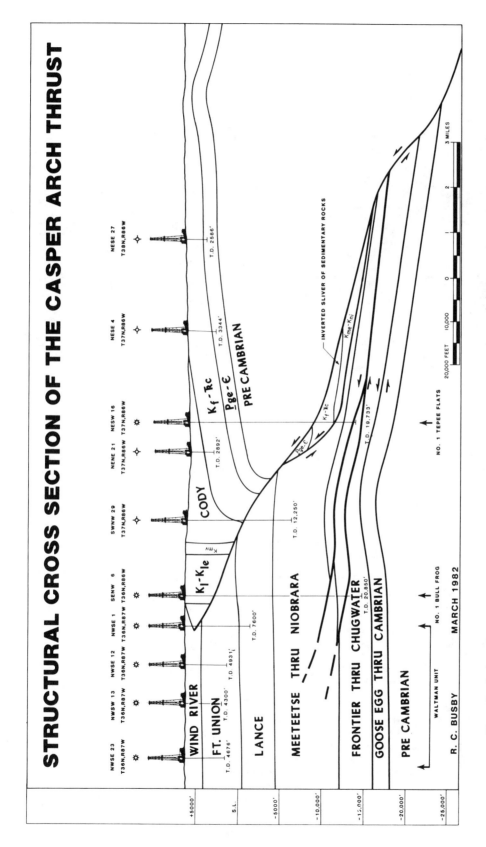

STRUCTURAL CROSS SECTION OF THE CASPER ARCH THRUST

Figure 2.   Basic geologic model of foreland mountain thrust in the Casper arch thrust fault area (after R.C. Busby, personal communication). Cross section located in the Waltman, Bullfrog, and Tepee Flats areas, Natrona County, Wyoming. Kl-Kle = Lance-Lewis, Kmv = Mesaverde.

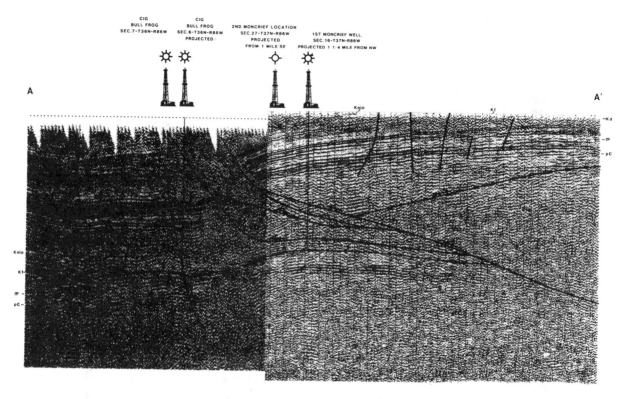

Figure 3. Seismic profile A-A' used for geologic modeling (see Fig. 1 for profile location). p€ = Precambrian, P = Tensleep, Kd = Dakota, Kf = Frontier, Knio = Niobrara.

relative amplitude. Ray tracing was performed on three of the models. A brief discussion of each of the models follows.

Basic Model

A basic depth model was constructed using seismic profile A-A' (Fig. 3) and the Moncrief No. 16-1 Tepee Flats well (sec. 16, T. 37 N., R. 86 W.). Projection of the Moncrief well into the seismic line confirmed the interpretation of a large thrust fault dipping approximately 20° northeast. Sonic velocities in the Moncrief well indicated the Precambrian to be almost a constant velocity of 19,000 ft/sec throughout its thickness of 8,898 ft. Sonic velocities decreased at three intervals 8,150 ft, 12,750 ft, and 14,000 ft, which are interpreted as major fracture or fault zones within the Precambrian section. The lower zone is a fault that shows up clearly on the seismic line (Fig. 3). Interestingly, the velocity contrast between the base of the crystalline Precambrian rocks and the underlying Triassic sedimentary rocks is only a few hundred ft/sec and, therefore, the thrust fault at the base of Precambrian does not necessarily show up as a strong reflector on the seismic data.

Sonic velocity information from wells in the Wind River Basin and on the Casper arch indicate the sedimentary section could be simply divided into three generalized velocity units: 1- Tertiary Wind River Formation to Upper Cretaceous Meeteetse Formation, about 13,000 ft thick with an interval velocity of 12,000 ft/sec in the basin: 2- Upper Cretaceous Meeteetse Formation to Upper Cretaceous Frontier Formation, about 6,500 ft thick with an interval velocity of 14,000 ft/sec in the basin and 10,500 ft/sec on the Casper arch; and 3- Upper Cretaceous Frontier Formation to Precambrian rocks, about 6,000 ft thick with an interval velocity of 16,500 ft/sec in the basin and 14,000

ft/sec on the arch. A velocity of 19,000 ft/sec was used in the model for the Precambrian rocks of the basement and the thrust.

The basic geologic depth model is a flat-lying sedimentary section underneath a thrust fault dipping 20° (Fig. 4). The resulting seismic time response dramatically indicates a velocity pull up of horizons creating apparent structure under the fault. About 1.100 seconds of time structure is created on the subthrust Precambrian basement reflection. The reflecting horizons show abrupt changes in dip which correspond to the overlying velocity changes in the thrust wedge. The abruptness of change in dip is an artifact of the simplified four layer model and occurs more smoothly on actual seismic sections because of the more gradual change in velocity with depth. However, the leading edge of Precambrian rock in the thrust is an abrupt velocity change and often causes a sharp pull up of reflecting horizons. Pull up begins in front of the true position of the leading edge of the wedge of Precambrian because of ray path bending (see ray path model discussion). The straight fault plane appears curved in the model because of the changing thickness of suprathrust rocks overlying the fault. The amplitude of the reflection at the fault plane is directly proportional to the velocity contrast across that boundary and decreases with depth as the underlying sedimentary rock velocity approaches the Precambrian rock velocity. The reflecting horizons below the fault do not appear to intersect the fault plane reflection because of the sharp bending of ray paths at that truncation boundary. This is caused by the velocity contrast and the steep dip geometries.

Steepened Fault Model

Many of the characteristics of the basic model are exaggerated when the angle of the fault plane is increased (Fig. 5). Most dramatic is the increase in the rate of pull up as the Precambrian rocks thicken more rapidly. The horizontal mispositioning of the fault plane reflection is evident on the unmigrated seismic time response. This effect occurs on the other models, but is more obvious as the angle of the fault increases. This effect also increases the gap between the fault reflection and the reflecting horizons below it.

Structure Below Fault

A 3,000 ft structural high was introduced on the Precambrian horizon of the geologic depth model (Fig. 6) to determine the affect of the thrust on the time response. The pull-up effect flattened the true structural dip on the right flank of the structure; consequently the existence of a large structure was almost completely masked on the seismic section. The crest of the small remaining time structure was shifted by velocity pull up back under the thrust, about 6,000 ft from its true depth position. Therefore, it is expected that crests of seismic time structures will shift toward the leading edge of the thrust when properly converted to depth.

Curved Upper Surface of the Precambrian Suprathrust

As the leading edge of the Precambrian suprathrust is curved more sharply, causing a more rapid rate of Precambrian thickness change, the affect of the velocity layers above the thrust becomes more pronounced (Fig. 7). The reflecting horizons below the thrust are sharply segmented in response to the "apparent" thickness increase caused by folding the higher velocity rocks in the hanging wall. The reflection from the fault is also very contorted in the time response compared to its uniform dip in the depth model. The leading edge of the fault under the Precambrian is almost flattened in the time response. As the dips on the leading edge of the thrust increase, the horizontal and vertical mispositioning of the seismic reflections is larger and more apparent.

Variation of Precambrian Thickness

Interpertation of several seismic lines and a magnetotelluric survey indicate the leading edge of the Precambrian varies in thickness. Figure 8 illustrates how variation in Precambrian thickness can create a non-existent time structure on the reflection sections. The time crest of the apparent structure corresponds to the thickest Precambrian rocks (on the lead edge) which cause the maximum pull-up effect. A sharp syncline with a bow-tie effect is caused by the wedge of low velocity rock on the leading edge of the upper plate of the thrust. This type of velocity distribution could also occur if the reverse fault relaxes, becoming a listric normal fault causing younger clastics to be deposited on the down-dropped block.

Recumbent Sliver

In the Moncrief Tepee Flats No. 16-1 well (sec. 16, T. 37 N., R. 86 W.) a recumbent sliver of Triassic to Upper Cretaceous rocks is present below the thrust (Fig. 9). Modeling this additional lens of high velocity rock exaggerates the velocity pull up of the horizons below the fault. Because the overall thickness of high velocity rocks increases at the leading edge of the thrust, the pull-up effect is very abrupt and appears as a fault on the seismic time response. In addition, unmigrated reflections occur across the discontinuity enhancing the appearance of a fault even though the horizons are flat in the depth model. This same effect could be created if the nose of the thrust was blunt.

Accreted Sliver Zone

This model demonstrates the affect of an accreted recumbent sliver zone all along the fault plane (Fig. 10). The abrupt pull up at the leading edge of the thrust is present as it was in the recumbent sliver model. However, the overall amount of time pull up is 1.000 second or slightly less than the 1.100 seconds in the recumbent sliver model. This is because the accreted sliver zone thickness has replaced some of the higher velocity Precambrian rock thickness.

RAY TRACE MODELS

Three ray trace models were prepared to study the affects model geometries have on the final seismic time responses (Figs. 11, 12, and 13). The actual acquisition parameters of 24-fold, 220 ft group interval, split-spread geometry with an overall length of 11,880 ft were built into the ray trace models. The modeling assumes reflection on the top of selected subthrust horizons and the ray path is plotted on the model only if the ray returns back to the surface within the defined spread length. Many of the rays return to the surface beyond the spread length and, therefore, would not be recorded.

The ray trace on the basic model shows the degree of ray path bending that takes place at several positions along the model (Fig. 11). The model indicates reflection paths on the top of the Precambrian basement horizon. The ray paths at the leading edge of the thrust illustrate the unmigrated position of data on the seismic section. The ray paths at the truncation of the Frontier Formation under the fault also demonstrate that the ray paths are strongly divergent and create the "null zone" or gap between the fault reflection and the horizon reflections that was visible on the previous seismic models.

The recumbent sliver model (Fig. 12) indicates reflection paths on the top of the Precambrian basement horizon. It is very similar to that of the basic model and shows the same mispositioning of reflections and blind zones.

The ray trace model with the varying Precambrian rock thickness (Fig. 13) is based on reflections on the Frontier Formation. The large velocity contrast at the base of the Precambrian wedge is the dominant influence on ray path bending. This model also shows how complex the ray paths can be on the leading edge of a steeply dipping thrust. The non-vertical nature of these ray paths is the cause of many diffractions and lack of good reflections in the seismic time response. Undoubtedly many of these same distortions occur on real seismic data and must be accounted for when preparing final depth maps.

SUMMARY OF MODELING RESULTS

A high velocity wedge of thrusted Precambrian rock causes an abrupt pull up of seismic reflections, creating apparent structure. The magnitude of the apparent seismic time structure is directly proportional to the thickness of the wedge of Precambrian rocks. The presence of a high velocity recumbent sliver under the thrust fault further exaggerates the velocity pull-up effect forming a sharp offset in underlying reflections creating diffractions that look like a fault on the time sections.

Seismic reflections from a flat uniformly dipping thrust may be very curved and contorted on the seismic section because of overlying layer velocity changes. In general, abrupt changes in dip of seismic reflections under the thrust can be a result of abrupt changes in overlying velocities, such as the edge of a recumbent sliver or discontinuities in the Precambrian. A reflection from the base of the Precambrian wedge will occur only where there is significant velocity contrast across that boundary. The amplitude of the thrust reflection will usually decrease with depth as the velocity of the subthrust sedimentary section approaches that of the Precambrian.

ROCKY MOUNTAIN ASSOCIATION OF GEOLOGISTS

BASIC PRECAMBRIAN WEDGE

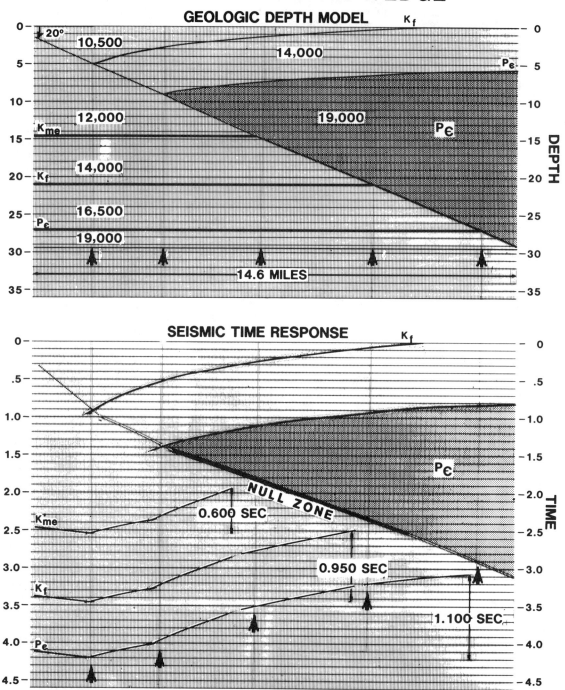

Figure 4. Basic model showing Meeteetse (Kme), Frontier (Kf), and Precambrian (p€) horizons with rock velocities in ft/sec. Depth scale is in seconds. Arrows indicate the positions of horizon truncations on the fault.

STEEP FAULT

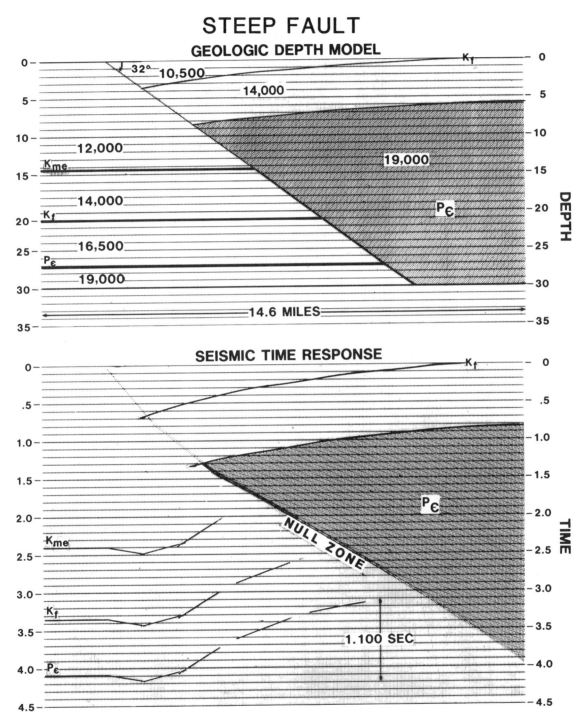

GEOLOGIC DEPTH MODEL

32°   10,500

14,000

12,000

K$_{me}$

14,000

K$_f$

16,500

P$_\in$

19,000

19,000

P$_\in$

14.6 MILES

DEPTH

SEISMIC TIME RESPONSE

K$_f$

P$_\in$

K$_{me}$

NULL ZONE

K$_f$

1.100 SEC

P$_\in$

TIME

Figure 5.   Steep fault model with fault dipping 32° showing total pull up of 1.100 seconds.

SUBTHRUST STRUCTURE

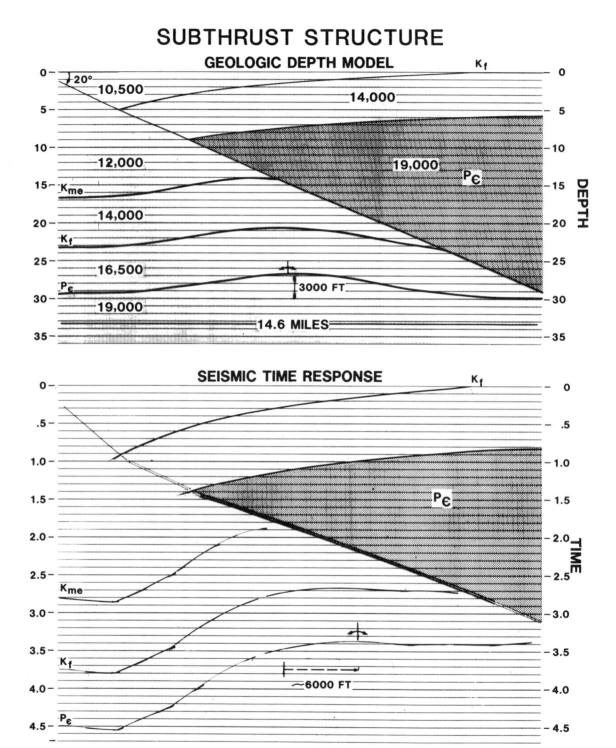

Figure 6. Subthrust structure model showing the relative shift between actual geologic structure and the seismic time structure.

ROCKY MOUNTAIN ASSOCIATION OF GEOLOGISTS

CURVED UPPER PLATE

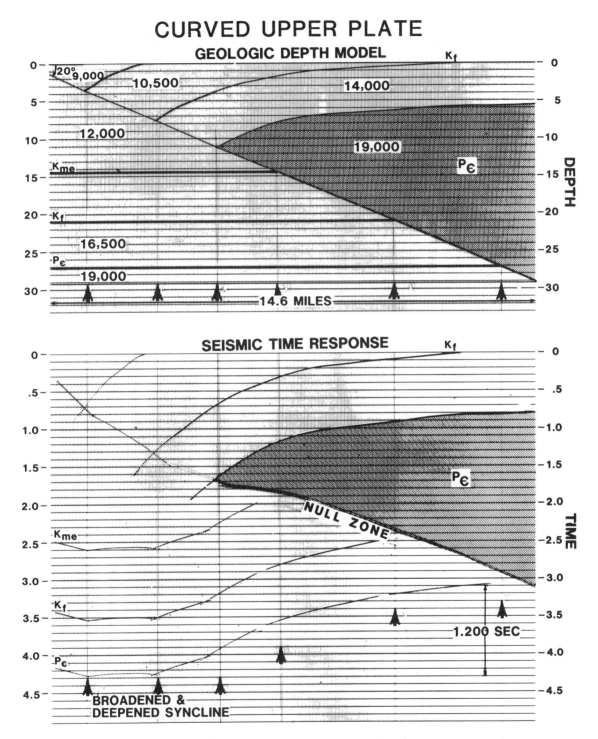

Figure 7. Curved upper plate model showing increased distortion of the fault plane reflection and increased mispositioning of reflectors from the steeply dipping leading edge of the thrust.

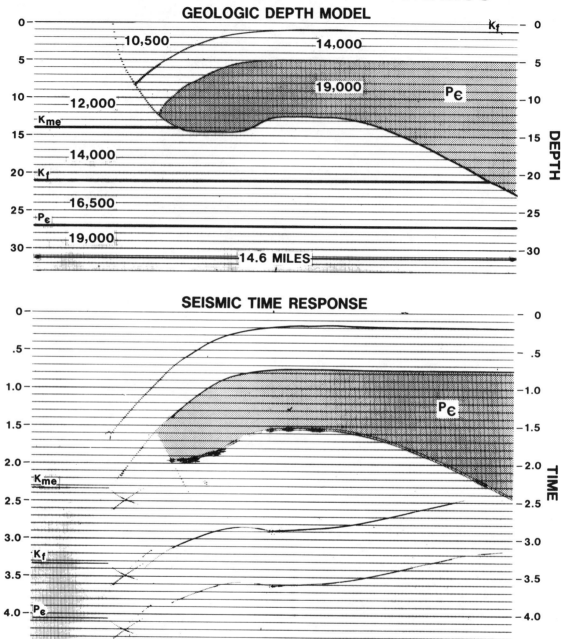

Figure 8. Varying Precambrian thickness model showing the effects of lateral variation in the wedge.

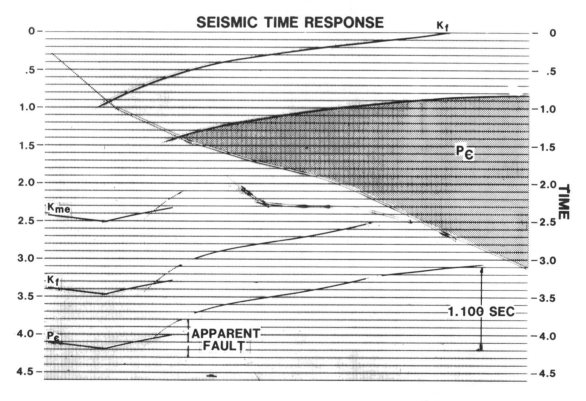

RECUMBENT SLIVER

GEOLOGIC DEPTH MODEL

K$_f$

10,500

14,000

12,000

19,000

K$_{me}$

16,000

Pϵ

14,000

K$_f$

Pϵ

16,500

19,000

14.6 MILES

DEPTH

SEISMIC TIME RESPONSE

K$_f$

Pϵ

K$_{me}$

K$_f$

1.100 SEC

Pϵ

APPARENT
FAULT

TIME

Figure 9. Recumbent silver model showing the increased pull up creating an apparent fault.

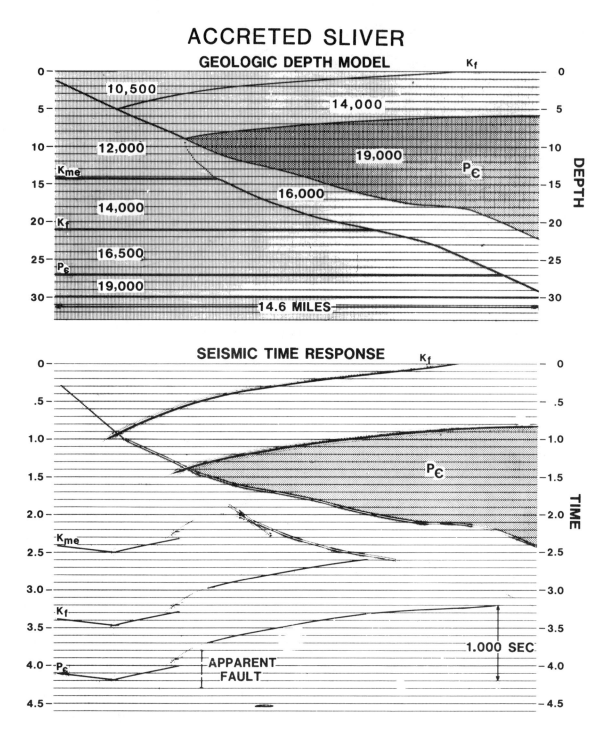

Figure 10. Accreted sliver zone model showing less pull up than the recumbent sliver model.

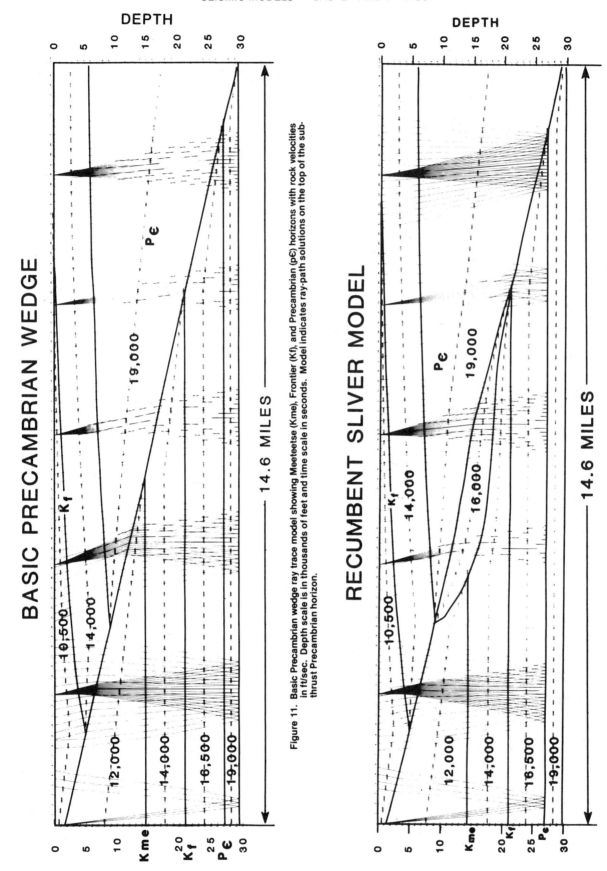

Figure 11. Basic Precambrian wedge ray trace model showing Meeteetse (Kme), Frontier (Kf), and Precambrian (pC) horizons with rock velocities in ft/sec. Depth scale is in thousands of feet and time scale in seconds. Model indicates ray-path solutions on the top of the subthrust Precambrian horizon.

Figure 12. Recumbent silver ray trace model showing ray-path solutions on the top of the subthrust Precambrian horizon.

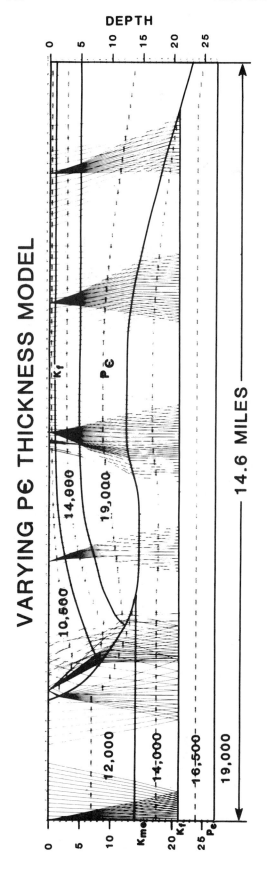

Figure 13. Varying Precambrian thickness ray trace model showing ray-path solutions on the top of the subthrust Frontier horizon.

Crests of seismic time structures under the thrust shift toward the leading edge of the thrust when converted to depth. Variations in the thickness of thrusted Precambrian can both create and destroy structure on the reflection time section. Real structural reversal can be completely flattened by velocity pull-up effect.

SEISMIC INTERPRETATION

Seismic data were interpreted for both the suprathrust (hanging wall) and the basin block subthrust (footwall) of the Casper arch thrust fault area. Five regional small-scale seismic profiles have been assembled by combining several seismic lines (Fig. 14). Displayed from north to south in Figures 3 and 14, profiles A-A′ through E-E′ present a series of cross sections of the west flank of the Casper arch and the southeastern arm of the Wind River Basin. Changes in the style of deformation including the geometry of the thrust fault, thickness of the suprathrust, and dips in the subthrust section have been mapped.

Suprathrust Casper Arch

The structural interpretation on the suprathrust of the Casper arch thrust fault is made from a combination of seismic data, well control, and mapped surface geology (Fig. 15). Because of the steep dips and structural complexity of the shallow section, seismic data is very poor and sometimes only dip segments can be mapped. Well control is helpful for recognizing faults, but the steep dips and monotonous character of non-marine Cretaceous Lance through Tertiary Wind River formations makes recognition of geologic intervals difficult or impossible. Therefore, a generalized shallow form-line map combining fragmentary bits of information from outcrop, seismic, and well control was constructed. The suprathrust block shows variation in structural style along the leading edge of the thrust (Fig. 15).

A series of anticlinal closures along the crest of the suprathrust plate has been mapped in the Precambrian time structure (Fig. 16). Most of the structures are broad features, such as the plunging nose of Notches and Okie fields and the anticlines of Powder River and Pine Mountain fields. However, South Casper Creek and Poison Spider fields are narrow asymmetrical structures near the leading edge of the suprathrust. North of Pine Mountain, Precambrian rocks dip steeply into the basin along the thrust front and are structurally very low.

Two major shear zones have been recognized by the offset of both outcrop patterns at the surface and structural trends on the suprathrust (Fig. 15). The shear zones are probably several miles wide and contain numerous strike-slip faults even though they have been mapped as single faults. The shear zones may be Precambrian zones of weakness that have been reactivated periodically, especially in post-Pennsylvanian — pre-Jurassic time when a few of the structures on the Casper arch were growing. Examples of some of these structures are Emerald field, Meadow Creek field (R.C. Busby, personal communication) and East Tisdale field (Eckelberg, 1958). The bending of the outcrop patterns of the Mesaverde through Tertiary Fort Union formations indicates shear zones have had movement at least through lower Eocene.

The northern shear zone trends northeast through Emerald field (Fig. 16). A distinct eastward bend of the normally N 40° W trending outcrop of Cretaceous Cody through Fort Union formations occurs along the shear zone in T. 36 N., R. 85 and 86 W. At the northeast end, the shear zone includes a northeast-trending fault that is the northern boundary of Emerald field. The relative vertical displacement on the shear zone changes from up on the north side in the Okie field area to up on the south side

ROCKY MOUNTAIN ASSOCIATION OF GEOLOGISTS

in the Emerald field and up on the north side on the north end of Big Sulphur Springs anticline in T.37N., R.83W. (R.C. Busby, personal communication, 1983). The shear is interpreted as having left-lateral movement.

The southern shear zone trends northeast from the southeast corner of T.35N., R.85W. where the normally N 40° W-trending outcrops of Cody through Fort Union formations are bent westward. The shear zone continues northeast along the north side of Pine Mountain before bending eastward and passing south of Gothberg field (T.36N., R.82W.). South of the shear zone, west flanks are steep on Laramide compressional folds, such as Immigrant Gap, Iron Creek, Oil Mountain, Poison Spider, South Casper Creek, and Pine Mountain anticlines. North of the shear zone, east flanks are steep on compressional folds of Laramide age, such as North Casper Creek, Gothberg, Big Sulphur Springs, and Notches anticlines, which also have high-angle reverse faults on the east flank upthrown on the west. This shear is interpreted as left lateral (R.C. Busby, personal communication, 1983) and is upthrown on the south side in this area.

Other smaller scale offsets are apparent on the Casper arch and include a zone between Notches and Okie fields, another between South Casper Creek and Poison Spider fields, and another between Oil Mountain and Iron Creek fields. These offsets are small compared to those on the two major shear zones.

Each of the plates separated by these shear zones may have responded differently to later Laramide compressional forces that caused folding on the Casper arch and thrusting along the Casper arch thrust fault. The middle block between the two shear zones appears to have been dropped down relative to the others. Regional seismic profiles B-B' and C-C' (Fig. 14) show the leading edge of the Precambrian rocks in the thrust has broken off and rotated along a high-angle reverse fault at the east side of the Notches and Powder River fields. This fault is a major back-limb thrust, antithetic to the Casper arch thrust fault with offset indicated by reflections within the Precambrian. These reflectors in the Precambrian rocks are probably fault gouge zones that demonstrate internal shearing within the thrust block.

The back-limb fault caused the leading edge of the thrust to rotate downward, and the overlying sedimentary section to slump down the steeply inclined Precambrian surface, creating the thrusted structures of Boone Dome and North Boone Dome (Fig. 14, profile C-C'). The décollement of the gravity slide structure is in the thick Cretaceous shale above the Dakota Formation. Most of the shallow deformation is limited to the middle shear zone block, especially at the south end where Precambrian rocks have dropped down to the lowest position on the thrust and formed the space for a series of thrusted folds (Fig. 14, profile C-C').

The seismic data reveal that the producing structures on the suprathrust are comprised of two different, but related, structural styles. The largest structures are basement involved, such as the drape-fold structures of Notches, Okie, and Powder River fields or the folded thrusts of Arminto, Pine Mountain, South Casper Creek, and Poison Spider fields. The northeast-trending shear-zone structure at Emerald field is another example of basement influence. A second style of structure is detached from basement, such as the fields of Boone Dome, North Boone Dome, and Lox, which are small thrusts on the leading edge of the large Casper arch thrust. These smaller features are not underlain by structures at the basement level and are second order features. Mapping on deeper horizons (Precambrian, Pennsylvania Tensleep, Cretaceous Dakota) shows basement influence, whereas structure maps on the Frontier and

shallower horizons show structures detached from basement.

Subthrust Basin Block

Well Control

Synthetic seismograms constructed from sonic logs of four deep wells were used to correlate geologic formations to the seismic data. They are:

1- Union Oil of California No. 1-K11-Hell's Half Acre NE, NW sec. 11, T.35N., R.86W., (TD 22,431 ft);
2- Union Oil of California No. 8-West Poison Spider SE, SE sec. 11, T.35N., R.86W., (TD 17,910 ft);
3- Colorado Interstate Gas No. 1-6-36-86-Bull Frog Unit SE, NW sec. 6, T.36N., R.86W., (TD 20,850 ft);
4- Moncrief Oil Co. 16-1-Tepee Flats Unit NE, SW sec. 16, T.37 N., R.86W., (TD 19,733 ft).

Data Correlation

Seismic profile A-A' (Fig. 3) near the Moncrief Tepee Flats Unit No. 16-1 well was used to project sedimentary section beneath the Casper arch thrust. The Tepee Flats well was used to establish Precambrian granite velocity (19,000 ft/sec), to recognize fault zones within the crystalline rock, to recognize the recumbent fault sliver, and to determine the reflection of the pay zone (Frontier Formation). With this information and some confidence in the correlation of the subthrust reflections, the remaining seismic sections were interpreted under the thrust.

The truncation of sedimentary rocks by fault contact against either Precambrian rocks or recumbent sedimentary rocks is not easily identifiable because the data is very spotty and discontinuous. The synthetic seismogram made from the Moncrief sonic log indicates stronger reflections originate from low velocity fault zones within the Precambrian than from the contact of Precambrian with rocks of the recumbent sliver or the Cretaceous basin block. Therefore, the contact at the base of the Precambrian wedge is probably not a strong reflector on the seismic sections (Fig. 3). This contact zone is best characterized as a "null zone" containing no reflections because: 1- the rocks are highly deformed within the zone; 2- the acoustic contrast is small; and 3- reflections are mispositioned horizontally and vertically on the seismic record (Fig. 5).

Depth Conversion — Frontier Formation

Under the thrust fault the effects of the anomalous velocities as determined by the modeling were compensated for in the depth conversion process. Depth conversions were made by adopting a sequential layer approach (Fig. 17). First, the depth to the top of the suprathrust Precambrian was obtained by using a velocity function for the sedimentary section on the Casper arch and then the isopach of the Precambrian wedge was added. Precambrian wedge thickness was determined by using a 19,000 ft/sec velocity and the appropriate Precambrian isochron value at each shot point. The recumbent fault sliver isopach was then added to give a total depth to the basal fault. The recumbent sliver isopach was determined using 16,000 ft/sec as a representative interval velocity. Finally, the total depth to the top of the subthrust Frontier was determined using a deep Wind River Basin velocity function for the seismic time difference (Δt) corresponding to the amount of sedimentary rocks between the basal fault and the top of the Frontier. This sequential layer method of depth conversion, as illustrated in an example calculation (Fig. 17), not only corrected for the velocity pull up created by the Precambrian wedge and recumbent sliver, but also compensated for the depth dependence of rock velocities in the deep Wind River Basin.

The elaborate procedure used in the sequential layer method

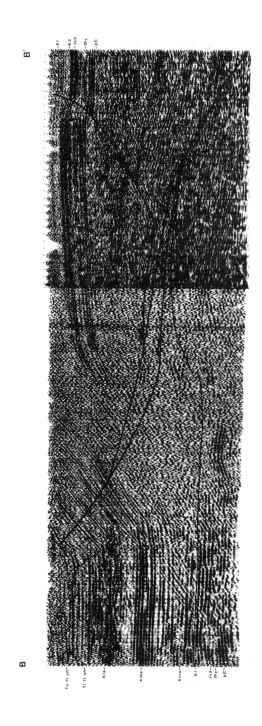

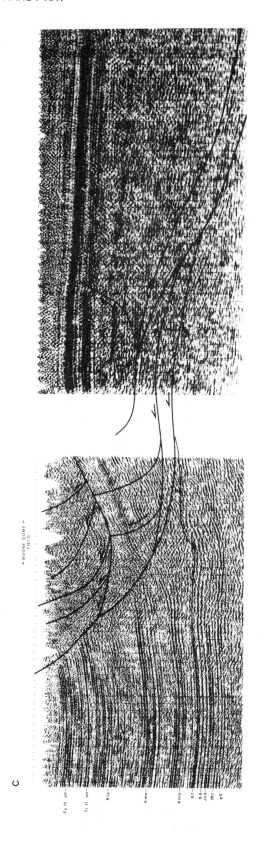

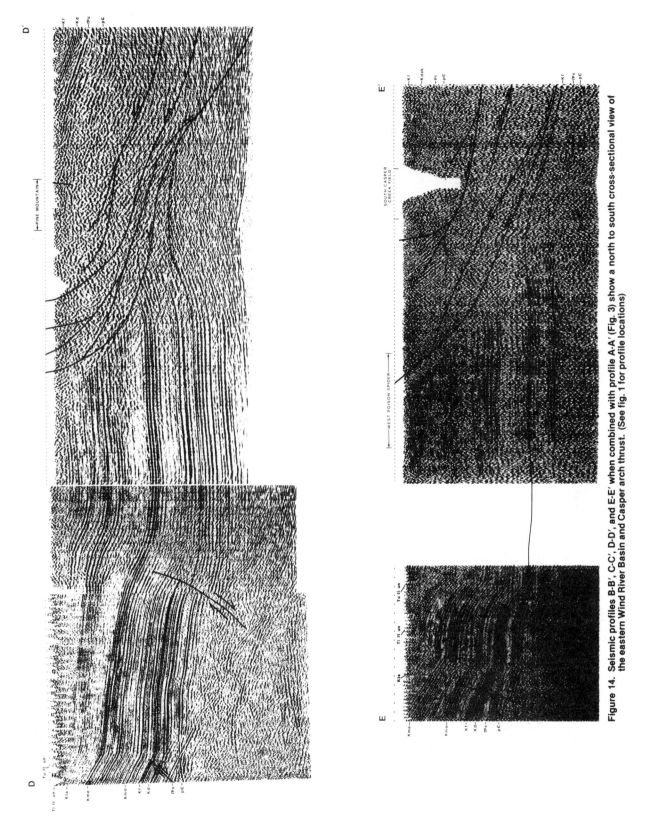

Figure 14.   Seismic profiles B-B', C-C', D-D', and E-E' when combined with profile A-A' (Fig. 3) show a north to south cross-sectional view of the eastern Wind River Basin and Casper arch thrust. (See fig. 1 for profile locations)

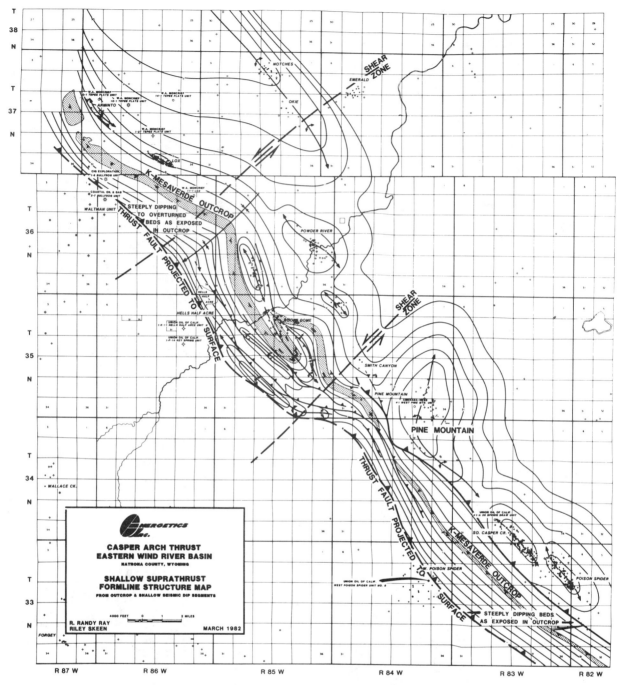

Figure 15. Shallow suprathrust form-line structure map.

to account for and eliminate the pull-up effects caused by velocity is a technically sound approach, however, it is not without risk. It is important to recognize that the largest differences in the depth conversion solutions are caused by relatively small errors in the interpretation of the Precambrian wedge isochron.

The histogram plot of velocities (Fig. 18) shows a large difference in velocity between Precambrian rocks (19,000 ft/sec) and Mesozoic/Paleozoic rocks (16,000 ft/sec) or Lower Cretaceous

rocks (14,000 ft/sec). The smallest percentage difference in velocity (15%) is between Lower Cretaceous rocks of the subthrust and Mesozoic/Paleozoic rocks of the recumbent sliver. By contrast, the largest percentage change in velocity (35%) is between Lower Cretaceous rocks and Precambrian rocks. Therefore, small changes in picking the Precambrian isochron cause larger differences in thickness than the same change in isochron of slower velocity rocks of the Lower Cretaceous or

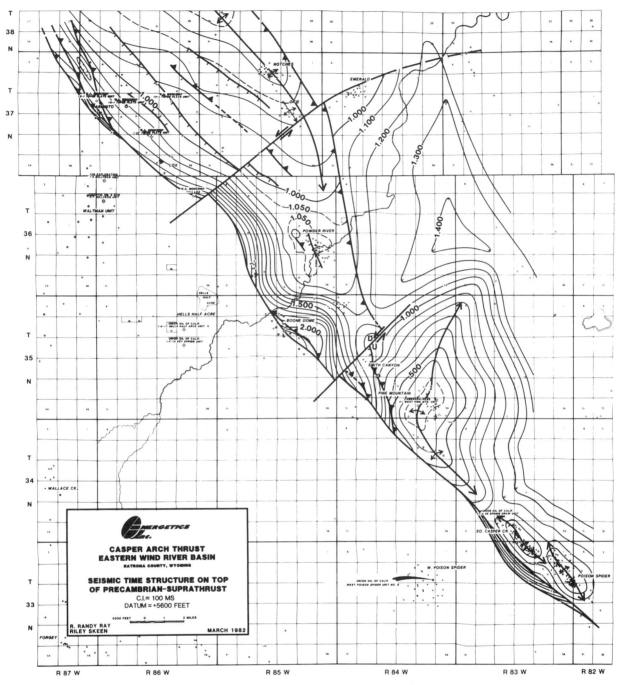

Figure 16. Seismic time structure on the top of the Precambrian suprathrust.

Mesozoic/Paleozoic.

Structure

Cretaceous Frontier Formation was chosen as the key mapping horizon and for discussion in this report because it is the producing zone at Tepee Flats. It is a fair to good subthrust reflector.

Several large structural anomalies were mapped throughout the study area in addition to the Tepee Flats anomaly. All of the time structures shifted to the west or toward the leading edge of the thrust when the velocity pull-up effects were compensated for as predicted by the original models.

The crest of the Tepee Flats structure, mapped in subsea depth, actually is approximately two miles southwest of the 16-1 discovery well. Once the velocity pull up is corrected, the true structural configuration of the anticline is better understood.

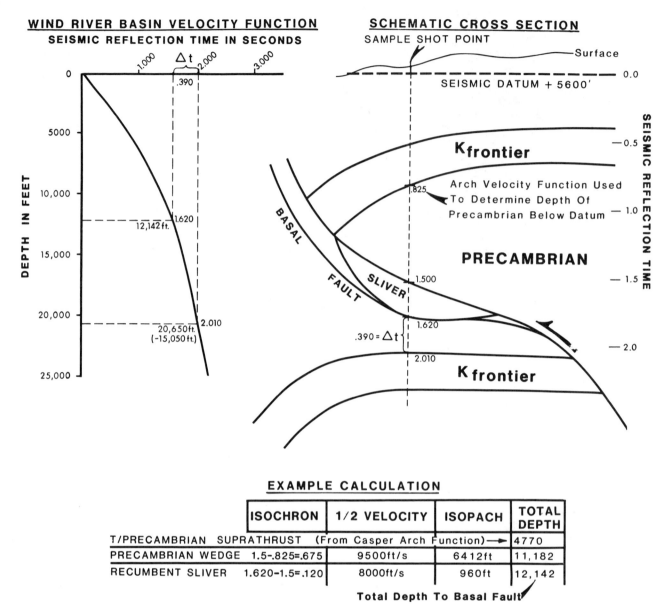

Figure 17. Schematic drawing showing components of the sequential layer depth conversion method.

The 16-1 well appears to be approximately 1,000 ft structurally low to the crest. Also, the Tepee Flats structure is interpreted as coalescing with the Bull Frog structure (Fig. 19). This large structure may encompass more than 10,000 acres and is on trend with the Madden structure, 20 mi to the west.

Casper Arch Thrust Fault

The Casper arch thrust, when observed from north to south in the project area, is interpreted to have less and less horizontal displacement. Consequently, less basin overlap was observed at the south end of the Casper arch. Interpretation of a seismic dip line approximately six miles northwest of the Moncrief 16-1 well indicates that sedimentary rocks extend eastward in the subthrust approximately 10 mi (Gries, 1983, fig. 17). Seis-

mic profile A-A', near the Tepee Flats discovery well, demonstrated approximately seven miles of thrust overlap (Fig. 3).

The Casper arch thrust steepens from about 20° near the Moncrief well to nearly 40° in the south part of the project area, just east of West Poison Spider field. Steepening of the fault plane and shortening of the horizontal transport distance is consistent with the present day configuration of the southern Wind River Basin. Regional seismic profiles D-D' and E-E' show that the basin is narrower to the south (Fig. 14). At the time of thrust movement from east to west, the Rattlesnake Hills (southwest of the Wind River Basin) acted as a buttress to lateral movement and compression along the Casper arch thrust. This buttress responded as a fulcrum or point of pivotal rotation. Hence, to the north, progressively greater horizontal transport occurred.

ROCKY MOUNTAIN ASSOCIATION OF GEOLOGISTS

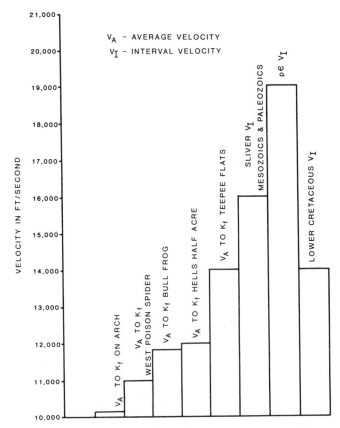

Figure 18. Histogram of rock velocities in the eastern Wind River Basin and Casper arch areas.

It is likely that the pre-existing basement lineaments (shear zones) also allowed for differing amounts of movement and different reactions to compression. As previously noted in the discussion on the "Suprathrust Casper arch", there appear to be three tectonic blocks, differentiated by these shear zones (Fig. 20).

The major zone of transport was along the basal fault between either Precambrian or recumbent Mesozoic/Paleozoic sedimentary rocks and Cretaceous basin subthrust sedimentary rocks (Fig. 21). The age of the major movement on the thrust is latest Laramide or younger (Keefer, 1970). However, seismic data and geometric limitations when attempting to restore the section palinspastically lead us to believe there might have been some pre-Lance movement on the thrust.

Faults in Precambrian Section

The eastward thickening wedge of Precambrian rock is probably one large wedge made up of several sheets, slivers, or imbricate blocks separated by faults or fault zones in the Precambrian. The sonic log in the Moncrief 16-1 well indicates several fault zones (Ray and others, 1983). Two of these zones show up as laterally continuous reflections on the seismic profile A-A' (Fig. 3).

Bedding Plane Faults

East-west horizontal compression, which caused the folding and subsequent faulting of the Casper arch, resulted in the Casper arch thrust cutting up section (ramping) through competent rocks. Seismic data indicates the thrust becomes a bed-ding-plane fault (décollement zone) in the incompetent, overpressured Cretaceous Niobrara shales. This décollement appears to have a maximum width of five to six miles, before the fault ramps up section again. The seismic interpretation of a décollement zone in or near the Niobrara is consistent with the faulting reported from associated deep wells. The deep wells are overpressured in or near the Niobrara forming a classical low friction zone for bedding plane faults.

False Fault

On the Frontier time structure map (Fig. 19), a reverse fault (thrusted east to west) is mapped in the basin block subthrust on the southwest flank of the Tepee Flats time structure and continues to the southeast. The offset in the reflection package is undeniable on seismic profile A-A' (Fig. 3). This fault is either small or non-existent as predicted by the recumbent sliver model (Fig. 9). However, there are many thrust faults in the CIG No. 1-6-36-86 Bullfrog Unit well which repeat the Frontier Formation. These are not interpreted to be basement involved, but instead are believed to be bedding-plane faults in front of the main Casper arch thrust (Fig. 2).

Precambrian Wedge

The isochron of the wedge of Precambrian rocks (Fig. 20) shows a fairly uniform increase in thickness to the east. The isochron contours parallel the leading edge of the fault from the north end to the south. The isochron is thicker at the south end of the area where the fault is steeper. The only significant departures from a smoothly thickening wedge are associated with the offsets at shear zones.

Recumbent Slivers

The highly variable thickness of the sedimentary rocks in the fault sliver is demonstrated in the Moncrief wells at Tepee Flats (Fig. 22). The Moncrief 16-1 and 27-1 wells are only two miles apart, and are slightly sub-parallel to strike, yet the 16-1 had 2,336 ft of sedimentary rocks in the recumbent section and the 27-1 is reported to have drilled in excess of 8,000 ft of recumbent section in the fault sliver. The overturned Mesozoic/Paleozoic rocks in the fault sliver are a remnant from a large recumbent fold.

Although, not easily identifiable on all of the seismic lines, the recumbent zone has been isolated and defined on a sliver isochron map (Fig. 22). The sliver isochron is highly variable and irregular in its geometric configuration. This variability could be anticipated because the rocks were highly deformed and shattered as they were emplaced by faulting.

Subthrust Section

The subcrop of the subthrust sedimentary section on the main transport fault plane is shown in figure 23. All of the potentially productive zones are expected to be present beneath Precambrian rocks including the Meeteetse, Frontier, Muddy, Lakota, Morrison, Sundance, and Tensleep formations. However, progressively more of the section is truncated to the east until hanging-wall Precambrian rocks are in thrust contact with Precambrian rocks of the footwall.

The sedimentary subcrops of the Niobrara and Frontier formations at the base of the fault flatten and have more separation as a result of the décollement in the Niobrara. This is particularly evident in the northern half of the area (Fig. 23). The lateral extent of the décollement zone coincides with the position of the back-limb thrust, which forms the fault on the east side of Notches and Powder River fields (Figs. 16 and 23). This coincidence suggests that rotation of the lead edge of the Precambrian thrust was limited to the area of bedding-plane

ROCKY MOUNTAIN ASSOCIATION OF GEOLOGISTS

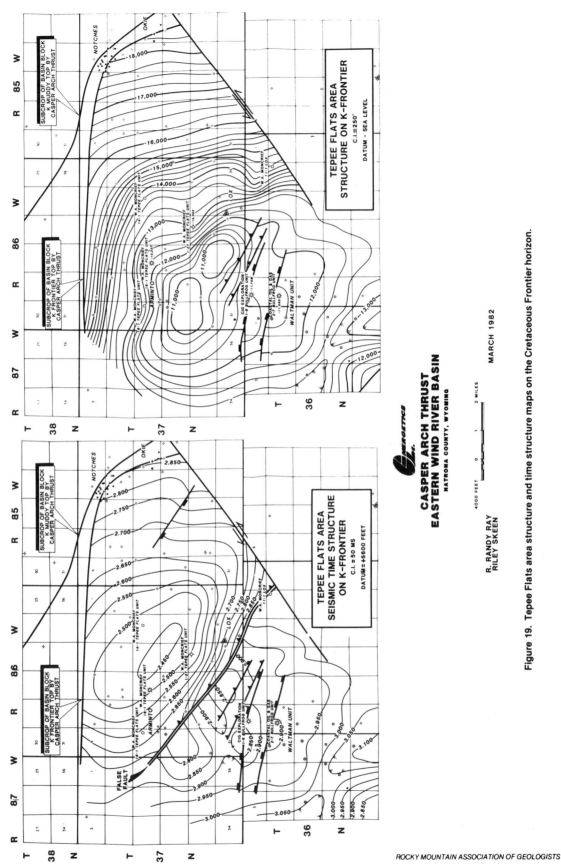

Figure 19. Tepee Flats area structure and time structure maps on the Cretaceous Frontier horizon.

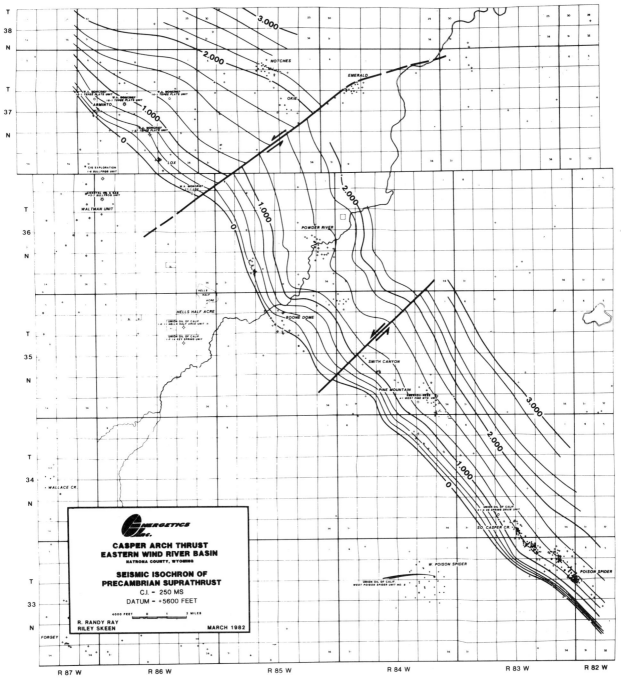

Figure 20. Seismic isochron of Precambrian suprathrust.

slippage.

Basin Development and Sedimentation

Isochron maps of the Lance, lower Fort Union, and upper Fort Union were made in the section southwest of the Casper arch thrust. Isochron synclinal thicks, located nearly parallel to and only a few miles east of the present day basin axis, indicate a shift of the basin depo-center westward through geologic time. This shift of the depo-center was probably caused by the compressional movement of the Casper arch thrust westward during the Laramide orogeny.

Isochron thinning during Lance deposition across West Poison Spider, Hell's Half Acre, and Bull Frog fields indicate these features have some Lance age development. Other structures in this area, such as Waltman field, have no Lance isochron thinning, indicating they are younger.

SUMMARY AND CONCLUSIONS

Generalized observations derived from interpreting seismic records in the Casper arch thrust study area are applicable to other Rocky Mountain foreland structures:

ROCKY MOUNTAIN ASSOCIATION OF GEOLOGISTS

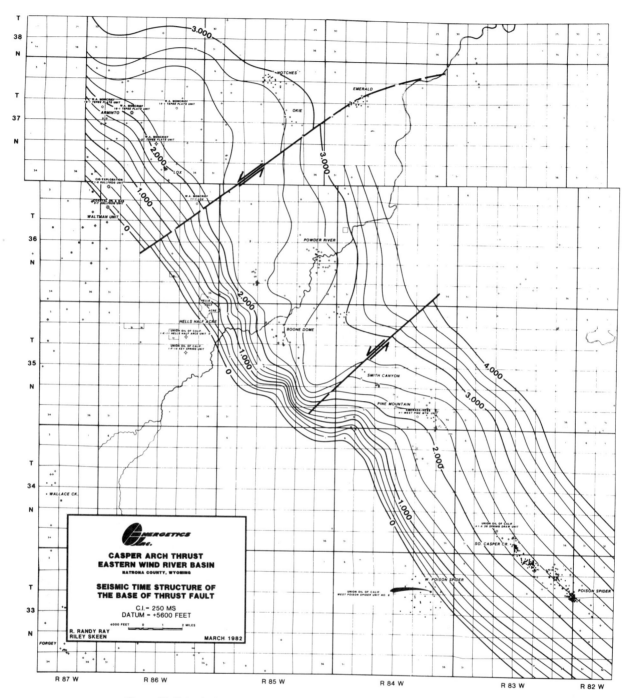

Figure 21. Seismic time structure on the base of the main transport thrust fault.

1- Matching the geologic depth model to the seismic response model and subsequently to seismic data proved critical to understanding and interpreting seismic sections with a higher degree of confidence.

2- The wedge of Precambrian rocks was verified as a major contribution of velocity pull up on seismic data.

3- Time structures observed on seismic data shift in depth toward the leading edge of the thrust, and the shift can be

in the range of a few hundred feet to a few miles.

4- Recumbent rocks in fault slivers underlying Precambrian rocks also affect subthrust seismic time response.

5- False faults are created on seismic records by rapid thickening of higher velocity rocks in the suprathrust resulting in large velocity pull ups.

6- Depth conversion using a sequential layer method provides a more accurate structural configuration than a time dis-

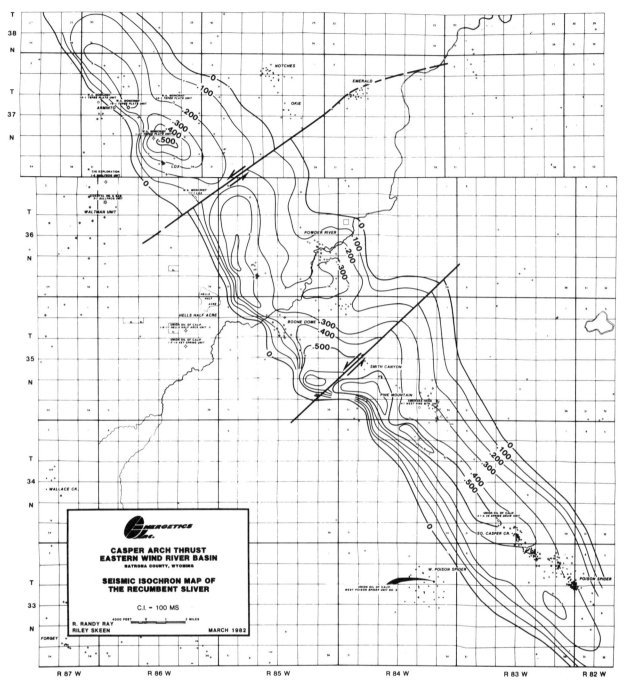

Figure 22. Seismic isochron map of the recumbent sliver.

play. This is accomplished by first migrating the seismic time structure map, and then calculating the thickness of sequential layers.

7- Ray trace modeling indicates ray path distortion at rock velocity boundaries. Focusing and de-focusing of the recorded reflection data results in blind zones to detection.

8- A "null zone", or zone of little or no reflection character, can be expected at the base of the transport fault, rather than a clear cut, sharp boundary corresponding to the trunca-

cation of strata against the fault.

9- Precambrian rocks thicken relatively uniformly along the width of a foreland thrust, but this does not preclude the possibilities of lateral thickness changes along the strike of the thrust.

10- The wedge of Precambrian rocks might be splintered, severely shattered, and thrusted.

11- The angle of the thrust plane may vary along the strike. The Casper arch thrust fault steepens from north to south, but does not become vertical.

ROCKY MOUNTAIN ASSOCIATION OF GEOLOGISTS

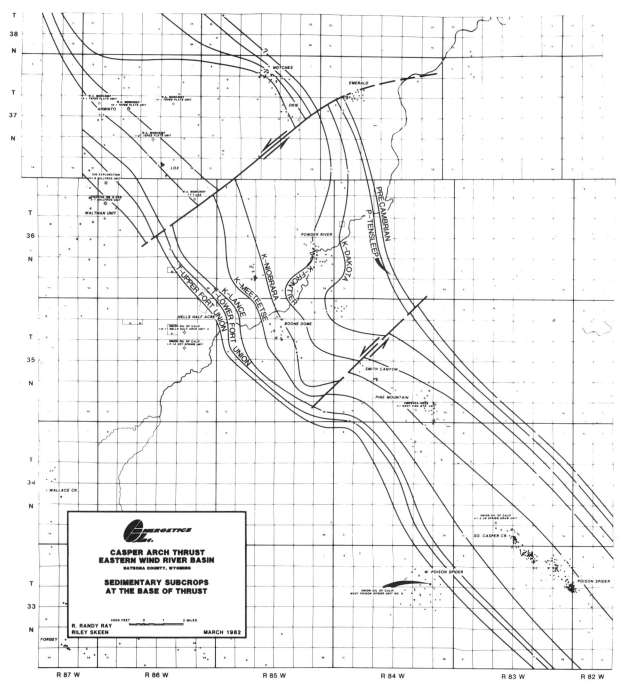

Figure 23. Sedimentary subcrops at the base of the main transport thrust.

REFERENCES

Eckelberg, D.J., 1958, Tisdale anticline: Wyoming Geol. Assoc. Guidebook, 13th Ann. Field Conf., p. 200-203.

Gries, Robbie, 1983, Oil and gas prospecting beneath the Precambrian of foreland thrust plates in the Rocky Mountains: Am. Assoc. Petroleum Geologists Bull., v. 67, p. 1-26.

Keefer, W.R., 1970, Structural geology of the Wind River Basin, Wyoming: U.S. Geol. Survey Prof. Paper 495-D, p. D1-D35.

Ray, R.R., Gries, Robbie, and Babcock, J.W., 1983, Acoustic velocities, synthetic seismograms, and lithologies of thrusted Precambrian rocks, Rocky Mountain foreland: This volume.

The American Association of Petroleum Geologists Bulletin
V. 67, No. 1 (January 1983), P. 1-28, 28 Figs., 3 Tables

Oil and Gas Prospecting Beneath Precambrian of Foreland Thrust Plates in Rocky Mountains[1]

ROBBIE GRIES[2]

ABSTRACT

Only 16 wells in the Rocky Mountain region have drilled through Precambrian rocks to test the 3 to 6 million acres of sedimentary rocks that are concealed and virtually unexplored beneath mountain-front thrusts. One recent test is a major gas discovery, another a development oil well, and over half of the unsuccessful tests had oil or gas shows. These wells have not only set up an exciting play, they have also helped define the structural geometry of the mountain-front thrusts, including the angle of the thrust, the amount of horizontal displacement, and the presence or absence of fault slivers containing overturned Mesozoic or Paleozoic rocks. Important for further geophysical exploration, these wells have provided vital data on seismic velocities in Precambrian rocks. Analysis of these data has stimulated further exploration along the fronts already drilled: in Wyoming, the Emigrant Trail thrust, the Washakie thrust, the Wind River thrust, the thrust at the north end of the Laramie Range, and the Casper arch; in Utah and Colorado, the Uncompahgre and Uinta uplifts.

The geologic success of these wells has encouraged leasing and seismic acquisition on every other mountain-front thrust in the Rockies. An unsuccessful attempt to drill through the Arlington thrust of the Medicine Bow Range will probably only momentarily daunt that play, and the attempted penetration of the Axial arch in Colorado has not condemned that area; in fact, another well is being drilled at this time. Untested areas that will be explored in the near future are: in Wyoming, the south flank of the Owl Creek Range, the southwest flank of the Gros Ventre Range, the east and west flanks of the Big Horn Mountains, the west flank of the Big Horn basin, the north flank of the Hanna basin; in Utah, the south flank of the Uinta Mountains; and in Colorado, the White River uplift, the north flank of North Park basin, and the Front Range.

[1]Manuscript received, February 11, 1982; accepted, July 23, 1982.
[2]Consulting geologist, Amarex, Inc., Denver, Colorado.
I thank Robert R. Berg, Texas A&M University, for encouraging me to write this paper; J. David Love, U.S. Geological Survey, for sharing ideas and information he has collected through years of work in Wyoming; and James A. Uhrlaub, Natomas North America, for encouraging me to think in an unconventional way while exploring for oil and gas in the Rockies and for introducing me to the mountain-front thrust play. R. E. Knight is thanked for helping with the drilling data on 14 of these wells. M. P. (Penny) Frush, editor, *Mountain Geologist*, deserves a large measure of gratitude for her work on the precursor to this paper published in the *Mountain Geologist*, January 1981.

INTRODUCTION

In the Rocky Mountain region, 3 to 6 million acres prospective for hydrocarbons are buried and overlain in thrust contact by Precambrian rocks. This vast subthrust area has been tested by only 16 wells (Fig. 1), which have drilled through Precambrian crystalline and metasedimentary rocks into underlying younger sedimentary strata. In January 1981, Moncrief Oil Co. announced the first major sub-Precambrian wildcat discovery when their 16-1 Tepee Flats well (Sec. 16, T37N, R86W, Natrona County, Wyoming) tested 11 MMCFGD. One other well, the True 31-6 Hakalo (Sec. 6, T32N, R75W, Converse County, Wyoming) drilled in 1971 was a development oil well producing 70 bbl of oil per day, and over half of the remaining unsuccessful wells have had hydrocarbon shows. In conjunction with more sophisticated seismic processing and geologic interpretations, these wells will be the basis for more successful exploration beneath several other mountain-front thrusts in the Rocky Mountain Foreland province.

Credit for this exciting play must go to the many people who for the last 35 years fought against great managerial prejudices to get wells drilled through Precambrian rocks and probably suffered more than the normal disappointment over a dry hole because of it. Their efforts, however, were not in vain because they gave the industry a solid foundation for exploration beneath these thrust plates and more important, gave the scientific world the much needed third dimension to begin to interpret the structure of the Rocky Mountains, a truly unique structural province.

The data presented in this paper (Table 1) are derived from a combination of scout cards, well logs, and conversations with geologists and engineers. These sources, while the best available, are naturally subject to error in reporting or interpretation. The wells described in this paper are limited to those that are known to have penetrated Precambrian rocks and exclude those that were drilled through the toe of these thrusts penetrating only sedimentary strata in the mountain front.

ROCKY MOUNTAIN FORELAND THRUSTS

The Rocky Mountain Foreland basin and uplift province is unique structurally as a system of tremendously high Precambrian-cored uplifts and adjacent deep basins, with vertical relief of 25,000 to 50,000 ft (7.6 to 15.2 km). This province extends from southwest Montana to northern New Mexico and from the western margin of the central Great Plains to the western Overthrust belt of southwestern Montana, western Wyoming, central Utah, and northern

1

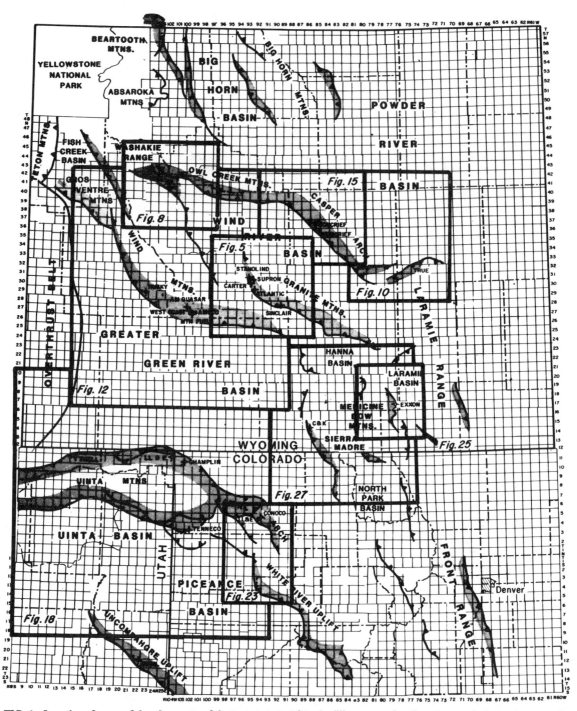

FIG. 1—Location of successful and unsuccessful attempts to test 3 to 6 million acres of sedimentary rocks beneath mountain-front thrusts. Stipple indicates approximate extent of subthrust sedimentary section. Blocks refer to other figures in paper.

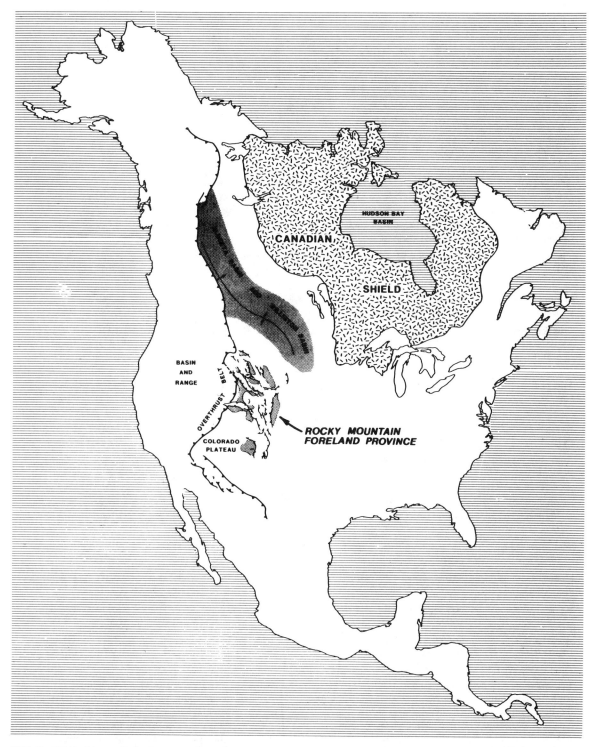

FIG. 2—Rocky Mountain Foreland province of Wyoming and Colorado is a system of basement-involved uplifts and adjacent depositional basins having vertical reliefs of 25,000 to 50,000 ft (7.6 to 15.2 km). It is in the west-central United States, adjacent to the western Overthrust belt.

Table 1. Well Data on 16 Wells that Penetrated Precambrian

Well Name and Operator	Location	Thrust	Spud Date	Compl. Date	Drlg. Time (mo.)	Top of Precambrian ft (m)	Precambrian Thickness	Formation Beneath Precambrian
Carter 1 Unit	Sec. 32, 30N, 93W Fremont Co., Wyo.	Emigrant Trail	1-12-50	6-7-50	6	1,200 (365)	1,900 (579)	Cambrian
Shell 1 Govt.	Sec. 9, 42N, 105W Fremont Co., Wyo.	E A	6-29-56	12-6-56	5	450 (137)	7,408 (2,258)	Ord.? or Dev.
Tennessee Gas 1 USA Margie Hicks-A	Sec. 3, 3N, 103W Moffat Co., Colo.	Uinta Mountain	9-22-60	1-28-61	4	6,020 (1,835)	1,973 (601)	Cambrian Ladore
Atlantic Refining 1 Ice Slough	Sec. 9, 29N, 93W Fremont Co., Wyo.	Emigrant Trail	11-3-60	12-18-60	1.5	1,092 (333)	2,056 (627)	Cambrian?
Sinclair 1 Cooper Creek	Sec. 19, 28N, 90W Fremont Co., Wyo.	Emigrant Trail	4-22-61	9-15-61	5	Surf.	2,087 (636)	Tertiary Wasatch
Shell 21X-9 Dahlgreen Creek	Sec. 9, 2N, 14E Summit Co., Utah	Uinta Mountain	11-17-67	10-8-68	10.3	2,940 (896)	8,790 (2,679)	Permian Phosphoria
Mountain Fuel 1 Dickey Springs	Sec. 24, 27N, 101W Fremont Co., Wyo.	Wind River	6-14-70	10-3-70	3.5	Surf.	4,290 (1,308)	Tertiary Wasatch
True & Rainbow Res. Shaffer Fed. 41-6	Sec. 6, 32N, 75W Converse Co., Wyo.	Northern Laramie Range	3-14-71	4-17-71	1	2,400 (732)	1,110 (335)	Cretaceous Cody
True 31-6 Hakalo	Sec. 6, 32N, 75W Converse Co., Wyo.	Northern Laramie Range	7-8-71	8-17-71	1.3	1,960 (597)	170 (52)	Cretaceous Cody
Husky 8-2 Federal	Sec. 2, 29N, 106W Sublette Co., Wyo.	Wind River	5-29-73	10-2-73	4	4,418 (1,347)	6,247 (1,904)	Ord.? or Dev.
American Quasar 1 Skinner Fed.	Sec. 32, 28N, 101W Fremont Co., Wyo.	Uinta Mountain	9-27-74	11-6-75	7	Surf.	10,090 (3,075)	Cretaceous Cody
Champlin Fed. 31-19 (Bear Springs)	Sec. 19, 11N, 102W Moffat Co., Colo.	Uinta Mountain	8-30-75	4-10-76	6.3	Surf.	7,100 (2,164)	Cretaceous Mancos
West Coast Oil 1 Skinner Fed.	Sec. 9, 27N, 101W Fremont Co., Wyo.	Wind River	9-12-76	1-22-77	4.3	3,600 (1,097)	1,400 427)	Cretaceous?
Mobil C-1 McCormick Fed.	Sec. 11, 21S, 22E Grand Co., Utah	Uncom- pahgre	2-3-77	12-1-77	9.5	3,600 (1,097)	14,000 (4,267)	Miss. or Dev.
Supron Energy 1 F-28-30-93	Sec. 28, 30N, 93W Fremont Co., Wyo.	Emigrant Trail	9-12-80	11-3-80	2	400 (122)	5,000 (1,524)	Permian Phosphoria
Moncrief 16-1 Tepee Flats	Sec. 16, 37N, 86W Natrona Co., Wyo.	Casper Arch	10-18-79	1-15-81	15	6,162 (1,878)	8,865 (2,702)	Triassic Chugwater

Arizona (Fig. 2). The Rocky Mountain foreland thrusts differ from the Overthrust belt thrusts in several ways: primarily, the Precambrian basement involvement in foreland thrusts and also the generally higher angle of the foreland thrusts compared to the low-angle, continuous sled-runner thrusts of the Overthrust belt (Fig. 3).

The dip of the fault plane on foreland thrust structures has long baffled geologists, and lack of information about this angle has hampered understanding of the foreland structural configuration and the causes of deformation.

Geologists mapping in the Rocky Mountain Foreland province have described the same mountain-front fault boundaries as everything from normal faults to low-angle thrust faults. In the last 20 years, arguments on the angle of the fault plane have divided into two basic camps (Fig. 4), those who thought the uplifts were thrust and/or fold thrusted as a result of crustal compression, as described by Berg (1962), and those who thought the uplifts were forced folds resulting

212

Rocks to Test Subthrust Sedimentary Section

Subthrust Fault Sliver Thickness ft (m)	Formation Beneath Fault Sliver	Formation at TD	TD ft (m)	Dip at TD	Shows
880 (268)	Cretaceous Thermopolis	Pennsylvanian Tensleep	6,591 (2,009)	13° NE	Oil cut and stain from Phosphoria cores.
820 (250)	Cretaceous Mowry	Jurassic Sundance	10,689 (3,258)	15° NE	Rec. 69 bbl gas-cut muddy water from Phosphoria and Tensleep in fault sliver.
370 (113)	Triassic Moenkopi	Pennsylvanian Weber	9,371 (2,856)	Low angle SE-SW	Sample shows in Weber, Moenkopi, Park City in fault sliver. Sample shows in Weber of subthrust.
520 (158)	Cretaceous Thermopolis	Pennsylvanian Tensleep	6,552 (1,997)	NE?	Dead & live oil in subthrust Phosphoria; recovered 660 ft slightly gas- and mud-cut water w/strong sulfur odor.
none	–	Cretaceous Frontier	12,224 (3,726)	20°? S	Oil shows from well samples in Cody; DST rec. 15 ft of mud.
2,083 (635)	Cretaceous Mesaverde	Jurassic Morrison	17,100 (5,212)	0°?	Sample shows in Phosphoria, Moenkopi, Thaynes, and Shinarump of fault sliver, and in subthrust Dakota, no tests in subthrust.
none	–	Cretaceous Mesaverde	12,282 (3,744)	7 to 10° N-NW	Trace of fluorescence in subthrust shales.
1,450 Overturned Cret. Cody (442)	Cretaceous Cody	Cretaceous Skull Creek	7,898 (2,407)	?	None reported.
2,590 (789)	Cretaceous Cody	Cretaceous Skull Creek	7,675 (2,339)	?	Sample shows in Frontier, Mowry, Muddy; completed for 70 BOPD in Muddy.
unknown	unknown	Pennsylvanian Amsden	12,944 (3,945)	?	None.
3,900 (1,189)	Cretaceous Mesaverde	Cretaceous Mesaverde	15,040 (4,584)	50° N-NW	Gas to surf. on Mesaverde (gaugeable), rec. 2,458 ft gas & very slightly oil-cut mud; minor sample shows in lower 5,000 ft of Precambrian.
none	–	Pennsylvanian Weber	13,810 (4,209)	15° NE	Sample shows in Precambrian, Moenkopi, and Phosphoria; recovered 465 ft gas-cut mud, 6,816 ft gas-cut water in Dakota test.
unknown	–	Cretaceous Mowry	9,700 (2,957)	45° N at 8,100 ft	None reported.
unknown	unknown	Miss. or Dev.	19,270 (5,873)	27° S-SW	None reported.
450 (137)	Triassic Chugwater	Pennsylvanian Tensleep	7,769 (2,368)	10° SW	Sample shows in Phosphoria.
2,373 (723)	Cretaceous Cody?	Jurassic Morrison	19,733 (6,015)	?	Tested 11 MCFGD from Frontier; producing 6 to 7 MCFGD.

from vertical uplift, as described by Stearns (1978). Drilling through the edges of these thrusts has produced the most concrete data concerning the angle of the thrust fault plane, and of the six mountain fronts drilled to date, the calculated angle of the thrust ranges from 20 to 45°.

Drilling and seismic exploration have also revealed many other characteristics of mountain-front thrusts not predicted by laboratory modeling, such as the following.

1. Subthrust structural highs generally are coincident with structural highs, anticlines, and Precambrian outcrops on the overthrust sheet (Berg, 1962). This does not preclude the presence of hydrocarbons in the deeper subthrust structures or stratigraphic traps. However, most subthrust tests to date have been drilled where major anticlinal trends in the basinal block intersect and can be projected beneath the overthrust block (e.g., where the Moxa arch and the Douglas Creek arch intersect the Uintas, the Rock Springs uplift intersects both the Uintas and the Wind River Range, and the Winkleman

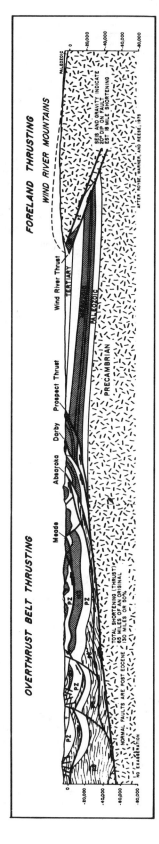

FIG. 3—Disturbed belt or Overthrust belt of western North America involves long continuous sled-runner type overlapping thrusts with 20 to 100 mi (32 to 161 km) of horizontal displacement and 5,000 to 10,000 ft (1.5 to 3.0 km) of vertical displacement, whereas foreland thrusts have 2 to 20 mi (3.2 to 32.2 km) of horizontal displacement and 25,000 to 50,000 ft (7.6 to 15.2 km) of vertical displacement. This cross section across Overthrust belt and Wind River foreland thrust demonstrates some structural differences in the two phenomena.

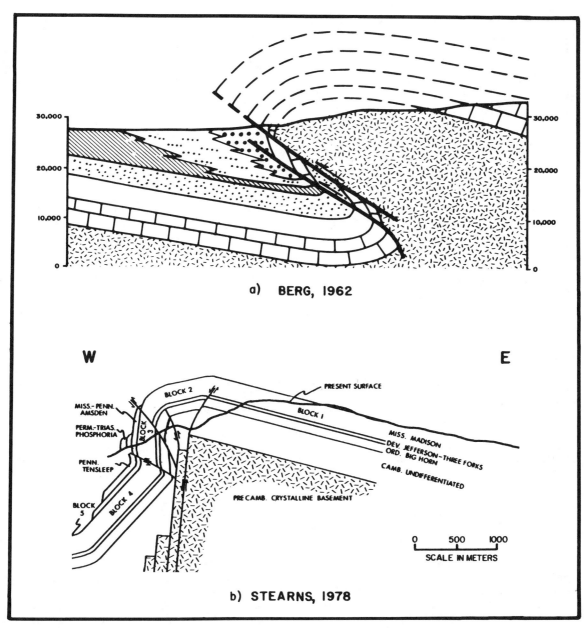

a) BERG, 1962

W   E

b) STEARNS, 1978

FIG. 4—Most disputed question concerning foreland structures has been whether or not they are thrust folds caused by compression, as described by Berg (1962), or forced folds caused by vertical uplift, as theorized by Stearns (1978).

dome trend intersects the Emigrant Trail thrust).

2. If Paleozoic rocks are present just beneath the Precambrian, they are usually in a fault sliver of overturned beds formed as a result of the fold-fault process (Berg, 1962) of mountain-front compressional thrusting. "Fault sliver" is a term used in this paper to describe the reverse fault-bounded sliver of sedimentary rocks commonly found beneath the Precambrian. A fault sliver usually has numerous other thrust faults within it and the sedimentary rocks within the sliver are usually overturned and highly deformed.

3. The amount of dip-slip thrust overlap on the mountain flank thrusts cannot yet be predicted along the structural trend (i.e., northwest-southeast-trending flanks of the ranges will not necessarily have more overlap than an east-west-trending mountain flank. It has been suggested that dip-slip thrust overlap at the ends or on the east-west-trending flanks of these ranges was minimal, and strike-slip movement was more prevalent (Stone, 1969; Brown, 1981; Sales, 1968). However, the known presence of subthrust sedimentary section 5 mi (8 km) behind the thrust edge in the American Quasar well on the south flank of the Wind River uplift and seismic data indicating 20 mi (32.2 km) or more of overthrust

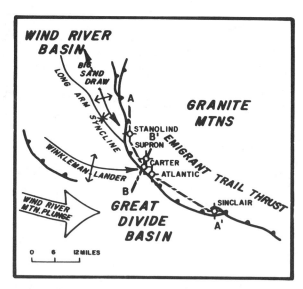

FIG. 5—Carter, Atlantic Refining, Sinclair, and Supron wells drilled through Precambrian granite into sedimentary rocks beneath Emigrant Trail thrust of Granite Mountains in central Wyoming. AA' is location of cross section of Figure 7; BB' is section of Figure 6.

must cause revision of this hypothesis to include significant thrust overlap on this particular east-west-trending flank. In addition, the evidence for significant thrust overlap on the south flank of the Owl Creek Range and Granite Mountains indicates a component of north-south compression on these east-west-trending mountain ranges (Gries, 1982). In a compressional model that includes significant north-south compression, the east-west-trending Uinta Mountains are no longer an enigma, as they have been with many previous foreland structural models.

4. The presence of overturned Paleozoic as opposed to upright Tertiary or Cretaceous strata beneath the Precambrian is also not yet predictable based on the part of the mountain flank being drilled. Where drilled, the Wind River and Emigrant Trail thrusts had overturned Paleozoic rocks beneath the Precambrian on the west flank and normal Tertiary and Cretaceous section beneath the Precambrian on the outer edge of the southern flank. These two similar cases alone might have been used to establish a pattern, except that drilling in the east-west-trending Uintas has revealed both overturned Paleozoic rocks and minimally disturbed Cretaceous rocks.

5. Three mountain fronts have had more than one well drilled through them. Four wells drilled through the flank of the Wind River thrust show three very different subthrust structural phenomena: (a) overturned fault sliver of Paleozoic rocks on the west flank; (b) normal Tertiary and older sedimentary rocks on the outer edge of the south flank; and (c) fault slivers of intensely folded Cretaceous rocks 5 mi (8 km) behind the south edge of the thrust. Similar sequences have been found beneath the Emigrant Trail (with four tests) and Uinta Mountain thrusts (with three tests).

6. Although movement on the foreland mountain uplifts was most active during the Laramide orogeny (Late Cretaceous through Eocene), there was episodic movement on some

of these uplifts during Precambrian, Paleozoic, and Mesozoic times. Hence, oil and gas prospecting should include analyses of hydrocarbon generation, migration, and entrapment prior to Laramide events.

Until many wells are drilled on all the foreland thrusts, we will be plagued with unanswered questions on both the geometry and causes of these thrusts.

Emigrant Trail Thrust

Granite Mountains, Central Wyoming

Carter Oil Co. drilled the first well (Sec. 32, T30N, R93W, Fremont County, Wyoming) through Precambrian rocks on a mountain-front thrust when they drilled through the Emigrant Trail thrust in central Wyoming (Fig. 5) in 1950. The well, Carter 1 Unit, penetrated 1,900 ft (579 m) of Precambrian granite before drilling out of the granite into an 880-ft (268 m) thick fault sliver of overturned Cambrian through Jurassic rocks (Berg, 1961). The well then drilled through a normal stratigraphic section from the Cretaceous Thermopolis Shale through Pennsylvanian Tensleep Sandstone. Regional mapping of the area shows that the major syncline of the Long Creek Arm of the Wind River basin slips beneath the Emigrant Trail thrust about 8 mi (12.9 km) north of the well. The 13° NE dip at total depth and the 5,925 ft (1,805.9 m) of fresh water recovered on a drill-stem test from subthrust Tensleep are indications that perhaps there was no stratigraphic or structural barrier in the Tensleep between the well and outcrops to the southwest (Fig. 6). A drill-stem test in the Lower Jurassic Sundance Formation also recovered 3,160 ft (963.2 m) of fresh water. The only shows in the subthrust section in this well were oil cut and stain from cores in the Permian Phosphoria Formation. Oil shows of this type are common in the Phosphoria and are probably not related to structural entrapment.

Stanolind drilled the 1 Scarlett Ranch Unit well (Sec. 22, T31N, R94W, Fremont County, Wyoming) 9 mi (14.5 km) north of the Carter well in 1956. It was located slightly northeast of the major synclinal axis (Fig. 5). The Stanolind well was spudded in Paleozoic rocks instead of Precambrian and was about 5,000 ft (1.5 km) structurally lower than the Carter well (Berg, 1962). Because of the greater thickness of Precambrian and younger rocks in the overthrust block (Fig. 7), the Stanolind well was still in Precambrian rocks at 8,000 ft (2.4 km). Another several hundred feet of drilling might have put them into subthrust Paleozoic rocks, on trend with the prolific Big Sand Draw-Beaver Creek trend that has produced over 175 million bbl of oil and 1 tcf gas to date. As the Stanolind well is north of the basinal axis, it could be in an optimum position for significant hydrocarbon accumulations in any existing structural or stratigraphic traps as an extension of the Big Sand Draw trend (Lowell, 1978). Several shallow tests drilled on the flanks of the Granite Mountains, which failed to penetrate the Precambrian, are described by Love (1970).

In 1960, Atlantic Refining drilled the 1 Ice Slough well (Sec. 9, T29N, R93W, Fremont County, Wyoming) 2 mi (3.2 km) south of the 1950 Carter test (Figs. 5, 7). The results were essentially the same as those of the Carter well, having drilled 2,056 ft (627 m) of granite before coming out in a fault

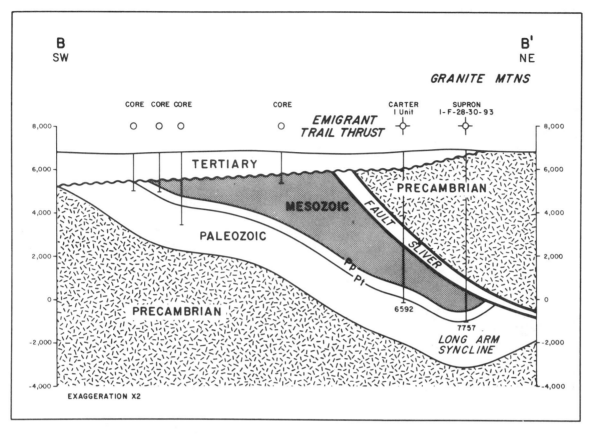

FIG. 6—Early wells drilled through Emigrant Trail thrust had no structural or stratigraphic barriers from outcrop to southwest. Location BB′ shown on Figure 5.

sliver of overturned Paleozoic and lower Mesozoic rocks and then through another fault and into a normal stratigraphic section of Thermopolis through Tensleep formations. Cores from the Phosphoria Formation contained live and dead oil in vugular porosity. A drill-stem test in the Phosphoria recovered 660 ft (201 m) of slightly gas- and mud-cut water with strong sulfur odor. Another test, just above the Tensleep, recovered 5,110 ft (1,557.5 m) of slightly muddy fresh water. The fresh water and the northeast dips in this zone again possibly indicate there is no structural barrier on this synclinal flank between the well and the outcrop. This well was abandoned. Both it and the Carter well were probably drilled on seismic velocity pullups (faster velocities in Precambrian rocks create unrealistic seismic anticlinal structures commonly referred to as "velocity pullups") on the southwest flank of the Long Arm syncline of the Wind River basin.

In 1980, Supron Energy drilled 6,200 ft (1.9 km) northeast of the Carter well (Sec. 28, T30N, R93W, Fremont County, Wyoming). The well drilled through 5,000 ft (1.5 km) of Precambrian rocks, which was 2,000 ft (610 m) more than in the Carter well (Fig. 6). This confirmed a 25 to 27° dip on the thrust plane, which was the angle predicted by Berg (1961) and Love (1970). Dips in the sedimentary section at the bottom of the hole were rumored to be about 10° SW indicating they may have been on the previously untested northeast

flank of the subthrust Long Arm syncline. Those dips also would indicate they were not on top of a structural closure; however, without shows in the Tensleep, Supron may not try to test updip.

In 1961, the Sinclair 1 Cooper Creek well (Sec. 19, T28N, R90W, Fremont County, Wyoming) was drilled through the Precambrian on the south flank of the Emigrant Trail thrust about 17 mi (27.4 km) southeast of the Carter well (Figs. 5, 7). A 2,087-ft (636 m) section of Precambrian granite was drilled before the well penetrated undisturbed Tertiary Wasatch Formation (Love, 1970). A normal Tertiary and Cretaceous section was drilled to a total depth of 12,224 ft (3.7 km) in the Frontier Formation, which had gentle south dip. Oil shows in sample cuttings from the Cretaceous Cody were reported, but one drill-stem test in the Cody recovered only 15 ft (4.6 m) of mud.

The subthrust play of the Emigrant Trail thrust can be divided into three parts (Fig. 7): (1) the northern end, with the prolific Big Sand Draw trend intersecting the thrust, (2) the central part characterized by being structurally highest in both the overthrust plate and the subthrust plate, and (3) the southeast flank, with characteristics of the Great Divide basin to the south including thick Tertiary and Cretaceous sections. The potential of this play in all three areas still remains inadequately tested.

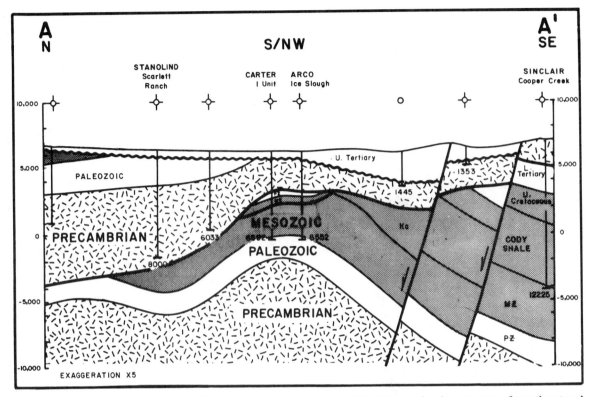

FIG. 7—North-southeast cross section AA′ along leading edge of Emigrant Trail thrust, showing structure of overthrust and subthrust blocks. For location see Figure 5.

E A Thrust

Washakie Range, Northwest Flank of Wind River Basin, Northwest Wyoming

A second mountain-flank thrust was tested in 1956 when Shell drilled the 1 Government well (Sec. 9, T42N, R105W, Fremont County, Wyoming) through Precambrian rocks near Dubois (Fig. 8). It drilled through 7,408 ft (2.3 km) of granite and schist in the E A thrust plate and then into a fault sliver of overturned, deformed, and sheared lower Paleozoic through Cretaceous rocks (Fig. 9). At 8,678 ft (2.6 km), they drilled out of the fault sliver and into a normal stratigraphic section of Cretaceous through Jurassic rocks, ending in the Jurassic Sundance Formation at a total depth of 10,689 ft (3.3 km). A drill-stem test in the overturned Tensleep and Phosphoria recovered 69 bbl of gas-cut muddy water.

This test well, like some of the early Emigrant Trail tests, probably failed because it was located on the wrong side of the major basinal axis (Fig. 8), with no stratigraphic or structural barriers between the well and the outcrop. The Wind River basin syncline can be projected beneath the E A thrust to the northeast of the well, which is consistent with the 15° NE dips found in the lower strata of the Shell test. The area northeast of the synclinal axis continues to offer good exploration potential. Both seismic and regional geologic studies indicate subthrust sedimentary sections may extend for some distance north-northeast of the Shell well (Fig. 9).

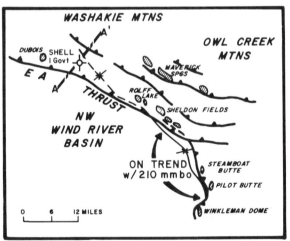

FIG. 8—Shell drilled through E A thrust of Washakie Mountain front into sedimentary rocks of northwest Wind River basin. AA′ is section shown in Figure 9.

North End of Laramie Range

South Flank, Powder River Basin, East-Central Wyoming

In 1971, a True Oil Co. development well in South Glenrock field (NE ¼, NE ¼, Sec. 6, T32N, R75W, Converse

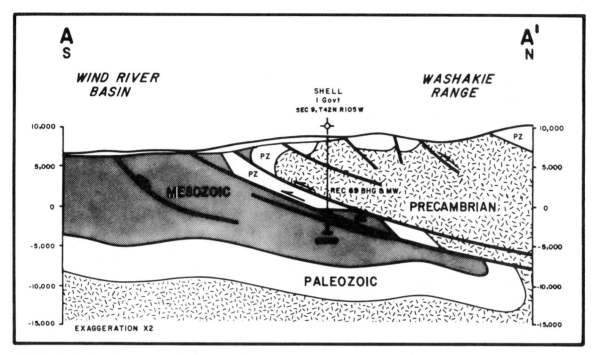

FIG. 9—Shell drilled over 7,400 ft (2.3 km) of crystalline basement, fault slivers of overturned Paleozoic and lower Mesozoic rocks, and finally into relatively undisturbed Cretaceous and Jurassic sections. Location of cross section shown on Figure 8.

County, Wyoming) drilled through 1,100 ft (335 m) of Precambrian rocks at the north end of the Laramie Range before penetrating a thrust and drilling into Cretaceous rocks (Fig. 10). In this well, the True and Rainbow Res. Shaffer Federal 41-6, 1,450 ft (442 m) of steeply dipping Cody Shale overlying normal older Cretaceous sandstones indicated probable overturning and/or thrust faulting in the Cody (Curry, 1972). Total depth was 7,898 ft (2.4 km) in the Cretaceous Skull Creek Shale. Drill-stem tests in the Cretaceous Muddy Sandstone recovered only 60 ft (18 m) of mud, and the well was abandoned. Later that year, an offset well (True 31-6 Hakalo, NW ¼, NE ¼, Sec. 6, T32N, R75W) a few hundred feet to the west, drilled through 170 ft (52 m) of Precambrian rocks before crossing the thrust and again drilling into disturbed and thicker Cretaceous Cody Shale (Fig. 11). Drilled to the Skull Creek, this well was completed as an oil well in the subthrust Muddy formation with a potential for 70 bbl of oil per day. This was the first sub-Precambrian production in the Rocky Mountain region.

Wind River Mountain Front

West-Central Wyoming

Four wells have been drilled through Precambrian rocks along the west and south flanks of the Wind River Mountains (Fig. 12). All were dry holes, but because they revealed subthrust structure, good reservoir and source rocks, shows of gas, and a low-angle thrust, exploration continues along this front.

In 1973, the Husky 8-2 Federal (Sec. 2, T29N, R106W,

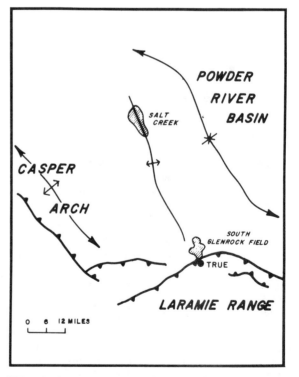

FIG. 10—Precambrian rocks in thrust on north flank of Laramie Range were penetrated by three True wells (Sec. 6, T32N, R75W, Converse County, Wyoming).

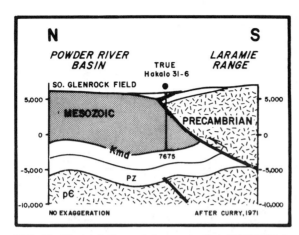

FIG. 11—First sub-Precambrian production (70 BOPD) in Rocky Mountains was from a development well drilled by True Oil in South Glenrock field.

Sublette County, Wyoming) drilled 4,418 ft (1.3 km) of Tertiary rocks, then 6,247 ft (1.9 km) of Precambrian crystalline rocks before entering a fault sliver with an overturned section starting with an undated black shale overlying Mississippian rocks of probably Osagian through Chesterian age, and finally into the Pennsylvanian Amsden Formation (Fig. 13). Because Husky's rig did not have the capacity to drill deeper, drilling stopped at 12,944 ft (3.9 km) in the Amsden, and the well was abandoned. The well never drilled through the fault sliver to test the normal Cretaceous and older section that extends 20 mi (32.2 km) or more back beneath the Wind River thrust, according to COCORP seismic interpretations (Smithson et al, 1979). There were no oil or gas shows reported in this well.

Three wells drilled through Precambrian rocks on the south flank of the Wind River thrust (Fig. 12). In 1970, Mountain Fuel spudded the 1 Dickey Springs well (Sec. 24, T27N, R101W, Fremont County, Wyoming), and penetrated 4,290 ft (1.3 km) of Precambrian rocks before drilling into the Tertiary Wasatch Formation (Fig. 14). Subsequent formation tops included Tertiary Fort Union Formation at 5,000 ft (1.5 km), Cretaceous Lance Formation at 7,622 ft (2.3 km), and Cretaceous Mesaverde Formation at 10,940 ft (3.3 km). Dips in the subthrust section were 7 to 10° NNW. Three drill-stem tests were run in the Mesaverde and recovered only minor amounts of mud before the well was abandoned.

American Quasar subsequently drilled the 1 Skinner Federal (Sec. 32, T28N, R101W, Fremont County, Wyoming), 5 mi (8 km) northwest of the Mountain Fuel well. After drilling about 10,100 ft (3.1 km) of Precambrian, the well penetrated what appeared to be Cretaceous Cody Shale, based on the presence of *Inoceramus* fragments and glauconite. This shale dipped about 30° N (Fig. 14). At 13,434 ft (4.1 km), a sandstone was penetrated possibly in the Cretaceous Frontier Formation, dipping about 50° N. Cutting another thrust fault at 14,000 ft (4.3 km), the well drilled into younger Cretaceous Mesaverde sandstones with numerous coal beds. At the total depth of 15,040 ft (4.6 km), dips were 50° NNW. Seven drill-stem tests were run across the fault at the top of the Mesaverde—five were misruns, one recovered 1,146 ft (349

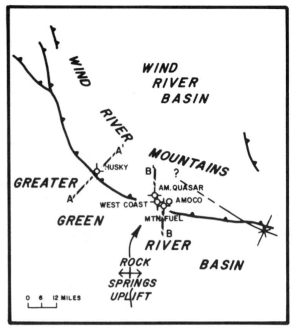

FIG. 12—Four wells drilled through Precambrian of Wind River Mountain thrust; Amoco is currently drilling a fifth. Section AA' is Figure 13; BB' is Figure 14.

m) of mud, and the last one gauged a maximum of 29.4 MCFGD and recovered 2,458 ft (749 m) of gas and very slightly oil-cut mud from the pipe. No attempts were made to complete the well before abandonment.

In 1976-77, West Coast Oil drilled the 9,700-ft (3.0 km) 1 Skinner Federal test 2 mi (3.2 km) south (Sec. 9, T27N, R101W, Fremont County, Wyoming), probably trying to get updip from the shows in the American Quasar well. The West Coast well drilled out of the Precambrian at about 5,000 ft (1.5 km) and into beds of probably Cretaceous age (Fig. 14). Near the bottom of the hole, Cretaceous shale was identified by J. D. Love (personal commun., 1980). Dips averaged 45° N at 8,100 ft (2.5 km) in this well. It is likely that they did not drill through the second deeper thrust fault penetrated in the American Quasar well at 14,000 ft (4.3 km). The highly disturbed Cretaceous rocks tested in the subthrust part of these wells may indicate that the undisturbed subthrust beneath fault slivers has not yet been tested under the Wind River thrust.

Amoco Production is drilling (Fig. 12) a proposed 22,500-ft (6.9 km) Mississippian Madison test about 5 mi (8 km) east of this area (Sec. 17, T27N, R100W) and has already drilled through the Precambrian rocks of the overthrust block.

Casper Arch

East Flank of Wind River Basin, Central Wyoming

The Moncrief 16-1 Tepee Flats well (Sec. 16, T37N, R86W, Natrona County, Wyoming), drilled in 1979-81, was

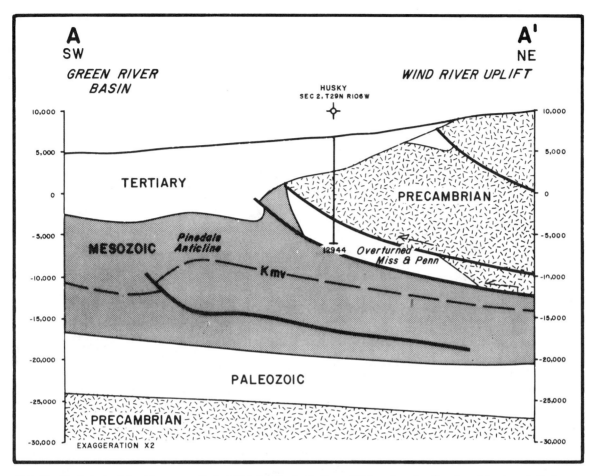

FIG. 13—Husky's rig did not have capacity to drill below the fault sliver of Paleozoic rocks; therefore did not reach Cretaceous rocks in subthrust. Location of cross section shown on Figure 12.

the first major and wildcat discovery beneath Precambrian rocks (Fig. 15) and had a reported initial potential of 11 MMCFGD. It produced 6 to 7 MMCFGD after being put on production. This discovery is to the mountain-front overthrust play what the discovery of Pineview field was to the Overthrust belt play in 1975 and likewise has set off leasing, geophysical shooting, and/or drilling through the Precambrian of every other mountain front in the Rocky Mountain foreland.

Moncrief Oil Co., on this intended 22,000-ft (6.7 km) Mississippian test, drilled 8,865 ft (2.7 km) of Precambrian rocks before entering subthrust sedimentary rocks at 15,027 ft (4.6 km) (Fig. 16). Just under the thrust it drilled into a fault sliver of overturned Triassic Chugwater Formation, through part of the Cretaceous section, and then at about 16,700 ft (5.1 km) drilled into a normal Cretaceous section above the Cretaceous Frontier Formation. After drilling down to Jurassic rocks at a total depth of 19,733 ft (6.0 km), where mechanical problems probably prevented drilling any deeper, the well was plugged back to the Frontier Formation and completed in a zone from 18,238 to 18,403 ft (5.6 km).

Seismic lines (Fig. 17) over the northwest Casper arch

(southern Big Horn Mountains) are remarkably similar to those shot over the Wind River thrust in demonstrating a low-angle (25 to 30°) dip on the thrust plane but no steepening with depth. This information is critical in developing new theories to explain the mechanics of forming the foreland thrusts, because most laboratory modeling has previously accepted a premise that these faults steepen with depth.

The lines over the Casper arch also demonstrate the presence of sedimentary rocks in a large structural block that is faulted higher than the rocks in the adjacent Wind River basin. This structural interpretation has been confirmed with the drilling of the Moncrief 16-1 well which penetrated Cretaceous Frontier rocks at 18,200 ft (5.5 km), whereas a few miles west in the basin syncline the Frontier Formation is at about 4.5 seconds or depths of 25,000 to 30,000 ft (7.6 to 9.1 km).

Moncrief is now drilling several other wells through the thrust. About 5 mi (8 km) southeast of their first well, the 11-1 Lox (Sec. 11, T36N, R86W) is scheduled to test the Mississippian Madison Formation at 21,000 ft (6.4 km). The 27-1 Tepee Flats well (Sec. 27, T37N, R86W) is currently being tested, after reaching a total depth of 21,382 ft (6.5 km). East

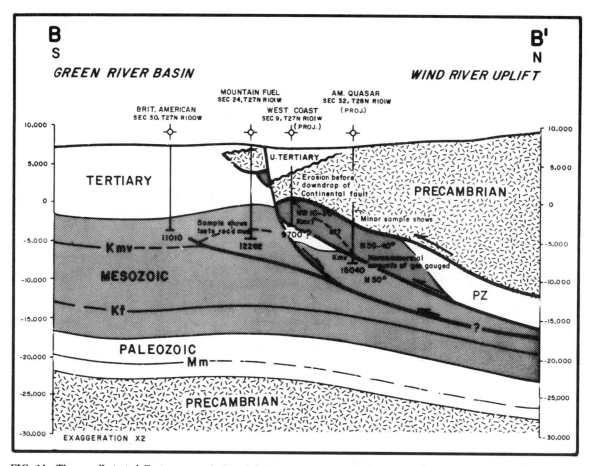

FIG. 14—Three wells tested Cretaceous rocks in subthrust on south end of Wind River Mountain thrust. See Figure 12 for location of section.

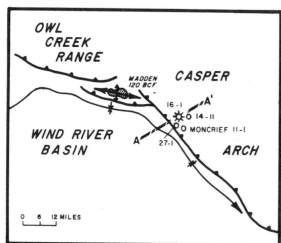

FIG. 15—Moncrief 16-1 Tepee Flats (Sec. 16, T37N, R86W, Natrona County, Wyoming) was first wildcat discovery below Precambrian rocks. It tested 11 MMCFGD from Cretaceous Frontier Formation. Two other tests to southeast are scheduled to test subthrust Mississippian formations. Another test east of discovery was spudded in early 1982. Section line AA' is Figure 16.

of the discovery well, the W. A. Moncrief 14-1 Tepee Flats (Sec. 14, T37N, R86W) recently was spudded to test subthrust Cretaceous rocks at 21,000 ft (6.4 km).

Uinta Mountain Uplift

Northwest Colorado, Northeast Utah

In 1960, Tennessee Gas and Oil Co. (T.G.& O.) drilled the 1-USA Margie Hicks-A well (Sec. 3, T3N, R103W, Moffat County, Colorado) north of Rangely field in northwest Colorado (Fig. 18). Rangely field has produced 615 million bbl of oil to date from the Pennsylvanian Weber Sandstone. Tennessee Gas' objective was a seismic structure in the subthrust, hoping for similar Weber potential. Interpretation of this well is primarily from R. B. Powers (oral commun., 1980), one of the geologists responsible for this play. T.G.& O. drilled into the Precambrian Uinta Mountain Group metasediments at 6,020 ft (1.8 km) (Fig. 19) and at 7,993 ft (2.4 km) into a fault sliver of overturned rocks ranging from Pennsylvanian Weber Sandstone through Triassic Moenkopi Formation. At 8,363 ft (2.5 km) they drilled out of the fault sliver and into a normal south-dipping stratigraphic section of Triassic Moenkopi Formation through Pennsylvanian Weber formation. Originally scheduled for 8,350 ft (2.5 km), drilling had

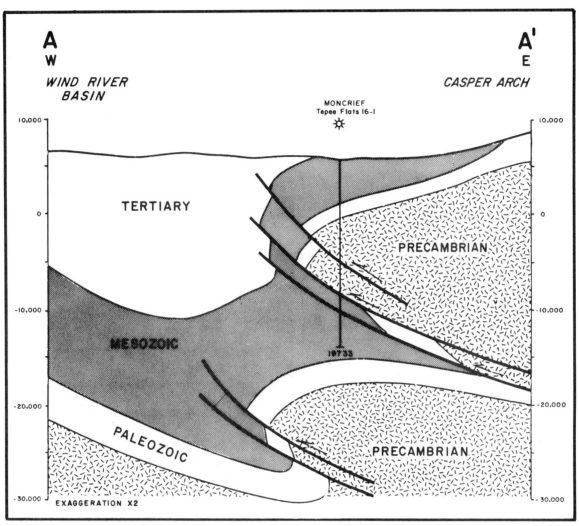

FIG. 16—Subthrust structure on Casper arch could be similar to Madden field northwest of Moncrief 16-1 Tepee Flats discovery well. This structure is an asymmetric fold related to compression of Owl Creek Range. Location of section shown on Figure 15.

to continue to 9,371 ft (2.9 km) to test the Weber objective, apparently because seismic velocity pullups affected the estimated target depth and structural interpretation. Tennessee Gas had moved its initial location 750 ft (229 m) south to allow the drill bit to drift updip (north) and also to avoid drilling a thicker Precambrian section. The Weber might have been topped structurally higher if the hole had actually drifted north. The bottom-hole location was about 1,300 ft (396 m) south of the estimated target spot. Dips of the fault planes were 30 and 32° N. All the beds in the fault sliver were north-dipping, and dips were at low angles to the southeast and southwest below the fault sliver. Numerous oil shows in well cuttings were described in this well. A drill-stem test of the overturned Weber in the fault sliver, in which there were scattered brown oil stains in samples and a drilling break, recovered only 590 ft (180 m) of muddy brackish water. Good oil shows were present in the Park City and Weber below the fault sliver, but no porosity was observed. Three drill-stem

tests in the Weber at the bottom of the hole recovered only mud or mud-cut salt water. Dips at bottom hole were generally southeast. The absence of porosity in the subthrust Weber Sandstone and other potential reservoirs is responsible for the lack of continued interest in the subthrust potential of this area.

In 1967, Shell drilled the northwest flank of the Uinta Mountain uplift (Fig. 18), reaching a total depth of 17,100 ft (5.2 km) in the 21X-9 Dahlgreen Creek Unit (Sec. 9, T2N, R14E, Summit County, Utah). It penetrated 8,790 ft (2.7 km) of Precambrian Uinta Mountain Group before drilling into a fault sliver with overturned Permian Phosphoria through Jurassic Nugget formations (Fig. 20). Another small fault sliver, with Triassic Shinarump section, was drilled before entering a normal section of Cretaceous Mesaverde through Jurassic Morrison. Dip was reported to be almost horizontal in the subthrust. Although no drill-stem tests were run in the subthrust, numerous shows were found in samples in the

Phosphoria and Moenkopi of the fault sliver and in the subthrust Dakota, which was topped at 16,790 ft (5.1 km).

Champlin drilled the northeast flank of the Uinta Mountains with the Federal 31-19 (Bear Springs) 1 well (Sec. 19, T11N, R102W, Moffat County, Colorado) in 1975-76 to a total depth of 13,810 ft (4.2 km) (Fig. 18). At 7,100 ft (2.2 km), they drilled into overturned Cretaceous Mancos Shale dipping to the south, and several hundred feet deeper, into Mancos dipping 13° NE (Fig. 21). That dip persisted to total

depth in the Pennsylvanian Weber Sandstone. Although oil shows in well cuttings began in the Precambrian, 1,500 ft (457 m) above the thrust, beneath the thrust oil shows were present only in the Triassic Moenkopi and Permian Phosphoria Formations. A drill-stem test in the Cretaceous Dakota Formation recovered 465 ft (142 m) of gas-cut mud and 6,816 ft (2,078 m) of gas-cut water. The Pennsylvanian Weber formation was porous, but apparently since no shows were seen, it was not tested. With north-northeast dip in the subthrust section, it is possible that a structural closure may exist south of this well, where topographic problems prevented seismic confirmation of critical south dip.

Louisiana Land and Exploration was scheduled to drill a 16,500 ft (5.0 km) Mississippian test on the north flank of the Uinta Mountains (Sec. 2, T2N, R25E, Daggett County, Utah), but had problems drilling fractured Precambrian rock and temporarily abandoned this attempt to test a subthrust section south of Clay Basin field in northeast Utah (Fig. 18).

Uncompahgre Mountain-Front Thrust

East-Central Utah

In 1977, Mobil drilled through the southwest flank of the Uncompahgre uplift. The Mobil C-1 McCormick Federal (Sec. 11, T21S, R22E, Grand County, Utah) drilled to a total depth of 19,270 ft (5.9 km) (Fig. 22), penetrating more Precambrian rocks, 14,000 ft (4.3 km), than any other well in the Rockies. At 17,600 ft (5.4 km), they drilled out of Precambrian and into very tight Mississippian or older dolomite, with beds dipping approximately 27° SSW at the bottom of the hole. From 17,600 to 19,100 ft (5.4 to 5.8 km) recorded dips were very erratic and possibly at higher angles. There were no shows reported in the subthrust and no tests were run. The well took 9 months to drill, including 3 months lost when pipe twisted off twice.

Mobil has drilled a second well (Sec. 30, T21S, R22E)

FIG. 18—Three wells have drilled through Precambrian metasediments of Uinta Mountain uplift, and fourth well operated by Louisiana Land & Exploration was attempted recently. AA' is Figure 19 section; BB' is Figure 20; CC' is Figure 21.

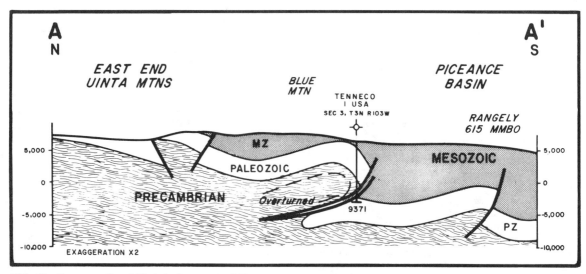

FIG. 19—Tennessee Gas drilled southeast end of Uinta Mountains through overturned Precambrian metasediments to test nonporous Weber Sandstone. Location of cross section shown on Figure 18.

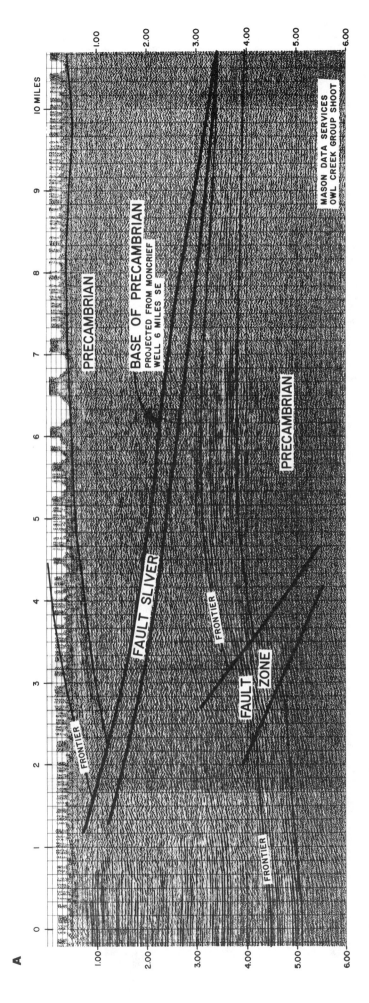

FIG. 17—(a) Fault flanking west side of Casper arch appears to be only 25 to 30° and does not steepen with depth. Base of Precambrian in overthrust is projected from about 6 mi (9.7 km) southeast at Moncrief well. More than 10 mi (16.1 km) of overthrow is shown on this line. Upthrown block beneath main Casper arch thrust is producing structure. (Line courtesy of Mason Data Services from their 1981 Owl Creek group shoot, not migrated.) (b) Same line, uninterpreted.

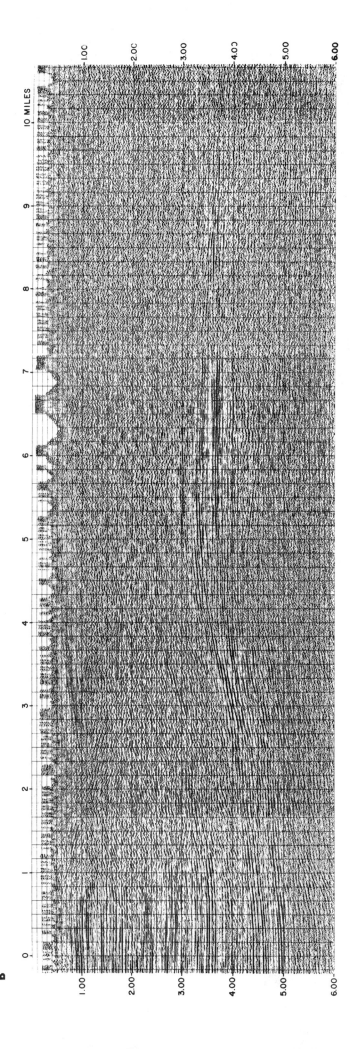

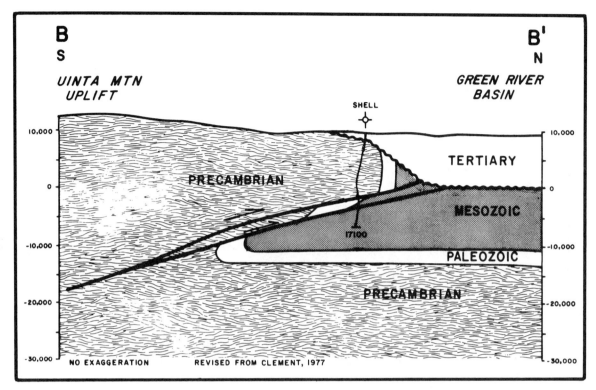

FIG. 20—Shell 21X-9 Dahlgreen Creek Unit well (Sec. 9, T2N, R14E, Summit County, Utah) had numerous oil shows in well-bore cuttings. See Figure 18 for location.

southwest of the first, but did not penetrate any Precambrian rocks in that test.

UNSUCCESSFUL ATTEMPTS TO PENETRATE SEDIMENTARY SECTION BENEATH PRECAMBRIAN

Since the Axial arch, Medicine Bow, and Sierra Madre plays are still active, three recently drilled wells in these areas deserve mention, despite their unsuccessful attempts to drill through Precambrian rocks.

Axial Arch

Northwest Colorado

In 1976, Conoco drilled the Axial 1-13 Federal (Sec. 13, T5N, R94W, Moffat County, Colorado; Fig. 23) on the Axial arch of northwest Colorado. After penetrating a normal sedimentary section of Cretaceous and older rocks, Precambrian rocks were drilled from 7,600 ft (7.3 km) to the total depth of 9,425 ft (2.9 km; Fig. 24). Their goal was a structure with four-way seismic closure located on the same structural block as the Wilson Creek-Danforth Hills trend to the west-southwest that has produced over 92 million bbl of oil. The thrust may be several hundred feet deeper, a speculation that still remains to be tested. The increased thickness of high-velocity Precambrian rocks found in this well will subtract from the amount of seismically calculated structural closure. Louisiana Land and Exploration was drilling a proposed

12,800 ft (3.9 km) Pennsylvanian Weber test about 1 mi (1.6 km) southwest of the Conoco well (Sec. 23, T5N, R94W) in 1981; however, it abandoned the well at 11,074 ft (3.4 km) and did not release any formation tops.

Arlington Fault

East Flank, Medicine Bow
Range, South-Central Wyoming

In the winter of 1979-80, Exxon (Fig. 25) spudded the 1 Medicine Bow (Sec. 10, T17N, R78W, Carbon County, Wyoming), and after drilling an unexpected 4,573 ft (1.4 km) of Precambrian rocks (Fig. 26), the well was plugged, leaving untested an interpreted seismic structure at the calculated depth of 8,500 ft (2.6 km). Application of faster velocities to the seismic data in the Precambrian may alter the anticipated structure, but it does not alter the seismic evidence for a sedimentary section extending some distance west of the east flank of the Medicine Bow Range, adjacent to the prolific oil-bearing Laramie basin.

East Flank, Washakie Basin,
South-Central Wyoming

In 1980, C&K Petroleum drilled the 1-36 Hartt Unit well (Sec. 36, T15N, R88W, Carbon Country, Wyoming) to a total depth of 5,856 ft (1.8 km) to test an interpreted seismic structure east of the Sierra Madre west-bounding fault (Fig.

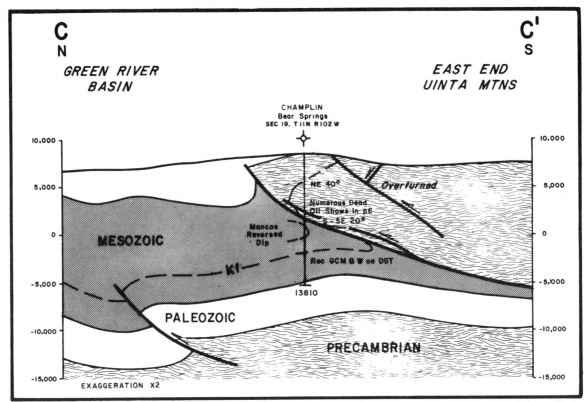

FIG. 21—Champlin drilled through 7,100 ft (2.2 km) of Precambrian metasediments into overturned Cretaceous Mancos Shale and eventually into porous, wet Weber sandstone that was dipping northeast 13°. See Figure 18 for location.

27). According to R. A. Andersen (oral commun., 1980), their geologists were surprised to drill from Tertiary cover into Precambrian at approximately 300 ft (91m). After drilling 700 ft (213m) of Precambrian schist, gneiss, and quartz veins, a thrust fault was cut and a thin section of lower

Triassic red beds, thin Permian beds, and a normal Pennsylvanian, Mississippian, and Cambrian section were penetrated. Precambrian crystalline rocks (phyllite, mica schist, and gabbro) were again penetrated at about 2,000 ft (610 m). The well was abandoned in Precambrian rock at 5,860 ft (1.8 km), a depth that was determined by the velocity survey to be the top of seismic reflections interpreted as sedimentary strata.

The rig was then moved to the 3-15 Hartt Unit location (Sec. 15, T15N, R88W) which was again interpreted to be in front of the mountain-front thrust. The second well penetrated several hundred feet of Tertiary cover before entering a Middle and Lower Triassic red-bed sequence. The rest of the section was nearly identical to the Permian through Cambrian section identified in the 1-36 well. Precambrian was penetrated at about 2,450 ft (747 m) and the well reached total depth at 4,175 ft (1.3 km), still in Precambrian. C&K is currently reevaluating seismic data with new velocity control before resuming operations in this area.

FUTURE OIL AND GAS EXPLORATION BENEATH MOUNTAIN FRONT THRUSTS

The wells that have drilled through Precambrian have refined and revised structural models and seismic interpretations of mountain-front thrusts, thereby better defining future prospective areas. By applying well data and velocity and dipmeter surveys to existing seismic data, better interpreta-

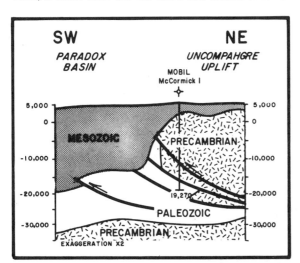

FIG. 22—Mobil C-1 McCormick (Sec. 11, T21S, R22E, Grand County, Utah) drilled through a record 14,000 ft (4.3 km) of Precambrian granite.

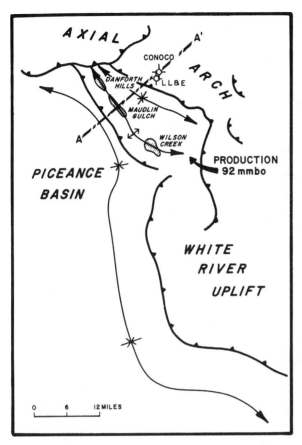

FIG. 23—Numerous seismic crews have been working on Axial arch and White River plateau of northwest Colorado, and future drilling can be expected despite 1976 failure of Conoco to drill through Precambrian and LL&E's abandoned attempt in 1982. AA' is location of cross section of Figure 24.

to whether it has had dominantly strike-slip movement, upthrust, or a significant element of compression and overthrust movement. The asymmetry of the range, the presence of the asymmetrically folded Madden structure just south of the Owl Creek Range, the proven thrusting in the Dubois area to the northwest, the numerous oil seeps on the flank of the mountain range, and recently acquired seismic data all indicate a subthrust province of untested sedimentary rocks beneath this mountain flank, and give substance to the theory of southward overthrusting. Several seismic surveys are currently under way on this mountain flank.

2. The northeast flank of the Beartooth Mountains near the Montana-Wyoming border, where the Big Horn basin synclinal axis and part of the basin's western flank are both overridden by the Beartooths, is an area untested for oil and gas potential. Any structures present would be highly prospective because few anticlines in the Big Horn basin are without hydrocarbon production.

3. The east and west flanks of the Big Horn Mountains have seismic evidence for 1 to 3 mi (1.6 to 4.8 km) of mountain-front thrusting over parts of the hydrocarbon-rich Big Horn and Powder River basins. The east flank also has evidence from well data and seismic data indicating several miles of horizontal overthrust, although sub-Precambrian rocks have not been penetrated by the drill bit.

4. The west flank of the Absaroka Mountains adjacent to the Fish Creek basin remains unexplored. Kansas-Nebraska Natural Gas (Sec. 24, T43N, R110W, Fremont County, Wyoming) drilled 2,500 ft (762 m) of Precambrian rock in this north-south-trending thrust before plugging and abandoning a proposed Paleozoic test. Interest in the subthrust potential here should increase with the recent discovery in the Fish Creek basin — the Exxon 1-2 Sohare-Federal (Sec. 34, T43N, R112W). This well only flowed about 0.5 MMCFGD from the Frontier and Muddy Formations, but as the first production in a relatively unexplored basin it may be an indication of larger unrealized potential for this area.

5. Data from seismic lines on the north flank of the Hanna basin in central Wyoming indicate several miles of unexplored sedimentary rocks beneath the north-bounding fault of the basin. Because of a rumored gas discovery 5 mi (8.0 km) south of the thrust flank in the Internorth, Inc., 1-30 (Sec. 30, T24N, R83W, Carbon County, Wyoming), the subthrust potential of the Ferris Mountains will probably be tested soon by additional seismic and drilling.

6. The subthrust of the south side of the Uinta uplift in Utah has not been drilled, although seismic activity has extended several miles north of the edge of the thrust. The proximity to the Altamont-Bluebell production makes this area highly prospective.

7. In Colorado, the west side of the White River uplift (Grand Hogback) has experienced extensive leasing and seismic activity in recent years, and drilling can be expected to commence there in the near future, as well as at several places along the Juniper Mountain-Axial arch trend. Recently, hundreds of miles of seismic line have been shot across the historic "block uplifts" of the Colorado Plateau region and this has prompted a needed reevaluation of the previous structural models, applying compressional concepts.

8. Gulf Oil is currently seismically exploring the Indepen-

tions are being made and new parameters for shooting are being developed to obtain greater resolution of thrust faults, subthrust sedimentary rocks, and structures. This has been a critical problem in the mountain-front play, as few companies will drill unless a structure is defined by seismic survey, and seismic records have often been insufficient for this purpose. There may be some areas in which good seismic records will never be obtained. Future exploration along the mountain flanks of the Rockies includes continued drilling along the thrusts that have already been drilled, with the benefit of the data already acquired.

Additional undrilled areas having subthrust potential for hydrocarbon discoveries include the following.

1. The Owl Creek Range on the north flank of the Wind River basin, is one of the largest unexplored mountain-front thrusts with the most evidence of subthrust hydrocarbon accumulations. Oil seeps have been described from Precambrian granite, Paleozoic rocks, Mesozoic rocks, and even from upper Eocene uranium sands (Love, oral commun., 1980) on the southern flank of these mountains. There has been some debate on the nature of this basin boundary, as

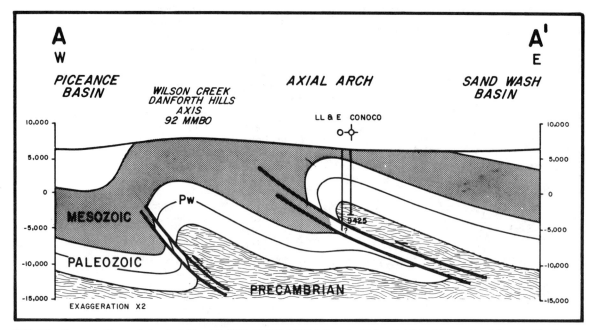

FIG. 24—Conoco's Precambrian test (Sec. 13, T5N, R94W, Moffat County, Colorado) failed to reach subthrust objective on Axial arch; Louisiana Land and Exploration attempted a Weber test 1 mi (1.6 km) southwest. See Figure 23 for location of section.

dence Mountain thrust on the North Park basin in north-central Colorado.

9. In light of new tectonic concepts of horizontal compression of Colorado mountain ranges and the presence of oil seeps in Precambrian rocks, the east flank of the Front Range will probably be more extensively explored and structurally defined through both seismic surveys and wildcat drilling.

GEOPHYSICAL WORK

Future plays will greatly depend on increased geophysical technology. The following aspects of this work will improve the chances of success.

1. Regional geology must be incorporated to avoid drilling the drained or flushed side of a major basinal syncline beneath the edge of a thrust.

2. Geophysical interpreters must include good velocity control and incorporate the limited geologic data for both the prospect area and for the entire mountain-flank thrusts. Interpreters are beginning to recognize the problems of recording, processing, and interpreting data from areas characterized by steeply dipping to overturned beds, numerous fault slivers, superimposed normal faults, a wide variety of Precambrian reflections, and large imbricate fault zones.

Velocities of both metasedimentary rocks and crystalline rocks in several of the wells that drilled through the Precambrian are shown in Table 2. Velocities were taken from sonic logs both near the top and near the base of the Precambrian where those data were available. The velocities in metasedimentary rocks seem to be slower than those in crystalline rocks. Additionally, velocities vary within crystalline rocks and are sometimes faster higher in the section that they are lower in the section.

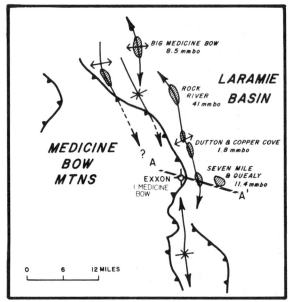

FIG. 25—Exxon drilled 4,573 ft (1.4 km) of unexpected Precambrian in their 1 Medicine Bow well. AA′ is shown on Figure 26.

3. The best interpretations of seismic information should involve extensive geologic extrapolations where data are commonly unreliable, such as the fault sliver zone. In addition, geologic modeling is recommended. This is being done by some companies where a model is proposed and known seismic parameters (velocities) are combined with the model

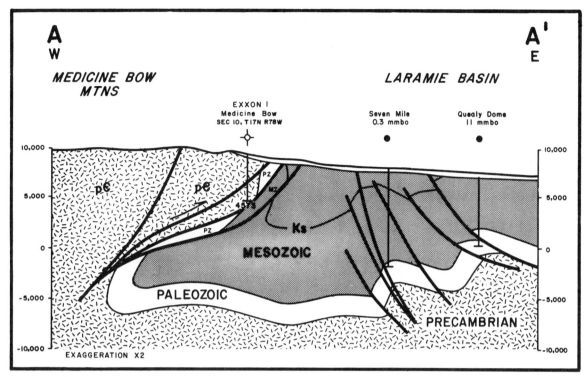

FIG. 26—Exxon's Precambrian test in Medicine Bow Range was unsuccessful in reaching Mesozoic sediments interpreted from seismic records. Location of section is shown on Figure 25.

to produce a synthetic seismic section in the lab. Comparisons with the real seismic sections are then made and adjustments applied to the geologic or geophysical parameters until the synthetic and real sections correspond to each other, and the final geologic model is the interpretation (R. G. Albertus, oral commun., 1980). As stated before, there may be some areas where good data are not attainable.

4. Seismic acquisition must include 1 to 3 mi (1.6 to 4.8 km) of additional data acquisition ("move out") beyond the expected width of the overthrow to record the deeper data and to acquire the necessary fold for optimum interpretation at the subthrust end of the line.

DRILLING RATES

All wells, except the Carter and Supron wells, drilled faster in Precambrian rocks, both crystalline and metasedimentary, than in the subthrust sedimentary rocks (Table 3). Usually more bits were used in the Precambrian than in an equivalent thickness of sedimentary rocks, therefore requiring more trip time. However, overall drilling progress for most wells drilling in Precambrian was equal to or better than drilling the subthrust sedimentary section. Typically, each bit averaged 160 to 200 ft (49 to 61 m) in the Precambrian, although the Shell 21X-9 Dahlgreen Creek well averaged only 43 ft (13 m) per bit. The fastest Precambrian drilling was American Quasar with an average of 422 ft per bit (129 m). Hard rock bits were used for drilling the Precambrian, and records show that diamond bits were used in the Precambrian only for a few feet in the Mobil C-2 McCor-

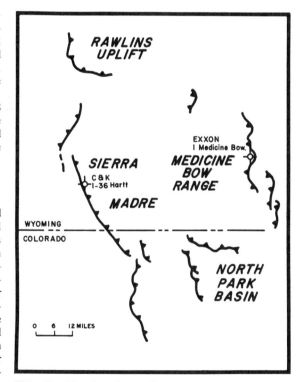

FIG. 27—C&K Petroleum drilled into complex crystalline and metamorphic rocks of Sierra Madre front in 1980.

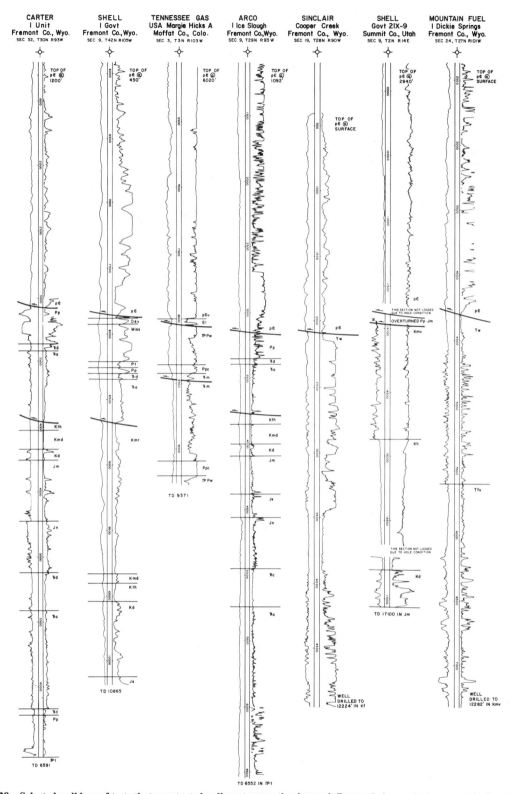

FIG. 28—Selected well logs of tests that penetrated sedimentary section beneath Precambrian rocks in mountain-front play of Rocky Mountain foreland.

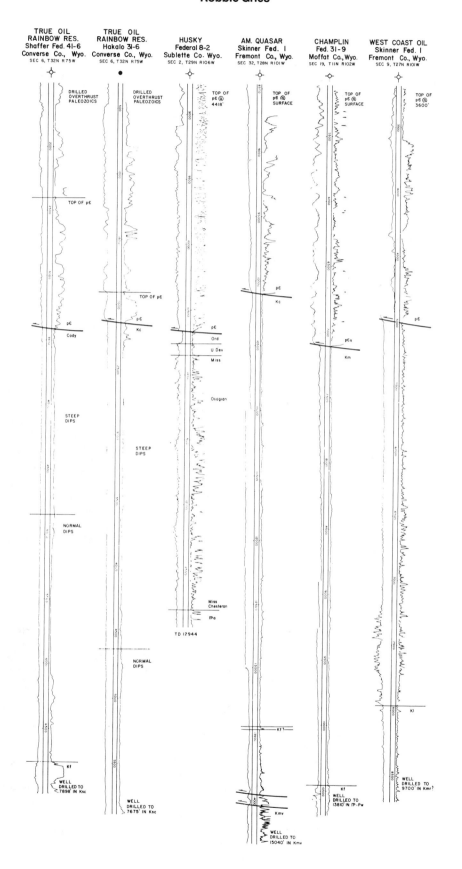

Table 2. Velocities Taken from Sonic Logs on Several Wells Drilled Through Precambrian of Rocky Mountains Foreland Thrusts

Well	Depth (ft)	Velocity on Sonic Logs (ft/sec)
CRYSTALLINE PRECAMBRIAN		
Supron 1 F-28-30-93 Sec. 28, T30N, R93W, Wyo.	1,000?- 5,400	18,000
American Quasar 1 Skinner Sec. 32, T28N, R101W, Wyo.	1,900 - 2,100 7,600 - 8,400 8,400 -10,089	18,500 19,000 18,000
West Coast 1 Skinner Fed. Sec. 9, T27N, R101W, Wyo.	3,600 - 3,800 4,800 - 5,000	18,000 20,500
Moncrief 16-1 Tepee Flats Sec. 16, T37N, R86W, Wyo.	7,500 - 7,800 14,800 -15,000	19,500 18,800
Exxon 1 Medicine Bow Sec. 10, T17N, R78W, Wyo.	1,700 - 1,900 5,050 - 2,230 4,000 - 4,250	19,800 17,500 18,600
METASEDIMENTARY PRECAMBRIAN		
Champlin 31-19 Sec. 19, T11N, R102W, Colo.	950 - 1,050 2,350 - 2,490 6,950 - 7,100	5,000 13,600 16,200
Conoco Axial 1-13 Sec. 13, T5N, R94W, Colo.	7,650 - 7,850 9,150 - 9,310	16,200 16,800

mick well when they lost a fish and had to sidetrack, and in some other wells when they were coring.

The first well to drill Precambrian, the Carter well in 1950, understandably had the most difficulty, taking 42 actual rotary days and 81 bits to drill 1,900 ft (579 m) of Precambrian crystalline rocks on the Emigrant Trail thrust. Drilling was twice as fast in the Paleozoic and Cretaceous sections beneath the thrust, where 4,691 ft (1.4 km) was drilled in 18 actual rotary days with 70 bits. The well had lost circulation problems and also had to fish frequently while drilling the Precambrian.

The next well, also drilled in crystalline Precambrian rocks, was the Shell 1 Government well drilled 6 years later on the E A thrust. It averaged 128 ft (39 m) per rotary day in Precambrian rocks and slowed down to 69 ft (21 m) per rotary day in the subthrust Paleozoic and Cretaceous rocks. It required 58 rotary days for drilling 7,408 ft (2,258 m) of Precambrian using 55 bits, while the subthrust drilling took 41 actual rotary days for 2,831 ft (863 m) and 68 bits, including one diamond bit used in the Mowry section.

Tennessee Gas drilled the first well in the Precambrian Uinta Mountain metasediments taking 22 rotary days to penetrate 1,973 ft (601 m) with 13 bits (90 ft, or 27 m, per rotary day), while the 1,376 ft (419 m) of subthrust Paleozoic and Triassic section required 20 days with 20 bits (69 ft, or 21 m, per rotary day). Tennessee Gas was stuck and had to fish or ream frequently in the Paleozoic rocks of the overthrust, but drilled fairly smoothly through the Precambrian.

Atlantic Refining Co., in 1960, a few miles from the Carter well, drilled 2,056 ft (627 m) of Precambrian in 8.5 rotary

days using 15 bits, an average of 242 ft (74 m) per rotary day. The subthrust drilled slower, penetrating 3,404 (1.0 km) ft of Paleozoic and Mesozoic section at 213 ft (65 m) per rotary day. Atlantic had to fish once in the Precambrian and twisted off and fished once in the subthrust. They used diamond bits in the Triassic Dinwoody and Permian Phosphoria subthrust sections.

Sinclair 1 Cooper Creek also took only 8 days to drill 2,087 ft (636 m) of Precambrian, averaging 327 ft (100 m) per rotary day, but drilling really slowed down in the Cretaceous subthrust section, averaging only 100 ft (30 m) per rotary day. They were down 13 days fishing or reaming, mostly in the bottom 2,000 ft (610 m) of the hole.

The Shell 21X-9 Dahlgreen well, the second to drill Precambrian metasediments on the Uinta Mountain uplift, took 76 actual rotary days to drill 8,790 ft (2.7 km) of Precambrian, averaging 116 ft (35 m) per rotary day and using 106 bits. The subthrust section was even slower, taking 62 days to drill 5,440 ft (17 km), an average of 86 ft (26 m) per rotary day and using 46 bits. They had constant hole problems, were reaming continually, and had difficulty logging the well. Hole problems may have prevented testing the shows that were seen in well cuttings.

Mountain Fuel, drilling the first test on the Wind River thrust, averaged 186 ft (57 m) per rotary day in the 4,290 ft (1.3 km) of Precambrian crystalline rocks. They averaged 180 ft (55 m) per rotary day in the 7,822 ft (2.4 km) of subthrust Tertiary and Cretaceous rocks. They used 24 bits in the Precambrian and 54 bits in the subthrust.

In 1973, when Husky drilled northwest of Mountain Fuel,

Table 3. Data Comparing Precambrian Drilling to Drilling in Sedimentary Section Beneath Thrusts

Year	Well	Precambrian Depths ft	Thickness ft	Days Act. Rot.	Feet per Rotary Day	No. of Bits Used	Bit Size	Sedimentary Subthrust	Thick ft	Days Act. Rot.	Feet per Rot. Day	No. of Bits	Bit Size
1950	Carter 1 Unit	1,200-3,100	1,900	42	45.2	81	9"	3,100-6,591 Paleozoic-Mesozoic	3,491	36	97	47	9"
1956	Shell 1 Govt.	450-7,858	7,408	58	128	9	17 1/2" 9"	7,858-10,689 Paleozoic-Mesozoic	2,831	41	69	68	9" 1 diam.
1960	Atlantic Refining 1 Ice Slough	1,092-3,148	2,056	8.5	242	15	9"	3,148-6,552 Paleozoic-Mesozoic	3,404	16	213	26	9" 2 diam.
1961	Tennessee Gas 1 USA Margie Hicks-A	6,020-7,995	1,975	22	90	13	9"	7,995-9,371 Paleozoic	1,376	20	69	20	9"
1961	Sinclair–Cooper Creek	0-2,087	2,087	8	260	6 13	13 3/4" 9"	2,087-10,224 Cretaceous	8,137	81	100	92 34	9" 7 7/8"
1968	Shell 21X-9 Dahlgreen Creek	2,970-11,760	8,790	76	116	16 90	12 1/4" 8 3/4"	11,760-17,100 Paleozoic-Mesozoic	5,340	62	86	32 14	8 3/4" 6"
1970	Mountain Fuel 1 Dickey Springs	0-4,290	4,290	22.8	186	4 20	12 1/4" 8 3/4"	4,290-12,272 Tertiary–Cretaceous	7,982	44.4	180	54	8 3/4"
1973	Husky 8-2 Federal	4,418-10,665	6,247	58	108	22 9	12 1/4" 8 3/4"	10,665-12,944 Paleozoic	2,279	23	99	10	8 3/4"
1975	American Quasar 1 Skinner Federal	0-10,090	10,090	57.2	176	8 18	12 1/4" 9 7/8"	10,090-15,040 Cretaceous	4,950	72	69	10 9	9 7/8" 6 1/2"
1975	Champlin Federal 31-19	0-7,100	7,100	64	111	30	12 1/4"	7,100-13,810 Cretaceous	6,710	61	110	12 31	12 1/4" 8 1/2"
1976	West Coast Oil Skinner Federal	3,600-5,000	1,400	12	117	10	7 7/8"	5,000-8,182 Cretaceous	3,182	29	110	8	7 7/8"
1977	Mobil C-1 McCormick Fed.	3,600-17,600	14,000	116	121	29 4	12 1/4" 9 1/2"	17,600-19,302 Paleozoic	1,702	25	68	12	9 1/2"
1980	Supron 1 F-28-30-93	400-5,400	5,000	31.7	156	21	8 3/4"	5,400-7,769 Triassic - Permian	2,369	14.3	166	21	8 3/4"
1981	Moncrief 16-1 Tepee Flats	6,162-15,027	8,865	50	177	29	12 1/4"	15,027-18,630 Triassic-Cretaceous	3,603	46	76	17 14	12 1/4" 9 1/2"

they averaged 108 ft (33 m) per rotary day in the Precambrian, penetrating 6,247 ft (1.9 km) in 58 days and using 30 bits. They then used 23 bits to drill the 2,179 ft (664 m) of Paleozoic rocks in the subthrust, with an average of 99 ft (30 m) per rotary day. This well was generally free of drilling problems.

American Quasar drilled 10,090 ft (3.1 km) of Precambrian rocks on this same mountain front in 57 rotary days (averaging 176 ft, or 54 m, per day) using 26 bits. In the subthrust Cretaceous section, they averaged only about 69 ft (21 m) per rotary day and used 19 bits. They moved the rig off at 13,010 ft (4.0 km) and then came back 5 months later and drilled down to 15,040 ft, (4.6 km), ran seven drill-stem tests, and plugged and abandoned the well.

Champlin, in 1975, used 30 bits drilling 7,100 ft (2.2 km)

of Precambrian Uinta Mountain metasediments averaging 111 ft (34 m) per rotary day. In the Cretaceous through Permian subthrust they used 43 bits to drill 6,710 ft (2.0 km), averaging 110 ft (34 m) per rotary day. They had to condition the hole several times, but generally experienced trouble-free drilling.

West Coast Oil drilled 1,400 ft (427 m) of Precambrian rock in 12 days, averaging 117 ft (36 m) per rotary day and using 10 bits on their Wind River thrust test in 1976. They averaged 110 ft (34 m) per rotary day in the Cretaceous subthrust and used 8 bits for 3,182 ft (970 m). Their drilling was also relatively trouble-free.

The first well drilled through the Precambrian crystalline rocks of the Uncompahgre uplift was the Mobil C-1 McCormick well in 1977, which drilled a record 14,000 ft (4.3 km) in 116 days, an average of 121 ft (37 m) per rotary day. They used 33 bits, including one diamond bit which was used when they twisted off in the granite and had to sidetrack. In the Paleozoic carbonates beneath the thrust, drilling slowed to an average of 68 ft (21 m) per rotary day for 1,702 ft (519 m), using 12 bits. They twisted off again in the subthrust, and essentially lost 3 months drilling time with hole problems.

Supron Energy drilled the latest well on the Emigrant Trail thrust (their 1 F-28-30-93) and drilled 5,000 ft (1.5 km) of Precambrian in 32 days at a rate of 156 ft (48 m) per rotary day, using 21 bits. A little faster were the lower Mesozoic and Paleozoic rocks in the subthrust, where drilling averaged 166 ft (51 m) per rotary day for 2,369 ft (722 m) and used only 7 bits.

Moncrief 16-1 Tepee Flats, on the Casper arch, recently averaged about 177 ft (54 m) per rotary day in 8,865 ft (2.7 km) of Precambrian crystalline rock, using 29 bits. Numerous drilling problems were experienced in the Precambrian, including twisting off once and having to whipstock. In the subthrust Triassic and Cretaceous rocks, from 15,143 to 18,630 ft (4.6 to 5.7 km), drilling averaged only about 76 ft (23 m) per rotary day, using 31 bits. Hole problems in the subthrust caused Moncrief to give up on its original objective at 21,000 ft (6.4 and plug back to the Frontier. They were unable to log the last 1,000 ft (300 m) of the hole.

CONCLUSIONS

The vast province of virtually unexplored subthrust mountain-flank acreage throughout the Rocky Mountain region contains enormous potential for hydrocarbon resources. Evolution of structural concepts and improvements in acquisition and interpretation of seismic data combined with well data from the 16 wells that have drilled through Precambrian make this one of the most daring and promising plays of the Rocky Mountains. Important to structural geology, continued seismic work and drilling will provide much needed

third-dimensional data to better interpret and theorize foreland structures.

REFERENCES CITED

Berg, R. R., 1961, Laramide tectonics of the Wind River Mountains, *in* Symposium on Late Cretaceous rocks: Wyoming Geol. Assoc., 16th Ann. Field Conf. Guidebook, p. 70-80.

——— 1962, Mountain flank thrusting in Rocky Mountain foreland, Wyoming and Colorado: AAPG Bull., v. 46, p. 704-707.

Brown, W. G., 1981, Surface and subsurface examples from the Wyoming foreland as evidences of a regional compressional origin for the Laramide orogeny, *in* D.W. Boyd, and J.S. Lillegraven, eds., Rocky Mountain foreland basement tectonics, Univ. Wyoming, Contributions to Geology, v. 19, p. 175-177.

Clement, J. H., 1977, Geological-geophysical illustrations of structural interpretations in Rocky Mountain basement tectonic terranes; lecture outline, illustrations and references: AAPG Structural Geology School, Vail, Colo., 15 p., (with credit to R. J. Carlson, 1968, for interpretive work).

Curry, W. H., and W. H. Curry, III, 1972, South Glenrock oil field, Wyoming: prediscovery thinking and post-discovery description, *in* Stratigraphic oil and gas fields—classification, exploration methods, and case histories: AAPG Mem. 16, 421 p.

Gries, R. R., 1982, North-south compression of Rocky Mountain foreland structures (abs.): AAPG Bull., v. 66, p. 574.

Love, J. D., 1970, Cenozoic geology of the Granite Mountain area, central Wyoming: U.S. Geol. Survey Prof. Paper 495C, 153 p.

Lowell, J. D., 1978, Petroleum potential of the footwall of the Emigrant Trail thrust, southwest Wind River basin, Wyoming (abs.): Wyoming Geol. Assoc., 30th Ann. Field Conf. Guidebook, p. 75.

Royse, F., Jr., M.A. Warner, and D.L. Reese, 1975, Thrust belt structural geometry and related stratigraphic problems Wyoming-Idaho-northern Utah *in* Symposium on deep drilling frontiers in the central Rocky Mountains: Rocky Mountain Assoc. Geologists, plate IV.

Sales, J. K., 1968, Crustal mechanics of Cordilleran foreland deformation: a regional and scale-model approach: AAPG Bull., v. 52, p. 2016-2044.

Stearns, D. W., 1978, Faulting and forced folding in the Rocky Mountain foreland; *in* V. Matthews, III, ed., Laramide folding associated with basement block faulting in the western United States: Geol. Soc. America Mem. 151, p. 1-36.

Stone, D. S., 1969, Wrench faulting and Rocky Mountain tectonics: Mtn. Geologist, v. 6, no. 2, p. 67-79.

Smithson, S. B., et al, 1979, Structure of Laramide Wind River uplift, from COCORP deep reflection data and from gravity data: Jour. Geophys. Research, v. 84, no. B11, p. 5955-5972.

STRIKE-SLIP DEFORMATION

Reprinted by permission of the Society of Economic Paleon-
tologists and Mineralogists from K. T. Biddle and N. Christie-
Blick, eds., *Strike-Slip Deformation, Basin Formation, and
Sedimentation*, SEPM Special Publication 37, 1985, p. 1-34.

DEFORMATION AND BASIN FORMATION ALONG STRIKE-SLIP FAULTS[1]

NICHOLAS CHRISTIE-BLICK,
*Department of Geological Sciences and
Lamont-Doherty Geological Observatory of Columbia University
Palisades, New York 10964;*

AND

KEVIN T. BIDDLE
*Exxon Production Research Company
P.O. Box 2189
Houston, Texas 77252-2189*

ABSTRACT: Significant advances during the decade 1975 to 1985 in understanding the geology of basins along strike-slip faults include
the following: (1) paleomagnetic and other evidence for very large magnitude strike slip in some orogenic belts; (2) abundant paleomagnetic evidence for the pervasive rotation of blocks about vertical axes within broad intracontinental transform boundaries; (3) greater
appreciation for the wide range of structural styles along strike-slip faults; (4) new models for the evolution of strike-slip basins; and
(5) a body of new geophysical and geological data for specific basins. In the light of this work, and as an introduction to the remainder
of the volume, the purpose of this paper is to summarize the major characteristics of and controls on structural patterns along strike-
slip faults, the processes and tectonic settings of basin formation, and distinctive stratigraphic characteristics of strike-slip basins.

Strike-slip faults are characterized by a linear or curvilinear principal displacement zone in map view, and in profile, by a subvertical
fault zone that ranges from braided to upward-diverging within the sedimentary cover. Many strike-slip faults, even those involving
crystalline basement rocks, may be detached within the middle to upper crust. Two prominent characteristics are the occurrence of en
echelon faults and folds, within or adjacent to the principal displacement zone, and the co-existence of faults with normal and reverse
separation. The main controls on the development of structural patterns along strike-slip faults are (1) the degree to which adjacent
blocks either converge or diverge during strike slip; (2) the magnitude of displacement; (3) the material properties of the sediments
and rocks being deformed; and (4) the configuration of pre-existing structures. Each of these tends to vary spatially, and, except for
the last, to change through time. It is therefore not surprising that structural patterns along strike-slip faults differ in detail from simple
predictions based on the instantaneous deformation of homogeneous materials. In the analysis of structural style, it is important to
attempt to separate structures of different ages, and especially to distinguish structures due to strike-slip deformation from those predating or post-dating that deformation. Distinctive aspects of structural style for strike-slip deformation on a regional scale include
evidence for simultaneous shortening and extension, and for random directions of vergence in associated thrusts and nappes.

Sedimentary basins form along strike-slip faults as a result of localized crustal extension, and, especially in zones of continental
convergence, of localized crustal shortening and flexural loading. A given basin may alternately experience both extension and shortening through variations in the motion of adjacent crustal blocks, or extension in one direction (or in one part of the basin) may be
accompanied by shortening in another direction (or in another part of the basin). The directions of extension and shortening also tend
to vary within a given basin, and to change through time; and the magnitude of extension may be depth-dependent. Theoretical studies
and observations from basins where strike-slip deformation has ceased suggest that many strike-slip basins experience very little thermally driven post-rift subsidence. Strike-slip basins are typically narrow (less than about 50 km wide), and they rapidly lose anomalous
heat by accentuated lateral as well as vertical conduction. Detached or thin-skinned basins also tend to be cooler after rifting has ended
than those resulting from the same amount of extension of the entire lithosphere. In some cases, subsidence may be arrested or its
record destroyed as a result of subsequent deformation. Subsidence due to extension, thermal contraction, or crustal loads is amplified
by sediment loading.

The location of depositional sites is determined by (1) crustal type and the configuration of pre-existing crustal structures; (2) variations
in the motion of lithospheric plates; and (3) the kinematic behavior of crustal blocks. The manner in which overall plate motion is
accommodated by discrete slip on major faults, and by the rotation and internal deformation of blocks between those faults is especially
important. Subsidence history cannot be determined with confidence from present fault geometry, which therefore provides a poor basis
for basin classification. Every basin is unique, and palinspastic reconstructions are useful even if difficult to undertake.

Distinctive aspects of the stratigraphic record along strike-slip faults include (1) geological mismatches within and at the boundaries
of basins; (2) a tendency for longitudinal as well as lateral basin asymmetry, owing to the migration of depocenters with time; (3)
evidence for episodic rapid subsidence, recorded by thick stratigraphic sections, and in some marine basins by rapid deepening; (4)
the occurrence of abrupt lateral facies changes and local unconformities; and (5) marked differences in stratigraphic thickness, facies
geometry, and occurrences of unconformities from one basin to another in the same region.

INTRODUCTION

Strike-slip deformation occurs where one crustal or lithospheric block moves laterally with respect to an adjacent
block. In reality, most "strike-slip" faults accommodate
oblique displacements along some segments or during part
of the time they are active; and most are associated with
an assemblage of related structures including both normal
and reverse faults. A component of oblique slip is also required for the formation of sedimentary basins. However,

the title of this paper and that of the volume were chosen
to emphasize tectonics and sedimentation in regions where
strike-slip deformation is prominent.

In this paper, we follow Mann et al. (1983) in using the
term "strike-slip basin" for any basin in which sedimentation is accompanied by significant strike slip. We acknowledge that at any given time some strike-slip basins
are hybrids associated with regional crustal extension or
shortening, and most are composite, influenced by varying
tectonic controls during their evolution (for various recent
perspectives about sedimentary basins in a plate-tectonic
framework see Green, 1977; Dickinson, 1978; Bally and
Snelson, 1980; Klemme, 1980; Bois et al., 1982; Dewey,

---

[1]Lamont-Doherty Geological Observatory Contribution No. 3910.

1982; Reading, 1982; Kingston et al., 1983a; Perrodon and Masse, 1984). Strike-slip basins also occur in a wide range of plate-tectonic settings including (1) intracontinental and intraoceanic transform zones; (2) divergent plate boundaries and extensional continental settings; and (3) convergent plate boundaries and contractional continental settings.

Among numerous articles on aspects of strike-slip deformation, basin formation, and sedimentation published in the last decade or so, several have been influential (Wilcox et al., 1973; Crowell, 1974a, b; Freund, 1974; Sylvester and Smith, 1976; Segall and Pollard, 1980; a collection of articles edited by Ballance and Reading, 1980a; Reading, 1980; Aydin and Nur, 1982a; Mann et al., 1983). Sylvester (1984) provides a compilation of classic papers on the mechanics, structural style, and displacement history of strike-slip faults, with emphasis on examples from California. In view of these existing summaries, we draw attention here to significant recent advances in understanding the large-scale characteristics of strike-slip faults and strike-slip basins, including several topics not discussed or only briefly mentioned in Ballance and Reading (1980a) and Sylvester (1984).

The remainder of this summary paper introduces two broad themes, which are elaborated in the articles that follow: (1) the characteristics of and controls on structural patterns along strike-slip faults; and (2) the processes and tectonic settings of basin formation, and distinctive aspects of the stratigraphic record of strike-slip basins.

PROGRESS DURING THE DECADE 1975–1985

Examples of advances in understanding the geology of strike-slip basins during the decade 1975 to 1985 include the following: (1) paleomagnetic and other evidence suggesting the very large magnitude of strike slip in some orogenic belts; (2) abundant paleomagnetic evidence where continents are intersected by diffuse transform plate boundaries for pervasive rotation of blocks about vertical axes, with implications for processes of deformation, basin formation and palinspastic reconstruction of strike-slip basins; (3) greater appreciation for the range of structural styles along strike-slip faults, both on the continents and in the ocean basins, and for the processes by which those styles arise; (4) new theoretical and empirical models for the evolution of strike-slip basins; and (5) new geophysical and geological data for many strike-slip basins, some of which are reported in this volume.

Large-Magnitude Strike Slip in Orogenic Belts

The role of strike slip in the evolution of orogenic belts and associated sedimentary basins has been recognized for several decades (references in Ballance and Reading, 1980b; and Sylvester, 1984), but the possible magnitude of such deformation may have been underestimated. A combination of paleomagnetic, faunal, and other geological data now indicate that in some complex orogens such as the North American Cordillera, individual elements of the tectonic collage (Helwig, 1974), termed terranes, have moved thousands not merely hundreds of kilometers with respect to each other along the trend of the orogen (Jones et al., 1977;

Irving, 1979; Beck, 1980; Coney et al., 1980; Irving et al., 1980; Champion et al., 1984; Eisbacher, 1985 this volume). Although part of this longitudinal displacement in the Cordillera pre-dates accretion, a significant component occurred after the terranes were sutured to North America. The Stikine Terrane, for example, appears to have been displaced northward by 13° to 20° since early Cretaceous time, or on the order of 1,500 km with respect to the North American craton (Jones et al., 1977; Irving et al., 1980; Chamberlain and Lambert, 1985; Eisbacher, 1985 this volume).

Large-scale strike-slip deformation also permits lateral tectonic escape in zones of continental convergence, such as between India and Eurasia (Molnar and Tapponnier, 1975), or on a smaller scale, in Turkey (Şengör et al., 1985 this volume). Molnar and Tapponnier (1975) estimated that about one third to half of the relative plate motion between India and Eurasia since the onset of continental collision in Eocene-Oligocene time (at least 1,500 km) could be accounted for by a comparable amount of strike-slip faulting in China and Mongolia.

We do not imply that such huge displacements characterize all orogenic belts, or that available paleomagnetic results are in every case without ambiguity, but only that the possibility of large-scale strike slip should be seriously entertained unless precluded by firm data. In the Appalachian-Caledonide orogen, for example, paleomagnetic studies initially suggested cumulative sinistral offset of as much as 2,000 km in late Paleozoic time (Kent and Opdyke, 1978, 1979; van der Voo et al., 1979; van der Voo and Scotese, 1981; Kent, 1982; van der Voo, 1982, 1983; Perroud et al., 1984). The timing, magnitude, and sense of displacement have recently been questioned on the basis of (1) new determinations of the Early Carboniferous pole for cratonic North America (Irving and Strong, 1984; Kent and Opdyke, 1985); (2) doubt about the age of the magnetization directions measured in some of the samples from eastern North America and Scotland (Donovan and Meyerhoff, 1982; Roy and Morris, 1983; Cisowski, 1984); and (3) the difficulty of finding appropriate faults on which to distribute the displacement (Ludman, 1981; Bradley, 1982; Donovan and Meyerhoff, 1982; Parnell, 1982; Winchester, 1982; Smith and Watson, 1983; Briden et al., 1984; Haszeldine, 1984). Where large displacements have occurred, however, even relatively young basins may have been dismembered and strung out over huge distances, and some geological mismatches may be resolved only by considering the history of an entire orogenic belt.

Rotations About Vertical Axes

The rotation of blocks about vertical axes and the bending of segments of orogenic belts have long been postulated on structural grounds (Carey, 1955, 1958; Albers, 1967; Freund, 1970; Garfunkel, 1974; Dibblee, 1977). Paleomagnetic data now confirm that such rotations tend to be pervasive in strike-slip regimes over a wide range of scales, especially where continents are intersected by diffuse transform plate boundaries (Figs. 1, 2), and the data suggest additional constraints on the timing of rotation and on the kinematics of deformation (Beck, 1980; Cox, 1980; Lu-

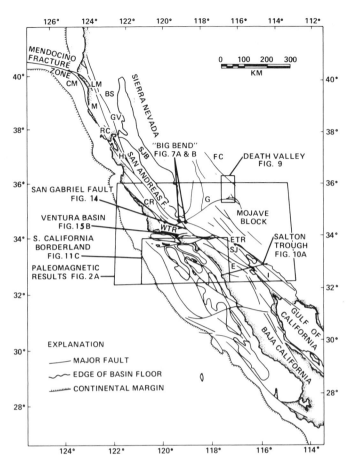

FIG. 1.—Major faults in the diffuse transform plate boundary of California and adjacent parts of Baja California, together with selected basins and elevated blocks mentioned in the text: BS, Bartlett Springs fault zone; CM, Cape Mendocino; CR, Coast Ranges; E, Elsinore fault; ETR, eastern Transverse Ranges; FC, Northern Death Valley-Furnace Creek fault zone; G, Garlock fault; GV, Green Valley fault zone; H, Hayward fault zone; I, Imperial fault; LM, Lake Mountain fault zone; M, Maacama fault zone; RC, Healdsburg-Rodgers Creek fault zone; SJ, San Jacinto fault; SJB, San Joaquin Basin; WTR, western Transverse Ranges (from King, 1969; Crowell, 1974a; Herd, 1978; Nilsen and McLaughlin, 1985 this volume). Map shows locations of areas and cross sections illustrated in Figures 2A, 7, 9, 10A, 11C, 14, 15B.

yendyk et al., 1980, 1985; Ron et al., 1984; Ron and Eyal, 1985).

Paleomagnetic evidence from Neogene rocks of southern California indicates that blocks such as the western Transverse Ranges (WTR in Fig. 1), bounded by east-striking left-slip faults, have experienced net clockwise rotations of between 35° and 90°, with sites near one major right-slip fault being rotated by more than 200° (near DB in Fig. 2A; Luyendyk et al., 1985). In contrast, the Mojave block, located between the San Andreas and Garlock faults, and characterized by northwest-striking right-slip faults, seems to have been rotated counterclockwise by about 15° ± 11° since 6 Ma (c in Fig. 2A; Morton and Hillhouse, 1985), but with large variations in declination over distances of 30 to 120 km from one sub-block to another. In the Cajon Pass region (f in Fig. 2A), there has been no significant rotation since 9.5 Ma (Weldon et al., 1984). These results generally support the tectonic models of Freund (1970), Garfunkel (1974), and Dibblee (1977), in which major crustal blocks

deform internally like a set of dominoes. Garfunkel (1974) suggested as much as 30° of rotation for the Mojave block, approximately twice that determined paleomagnetically, but he probably overestimated by a factor of two to three the magnitude of displacement on the strike-slip faults (Dokka, 1983).

The dimensions of the rotating blocks are uncertain for three reasons: (1) As noted above, blocks tend to deform internally, producing dispersion in declination anomalies; (2) In some cases, rotation seems to have occurred during deposition or eruption of the sedimentary and volcanic rocks studied, so that there is a systematic variation in the magnitude of rotation with age (Luyendyk et al., 1985); (3) Similar rotations may characterize blocks that were actually behaving independently. It is also likely that the boundaries of crustal fragments have changed through time, as displacement occurred sequentially along different strike-slip faults. Some blocks may have undergone, at different times, both clockwise and counterclockwise rotation. A possible example is the San Gabriel block between the San Gabriel and San Andreas faults (SB and e in Fig. 2A), although available data are inconclusive (Ensley and Verosub, 1982; Luyendyk et al., 1985). Small blocks can rotate at an alarming speed. Plio-Pleistocene sediments of the Vallecito-Fish Creek Basin in the western Imperial Valley (d in Fig. 2A) have been rotated 35° since 0.9 Ma (Opdyke et al., 1977; Johnson et al., 1983), and there is thus no assurance that rotation accumulates at a uniform rate, any more than does displacement.

Paleomagnetic and structural data from northern Israel indicate that Neogene intraplate deformation was accommodated by block rotation and strike-slip deformation similar to but on a smaller scale than that documented in California (Fig. 2B; Ron et al., 1984; Ron and Eyal, 1985). Domains of left-slip faults (Galilee and Carmel regions) have rotated clockwise by 23° to 35° since the Cretaceous, and domains of right-slip faults (Galilee and Tiberias regions), counterclockwise by 23° to 53° since the Cretaceous and Miocene, respectively. Sites associated with east-striking normal faults in Galilee yield the expected Cretaceous direction. Strike slip on individual faults west of the Dead Sea fault zone (Sea of Galilee, Fig. 2B) is measurable in hundreds of meters, beginning in late Miocene to early Pliocene time (Ron and Eyal, 1985). The normal faults are predominantly of post-middle Pliocene age.

The significance of these paleomagnetic results is that strike-slip basins can no longer be viewed solely in terms of bends, oversteps or junctions along strike-slip faults, to list some popular models; rotations may play an important role in basin formation. In addition, we should expect facies and paleogeographic elements not only to be offset along strike-slip faults, but also to be systematically misaligned. A further implication of block rotations, discussed below, is that the blocks and bounding faults are detached at some level in the crust or upper mantle (Terres and Sylvester, 1981; Dewey and Pindell, 1985).

Structural Style of Strike-Slip Faults

There is growing appreciation for the wide range of structural styles along strike-slip faults, both on the conti-

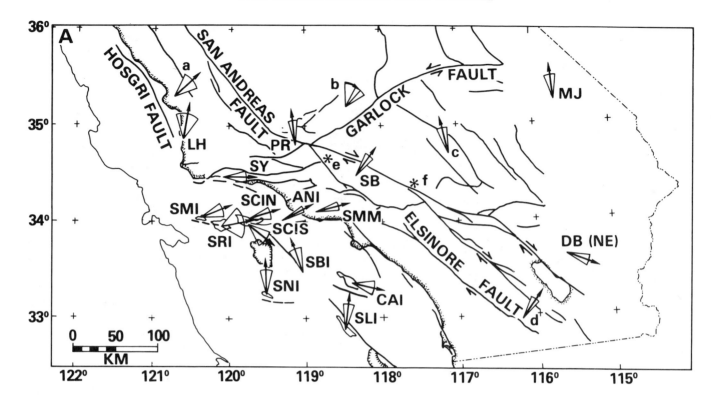

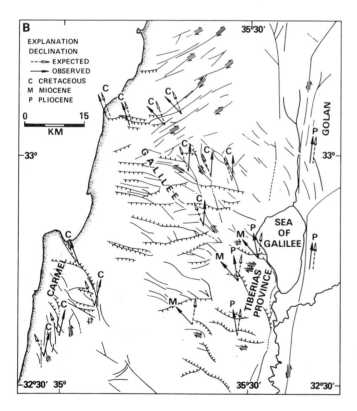

FIG. 2.—A) Paleomagnetic declinations measured in rocks of Neogene age (older than 13 Ma, except where indicated) and Quaternary sediments (site d) in southern California (see Fig. 1 for location). For each site, except e and f, the mean declination is shown along with the 95% confidence limit on the mean. Most of the data and the figure are from Luyendyk et al. (1985). Additional published data; a, Greenhaus and Cox (1979); b, apparent rotation between about 80 and 20 Ma (in comparison with the late Cretaceous pole of Irving, 1979; Kanter and McWilliams, 1981); c, mean counterclockwise rotation of 15° ± 11° since 6 Ma, but with large variations in declination from one sub-block to another (Morton and Hillhouse, 1985); d, 35° clockwise rotation since 0.9 Ma (Opdyke et al., 1977; Johnson et al., 1983); e, no major rotation since 8.5 Ma; the data can be interpreted in terms of clockwise rotation prior to 7.5 Ma, and counterclockwise rotation after that time (Ensley and Verosub, 1982); f, no significant rotation since 9.5 Ma (Weldon et al., 1984). Abbreviations for sites: ANI, Anacapa Island; CAI, Catalina Island; DB (NE) northeastern Diligencia Basin; LH, Lions Head; MJ, Mojave; PR, Plush Ranch Formation, Lockwood Valley; SB, Soledad Basin; SBI, Santa Barbara Island; SCIN, north Santa Cruz Island; SCIS, south Santa Cruz Island; SLI, San Clemente Island; SMI, San Miguel Island; SMM, Santa Monica Mountains; SNI, San Nicolas Island; SRI, Santa Rosa Island; SY, Santa Ynez Range; a, Morro Rock-Islay Hill Complex; b, southern Sierra Nevada; c, Mojave block; d, Vallecito-Fish Creek Basin; e, Ridge Basin; f, Cajon Pass. Faults are from Jennings (1975).

B) Paleomagnetic declinations measured in rocks of Cretaceous, Miocene, and Pliocene age in northern Israel; strike-slip faults, with sense of slip; and normal faults, indicated by hachures on downthrown side (simplified from Ron et al. 1984; Ron and Eyal, 1985). Domains associated with left-slip faults have undergone clockwise rotation, whereas those associated with right-slip faults have experienced counterclockwise rotation; domains associated predominantly with normal faults yield the expected declination.

nents and in the ocean basins, and for the processes by which those styles arise. Wilcox et al. (1973) recognized that one of the main controls on structural style along continental strike-slip faults cutting appreciable thicknesses of sediment is the degree to which adjacent blocks either converge or diverge during deformation. Little attention has been paid subsequently to the structures associated with divergence (see Harding et al., 1985 this volume). In addition, although experimental models, such as those of Wilcox et al. (1973), constitute a useful point of departure for analyzing the development of structures along strike-slip faults, they cannot account adequately for the considerably larger length scale, lower strain rate, greater strength, complex variations in block motion, syn-deformational sedimentation, or material heterogeneity inherent in natural examples (for further discussion, see Hubbert, 1937).

Since transform faults were first recognized in the ocean basins (Wilson, 1965; Sykes, 1967), a wealth of information has accumulated about their physiographic and structural characteristics, particularly from investigations with deep-towed echo-sounding and side-scan sonar instruments such as Deep Tow, GLORIA, Sea Beam, and Sea MARC 1 (Lonsdale, 1978; Macdonald et al., 1979; Searle, 1979, 1983; Bonatti and Crane, 1984; Fox and Gallo, 1984), the deep-towed ANGUS camera (Karson and Dick, 1983), and the deep-sea submersible ALVIN (Choukroune et al., 1978; Karson and Dick, 1983). The gross crustal structure across fracture zones has been determined by a variety of geophysical means, including seismic refraction experiments (e.g., Detrick and Purdy, 1980; Sinha and Louden, 1983), and multichannel seismic reflection profiles (e.g., Mutter and Detrick, 1984). Fossil oceanic fracture zones have also been studied in ophiolites such as the Coastal Complex of Newfoundland (Karson, 1984). Sediment accumulations in oceanic transform zones are generally thinner than about 1 km, except in areas such as the northern Gulf of California, where there is an abundant supply of terrigenous detritus (Crowell, 1981a; Kelts, 1981). The main structural features of the transforms are therefore expressed largely in igneous and metaigneous rocks, and they differ in detail from the styles of many continental strike-slip faults.

The morphology of oceanic transforms is in part a function of the offset and spreading rate of associated ridges (Fox and Gallo, 1984), and of changes in spreading direction (Menard and Atwater, 1969; Macdonald et al., 1979; Bonatti and Crane, 1984). For example, slow-slipping transforms (full rate of 1.5 to 5 cm/yr), such as those of the North Atlantic, are characterized by prominent linear topography and aligned closed basins oriented transverse to offset ridges. The relief increases from about 1,500 m for small-offset (<30 km) transforms to several thousand meters for those with large offset (>100 km), and gradients of valley walls are typically 20° to 30°, with scarps locally near vertical (Fox and Gallo, 1984). Ridge-flank topography tends to deviate in the direction of strike-slip within a few kilometers of the transform. In contrast, many fast-slipping transforms (full rate of 12 to 18 cm/yr), such as those of the eastern Pacific, are characterized by broad zones (7 to 150 km wide), composed of numerous small-offset transform segments, linked by oblique extensional domains. Relief ranges from several hundred to a few thousand meters (Fox and Gallo, 1984). The pronounced topography of some transforms may be an expression of changes in plate motion, and of convergent strike-slip deformation. For example, near its eastern intersection with the mid-Atlantic ridge, the Romanche fracture zone shallows to little more than 1,000 m below sea level, and the presence of reefal limestone shows that it was locally emergent at about 5 Ma (Bonatti and Crane, 1984). About 400 km farther east, there is evidence from a seismic reflection profile for reverse faulting and folding on the north flank of the fracture zone (Lehner and Bakker, 1983).

The investigation of oceanic transform faults is currently a major frontier in the geological and geophysical sciences, but in view of the emphasis of this volume on strike-slip deformation and basin formation in continental settings, in the remainder of this paper we focus primarily on continental examples.

Models for Strike-Slip Basins

Models for strike-slip basins are of two kinds: (1) theoretical models, derived from relatively simple assumptions about the thermal and mechanical properties of the lithosphere, but which can be compared with natural basins (Rodgers, 1980; Segall and Pollard, 1980; and in this volume, Aydin and Nur, 1985; Guiraud and Seguret, 1985; Pitman and Andrews, 1985; Royden, 1985); and (2) empirical models, which represent a distillation of geological and geophysical data from several different basins (Crowell, 1974a, b; Aydin and Nur, 1982a; Mann et al., 1983; and in this volume, Nilsen and McLaughlin, 1985; Şengör et al., 1985).

One theoretical approach for investigating the development of a pull-apart basin (Fig. 11A, B below; Burchfiel and Stewart, 1966; Crowell, 1974a, b), the simplest type of basin along a strike-slip fault, is to model the state of stress, secondary fracturing, and vertical displacements produced by infinitesimal lateral displacements on overstepping discontinuities. It is assumed for the sake of simplicity that the discontinuities are planar, vertical, and parallel, that the crust is composed of an isotropic, homogeneous, linear elastic material, and that the far-field stress is spatially uniform (Rodgers, 1980; Segall and Pollard, 1980; based on earlier studies by Chinnery, 1961, 1963; Weertman, 1965). Although such models reproduce the first-order characteristics of pull-apart basins located between overstepping strike-slip faults, they do not account satisfactorily for protracted deformation, material heterogeneity, inelastic behavior, variations in the state of stress with depth, or changes in fault geometry with depth; and currently available models are not applicable to more complex geometry and kinematic history.

Another approach (e.g., Pitman and Andrews, 1985 this volume; Royden, 1985 this volume) is to consider the subsidence and thermal history of pull-apart basins using models similar to the stretching model of McKenzie (1978), but incorporating such effects as finite rifting times, accentuated lateral heat flow from narrow basins, and depth-dependent extension (Jarvis and McKenzie, 1980; Royden and

Keen, 1980; Steckler, 1981; Steckler and Watts, 1982; Cochran, 1983). In the McKenzie model, subsidence results from instantaneous, uniform extension of the lithosphere (rifting), and from subsequent cooling by vertical conduction of heat. Theoretical studies and observations from basins where strike-slip deformation has ceased suggest that many strike-slip basins experience very little thermally driven post-rift subsidence. Basins along strike-slip faults are typically narrow (less than about 50 km wide; Aydin and Nur, 1982a; and in this volume, Aspler and Donaldson, 1985; Cemen et al., 1985; Guiraud and Seguret, 1985; Johnson, 1985; Link et al., 1985; Manspeizer, 1985; Nilsen and McLaughlin, 1985; Royden, 1985; Şengör et al., 1985), and much of the thermal anomaly decays during the rifting stage (Pitman and Andrews, 1985 this volume). In some cases, the absence of evidence for post-rift thermal subsidence may also be due to subsequent deformation of the basin (Reading, 1980; Mann et al., 1983; Nilsen and McLaughlin, 1985 this volume). An example of a large pull-apart basin that may have experienced some thermal subsidence after extension ceased is the Magdalen Basin (Carboniferous), approximately 100 km by 200 km, and situated in the Gulf of St. Lawrence between Newfoundland and New Brunswick (Bradley, 1982; Mann et al., 1983). From an analysis of subsidence and heat-flow data for the Vienna Basin (Miocene), Royden (1985 this volume) has shown that extension was confined to shallow crustal levels, and that the basin formed as a pull-apart between tear faults of the Carpathian nappes (Fig. 10B below). Detached or thin-skinned basins such as the Vienna Basin, also tend to be cooler during post-rift subsidence than those produced by the same amount of extension of the entire lithosphere.

Theoretical models are useful for analyzing processes of basin formation along strike-slip faults, although they are difficult to evaluate owing to the complexities of the real world. Examples of uncertainties are as follows: (1) lithospheric and crustal thickness prior to strike-slip deformation; (2) for predominantly extensional basins, the magnitude of extension, and the variation of extension with depth and from one part of a basin to another; (3) the relative contributions of lithospheric stretching and igneous intrusion in accommodating extension; (4) the manner in which complex fault geometry evolved, and the possible role in basin subsidence of flexural loading due to crustal shortening; (5) for starved basins, poor paleobathymetry; and for non-marine basins, the lack of a suitable datum from which to measure subsidence; (6) age control, particularly in non-marine basins; (7) the degree to which lithification occurred by physical compaction or by cementation of externally derived minerals; (8) the magnitudes of background heat flow and of heat production within the basin sediments; and (9) the relative contributions of vertical and lateral heat conduction, and of fluid motion in thermal history. In spite of these uncertainties, however, modelling studies have great potential when applied to areas for which there is abundant geological and geophysical data.

Recent empirical models for strike-slip basins (e.g., Aydin and Nur, 1982a; Mann et al., 1983) emphasize the mechanisms by which basins evolve along a single strike-slip fault or at fault oversteps, but no attempt has been made to improve on the qualitative models of Crowell (1974a, b) for more complicated basins involving intersecting strike-slip faults and significant block rotation. In spite of notable improvements in palinspastic reconstructions for orogenic belts, such restorations are commonly difficult to undertake in individual basins, which may have been deformed and dislocated by subsequent strike slip.

Geophysical, Stratigraphic, and Sedimentological Data From Strike-Slip Basins

Seismic-reflection profiles provide an important tool for basin analysis, augmenting seismic refraction, gravity, and magnetic data, which suggest only the broadest features of basin geometry. Where outcrops and wells or boreholes are few, reflection data, calibrated by available information from wells, are indispensible in developing time-stratigraphy and determining the internal geometry of the basin fill (Vail et al., 1977), and they can provide important insights into the large-scale relations between structural evolution and sedimentation. During the past decade, there has been tremendous progress in techniques for the acquisition and processing of multichannel seismic data, considerable amounts of which have been obtained by industry and academic institutions in areas affected by strike-slip deformation. Unfortunately, much of this information remains relatively inaccessible to a large part of the scientific community. One exception is a series of three volumes edited by A. W. Bally (1983) that includes several seismic profiles across strike-slip zones (D'Onfro and Glagola, 1983; Harding, 1983; Harding et al., 1983; Lehner and Bakker, 1983; Roberts, 1983). In the future, data such as these, when coupled with standard field observations, should significantly improve our understanding of the development of strike-slip basins.

Deep seismic-reflection studies, such as those undertaken in a variety of tectonic settings in the United States by COCORP and CALCRUST, in Canada by LITHOPROBE, and in Europe by BIRPS, should also provide important information about the large-scale characteristics of crustal structure associated with strike slip. By running the seismic recorders for longer than in conventional seismic experiments, it is possible to obtain images of geological structure through the crust to the Moho, and into the upper mantle (White, 1985). Direct sampling of the deep crust may also be possible through programs such as Continental Scientific Drilling.

Continued refinement of the correlation between biostratigraphy, magnetostratigraphy and the numerical timescale (e.g., Armstrong, 1978; Harland et al., 1982; Palmer, 1983; Salvador, 1985) permits improved resolution of times, durations, and rates of sedimentation. Of special significance for the analysis of strike-slip basins in complex continental settings is the more precise correlation obtainable between non-marine, marginal marine and marine successions (e.g., Royden, 1985; Şengör et al., 1985, both in this volume).

Modern facies analysis has drawn attention to the numerous factors influencing basin fill, such as basin geometry and subsidence history, climate, sediment source, drainage patterns, transport and depositional mechanisms, depositional setting, and geological age. Examples of facies analysis in strike-slip basins are numerous papers in the volumes edited by Ballance and Reading (1980a) and by Crowell and Link (1982); and articles in this volume by Aspler and Donaldson (1985), Johnson (1985), Link et al. (1985), and Nilsen and McLaughlin (1985).

STRUCTURES ALONG STRIKE-SLIP FAULTS

Major Characteristics

Strike-slip faults are characterized by a linear or curvilinear principal displacement zone in map view (Fig. 3), because significant lateral displacement cannot be accommodated where there are discontinuities or abrupt changes in fault orientation without pervasive deformation within one or both of the juxtaposed blocks. For example, angular deviations of as little as 3° between 12 to 13 km-long fault segments along the southern San Andreas fault zone may be responsible for young topographic features, as well as for the spatial distribution of aseismic triggered slip (Bilham and Williams, 1985). As viewed in profile, and in places in outcrop, most prominent strike-slip faults involve igneous and metamorphic basement rocks as well as supracrustal sediments and sedimentary rocks. Such faults are commonly termed "wrench faults," particularly in the literature of petroleum geology (e.g., Kennedy, 1946; Anderson, 1951; Moody and Hill, 1956; Wilcox et al., 1973;

Harding et al., 1985 this volume). Typically, they consist of a relatively narrow, sub-vertical principal displacement zone at depth, and within the sedimentary cover, of braided splays that diverge and rejoin both upwards and laterally (Fig. 4). Arrays of upward-diverging fault splays are known as "flower structures" (attributed to R. F. Gregory by Harding and Lowell, 1979), or less commonly, "palm tree structures" (terminology of A. G. Sylvester and R. R. Smith; Sylvester, 1984). Some strike-slip faults terminate at depth (or upward) against low-angle detachments that may be located entirely within the sedimentary section or involve basement rocks as well. Examples are high-angle to low-angle tear faults and lateral ramps of foreland thrust and fold belts (Dahlstrom, 1970; Butler, 1982; Royden et al., 1982; Royden, 1985 this volume), and tear faults associated with pronounced regional extension, as in the Basin and Range Province of the western United States (Wright and Troxel, 1970; Davis and Burchfiel, 1973; Guth, 1981; Wernicke et al., 1982; Stewart, 1983; Cheadle et al., 1985).

The distinction between wrench faults and tear faults according to whether they are "thick-skinned" or "thin-skinned" (e.g., Sylvester, 1984) is somewhat arbitrary. Several authors have suggested that crustal blocks and associated wrench faults in central and southern California are decoupled near the base of the seismogenic crust (10–15 km) from a deeper aseismic shear zone that accommodates mo-

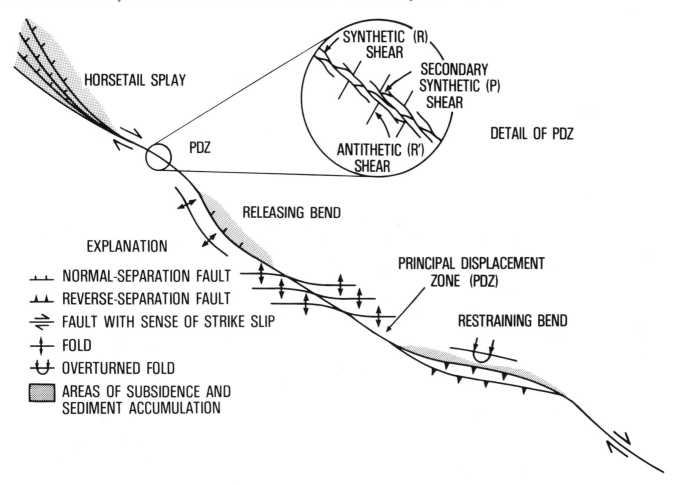

FIG. 3.—The spatial arrangement, in map view, of structures associated with an idealized right-slip fault.

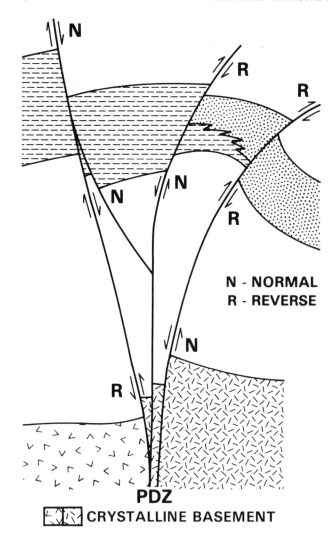

MAJOR CHARACTERISTICS

- **BASEMENT - INVOLVED**
- **PDZ SUB-VERTICAL AT DEPTH**
- **UPWARD DIVERGING & REJOINING SPLAYS**

JUXTAPOSED ROCKS

- **CONTRASTING BASEMENT TYPE**
- **ABRUPT VARIATIONS IN THICKNESS & FACIES IN A SINGLE STRATI-GRAPHIC UNIT**

SEPARATION IN ONE PROFILE

- **NORMAL- & REVERSE-SEPARATION FAULTS IN SAME PROFILE**
- **VARIABLE MAGNITUDE & SENSE OF SEPARATION FOR DIFFERENT HORIZONS OFFSET BY THE SAME FAULT**

SUCCESSIVE PROFILES

- **INCONSISTENT DIP DIRECTION ON A SINGLE FAULT**
- **VARIABLE MAGNITUDE & SENSE OF SEPARATION FOR A GIVEN HORIZON ON A SINGLE FAULT**
- **VARIABLE PROPORTIONS OF NORMAL- & REVERSE-SEPARATION FAULTS**

N - NORMAL
R - REVERSE

PDZ

CRYSTALLINE BASEMENT

 **TIME-STRATIGRAPHIC UNIT WITH VARIABLE SEDIMENTARY FACIES**

FIG. 4.—The major characteristics, in transverse profile, of an idealized strike-slip fault.

tion between the Pacific and North American plates (e.g., Hadley and Kanamori, 1977; Lachenbruch and Sass, 1980; Yeats, 1981; Hill, 1982; Crouch et al., 1984; Turcotte et al., 1984; Nicholson et al., 1985a, b; Webb and Kanamori, 1985). Recent COCORP seismic reflection profiling in California suggests that the Garlock fault, long recognized as both a major wrench zone (Moody and Hill, 1956) and a tear fault bounding an extensional allochthon (Davis and Burchfiel, 1973; Burchfiel et al., 1983), may terminate downwards against a mid-crustal (9–21 km), low-angle reflecting horizon (Cheadle et al., 1985).

A prominent feature of many strike-slip faults is the occurrence of "en echelon" faults and folds within and adjacent to the principal displacement zone (Figs. 3, 5). The term en echelon refers to a stepped arrangement of relatively short, consistently overlapping or underlapping structural elements that are approximately parallel to each other, but oblique to the linear zone in which they occur (Biddle and Christie-Blick, 1985a this volume; modified from Goguel, 1948, p. 435; Cloos, 1955; Campbell, 1958; Harding and Lowell, 1979). En echelon has also been used for oblique

elements extending tens or even hundreds of kilometers from a principal displacement zone (e.g., Wilcox et al., 1973), where it is doubtful that they have much to do with strike-slip deformation (see Harding et al., 1985 this volume), and in several classic papers on thrust and fold belts for inconsistently overlapping elements arranged parallel to a zone of deformation (e.g., Rodgers, 1963; Gwinn, 1964; Armstrong, 1968; Fitzgerald, 1968; Dahlstrom, 1970). We prefer to describe the latter as a "relay" arrangement (Harding and Lowell, 1979), because it is geometrically different and because it characterizes distinctly different tectonic regimes, those associated with regional extension or with regional shortening rather than with strike slip (Harding and Lowell, 1979; Harding, 1984).

In strike-slip regimes, we also distinguish between en echelon arrangements of structures along a given principal displacement zone, and oversteps between different segments of the principal displacement zone ("en relais" of Harris and Cobbold, 1984; Biddle and Christie-Blick, 1985a this volume). Solitary oversteps (Guiraud and Seguret, 1985 this volume) and many multiple oversteps (Biddle and

Christie-Blick, 1985a this volume) do not constitute a linear zone and by our definition are not en echelon. This is not simply a matter of scale or semantics. Oversteps between different segments and en echelon structures (our usage) appear to have different origins (Aydin and Nur, 1985 this volume; see discussion below). We note, however, that there is no general agreement about such distinctions.

Idealized en echelon arrangements, such as those shown in Figure 3, are reproduced most closely in model studies involving clay, unconsolidated sand, sheets of paraffin wax, Plasticine, or wet tissue paper (Riedel, 1929; Cloos, 1955; Pavoni, 1961; Emmons, 1969; Morgenstern and Tchalenko, 1967; Tchalenko, 1970; Wilson, 1970; Lowell, 1972; Wilcox et al., 1973; Courtillot et al., 1974; Freund, 1974; Mandl et al., 1977; Graham, 1978; Groshong and Rodgers, 1978; Rixon, 1978; Gamond, 1983; Odonne and Vialon, 1983; Harris and Cobbold, 1984), in the experimental deformation of homogeneous rock samples under confining pressure (Logan et al., 1979; Bartlett et al., 1981), and in the deformation of alluvium during large earthquakes (Tchalenko, 1970; Tchalenko and Ambraseys, 1970; Clark, 1972, 1973; Sharp, 1976, 1977; Philip and Megard, 1977). Five sets of fractures are commonly observed (Fig. 5): (1) synthetic strike-slip faults or Riedel (R) shears; (2) antithetic strike-slip faults or conjugate Riedel (R') shears; (3) secondary synthetic faults or P shears; (4) extension or tension fractures (see Biddle and Christie-Blick, 1985a this volume); and (5) faults parallel to the principal displacement zone, or Y shears of Bartlett et al. (1981). In experimental deformation of Indiana limestone, Bartlett et al. (1981) have also described what they call X shears (not shown in Fig. 5). These are symmetrical with R' in relation to the principal displacement zone. The sense of offset for R and P shears is the same as that of the principal displacement zone, whereas R' and X shears have the opposite sense

of offset. The approximate orientation at which faults and associated folds develop under simple conditions is indicated in Figure 5. For the right-slip system illustrated, folds, P shears and X shears are right-handed (Campbell, 1958; Wilcox et al., 1973; Biddle and Christie-Blick, 1985a this volume), whereas R shears, R' shears and extension (tension) fractures are left-handed. As discussed below, however, geological examples tend to be more complicated, and even in the case of Holocene deformation (Fig. 7 below), observed arrangements of structures do not necessarily conform to those predicted by models or experiments. This is because rocks are heterogeneous, because structures develop sequentially rather than instantaneously, and because early-formed structures tend to be rotated during protracted deformation. The structural style is also affected by even a small component of extension or shortening across the principal displacement zone (Wilcox et al., 1973; Harding et al., 1985 this volume), a circumstance that favors fault-parallel folds rather than en echelon ones.

In addition to the occurrence of flower structures, strike-slip faults exhibit several characteristics in profile that result in part from the segmentation of wedge-shaped sediment or rock bodies with laterally variable facies, and in part from a component of convergence or divergence across different fault strands (Fig. 4). Examples are (1) the presence in a given profile across the principal displacement zone of both normal- and reverse-separation faults, and of variable proportions of normal and reverse faults in different profiles; (2) the tendency in a single profile for the magnitude and sense of separation of a given fault splay to vary from one horizon to another; and (3) the tendency in successive profiles for a given fault to dip alternately in one direction and then in the opposite direction, and to display variable separation (both magnitude and sense) for a single horizon.

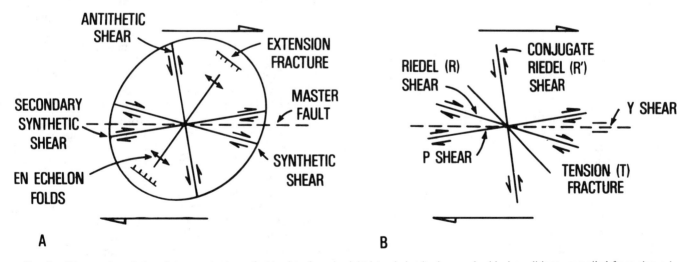

A   B

Fig. 5.—The angular relations between structures that tend to form in right-lateral simple shear under ideal conditions, compiled from clay-cake models and from geological examples. Arrangements of structures along left-slip faults may be determined by viewing the figure in reverse image. A) Terminology largely from Wilcox et al. (1973), superimposed on a strain ellipse for the overall deformation. B) Reidel shear terminology, modified from Tchalenko and Ambraseys (1970) and Bartlett et al. (1981). Extension fractures form when effective stresses are tensile (i.e., when pore-fluid pressure exceeds lithostatic pressure); tension fractures form when lithostatic loads become negative (J. T. Engelder, personal commun., 1985). In geological examples, faults with normal separation tend to develop parallel to the orientation of the extension and tension fractures in A and B; faults with reverse separation tend to develop parallel to the orientation of the fold in A. See text for further discussion of structures along strike-slip faults.

On a regional scale, distinctive aspects of strike-slip structural style are evidence for simultaneous shortening and extension, and for random directions of vergence in associated thrusts and nappes (e.g., Heward and Reading, 1980; Miall, 1985 this volume).

Controls on Structural Patterns

The main controls on the development of structural patterns along individual continental strike-slip faults are (1) the degree to which adjacent blocks either converge or diverge during strike-slip; (2) the magnitude of displacement; (3) the material properties of the sediments and rocks being deformed; and (4) the configuration of pre-existing structures. Each of these factors tends to vary along any given fault, and, except for the last, to change through time. Individual faults are commonly elements of broader regions of deformation, particularly along major intracontinental transform zones, and different faults may be active episodically at different times as a result of block rotation or the reorganization of block boundaries. Structural style may also be influenced directly by the rotation of small blocks that can produce segments of relative convergence and divergence even along straight strike-slip faults (Fig. 6), where adjacent blocks are for the most part neither converging nor diverging. The material properties of the sediments and rocks being deformed tend to change with time where, for example, strike slip is accompanied by sedimentation, so that sediments formerly near the surface become progressively more deeply buried; or where strike slip is accompanied by uplift, so that once buried strata are brought closer to the surface (as in the familiar mechanism of Karig, 1980, for raising "knockers" from great depths in accretionary prisms). Pre-existing structures can markedly influence the location and orientation of strike-slip faults, or simply complicate the overall structural pattern without being reactivated during strike-slip deformation.

Convergent, Divergent and Simple Strike-Slip Faults.—The terms "convergent wrench fault" and "divergent wrench fault" were introduced by Wilcox et al. (1973) to describe basement-involved strike-slip fault zones that, judging from the proportions of reverse-separation and normal-separation faults along and adjacent to the principal displacement zone, are thought to involve a significant component of shortening or extension, respectively. Convergent and divergent thus imply much the same as transpressional and transtensional of Harland (1971), although they refer to kinematics rather than to stress, and are for this reason preferred. We note that the use of these terms in the context of deformational process, rather than for description alone, presupposes that strike slip and transverse shortening or extension occurred simultaneously, not sequentially. In some cases, it may be difficult to make such distinctions. Faults along which there is no evidence for preferential convergence or divergence were described by Wilcox et al. (1973) as "simple parallel wrench faults," by Harding and Lowell (1979) as "side-by-side wrench faults," and by Mann et al. (1983) as "slip-parallel faults." Here, we use the term "simple strike-slip fault" in order to avoid confusion with a geometrical pattern in which a number of faults are actually parallel to

each other or arranged side-by-side, and because for any fault, slip is by definition parallel with the fault surface (see Biddle and Christie-Blick, 1985a this volume).

In comparison with simple strike slip, convergent strike slip leads to the development not only of abundant reverse faults, including low-angle thrust faults, but also of folds that are arranged both en echelon and parallel to the principal displacement zone (Fig. 3; Lowell, 1972; Wilcox et al., 1973; Sylvester and Smith, 1976; Lewis, 1980; and in this volume, Anadón et al., 1985; Mann, et al., 1985; Şengör et al., 1985; Steel et al., 1985). In cases of pronounced convergence, as in the western Transverse and Coast Ranges of California (WTR and CR in Fig. 1), the structural style becomes similar to that of a thrust and fold belt (Nardin and Henyey, 1978; Suppe, 1978; Yeats, 1981, 1983; Crouch et al., 1984; Davis and Lagoe, 1984; Wentworth et al., 1984; Namson et al., 1985). In contrast, along divergent strike-slip faults, folds are less well developed, and commonly consist of flexures arranged parallel rather than oblique to the principal displacement zone, although both types are known (Fig. 3; Wilcox et al., 1973; Nelson and Krausse, 1981; Harding et al., 1985 this volume).

Flower structures develop along both convergent strike-slip faults (Allen, 1957, 1965; Wilcox et al., 1973; Sylvester and Smith, 1976; Harding and Lowell, 1979; Harding et al., 1983; Harding, 1985), where they are associated with prominent antiforms and known as positive flower structures, and along divergent faults, where they are associated with synforms and termed negative (D'Onfro and Glagola, 1983; Harding, 1983, 1985; Harding et al., 1985 this volume). The occurrence of upward-diverging fault splays is thus not due essentially to convergence but to the propagation of faults upward through the sedimentary cover toward a free surface (Allen, 1965; Sylvester and Smith, 1976), an interpretation confirmed in experimental studies with unconsolidated sand (Emmons, 1969) and with homogeneous rock samples under confining pressure (Bartlett et al., 1981).

Strike-slip faults may be characterized by convergence or divergence for considerable distances (Namson et al., 1985; Harding et al., 1985 this volume), or only locally at restraining and releasing bends, fault junctions, and oversteps (Crowell, 1974b; Aydin and Nur, 1982b; Mann et al., 1985 this volume). Shortening and extension can occur simultaneously at oversteps associated with block rotation. A small-scale example is provided by a set of left-stepping en echelon R shears formed in a newly seeded carrot field during the Imperial Valley, California, earthquake of 15 October, 1979 (Fig. 6A; Terres and Sylvester, 1981). Elongate blocks of soil separated along furrows, became detached from the less rigid subsoil (a "Riedel flake" of Dewey, 1982), and rotated as much as 70°. The complex pattern of fractures, faults, folds and buckles, and gaps is related to motion on individual blocks. Şengör et al. (1985 this volume) describe an analogous rotated flake more than 70 km long, along the North Anatolian fault. Block rotation may also produce local convergence and divergence along relatively straight fault segments (Fig. 6B; Dibblee, 1977; Nicholson et al., 1985a, b). From an analysis of earthquake hypocentral locations and first-motion studies near the intersection of the San Andreas and San Jacinto faults, California, Ni-

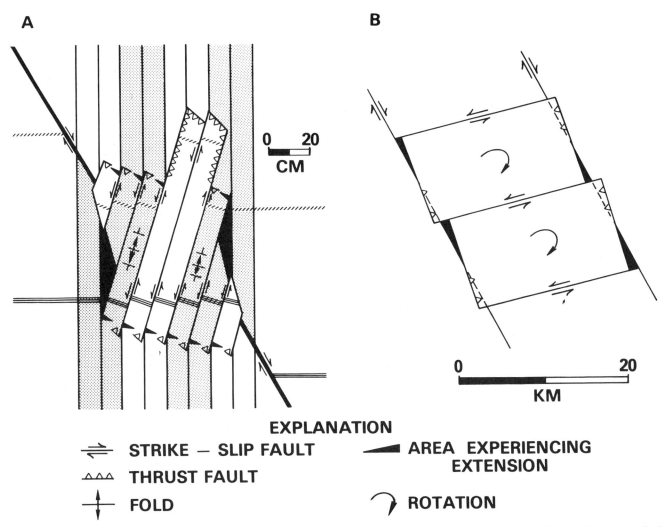

A

B

0 20
CM

0 20
KM

EXPLANATION

⇌ **STRIKE – SLIP FAULT**

◣ **AREA EXPERIENCING EXTENSION**

△△△ **THRUST FAULT**

╪ **FOLD**

↻ **ROTATION**

Fɪɢ. 6.—A) Geometric model illustrating the development of fractures, faults, folds and buckles, and gaps observed in a surface rupture in the Imperial Valley, California (I in Fig. 1), following the earthquake of 15 October, 1979 (from Terres and Sylvester, 1981). The spaced parallel ruling represents the furrows in a newly seeded carrot field.

B) A model for the rotation of blocks near the intersection of the San Andreas and San Jacinto faults, California, from earthquake hypocentral locations and first-motion studies (from Nicholson et al., 1985a, b; and similar to a figure of Dibblee, 1977). During a large earthquake, there is right slip on one of the major bounding faults; during the inter-seismic interval, the major faults are locked, and right-lateral motion is taken up by clockwise rotation of small blocks, and by minor left-slip faults. Incompatible motion at the corners of blocks leads to the development of normal and reverse fault segments, which alternate along the major right-slip faults. If such a non-elastic process were to continue on a geological timescale, it would produce complex wedge-shaped basins in areas subject to extension, and crustal flakes detached above thrust faults merging with the right-slip faults.

cholson et al. (1985a, b) have shown that incompatible motion at the corners of blocks leads to the development of normal- and reverse-fault segments, which alternate along the major right-slip faults.

Displacement Magnitude.—Experimental work, sub-surface studies and field observations indicate that structural style along strike-slip faults is influenced qualitatively by the magnitude of displacement (Wilcox et al., 1973; Harding, 1974; Harding and Lowell, 1979; Bartlett et al., 1981). In experiments, folds and shear fractures tend to form sequentially, although not in the same abundance or in the same order for all materials under all conditions (Wilcox et al., 1973; Bartlett et al., 1981). In clay models, for example, folds and Riedel shears develop first, followed by P shears, and finally by the development of a relatively

narrow through-going principal displacement zone (Morgenstern and Tchalenko, 1967; Tchalenko, 1970; Wilcox et al., 1973). In experiments with rock samples under confining pressure, R and P shears develop concurrently, and R shears tend to propagate toward the orientation of the principal displacement zone with increasing displacement (Bartlett et al., 1981). R′ and X shears form at this latter stage. The simultaneous development of both R and P shears cannot be explained by the Coulomb-Mohr failure criterion. The width of the zone of deformation increases rapidly during the initial development of folds and fractures, but tends to stabilize quickly because weakening of the sheared material leads to the concentration of subsequent deformation (Odonne and Vialon, 1983). Early-formed structures tend to rotate during progressive deformation (Tchalenko, 1970;

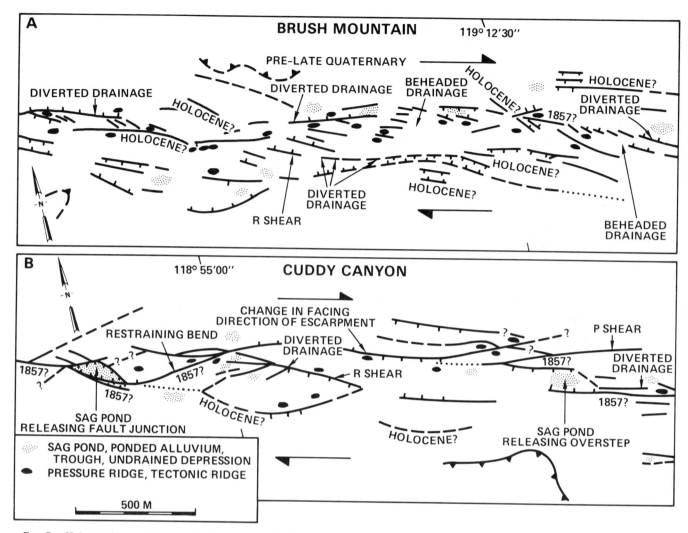

FIG. 7.—Holocene faults and tectonic landforms along the "big bend" of the San Andreas fault, California, A) south of Brush Mountain, and B) in Cuddy Canyon, east of Frazier Park (simplified from T. Davis and E. Duebendorfer, unpublished mapping, 1981, with permission from the authors; see Fig. 1 for location). Hachures on topographically lower side of fault traces; sawteeth on upper plate of thrust faults. The maps indicate the structural complexity that can develop in a few thousand years along a major strike-slip fault. The sediments at the surface are largely of Neogene and Quaternary age (A) and Quaternary age (B). The stipple pattern indicates the sites of basins (sag ponds, ponded alluvium, troughs, undrained depressions); elliptical black dots indicate elevated blocks (pressure ridges, tectonic ridges). Except for a few obvious examples of basins at releasing fault junctions or releasing oversteps, at the scale of these examples, there is no obvious relation between sites of sedimentation and fault geometry.

Wilcox et al., 1973; Rixon, 1978; Odonne and Vialon, 1983), particularly R′ shears, initially oriented at a high angle to the principal displacement zone. Thus the final orientations of folds and faults depend on the magnitude of displacement and on the time at which these structures formed during the deformation history.

Many of these features documented in experiments, can also be observed in geological examples. For instance, strike-slip faults with small lateral displacement in the basement are generally expressed within the sedimentary cover by discontinuous faults and folds (e.g., the Newport-Inglewood fault zone of the Los Angeles Basin with right slip of about 200 to 750 m; Harding, 1974). At the same scale, faults with large displacement (such as the San Andreas) consist of a through-going principal displacement zone (see Fig. 5 of Harding and Lowell, 1979). One difference, however, between experiments and natural examples is that in

nature, deformation is commonly accompanied by sedimentation, so that younger sediments record less offset than older ones. Faults within Holocene alluvium along the San Andreas fault zone tend to be discontinuous (Fig. 7; from T. Davis and E. Duebendorfer, unpublished mapping, 1981). Figure 7 also illustrates the rate at which structural complexities can arise. Given an immensely longer geological timescale, and the fact that rocks and sediments are heterogeneous even before deformation, we should not be surprised if the orientation and geometry of observed structures along a given strike-slip fault depart significantly from those predicted by idealized strain ellipse summaries (Fig. 5). This is of considerable importance to the petroleum geologist constructing structure maps for a prospective exploration target from limited data. Idealized models are useful, but they should be applied with caution. That such models commonly "match" geological observations is in part due to the large number of different fault orientations available (Fig. 5), and interpretations based on such comparisons are not necessarily correct. In addition, the absence of a simple strike-slip structural style does not eliminate the possibility that strike slip played a major role in the deformation.

Material Properties.—The character of a given strike-slip fault zone is also a function of the lithology of the juxtaposed rocks, the confining pressure, fluid pressure, and temperature conditions at which deformation occurred, and the rates of strain and recovery (Donath, 1970; Mandl et al., 1977; Sibson, 1977; Logan, 1979; Logan et al., 1979; Ramsay, 1980; Bartlett et al., 1981; Wise et al., 1984). These factors are likely to change during continuing strike-slip deformation as a result of lateral variations in depositional facies, or through sedimentation and burial, uplift and erosion, changes in provenance, or changes in heat flow.

Though difficult to predict, fault-zone characteristics are of significance to petroleum geologists because they influence the tendency of faults to behave as either conduits or barriers to fluid migration. Depending on the conditions of deformation, the accumulation of displacement along a fault may be accompanied by the development of gouge, which can act as a barrier to fluid migration (e.g., Pittman, 1981). However, displacement may also promote fracturing and additional avenues for leakage. As with other faults, strike-slip faults may juxtapose rocks with significantly different permeability, thus forming either seals or conduits for the migration of petroleum across the fault surface (see Downey, 1984, for a general treatment of hydrocarbon seals).

Pre-existing Structures.—In regions where strike-slip faults are present, pre-existing structures are of two kinds, "essential" and "incidental." As applied to the analysis of strike-slip deformation, essential structures are defined here as those which significantly influence the location and orientation of faults and folds during strike slip. Incidental structures are inherited and contribute little to that deformation. Both types of structure are elements of the overall structural pattern of the region, but incidental ones should be excluded from the analysis strike-slip structural style.

Essential structures commonly influence patterns of strike-slip deformation on the continents. The process can be observed today in the Upper Rhine graben (Fig. 8), formed by extension in middle Eocene to early Miocene time in response to Alpine deformation (Illies, 1975; Illies and Greiner, 1978; Şengör et al., 1978). In Pliocene to Holocene time, normal faults parallel to the graben were reactivated as strike-slip faults during uplift of the Alps. Earthquake first-motion studies suggest that much of the current deformation is due to left slip (Fig. 8; Ahorner, 1975), and this sense of displacement is consistent with measurements of in-situ stress (Illies and Greiner, 1978). Other examples of essential structures are discussed in this volume by Cemen et al. (1985), Harding et al. (1985), Mann et al. (1985), Royden (1985), Şengör et al. (1985), and Zalan et al. (1985). In the oceans, the location of some major transform faults is controlled by weaknesses within the continents prior to continental separation (e.g., the Atlantic equatorial megashear zone; Bonatti and Crane, 1984; Zalan et al., 1985 this volume).

Some basement-involved folds in the Death Valley area, southeastern California, illustrate the concept of incidental pre-existing structures (Figs. 1, 9). Central Death Valley developed in late Miocene through Holocene time as an east-tilted half graben bounded to the north and south by northwest-striking right-slip faults (the northern Death Valley–Furnace Creek fault zone and the southern Death Valley fault zone; Stewart, 1983). Although the type example of a pull-apart basin (Fig. 9B; Burchfiel and Stewart, 1966; Mann et al., 1983), Death Valley is not a particularly good example, because it is only one of many grabens in the Basin and Range Province that formed in response to late Cenozoic regional extension, not as a result of a bend or an overstep in a strike-slip fault (see reviews of the Basin and Range Province by Stewart, 1978; Eaton, 1982; and Fig. 1 of Cemen et al., 1985 this volume). The strike-slip faults appear to be tear faults within an extensional allochthon separating areas that experienced different amounts of extension (Wright and Troxel, 1970; Wernicke et al., 1982; Stewart, 1983). There is no evidence for strike slip within and parallel to the *central* segment of Death Valley (Wright and Troxel, 1970), although the basin undoubtedly opened obliquely, as shown in Figure 9B, and is at present extending in an approximately northwest-southeast direction (Sbar, 1982). Such a strike-slip fault was shown on an earlier map by Hill and Troxel (1966; trend S_1 in the strain ellipse insert of Fig. 9A), and reproduced by Mann et al. (1983). Evidence cited to support that interpretation, in addition to the presence of oblique striae on fault surfaces in the Black Mountains, consists of spectacular northwest-plunging folds within the basement (trend A in Fig. 9A). Hill and Troxel (1966) described the folds as en echelon, but regional arguments, including evidence for the pressure and temperature conditions at the time of folding, suggest that they are probably incidental oblique structures predating strike-slip deformation. Folds with similar orientation are present in metamorphic rocks of both the Panamint Range, west of central Death Valley, and in the Funeral Mountains, north of (and approximately parallel to) the northern Death Valley-Furnace Creek fault zone. Metamorphism in the Funeral Mountains occurred, probably in late Mesozoic time, at a temperature of 600° to 700° C, and at a pressure of 7.2 to 9.6 kb (Labotka, 1980). In the Panamint Range, folding occurred at 70 to 80 Ma during retrograde metamorphism, and at a temperature of about 450° C, following prograde metamorphism at 400° to 700° C (Labotka, 1981; Labotka and Warasila, 1983). These metamorphic conditions and the age of folding inferred in the Panamint Range are incompatible with the Miocene and younger tectonic denudation that accompanied strike-slip deformation in the Death Valley area (see Cemen et al., 1985 this volume). Although low-angle normal faults follow the top of the basement, the folds themselves are incidental structures.

BASINS ALONG STRIKE-SLIP FAULTS

Some basins along strike-slip faults developed as a direct response to strike-slip deformation. Others owe their origin to a different tectonic regime, in which strike slip played only a subsidiary role in basin development. Examples of

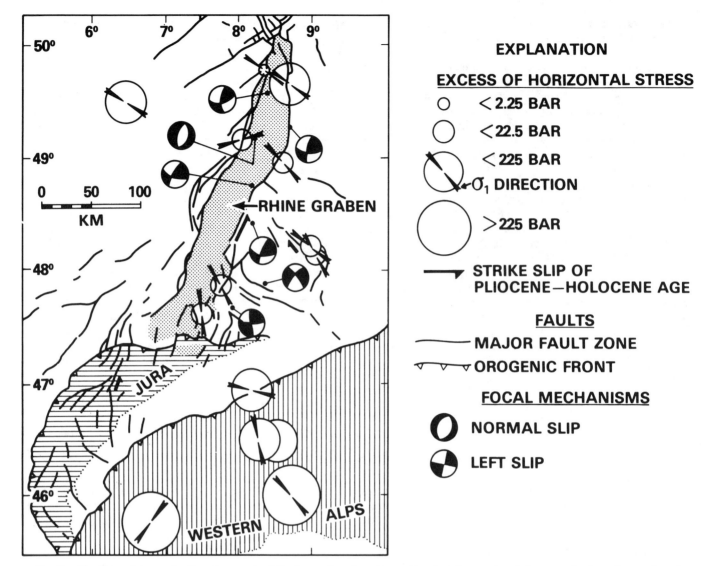

FIG. 8.—The Rhine graben and adjacent segments of the Jura and western Alps; with selected examples of the excess horizontal stress from in situ determinations; and earthquake focal mechanisms (modified from Ahorner, 1975; Illies and Greiner, 1978). Normal faults of Eocene to Miocene age were reactivated as left-slip faults in Pliocene to Holocene time, during uplift of the Alps.

the latter are (1) some intracontinental grabens, foreland basins, and forearc basins associated with strike-slip faults, which though active during sedimentation, had little influence on sedimentation patterns; and (2) various basins that were subsequently reactivated by strike-slip deformation. Basins of these types are considered here only to the extent that the geology is related to strike-slip deformation.

One of the most obvious characteristics of a strike-slip basin is its present geometry in map view, particularly the geometry of any bounding faults, and this has been a useful point of departure for various classification schemes and for most discussions about processes of basin formation (Carey, 1958; Kingma, 1958; Lensen, 1958; Quennell, 1958; Burchfiel and Stewart, 1966; Clayton, 1966; Belt, 1968; Freund, 1971; Crowell, 1974a, b, 1976; Ballance, 1980; Aydin and Nur, 1982a, b; Burke et al., 1982; Crowell and Link, 1982; Fralick, 1982; Mann and Burke, 1982; Mann et al., 1983; Mann and Bradley, 1984; see glossary in Bid-

dle and Christie-Blick, 1985a this volume). However, with the exception of very young sedimentary accumulations, the geometry, location, and perhaps orientation of a given basin have undoubtedly changed with time, and some prominent faults may be younger than much of the sedimentary fill. In many cases, present geometry may thus provide only limited information about either the kinematic history or the ultimate controls on basin evolution, and the utility of some of the classification schemes is questionable.

In this section, we focus on processes of basin formation, on the tectonic setting of depositional sites, and on certain distinctive characteristics of the stratigraphic record along strike-slip faults. We emphasize as others have before (Bally and Snelson, 1980) that every basin has a unique history, and that simple models may provide only a superficial summary of basin development. No strike-slip basin can be considered thoroughly understood unless its history can be reconstructed by a series of well-constrained palinspastic

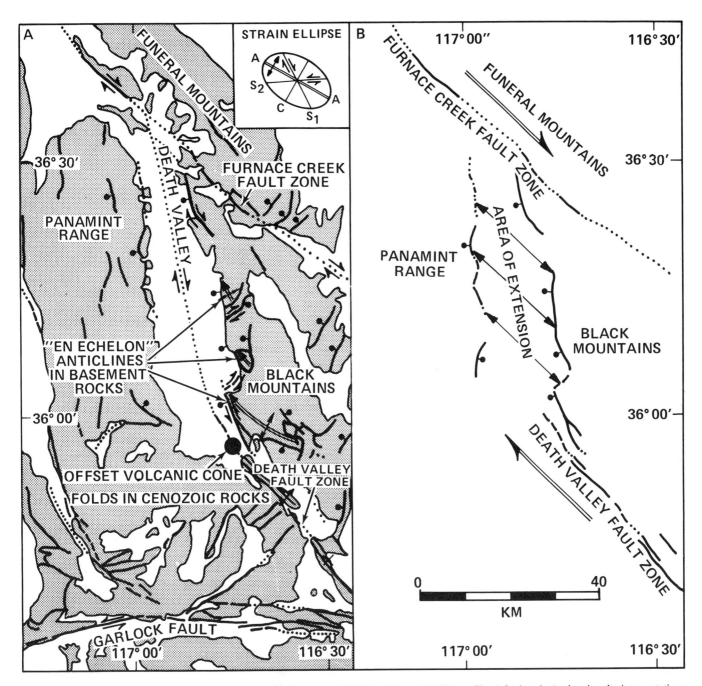

Fig. 9.—Major fault zones in the Death Valley area, California (modified from Stewart, 1983; see Fig. 1 for location), showing the interpretations of A) Hill and Troxel (1966), and B) Burchfiel and Stewart (1966). Shading (A) indicates outcrops of Proterozoic to Tertiary sedimentary and volcanic rocks; unshaded area represents Quaternary alluvial deposits. Evidence for strike slip along the northern Death Valley-Furnace Creek and southern Death Valley fault zones includes en echelon folds in Cenozoic rocks and an offset volcanic cone, together with regional stratigraphic arguments.

A) A buried strike-slip fault is inferred in the central north-trending segment of Death Valley on the basis of oblique striae on fault surfaces in the Black Mountains, and of "en echelon" anticlines in basement rocks (Hill and Troxel, 1966). The insert compares the orientations of observed structures with an idealized strain ellipse for the overall deformation; right slip inferred parallel to direction C is incompatible with orientations summarized in Figure 5.

B) Death Valley interpreted as a pull-apart along an oblique segment of a strike-slip fault system (Burchfiel and Stewart, 1966). Indicators of crustal stress and regional seismicity indicate continued extension in an approximately northwest-southeast direction parallel with the Furnace Creek and southern Death Valley fault zones (Sbar, 1982). See text for further explanation.

maps. Many of the examples discussed are taken from California, a classic region with which we are most familiar, but where appropriate, we draw attention to articles in this volume dealing with strike-slip basins in other parts of the world.

Processes of Basin Formation

Sedimentary basins form along strike-slip faults as a result of localized crustal extension, and especially in zones of continental convergence, of localized crustal shortening. Individual basins vary greatly in size (e.g., Aydin and Nur, 1982a; Mann et al., 1983), but they tend to be smaller than those produced by regional extension (many intracontinental grabens) or regional shortening (foreland and forearc basins). In addition, a given basin may alternately experience both extension and shortening on a timescale of thousands to millions of years, through variations in the motion of adjacent crustal blocks (e.g., the Ventura Basin, California, Fig. 15 below; and in this volume, Miall, 1985; Nilsen and McLaughlin, 1985; Steel et al., 1985); or extension in one direction (or in one part of the basin) may be accompanied by shortening in another direction (or in another part of the basin). A possible example of the latter is the occurrence of positive flower structures in the Mecca Hills on the northeastern side of the Salton Trough, southern California, a basin that can be related to extension in a northwestward direction in the overstep between the San Andreas and Imperial faults (Fig. 1; Sylvester and Smith, 1976; Crowell, 1981b; Fig. 6 of Harding et al., 1985 this volume). The directions of extension and shortening also tend to vary within a given basin and to change through time, especially where crustal blocks rotate (as seen on a small scale in Fig. 6 A), where there are significant differences in the rate of internal strain of adjacent crustal blocks (see Figs. 10, 11, and 12 of Şengör et al., 1985 this volume), or when there are changes in lithospheric plate motion.

In the simplest strike-slip basins (e.g., the "pull-apart hole" and "sharp pull-apart basin" of Crowell, 1974a, b), the bounding blocks are torsionally rigid and deform only at their edges, and subsidence is due to extension only in a direction parallel to the regional strike of the fault(s). Examples include many small pull-apart basins (Aydin and Nur, 1982a; Mann et al., 1983), and some detached or thin-skinned basins such as the Vienna Basin (Fig. 10B; Royden, 1985 this volume). In the vicinity of the Salton Trough, however, the crustal structure inferred from seismic refraction and gravity data (Fuis et al., 1984) suggests that significant crustal thinning has occurred outside the overstep between the San Andreas and Imperial faults (Fig. 10A). The interpretation of the gravity data is not unique, but the relatively flat gravity profile across the Salton Trough requires that the upper surface of the subbasement (lined area) largely mirror the contact between the sedimentary rocks and basement (Fuis et al., 1984). The subbasement was modelled with a density of 3.1 gm/cm^3, and refraction data suggest a P-wave velocity of greater than or equal to 7.2 km/s, consistent with significant amounts of mafic igneous rocks intruded into the lower part of the crust. The depth to the top of the subbasement decreases abruptly from 16

km at the Salton Sea to approximately 10 km at the U.S.-Mexico border, about 30 km to the south, indicating that the gross crustal structure is for the most part related to the opening of the Gulf of California, and is not simply inherited from an earlier phase of regional extension.

The profile shown in Figure 10A is probably characteristic of junctures between continental transform systems and divergent plate boundaries, but some strike-slip basins are detached. Examples are known from (1) areas of pronounced regional shortening, such as the Vienna Basin, which formed adjacent to tear faults of the Carpathian nappes (Fig. 10B; Royden, 1985 this volume), and the St. George Basin, located in the forearc of the Bering Sea, Alaska (Marlow and Cooper, 1980); and (2) areas subject to marked regional extension, such as the West Anatolian extensional province of Turkey (Fig. 18 of Şengör et al., 1985 this volume), and the Basin and Range Province of the western United States (Cemen et al., 1985; Link et al., 1985, both in this volume). Detached strike-slip basins may also prove to be relatively common in intracontinental transform zones, particularly those located along former convergent plate boundaries. In central and southern California, for example, the presence of mid-crustal detachments is suggested by (1) earthquakes with low-angle nodal planes, and the alignment of earthquake hypocenters parallel to the gently dipping base of the seismogenic crust (Nicholson et al., 1985a, b; Webb and Kanamori, 1985); (2) a pronounced change, near the junction of the San Andreas and San Jacinto faults (SJ in Fig. 1), in the patterns of seismicity with depth, with shallow seismicity suggesting the rotation of small crustal blocks (Fig. 6B; Nicholson et al., 1985a, b); (3) regional patterns of earthquake travel-time residuals (Hadley and Kanamori, 1977); (4) the distribution of upper crustal seismic velocities (Hearn and Clayton, 1984; Nicholson et al., 1985a); (5) geodetic constraints on the effective elastic thickness of the upper crust (Turcotte et al., 1984); (6) paleomagnetically determined patterns of large-scale block rotation (Fig. 2A; Luyendyk et al., 1985); (7) deep seismic-reflection studies that reveal the presence of low-angle reflecting horizons (Cheadle et al., 1985); and (8) evidence from surface geology and shallow seismic reflection profiles for numerous Neogene and Quaternary low-angle faults (Yeats, 1981, 1983; Crouch et al., 1984; Wentworth et al., 1984; Namson et al., 1985). We note that it is not yet clear how much of the California margin is detached, although the scale probably exceeds that of individual basins. Manspeizer (1985 this volume) interprets the Dead Sea Basin as detached on a smaller scale above listric normal faults bounded by the overstepping strands of the Dead Sea fault zone (his Fig. 13, after a concept of K. Arbenz, and reproduced on the cover of this book).

As discussed above, lithospheric or crustal extension produces a thermal anomaly, and subsequent cooling leads to additional subsidence (McKenzie, 1978; Royden and Keen, 1980; Steckler and Watts, 1982). In narrow basins, and in those for which extension is not "instantaneous," but occurs over an interval of more than about 10 m.y., a significant fraction of the thermal anomaly decays during the rifting stage, increasing the amount of syn-rift subsidence at the expense of the post-rift (Jarvis and McKenzie, 1980;

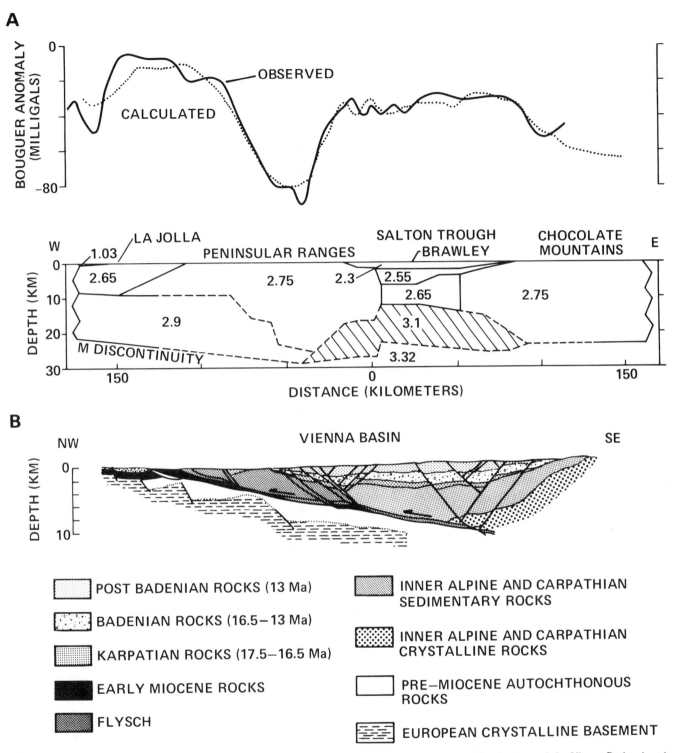

FIG. 10.—A comparison of thick-skinned and thin-skinned pull-apart basins: the Salton Trough, California (A), and the Vienna Basin, Austria and Czechoslovakia (B).

A) Cross section of southern California from La Jolla to the Chocolate Mountains (from Fuis et al., 1984, see Fig. 1 for location). The observed gravity anomaly is compared with the anomaly calculated from the model (densities in gm/cm³). Solid boundaries are those controlled by seismic refraction data; dashed lines indicate boundaries adjusted to fit the gravity data. Sub-basement (lined area, density 3.1 gm/cm³) beneath the Salton Trough provides most of the gravitational compensation for sedimentary rocks (densities 2.3 and 2.55 gm/cm³) and inferred metasedimentary rocks (density 2.65 gm/cm³). The San Andreas and Imperial faults are located near the east and west edges of the block with density of 2.65 gm/cm³.

B) Cross section of the Miocene Vienna Basin, a detached pull-apart basin superimposed partly on nappes of the outer Carpathian flysch belt, and partly on nappes of the inner Carpathians (section 3 from Fig. 6 of Royden, 1985 this volume). Tertiary thrusts are indicated by arrows; Miocene normal faults displace Miocene rocks at the surface; normal faults confined to the autochthon are mainly Jurassic syn-sedimentary faults associated with Mesozoic rifting.

Steckler, 1981; Cochran, 1983; Pitman and Andrews, 1985 this volume). The post-rift thermal anomaly is also less for detached basins than for those produced by an equivalent amount of extension of the entire lithosphere, and detached basins thus experience reduced post-rift subsidence. Extensional strike-slip basins tend to be short-lived, but the rifting stage is preferentially represented in the stratigraphic record because most such basins are small, in some cases detached, and in many cases subject to subsequent uplift and erosion.

Another potentially important mechanism for basin subsidence along strike-slip faults, in addition to crustal extension, is loading due to the local convergence of crustal blocks (e.g., the Ventura Basin of southern California; Fig. 15 below; Burke et al., 1982; Yeats, 1983). Similar basins bounded by major thrust faults are discussed by Şengör et al. (1985 this volume) and by Steel et al. (1985 this volume). As in foreland basins (Beaumont, 1981), patterns of subsidence are likely to be influenced by lithospheric flexure, a mechanism of regional isostatic compensation in which loads are supported by broad deflection of the lithosphere as a result of lithospheric rigidity (Forsyth, 1979; Watts, 1983), but to our knowledge, no attempt has yet been made to model flexure in strike-slip basins. Some strike-slip basins experience both extension and shortening (e.g., a pull-apart basin adjacent to a convergent strike-slip fault). Flexural effects are reduced when the lithosphere is warm and relatively weak, and we presume that in such circumstances subsidence induced by loading would be localized.

Subsidence due to extension, thermal contraction or crustal loads is amplified by sediment loading (Steckler and Watts, 1982). Simple isostatic considerations indicate that for typical crustal and sediment densities, sediment loading accounts for about half the total subsidence. Loading by water is important only to the extent that sea level or lake level varies.

Although it is relatively simple to construct models of the processes outlined above for hypothetical basins, it is clearly difficult to unravel the processes from the geological record of a given strike-slip basin. A fruitful avenue of research will be to use a combination of geological and geophysical data from well studied basins to test these theoretical concepts.

Tectonic Setting of Depositional Sites

The location of depositional sites along strike-slip faults is controlled by several factors operating at a variety of length scales and time scales. Along intracontinental transform zones, such as in California, and the Dead Sea fault zone, these include (1) crustal type and the configuration of pre-existing crustal structures, especially the distribution, orientation and dimensions of zones of weakness, such as faults; (2) variations in the overall motion of adjacent lithospheric plates; and (3) the kinematic behavior of crustal blocks within the transform zone.

Pre-existing Crustal Structure.—Pre-existing crustal structure influences the location of major strike-slip faults, and the way in which crustal blocks between those faults move and deform to produce basins. The San Andreas fault, for

example, originated in southern California by about 8 Ma, with most of the 320 km of right slip accumulating since about 5.5 Ma during oblique opening of the Gulf of California (Fig. 1; Karig and Jensky, 1972; Crowell, 1981b; Curray and Moore, 1984). Strike-slip faulting is only the latest event in a long geologic history, extending back into Proterozoic time, and including the assembly, amalgamation and accretion of various suspect terranes to North America during Mesozoic time; Mesozoic to Miocene arc magmatism, associated forearc and perhaps back-arc sedimentation; and Oligocene to Miocene Basin and Range extension (Hamilton, 1978; Crowell, 1981a, b; Dokka and Merriam, 1982; Champion et al., 1984). The present Gulf is the result of seafloor spreading beginning in latest Miocene time, but possibly along the line of a somewhat older Miocene tectonic depression that has been termed the proto-Gulf of California (Karig and Jensky, 1972; Terres and Crowell, 1979; Crowell, 1981a, b; but see Curray and Moore, 1984, for a modified interpretation). With local exceptions, such as in the vicinity of the Transverse Ranges (ETR and WTR in Fig. 1), the San Andreas fault and related transform faults to the south in the Gulf of California approximately parallel the boundaries of older tectonic elements, and occupy a position along the eastern edge of the Cretaceous Peninsular Ranges and Baja California Batholith (Hamilton, 1978).

Vink et al. (1984) suggested that the Gulf of California and the associated transform system formed just within the continent because continental lithosphere is weaker than oceanic lithosphere, and that rifting close to a continent-ocean boundary invariably follows a continental pathway. A similar model has been proposed by Steckler and ten Brink (1985) to explain the development in mid-Miocene time of the Dead Sea fault, at the expense of extension in the Gulf of Suez. According to their model, the hinge zone along the Mediterranean margin between thinned and unthinned continental crust acted as a barrier to continued northward propagation of the Red Sea rift.

The distribution of Neogene basins in California is probably influenced to a certain extent by crustal composition and thickness, and by the location of older basins (Blake et al., 1978). For example, the San Joaquin Basin (SJB in Fig. 1), located along the eastern side of the San Andreas Fault in central California, is superimposed on a forearc basin of late Jurassic to mid-Miocene age, which in turn overlies the boundary between Sierran-arc and ophiolitic basement (Blake et al., 1978; Hamilton, 1978; Bartow, 1984; Bartow and McDougall, 1984; Namson et al., 1985).

Relative Plate Motion.—Another factor that influences the development of sedimentary basins along a continental transform zone is the overall relative motion of the adjacent lithospheric plates (Crowell, 1974b; Mann et al., 1983). Such motion may be strictly transform, or it may involve a component of either convergence or divergence (Harland, 1971). In practice, however, it is generally difficult to predict the behavior of individual blocks solely on this basis, owing to the complex jostling that takes place in a broad transform plate boundary, even when there is little variation in relative plate movement. The character of the plate boundary is also sensitive to the migration of any unstable triple junctions, and to changes in the position of the instantaneous relative-motion pole. Either of these eventualities may be associated with the reorganization of plate-boundary geometry, and with the transfer of slices of one plate to another (Crowell, 1979).

Mann et al. (1983) argued that the gross distribution of regions of extension and shortening along the San Andreas and Dead Sea transform systems can be explained by comparing the orientation of the major faults with the interplate slip lines suggested by Minster et al. (1974). According to Mann et al., active strike-slip basins occur preferentially where the principal displacement zone is divergent with respect to overall plate motion (e.g., the Wagner and Delfin Basins of the northern Gulf of California; and the Dead Sea Basin); push-up blocks (the Transverse Ranges, California, and the Lebanon Ranges) occur where the principal displacement zone is convergent. Such statements are probably valid where most of the plate motion is taken up along a single fault (e.g., Dead Sea fault zone), but less certain where the motion is accommodated by a number of major faults, and especially where crustal blocks rotate or deform internally.

Estimates of relative motion between the Pacific and North American plates, and of Quaternary slip rates for strike-slip faults in California suggest that there should be pronounced shortening across the Transverse Ranges of California (ETR and WTR in Fig. 1), an observation qualitatively supported by abundant geological observations (Nardin and Henyey, 1978; Jackson and Yeats, 1982; Yeats, 1983; Crouch et al., 1984). About half to two thirds of the Pacific-North American relative motion (56 mm/yr; Minster et al., 1974; Minster and Jordan, 1978, 1984) occurs on the San Andreas fault (Sieh and Jahns, 1984; Weldon and Humphreys, 1985). Recent estimates of the slip rate are 25 mm/yr during the Quaternary in southern California (Weldon and Sieh, 1985), and 34 mm/yr during the Holocene in central California (Sieh and Jahns, 1984). The remaining displacement is thought to be taken up by the San Jacinto and Elsinore faults (about 10 and 1 mm/yr, respectively, for Quaternary time; Sharp, 1981; Ziony and Yerkes, 1984), and by offshore faults such as the San Gregorio-Hosgri (6 to 13 mm/yr for the late Pleistocene and Holocene; Weber and Lajoie, 1977; fault not shown in Fig. 1), assuming that Quaternary plate motion has been much the same as the average motion for the past 5 to 10 m.y. In assessing these data, Weldon and Humphreys (1985) have concluded that the magnitude of Quaternary shortening across the Transverse Ranges is considerably smaller than expected. They suggest that plate motion may have been taken up not merely by translation on the major faults but in part by counterclockwise rotation about a pole 650 km southwest of the "big bend" of the San Andreas fault (Fig. 1), an interpretation that contrasts with paleomagnetic evidence for predominantly clockwise rotations west of the San Andreas on a longer time scale (Fig. 2A; Luyendyk et al., 1985). The rates of both strike-slip faulting and regional shortening are uncertain (see Bird and Rosenstock, 1984, for a different model), but available data indicate only a qualitative relation between the orientation of interplate slip lines and patterns of uplift and subsidence, even for tectonic features as prominent as the Transverse Ranges. In zones of continental convergence, such as Turkey (Şengör et al., 1985 this volume), patterns of deformation are even more complicated.

The Neogene history of California also provides examples of plate boundary reorganization resulting from the migration of unstable triple junctions and changes in the position of the instantaneous relative-motion pole. The transform plate margin originated in California at about 30 Ma, following the impingement of the Pacific plate against the North American plate (Atwater, 1970; Atwater and Molnar, 1973). The transform system lengthened, probably intermittently, by northward motion of the northern trench-transform-transform triple junction and by generally southward motion of the southern transform-trench-ridge triple junction. At the same time, perhaps because of irregularities along the plate boundary (Crowell, 1979), slices of the North American plate were incorporated within the evolving transform system and offset differentially along the plate margin (Crouch, 1979). By about 5.5 Ma, the southern triple junction reached the mouth of the proto-Gulf of California, and much of the transform motion was taken up by the San Andreas fault, effectively transferring the Peninsular Ranges and Baja California to the Pacific plate (Curray and Moore, 1984). This reorganization of the plate boundary may be related to a small change in the relative plate motion determined by Page and Engebretson (1984). Another consequence of the inferred change in plate motion may be the onset in Pliocene and Pleistocene time of the shortening, described above, across the Transverse Ranges and Coast Ranges of California, and an acceleration in the subsidence rates for several basins (Yeats, 1978). In the Ventura Basin, for example, where considerable geochronological precision is possible, subsidence rates are estimated to have increased from approximately 250 m/m.y. at about 4 Ma to between 2,000 and 4,000 m/m.y. in the past million years. Note that these figures of Yeats (1978) incorporate inferred changes in water depth, but no corrections for compaction, loading or eustatic changes in sea level.

A modern analogue for the geological complexities at a trench-transform-transform triple junction, and a mechanism by which tectonic slices are transferred from one plate to another, has been described by Herd (1978). In the vicinity of the Cape Mendocino triple junction, the San Andreas fault appears to split into two subparallel fault zones, approximately 70 to 100 km apart. The western zone, the San Andreas proper, terminates at Cape Mendocino (CM in Fig. 1). The eastern zone, consisting of the right-stepping Hayward, Healdsburg-Rodgers Creek, Maacama and Lake Mountain fault zones (H, RC, M, and LM in Fig. 1), appears to be very youthful, and extends northward onto the continental shelf about 150 km north of Cape Mendocino.

Kinematic Behavior of Crustal Blocks.—The kinematic behavior of fault-bounded crustal blocks within a transform zone has long been considered the principal control on the development of strike-slip basins (Lensen, 1958; Kingma, 1958; Quennell, 1958; Crowell, 1974a, b, 1976; Reading, 1980; Aydin and Nur, 1982a; Mann et al., 1983). The general idea is that subsidence tends to occur where strike slip is accompanied by a component of divergence, as a result, for example, of a bend or an overstep in the fault trace ("pull-apart basin" of Burchfiel and Stewart, 1966) or through extension near a fault junction ("fault-wedge basin" of Crowell, 1974b). Uplift occurs where there is a component of convergence, although an overridden block may be depressed by the overriding one. Examples of strike-slip basins with different geometry and of different size are illustrated in Figure 11. Crowell (1974b) described bends associated predominantly with stretching and subsidence or with shortening and uplift as "releasing" and "restraining" bends, respectively, and here, we extend the use of the terms

"releasing" and "restraining" to kinematically equivalent oversteps and fault junctions. As recognized by Crowell (1974a, b), the kinematic behavior of individual crustal blocks is superimposed on broader patterns of plate interaction, and in complexly braided fault systems, many basins experience episodic subsidence as a result of changes in fault geometry and/or block motion. In some cases, therefore, the stratigraphic record of a strike-slip basin may not be related in a simple way to the present configuration of faults, and it may even be difficult to predict contemporary patterns of subsidence and uplift from fault geometry.

Consider, for example, basins and horsts associated with fault junctions. Some of the numerous possible patterns of subsidence and uplift at a simple fault junction are illustrated in Figure 12, modified from Figure 11 of Crowell (1974b). As originally discussed by Crowell, and reproduced in subsequent summary articles (e.g., Bally and Snelson, 1980; Reading, 1980, 1982; Freund, 1982), the wedge between the faults shown in Figure 12A is said to be "compressed and elevated" where the "faults converge," and "extended" where the "faults diverge." It is clear, however, that faults that converge in one direction diverge in the opposite direction. Assuming deformation only along the edges of blocks, and rotation only to the extent required by fault curvature, uplift and subsidence are related to the orientation (dip and strike) of the faults with respect to the overall slip vectors of the blocks (horizontal in Fig. 12B), and to the amount of extension or shortening associated with each fault. Although distinct in cross section, note that similar map-view configurations of basins and horsts can arise at either releasing or restraining junctions. More complicated arrangements can be envisaged at junctions with both releasing and restraining characteristics, and where blocks are internally deformed or rotated (see Fig. 6). Segmentation of the block between the branching faults may produce grabens within a horst or horsts within a graben, as in the Dead Sea fault zone (Garfunkel, 1981). In general, we expect basins and horsts to evolve continuously by a combination of fault slip and rotations about both vertical and horizontal axes, but there are also discontinuous changes due to episodic slip and rotation, and to the propagation of new faults. These are needed from time to time to eliminate tectonic knots, or complexities that inhibit further strike-slip deformation. Thus basins with thick sedimentary accumulations may at times experience uplift, and horsts with little preserved sediment may subside. Many strike-slip basins are actually slices of basins offset along one or more younger strike-slip faults.

An important aspect of strike-slip deformation and basin formation, but one for which documentation is acquired with difficulty, is the manner in which complexities such as fault bends, oversteps and junctions arise or are subsequently removed. Some curvature in strike-slip faults is attributable to a near pole of plate rotation (e.g., the Dead Sea fault zone; Garfunkel, 1981), and some bends may result from crustal heterogeneity or from local variations in the distribution of stress influencing the path of fault propagation. In eastern Jamaica, for example, a prominent right-hand restraining bend between the left-lateral east-striking Plan-

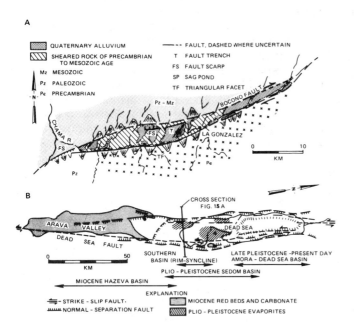

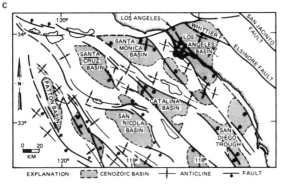

FIG. 11.—A comparison in map view of strike-slip basins of different ages, geometry, and scale.

A) The La González Basin, a lazy Z-shaped pull-apart basin of Pliocene (?) to Quaternary age along the Boconó fault zone, Venezuela (from Schubert, 1980).

B) The Dead Sea Rift, a rhomboidal pull-apart basin of Miocene to Holocene age (from Zak and Freund, 1981). Since Miocene time, the depocenter has migrated northward from the Arava Valley to the site of the present Dead Sea. For cross section, see Figure 15 A.

C) Selected faults, anticlines, and Cenozoic strike-slip basins in the southern California borderland (modified from Moore, 1969; Junger, 1976; Howell et al., 1980; see Figs. 1 and 2A for location). Many of these basins differ from the La González and Dead Sea Basins in being bounded by strike-slip faults of different orientations in a broad transform zone.

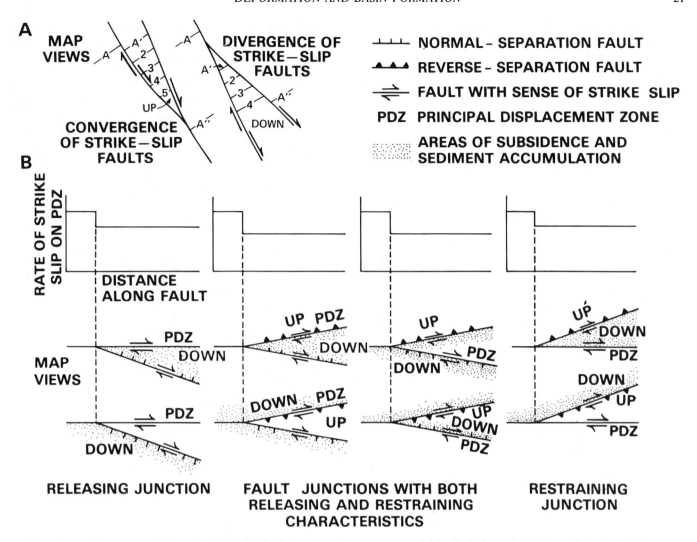

FIG. 12.—A) Sketch maps showing uplift of the tip of a fault wedge with convergence of right-slip faults, and subsidence of the tip with divergence (from Crowell, 1974b).

B) An alternative version of A, showing a range of possible results of slip along a bifurcating right-slip fault. See text for futher explanation.

tain Garden and Duanvale fault zones appears to have nucleated in Miocene time on northwest-striking normal faults that bounded a Paleogene graben (Fig. 2 of Mann et al., 1985 this volume). Other bends in strike-slip faults are due to the deformation of initially straight faults, as a result of (1) incompatible slip at a fault junction; (2) rotations within one or more adjacent blocks; or (3) intersection of a strike-slip fault with a zone of greater extensional or convergent strain. The development of the "big bend" of the San Andreas fault (Fig. 1) was attributed by Bohannon and Howell (1982) to incompatible displacement in late Cenozoic time on the San Andreas and Garlock faults (320 km of right slip, and 65 km of left slip, respectively; Smith, 1962; Smith and Ketner, 1970; Crowell, 1981b). As discussed above, this deformation was accompanied by counterclockwise rotation of the Mojave block between the faults (Fig. 2A; Morton and Hillhouse, 1985). Bends may also develop as a result of small-scale block rotation of the sort documented by Nicholson et al. (1985a, b; Fig. 6B). Şengör et al. (1985 this volume) describe fault bends due to the intersection of

strike-slip faults with zones of accentuated extensional strain at the western end of the North Anatolian fault, and of accentuated convergent strain near the junction of the North Anatolian and East Anatolian faults in eastern Turkey (see their Figs. 10, 11, and 12).

Oversteps, and branching and braiding are fundamental features of many strike-slip fault zones and fault systems, and they develop by a number of mechanisms: (1) bending of initially straight faults; (2) direct and indirect interaction between faults; (3) segmentation of curved faults; (4) faulting within a weak zone oblique to possible failure planes; and (5) reactivation of pre-existing extension fractures (Crowell, 1974b; Freund, 1974; Segall and Pollard, 1980; 1983; Mann et al., 1983; Aydin and Nur, 1982a, 1985 this volume). Mann et al. (1983) proposed that pull-aparts evolve from incipient to mature ("extremely developed") basins through a sequence of closely related states. According to them, basins tend to form at releasing bends, and develop by way of spindle-shaped and "lazy S" (or "lazy Z") basins such as the La González Basin, Venezuela (Fig. 11A;

Schubert, 1980), to rhomboidal basins such as the Dead Sea Rift (Fig. 11B; Zak and Freund, 1981; Manspeizer, 1985 this volume), and eventually, in some cases, to long narrow troughs floored by oceanic crust (e.g., the Cayman Trough of the northern Caribbean). In this model, oversteps arise by the propagation of secondary strike-slip faults in the vicinity of the releasing bend. A possible example of this branching process is the junction between the San Gabriel and San Andreas faults, the site of the Ridge Basin (Fig. 1). Between about 12 and 5 Ma, 13 km of sediment was deposited at a right bend along the San Gabriel fault, which at the same time experienced as much as 60 km of right slip (Crowell, 1982; Nilsen and McLaughlin, 1985 this volume). Beginning between 5 and 6 Ma, the San Gabriel fault ceased to be active, and strike-slip deformation was taken over by the San Andreas fault.

The interaction of parallel faults propagating from opposite directions is inherent in the models of Rodgers (1980) and Segall and Pollard (1980) for the evolution of pull-apart basins, and a possible example, the Soria Basin of northern Spain, is discussed in this volume by Guiraud and Seguret (1985). Theoretical studies indicate that significant interaction should occur between strike-slip faults if they are separated by less than twice the depth of faulting (Segall and Pollard, 1980). For strike-slip faults in California, where seismicity is observed to depths of 10 to 15 km, interaction is expected if faults are closer than 20 to 30 km. The other mechanisms listed above for generating oversteps are discussed elsewhere in this volume by Aydin and Nur (1985), and only briefly mentioned here. A consistent sense of overstepping along some curved fault zones suggests a relation between the sense of step and curvature (Aydin and Nur, 1985 this volume). An example of overstepping faults within an inappropriately oriented weak zone is the series of transform faults in the Gulf of California (Fig. 2 of Mann et al., 1983). The reactivation of pre-existing extension fractures appears to be largely a small-scale phenomenon (Segall and Pollard, 1983).

In the light of the foregoing discussion of the kinematics of strike-slip basins, we here consider the interpretation of one of the key geological elements needed to undertake a palinspastic reconstruction: piercing points of known age with which to derive the displacement history of the major strike-slip faults (e.g., Crowell, 1962, 1982, for offsets across the San Andreas and San Gabriel faults, California; and Freund et al., 1970, for offsets across the Dead Sea fault). Owing to the paucity of suitable geologic "lines," such reconstructions are commonly difficult. In the simplest case, the time of earliest movement on a strike-slip fault is given approximately by the age of the youngest rocks offset by the maximum amount (T_1 in Fig. 13A). For the San Andreas fault in southern California, this is late Miocene (Crowell, 1981a). The timing for the Dead Sea fault is less well constrained as approximately mid-Miocene, but definitely younger than basaltic dikes dated as about 20 Ma (Garfunkel, 1981), and the nature of this sort of uncertainty is shown diagrammatically in Figure 13A. Another difficulty arises if displacement varies along the fault as well as increasing with time. Strike-slip faults commonly branch and intersect, and in places accommodate differential ex-

tension or shortening. Figure 13 illustrates alternative interpretations of the displacement history of a hypothetical fault, given piercing points of known age (T_0, T_1, T_2, T_3), known displacement magnitude (D_0, D_1, D_2, D_3), and known location along one side of the fault (L_0, L_1, L_2, L_3). In one interpretation (Fig. 13A), the fault moves episodically over a long interval, beginning between times T_1 and T_2, and ending at T_4. For rocks of a given age, the magnitude of displacement is the same at all points along the fault. In the second interpretation (Fig. 13B), faulting is confined to a short interval between times T_3 and T_4. Piercing points of different ages indicate different magnitudes of displacement, because displacement varies along the fault. In reality, a given fault may exhibit displacement that is both variable and episodic, with the rate of movement changing from one interval of geologic time to another (not shown in Fig. 13). A given data set may also be incompatible with the extreme interpretations presented. The point of the illustration, however, is that in areas of complex deformation it is important to interpret the offset history of a given strike-slip fault in a regional plate-tectonic context, incorporating other stratigraphic clues to the timing of faulting.

Distinctive Aspects of the Stratigraphic Record

Strike-slip basins are present in many different plate tectonic settings, and they are filled with sediments deposited in a variety of marine and non-marine environments, subject to a range of climatic conditions. In spite of these obvious differences, however, certain aspects of the stratigraphic record appear to be distinctive. These are (1) geological mismatches within and at the boundaries of basins, that is, features which document the occurrence of strike slip; (2) a tendency for longitudinal as well as lateral basin asymmetry, owing to the migration of depocenters with time; (3) evidence for episodic rapid subsidence, recorded by thick stratigraphic sections, and in some marine basins by rapid deepening; (4) the development of pronounced topographic relief, which is associated with abrupt lateral facies changes and local unconformities at basin margins; and (5) marked differences in stratigraphic thickness, facies geometry, and the occurrence of unconformities from one basin to another in the same region.

Geological Mismatches.—A geological mismatch occurs where rocks juxtaposed by a fault require a considerable amount of displacement to have taken place on the fault or on another structure cut by the fault. Such mismatches occur at sutures, in thrust and fold belts, and across some low-angle normal faults in extensional allochthons. They are also common in regions deformed by major strike-slip faults. The segment of the San Gabriel fault between the southern part of the Ridge Basin and the eastern Ventura Basin is a good example (Fig. 14; Crowell, 1982). Lateral facies relations, clast-size trends, and paleocurrent data suggest that conglomerate of the upper Miocene Modelo Formation of the Ventura Basin was derived from a nearby source to the northeast. The conglomerate consists of distinctive clasts of gabbro, norite, anorthosite, and gneiss as large as 1.5 m in diameter for which no nearby source is known across the

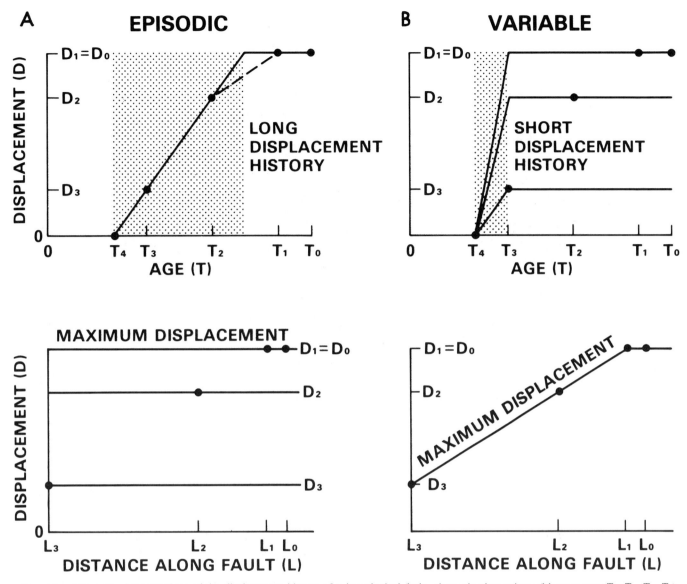

FIG. 13.—Alternative interpretations of the displacement history of a hypothetical fault, given piercing points of known age (T_0, T_1, T_2, T_3), known displacement magnitude (D_0, D_1, D_2, D_3), and known location along one side of the fault (L_0, L_1, L_2, L_3).

A) The fault moves episodically over a long interval, beginning between time T_1 and time T_2, and ending at T_4. For rocks of a given age, the magnitude of displacement is the same at all points along the fault. Note that by connecting displacement-age pairs, one might infer (perhaps incorrectly) that faulting began as early as time T_1, and that the slip rate increased after T_2 (dashed line).

B) Movement along the fault is confined to a short interval between times T_1 and T_4. Piercing points of different ages indicate different magnitudes of displacement, because displacement varies along the fault.

San Gabriel fault. Instead, that region is underlain by thick Miocene and older sedimentary rocks overlying a basement terrane quite different from that represented in the Modelo clasts. Immediately across the fault from the Modelo conglomerate, the Violin Breccia, also upper Miocene and also marine, was derived from a predominantly gneissic source to the southwest, where the basement is still largely covered by Miocene and older sedimentary rocks. The mismatch between sediments and suitable source rocks on both sides of the San Gabriel fault can be resolved, however, by removing between 35 and 60 km of right slip (Crowell, 1982). We emphasize that although evidence of this sort is important for documenting the magnitude or even occur-

rence of lateral offsets, the presence of similar geology on opposite sides of a fault at a particular locality does not necessarily preclude strike-slip deformation, a phenomenon that Crowell (1962) termed regional trace slip.
Basin Asymmetry.—Many sedimentary basins are asymmetrical, especially if faults occur preferentially along one side, and as in the case of grabens formed by regional extension (Harding, 1984), the sense of asymmetry in strike-slip basins may change from one profile to another (e.g., the Gulf of Elat; Ben-Avraham et al., 1979). The faults bounding strike-slip basins may be characterized by either normal separation (e.g., the Dead Sea Rift; Fig. 15A) or reverse separation (e.g., the Ventura Basin, California; Fig.

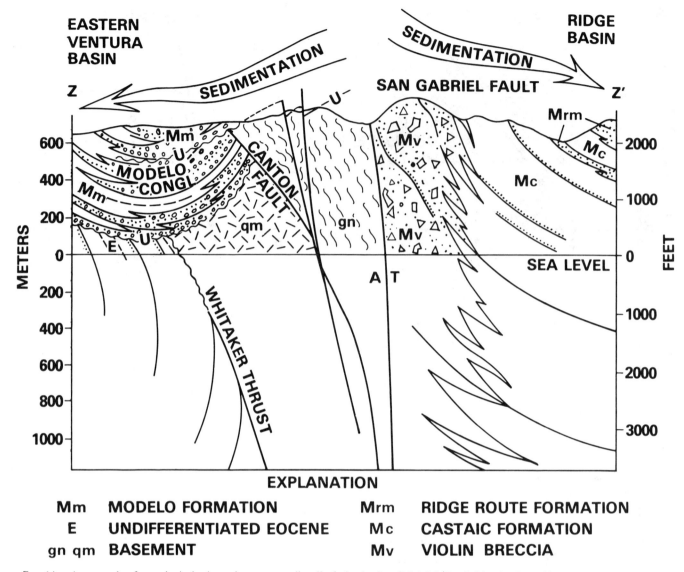

Fig. 14.—An example of a geological mismatch across a strike-slip fault: the San Gabriel fault, California (from Crowell, 1982; see Fig. 1 for location). The Modelo Conglomerate, derived from the northeast, is faulted against the Violin Breccia, derived from the southwest. T, displacement toward the observer; A, displacement away from the observer.

15B), or as discussed in the section on structural style, both normal and reverse faults may be present in the same basin (see Nilsen and McLaughlin, 1985 this volume). A particularly distinctive feature of strike-slip basins is the tendency for longitudinal as well as lateral asymmetry. The depocenter of the Dead Sea Basin, for example, has migrated northward more than 100 km from the site of the Arava Valley in the Miocene to the present Dead Sea (Fig. 11B; Zak and Freund, 1981; Manspeizer, 1985 this volume). Other basins with longitudinal asymmetry described in this volume are the Ridge Basin, California, and Hornelen Basin, Norway (Nilsen and McLaughlin, 1985), the Soria Basin of northern Spain (Guiraud and Seguret, 1985), the Nonacho Basin of Canada (Aspler and Donaldson, 1985), and possibly the Central Basin of Spitsbergen (Steel et al., 1985). *Episodic Rapid Subsidence.*—Strike-slip basins are characterized by extremely rapid rates of subsidence (Fig. 16),

even more rapid than many grabens and foreland basins, and where there is an abundant sediment supply, by very thick stratigraphic sections in comparison with lateral basin dimensions (S. Y. Johnson, 1985; Nilsen and McLaughlin, 1985, both in this volume). For example, about 13 km of sediment accumulated in the Ridge Basin in only 7 m.y. (Crowell and Link, 1982), and 5 km of sediment was deposited in the Vallecito-Fish Creek Basin in about 3.4 m.y. (N. M. Johnson et al., 1983). The Ventura Basin subsided nearly 4 km in the past 1 m.y. (Fig. 15B; Yeats, 1978). Marine basins and some deep lakes tend to become temporarily starved of sediment, a situation that promotes the accumulation of fine-grained organic-rich sediments suitable for the generation of petroleum (Graham et al., 1985; Link et al., 1985 this volume). Depending on local patterns of deformation, however, the subsidence in strike-slip basins is also episodic, and may end abruptly. The Vallecito-

Fish Creek Basin has been uplifted more than 5 km in the past 0.9 m.y. (N. M. Johnson et al., 1983).

Local Facies Changes and Unconformities.—Although not diagnostic of a strike-slip setting, many strike-slip basins form adjacent to uplifted blocks with pronounced topographic relief. As described in many of the papers that follow, this leads to very coarse sedimentary facies along some basin margins and to abrupt lateral facies changes. Local vertical movements of blocks result in localized unconformities.

Contrasts Between Basins.—Again, because of local tectonic controls, patterns of sedimentation vary markedly from one basin to another within the same region (see the description of Eocene sedimentation in Washington by Johnson, 1985 this volume). In the case of basins for which original geometry has been obscured by subsequent defor-

A

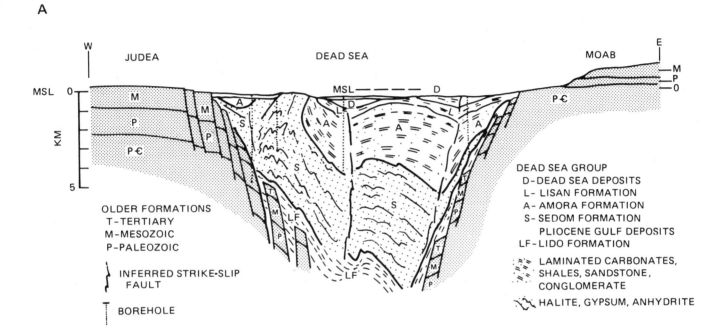

B

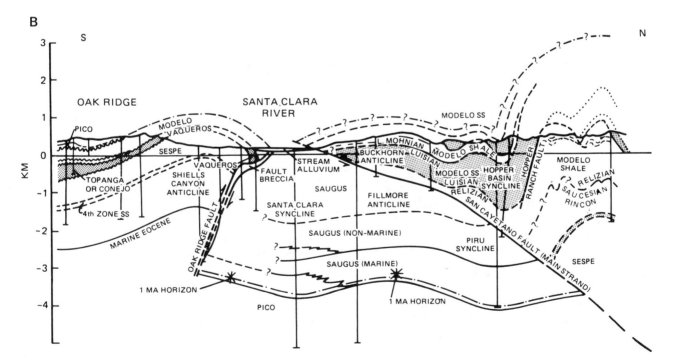

FIG. 15.—A comparison of strike-slip basins in profile.
A) The Dead Sea Rift, bounded by faults with normal separation (from Zak and Freund, 1981; see Fig. 11B for location).
B) The Ventura Basin, California, bounded by faults with reverse separation (from Yeats, 1983; see Fig. 1 for location).

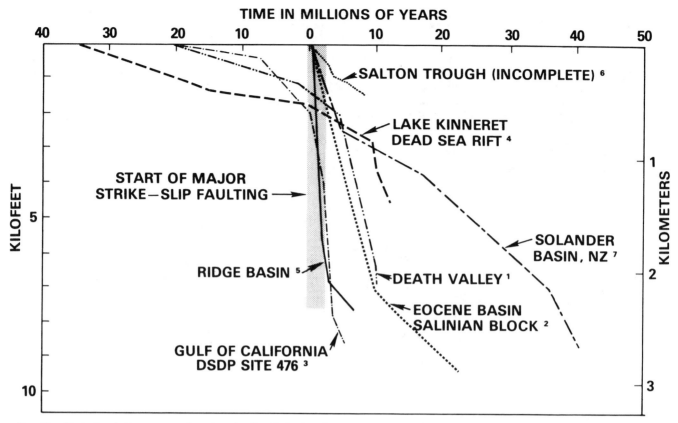

FIG. 16.—Tectonic subsidence curves for selected strike-slip basins. Tectonic subsidence is calculated by correcting cumulative stratigraphic thicknesses for the effects of sediment loading, sediment compaction, and water-depth changes. For each basin, the patterned area in the vicinity of the origin represents the approximate time of onset of major strike-slip deformation. Sources of data: 1, Hunt and Mabey (1966); 2, Graham (1976); 3, Curray et al. (1982); 4, Ben-Avraham et al. (1981); 5, Crowell and Link (1982), and papers therein; 6, Kerr et al. (1979); 7, Norris and Carter (1980).

mation (e.g., many Paleozoic and Precambrian examples), regional stratigraphic comparisons may provide some of the more important clues that sedimentation was accompanied by strike-slip deformation (Heward and Reading, 1980; Aspler and Donaldson, 1985 this volume).

ECONOMIC IMPORTANCE

Oil and gas are the most important resource exploited in strike-slip basins. Huff (1980) estimated that as much as 133 billion barrels has been discovered, but we regard this figure as rather high. The Dead Sea, known as Lake Asphaltitis to the ancient Greeks, was mentioned by the Latin authors Josephus and Tacitus as a source of asphalt used for calking ships and preparing medicines (Nissenbaum, 1978). In modern times, some of the most productive basins have been in California. About 10 billion barrels of oil have been produced from the Los Angeles, Ventura, Santa Maria, Salinas, and Cuyama Basins alone (Taylor, 1976). Current production from these basins is about 600,000 barrels of oil per day, and the ultimate recovery is estimated to be as much as 20 billion barrels of oil equivalent. The Point Arguello oil field, discovered in the offshore Santa Maria Basin in 1981, is the largest U.S. discovery since Prudhoe Bay, Alaska, with between 2.2 and 2.5 billion barrels of oil in place, and total recoverable reserves of be-

tween 300 and 500 million barrels (Crain et al., 1985). Other northern California offshore basins are promising (Crouch and Bachman, 1985). Additional reserves may be discovered in provinces affected by strike-slip deformation elsewhere in the world, such as the southern Caribbean region (Leonard, 1983), the Bering Sea (Marlow, 1979; Fisher, 1982), the Dead Sea (Wilson et al., 1983), and parts of China (Li Desheng, 1984). Individual strike-slip basins range from very rich to non-productive, depending on such factors as the presence of source rocks, thermal history and maturation, migration potential, reservoir quality and distribution, occurrence of traps and seal, and preservation of trapped hydrocarbons (Kingston et al., 1983b; Biddle and Christie-Blick, 1985b). Most important is the timing of maturation, migration and trap formation, because strike-slip basins tend to be short-lived. Examples of commodities, other than petroleum, found in strike-slip basins are ground water, coal and lignite, geothermal energy, and subsurface brines.

CONCLUSIONS

(1) Structural patterns along strike-slip faults differ in detail from simple predictions. The geometry of faults and folds along a given fault zone is generally a result of protracted, episodic deformation of heterogeneous sediments

and sedimentary rocks, involving the rotation of crustal blocks at a variety of scales, as well as strike slip, together with varying degrees of convergence and divergence, all superimposed on a pre-existing structural grain. On a regional scale, distinctive aspects of the structural style are the occurrence of en echelon structures, evidence for simultaneous shortening and extension, and random directions of vergence in associated thrusts and nappes.

(2) Basins form along strike-slip faults as a result of localized crustal extension and/or localized crustal shortening. The main processes leading to subsidence are mechanical thinning, thermal contraction, and loading due to the convergence of crustal blocks, amplified by the effects of sediment loading. The rifting stage is preferentially represented in the stratigraphic record of many strike-slip basins, because they are typically narrow, and in some cases detached (and therefore subject to less post-rift thermal subsidence), and many basins experience subsequent uplift and erosion. The importance of lithospheric flexure depends on the amount of extension involved in basin formation. Flexural effects are probably reduced during rifting.

(3) The location of depositional sites along strike-slip faults is controlled by crustal type and the configuration of pre-existing crustal structures, variations in the motion of adjacent lithospheric plates, and the kinematic behavior of crustal blocks. A key factor influencing the development of sedimentary basins is the manner in which overall plate motion is accommodated by discrete slip on major faults, and by the rotation and internal deformation of blocks. Subsidence history cannot be determined with confidence from present fault geometry, which therefore provides a poor basis for basin classification.

(4) Distinctive aspects of the stratigraphic record along strike-slip faults are geological mismatches; a tendency for basins to be asymmetrical both longitudinally and laterally; thick stratigraphic sections representing short intervals of time; the occurrence of abrupt lateral facies changes and local unconformities; and marked differences in stratigraphic thickness, facies geometry, and occurrence of unconformities from one basin to another in the same region.

(5) A major frontier for research in strike-slip basins is that of integrated geophysical, geological and modelling studies in a variety of plate-tectonic settings. The geophysical work should include standard seismic reflection, seismic refraction, gravity, magnetic and heat flow measurements, together with paleomagnetic studies to establish the magnitude and timing of rotations, and the boundaries of the blocks experiencing rotation. The geological work should include structural, stratigraphic and sedimentological studies using outcrop, borehole and seismic reflection data, together with investigations of diagenesis and paleothermometry. Modelling should be directed at using all available data in quantitative tests of our notions of the processes that control the development of strike-slip basins.

ACKNOWLEDGMENTS

This paper is an outgrowth of work conducted for Exxon Production Research Company, and we are grateful to that company for permission to publish and for support in the preparation of the manuscript. Christie-Blick acknowledges additional logistical support from an ARCO Foundation Fellowship. For stimulating discussions at various times about deformation and basins along strike-slip faults and for preprints of articles, we thank numerous geoscientists, but especially Tod Harding and Richard Vierbuchen (Exxon); Roger Bilham, Gerard Bond, Kathy Crane, Dennis Kent, Michelle Kominz, Craig Nicholson, Leonardo Seeber, Michael Steckler, Anthony Watts, and Patrick Williams (Lamont-Doherty Geological Observatory); John Crowell, Bruce Luyendyk and Arthur Sylvester (University of California, Santa Barbara); James Helwig and Jay Namson (ARCO); Ray Weldon (Caltech); M. J. Cheadle (University of Cambridge); William Dickinson (University of Arizona); and all the participants of the 1984 Research Symposium of the Society of Economic Paleontologists and Mineralogists on Strike-Slip Deformation, Basin Formation, and Sedimentation. Thom Davis and Ernie Duebendorfer kindly allowed us to use their unpublished mapping of neotectonic features along the San Andreas Fault to construct Figure 7. The manuscript benefitted from constructive reviews by Gerard Bond, Kevin Burke, Kathy Crane, James Helwig, Dennis Kent, Craig Nicholson, Harold Reading, and Arthur Sylvester.

REFERENCES

AHORNER, L., 1975, Present-day stress field and seismotectonic block movements along major fault zones in central Europe: Tectonophysics, v. 19, p. 233–249.

ALBERS, J. P., 1967, Belt of sigmoidal bending and right-lateral faulting in the western Great Basin: Geological Society of America Bulletin, v. 78, p. 143–156.

ALLEN, C. R., 1957, San Andreas fault zone in San Gorgonio Pass, southern California: Geological Society of America Bulletin, v. 68, p. 315–350.

———, 1965, Transcurrent faults in continental areas, *in* A Symposium on Continental Drift: Philosophical Transactions of Royal Society of London, Series A, v. 258, p. 82–89.

ANADÓN, P., CABRERA, L., GUIMERÀ, J., AND SANTANACH, P., 1985, Paleogene strike-slip deformation and sedimentation along the southeastern margin of the Ebro Basin, *in* Biddle K. T., and Christie-Blick, N., eds., Strike-Slip Deformation, Basin Formation, and Sedimentation: Society of Economic Paleontologists and Mineralogists Special Publication No. 37, p. 303–318.

ANDERSON, E. M., 1951, The dynamics of faulting and dyke formation with applications to Britain: Edinburgh, Oliver and Boyd, 2nd edition, 206 p.

ARMSTRONG, R. L., 1968, Sevier orogenic belt in Nevada and Utah: Geological Society of America Bulletin, v. 79, p. 429–458.

———, 1978, Pre-Cenozoic Phanerozoic time scale—Computer file of critical dates and consequences of new and in-progress decay-constant revisions, *in* Cohee, G. V., Glaessner, M. F., and Hedberg, H. D., eds., Contributions to the Geologic Time Scale: American Association of Petroleum Geologists Studies in Geology No. 6, p. 73–91.

ASPLER, L. B., AND DONALDSON, J. A., 1985, The Nonacho Basin (Early Proterozoic), Northwest Territories, Canada: Sedimentation and deformation in a strike-slip setting, *in* Biddle K. T., and Christie-Blick, N., eds., Strike-Slip Deformation, Basin Formation, and Sedimentation: Society of Economic Paleontologists and Mineralogists Special Publication No. 37, p. 193–209.

ATWATER, T., 1970, Implications of plate tectonics for the Cenozoic tectonic evolution of western North America: Geological Society of America Bulletin, v. 81, p. 3513–3536.

ATWATER, T., AND MOLNAR, P., 1973, Relative motion of the Pacific and North American plates deduced from sea-floor spreading in the Atlantic, Indian, and South Pacific Oceans, *in* Kovach, R. L., et al., eds., Proceedings, Tectonic Problems of the San Andreas Fault System:

Stanford University Publications in Geological Sciences, v. 13, p. 136–148.

AYDIN, A., AND NUR, A., 1982a, Evolution of pull-apart basins and their scale independence: Tectonics, v. 1, p. 91–105.

———, 1982b, Evolution of pull-apart basins and push-up ranges: Pacific Petroleum Geologists Newsletter: American Association of Petroleum Geologists, Pacific Section, Nov. 1982, p. 2–4.

———, 1985, The types and role of stepovers in strike-slip tectonics, *in* Biddle K. T., and Christie-Blick, N., eds., Strike-Slip Deformation, Basin Formation, and Sedimentation: Society of Economic Paleontologists and Mineralogists Special Publication No. 37, p. 35–44.

BALLANCE, P. F., 1980, Models of sediment distribution in non-marine and shallow marine environments in oblique-slip fault zones, *in* Ballance, P. F., and Reading, H. G., eds., Sedimentation in Oblique-Slip Mobile Zones: International Association of Sedimentologists Special Publication No. 4, p. 229–236.

BALLANCE, P. F., AND READING, H. G., eds., 1980a, Sedimentation in Oblique-Slip Mobile Zones: International Association of Sedimentologists Special Publication No. 4, 265 p.

———, 1980b, Sedimentation in oblique-slip mobile zones: an introduction, *in* Ballance, P. F., and Reading, H. G., eds., Sedimentation in Oblique-Slip Mobile Zones: International Association of Sedimentologists Special Publication No. 4, p. 1–5.

BALLY, A. W., ed., 1983, Seismic Expression of Structural Styles: American Association of Petroleum Geologists Studies in Geology Series, No. 15, 3 volumes.

BALLY, A. W., AND SNELSON, S., 1980, Realms of subsidence, *in* Miall, A. D., ed., Facts and Principles of World Petroleum Occurrence: Canadian Society of Petroleum Geologists Memoir 6, p. 9–94.

BARTLETT, W. L., FRIEDMAN, M., AND LOGAN, J. M., 1981, Experimental folding and faulting of rocks under confining pressure. Part IX. Wrench faults in limestone layers: Tectonophysics, v. 79, p. 255–277.

BARTOW, J. A., 1984, Geologic map and cross sections of the southeastern margin of the San Joaquin Valley, California: United States Geological Survey Miscellaneous Investigations Series Map I–1496, Scale 1:125,000.

BARTOW, J. A., AND MCDOUGALL, K., 1984, Tertiary stratigraphy of the southeastern San Joaquin Valley, California: United States Geological Survey Bulletin 1529-J, 41 p.

BEAUMONT, C., 1981, Foreland basins: Geophysical Journal of Royal Astronomical Society, v. 65, p. 291–329.

BECK, M. E., JR., 1980, Paleomagnetic record of plate-margin tectonic processes along the western margin of North America: Journal of Geophysical Research, v. 85, p. 7115–7131.

BELT, E. S., 1968, Post-Acadian rifts and related facies, eastern Canada, *in* Zen, E-An, White, W. S., Hadley, J. B., and Thompson, J. B., Jr., eds., Studies of Appalachian Geology, Northern and Maritime: New York, Wiley Interscience, p. 95–113.

BEN-AVRAHAM, Z., ALMAGOR, G., AND GARFUNKEL, Z., 1979, Sediments and structure of the Gulf of Elat (Aqaba)—northern Red Sea: Sedimentary Geology, v. 23, p. 239–267.

BEN-AVRAHAM, Z., GINZBURG, A., AND YUVAL, Z., 1981, Seismic reflection and refraction investigations of Lake Kinneret—central Jordan Valley, Israel: Tectonophysics, v. 80, p. 165–181.

BIDDLE, K. T., AND CHRISTIE-BLICK, N., 1985a, Glossary—Strike-slip deformation, basin formation, and sedimentation, *in* Biddle, K. T., and Christie-Blick, N., eds., Strike-Slip Deformation, Basin Formation, and Sedimentation: Society of Economic Paleontologists and Mineralogists Special Publication No. 37, p. 375–386.

BIDDLE, K. T., AND CHRISTIE-BLICK, N. H., 1985b, Basin formation, structural traps, and controls on hydrocarbon occurrence along wrench-fault zones: Offshore Technology Conference Paper OTC 4872, p. 291–295.

BILHAM, R., AND WILLIAMS, P., 1985, Sawtooth segmentation and deformation processes on the southern San Andreas fault, California: Geophysical Research Letters, v. 12, p. 557–560.

BIRD, P., AND ROSENSTOCK, R. W., 1984, Kinematics of present crust and mantle flow in southern California: Geological Society of America Bulletin, v. 95, p. 946–957.

BLAKE, M. C., JR., CAMPBELL, R. H., DIBBLEE, T. W., JR., HOWELL, D. G., NILSEN, T. H., NORMARK, W. R., VEDDER, J. C., AND SILVER, E. A., 1978, Neogene basin formation in relation to plate-tectonic evolution of San Andreas fault system, California: American Association

of Petroleum Geologists Bulletin, v. 62, p. 344–372.

BOHANNON, R. G., AND HOWELL, D. G., 1982, Kinematic evolution of the junction of the San Andreas, Garlock, and Big Pine faults, California: Geology, v. 10, p. 358–363.

BOIS, C., BOUCHE, P., AND PELET, R., 1982, Global geologic history and distribution of hydrocarbon reserves: American Association of Petroleum Geologists Bulletin, v. 66, p. 1248–1270.

BONATTI, E., AND CRANE, K., 1984, Oceanic fracture zones: Scientific American, v. 250, p. 40–51.

BRADLEY, D. C., 1982, Subsidence in Late Paleozoic basins in the northern Appalachians: Tectonics, v. 1, p. 107–123.

BRIDEN, J. C., TURNELL, H. B., AND WATTS, D. R., 1984, British paleomagnetism, Iapetus Ocean, and the Great Glen Fault: Geology, v. 12, p. 428–431.

BURCHFIEL, B. C., AND STEWART, J. H., 1966, "Pull-apart" origin of the central segment of Death Valley, California: Geological Society of America Bulletin, v. 77, p. 439–442.

BURCHFIEL, B. C., WALKER, D., DAVIS, G. A., AND WERNICKE, B., 1983, Kingston Range and related detachment faults—a major "breakaway" zone in the southern Great Basin: Geological Society of America Abstracts with Programs, v. 15, p. 536.

BURKE, K., MANN, P., AND KIDD, W., 1982, What is a ramp valley?: 11th International Congress on Sedimentology, Hamilton, Ontario, International Association of Sedimentologists, Abstracts of Papers, p. 40.

BUTLER, R. W. H., 1982, The terminology of structures in thrust belts: Journal of Structural Geology, v. 4, p. 239–245.

CAMPBELL, J. D., 1958, En echelon folding: Economic Geology, v. 53, p. 448–472.

CAREY, S. W., 1955, The orocline concept in geotectonics: Royal Society of Tasmania Proceedings, v. 89, p. 255–288.

———, 1958, A tectonic approach to continental drift, *in* Carey, S. W., convenor, Continental Drift: A Symposium: Hobart, University of Tasmania, p. 177–355.

CEMEN, I., WRIGHT, L. A., DRAKE, R. E., AND JOHNSON, F. C., 1985, Cenozoic sedimentation and sequence of deformational events at the southeastern end of the Furnace Creek strike-slip fault zone, Death Valley region, California, *in* Biddle, K. T., and Christie-Blick, N., eds., Strike-Slip Deformation, Basin Formation, and Sedimentation: Society of Economic Paleontologists and Mineralogists Special Publication No. 37, p. 127–141.

CHAMBERLAIN, V. E., AND LAMBERT, R. St J., 1985, Cordilleria, a newly defined Canadian microcontinent: Nature, v. 314, p. 707–713.

CHAMPION, D. E., HOWELL, D. G., AND GROMME, C. S., 1984, Paleomagnetic and geologic data indicating 2500 km of northward displacement for the Salinian and related terranes, California: Journal of Geophysical Research, v. 89, p. 7736–7752.

CHEADLE, M. J., CZUCHRA, B. L., BYRNE, T., ANDO, C. J., OLIVER, J. E., BROWN, L. D., KAUFMAN, S., MALIN, P. E., AND PHINNEY, R. A., 1985, The deep crustal structure of the Mojave Desert, California, from COCORP seismic reflection data: Tectonics, in press.

CHINNERY, M. A., 1961, The deformation of the ground around surface faults: Seismological Society of America Bulletin, v. 51, p. 355–372.

———, 1963, The stress changes that accompany strike-slip faulting: Seismological Society of America Bulletin, v. 53, p. 921–932.

CHOUKROUNE, P., FRANCHETEAU, J., AND LE PICHON, X., 1978, In situ structural observations along Transform Fault A in the FAMOUS area Mid-Atlantic Ridge: Geological Society of America Bulletin, v. 89, p. 1013–1029.

CISOWSKI, S. M., 1984, Evidence for early Tertiary remagnetization of Devonian rocks from the Orcadian basin, northern Scotland, and associated transcurrent fault motion: Geology, v. 12, p. 369–372.

CLARK, M. M., 1972, Surface rupture along the Coyote Creek Fault, *in* The Borrego Mountain Earthquake of April 9, 1968: United States Geological Survey Professional Paper 787, p. 55–86.

———, 1973, Map showing recently active breaks along the Garlock and associated faults, California: United States Geological Survey Miscellaneous Investigations Series Map I–741.

CLAYTON, L., 1966, Tectonic depressions along the Hope Fault, a transcurrent fault in North Canterbury, New Zealand: New Zealand Journal of Geology and Geophysics, v. 9, p. 95–104.

CLOOS, E., 1955, Experimental analysis of fracture patterns: Geological Society of America Bulletin, v. 66, p. 241–256.

COCHRAN, J. R., 1983, Effects of finite rifting times on the development of sedimentary basins: Earth and Planetary Science Letters, v. 66, p. 289–302.

CONEY, P. J., JONES, D. L., AND MONGER, J. W. H., 1980, Cordilleran suspect terranes: Nature, v. 288, p. 329–333.

COURTILLOT, V., TAPPONIER, P., AND VARET, J., 1974, Surface features associated with transform faults—A comparison between observed examples and an experimental model: Tectonophysics, v. 24, p. 317–329.

COX, A., 1980, Rotation of microplates in western North America, *in* Strangway, D. W., ed., The Continental Crust and its Mineral Deposits: Geological Association of Canada Special Paper 20, p. 305–321.

CRAIN, W. E., MERO, W. E., AND PATTERSON, D., 1985, Geology of the Point Arguello discovery: American Association of Petroleum Geologists Bulletin, v. 69, p. 537–545.

CROUCH, J. K., 1979, Neogene tectonic evolution of the California continental borderland and western Transverse Ranges: Geological Society of America Bulletin, Part I, v. 90, p. 338–345.

CROUCH, J. K., BACHMAN, S. B., AND SHAY, J. T., 1984, Post-Miocene compressional tectonics along the central California margin, *in* Crouch, J. K., and Bachman, S. B., eds., Tectonics and Sedimentation Along the California Margin: Society of Economic Paleontologists and Mineralogists, Pacific Section, v. 38, p. 37–54.

CROUCH, J., AND BACHMAN, S., 1985, California basins gain attention: American Association of Petroleum Geologists Explorer, April, p. 1, 8–9.

CROWELL, J. C., 1962, Displacement along the San Andreas fault, California: Geological Society of America Special Paper No. 71, 61 p.

———, 1974a, Sedimentation along the San Andreas fault, California, *in* Dott, R. H., Jr., and Shaver, R. H., eds., Modern and Ancient Geosynclinal Sedimentation: Society of Economic Paleontologists and Mineralogists Special Publication No. 19, p. 292–303.

———, 1974b, Origin of late Cenozoic basins in southern California, *in* Dickinson, W. R., ed., Tectonics and Sedimentation: Society of Economic Paleontologists and Mineralogists Special Publication No. 22, p. 190–204.

———, 1976, Implications of crustal stretching and shortening of coastal Ventura Basin, California, *in* Howell, D. G., ed., Aspects of the Geologic History of the California Continental Borderland: American Association of Petroleum Geologists, Pacific Section, Miscellaneous Publication 24, p. 365–382.

———, 1979, The San Andreas fault system through time: Journal of Geological Society of London, v. 136, p. 293–302.

———, 1981a, An outline of the tectonic history of southeastern California, *in* Ernst, W. G., ed., The Geotectonic Development of California (Rubey Volume 1): Englewood Cliffs, New Jersey, Prentice-Hall, p. 584–600.

———, 1981b, Juncture of San Andreas transform system and Gulf of California rift: Oceanologica Acta, Proceedings, 26th International Geological Congress, Paris, Geology of Continental Margins Symposium, p. 137–141.

———, 1982, The tectonics of Ridge Basin, southern California, *in* Crowell, J. C., and Link, M. H., eds., Geologic History of Ridge Basin, Southern California: Society of Economic Paleontologists and Mineralogists, Pacific Section, p. 25–42.

CROWELL, J. C., AND LINK, M. H., eds., 1982, Geologic History of Ridge Basin, Southern California: Society of Economic Paleontologists and Mineralogists, Pacific Section, 304 p.

CURRAY, J. R., AND MOORE, D. G., 1984, Geologic history of the mouth of the Gulf of California, *in* Crouch, J. K., and Bachman, S. B., eds., Tectonics and Sedimentation Along the California Continental Margin: Society of Economic Paleontologists and Mineralogists, Pacific Section, v. 38, p. 17–36.

CURRAY, J. R., AND MOORE, D. G., et al., 1982, Baja California passive margin transect: Sites 474, 475 and 476, *in* Curray, J. R., Moore, D. G., et al., Initial Reports of Deep Sea Drilling Project, v. 64, Part 1, p. 35–210.

DAHLSTROM, C. D. A., 1970, Structural geology in the eastern margin of the Canadian Rocky Mountains: Bulletin of Canadian Petroleum Geology, v. 18, p. 332–406.

DAVIS, G. A., AND BURCHFIEL, B. C., 1973, Garlock fault: an intracontinental transform structure, southern California: Geological Society of America Bulletin, v. 84, p. 1407–1422.

DAVIS, T., AND LAGOE, M., 1984, Cenozoic structural development of the north-central Transverse Ranges and southern margin of the San Joaquin Valley: Geological Society of America Abstracts with Programs, v. 16, p. 484.

DETRICK, R. S., JR., AND PURDY, G. M., 1980, The crustal structure of the Kane fracture zone from seismic refraction studies: Journal of Geophysical Research, v. 85, p. 3759–3777.

DEWEY, J. F., 1982, Plate tectonics and the evolution of the British Isles: Journal of Geological Society of London, v. 139, p. 371–412.

DEWEY, J. F., AND PINDELL, J. L., 1985, Neogene block tectonics of eastern Turkey and northern South America: Continental applications of the finite difference method: Tectonics, v. 4, p. 71–83.

DIBBLEE, T. W., JR., 1977, Strike-slip tectonics of the San Andreas fault and its role in Cenozoic basin evolvement, *in* Late Mesozoic and Cenozoic Sedimentation and Tectonics in California: San Joaquin Geological Society Short Course, p. 26–38.

DICKINSON, W. R., 1978, Plate tectonic evolution of sedimentary basins, *in* Dickinson, W. R., and Yarborough, H., eds., Plate tectonics and hydrocarbon accumulation: American Association of Petroleum Geologists Continuing Education Course Note Series No. 1, p. 1–62.

D'ONFRO, P., AND GLAGOLA, P., 1983, Wrench fault, southeast Asia, *in* Bally, A. W., ed., Seismic Expression of Structural Styles, American Association of Petroleum Geologists Studies in Geology Series 15, v. 3, p. 4.2–9 to 4.2–12.

DOKKA, R. K., 1983, Displacements on late Cenozoic strike-slip faults of the central Mojave Desert, California: Geology, v. 11, p. 305–308.

DOKKA, R. K., AND MERRIAM, R. H., 1982, Late Cenozoic extension of northeastern Baja California, Mexico: Geological Society of America Bulletin, v. 93, p. 371–378.

DONATH, F. A., 1970, Some information squeezed out of rock: American Scientist, v. 58, p. 54–72.

DONOVAN, R. N., AND MEYERHOFF, A. A., 1982, Comment on 'Paleomagnetic evidence for a large (~2,000 km) sinistral offset along the Great Glen Fault during Carboniferous time:' Geology, v. 10, p. 604–605.

DOWNEY, M. W., 1984, Evaluating seals for hydrocarbon accumulations: American Association of Petroleum Geologists Bulletin, v. 68, p. 1752–1763.

EATON, G. P., 1982, The Basin and Range Province: Origin and tectonic significance: Annual Review of Earth and Planetary Sciences, v. 10, p. 409–440.

EISBACHER, G. H., 1985, Pericollisional strike-slip faults and syn-orogenic basins, Canadian Cordillera, *in* Biddle, K. T., and Christie-Blick, N., eds., 1985, Strike-Slip Deformation, Basin Formation, and Sedimentation: Society of Economic Paleontologists and Mineralogists Special Publication No. 37, p. 265–282.

EMMONS, R. C., 1969, Strike-slip rupture patterns in sand models: Tectonophysics, v. 7, p. 71–87.

ENSLEY, R. A., AND VEROSUB, K. L., 1982, Biostratigraphy and magnetostratigraphy of southern Ridge Basin, central Transverse Ranges, California, *in* Crowell, J. C., and Link, M. H., eds., Geologic History of Ridge Basin, Southern California: Society of Economic Paleontologists and Mineralogists, Pacific Section, p. 13–24.

FISHER, M. A., 1982, Petroleum geology of Norton Basin, Alaska: American Association of Petroleum Geologists Bulletin, v. 66, p. 286–301.

FITZGERALD, E. L., 1968, Structure of British Columbia foothills: American Association of Petroleum Geologists Bulletin, v. 52, p. 641–664.

FORSYTH, D. W., 1979, Lithospheric flexure: Reviews of Geophysics and Space Physics, v. 17, p. 1109–1114.

FOX, P. J., AND GALLO, D. G., 1984, A tectonic model for ridge-transform-ridge plate boundaries: Implications for the structure of oceanic lithosphere: Tectonophysics, v. 104, p. 205–242.

FRALICK, P. W., 1982, Wrench basin type inferred from sedimentation style: Examples from the Upper Paleozoic Cumberland Basin, Maritime, Canada: 11th International Congress on Sedimentology, Hamilton, Ontario, International Association of Sedimentologists, Abstracts of Papers, p. 38.

FREUND, R., 1970, Rotation of strike slip faults in Sistan, southeast Iran: Journal of Geology, v. 78, p. 188–200.

———, 1971, The Hope fault, a strike-slip fault in New Zealand: New Zealand Geological Survey Bulletin, v. 86, p. 1–49.

———, 1974, Kinematics of transform and transcurrent faults: Tectonophysics, v. 21, p. 93–134.

————, 1982, The role of shear in rifting, *in* Pálmason, G., ed., Continental and Oceanic Rifts: American Geophysical Union Geodynamics Series, v. 8, p. 33–39.

FREUND, R., GARFUNKEL, Z., ZAK, I., GOLDBERG, M., WEISSBROD, T., AND DERIN, B., 1970, The shear along the Dead Sea rift: Philosophical Transactions of Royal Society of London, Series A, v. 267, p. 107–130.

FUIS, G. S., MOONEY, W. D., HEALY, J. H., MCMECHAN, G. A., AND LUTTER, W. J., 1984, A seismic refraction survey of the Imperial Valley region, California: Journal of Geophysical Research, v. 89, p. 1165–1189.

GAMOND, J. F., 1983, Displacement features associated with fault zones: a comparison between observed examples and experimental models: Journal of Structural Geology, v. 5, p. 33–45.

GARFUNKEL, Z., 1974, Model for the late Cenozoic tectonic history of the Mojave Desert, California, and for its relation to adjacent regions: Geological Society of America Bulletin, v. 85, p. 1931–1944.

————, 1981, Internal structure of the Dead Sea leaky transform (rift) in relation to plate kinematics: Tectonophysics, v. 80, p. 81–108.

GOGUEL, J., 1948, Introduction à l'étude mécanique des déformations de l'écorce terrestre: Mémoires, Carte Géologique Détaillée de la France, 2nd edition, 530 p.

GRAHAM, R. H., 1978, Wrench faults, arcuate fold patterns and deformation in the southern French Alps: Proceedings of Geological Association, v. 89, p. 125–142.

GRAHAM, S. A., 1976, Tertiary sedimentary tectonics of the central Salinian block of California [unpubl. Ph.D. dissertation]: Stanford, Stanford University, 510 p.

GRAHAM, S. A., AND WILLIAMS, L. A., 1985, Tectonic, depositional, and diagenetic history of Monterey Formation (Miocene), central San Joaquin Basin, California: American Association of Petroleum Geologists Bulletin, v. 69, p. 385–411.

GREEN, A. R., 1977, The evolution of the Earth's crust and sedimentary basin development: Offshore Technology Conference Paper OTC 2885, p. 67–72.

GREENHAUS, M. R., AND COX, A., 1979, Paleomagnetism of the Morro Rock-Islay Hill Complex as evidence for crustal block rotations in central coastal California: Journal of Geophysical Research, v. 84, p. 2393–2400.

GROSHONG, R. H., AND RODGERS, D. A., 1978, Left-lateral strike-slip fault model, *in* Wickham, J., and Denison, R., eds., Structural Style of Arbuckle Region: Geological Society of America Field Trip No. 3.

GUIRAUD, M., AND SEGURET, M., 1985, A releasing solitary overstep model for the late Jurassic—early Cretaceous (Wealdian) Soria strike-slip basin (northern Spain), *in* Biddle, K. T., and Christie-Blick, N., eds., Strike-Slip Deformation, Basin Formation, and Sedimentation: Society of Economic Paleontologists and Mineralogists Special Publication No. 37, p. 159–175.

GUTH, P. L., 1981, Tertiary extension north of the Las Vegas Valley shear zone, Sheep and Desert Ranges, Clark County, Nevada: Geological Society of America Bulletin, Part I, v. 92, p. 763–771.

GWINN, V. E., 1964, Thin-skinned tectonics in the Plateau and northwestern Valley and Ridge Provinces of the central Appalachians: Geological Society of America Bulletin, v. 75, p. 863–900.

HADLEY, D., AND KANAMORI, H., 1977, Seismic structure of the Transverse Ranges, California: Geological Society of America Bulletin, v. 88, p. 1469–1478.

HAMILTON, W., 1978, Mesozoic tectonics of the western United States, *in* Howell, D. G., and McDougall, K. A., eds., Mesozoic Paleogeography of the Western United States: Pacific Coast Paleogeography Symposium 2, Society of Economic Paleontologists and Mineralogists, Pacific Section, p. 33–70.

HARDING, T. P., 1974, Petroleum traps associated with wrench faults: American Association of Petroleum Geologists Bulletin, v. 58, p. 1290–1304.

————, 1983, Divergent wrench fault and negative flower structure, Andaman Sea, *in* Bally, A. W., ed., Seismic Expression of Structural Styles: American Association of Petroleum Geologists Studies in Geology, Series 15, v. 3, p. 4.2–1 to 4.2–8.

————, 1984, Graben hydrocarbon occurrences and structural style: American Association of Petroleum Geologists Bulletin, v. 68, p. 333–362.

————, 1985, Seismic characteristics and identification of negative flower structures, positive flower structures, and positive structural inversion: American Association of Petroleum Geologists Bulletin, v. 69, p. 582–600.

HARDING, T. P., AND LOWELL, J. D., 1979, Structural styles, their plate-tectonic habitats, and hydrocarbon traps in petroleum provinces: American Association of Petroleum Geologists Bulletin, v. 63, p. 1016–1058.

HARDING, T. P., GREGORY, R. F., AND STEPHENS, L. H., 1983, Convergent wrench fault and positive flower structure, Ardmore Basin, Oklahoma, *in* Bally, A. W., ed., Seismic Expression of Structural Styles: American Association of Petroleum Geologists Studies in Geology, Series 15, v. 3, p. 4.2–13 to 4.2–17.

HARDING, T. P., VIERBUCHEN, R. C., AND CHRISTIE-BLICK, N., 1985, Structural styles, plate-tectonic settings, and hydrocarbon traps of divergent (transtensional) wrench faults, *in* Biddle, K. T., and Christie-Blick, N., eds., Strike-Slip Deformation, Basin Formation, and Sedimentation: Society of Economic Paleontologists and Mineralogists Special Publication no. 37, p. 51–77.

HARLAND, W. B., 1971, Tectonic transpression in Caledonian Spitsbergen: Geological Magazine, v. 108, p. 27–42.

HARLAND, W. B., COX, A. V., LLEWELLYN, P. G., PICTON, C. A. G., SMITH, A. G., AND WALTERS, R., 1982, A Geologic Time Scale: Cambridge, Cambridge University Press, 131 p.

HARRIS, L. B., AND COBBOLD, P. R., 1984, Development of conjugate shear bands during simple shearing: Journal of Structural Geology, v. 7, p. 37–44.

HASZELDINE, R. S., 1984, Carboniferous North Atlantic paleogeography: stratigraphic evidence for rifting, not megashear or subduction: Geological Magazine, v. 121, p. 443–463.

HEARN, T. M., AND CLAYTON, R. W., 1984, Crustal structure and tectonics in Southern California (abs.): EOS, v. 65, p. 992.

HERD, D. G., 1978, Intracontinental plate boundary east of Cape Mendocino, California: Geology, v. 6, p. 721–725.

HELWIG, J., 1974, Eugeosynclinal basement and a collage concept of orogenic belts, *in* Dott, R. H., Jr., and Shaver, R. H., eds., Modern and Ancient Geosynclinal Sedimentation: Society of Economic Paleontologists and Mineralogists Special Publication No. 19, p. 359–376.

HEWARD, A. P., AND READING, H. G., 1980, Deposits associated with a Hercynian to late Hercynian continental strike-slip system, Cantabrian Mountains, northern Spain, *in* Ballance, P. F., and Reading, H. G., eds., Sedimentation in Oblique-Slip Mobile Zones: International Association of Sedimentologists Special Publication No. 4, p. 105–125.

HILL, D. P., 1982, Contemporary block tectonics: California and Nevada: Journal of Geophysical Research, v. 87, p. 5433–5450.

HILL, M. L., AND TROXEL, B. W., 1966, Tectonics of Death Valley region, California: Geological Society of America Bulletin, v. 77, p. 435–438.

HOWELL, D. G., CROUCH, J. K., GREENE, H. G., MCCULLOCH, D. S., AND VEDDER, J. G., 1980, Basin development along the late Mesozoic and Cainozoic California margin: a plate tectonic margin of subduction, oblique subduction and transform tectonics, *in* Ballance, P. F., and Reading, H. G., eds., Sedimentation in Oblique-Slip Mobile Zones: International Association of Sedimentologists Special Publication No. 4, p. 43–62.

HUBBERT, M. K., 1937, Theory of scale models as applied to the study of geologic structures: Geological Society of America Bulletin, v. 48, p. 1459–1520.

HUFF, K. F., 1980, Frontiers of world exploration, *in* Miall, A. D., ed., Facts and Principles of World Petroleum Occurrence: Canadian Society of Petroleum Geologists Memoir 6, p. 343–362.

HUNT, C. B., AND MABEY, D. R., 1966, Stratigraphy and structure, Death Valley, California: United States Geological Survey Professional Paper 494-A, 162 p.

ILLIES, J. H., 1975, Intraplate tectonics in stable Europe as related to plate tectonics in the Alpine System: Geologische Rundschau, v. 64, p. 677–699.

ILLIES, J. H., AND GREINER, G., 1978, Rhinegraben and the Alpine System: Geological Society of America Bulletin, v. 89, p. 770–782.

IRVING, E., 1979, Paleopoles and paleolatitudes of North America and speculations about displaced terrains: Canadian Journal of Earth Sciences, v. 16, p. 669–694.

IRVING, E., AND STRONG, D. F., 1984, Palaeomagnetism of the Early Carboniferous Deer Lake Group, western Newfoundland: no evidence for mid-Carboniferous displacement of "Acadia:" Earth and Planetary Science Letters, v. 69, p. 379–390.

IRVING, E., MONGER, J. W. H., AND YOLE, R. W., 1980, New paleomagnetic evidence for displaced terranes in British Columbia, *in* Strangway, D. W., ed., The Continental Crust and its Mineral Deposits: Geological Association of Canada Special Paper 20, p. 441–456.

JACKSON, P. A., AND YEATS, R. S., 1982, Structural evolution of Carpinteria Basin, western Transverse Ranges, California: American Association of Petroleum Geologists Bulletin, v. 66, p. 805–829.

JARVIS, G. T., AND MCKENZIE, D. P., 1980, Sedimentary basin formation with finite extension rates: Earth and Planetary Science Letters, v. 48, p. 42–52.

JENNINGS, C. W., compiler, 1975, Preliminary fault and geologic map of southern California, *in* Crowell, J. C., ed, San Andreas Fault in Southern California: California Division of Mines and Geology Special Report 118, Scale 1:750,000.

JOHNSON, N. M., OFFICER, C. B., OPDYKE, N. D., WOODWARD, G. D., ZEITLER, P. K., AND LINDSAY, E. H., 1983, rates of late Cenozoic tectonism in the Vallecito-Fish Creek basin, western Imperial Valley, California: Geology, v. 11, p. 664–667.

JOHNSON, S. Y., 1985, Eocene strike-slip faulting and nonmarine basin formation in Washington, *in* Biddle, K. T., and Christie-Blick, N., eds., Strike-Slip Deformation, Basin Formation, and Sedimentation: Society of Economic Paleontologists and Mineralogists Special Publication No. 37, p. 283–302.

JONES, D. L., SILBERLING, N. J,, AND HILLHOUSE, J., 1977, Wrangellia—A displaced terrane in northwestern North America: Canadian Journal of Earth Sciences, v. 14, p. 2565–2577.

JUNGER, A., 1976, Tectonics of the southern California borderland, *in* Howell, D. G., ed., Aspects of the Geologic History of the California Continental Borderland: American Association of Petroleum Geologists, Pacific Section, Miscellaneous Publication 24, p. 486–498.

KANTER, L. R., AND MCWILLIAMS, M. O., 1982, Rotation of the southernmost Sierra Nevada, California: Journal of Geophysical Research, v. 87, p. 3819–3830.

KARIG, D. E., 1980, Material transport within accretionary prisms and the "knocker" problem: Journal of Geology, v. 88, p. 27–39.

KARIG, D. E., AND JENSKY, W., 1972, The proto-Gulf of California: Earth and Planetary Science Letters, v. 17, p. 169–174.

KARSON, J. A., 1984, Variations in structure and petrology in the Coastal Complex, Newfoundland: anatomy of an oceanic fracture zone, *in* Gass, I. G., Lippard, S. J., and Shelton, A. W., eds., Ophiolites and Oceanic Lithosphere: Geological Society of London, Special Publication No. 13, p. 131–144.

KARSON, J. A., AND DICK, H. J. B., 1983, Tectonics of ridge-transform intersections at the Kane fracture zone: Marine Geophysical Researches, v. 6, p. 51–98.

KELTS, K., 1981, A comparison of some aspects of sedimentation and translational tectonics from the Gulf of California and the Mesozoic Tethys, northern Penninic margin: Eclogae Geologicae Helvetiae, v. 74, p. 317–338.

KENNEDY, W. Q., 1946, The Great Glen Fault: Quarterly Journal of Geological Society of London, v. 102, p. 41–76.

KENT, D. V., 1982, Paleomagnetic evidence for post-Devonian displacement of the Avalon platform (Newfoundland): Journal of Geophysical Research, v. 87, p. 8709–8716.

KENT, D. V., AND OPDYKE, N. D., 1978, Paleomagnetism of the Devonian Catskill red beds: Evidence for motion of the coastal New England–Canadian Maritime region relative to cratonic North America: Journal of Geophysical Research, v. 83, p. 4441–4450.

———, 1979, The Early Carboniferous paleomagnetic field of North America and its bearing on tectonics of the northern Appalachians: Earth and Planetary Science Letters, v. 44, p. 365–372.

———, 1985, Multicomponent magnetizations from the Mississippian Mauch Chunk Formation of the central Appalachians and their tectonic implications: Journal of Geophysical Research, v. 90, p. 5371–5383.

KERR, D. R., PAPPAJOHN, S., AND PETERSON, G. L., 1979, Neogene stratigraphic section at Split Mountain, eastern San Diego County, California, *in* Crowell, J. C., and Sylvester, A. G., eds., Tectonics of the

Juncture Between the San Andreas Fault System and the Salton Trough, Southeastern California: Santa Barbara, California, Department of Geological Sciences, University of California, p. 111–123.

KING, P. B., 1969, Tectonic map of North America: United States Geological Survey, Scale 1:5,000,000.

KINGMA, J. T., 1958, Possible origin of piercement structures, local unconformities, and secondary basins in the Eastern Geosyncline, New Zealand: New Zealand Journal of Geology and Geophysics, v. 1, p. 269–274.

KINGSTON, D. R., DISHROON, C. P., AND WILLIAMS, P. A., 1983a, Global basin classification system: American Association of Petroleum Geologists Bulletin, v. 67, p. 2175–2193.

———, 1983b, Hydrocarbon plays and global basin classification: American Association of Petroleum Geologists Bulletin, v. 67, p. 2194–2198.

KLEMME, H. D., 1980, Petroleum basins—classifications and characteristics: Journal of Petroleum Geology, v. 3, p. 187–207.

LABOTKA, T. C., 1980, Petrology of a medium-pressure regional metamorphic terrane, Funeral Mountains, California: American Mineralogist, v. 65, p. 670–689.

———, 1981, Petrology of an andalusite-type regional metamorphic terrane, Panamint Mountains, California: Journal of Petrology, v. 22, p. 261–296.

LABOTKA, T. C., AND WARASILA, R., 1983, Ages of metamorphism in the central Panamint Mountains, California: Geological Society of America Abstracts with Programs, v. 15, p. 437.

LACHENBRUCH, A. H., AND SASS, J. H., 1980, Heat flow and energetics of the San Andreas fault zone: Journal of Geophysical Research, v. 85, p. 6185–6222.

LEHNER, P., AND BAKKER, G., 1983, Equatorial fracture zone (Romanche fracture), *in* Bally, A. W., ed., Seismic Expression of Structural Styles: American Association of Petroleum Geologists Studies in Geology, Series 15, v. 3, p. 4.2–25 to 4.2–29.

LENSEN, G. J., 1958, A method of graben and horst formation: Journal of Geology, v. 66, p. 579–587.

LEONARD, R., 1983, Geology and hydrocarbon accumulations, Columbus Basin, offshore Trinidad: American Association of Petroleum Geologists, v. 67, p. 1081–1093.

LEWIS, K. B., 1980, Quaternary sedimentation on the Hikurangi oblique-subduction and transform margin, New Zealand, *in* Ballance, P. F., and Reading, H. G., eds., Sedimentation in Oblique-Slip Mobile Zones: International Association of Sedimentologists Special Publication No. 4, p. 171–189.

LI DESHENG, 1984, Geologic evolution of petroliferous basins on continental shelf of China: American Association of Petroleum Geologists Bulletin, v. 68, p. 993–1003.

LINK, M. H., ROBERTS, M. T., AND NEWTON, M. S., 1985, Walker Lake Basin, Nevada: An example of late Tertiary (?) to Recent sedimentation in a basin adjacent to an active strike-slip fault, *in* Biddle, K. T., and Christie-Blick, N., eds., Strike-Slip Deformation, Basin Formation, and Sedimentation: Society of Economic Paleontologists and Mineralogists Special Publication No. 37, p. 105–125.

LOGAN, J. M., 1979, Brittle phenomena: Reviews of Geophysics and Space Physics, v. 17, p. 1121–1132.

LOGAN, J. M., FRIEDMAN, M., HIGGS, N. G., DENGO, C., AND SHIMAMOTO, T., 1979, Experimental studies of simulated gouge and their application to studies of natural fault zones: United States Geological Survey Open-File Report 79-1239, p. 305–343.

LONSDALE, P., 1978, Near-bottom reconnaissance of a fast-slipping transform zone at the Pacific-Nazca plate boundary: Journal of Geology, v. 86, p. 451–472.

LOWELL, J. D., 1972, Spitsbergen Tertiary orogenic belt and the Spitsbergen fracture zone: Geological Society of America Bulletin, v. 83, p. 3091–3102.

LUDMAN, A., 1981, Significance of transcurrent faulting in eastern Maine and location of the suture between Avalonia and North America: American Journal of Science, v. 281, p. 463–483.

LUYENDYK, B. P., KAMERLING, M. J., AND TERRES, R., 1980, Geometric model for Neogene crustal rotations in southern California: Geological Society of America Bulletin, Part I, v. 91, p. 211–217.

LUYENDYK, B. P., KAMERLING, M. J., TERRES, R. R., AND HORNAFIUS, J. S., 1985, Simple shear of southern California during Neogene time suggested by paleomagnetic declinations: Journal of Geophysical Research, in press.

MACDONALD, K. C., KASTENS, K., SPIESS, F. N., AND MILLER, S. P., 1979, Deep tow studies of the Tamayo transform fault: Marine Geophysical Researches, v. 4, p. 37–70.

MANDL, G., DE JONG, L. N. G., AND MALTHA, A., 1977, Shear zones in granular material: Rock Mechanics, v. 9, p. 95–144.

MANN, P., AND BRADLEY, D., 1984, Comparison of basin types in active and ancient strike-slip zones (abs.): American Association of Petroleum Geologists Bulletin, v. 68, p. 503.

MANN, P., AND BURKE, K., 1982, Basin formation at intersections of conjugate strike-slip faults: examples from southern Haiti: Geological Society of America Abstracts with Programs, v. 14, p. 555.

MANN, P., HEMPTON, M. R., BRADLEY, D. C., AND BURKE, K., 1983, Development of pull-apart basins: Journal of Geology, v. 91, p. 529–554.

MANN, P., DRAPER, G., AND BURKE, K., 1985, Neotectonics of a strike-slip restraining bend system, Jamaica, in Biddle, K. T., and Christie-Blick, N., eds., Strike-Slip Deformation, Basin Formation, and Sedimentation: Society of Economic Paleontologists and Mineralogists Special Publication No. 37, p. 211–226.

MANSPEIZER, W., 1985, The Dead Sea Rift: Impact of climate and tectonism on Pleistocene and Holocene sedimentation, in Biddle, K. T., and Christie-Blick, N., eds., Strike-slip Deformation, Basin Formation, and Sedimentation: Society of Economic Paleontologists and Mineralogists Special Publication No. 37, p. 143–158.

MARLOW, M. S., 1979, Hydrocarbon prospects in Navarin basin province, northwest Bering Sea shelf: Oil and Gas Journal, October 29, v. 77, p. 190–196.

MARLOW, M. S., AND COOPER, A. K., 1980, Mesozoic and Cenozoic structural trends under southern Bering Sea shelf: American Association of Petroleum Geologists Bulletin, v. 64, p. 2139–2155.

McKENZIE, D., 1978, Some remarks on the development of sedimentary basins: Earth and Planetary Science Letters, v. 40, p. 25–32.

MENARD, H. W., AND ATWATER, T., 1969, Origin of fracture zone topography: Nature, v. 222, 1037–1040.

MIALL, A. D., 1985, Stratigraphic and structural predictions from a plate-tectonic model of an oblique-slip orogen: The Eureka Sound Formation (Campanian-Oligocene), northeast Canadian Arctic islands, in Biddle, K. T., and Christie-Blick, N., eds., Strike-Slip Deformation, Basin Formation, and Sedimentation: Society of Economic Paleontologists and Mineralogists Special Publication No. 37, p. 000–000.

MINSTER, J. B. AND JORDAN, T. H., 1978, Present-day plate motions: Journal of Geophysical Research, v. 83, p. 5331–5354.

———, 1984, Vector constraints on Quaternary deformation of the western United States east and west of the San Andreas fault, in Crouch, J. K., and Bachman, S. B., eds., Tectonics and Sedimentation Along the California margin: Society of Economic Paleontologists and Mineralogists, Pacific Section, v. 38, p. 1–16.

MINSTER, J. B., JORDAN, T. H., MOLNAR, P., AND HAINES, E., 1974, Numerical modelling of instantaneous plate tectonics: Geophysical Journal of Royal Astronomical Society, v. 36, p. 541–576.

MOLNAR, P., AND TAPPONNIER, P., 1975, Cenozoic tectonics of Asia: Effects of a continental collision: Science, v. 189, p. 419–426.

MOODY, J. D., AND HILL, M. J., 1956, Wrench-fault tectonics: Geological Society of America Bulletin, v. 67, p. 1207–1246.

MOORE, D. G., 1969, Reflection profiling studies of the California continental Borderland: Structure and Quaternary turbidite basins: Geological Society of America Special Paper 107, 138 p.

MORGENSTERN, N. R., AND TCHALENKO, J. S., 1967, Microscopic structures in kaolin subjected to direct shear: Géotechnique, v. 17, p. 309–328.

MORTON, J. L., AND HILLHOUSE, J. W., 1985, Paleomagnetism and K-Ar ages of Miocene basaltic rocks in the western Mojave Desert, California: manuscript.

MUTTER, J. C., AND DETRICK, R. S., 1984, Multichannel seismic evidence for anomalously thin crust at Blake Spur fracture zone: Geology, v. 12, p. 534–537.

NAMSON, J. S., DAVIS, T. L., AND LAGOE, M. B., 1985, Tectonic history and thrust-fold deformation style of seismically active structures near Coalinga, California: United States Geological Survey Professional Paper, submitted.

NARDIN, T. R., AND HENYEY, T. L., 1978, Pliocene-Pleistocene diastrophism of Santa Monica and San Pedro shelves, California continental borderland: American Association of Petroleum Geologists Bulletin, v. 62, p. 247–272.

NELSON, W. J., AND KRAUSSE, H.-F., 1981, The Cottage Grove fault system in southern Illinois: Illinois Institute of Natural Resources, State Geological Survey Division, Circular 522, 65 p.

NICHOLSON, C., SEEBER, L., WILLIAMS, P., AND SYKES, L. R., 1985a, Seismicity and fault kinematics through the eastern Transverse Ranges, California: Block rotation, strike-slip faulting and shallow-angle thrusts: Journal of Geophysical Research, in press.

NICHOLSON, C., SEEBER, L., WILLIAMS, P. L., AND SYKES, L. R., 1985b, Seismic deformation along the southern San Andreas fault, California: Implications for conjugate slip rotational block tectonics: Tectonics, submitted.

NILSEN, T. H., AND McLAUGHLIN, R. J., 1985, Comparison of tectonic framework and depositional patterns of the Hornelen strike-slip basin of Norway and the Ridge and Little Sulphur Creek strike-slip basins of California, in Biddle, K. T., and Christie-Blick, N. eds., Strike-Slip Deformation, Basin Formation, and Sedimentation: Society of Economic Paleontologists and Mineralogists Special Publication No. 37, p. 80–103.

NISSENBAUM, A., 1978, Dead Sea Asphalts—Historical aspects: American Association of Petroleum Geologists Bulletin, v. 62, p. 837–844.

NORRIS, R. J., AND CARTER, R. M., 1980, Offshore sedimentary basins at the southern end of the Alpine fault, New Zealand, in Ballance, P. F., and Reading, H. G., eds., Sedimentation in Oblique-Slip Mobile Zones: International Association of Sedimentologists Special Publication No. 4, p. 237–265.

ODONNE, F., AND VIALON, P., 1983, Analogue models of folds above a wrench fault: Tectonophysics, v. 99, p. 31–46.

OPDYKE, N. D., LINDSAY, E. H., JOHNSON, N. M., AND DOWNS, T., 1977, The paleomagnetism and magnetic polarity stratigraphy of the mammal-bearing section of Anza Borrego State Park, California: Quaternary Research, v. 7, p. 316–329.

PAGE, B. M., AND ENGEBRETSON, D. C., 1984, Correlation between the geologic record and computed plate motions for central California: Tectonics, v. 3, p. 133–155.

PALMER, A. R., 1983, The Decade of North American Geology 1983 time scale: Geology, v. 11, p. 503–504.

PARNELL, J. T., 1982, Comment on 'Paleomagnetic evidence for a large (2,000 km) sinistral offset along the Great Glen fault during Carboniferous time:' Geology, v. 10, p. 605.

PAVONI, N., 1961, Die Nordanatolische Horizontalverschiebung: Geologische Rundschau, v. 51, p. 122–139.

PERRODON, A., AND MASSE, P., 1984, Subsidence, sedimentation and petroleum systems: Journal of Petroleum Geology, v. 7, p. 5–26.

PERROUD, H., VAN DER VOO, R., AND BONHOMMET, N., 1984, Paleozoic evolution of the Armorica plate on the basis of paleomagnetic data: Geology, v. 12, p. 579–582.

PHILIP, H., AND MEGARD, F., 1977, Structural analysis of the superficial deformation of the 1969 Pariahuanca earthquakes (central Peru): Tectonophysics, v. 38, p. 259–278.

PITMAN, W. C., III, AND ANDREWS, J. A., 1985, Subsidence and thermal history of small pull-apart basins, in Biddle, K. T., and Christie-Blick, N., eds., Strike-Slip Deformation, Basin Formation, and Sedimentation: Society of Economic Paleontologists and Mineralogists Special Publication No. 37, p. 45–49.

PITTMAN, E. D., 1981, Effect of fault-related granulation on porosity and permeability of quartz sandstones, Simpson Group (Ordovician), Oklahoma: American Association of Petroleum Geologists Bulletin, v. 65, p. 2381–2387.

QUENNELL, A. M., 1958, The structural and geomorphic evolution of the Dead Sea Rift: Quarterly Journal of Geological Society of London, v. 114, p. 1–24.

RAMSAY, J. G., 1980, Shear zone geometry: a review: Journal of Structural Geology, v. 2, p. 83–99.

READING, H. G., 1980, Characteristics and recognition of strike-slip fault systems, in Ballance, P. F., and Reading, H. G., eds., Sedimentation in Oblique-Slip Mobiles Zones: International Association of Sedimentologists Special Publication No. 4, p. 7–26.

———, 1982, Sedimentary basins and global tectonics: Proceedings of

Geological Association, v. 93, p. 321–350.

RIEDEL, W., 1929, Zur Mechanik geologischer Brucherscheinungen: Zentrablatt für Mineralogie, Geologie und Pälaeontologie, v. 1929 B, p. 354–368.

RIXON, L. K., 1978, Clay modelling of the Fitzroy Graben: Bureau of Mineral Resources Journal of Australian Geology and Geophysics, v. 3, p. 71–76.

ROBERTS, M. T., 1983, Seismic example of complex faulting from northwest shelf of Palawan, Phillipines, in Bally, A. W., ed., Seismic Expression of Structural Styles: American Association of Petroleum Geologists Studies in Geology, Series 15, v. 3, p. 4.2–18 to 4.2–24.

RODGERS, D. A., 1980, Analysis of pull-apart basin development produced by en echelon strike-slip faults, in Ballance, P. F., and Reading, H. G., eds., Sedimentation in Oblique-Slip Mobile Zones: International Association of Sedimentologists Special Publication No. 4, p. 27–41.

RODGERS, J., 1963, Mechanics of Appalachian foreland folding in Pennsylvania and West Virginia: American Association of Petroleum Geologists Bulletin, v. 47, p. 1527–1536.

RON, H., FREUND, R., GARFUNKEL, Z., AND NUR, A., 1984, Block rotation by strike-slip faulting: structural and paleomagnetic evidence: Journal of Geophysical Research, v. 89, p. 6256–6270.

RON, H., AND EYAL, Y., 1985, Intraplate deformation by block rotation and mesostructures along the Dead Sea transform, northern Israel: Tectonics, v. 4, p. 85–105.

ROY, J. L., AND MORRIS, W. A., 1983, A review of paleomagnetic results from the Carboniferous of North America; the concept of Carboniferous geomagnetic field horizon markers: Earth and Planetary Science Letters, v. 65, p. 167–181.

ROYDEN, L. H., 1985, The Vienna Basin: A thin-skinned pull-apart basin, in Biddle, K. T., and Christie-Blick, N., eds., Strike-Slip Deformation, Basin Formation, and Sedimentation: Society of Economic Paleontologists and Mineralogists Special Publication No. 37, p. 319–338.

ROYDEN, L., AND KEEN, C. E., 1980, Rifting processes and thermal evolution of the continental margin of eastern Canada determined from subsidence curves: Earth and Planetary Science Letters, v. 51, p. 343–361.

ROYDEN, L. H., HORVÁTH, F., AND BURCHFIEL, B. C., 1982, Transform faulting, extension, and subduction in the Carpathian Pannonian region: Geological Society of America Bulletin, v. 93, p. 717–725.

SALVADOR, A., 1985, Chronostratigraphic and geochronometric scales in COSUNA stratsigraphic correlation charts of the United States: American Association of Petroleum Geologists Bulletin, v. 69, p. 181–189.

SBAR, M. L., 1982, Delineation and interpretation of seismotectonic domains in western North America: Journal of Geophysical Research, v. 87, p. 3919–3928.

SCHUBERT, C., 1980, Late-Cenozoic pull-apart basins, Boconó fault zone, Venezuelan Andes: Journal of Structural Geology, v. 2, p. 463–468.

SEARLE, R. C., 1979, Side-scan sonar studies of North Atlantic fracture zones: Journal of Geological Society of London, v. 136, p. 283–292.

———, 1983, Multiple, closely spaced transform faults in fast-slipping fracture zones: Geology, v. 11, p. 607–610.

SEGALL, P., AND POLLARD, D. D., 1980, Mechanics of discontinuous faults: Journal of Geophysical Research, v. 85, p. 4337–4350.

———, 1983, Nucleation and growth of strike slip faults in granite: Journal of Geophysical Research, v. 88, p. 555–568.

SENGÖR, A. M. C., BURKE, K., AND DEWEY, J. F., 1978, Rifts at high angles to orogenic belts: Tests for their origin and the Upper Rhine Graben as an example: American Journal of Science, v. 278, p. 24–40.

ŞENGÖR, A. M. C., GÖRÜR, N., AND SAROGLU, F., 1985, Strike-slip faulting and related basin formation in zones of tectonic escape: Turkey as a case study, in Biddle, K. T., and Christie-Blick, N., eds., Strike-Slip Deformation, Basin Formation, and Sedimentation: Society of Economic Paleontologists and Mineralogists Special Publication No. 37, p. 227–264.

SHARP, R. V., 1976, Surface faulting in Imperial Valley during the earthquake swarm of January–February, 1975, Seismological Society of America Bulletin, v. 66, p.1145–1154.

———, 1977, Map showing the Holocene surface expression of the Brawley Fault, Imperial County, California: United States Geological

Survey Miscellaneous Field Studies Map MF-838.

———, 1981, Variable rates of late Quaternary strike slip on the San Jacinto fault zone, southern California: Journal of Geophysical Research, v. 86, p. 1754–1762.

SIBSON, R. H., 1977, Fault rocks and fault mechanisms: Journal of Geological Society of London, v. 133, p. 191–213.

SIEH, K. E., AND JAHNS, R. H., 1984, Holocene activity of the San Andreas fault at Wallace Creek, California: Geological Society of America Bulletin, v. 95, p. 883–896.

SINHA, M. C., AND LOUDEN, K. E., 1983, The Oceanographer fracture zone–I. Crustal structure from seismic refraction studies: Geophysical Journal of Royal Astronomical Society, v. 75, p. 713–736.

SMITH, D. I., AND WATSON, J., 1983, Scale and timing of movements on the Great Glen fault, Scotland: Geology, v. 11, 523–526.

SMITH, G. I., 1962, Large lateral displacement on Garlock fault, California, as measured from offset dike swarm: American Association of Petroleum Geologists Bulletin, v. 46, p.85–104.

SMITH, G. I., AND KETNER, K. B., 1970, Lateral displacement on the Garlock fault, southeastern California, suggested by offset sections of similar metasedimentary rocks: United States Geological Survey Professional Paper 700-D, p. 1–9.

STECKLER, M. S., 1981, Thermal and mechanical evolution of Atlantic-type margins [unpubl. Ph.D. thesis]: New York, Columbia University, 261 p.

STECKLER, M. S., AND TEN BRINK, U. S., 1985, Replacement of the Gulf of Suez rift by the Dead Sea transform: The role of the hinge zones in rifting (abs.): EOS, v. 66, p. 364.

STECKLER, M. S., AND WATTS, A. B., 1982, Subsidence history and tectonic evolution of Atlantic-type continental margins, in Scrutton, R. A., ed., Dynamics of Passive Margins: American Geophysical Union Geodynamics Series, v. 6, p. 184–196.

STEEL, R., GJELBERG, J., HELLAND-HANSEN, W., KLEINSPEHN, K., NØTTVEDT, A., AND RYE-LARSEN, M., 1985, The Tertiary strike-slip basins and orogenic belt of Spitsbergen, in Biddle, K. T., and Christie-Blick, N., eds., Strike-Slip Deformation, Basin Formation, and Sedimentation: Society of Economic Paleontologists and Mineralogists Special Publication No. 37, p. 227–264.

STEWART, J. H., 1978, Basin-range structure in western North America: A review, in Smith, R. B., and Eaton, G. P., eds., Cenozoic tectonics and regional geophysics of the western Cordillera: Geological Society of America Memoir 152, p. 1–31.

———, 1983, Extensional tectonics in the Death Valley area, California: Transport of the Panamint Range structural block 80 km northwestward: Geology, v. 11, p. 153–157.

SUPPE, 1978, Cross section of southern part of northern Coast Ranges and Sacramento Valley, California: Geological Society of America Map and Chart Series, MC-28B.

SYKES, L. R., 1967, Mechanism of earthquakes and nature of faulting on the mid-oceanic ridges: Journal of Geophysical Research, v. 72, p. 2131–2153.

SYLVESTER, A. G., compiler, 1984, Wrench fault tectonics: American Association of Petroleum Geologists Reprint Series, No. 28, 374 p.

SYLVESTER, A. G., AND SMITH, R. R., 1976, Tectonic transpression and basement-controlled deformation in San Andreas fault zone, Salton Trough, California: American Association of Petroleum Geologists Bulletin, v. 60, p. 2081–2102.

TAYLOR, J. C., 1976, Geologic appraisal of the petroleum potential of offshore southern California: the borderland compared to onshore coastal basins: United States Geological Survey Circular 730, 43 p.

TCHALENKO, J. S., 1970, Similarities between shear zones of different magnitudes: Geological Society of America Bulletin, v. 81, p. 1625–1640.

TCHALENKO, J. S., AND AMBRASEYS, N. N., 1970, Structural analyses of the Dasht-e Bayaz (Iran) earthquake fractures: Geological Society of America Bulletin, v. 81, p. 41–60.

TERRES, R., AND CROWELL, J. C., 1979, Plate tectonic framework of the San Andreas-Salton Trough juncture, in Crowell, J. C., and Sylvester, A. G., eds., Tectonics of the Juncture Between the San Andreas Fault System and the Salton Trough, Southeastern California: Santa Barbara, California, Department of Geological Sciences, University of California, p. 15–25.

TERRES, R. R., AND SYLVESTER, A. G., 1981, Kinematic analysis of rotated fractures and blocks in simple shear: Seismological Society of American Bulletin, v. 71, p. 1593–1605.

TURCOTTE, D. L., LIU, J. Y., AND KULHAWY, F. H., 1984, The role of an intracrustal asthenosphere on the behavior of major strike-slip faults: Journal of Geophysical Research, v. 89, p. 5801–5816.

VAIL, P. R., MITCHUM, R. M., JR., TODD, R. G., WIDMIER, J. M., THOMPSON, S., JR., SANGREE, J. B., BUBB, J. N., AND HATELID, W. G., 1977, Seismic stratigraphy and global changes of sea level, *in* Payton, C. E., ed., Seismic Stratigraphy—Applications to Hydrocarbon Exploration: American Association of Petroleum Geologists Memoir 26, p. 49–212.

VAN DER VOO, R., 1982, Pre-Mesozoic paleomagnetism and plate tectonics: Annual Review of Earth and Planetary Sciences, v. 10, p. 191–220.

———, 1983, Paleomagnetic constraints on the assembly of the Old Red continent: Tectonophysics, v. 91, p. 271–283.

VAN DER VOO, R., AND SCOTESE, C. R., 1981, Paleomagnetic evidence for a large (~2,000 km) sinistral offset along the Great Glen Fault during Carboniferous time: Geology, v. 9, p. 583–589.

VAN DER VOO, R., FRENCH, A. N., AND FRENCH, R. B., 1979, A paleomagnetic pole position from the folded Upper Devonian Catskill red beds, and its tectonic implications: Geology, v. 7, p. 345–348.

VINK, G. E., MORGAN, W. J., AND ZHAO, W.-L., 1984, Preferential rifting of continents: A source of displaced terranes: Journal of Geophysical Research, v. 89, p. 10,072–10,076.

WATTS, A. B., 1983, The strength of the Earth's crust: Marine Technology Society Journal, v. 17, p. 5–17.

WEBB, T. H., AND KANAMORI, H., 1985, Earthquake focal mechanisms in the eastern Transverse Ranges and San Emigdio Mountains, southern California, and evidence for a regional decollement: Seismological Society of America Bulletin, v. 75, p. 737–757.

WEBER, G. E., AND LAJOIE, K. R., 1977, Late Pleistocene and Holocene tectonics of the San Gregorio fault zone between Moss Beach and Point Ano Nuevo, San Mateo County, California: Geological Society of America Abstracts with Programs, v. 9, p. 524.

WEERTMAN, J., 1965, Relationship between displacements on a free surface and the stress on a fault: Seismological Society of America Bulletin, v. 55, p. 945–953.

WELDON, R., AND HUMPHREYS, G., 1985, A kinematic model of southern California: Tectonics, in press.

WELDON, R. J., II, AND SIEH, K. E., 1985, Holocene rate of slip and tentative recurrence interval for large earthquakes on the San Andreas fault, Cajon Pass, southern California: Geological Society of American Bulletin, v. 96, p. 793–812.

WELDON, R. J., WINSTON, D. S., KIRSCHVINK, J. L., AND BURBANK, D. W., 1984, Magnetic stratigraphy of the Crowder Formation, Cajon Pass, southern California: Geological Society of America Abstracts with Programs, v. 16, p. 689.

WENTWORTH, C. M., BLAKE, M. C., JR., JONES, D. L., WALTER, A. W., AND ZOBACK, M. D., 1984, Tectonic wedging associated with emplacement of the Franciscan assemblage, California Coast Ranges, *in* Blake, M. C., Jr., ed., Franciscan Geology of Northern California: Society of Economic Paleontologists and Mineralogists, Pacific Section, v. 43, p. 163–173.

WERNICKE, B., SPENCER, J. E., BURCHFIEL, B. C., AND GUTH, P. L., 1982, Magnitude of crustal extension in the southern Great Basin: Geology, v. 10, p. 499–502.

WHITE, R. S., 1985, Seismic reflection profiling comes of age: Geological Magazine, v. 122, p. 199–201.

WILCOX, R. E., HARDING, T. P., AND SEELY, D. R., 1973, Basic wrench tectonics: American Association of Petroleum Geologists Bulletin, v. 57, p. 74–96.

WILSON, G., 1970, Wrench movements in the Aristarchus region of the Moon: Proceedings of Geological Association, v. 81, p. 595–608.

WILSON, J. E., KASHAI, E. L., AND CROKER, P., 1983, Hydrocarbon potential of Dead Sea Rift Valley: Oil and Gas Journal, v. 81, No. 25, June 20, p. 147–154.

WILSON, J. T., 1965, A new class of faults and their bearing on continental drift: Nature, v. 207, p. 343–347.

WINCHESTER, J. A., 1982, Comment on 'Paleomagnetic evidence for a large (~2,000 km) sinistral offset along the Great Glen fault during Carboniferous time:' Geology, v. 10, p. 487–488.

WISE, D. U., DUNN, D. E., ENGELDER, J. T., GEISER, P. A., HATCHER, R. D., KISH, S. A., ODOM, A. L., AND SCHAMEL, S., 1984, Fault-related rocks: Suggestions for terminology: Geology, v. 12, p. 391–394.

WRIGHT, L. A., AND TROXEL, B. W., 1970, Summary of regional evidence for right-lateral displacement in the western Great Basin: Discussion: Geological Society of America Bulletin, v. 81, p. 2167–2173.

YEATS, R. S., 1978, Neogene acceleration of subsidence rates in southern California: Geology, v. 6, p. 456–460.

———, 1981, Quaternary flake tectonics of the California Transverse Ranges: Geology, v. 9, p. 16–20.

———, 1983, Large-scale Quaternary detachments in Ventura Basin, southern California: Journal of Geophysical Research, v. 88, p. 569–583.

ZALAN, P. V., NELSON, E. P., WARME, J. E., AND DAVIS, T. L., 1985, The Piaui Basin: Rifting and wrenching in an equatorial Atlantic transform basin, *in* Biddle, K. T., and Christie-Blick, N., eds., Strike-Slip Deformation, Basin Formation, and Sedimentation: Society of Economic Paleontologists and Mineralogists Special Publication No. 37, p. 000–000.

ZAK, I., AND FREUND, R., 1981, Asymmetry and basin migration in the Dead Sea rift, *in* Freund, R., and Garfunkel, Z., eds., The Dead Sea Rift: Tectonophysics, v. 80, p.27–38.

ZIONY, J. I., AND YERKES, R. F., 1984, Fault slip-rate estimation for the Los Angeles region: Challenges and opportunities (abs.): Seismological Society of America, Eastern Section, Earthquake Notes, v. 55, No. 1, p. 8.

Reprinted by permission of the Society of Economic Paleontologists and Mineralogists from K. T. Biddle and N. Christie-Blick, eds., *Strike-Slip Deformation, Basin Formation, and Sedimentation*, SEPM Special Publication 37, 1985, p. 375-386.

GLOSSARY—STRIKE-SLIP DEFORMATION, BASIN FORMATION, AND SEDIMENTATION[1]

KEVIN T. BIDDLE

Exxon Production Research Company, P. O. Box 2189, Houston, Texas 77252-2189;

AND

NICHOLAS CHRISTIE-BLICK

Department of Geological Sciences and Lamont-Doherty Geological Observatory of Columbia University, Palisades, New York 10964

INTRODUCTION

Many of the geological terms having to do with strike-slip deformation, basin formation, and sedimentation are used in a variety of ways by different authors (e.g., pull-apart basin), or they are synonymous with other words (e.g., left-lateral, sinistral). Rather than enforcing a rigorously uniform terminology in this book, we decided to set down our preferred definitions in a glossary, and where appropriate to indicate alternative usage. In selecting terms for definition, we have tried to steer a course between being overly encyclopedic and providing a list useful to those having little familiarity with the geology of strike-slip basins, especially those described in this volume. Some words (e.g., cycle) have additional meanings in the geological sciences not included here, and this glossary should therefore be used in the context of strike-slip basins. The references cited are those from which we obtained definitions, or which illustrate the concept embodied by a particular term. We have not attempted to provide original references for every term, especially for those long used in the geological literature.

THE GLOSSARY

Anastomosing—Pertaining to a network of branching and rejoining surfaces or surface traces. Commonly used to describe braided fault systems.

Antithetic fault—Originally defined by H. Cloos (1928, 1936) to describe faults that dip in a direction opposite to the dip of the rocks displaced, and that rotate fault-bounded blocks so that the net slip on each fault is greater than it would be without rotation (Dennis, 1967, p. 3). Many authors now use the term to describe faults that (1) are subsidiary to a major fault and have less displacement than that fault, (2) formed in the same stress regime as the major fault with which they are associated, (3) are oriented at a high angle to the major fault (in map view for strike-slip faults, in cross-sectional view for normal faults), and (4) for strike-slip faults, have a sense of displacement opposite that of the major fault, or for normal faults, dip in the opposite direction. Antithetic strike-up faults compose the R' set of Reidel shears formed in simple shear (Fig. 1).

Aulacogen—A term introduced by Shatski (1946a, b) to describe narrow, elongate sedimentary basins that extend into cratons from either a geosyncline or a mountain belt that formed from a geosyncline (for a discussion of genesis in terms of plate tectonics, see Hoffman et al., 1974).

Basin—(1) A site of pronounced sediment accumulation; (2) a relatively thick accumulation of sedimentary rock (for a discussion of the history and usage of the word, see Dennis, 1967, p. 9; Bates and Jackson, 1980, p. 55).

Bubnoff curve—A plot of subsidence versus time (Fischer, 1974).

Bubnoff unit—A standard measure of geologic rates, such as subsidence rates, defined as 1 m/m.y. (Fischer, 1969; Bates and Jackson, 1980, p. 84).

Burial history curve—A plot, for a given location, of the cumulative thickness of sediments overlying a surface versus time (Philipp, 1961; van Hinte, 1978).

Closing bend—See restraining bend.

Compaction—(1) The reduction in bulk volume or thickness of, or the pore space within, a body of sediment in response to the increasing weight of a superimposed load; (2) the physical process by which fine-grained sediment is converted to consolidated rock (modified from Bates and Jackson, 1980, p. 127).

Compression—(1) A system of stresses that tends to shorten or decrease the volume of a substance (preferred definition, modified from Bates and Jackson, 1980, p. 130). Uniaxial compression involves one nonzero principal stress, which is compressive; in triaxial compression, all three principal stresses are nonzero (Means, 1976, p. 80). It is also possible for a compressive principal stress to occur with one or more tensile principal stresses. (2) A state of strain in which material lines become shorter under compressive stress (J. T. Engelder, personal commun., 1985; Aydin and Nur, 1985 this volume; see contraction).

Compressional bend—See restraining bend.

Compressional overstep—See restraining overstep.

Conjugate Riedel shear—Synonymous with R' Riedel or antithetic shear (Fig. 1). See Riedel shear, synthetic and antithetic faults.

Contraction—A strain involving (1) a reduction in volume (e.g., thermal contraction), or (2) a reduction of length (e.g., contraction fault of Norris, 1958; McClay, 1981). Contraction has been gaining popularity as the general strain term associated with compressive stress, much like the relationship between the stress term tension and the strain

---

[1]Lamont-Doherty Geological Observatory Contribution No. 3913.

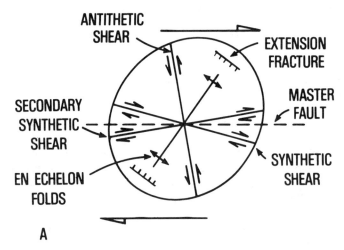

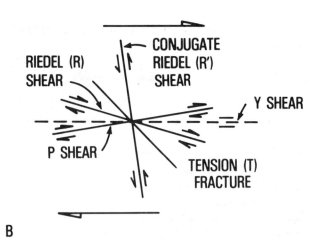

A   B

Fig. 1.—The angular relations between structures that tend to form in right-lateral simple shear under ideal conditions, compiled from clay-cake models and from geological examples. Arrangements for structures along left-slip faults may be determined by viewing the figure in reverse image. A) Terminology largely from Wilcox et al. (1973), superimposed on a strain ellipse for the overall deformation. B) Riedel shear terminology, modified from Tchalenko and Ambraseys (1970) and Bartlett et al. (1981). See glossary for definitions of terms.

term extension. The word shortening, as used by Hobbs et al. (1976, p. 27), may be a better choice for general use because it does not imply a volume change.

Convergent bend—A bend in a strike-slip fault that results in overall crustal shortening in the vicinity of the bend (synonymous with restraining bend of Crowell, 1974a).

Convergent overstep—See restraining overstep.

Convergent (transpressional) strike-slip or wrench fault—A strike-slip or wrench fault along which strike-slip deformation is accompanied by a component of shortening transverse to the fault (Wilcox et al., 1973).

Cycle—(1) An interval of time during which one series of recurrent events is completed (preferred definition); (2) a sequence of sediment or rock units repeated in a succession (for additional discussion, see Bates and Jackson, 1980, p. 156).

Depositional sequence—A stratigraphic unit composed of a relatively conformable succession of genetically related strata and bounded at its top and base by unconformities or their correlative conformities (Mitchum, 1977, p. 206).

Dextral—Pertaining to the right (e.g., dextral slip is right slip).

Dip—The acute angle between an inclined surface and the horizontal, measured in a vertical plane perpendicular to strike.

Dip separation—Separation measured parallel to the dip of a fault (modified from Crowell, 1959, p. 2662; Bates and Jackson, 1980, p. 177). See separation.

Dip slip—The component of slip measured parallel to the dip of a fault (Crowell, 1959, p. 2655; Bates and Jackson, 1980, p. 177). See slip.

Dip-slip fault—A fault along which most of the displacement is accomplished by dip slip (modified from Bates and Jackson, 1980, p. 177).

Divergent bend—A bend in a strike-slip fault that results in overall crustal extension in the vicinity of the bend (synonymous with releasing bend of Crowell, 1974a).

Divergent overstep—See releasing overstep.

Divergent (transtensional) strike-slip or wrench fault—A strike-slip or wrench fault along which strike-slip deformation is accompanied by a component of extension transverse to the fault (Wilcox et al., 1973; Harding et al., 1985 this volume).

Downlap—A base-discordant relation in which initially inclined strata terminate downdip against an initially horizontal or inclined surface (Mitchum, 1977, p. 206).

Drag fold—(1) A fold produced by movement along a fault (see normal drag and reverse drag). In this context, the term is somewhat misleading because folding commonly initiates before faulting (Hobbs et al., 1976, p. 306). (2) A minor fold formed in a less competent bed between more competent beds by movement of the competent beds in opposite directions relative to one another (Bates and Jackson, 1980, p. 186).

Drape fold—A fold in a sedimentary layer that conforms passively to the configuration of underlying structures (Friedman et al., 1976, p. 1049). A fold formed by differential compaction is an example of a drape fold.

Dynamic analysis—The study of kinematics and kinetics that relates strains to the evolution of stress fields.

Echelon—Step (e.g., echelon faults of Clayton, 1966, and of Segall and Pollard, 1980, meaning overstepping faults).

En echelon—(1) A stepped arrangement of relatively short, consistently overlapping or underlapping structural elements such as faults or folds that are approximately parallel to each other but oblique to the linear or relatively narrow zone in which they occur (preferred definition, modified from Campbell, 1958; Harding and Lowell, 1979; see Fig. 2). En echelon arrangements can occur in both map view and cross section (Shelton, 1984; Aydin and Nur, 1985 this volume). (2) Any stepped arrangement of two or more overlapping or underlapping structural elements such as faults or folds that are approximately parallel to each other and to the zone in which they occur, without reference to whether the sense of overstep is consistent or inconsistent (e.g., for strike-slip deformation, D. A. Rodgers, 1980; Aydin and Nur, 1985 this volume; for thrust and fold belts, J. Rodgers, 1963; Armstrong, 1968; Dahlstrom, 1970). See oblique, relay pattern.

Extension—A strain involving an increase in length.

Extension fault—A fault that results in lengthening of an arbitrary datum, commonly but not necessarily bedding

(synonymous with one usage of normal fault; Suppe 1985, p. 269). The term may be applied to faults of any dip (Christie-Blick, 1983).

Extension fracture—A mode I crack, or one that shows no motion in the plane of the crack (Lawn and Wilshaw, 1975, p. 52; J. T. Engelder, personal commun., 1985). Extension fractures form when effective stresses are tensile (i.e., when pore-fluid pressure exceeds lithostatic pressure). Partly synonymous with T fracture of Tchalenko and Ambraseys (1970). See tension fracture. In strike-slip systems, extension fractures and tension fractures form in response to simple shear at about 45° to the master fault (Fig. 1).

Extensional bend—See releasing bend.

Extensional overstep—See releasing overstep.

External rotation—A change in the orientation of structural features during deformation with reference to coor-

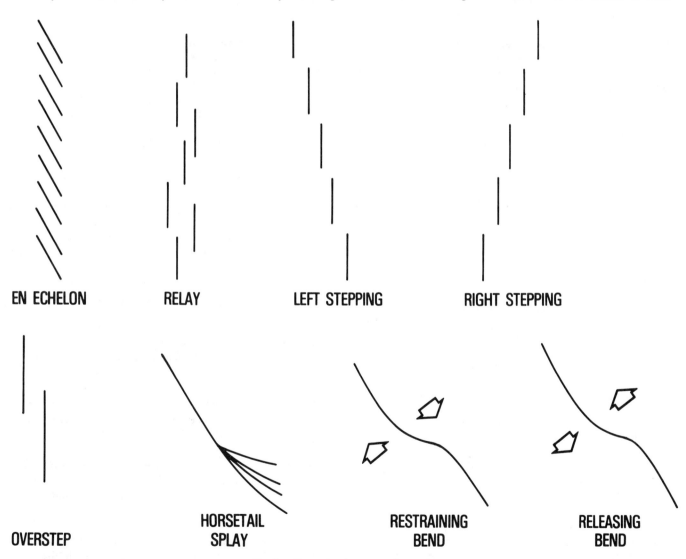

EN ECHELON   RELAY   LEFT STEPPING   RIGHT STEPPING

OVERSTEP   HORSETAIL SPLAY   RESTRAINING BEND   RELEASING BEND

FIG. 2.—Some simple structural patterns.

dinate axes external to the deformed body (Bates and Jackson, 1980, p. 218).

Facies—(1) Laterally or sequentially associated bodies of sediment or sedimentary rock distinguished on the basis of objective lithologic and paleontologic characteristics that reflect the processes and environments of deposition and/or diagenesis; (2) distinctive, adjacent, coeval rock units (White, 1980); (3) the lithologic and paleontologic characteristics of a body of sediment or sedimentary rock that reflect the processes and environments of deposition (original sense of Gressly, 1838). See Walker (1984) for further discussion and references to numerous reviews.

Fault-angle depression—A subsiding depression parallel to the trace of an oblique-slip fault (Ballance, 1980, p. 232).

Fault-flank depression—A depression between subsidiary folds of a strike-slip fault system (Crowell, 1976).

Fault-slice ridge—A linear topographic high associated with a fault-bounded uplifted block within a fault zone (Crowell, 1974b; synonymous with pressure ridge).

Fault splay—A subsidiary fault that merges with and is genetically related to a more prominent fault. Fault splays are common near the termination of a major strike-slip fault (Fig. 2), unless this is at an intersection with another strike-slip fault.

Fault strand—An individual fault of a set of closely spaced, parallel or subparallel faults of a fault system.

Fault-wedge basin—A basin formed by extension at a releasing junction between two predominantly strike-slip faults having the same sense of offset (Crowell, 1974a; synonymous with wedge graben of Freund, 1982). See releasing fault junction.

Flexure—(1) A fold produced by a force couple applied parallel to the direction of deflection (Suppe, 1985, p. 360); (2) a mechanism of regional isostatic compensation in which loads are supported by broad deflection of the lithosphere as a result of lithospheric rigidity (Watts and Ryan, 1976; Watts, 1983; M. S. Steckler, personal commun., 1985).

Flower structure—An array of upward-diverging fault splays within a strike-slip zone (attributed to R. F. Gregory by Harding and Lowell, 1979; see positive flower structure and negative flower structure; synonymous with palm-tree structure of Sylvester, 1984, but preferred for reasons of precedence).

Forced fold—A fold whose overall shape and trend are dominated by the shape and trend of an underlying forcing member (Stearns, 1978).

Foreland—A more-or-less stable area underlain by continental crust, and adjacent to an orogenic belt, toward which rocks of the belt were tectonically transported (Bates and Jackson, 1980, p.241).

Fracture zone—An extension of a transform fault beyond its intersection with an oceanic ridge. Fracture zones are characterized by dip slip, especially where juxtaposed oceanic crust is of markedly different age, and usually they do not experience strike slip (see Freund, 1974; Fox and Gallo, 1984).

Graben—An elongate, relatively depressed block bounded by normal faults (Bates and Jackson, 1980, p. 268).

Heat flow—The product of a thermal gradient and the thermal conductivity of the material across which the thermal gradient is measured.

Horsetail splay—One of a set of curved fault splays near the end of a strike-slip fault that merge with that fault. The set forms an array that crudely resembles a horse's tail (Figs. 2, 3).

Internal rotation—A change in the orientation of structural features during deformation with reference to coordinate axes internal to the deformed body (Bates and Jackson, 1980, p. 322).

Kinematic analysis—The analysis of a movement pattern based on displacement without reference to force or stress (modified from Spencer, 1977, p. 39).

Leaky transform—A transform plate boundary characterized by significant volcanism and/or intrusion along its length. See Garfunkel (1981) for a continental example. See transform fault, transform margin.

Left-hand overstep or stepover—An overstep (stepover) in which one fault or fold segment occurs to the left of the adjacent segment from which it is being viewed (Campbell, 1958; Wilcox et al., 1973). See left-stepping, overstep, stepover. For oversteps in cross section, it is necessary to specify the direction from which the overstep is being viewed.

Left-lateral—Refers to an offset along a fault in map view, in which the far side is apparently displaced to the left with respect to the near side.

Left separation—Strike separation in which the far side of a fault is apparently displaced to the left with respect to the near side. See separation, strike separation.

Left slip—The component of slip measured parallel to the strike of a fault in which the far side of the fault is displaced to the left with respect to the near side. See slip.

Left-stepping—Refers to an overstep in which one fault or fold segment occurs to the left of the adjacent segment from which it is being viewed (Fig. 2). See left-hand overstep.

Lineament—A linear topographic feature of regional extent that is thought to reflect crustal structure (Hobbs et al., 1976, p. 267).

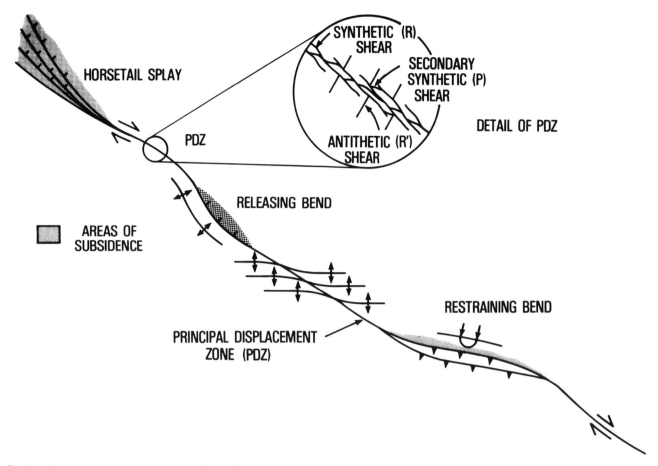

FIG. 3.—The spatial arrangement, in map view, of structures associated with an idealized right-slip fault. See glossary for definitions of terms.

Listric fault—A curved, generally concave-upward, downward-flattening fault (see Bally et al., 1981). Listric faults may be characterized by normal or reverse separation.

Lithosphere—The outer shell of the Earth consisting of crust and upper mantle, and characterized by strength relative to the underlying asthenosphere for deformation at geologic rates (Bates and Jackson, 1980, p. 364; Watts, 1983). The base of the lithosphere can be defined by a number of different properties that reflect rheology, such as temperature, seismic velocity and degree of seismic attenuation.

Marginal basin—A semi-isolated basin lying behind the volcanic chain of an island arc system (Karig, 1971, p. 2542).

Master fault—A major fault in a fault system (Wilcox et al., 1973; Rodgers, 1980; nearly synonymous with principal displacement zone of Tchalenko and Ambraseys, 1970).

Megashear—A strike-slip fault with horizontal displacement that significantly exceeds the thickness of the crust (Carey, 1958, 1976, p. 85).

Multiple overstep—A series of discontinuities between approximately parallel overlapping or underlapping strike-slip faults (new term). See overstep, overlap, and underlapping faults.

Negative flower structure—A flower structure in which the upward-diverging fault splays are predominantly of normal separation and commonly associated with a prominent synformal structure or structures in strata above, or cut by, the faults (Harding, 1983, 1985; Harding et al., 1985 this volume). See flower structure.

Net slip—The displacement vector connecting formerly adjacent points on opposite sides of a fault (modified from Hobbs et al., 1976, p. 300).

Normal drag—Folding near a fault resulting from resistance to slip along the fault. Folded strata are convex toward the slip direction on both sides of the fault. See drag fold.

Normal fault—(1) A fault with normal separation across which the hanging wall is apparently lowered with respect to the footwall (preferred definition; Hill, 1959); (2) a fault generated by normal slip (see Gill, 1941, 1971). The term may be applied to faults of any dip (Christie-Blick, 1983).

Normal separation—Separation measured parallel to the dip of a fault across which the hanging wall is apparently lowered with respect to the footwall. See separation.

Normal slip—The component of slip measured parallel to the dip of a fault across which the hanging wall is lowered with respect to the footwall. See slip.

Oblique—Not parallel; intersecting at an acute angle. En echelon elements are oblique to the zone in which they occur, but not all oblique elements are en echelon. See en echelon.

Oblique slip—The relative displacement of formerly adjacent points on opposite sides of a fault, involving components of both dip slip and strike slip. See slip, dip slip, strike slip.

Oblique-slip fault—A fault along which displacement is accomplished by a combination of strike slip and dip slip. See slip-oblique fault.

Oblique subduction—The relative displacement of one lithospheric plate beneath another plate such that in map view the displacement vector is oblique to the plate boundary.

Onlap—A base-discordant stratigraphic relation in which initially horizontal or inclined strata terminate updip against an initially inclined surface (modified from Mitchum, 1977, p. 208).

Opening bend—See releasing bend.

Orogeny—Profound deformation of rock bodies along restricted zones and within a limited time interval (Dennis, 1967, p. 112).

Overlap—(1) The distance between the ends of overlapping parallel faults, measured parallel to the faults (Rodgers, 1980; Mann et al., 1983; Aydin and Nur, 1985 this volume; generally applied to strike-slip faults in map view; nearly synonymous with separation of Segall and Pollard, 1980); (2) a relation between two superimposed stratigraphic units onlapping a given surface, in which the upper unit extends beyond the line of pinch-out in the lower unit.

Overstep—(1) a discontinuity between two approximately parallel overlapping or underlapping faults (Fig. 2; synonymous with stepover of Aydin and Nur, 1982a, b; 1985 this volume). Oversteps can occur in both map view and cross section, and on both strike-slip and dip-slip faults (Aydin and Nur, 1985 this volume), but the term is commonly applied to strike-slip faults in map view. See solitary overstep, multiple overstep, releasing overstep, and restraining overstep. (2) A stratigraphic relation in which one or more stratigraphic units unconformably overlie the eroded edge of older, generally tilted or folded sedimentary rocks.

Palm-tree structure—An array of upward-diverging fault splays within a strike-slip zone (nomenclature of A. G. Sylvester and R. R. Smith; Sylvester, 1984; synonymous with flower structure of Harding and Lowell, 1979, a term that has precedence).

P shear—One of a set of faults that develop in simple shear generally after the formation of Riedel shears. P shears have the same sense of displacement as Riedel R shears, and form at an angle to the principal displacement zone that is of about the same magnitude but of opposite sign (Skempton, 1966; Tchalenko and Ambraseys, 1970; Figs. 1, 3). Synonymous with secondary synthetic strike-slip fault.

Piercing points—The points of intersection of formerly contiguous linear features (real or constructed) on opposite sides of a fault, by means of which net slip on the fault may be determined (Crowell, 1959, p. 2656). Examples of such linear features are the pinchout line of a sedimentary wedge, offset streams, and facies boundaries used in conjunction with structure contours.

Plunge—The acute angle between an inclined line and the horizontal, measured in a vertical plane containing the line.

Pop-up—A relatively uplifted block between thrusts verging in opposite directions. Originally applied to structures in thrust and fold belts (Butler, 1982).

Positive flower structure—A flower structure in which the upward-diverging fault splays are predominantly of reverse separation and commonly associated with a prominent antiformal structure or structures in strata above, or cut by, the faults (Harding et al., 1983; Harding, 1985). See flower structure.

Pressure ridge—A linear topographic high associated with a fault-bounded uplifted block within a fault zone (Tchalenko and Ambraseys, 1970; synonymous with fault-slice ridge).

Principal displacement zone—A relatively narrow zone that accounts for most of the slip on a given fault (Tchalenko and Ambraseys, 1970, p. 43; Fig. 3). See master fault.

Progradation—The outward building of sediment in the direction of transport, generally, but not exclusively, from a shoreline toward a body of water.

Pull-apart basin—(1) A basin formed by crustal extension at a releasing bend or releasing overstep along a strike-slip fault zone (preferred definition; Burchfiel and Stewart, 1966; Crowell, 1974b; Mann et al., 1983; nearly synonymous with rhomb graben); (2) any basin resulting from crustal extension (Klemme, 1980; Bois et al., 1982).

Pure shear—A homogeneous deformation involving either a plane strain or a general strain in which lines of particles that are parallel to the principal axes of the strain ellipsoid have the same orientation before and after deformation (Hobbs et al., 1976, p. 28). Also referred to as an irrotational deformation or strain. See shear, simple shear.

Push-up—A block elevated by crustal shortening at a restraining bend or restraining overstep along a strike-slip fault zone (Aydin and Nur, 1982a; Mann et al., 1983). See rhomb horst.

Ramp valley—A topographic basin bounded by reverse faults (Willis, 1928, p. 493; Burke et al., 1982). Not all ramp valleys are related to strike-slip deformation.

Regression—A seaward retreat of a shoreline, generally expressed as a seaward migration of shallow-marine facies (modified from Mitchum, 1977, p. 209).

Relay pattern—A shingled arrangement of inconsistently overlapping or underlapping structural elements such as faults or folds that are approximately parallel to each other and to the elongate zone in which they occur (modified from Harding and Lowell, 1979; Fig. 2). Many authors do not distinguish between en echelon and relay arrangements. See Christie-Blick and Biddle (1985 this volume).

Releasing bend—A bend in a strike-slip fault associated with overall crustal extension in the vicinity of the bend (Figs. 2, 3; Crowell, 1974a; synonymous with divergent bend).

Releasing fault junction—A junction between two strike-slip faults associated with overall crustal extension and basin formation between the faults (Christie-Blick and Biddle, 1985 this volume). See fault-wedge basin, wedge graben.

Releasing overstep—A right overstep between right-slip faults or a left overstep between left-slip faults associated with overall crustal extension and basin formation between the faults (Christie-Blick and Biddle, 1985 this volume).

Restraining bend—A bend in a strike-slip fault associated with overall crustal shortening and uplift in the vicinity of the bend (Figs. 2, 3; Crowell, 1974a; synonymous with convergent bend).

Restraining fault junction—A junction between two strike-slip faults associated with overall crustal shortening and uplift between the faults (Christie-Blick and Biddle, 1985 this volume).

Restraining overstep—A right overstep between left-slip faults or a left overstep between right-slip faults associated with overall crustal shortening and uplift between the faults (Christie-Blick and Biddle, 1985 this volume).

Retrogradation—The landward backstepping of sedimentary units usually but not exclusively from a shoreline, as expressed by a landward migration of facies belts.

Reverse drag—Deformation along a fault that creates a fold or set of folds whose curvature is opposite that which would be formed by normal drag folding. Reverse drag is a common feature of listric normal faults where hanging-wall folds are concave toward the slip direction.

Reverse fault—(1) A fault with reverse separation across which the hanging wall is apparently elevated with respect to the footwall (preferred definition; Hill, 1959); (2) a fault generated by reverse slip (see Gill, 1941, 1971). The term may be applied to faults of any dip (Christie-Blick, 1983).

Reverse separation—Separation measured parallel to the dip of a fault across which the hanging wall is apparently elevated with respect to the footwall. See separation.

Reverse slip—The component of slip measured parallel to the dip of a fault across which the hanging wall is elevated with respect to the footwall. See slip.

Rhombochasm—A parallel-sided gap in sialic (continental) crust occupied by simatic (oceanic) crust (modified from Carey, 1976, p. 81). One of S. W. Carey's type examples is the Gulf of California. For original definition and discussion, see Carey (1958).

Rhomb graben—A basin formed by crustal extension at a releasing bend or releasing overstep in a strike-slip fault zone (Freund, 1971; Aydin and Nur, 1982b; synonymous with pull-apart basin of Burchfiel and Stewart, 1966, and Crowell, 1974a, b, particularly sharp pull-aparts, or ones that are angular in map view).

Rhomb horst—A block elevated by crustal shortening at a restraining bend or restraining overstep in a strike-slip fault zone (Aydin and Nur, 1982b; nearly synonymous with push-up of Aydin and Nur, 1982a, and Mann et al., 1983, particularly those that are angular in map view).

Riedel shear—In simple shear, two sets of shear fractures tend to form, oriented at $\phi/2$ and $90°-\phi/2$ to the principal displacement zone (where ϕ is the internal coefficient of friction commonly taken to be about 30°). Shear fractures oriented at $\phi/2$ are called R shears whereas those formed at $90°-\phi/2$ are termed R′ shears (modified from Tchalenko and Ambraseys, 1970). See synthetic and antithetic faults (Figs. 1, 3).

Right-hand overstep or stepover—An overstep (stepover) in which one fault or fold segment occurs to the right of the adjacent segment from which it is being viewed (Campbell, 1958; Wilcox et al., 1973). See right-stepping, overstep, stepover. For oversteps in cross section, it is necessary to specify the direction from which the overstep is being viewed.

Right-lateral—Refers to an offset along a fault in map view, in which the far side is apparently displaced to the right with respect to the near side.

Right separation—Strike separation in which the far side of a fault is apparently displaced to the right with respect to the near side. See separation, strike separation.

Right slip—The component of slip measured parallel to the strike of a fault in which the far side of the fault is displaced to the right with respect to the near side. See slip.

Right-stepping—Refers to an overstep in which one fault or fold segment occurs to the right of the adjacent segment from which it is being viewed (Fig. 2). See right-hand overstep.

Rotation—Motion in which the path of a point in the moving object defines an arc around a specified axis.

Rotational strain—Strain in which the orientation of the strain axes is different before and after deformation (Bates and Jackson, 1980, p. 546). See simple shear.

Secondary synthetic fault—One of a set of faults that develop in simple shear, generally after the formation of synthetic faults (Riedel shears). Secondary synthetic faults have the same sense of displacement as the synthetic faults and form an angle to the principal displacement zone that is of about the same magnitude as the synthetic faults but of opposite sign (Figs. 1, 3). Synonymous with P shear. See synthetic and antithetic fault, Riedel shear.

Separation—(1) The apparent displacement of formerly contiguous surfaces on opposite sides of a fault, measured in any given direction (modified from Reid et al., 1913, p. 169; Crowell, 1959, p. 2661); (2) the perpendicular distance between overlapping parallel strike-slip faults (Rodgers, 1980; Mann et al., 1983; Aydin and Nur, 1985 this volume); (3) the distance between overstepping parallel strike-slip faults (either overlapping or underlapping), measured parallel to the faults (Segall and Pollard, 1980; nearly synonymous with overlap of Rodgers, 1980).

Shear—A strain resulting from stresses that cause, or tend to cause, parts of a body to move relatively to each other in a direction parallel to their plane of contact (modified from Bates and Jackson, 1980, p. 575). See pure shear, simple shear.

Simple shear—A constant volume, homogeneous deformation involving plane strain, in which a single family of parallel material planes is undistorted in the deformed state and parallel to the same family of planes in the undeformed state (Hobbs et al., 1976, p. 29). Also referred to as a rotational deformation or strain. See shear, pure shear.

Simple strike-slip or wrench fault—A strike-slip or wrench fault along which adjacent blocks move laterally with no component of shortening or extension transverse to the fault (Christie-Blick and Biddle, 1985 this volume; synonymous with simple parallel strike-slip or wrench fault of Wilcox et al., 1973; and with slip-parallel fault of Mann et al., 1983).

Sinistral—Pertaining to the left (e.g., sinistral slip is left slip).

Slickenside—A polished or smoothly striated surface on either side of a fault that results from motion along the fault (Bates and Jackson, 1980, p. 587).

Slip—The relative displacement of formerly adjacent points on opposite sides of a fault measured along the fault surface (modified from Reid et al., 1913, p. 168; Crowell, 1959, p. 2655).

Slip-oblique fault—A strike-slip fault along which strike-slip deformation is accompanied by a component of either shortening or extension transverse to the fault (Mann et al., 1983; includes convergent and divergent strike-slip or wrench faults of Wilcox et al., 1973; and transpressional and transtensional faults of Harland, 1971; nearly synonymous with oblique-slip fault).

Slip-parallel fault—A fault that strikes parallel to the azimuth of the slip direction (Mann et al., 1983; synonymous with simple parallel strike-slip or wrench fault of Wilcox et al., 1973; and with simple strike-slip fault of Christie-Blick and Biddle, 1985 this volume).

Solitary overstep—An isolated discontinuity, between two approximately parallel overlapping or underlapping faults (Guiraud and Seguret, 1985 this volume; Fig. 2).

Sphenochasm—A triangular gap of oceanic crust separating two continental blocks with fault margins converging to a point, and interpreted as having originated by the rotation of one block with respect to the other (modified from Carey, 1976, p. 81). The Bay of Biscay is one of S. W. Carey's examples of a sphenochasm. See Carey (1958) for original definition.

Splay—Generally synonymous with fault splay, a subsidiary fault that merges with, and is genetically related to, a more prominent fault.

Stepover—A discontinuity between two approximately parallel overlapping or underlapping faults (Aydin and Nur, 1982a, b; 1985 this volume; Aydin and Page, 1984; synonymous with overstep). Stepovers can occur in both map view and cross section, and on both strike-slip and dip-slip faults (Aydin and Nur, 1985 this volume), but the term is commonly applied to discontinuities on strike-slip faults in map view.

Stratigraphic separation—The stratigraphic thickness either cut out or repeated by a fault (modified from Crowell, 1959, p. 2663).

Strand—See fault strand.

Strike—The azimuth of the line of intersection of an inclined surface with a horizontal plane.

Strike separation—Separation measured parallel to the strike of a fault (modified from Crowell, 1959, p. 2662; Bates and Jackson, 1980, p. 618). See separation.

Strike slip—The component of slip measured parallel to the strike of a fault (Crowell, 1959, p. 2655; Bates and Jackson, 1980, p. 618). See slip.

Strike-slip fault—A fault along which most of the displacement is accomplished by strike slip (modified from Bates and Jackson, 1980, p. 618).

Strike-slip basin—Any basin in which sedimentation is accompanied by significant strike slip (modified from Mann et al., 1983).

Subsidence—The depression of an area of the Earth's crust with respect to surrounding areas.

Synthetic fault—Originally defined by H. Cloos (1928, 1936) to describe faults that dip in the same direction as the rocks displaced and that rotate fault-bounded blocks so that the net slip on each fault is less than it would be without rotation (Dennis, 1967, p. 148–149). Many authors now use the term to describe faults that (1) are subsidiary to a major fault and have less displacement than that fault; (2) formed in the same stress regime as the major fault with which they are associated; (3) are oriented at a low angle to the major fault (in map view for strike-slip faults, in cross-sectional view for normal faults); and (4) for strike-slip faults, have the same sense of displacement as the major fault with which they are associated, or for normal faults, dip in the same direction. The R set of Riedel shears and the P shears of Tchalenko and Ambraseys (1970) are synthetic faults (Figs. 1, 3).

Tear fault—A strike-slip or oblique-slip fault within or bounding an allochthon produced by either regional extension or regional shortening. Tear faults accommodate differential displacement within a given allochthon, or between the allochthon and adjacent structural units.

Tectonic depression—(1) Any structurally produced topographic low; (2) a topographic low produced by strike-slip deformation (Clayton, 1966).

Tectonic subsidence—That part of the subsidence at a given point in a sedimentary basin caused by a tectonic driving mechanism. Tectonic subsidence is calculated by removing the component of subsidence produced by non-tectonic processes such as sediment loading, sediment compaction, and water-depth changes (Watts and Ryan, 1976; Steckler and Watts, 1978; Keen, 1979; Bond and Kominz, 1984).

Tension—A system of stresses that tends to lengthen or increase the volume of a substance. Uniaxial tension involves one nonzero principal stress, which is tensile; in general tension, two principal stresses are tensile (Means, 1976, p. 79). It is possible for a tensile principal stress to occur with one or more compressive principal stresses.

Tension (T) fracture—A mode I crack that forms when lithostatic loads become negative (Lawn and Wilshaw, 1975, p. 52; J. T. Engelder, personal commun., 1985). See extension fracture. In strike-slip systems, extension fractures and tension fractures form in response to simple shear at about 45° to the master fault (Fig. 1; Tchalenko and Ambraseys, 1970).

Thermal subsidence—That part of the tectonic subsidence at a given point in a sedimentary basin caused by thermal contraction (Sleep, 1971; Parsons and Sclater, 1977). See tectonic subsidence.

Thrust fault—A map-scale contraction fault that shortens an arbitrary datum, commonly but not necessarily bedding (McClay, 1981). The term may be applied to faults of any dip, although thrust faults tend to dip less than 30° during active slip.

Trace slip—The component of slip measured parallel with the trace of a bed, vein, or other surface on the fault plane (Reid et al., 1913, p. 170; Beckwith, 1941, p. 2182).

Transcurrent fault—(1) A strike-slip fault, typically subvertical at depth and commonly involving igneous and metamorphic basement as well as supracrustal sediments and sedimentary rocks (see Moody and Hill, 1956; Freund, 1974; nearly synonymous with wrench fault); (2) a long, subvertical strike-slip fault that cuts strata approximately perpendicular to strike (original sense of Geikie, 1905, p. 169; modified from Dennis, 1967; p. 57).

Transform fault—A strike-slip fault that acts as a lithospheric plate boundary and terminates at both ends against major tectonic features (such as oceanic ridges, subduction zones, or rarely, other transform faults) that are also plate boundaries (Wilson, 1965; Freund, 1974).

Transform margin—A plate margin formed by a transform fault or system of transform faults and dominated by strike-slip deformation.

Transgression—A landward movement of a shoreline, generally expressed as a landward migration of shallow-marine facies (modified from Mitchum, 1977, p. 211).

Transtension—A system of stresses that operates in zones of oblique extension (modified from Harland, 1971). See divergent strike-slip or wrench fault, transpression.

Transpression—A system of stresses that operates in zones of oblique shortening (modified from Harland, 1971; Sylvester and Smith, 1976). See convergent strike-slip or wrench fault, transtension.

Trend—The azimuth of an inclined or horizontal line.

Unconformity—A buried surface of erosion or non-deposition (modified from J. C. Crowell, personal commun., 1975).

Underlapping faults—Approximately parallel faults that overstep without overlapping (applied to oceanic ridge segments by Pollard and Aydin, 1984).

Wedge graben—A basin formed by extension at a releasing junction between two predominantly strike-slip faults having the same sense of offset (Freund, 1982; synonymous with fault-wedge basin of Crowell, 1974a). See releasing fault junction.

Wrench fault—A strike-slip fault, typically sub-vertical at depth, involving igneous and metamorphic basement rocks as well as supracrustal sediments and sedimentary rocks (modified from Moody and Hill, 1956, p. 1208; Wilcox et al., 1973; nearly synonymous with transcurrent fault).

Y-shear—A fault that forms in response to simple shear, and as deformation continues gradually accommodates most of the movement along the principal displacement zone (Bartlett et al., 1981).

ACKNOWLEDGMENTS

In compiling this glossary we have been inspired by J. C. Crowell's insistence on the value of precise terminology for the effective communication of both observations and ideas. The manuscript was reviewed at various stages by A. Aydin, J. T. Engelder, A. M. Grunow, D. R. Seely, and C. C. Wielchowsky. We also thank I. W. D. Dalziel, K. A. Kastens, C. Nicholson, and M. S. Steckler for helpful suggestions. Remaining errors or omissions are, however, the responsibility of the authors. Logistical support was provided by Exxon Production Research Company, and by an ARCO Foundation Fellowship to Christie-Blick.

REFERENCES

ARMSTRONG, R. L., 1968, Sevier orogenic belt in Nevada and Utah: Geological Society of America Bulletin, v. 79, p. 429–458.
AYDIN, A., AND NUR, A., 1982a, Evolution of pull-apart basins and push-up ranges: Pacific Petroleum Geologists Newsletter, American Association of Petroleum Geologists, Pacific Section, Nov. 1982, p. 2–4.
———, 1982b, Evolution of pull-apart basins and their scale independence: Tectonics, v. 1, p. 91–105.
———, 1985, The types and role of stepovers in strike-slip tectonics, in Biddle, K. T., and Christie-Blick, N., Strike-Slip Deformation, Basin Formation, and Sedimentation: Society of Economic Paleontologists and Mineralogists Special Publication No. 37, p. 35–44.
AYDIN, A., AND PAGE, B. M., 1984, Diverse Pliocene-Quaternary tectonics in a transform environment, San Francisco Bay region, California: Geological Society of America Bulletin, v. 95, p. 1303–1317.
BALLANCE, P. F., 1980, Models of sediment distribution in non-marine and shallow marine environments in oblique-slip fault zones, in Ballance, P. F., and Reading, H. G., eds., Sedimentation in Oblique-Slip Mobile Zones: International Association of Sedimentologists Special Publication No. 4, p. 229–236.
BALLY, A. W., BERNOULLI, D., DAVIS, G. A., AND MONTADERT, L., 1981, Listric normal faults: Oceanologica Acta, Proceedings, 26th International Geological Congress, Geology of Continental Margins Symposium, Paris, p. 87–101.
BARTLETT, W. L., FRIEDMAN, M., AND LOGAN, J. M., 1981, Experimental folding and faulting of rocks under confining pressure. Part IX. Wrench faults in limestone layers: Tectonophysics, v. 79, p. 255–277.
BATES, R. L., AND JACKSON, J. A., 1980, eds., Glossary of Geology: American Geological Institute, 749 p.
BECKWITH, R. H., 1941, Trace-slip faults: American Association of Petroleum Geologists Bulletin, v. 25, p. 2181–2193.
BOIS, C., BOUCHE, P., AND PELET, R., 1982, Global geologic history and distribution of hydrocarbon reserves: American Association of Petroleum Geologists Bulletin, v. 66, p. 1248–1270.
BOND, G. C., AND KOMINZ, M. A., 1984, Construction of tectonic subsidence curves for the early Paleozoic miogeocline, southern Canadian

Rocky Mountains: Implications for subsidence mechanisms, age of breakup, and crustal thinning: Geological Society of America Bulletin, v. 95, p. 155–173.
BURCHFIEL, B. C., AND STEWART, J. H., 1966, "Pull-apart" origin of the central segment of Death Valley, California: Geological Society of America Bulletin, v. 77, p. 439–442.
BURKE, K., MANN, P., AND KIDD, W., 1982, What is a ramp valley?: 11th International Congress on Sedimentology, Hamilton, Ontario, International Association of Sedimentologists, Abstracts of Papers, p. 40.
BUTLER, R. W. H., 1982, The terminology of structures in thrust belts: Journal of Structural Geology, v. 4, p. 239–245.
CAMPBELL, J. D., 1958, En echelon folding: Economic Geology, v. 53, p. 448–472.
CAREY, S. W., 1958, A tectonic approach to continental drift, in Carey, S. W., convenor, Continental Drift: a Symposium: Hobart, University of Tasmania, 177–355.
———, 1976, The Expanding Earth: Amsterdam, Elsevier Scientific Publishing Company, Developments in Geotectonics 10, 488 p.
CHRISTIE-BLICK, N., 1983, Structural geology of the southern Sheeprock Mountains, Utah: Regional significance, in Miller, D. M., Todd, V. R., and Howard, K. A., eds., Tectonic and Stratigraphic Studies in the Eastern Great Basin: Geological Society of American Memoir 157, p. 101–124.
CHRISTIE-BLICK N., AND BIDDLE, K. T., 1985, Deformation and basin formation along strike-slip faults, in Biddle, K. T., and Christie-Blick, N., eds., Strike-Slip Deformation, Basin Formation, and Sedimentation: Society of Economic Paleontologists and Mineralogists Special Publication 37, p. 1–34.
CLAYTON, L., 1966, Tectonic depressions along the Hope Fault, a transcurrent fault in North Canterbury, New Zealand: New Zealand Journal of Geology and Geophysics, v. 9, p. 95–104.
CLOOS, H., 1928, Über antithetische Bewegungen: Geologische Rundschau, v. 19, p. 246–251.
———, 1936, Einführung in die Geologie: Berlin, Borntraeger, 503 p.
CROWELL, J. C., 1959, Problems of fault nomenclature: American Association of Petroleum Geologists Bulletin, v. 43, p. 2653–2674.
———, 1974a, Origin of late Cenozoic basins in southern California, in Dickinson, W. R., ed., Tectonics and Sedimentation: Society of Economic Paleontologists and Mineralogists Special Publication No. 22, p. 190–204.
———, 1974b, Sedimentation along the San Andreas fault, California, in Dott, R. H., Jr., and Shaver, R. H., eds., Modern and Ancient Geosynclinal Sedimentation: Society of Economic Paleontologists and Mineralogists Special Publication No. 19, p. 292–303.
———, 1976, Implications of crustal stretching and shortening of coastal Ventura Basin, California, in Howell, D. G., ed., Aspects of the Geologic History of the California Continental Borderland: American Association of Petroleum Geologists, Pacific Section, Miscellaneous Publication 24, p. 365–382.
DAHLSTROM, C. D. A., 1970, Structural geology in the eastern margin of the Canadian Rocky Mountains: Bulletin of Canadian Petroleum Geology, v. 18, p. 332–406.
DENNIS, J. G., ed., 1967, International Tectonic Dictionary: American Association of Petroleum Geologists Memoir 7, 196 p.
FISCHER, A. G., 1969, Geologic time-distance rates: the Bubnoff unit: Geological Society of America Bulletin, v. 80, p. 549–551.
———, 1974, Origin and growth of basins, in Fischer, A. G., and Judson, S., ed., Petroleum and Global Tectonics: Princeton, New Jersey, Princeton University Press, p. 47–82.
FOX, P. J., AND GALLO, D. G., 1984, A tectonic model for ridge-trans-form-ridge plate boundaries: Implications for the structure of oceanic lithosphere: Tectonophysics, v. 104, p. 205–242.
FREUND, R., 1971, The Hope Fault, a strike-slip fault in New Zealand: New Zealand Geological Survey Bulletin, v. 86, p. 1–49.
———, 1974, Kinematics of transform and transcurrent faults: Tectonophysics, v. 21, p. 93–134.
———, 1982, The role of shear in rifting, in Pálmason, G., ed., Continental and Oceanic Rifts: American Geophysical Union Geodynamics Series, v. 8, p. 33–39.
FRIEDMAN, M., HANDIN, J., LOGAN, J. M., MIN, K. D., AND STEARNS, D. W., 1976, Experimental folding of rocks under confining pressure: Part III. Faulted drape folds in multilithologic layered specimens: Geolog-

ical Society of America Bulletin, v. 87, p. 1049–1066.

GARFUNKEL, Z., 1981, Internal structure of the Dead Sea leaky transform (rift) in relation to plate kinematics: Tectonophysics, v. 80, p. 81–108.

GEIKIE, J., 1905, Structural and Field Geology: Edinburgh, Oliver and Boyd, 435 p.

GILL, J. E., 1941, Fault nomenclature: Royal Society of Canada Transactions, v. 35, p. 71–85.

———, 1971, Continued confusion in the classification of faults: Geological Society of America Bulletin, v. 82, p. 1389–1392.

GRESSLY, A., 1838, Observations géologique sur le Jura Soleurois: Nouveaux Mémoires de la Société Helvétique des Sciences Naturelles, v. 2, p. 1–112.

GUIRAUD, M., AND SEGURET, M., 1985, A releasing solitary overstep model for the late Jurassic—early Cretaceous (Wealdian) Soria strike-slip basin (northern Spain), in Biddle, K. T., and Christie-Blick, N., eds., Strike-Slip Deformation, Basin Formation, and Sedimentation: Society of Economic Paleontologists and Mineralogists Special Publication No. 37, p. 159–177.

HARDING, T. P., 1983, Divergent wrench fault and negative flower structure, Andaman Sea, in Bally, A. W., ed., Seismic Expression of Structural Styles, v. 3: American Association of Petroleum Geologists Studies in Geology Series 15, v. 3, p. 4.2-1 to 4.2-8.

———, 1985, Seismic characteristics and identification of negative flower structures, positive flower structures, and positive structural inversion: American Association Petroleum Geologists Bulletin, v. 69, p. 582–600.

HARDING, T. P., AND LOWELL, J. D., 1979, Structural styles, their plate-tectonic habitats, and hydrocarbon traps in petroleum provinces: American Association of Petroleum Geologists Bulletin, v. 63, p. 1016–1058.

HARDING, T. P., GREGORY, R. F., AND STEPHENS, L. H., 1983, Convergent wrench fault and positive flower structure, Ardmore Basin, Oklahoma, in Bally, A. W., ed., Seismic Expression of Structural Styles, v. 3, American Association of Petroleum Geologists Studies in Geology, Series 15, p. 4.2-13 to 4.2-17.

HARDING, T. P., VIERBUCHEN, R. C., AND CHRISTIE-BLICK, N., 1985, Structural styles, plate-tectonic settings, and hydrocarbon traps of divergent (transtensional) wrench faults, in Biddle, K. T., and Christie-Blick, N., eds., Strike-Slip Deformation, Basin Formation, and Sedimentation: Society of Economic Paleontologists and Mineralogists Special Publication No. 37, p. 51–78.

HARLAND, W. B., 1971, Tectonic transpression in Caledonian Spitsbergen: Geological Magazine, v. 108, p. 27–42.

HILL, M. L., 1959, Dual classification of faults: American Association of Petroleum Geologists Bulletin, v. 43, p. 217–237.

HOBBS, B. E., MEANS, W. D., AND WILLIAMS, P. F., 1976, An Outline of Structural Geology: New York, John Wiley and Sons, 571 p.

HOFFMAN, P., DEWEY, J. F., AND BURKE, K., 1974, Aulacogens and their genetic relation to geosynclines, with a Proterozoic example from Great Slave Lake, Canada, in Dott, R. H., Jr., and Shaver, R. H., eds., Modern and Ancient Geosynclinal Sedimentation: Society of Economic Paleontologists and Mineralogists Special Publication No. 19, p. 38–55.

KARIG, D. E., 1971, Origin and development of marginal basins in the western Pacific: Journal of Geophysical Research, v. 76, p. 2542–2561.

KEEN, C. E., 1979, Thermal history and subsidence of rifted continental margins—Evidence from wells on the Nova Scotian and Labrador Shelves: Canadian Journal of Earth Sciences, v. 16, p. 505–522.

KLEMME, H. D., 1980, Petroleum basins—classifications and characteristics: Journal of Petroleum Geology, v. 3, p. 187–207.

LAWN, B. R., AND WILSHAW, T. R., 1975, Fracture of Brittle Solids: New York, Cambridge University Press, 204 p.

MANN, P., HEMPTON, M. R., BRADLEY, D. C., AND BURKE, K., 1983, Development of pull-apart basins: Journal of Geology, v. 91, p. 529–554.

McCLAY, K. R., 1981, What is a thrust? What is a nappe?, in McClay, K. R., and Price, N. J., eds., Thrust and Nappe Tectonics: Geological Society of London Special Publication No. 9, p. 7–9.

MEANS, W. D., 1976, Stress and Strain: New York, Springer-Verlag, 339 p.

MITCHUM, R. M., JR., 1977, Seismic stratigraphy and global changes of sea level, Part 11: Glossary of terms used in seismic stratigraphy, in Payton, C. E., ed., Seismic Stratigraphy—Applications to Hydrocarbon Exploration: American Association of Petroleum Geologists Memoir 26, p. 205–212.

MOODY, J. D., AND HILL, M. L., 1956, Wrench-fault tectonics: Geological Society of America Bulletin, v. 67, p. 1207–1246.

NORRIS, D. K., 1958, Structural conditions in Canadian coal mines: Geological Survey of Canada Bulletin 44, 54 p.

PARSONS, B., AND SCLATER, J. G., 1977, An analysis of the variation of ocean floor bathymetry and heat flow with age: Journal of Geophysical Research, v. 82, p. 803–827.

PHILIPP, W., 1961, Struktur- und Lagerstattengeschichte des Erdolfeldes Eldingen: Deutsche Geologische Gesellschaft Zeitschrift, v. 112, p. 414–482.

POLLARD, D. D., AND AYDIN, A., 1984, Propagation and linkage of oceanic ridge segments: Journal of Geophysical Research, v. 89, p. 10,017–10,028.

REID, H. F., et al., 1913, Report of the Committee on the Nomenclature of Faults: Geological Society of America Bulletin, v. 24, p. 163–186.

RODGERS, D. A., 1980, Analysis of pull-apart basin development produced by en echelon strike-slip faults, in Ballance, P. F., and Reading, H. G., eds., Sedimentation in Oblique-Slip Mobile Zones: International Association of Sedimentologists Special Publication No. 4, p. 27–41.

RODGERS, J., 1963, Mechanics of Appalachian foreland folding in Pennsylvania and West Virginia: American Association of Petroleum Geologists Bulletin, v. 47, p. 1527–1536.

SEGALL, P., AND POLLARD, D. D., 1980, Mechanics of discontinuous faults: Journal of Geophysical Research, v. 85, p. 4337–4350

SHATSKI, N. S., 1946a, Basic features of the structures and development of the East European platform. Comparative tectonics of ancient platforms: Izvestiya Akademii Nauk SSSR, Ser. Geol. No. 1, p. 5–62.

———, 1946b, The Great Donets Basin and the Wichita System. Comparative tectonics of ancient platforms: Izvestiya Akademii Nauk SSSR, Ser. Geol. No. 6, p. 57–90.

SHELTON, J. W., 1984, Listric normal faults: An illustrated summary: American Association of Petroleum Geologists Bulletin, v. 68, p. 801–815.

SLEEP, N. H., 1971, Thermal effects of the formation of Atlantic continental margins by continental breakup: Royal Astronomical Society Geophysical Journal, v. 24, p. 325–350.

SKEMPTON, A. W., 1966, Some observations on tectonic shear zones: 1st Congress of International Society of Rock Mechanics, Lisbon, Proceedings, v. 1, p. 329–335.

SPENCER, E. W., 1977, Introduction to the Structure of the Earth: New York, McGraw-Hill Book Company, 2nd edition, 640 p.

STEARNS, D. W., 1978, Faulting and forced folding in the Rocky Mountains foreland, in Matthews, V., III, ed., Laramide Folding Associated With Basement Block Faulting in the Western United States: Geological Society of America Memoir 151, p. 1–37.

STECKLER, M. S., AND WATTS, A. B., 1978, Subsidence of the Atlantic-type continental margin off New York: Earth and Planetary Science Letters, v. 41, p. 1–13.

SUPPE, J., 1985, Principles of Structural Geology: Englewood Cliffs, New Jersey, Prentice-Hall, 537 p.

SYLVESTER, A. G., compiler, 1984, Wrench Fault Tectonics: American Association of Petroleum Geologists Reprint Series, No. 28, 374 p.

SYLVESTER, A. G., AND SMITH, R. R., 1976, Tectonic transpression and basement-controlled deformation in the San Andreas fault zone, Salton Trough, California: American Association of Petroleum Geologists Bulletin, v. 60, p. 2081–2102.

TCHALENKO, J. S., AND AMBRASEYS, N. N., 1970, Structural analysis of the Dasht-e Baÿaz (Iran) earthquake fractures: Geological Society of America Bulletin, v. 81, p. 41–60.

VAN HINTE, J. E., 1978, Geohistory analysis—application of micropaleontology in exploration geology: American Association of Petroleum Geologists Bulletin, v. 62, p. 201–222.

WALKER, R. G., 1984, General introduction: Facies, facies sequences and facies models, in Walker, R. G., ed., Facies Models: Geoscience Canada Reprint Series 1, 2nd edition, p. 1–9.

WATTS, A. B., 1983, The strength of the Earth's crust: Marine Technology Society Journal, v. 17, p. 5–17.

WATTS, A. B., AND RYAN, W. B. F., 1976, Flexure of the lithosphere and continental margin basins: Tectonophysics, v. 36, p. 25–44.

WHITE, D. A., 1980, Assessing oil and gas plays in facies-cycle wedges: American Association of Petroleum Geologists Bulletin, v. 64, p. 1158–1178.

WILCOX, R. E., HARDING, T. P., AND SEELY, D. R., 1973, Basic wrench tectonics: American Association of Petroleum Geologists Bulletin, v. 57, p. 74–96.

WILLIS, Bailey, 1928, Dead Sea problem: rift valley or ramp valley?: Geological Society of America Bulletin, v. 39, p. 490–542.

WILSON, J. T., 1965, A new class of faults and their bearing on continental drift: Nature, v. 207, p. 343–347.

Reprinted by permission of the Society of Economic Paleontologists and Mineralogists from W. R. Dickinson, ed., *Tectonics and Sedimentation*, SEPM Special Publication 22, 1974, p. 190-204.

ORIGIN OF LATE CENOZOIC BASINS IN SOUTHERN CALIFORNIA[1]

JOHN C. CROWELL

University of California, Santa Barbara, California

ABSTRACT

Several sedimentary basins in southern California, within and south of the Transverse Ranges, display a history suggestive of a pull-apart or a tipped-wedge origin. Beginning in the Miocene, these basins apparently originated along the soft and splintered margins of the Pacific and Americas plates. Basin walls were formed by both transform faults and by crustal stretching and dip-slip faulting. Basin floors developed on stretched and attenuated marginal rocks, and some floors grew as a complex of volcanic rocks and sediments. As basins enlarged, high-standing blocks are pictured as stretching and separating laterally from terranes that were originally adjacent. Older rocks exposed around basin margins therefore cannot always be extrapolated to depth beneath the basins.

Support for such speculative models comes from accumulating understanding of the modern Salton trough. This narrow graben is now being pulled apart obliquely, with faults of the San Andreas system serving as transforms. With widening, the walls sag and stretch, and margins are inundated by sedimentation that goes on hand in hand with deformation and with volcanism within the basin. The Los Angeles basin apparently started to form as an irregular pull-apart hole in the early midMiocene, and basin-floor volcanism accompanied subsequent voluminous sedimentation. Great thicknesses of Miocene beds and volcanic rocks in the western Santa Monica Mountains probably constitute the displaced northern part of the Los Angeles basin, and were laid down adjacent to high ground from which sediments and large detachment slabs were carried into the growing depression. Basins and intervening banks and ridges in the California Borderland may have originated in a broad right-slip regime where strike-slip faults converge and diverge in plan view to slice the terrane into wedge-shaped segments. Displacement along converging and diverging strike-slip faults bounding such wedges results in shortening and elevation, or in stretching and subsidence, respectively.

INTRODUCTION

As knowledge of the geologic history of late Cenozoic sedimentary basins in southern California has accumulated during the past half century, several genetic models have been proposed to explain their origin. At first the great thicknesses of sediments found in deep basins were credited to vertical tectonics only, in which subcrustal processes brought about subsidence of basins concomitantly with the elevation of adjacent highlands. Erosional debris from the highlands was pictured as washing across the basin margins and directly into the contiguous basins. Only slowly did the concept gather acceptance that major strike slip was significant, and was superimposed upon this pattern of vertical tectonics. For example, Eaton (1926) and Ferguson and Willis (1924) noted that strike slip was primarily responsible for the folds along the Newport-Inglewood zone in the Los Angeles basin, and Vickery (1925) interpreted the pattern of faults and folds east of the San Francisco Bay area in terms of strike slip. In the early fifties, rock sequences offset by many tens of kilometers on the San Andreas fault were recognized by Hill and Dibblee (1953), and strike slip of conglomerates

from their source areas was shown to be about 30 kilometers on the San Gabriel fault by Crowell (1952). During the two decades since then many workers have demonstrated great strike-slip components on several California faults, including those associated with major basins. As the data have come in, however, it has grown increasingly clear that other faults have essentially no component of strike slip, and that vertical tectonic movements involving steep flexures at basin margins, normal-slip faults, thrust-slip faults and detachment faults are also common. The record shows as well that deformation has been nearly continuous in southern California as a whole since early in the mid-Tertiary, and that this deformation has not always followed the same pattern.

California at present is being deformed as part of a broad transform zone, the sliced and segmented boundary between the Pacific plate and the Americas plate (Atwater, 1970). The origin of several modern basins, such as those within the Salton Trough and the Gulf of California, is related to their position at or near this plate boundary. Similar origins can be recognized for some more ancient basins. The in-

[1] Studies of the tectonics of southern California have recently been supported by the University of California, Santa Barbara, and the U.S. National Science Foundation, Grant GA 30901. I am also grateful to many students and colleagues for numerous discussions and comments and especially to Arne Junger for suggesting a diagram similar to that of Figure 9.

190

terpretation of the ocean-floor record including magnetic anomalies west of California reveals a history of major plate interaction across the region back into pre-Tertiary times, but detailed explanations of this interaction before mid-Tertiary are still inconclusive. According to Atwater (1970, fig. 17), this interaction since the mid-Tertiary has included long episodes of strike slip, and right slip has predominated in the vicinity of southern California for about 25 million years. The Americas plate has moved about 1500 km relative to the Pacific plate during this interval.

Only about 300 km of post-Oligocene right slip on the San Andreas fault is now recognized in southern California, however, leaving a difference of 1200 km or so in order to match interpretations of the land record with those from magnetic anomalies on the sea floor. This difference can most easily be accounted for by considering that other faults on land, such as those in the Great Basin and splays of the San Andreas in southern California and the adjacent borderland, took up the difference. In particular, a major right-slip fault may have coincided with the western edge of the continent where it joins the deep Pacific floor at the base of the Patton Escarpment (fig. 1). Despite the fact that right slip of the order needed to match the sea-floor interpretation is still not recognized on faults in southern California, in the borderland, or in the Mojave and Colorado Deserts, the idea that these regions are part of a broad transform zone attracts investigation. In this paper we will therefore accept as a premise the concept that southern California and its borderland have been very mobile laterally during the late Cenozoic as part of a broad and complicated transform zone, but without committing ourselves to the magnitude of total right slip across the soft and broad boundary between plates. We will search for models of basin origin and consider ways to recognize or test them.

BASIN GEOMETRY IN A TRANSFORM REGIME

Several types of basins can be envisaged theoretically along a transform boundary between major tectonic plates, and especially if the boundary is a complex zone of branching faults. Some terranes may be uplifted to make source areas, and others depressed to form basins (Crowell, 1974). If the strike-slip zone is distant from land and cuts the ocean floor only, high-standing blocks along the oceanic transform may not rise into the zone of erosion, so that the associated depressions receive little sediment from them. Near continents,

however, and especially along continental transforms, large volumes of sediment may be washed directly to nearby basins. Southern California during the late Cenozoic seems to fit the latter circumstances so that the following geometrical discussion starts with the assumption that the transform zone cuts continental crust. Moreover, in using the term 'transform,' emphasis is placed upon the plate-tectonic concept of major crustal plates moving laterally along a strike-slip zone and upon the relation of the strike-slip zone to spreading centers and subduction zones in order to account for lateral displacements of hundreds of kilometers (Wilson, 1965; Vine and Wilson, 1965). In terms of the geometry of rock units, however, such transform faults are major strike-slip fault zones with nearly vertical fault surfaces and long extent. In this sense, they are synonymous with "wrench faults" or "transcurrent faults." In the examples figured below, right slip rather than left slip is illustrated inasmuch as the San Andreas is a right-slip system.

STRIKE SLIP ALONG A STRAIGHT FAULT

If continental terrane with subdued or near-flat topography is cut by a long and vertical strike-slip fault, no differential elevation or subsidence will result from the deformation (fig. 2). The blocks merely slide by each other. With such a simple system there is little likelihood of fault-branching, and fault zones are straight, narrow, and relatively unbraided. Such a situation seems to prevail today along the straight stretch of the San Andreas between the central Temblor Range and the Gabilan Range (fig. 1, TR and GR). This stretch includes Parkfield and the part of the fault zone exhibiting creep and frequent small earthquakes (Brown and others, 1967).

STRIKE SLIP ALONG A FAULT WITH A GENTLE DOUBLE BEND

Displacement of adjacent blocks along a single dominating strike-slip fault with a gentle double curve displays one of two different geometries. On the one hand, the bend may be in the direction to free or release the blocks as they glide by each other, or on the other, to lock or restrain them. With right slip, for example, if the fault trace curves to the right (clockwise) in looking along the fault toward a displaced feature, the bend will be a *releasing double bend*, and if to the left (counterclockwise), a *restraining double bend* (fig. 3). With left slip, in contrast, a releasing bend curves to the left, and a restraining bend to

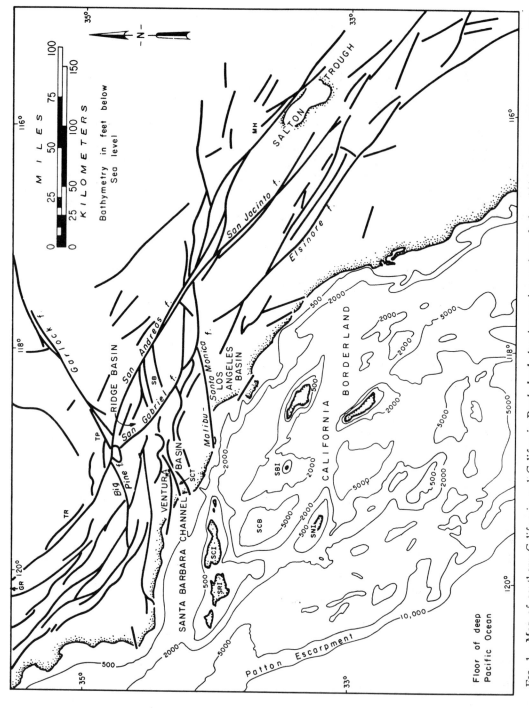

Fig. 1.—Map of southern California and California Borderland showing major onshore faults. Abbreviations: TP, Tejon Pass; SB, Soledad basin; SCT, Santa Clara trough; SRI, Santa Rosa Island; SCI, Santa Cruz Island; SCB, Santa Cruz basin; SBI, Santa Barbara Island; SNI, San Nicolas Island; GR, Gabilan Range; TR, Temblor Range; MH, Mecca Hills.

the right [Note mnemonically that if the two words are the same, a releasing bend results; if different, a restraining bend]. In this discussion the strike-slip fault is visualized as long and extensive, and the bends are relatively local departures in trend. If so, the bends are double in that one curvature takes the fault away from the regional trend and another brings it back into alignment. The direction of shift or strike slip of one block with respect to the other is defined by the strike of the fault on an extensive regional basis. If the fault is considered as a transform fault, this is the direction of relative motion between the major lithospheric plates.

If we consider the blocks as less rigid; that is, as relatively soft and deformable, movement along a fault with a double bend will cause shortening or crowding of the crustal rocks within concavities, and stretching at convexities (fig. 4), in the edges of adjacent blocks. Inasmuch as the crowding and stretching is relieved most easily at the terrane surface, shortening results in elevation of the ground surface, and stretching in subsidence. The maximum effect of these processes occurs near the strike-slip fault and at the point of maximum curvature. As the displacement continues, the centers of elevation or subsidence may move through time. Or on the other hand, one block can remain fixed in shape so that the same terrane continues to be elevated or depressed for long periods as the other block slides by and bends around it. The possibilities range through a continuum from one block remaining fixed in shape as the other participates in all of the bending, to both sharing equally in the bending. And on a single fault this style may change through time.

Ridge Basin, sited adjacent to the Pliocene major strand of the San Andreas fault in the central Transverse Ranges, apparently formed in such a setting (Crowell, 1954; 1962; in press). About 12,000 m (40,000 ft) of both

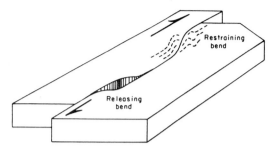

FIG. 3.—Right slip on fault with marked double bends results in pull-aparts at releasing bends and deformation and uplift at restraining bends.

marine and nonmarine sediments accumulated within a narrow basin at the stretched and depressed margin of the Americas Plate as it moved alongside a restricted source area (fig. 4). The lateral movement of the depositional site allowed the accumulation of the vast stratigraphic thickness of sediments by a gradual northwestward overlapping of older strata by younger, but without breaks or unconformities along the axis of the trough as the depocenter migrated. Older strata were carried away laterally as younger ones were deposited opposite the restricted source area. In this case, the uplifted source remained fixed in shape.

STRIKE SLIP ALONG A FAULT WITH A SHARP BEND

A long and straight strike-slip fault with a sharp double bend that sidesteps the fault trend for a relatively short distance can exhibit again two basically different situations. A sharp restraining double bend results in overlap and elevation at the bends (fig. 5), and a releasing double bend results in a pull-apart and subsidence (fig. 6). At restraining bends, geologic structures display oblique shortening, and at releasing bends, stretching or extension.

Restraining double bends bring about overlap

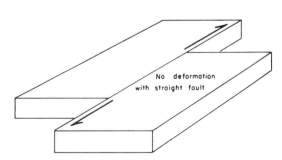

FIG. 2.—Strike slip on straight fault results in no deformation of crustal blocks.

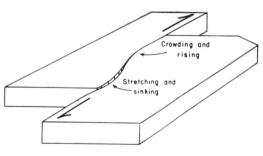

FIG. 4.—Right slip on gently curved fault results in crowding and uplift within convexities of deformable lithospheric plates and stretching and sagging within concavities.

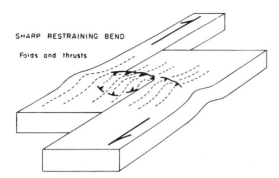

FIG. 5.—Severe deformation at sharp restraining bend results in folds and thrust faults.

of one block over the other, so that, under gravity, the edge of one block is depressed to form a shallow site for sedimentation adjacent to an elevated source area (Crowell, 1974, fig. 6). An example may be the depositional site of the wedge of continental sediments now accumulating on the San Bernardino Plain southwest and adjacent to the San Andreas, which at this place marks the boundary of the San Bernardino Mountains.

In contrast, releasing double bends form deep and narrow sedimentary basins at the pull-apart. These range in scale from small sag ponds within a restricted strike-slip zone floored by local country rock to true rhombochasms (Carey, 1958), such as those in the Gulf of California, floored by new lava above a spreading center or diapiric volcanic complex. Large and complex pull-aparts are treated more fully below.

BRANCHING AND BRAIDED STRIKE-SLIP ZONES

Strike-slip systems, such as those in California, consists of long and straight master faults with lesser branching faults or splays leading off from them. In addition, many of the major faults within the system are fault zones several kilometers wide containing fault slices and folds (Saul, 1967). Major splays lead away

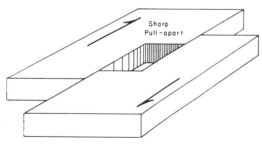

FIG. 6.—A sharp pull-apart on a right-slip fault.

from places where there is a slight change in strike of the master fault. In fact, where the Sunol-Calaveras system extends northward and away from the San Andreas near Hollister, the San Andreas northwest of the juncture is not now as active seismically and a crustal wedge may be forming (Burford and Savage, 1972). The inference is that the local northwestern part of the San Andreas is in the process of being supplanted as the major transform break by the Sunol-Calaveras in response to adjustments between the Pacific and Americas plates.

The splay may either continue on the new trend, or rejoin the original master fault to make a wedge or slice (fig. 7). Wedges range in size up to more than a hundred kilometers in length, such as the one between the San Gabriel and San Andreas faults within the Transverse Ranges. If huge wedges of this type are tipped longitudinally as displacement on the transform system continues, one end may subside to form a depositional site, and the other may elevate to provide a source area. Some further generalizations concerning broad zones of slices and wedges are discussed below.

Braided fault zones consisting of anastomosing faults and obliquely trending folds, such as the Sunol-Calaveras fault zone (Saul, 1967), may display local complexities that have only obscure kinematic relations to strike-slip origin. Individual faults exhibit dip slip and dip separations. Some originated as wedges within the zone were squeezed upward and others, during sagging of wedges downward. Clay-model experiments, such as those illustrated by Wilcox and others (1973), show these complicated patterns very well. Rocks on a regional scale

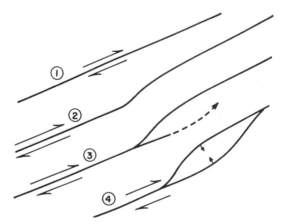

FIG. 7.—Diagrammatic map showing progressive development of fault splays and wedges on a right-slip fault. Straight fault (1) gradually develops a bend through time (2 and 3) and eventually forms a fault wedge (4).

are also weak, similar to clay, so that when a crustal block is pressed upward beside a nearly vertical fault, the hanging well sags outward and downward; the field geologist will correctly map it as a thrust fault. Such an origin for local complexities has been described, especially in New Zealand where horsts and grabens are associated with strike-slip zones (Kingma 1958; Lensen, 1958). Similar braided zones are recognizable in Maritime Canada (Belt, 1968), northwestern Scotland (Kennedy, 1946; Dearnley, 1962), along the Dead Sea fault zone of the Levant (Quennell, 1958; Freund, 1965), and along the north coast of South America in Colombia, Venezuela, and Trinidad (Malfait and Dinkelman, 1972; Wilcox and others, 1973; Crowell, 1974).

PULL-APART BASINS

At the present time the Gulf of California and Salton Trough are widening and lengthening as continental terrane to the west moves obliquely away from the mainland part of North America (Wegener, 1924; Carey, 1958, Fig. 42; Hamilton, 1961). This process is envisaged as the result of sea-floor spreading along the segmented and offset parts of the East Pacific Rise as it enters the Gulf of California at its southern end (Larson and others, 1968; Moore and Buffington, 1968; Larson, 1972; Larson and others, 1972). Geologic studies and geophysical surveys at the head of the Gulf and within the Salton Trough suggest that the Salton Trough lies above a series of spreading centers or diapiric masses with volcanic rocks at depth, and continental rocks, if any, are attenuated and fragmented near the center of the structure (Elders and others, 1972; Sumner, 1972; Henyey and Bischoff, 1973; Karig and Jensky, 1973). It is therefore probably a true rhombochasm, or chain of them, that has opened while abundant sediment has flooded into the widening hole (Crowell, 1974). Details of the structure along the northeastern border of the Salton Trough, for example, fit reasonably well the idea that they are the consequence of right slip at a steep basin margin with high-standing continental rocks on the northeast and deep quasi-oceanic crust on the southwest. In the Mecca Hills (fig. 1, MH) the deformed sedimentary section of Neogene age thickens rapidly toward the trough and the San Andreas zone there consists of braided faults of several ages, with complex folds and thrusts arranged between them (Dibblee, 1954; Hays, 1957; Crowell, 1962).

According to a pull-apart model, prisms of the oldest sediments lying upon the original basin floor should occur around the margins

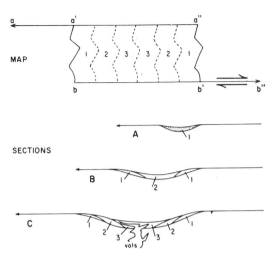

Fig. 8.—Sketch map and sections of pull-apart basin; see text for discussion.

only, but drilling and geophysical studies are not yet complete enough for confirmation. In the paragraphs below I will consider first some simple theoretical models and then geological processes and circumstances that complicate and modify them. The latter include especially the relative rates of sedimentation in comparison with rates of subsidence and deformation, changes in orientations and rates of deformation through time, and the strength or weakness of the crust and the magnitudes and arrangements of heterogeneities within it.

The simplest pull-apart model consists of side-stepped parallel transforms (a-a'-a" and b-b'-b" in fig. 8) sited above a volcanic center. The transform margins (a-a" and b-b") of the basin appear straight and parallel in map view, but the pull-apart margins (a'-b and a"-b') can have any shape and are here drawn crookedly to emphasize their critical geometric distinction from the transforms. When the pull-apart is born, a'-b and a"-b' fit together, but as the hole opens, the pull-apart walls sag independently. The walls extend and stretch, so that in time their structures may be very different and fail to match, although details in pre-existing rocks will still correlate. The walls along the transform margins of the pull-apart tend to sag also but continued strike-slip on them slices off segments out of line, and a complex braided zone results.

The first sediments laid down within the growing basin occupy a position as in figure 8, stage A, and the later layers are shown in stages B and C. As the hole widens, the margins are first stretched but in time lava comes up from below (stage C) so that down-dip near the center of the basin lavas and shallow intrusions

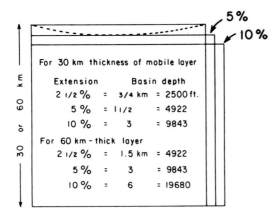

FIG. 9.—Diagram showing crustal cross section, 30 or 60 km thick, with 5 or 10% extension; see text for discussion.

may document extension. The emplacement of such lavas over active spreading centers or over diapirs from the mantle is suggested by gravity and heat-flow data and by the location of geothermal areas and young volcanics in the Salton Trough (Elders and others, 1972). It is also suggested by the distribution of volcanic rocks within the deeper and older parts of the Los Angeles basin, now displaced laterally, uplifted and eroded to view in the western Santa Monica Mountains.

Inasmuch as the crust is weak and easily folded, faulted, stretched, and compressed on a regional scale, some basins probably form in a broad transform system without volcanic flows and intrusions at depth. Most of the basins in the southern California region, including the Gulf of California and the continental borderland, range between 50 and 100 km in a northwest-southeast dimension. Inferences from the depths of earthquake foci along the San Andreas system (Eaton, 1967) and from plate-tectonic models (e.g., Isacks and others, 1968) suggest that although creep takes place below a depth of about 20 km, there is a near-uncoupling of the crust from the upper mantle at or within the low-velocity zone. Under stable cratonic crusts, the thickness of the lithosphere or depth to this zone of uncoupling is of the order of 110 km, under the central Great Basin about 20 km, and in the oceans away from ridges about 75 km (Walcott, 1970).

It may be of interest to estimate the amount of crustal extension needed to form a basin such as those in the California Borderland. For a rough and oversimplified two-dimensional model, if we assume that the lithosphere in southern California has a minimum thickness of

30 km beneath fragmented sialic crustal blocks and a maximum thickness of 60 km and that the average basin has an average depth across the block of 3 km and a northwest-southeast dimension of 60 km, then the extension needed to form the basin is 5 percent for the 60 km lithospheric thickness and 10 percent for the 30 km thickness (fig. 9).

Unfortunately, as yet we have no clear picture whether stretching or "necking down" of the lithosphere as much as 5 or 10 percent is reasonable. We can conclude, however, that stretching of crustal blocks of a few percent may begin the pull-apart process and start the formation of the basin before rupture of the basin floor and the entry of volcanics from below. In fact, in a broad weak transform zone, such as may prevail across the full width of southern California and its borderland, the volcanics may arise irregularly and diapirically into the growing gap after the breaking point is reached. At the same time the basin is being filled from the top by sedimentation.

In figure 8, the sidestepped transforms are shown as ending at a″ and b, but as the hole enlarges, complications at these corners are to be expected, such as continued minor growth along extensions of the transforms. Moreover, the angle between the transform direction and the pull-apart scarps (a′-b and a″-b′) might be very much less than shown in figure 8 so that in the field it would be difficult to locate the basin corners precisely. As the basin deepens, the pull-apart margins may stretch and subside through time, so that successively younger stratigraphic units lap farther and farther basinward leaving behind a record of minor unconformities. If the center of the basin has stretched to the point of rupture, and then has been intruded by volcanics to make a new floor, strata deposited earlier within the basin may be confined to the attenuated margins only and a deep well drilled in the basin center would not penetrate them. Instead, it would pass through younger sediments and into lava flows and associated volcaniclastic rocks. Below these it would drill through fragments only of the earlier basin fillings, and of the basin floor, and finally into diapiric masses of hypabyssal volcanics and volcanic feeder complexes. Such a model at depth suggests that it is unwise to extrapolate strata at the pull-apart margins very far down dip into some basins. The floors of true rhombochasms, for example, consist of new volcanic rocks, and lack the older rocks exposed around their margins, except perhaps as isolated blocks or "floaters." A summary sketch map is shown in figure 10, on which are portrayed possible features associated with pull-apart basins, but no single basin would be expected to possess them all.

FAULT-WEDGE BASINS

In regions such as southern California, and especially within its borderland, rhomboid and lens-shaped basins are associated with similarly shaped high-standing banks and ridges. Such a fragmented portion of the continental plate can be visualized theoretically as forming within a strike-slip regime if the major strike-slip faults converge and diverge in map view. For example, in a right-slip system where two major right-slip faults converge, assuming concurrent or intermittently alternating movement on each, the wedge between the faults will be compressed and elevated where the faults diverge, the block is extended and terrane subsides (fig. 11). Many faults in a broad and anastomosing system probably do not all move at the same time; those that predominate become straighter and longer, whereas some early faults are bent and rotated out of an orientation conducive to easy slip.

Ideas developed by Lensen (1958) to explain horsts and grabens and changing fault dips along strike-slip fault zones in New Zealand can be modified to apply to broad transform

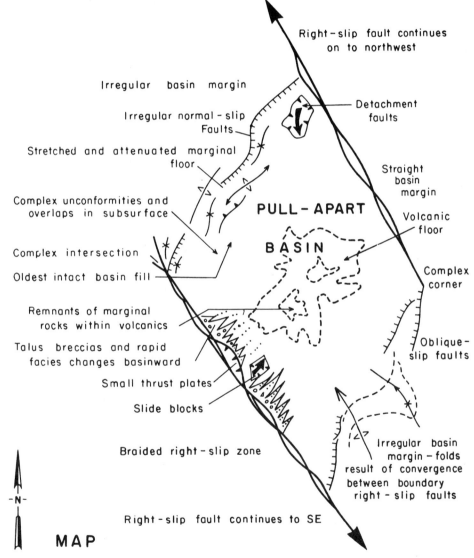

FIG. 10.—Sketch map of idealized pull-apart basin; see text for discussion.

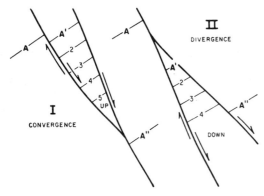

FIG. 11.—Sketch maps showing uplift of tip of fault wedge with convergence of right-slip faults, and subsidence of tip with divergence; see text for discussion.

borders in plate tectonics. If Lensen's concept of the principal horizontal stress is replaced by the concept of the direction of relative motion between two major plates, many of his geometric relations apply. The plate-motion direction will lie at 45° to this direction of principal horizontal stress. Ideally, in plate-tectonic theory, rigid plates glide by each other along a straight transform and the relative motion between the two plates is purely horizontal on a single vertical fault surface. At the present time this direction of relative motion is determined by measurements of the first motions of earthquakes, and by noting the orientation of the long and straight and predominating strike-slip faults that are clearly active.

In a boundary region of weak plate edges between rigid plates, however, braided zones apparently develop. In these systems some faults lie parallel to the direction of relative plate movement whereas others lie at an angle, usually a small angle. In general, those parallel to the plate-movement direction predominate and grow longer; those at an angle may rotate even farther out of alignment. The long predominating faults develop nearly vertical dips; those rotated out of alignment develop dips that depart considerably from the vertical. Strike-slip faults curving or bending away from the plate-movement direction gradually change from pure strike slip upon a vertical fault surface to those with first gentle oblique slip and then steep oblique slip as the fault strike bends more and more. If the curvature carries the fault into a region of extension, the fault becomes a normal oblique-slip fault; if into a region of compression, a reverse oblique-slip fault.

Such geometric relations are especially easy to envisage along braided zones of single major

strike-slip faults where all slices occur in surficial rocks. Along such master faults, the faults bounding the slices converge at depth to rejoin the strike of the through-going fault. In regions such as the California Borderland, however, the major faults do not meet at depth, but presumably end in the lower crust in approaching the low-velocity zone. Such a system, including pull-apart basins, highs owing to convergence, and lows owing to divergence between major right-slip faults, is shown in figure 12.

SOUTHERN CALIFORNIA BASINS

Salton Trough, an example of a pull-apart basin, and Ridge Basin, an example of a basin formed by the sagging of weak crust as plates move around a double bend, have been mentioned briefly above. It remains to comment concerning the applicability of models described above to other basins in southern California but some complicating factors need emphasis first. To begin with, the tectonic style across southern California has changed from place to place during the last 25 my so that older basins originated under different tectonic schemes from modern ones. Salton Trough, for example, originated during the past 4 to 6 my and all deformation of Pliocene and Quaternary age can probably be related somehow to this opening. The San Andreas fault in the Salton region, however, is older than this event and was associated with an earlier elongate basin or proto-Gulf (Crowell, 1971; Karig and Jensky, 1973).

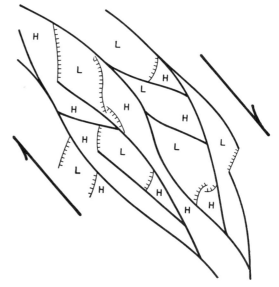

FIG. 12.—Sketch map of region in a transform regime, showing pull-apart basins and tipped fault wedges where right-slip faults converge or diverge; see text for discussion.

During the Quaternary the Transverse Ranges were deformed and uplifted as shown by the age of recent movements on dip-slip faults bounding the range and the age of folding (Crowell, 1971; Dibblee, 1971; Ehlig, in press). In fact, the San Fernando Earthquake of 1971 demonstrates that this deformation is still in progress (Grantz and others, 1971). Moreover, the major bend in the San Andreas fault near Tejon Pass (fig. 1, TP) and the origin or rejuvenation or reorientation of many structures in the region are perhaps all associated with the relative northwestward movement of the Peninsular Ranges toward the Transverse Ranges that accompanied the opening of the Gulf of California. Earlier events in southern California included major left slip in an east-west direction on the combined Malibu Coastal-Santa Monica fault system and Whittier-Elsinore fault system of as much as 90 km in the latest Miocene or early Pliocene (Yerkes and Campbell, 1971; Campbell and Yerkes, 1971), displacement on faults associated with opening of the Los Angeles basin in the early mid-Miocene, and the formation of steep and narrow grabens in the Soledad basin (fig. 1, SB) in the early Miocene (Jahns and Muehlberger, 1954; Crowell, 1968). Deformation in southern California has therefore been locally severe and intermittently continuous during the late Cenozoic as sedimentation and erosion have gone on concomitantly. Older stratigraphic units and structures have been modified by these processes so that we may not easily identify the original basin geometry. Careful palinspastic work, in which the results of younger deformations are first removed, must precede analysis in detail of basin tectonics in the past. This sound procedure, however, is difficult and imposing, and is well beyond the scope of this brief paper.

THE LOS ANGELES BASIN

One of the regions of southern California that displays this type of complicated history is the Los Angeles basin (fig. 13). It contains over 6750 m (22,000 ft) of Miocene and Pliocene sediments in its deepest part and has yielded nearly 6 billion barrels of petroleum to date (Gardett, 1971, table 1). The basin was completely filled with sediments by some time in the Pleistocene and is now being deformed, especially along the Newport-Inglewood trend (Harding, 1973; Yeats, 1973; Hill, 1971). Faulting and folding along this trend fit neatly into the concept of deformation along a system of simple right shear in a thick sedimentary section overlying a major fault at depth (Eaton, 1926; Platt and Stuart, in press). This buried fault, however, was primarily active before

deposition of the upper Miocene and Pliocene sediments, although it was probably active and instrumental in demarcating the western margin of the basin in early mid-Miocene times. The San Onofre Breccia, for example, was laid down to the east of a fault scarp along this trend during the early part of the middle Miocene; the debris came from a schist terrane, probably exposed to subaerial erosion (Woodford, 1925; Stuart, 1973), that bordered the basin on the west. During the later Miocene, however, the fault was overlapped and marine beds transgressed southwestward.

The southeastern margin of the Los Angeles basin is formed by deformed Cretaceous and lower Tertiary strata overlying Mesozoic sialic basement. Facies trends, particularly in Paleocene beds, show that the paleocontours extended approximately in a north-south direction, with marine waters deepening westward. Nonmarine beds succeeded these in the Oligocene and early Miocene (Sespe and Vaqueros formations) but with approximately the same depositional trends. Similar stratigraphic successions older than middle Miocene are recognizable in terranes surrounding the Los Angeles basin and were disrupted by tectonic processes involved in the origin of the basin (Yerkes and Campbell, 1971; Campbell and Yerkes, 1971). Marked sedimentation within the deepening Los Angeles basin ensued in the middle Miocene, and was quickly followed by both faulting and volcanism in this southeastern region. In the San Joaquin Hills (fig. 13, SJH), for example, irregular north-trending faults displace middle Miocene strata but in turn are intruded by middle Miocene volcanics (Vedder and others, 1957). In addition, some of these faults are also overlapped by upper Miocene beds (Monterey Formation). The continuity and complexity of faulting and deformation along this flank of the Los Angeles basin is emphasized by the fact that some faults of similar trend are clearly younger and truncate the volcanics and Monterey beds. Volcanic rocks are present nearly everywhere in the central part of the Los Angeles basin (Eaton, 1958; Yerkes & others, 1965, fig. 9), and imply a nearly hydrostatic uprising of lava to a compensating level within the basin. Only locally were volcanic rocks extruded at and beyond the basin margins, as in the Glendora region (Shelton, 1955; fig. 13, GV).

During the middle Miocene the Los Angeles basin extended northward beyond the limits of the present Los Angeles Plain. The western Santa Monica Mountains, for example, contain very thick sequences of flyschlike strata of this age that extend downward into the lower Miocene. The beds were probably deposited in fairly

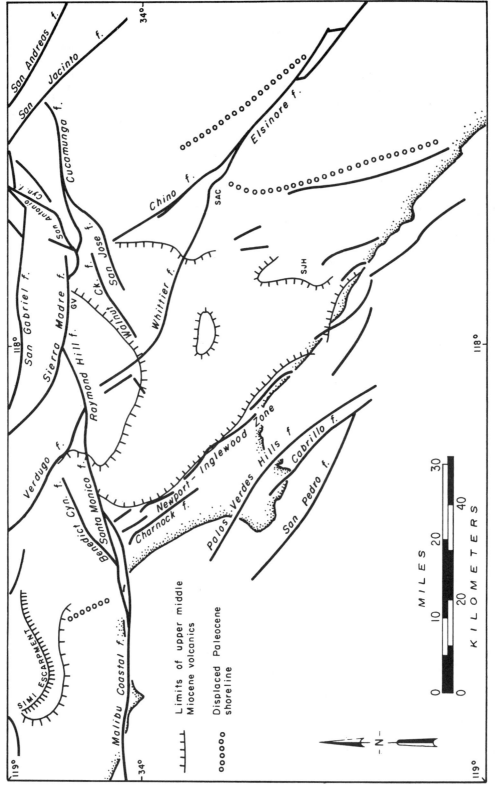

FIG. 13.—Sketch map of the Los Angeles basin region. Abbreviations: GV, Glendora volcanics; SAC, Santa Ana Canyon; SJH, San Joaquin Hills.

deep water with a nearby schist landmass on the southwest (Woodford and Bailey, 1928; Durrell, 1954) and with a coast to the north near the Simi Escarpment (fig. 13). Within this trough the lower and middle Miocene beds reach a total thickness of about 3800 m (12,500 ft) and are intercalated with middle Miocene volcanics aggregateing an aditional 3650 m (12,000 ft) approximately. The middle Miocene section is overlain by about 1400 m (4500 ft) of upper Miocene sandstone and shale. This very thick sequence of clastic sediments, now deformed and uplifted, thins northward very rapidly toward the Simi Escarpment. According to Campbell and others (1966), detachment blocks from the region of this escarpment slid southward into the deep basin during the mid-Miocene. Since these events the whole region has been displaced relatively westward with respect to the main part of the Los Angeles basin by left slip on the Malibu Coastal-Santa Monica fault system (Yerkes and Campbell, 1971; Campbell and Yerkes, 1971). Even more recently, during the Pleistocene, the Santa Monica Mountains were uplifted and deeply eroded. Upper lower and lower middle Miocene sandstone and shale, here considered as laid down in a deep part of the original Los Angeles basin, are now exposed to view along portions of the Malibu Coast. Dike and still complexes with associated flows document extension and perhaps afford an arrested view of the floor of a widening pull-apart basin.

During the Pliocene, the Los Angeles basin deepened but became considerably restricted. Thick sediments accumulated south of the present location of the Santa Monica Mountains and southwest of the Whittier fault zone (fig. 13.) This fault apparently originated near the end of the Miocene because middle Miocene beds are displaced nearly as much as Pliocene (Woodford and others, 1954, p. 75). It apparently formed the northeastern boundary of the Pliocene basin, but may have terminated at the southeast in a "corner" near Santa Ana Canyon (fig. 13, SAC). On the west, Pliocene beds overlapped the Newport-Inglewood fault zone and on the plunging southeastern margin of the basin overlapped as well upon deformed older strata. Subsidence, presumably accompanying extension and stretching and attenuation of crustal rocks, allowed the accumulation of thick sediments in the center of the Los Angeles basin. No volcanics are known to have invaded the basin floor during the Pliocene, either because extension was not sufficient to require it or because a hot spreading center did not then lie beneath it.

By the Pleistocene the Los Angeles basin had filled with sediment and shortly thereafter the beds were deformed, especially along the northeastern margin, demarcated primarily by the Whittier fault zone, and over the buried Newport-Inglewood fault zone (Harding, 1973; Yeats, 1973). The Elsinore fault system originated as a major strike-slip fault late in the Pliocene or in the Pleistocene (Gray, 1961), perhaps associated with the relative northwestward movement of the Peninsular Ranges as the Gulf of California opened. In addition, the San Jacinto fault to the east, with a total right slip of about 24 km (Sharp, 1967), may have resulted from the same movements. The southeastern part of the Elsinore fault, however, apparently has a right slip of only about 5 km, judging from the offset of an ancient cataclastic zone (R. V. Sharp, p. 22, in Lamar, 1972). The Elsinore fault is therefore probably younger than much of the movement on the Whittier fault with which it is now connected, and both faults are younger than the original opening of the Los Angeles basin in the early middle Miocene.

Bouguer gravity anomalies over the Los Angeles basin now show a broad and deep depression corresponding roughly with the thickest section of upper Miocene and Pliocene strata (McCulloh, 1960). The gravity measurements apparently reveal the thick mass of light sediments in the basin and not a high-standing diapir or volcanic complex derived from the upper mantle similar to that inferred to lie beneath the Salton Trough from high values of gravity and other data in that region (Elders and others, 1972, Fig. 3). More complete deep seismic and other geophysical data are needed before we can understand the structure at depth below the Los Angeles region, and before we can reconstruct the deeper structures formed by tectonic events in the late Cenozoic leading up to the present.

The history of the Los Angeles basin as we now understand it includes birth accompanied by rapid deepening during the early middle Miocene followed by thick sedimentation and volcanism, especially in the central part. Detachment faulting took place along the northern margin in the mid-Miocene. Truncation and left displacement of the irregular northern portion of the basin occurred in the late Miocene or after, and were accompanied by death of the Newport-Inglewood zone and nearly simultaneous birth of the Whittier fault zone. Thick sedimentation in the restricted Los Angeles basin followed during the Pliocene. The history ends with Pleistocene filling and northwest-southeast compression manifested by birth of the Elsinore fault and complex right-slip reactivation along the Whittier fault zone. Although it is speculative, perhaps the original basin was born as a pull-apart over a "hot spot"; this may have been associated with the passage of the East Pacific Rise along the coast, or with a local hot "plume.'" Later events in the Los Angeles basin, however, are less clearly related to a pull-apart origin.

OTHER BASINS

The California Borderland consists of a series of irregular topographic highs, some surmounted by islands, separated by basins (fig. 1). Those basins near the coast and accessible to debris from rivers in general contain more sediment than those far offshore and relatively isolated from land sources (Emery, 1960, p. 53; Moore, 1969, pl. 14). Extrapolation seaward of the land geology and geologic history into this region suggests a complex history similar to that just reviewed for the Los Angeles basin. Data to document the complicated history are however missing from the published record. In general, the pattern of escarpments, the shapes of banks, and ridges and islands, and the configuration of basins, suggests a pull-apart origin for some of the basins. This pattern is similar to that shown here in figure 12, consisting of diverging strike-slip faults in a broad and soft transform zone. The shape of the Santa Cruz basin (fig. 1, SCB) suggests that it is a pull apart formed when the ridge surmounted by Santa Barbara Island (fig. 1, SBI) was stretched away from the ridge connecting Santa Rosa Island and San Nicolas Island (fig. 1, SRI and SNI). Although other analogies between the style of deformation portrayed on figure 13 and the California Borderland can be recognized, it is not profitable to speculate further here in the absence of specific information. Nonetheless, such a pull-apart origin for some basins, associated with convergence and divergence along major strike-slip faults accompanied by attendant rising and tipping of fault-bounded wedges, should be entertained by geologists and geophysicists investigating the region.

The Ventura basin, and its offshore extension beneath the Santa Barbara Channel (fig. 1), may also have subsided in a regime of stretching. During the Pliocene, for example, the Santa Clara trough (fig. 1, SCT) received more than 4600 m (15,000 ft) of sediments that were then severely compressed in a north-south direction during the Pleistocene (Nagle and Parker, 1971, Fig. 12). In the absence of known Pliocene volcanism at depth, however, there is

little reason to suggest that the trough formed as a true sphenochasm or rhombochasm (Carey, 1958).

SUMMARY AND RECOMMENDATIONS

Understanding of the origin of the Salton trough and other basins along the San Andreas transform, now active, leads to the speculation that more ancient basins in southern California may have originated in a similar way. The geologic record is so complicated, however, that in this paper we have focused attention on what we ought to look for in order to find analogies rather than on documentation. In a strike-slip regime, pull-apart basins or rhombochasms will display straight margins where they are parallel to the transform direction, and irregular borders along the pull-apart margins. The transform margins although generally straight will be complicated in detail, and will consist of braided zones, slices, thrust blocks, and detachment faults where high terranes are structurally unsupported against low terranes. Pull-apart margins may display similar features, but even more irregularly. If pull-aparts stretch enough, or are sited above hot spreading centers or diapirs of magma from the upper mantle, volcanic rocks may enter their floors as dikes, sills, irregular bodies, and flows. Under such circumstances the floor of the basin may consist of volcanic rocks and young sediments deposited in the depression, and old or marginal rocks may be absent or present only as isolated blocks or "floaters." Structure and strata exposed around the margins of such pull-aparts cannot be extrapolated to the basin floor with confidence.

A system of anatomosing transforms, converging and diverging in map view, may give rise on a regional scale to wedge-shaped basins separated by uplands. In a soft crust, convergence of faults will result in squeezing of the terrane between the faults and in its deformation and uplift owing to isostatic compensation. Divergence of strike-slip faults will result in stretching and sagging, and the development of down-tipped triangular basins and pull-apart basins. Many of the latter will be rhombic in shape, and some may have floors composed of volcanics intermixed with infilled sediments with few or no remnants of previously existing crustal rocks. Inasmuch as much stretching and sagging and squeezing and uplift goes on hand in hand with sedimentation, complicated unconformities and overlaps are to be expected within the basins, and especially around their margins. Kinematics interpreted from local structures may therefore reveal only remotely their connection to a broad strike-slip regime.

REFERENCES CITED

ATWATER, TANYA, 1970, Implications of plate tectonics for the Cenozoic tectonic evolution of western North America: Geol. Soc. America Bull., v. 81, p. 3513–3536.

BELT, E. S., 1968, Post-Acadian rifts and related facies, eastern Canada, *in* Zen, E-An, White, W. S., Hadley, J. B., and Thompson, J. B., Jr. (eds.), Studies of Appalachian geology, northern and maritime: Wiley Interscience, N.Y., p. 95–113.

BROWN, R. D., JR., VEDDER, J. G., WALLACE, R. E., ROTH, E. F., YERKES, R. F., CASTLE, R. O., WAANANEN, A. O., PAGE, R. W., AND EATON, J. P., 1967, The Parkfield-Cholame, California earthquakes of June-August, 1966: U.S. Geol. Survey Prof. Paper 579, 66 p.

BURFORD, R. O., AND SAVAGE, J. C., 1972, Tectonic evolution of a crustal wedge caught within a transform fault system (abs.): Geol. Soc. America Abstracts with Programs, v. 4, p. 134.

CAMPBELL, R. H., AND YERKES, R. F., 1971, Cenozoic evolution of the Santa Monica Mountains-Los Angeles basin area: II. Relation to plate tectonics of the northeast Pacific Ocean: *ibid.,* v. 3, p. 92.

CAMPBELL, R. H., YERKES, R. F., AND WENTWORTH, C. M., 1966, Detachment faults in the central Santa Monica Mountains, California: U.S. Geol. Survey Prof. Paper 550-C, p. C1–C11.

CAREY, S. W., 1958, The tectonic approach to continental drift, *in* Carey, S. W. (ed.), Continental drift: Univ. Tasmania Geology Dept. Symposium No. 2, p. 177–355.

CROWELL, J. C., 1952, Probable large lateral displacement on the San Gabriel fault, southern California: Am. Assoc. Petroleum Geologists Bull., v. 36, p. 2026–2035.

———, 1954, Geology of the Ridge basin area, Los Angeles and Ventura Counties, California: Calif. Division Mines Bull. 170, map sheet 7.

———, 1962, Displacement along the San Andreas fault, California: Geol. Soc. America Special Paper 71, 61 p.

———, 1968, Movement histories of faults in the Transverse Ranges and speculations on the tectonic history of California, *in* Dickinson, W. R. and Grantz, Arthur (eds.), Proceedings of conference on geologic problems of San Andreas fault system: Stanford Univ. Pub. Geol. Sci., v. 12, p. 323–341.

———, 1971, Tectonic problems of the Transverse Ranges, California: Geol. Soc. America Abstracts with Programs, v. 3, p. 106.

———, 1974, Sedimentation along the San Andreas fault, California, *in* Dott, R. H., and Shaver, R. H. (eds.), Modern and Ancient geosynclinal sedimentation: Soc. Econ. Paleontologists and Mineralogists Special Pub. 19, p. 292–303.

DEARNLEY, RAYMOND, 1962, An outline of the Lewisian complex of the Outer Hebrides in relation to that of the Scottish mainland: Quart. Jour. Geol. Soc. London, v. 118, p. 143–176.

DIBBLEE, T. W., JR., 1954, Geology of the Imperial Valley Region, California: Calif. Division Mines Bull. 170, p. 21–28.

———, 1971, Geologic environment and tectonic development of the San Bernardino Mountains, California: Geol. Soc. America Abstracts with Programs, v. 3, p. 109–110.

DURRELL, CORDELL, 1954, Geology of the Santa Monica Mountains, Los Angeles and Ventura Counties, California: Calif. Division Mines Bull. 170, map sheet 8.

EATON, G. P., 1958, Miocene volcanic activity in the Los Angeles basin, California, *in* Higgins, J. W. (ed.), A guide to the geology and oil fields of the Los Angeles and Ventura region: Pacific Sec. Am. Assoc. Petroleum Geologists, Los Angeles, p. 55–58.

EATON, J. E., 1926, A contribution to the geology of Los Angeles basin, California: Am. Assoc. Petroleum Geologists Bull., v. 10, p. 753–767.

EATON, J. P., 1967, Instrumental seismic studies, Parkfield-Cholame, California earthquake of June-August 1966: U.S. Geol. Survey Prof. Paper 579, p. 57–65.

EHLIG, P. L., in press, Geologic framework of the San Gabriel Mountains: Calif. Division Mines and Geology Report on San Fernando, California earthquake of 1971.

ELDERS, W. A., REX, R. W., MEIDAV, TSVI, ROBINSON, P. T., AND BIEHLER, SHAWN, 1972, Crustal spreading in southern California: Science, v. 178, p. 15–24.

EMERY, K. O., 1960, The sea off southern California: Wiley, N.Y., 366 p.

FERGUSON, R. N., AND WILLIS, C. G., 1924, Dynamics of oil-field structure in southern California: Am. Assoc. Petroleum Geologists Bull., v. 8, p. 576–583.

FREUND, RAPHAEL, 1965, A model of the structural development of Israel and adjacent areas since upper Cretaceous times: Geol. Mag., v. 102, p. 189–205.

GARDETT, P. H., 1971, Petroleum potential of Los Angeles Basin, California, *in* Cram, I. H. (ed.), Future petroleum provinces of the United States—their geology and potential: Am. Assoc. Petroleum Geologists Mem. 15, v. 1, p. 298–308.

GRANTZ, ARTHUR AND OTHERS, 1971, The San Fernando, California earthquake of February 9, 1971: U.S. Geol. Survey Prof. Paper 733, 254 p.

GRAY, C. H., JR., 1961, Geology and mineral resources of the Corona South Quadrangle, California: Calif. Division Mines Bull. 178, 120 p.

HAMILTON, WARREN, 1961, Origin of the Gulf of California: Geol. Soc. America Bull., v. 72, p. 1307–1318.

HARDING, T. P., 1973, Newport-Inglewood trend, California: Am. Assoc. Petroleum Geologists Bull., v. 57, p. 97–116.

HAYS, W. H., 1957, Geology of part of the Cottonwood Springs quadrangle, Riverside County, California (Ph.D. thesis): Yale Univ., New Haven, 324 p.

HENYEY, T. L., AND BISCHOFF, J. L., 1973, Tectonic elements of the northern part of the Gulf of California: Geol. Soc. America Bull., v. 84, p. 315–330.

HILL, M. L., 1971, Newport-Inglewood zone and Mesozoic subduction, California: *ibid.,* v. 82, p. 2957–2962.

———, AND DIBBLEE, T. W., JR., 1953, San Andreas, Garlock and Big Pine faults, California: *ibid.,* v. 64, p. 443–458.

ISACKS, BRYAN, OLIVER, JACK, AND SYKES, L. R., 1968, Seismology and the new global tectonics: Jour. Geophys. Research, v. 73, p. 5855–5899.

JAHNS, R. H. AND MUEHLBERGER, W. R., 1954, Geology of Soledad Basin, Los Angeles County: Calif. Division Mines Bull. 170, Map Sheet 6.

KARIG, D. E., AND JENSKY, WALLACE, 1972, The proto-Gulf of California: Earth and Planetary Sci. Letters, v. 17, p. 169–174.

KENNEDY, W. Q., 1946, The Great Glen fault: Quart. Jour. Geol. Soc. London, v. 102, p. 41–76.

KINGMA, J. T., 1958, Possible origin of piercement structures, local uncomformities, and secondary basins in the Eastern Geosyncline, New Zealand: New Zealand Jour. Geology and Geophysics, v. 1, p. 269–274.

LAMAR, D. L., 1972, Microseismicity and recent tectonic activity in Whittier fault area, California: U.S. Geol. Survey National Center for Earthquake Research Final Technical Report, 44 p.

LARSON, P. A., MUDIE, J. D., AND LARSON, R. L., 1972, Magnetic anomalies and fracture-zone trends in the Gulf of California: Geol. Soc. America Bull., v. 83, p. 3361–3368.

LARSON, R. L., 1972, Bathymetry, magnetic anomalies, and plate tectonic history of the mouth of the Gulf of California: *ibid.*, v. 83, p. 3345–3360.

LARSON, R. L., MENARD, H. W., AND SMITH, S. M., 1968, Gulf of California, a result of ocean-floor spreading and transform faulting: Science, v. 161, p. 781–784.

LENSEN, G. J., 1958, A method of horst and graben formation: Jour. Geology, v. 66, p. 579–587.

MALFAIT, B. T., AND DINKELMAN, M. G., 1972, Circum-Caribbean tectonic and igneous activity and the evolution of the Caribbean Plate: Geol. Soc. America Bull., v. 83, p. 251–272.

McCULLOH, T. H., 1960, Gravity variations and the geology of the Los Angeles basin, California: U.S. Geol. Survey Prof. Paper 400-B, p. B320–B325.

MOORE, D. G., 1969, Reflection profiling studies of the California continental borderland: Geol. Soc. America Special Paper 107, 138 p.

———, AND BUFFINGTON, E. C., 1968, Transform faulting and growth of the Gulf of California since the late Pliocene: Science, v. 161, p. 1238–1241.

NAGLE, H. E. AND PARKER, E. S., 1971, Future oil and gas potential of onshore Ventura basin, California: *in* Cram, I. H. (ed.), Future Petroleum Provinces of the United States—their geology and potential: Am. Assoc. Petroleum Geologists, Mem. 15, v. 1, p. 254–297.

PLATT, J. P., AND STUART, C. J., in press, Newport-Inglewood fault zone, Los Angeles Basin, California (Discussion): *ibid.*, Bull.

QUENNELL, A. M., 1958, The structural and geomorphic evolution of the Dead Sea rift: Quart. Jour. Geol. Soc. London, v. 114, p. 1–24.

SAUL, R. B., 1967, The Calaveras fault zone: Calif. Division Mines and Geology Mineral Information Service, v. 20, p. 33–37.

SHARP, R. V., 1967, San Jacinto fault zone in the Peninsular Ranges of southern California: Geol. Soc. America Bull., v. 78, p. 705–730.

SHELTON, J. S., 1955, Glendora volcanic rocks, Los Angeles basin, California: *ibid.*, v. 66, p. 45–89.

STUART, C. J., 1973, Stratigraphy of the San Onofre Breccia, Laguna Beach area, California: *ibid.*, Abstracts with Programs, v. 5, p. 112.

SUMNER, J. R., 1972, Tectonic significance of gravity and aeromagnetic investigations at the head of the Gulf of California: *ibid.*, Bull., v. 83, p. 3103–3120.

VICKERY, F. P., 1925, The structural dynamics of the Livermore region: Jour. Geology, v. 33, p. 608–628.

VEDDER, J. G., YERKES, R. F., AND SCHOELLHAMER, J. E., 1957, Geologic map of the San Joaquin Hills-San Capistrano area, Orange County, California: U.S. Geol. Survey Oil and Gas Inv. Map OM-193.

VINE, F. J., AND WILSON, J. T., 1965, Magnetic anomalies over a young oceanic ridge off Vancouver Island: Science, v. 150, p. 485–489.

WALCOTT, R. I., 1970, Flexural rigidity, thickness, and viscosity of the lithosphere: Jour. Geophys. Research, v. 75, p. 3941–3954.

WEGENER, ALFRED, 1924, The origin of continents and oceans: Dover, N.Y. [1966 reprint], 246 p.

WILCOX, R. E., HARDING, T. P., AND SEELY, D. R., 1973, Basic wrench tectonics: Am. Assoc. Petroleum Geologists Bull., v. 57, p. 74–96.

WILSON, J. T., 1965, A new class of faults and their bearing on continental drift: Nature, v. 207, p. 343–347.

WOODFORD, A. O., 1925, The San Onofre Breccia, its nature and origin: Univ. Calif. Pub. Dept. Geol. Sci. Bull., v. 15, p. 159–280.

———, AND BAILEY, T. L., 1928, Northwestern continuation of the San Onofre Breccia: *ibid.*, v. 17, p. 187–191.

———, SCHOELLHAMER, J. E., VEDDER, J. G., AND YERKES, R. F., 1954, Geology of the Los Angeles basin, California: Calif. Division Mines Bull. 170, p. 65–81.

YEATS, R. S., 1973, Newport-Inglewood fault zone, Los Angeles basin, California. Am. Assoc. Petroleum Geologists Bull., v. 57, p. 117–135.

YERKES, R. F., AND CAMPBELL, R. H., 1971, Cenozoic evolution of the Santa Monica Mountains-Los Angeles basin area: I. Constraints on tectonic models: Geol. Soc. America Abstracts with Programs, v. 3, p. 222.

YERKES, R. F., McCULLOH, T. H., SCHOELLHAMER, J. E., AND VEDDER, J. G., 1965, Geology of the Los Angeles basin, California—an introduction: U.S. Geol. Survey Prof. Paper 420-A, 57 p.

The American Association of Petroleum Geologists Bulletin
V. 57, No. 1 (January 1973), P. 74–96, 16 Figs., 1 Table

Basic Wrench Tectonics[1]

RONALD E. WILCOX,[2] T. P. HARDING,[3] and D. R. SEELY[4]

Houston, Texas 77001

Abstract *En échelon* structures which may trap oil and gas develop in a systematic pattern along wrench zones in sedimentary basins. Laboratory clay models simulate the formation of *en échelon* folds and faults caused by wrenching. Folds form early in the deformation and are accompanied or followed by conjugate strike-slip, reverse, or normal faulting. Deformation may cease at any stage or may continue until strike slip along the wrench zone produces a wrench fault and separation of the severed parts of early structures. Oblique movements of fault blocks on opposite sides of a wrench fault cause divergence or convergence and enhancement, respectively, of extensional or compressional structures. Basins form in areas of extension and are filled with sediment, whereas upthrust blocks emerge in areas of compression and become sediment sources. The combined effects of wrenching in a petroliferous basin are to increase its prospectiveness for major hydrocarbon reserves.

INTRODUCTION

Wrench faults (Kennedy, 1946; Anderson, 1951) are high-angle strike-slip faults of great linear extent along which strike-slip may be tens of miles or considerably more. Basement invariably is involved in the deformation and a wrench zone is a swath of terrane deformed by wrenching prior to and concurrently with strike-slip along the throughgoing wrench fault. The term "wrench fault" has no genetic connotation.

En échelon folds are the most important structures of potential value for trapping hydrocarbons in most wrench zones. They are also useful for recognition of wrench zones (Figs. 1-5). A single *en échelon* fold can be depicted as an ellipse (Fig. 6), which represents the deformation of a circle in the wrench zone, with the longer ellipse axis *(A-A')* parallel with the fold axis. Other structural straps can be formed by faulting or a combination of faulting and folding. Four types of fractures can form during wrench deformation, and if the wrenching continues, any one or all of these fractures can become faults. In Figure 6, the fracture directions are shown as *X-X'* (the strike of the primary wrench fault or wrench zone), *C-C'* and *D-D'* (*en échelon* conjugate shear joints or strike-slip faults), and *B-B'* (*en échelon* tension joints or normal faults). The development and interrelations of these faults and the *en échelon* folds are the main subjects of this paper. In a previous paper Moody and Hill

(1956) have treated aspects of wrench tectonics, particularly as these pertain to proposed systems and patterns of sets of wrench faults.

Prolific reserves of hydrocarbons have been trapped in wrench structures, mainly in *en échelon* folds and faulted folds. Some of the largest and best known of these structures are anticlinal traps in the Los Angeles basin (Fig. 4) and the west side of the San Joaquin Valley (Fig. 5), California (see also Harding, 1973, the following paper in this issue).

Clay models that illustrate the mechanics and development of this structural style represent broad basins filled with structurally homogeneous sediments whose total thickness is small compared with the size of the basin. The models also aid in prediction of traps by providing visual examples of the three-dimensional relations between structural elements in wrench zones.

MECHANICS OF WRENCHING

Wrench faults form in response to horizontal shear couples within the earth's crust, and they can be simulated in clay models by moving tin sheets beneath a clay cake (Cloos, 1955). Simple wrenching results from the movements of the crustal blocks or tin sheets in opposite directions parallel with their adjacent edges. As a consequence of such parallel displacements, compressional and tensional stresses are generated in the overlying sediments or clay. If, instead of moving exactly parallel with the wrench fault, the basement blocks or the tin sheets converge or diverge slightly, the compressional or tensional stresses, respectively, that result from the basic wrench are enhanced. These important special cases of convergent and divergent wrenching are discussed after analysis of the more general case of simple parallel wrenching.

[1] Manuscript received, February 18, 1972; accepted, April 17, 1972.

[2] Esso Production Research Co.; present address: P.O. Box 1230, Bellaire, Texas 77401.

[3] Humble Oil & Refining Co.

[4] Esso Production Research Co.

We are grateful to E. Cloos, consultant to Esso Production Research Co., for contributing helpful suggestions during the course of this work. Our appreciation also is extended to J. Crowell for stimulating discussions.

74

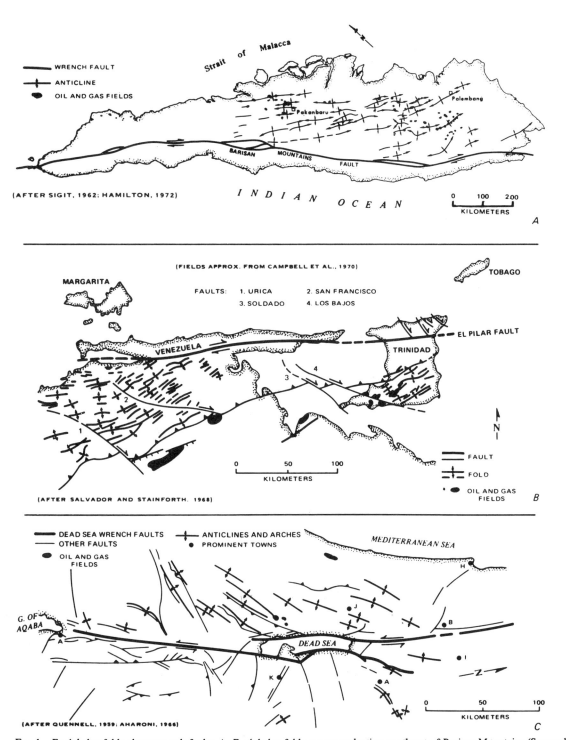

Fig. 1—*En échelon* folds along wrench faults. **A.** *En échelon* folds, some productive, northeast of Barisan Mountains (Semangko) fault in Central and South Sumatra basins, Sumatra. Oblique convergent subduction along adjacent Java trench is additional factor in deformation here. **B.** El Pilar fault and associated faults and *en échelon* folds in eastern Venezuela and Trinidad; note production from folds near Los Bajos fault, southwestern Trinidad. **C.** Dead Sea rift, Israel and Jordan; note location of Dead Sea between overlapping ends of major wrenches. Some *en échelon* folds are bounded by thrusts and several are marginally productive.

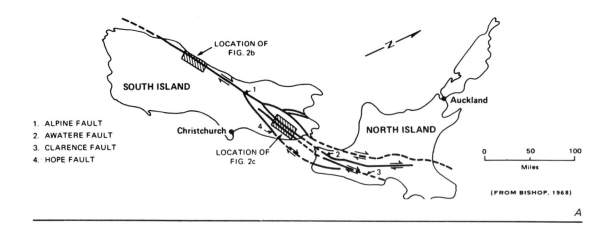

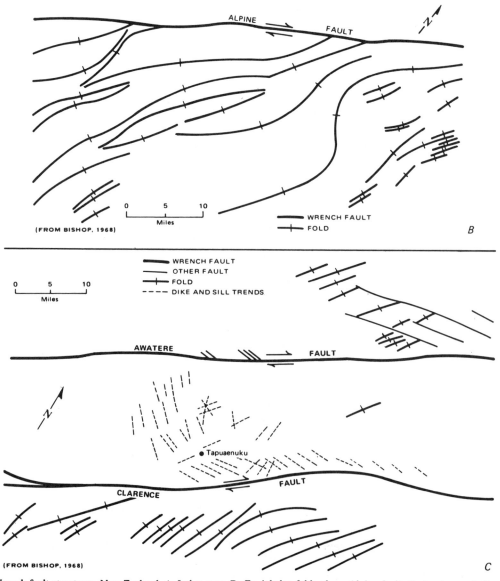

Fig. 2—Wrench-fault structures, New Zealand. **A.** Index map. **B.** *En échelon* folds along Alpine fault. **C.** Awatere and Clarence wrench faults and associated *en échelon* folds, dikes and sills, and subsidiary faults.

Simple Parallel Wrenching

Simple parallel wrenching is a special case of simple shear, which is one kind of finite homogeneous strain (Jaeger and Cook, 1969; Ramsay, 1967). The shear angle (ψ, Fig. 6) increases with increasing simple shear. In some crustal deformation and in clay models the initial deformations are plastic and involve folding. These are followed by a combination of plastic distortion and fracturing. As deformation proceeds, displacement along the wrench zone increases, and the zone of principal shear narrows. Finally, all of the slip occurs along a few closely spaced faults or along one throughgoing wrench fault, and subsequent deformations within either fault block are more or less independent of each other.

On a wrench model (Figs. 7, 8), it is convenient to mark the clay surface with a circle and to note how its shape changes during deformation. Points moving closer together mark compression, and points moving apart denote extension. The original circles (Fig. 7A) on the clay are aligned along the edge of the underlying tin sheet and deform into *en échelon* ellipses during the plastic phase of strain (Fig. 7B). Straight lines on the clay (Fig. 7A, normal to the line of circles) are trowel marks that become bent during deformation (Figs. 7B–C, 8D–F). Maximum compression and extension are parallel with the minor and major strain ellipse axes, respectively, and neither of these directions is parallel with or perpendicular to the shear direction imposed on the model, i.e., the strike of the wrench zone defined by the parallel edges of the tin sheets and the line of circles. It follows from the *en échelon* arrangement of ellipses (Fig. 7B) that all structures associated with each ellipse (Fig. 6) may be repeated along the wrench zone. This *en échelon* repetition of folds and faults is an important diagnostic feature of wrench zones (Figs. 1–5). (The size and spacing of circles/ellipses on the models is arbitrary; the spacing of folds and faults in the model wrench zones is determined by various characteristics of each model.)

The clay models of wrenching are all basically alike. The model in Figures 7 and 8 has left-lateral displacement, whereas the models in Figures 9 and 10 are right-lateral wrenches. (By convention, the sense of fault displacement is described by assuming that the block toward the observer is fixed, and the block across the wrench fault from the observer moves to his right or left.) Various structures form on each model, however, depending on the thickness and nature of the wet-clay cake, on the rate of deformation, on any special conditions built into the model, and to a certain degree, on chance. Included in the "chance" aspect that helps to determine the final model structures are, for example, slight inhomogeneities in the texture of the clay and the presence of hidden bubbles beneath the clay surface.

By analogy, the explorationist is faced with a host of unknown (chance) factors in interpreting wrench zones. Some of the more obvious factors are the effects of nonuniform stratigraphy (both thickness and composition), variable rates of deformation, and different directions of movement between crustal blocks during one stage of deformation or during succeeding stages. In spite of these inherent complexities in both nature and the models, however, the overall pattern of wrenching has key elements that are repeated, and the presence of any one or more structures of the basic pattern serves as a clue for recognizing this structural style and its associated prospective structures.

The structures of the basic wrench-tectonic patterns are *en échelon* folds, *en échelon* conjugate strike-slip faults, the main wrench fault or wrench-fault zone, and *en échelon* normal faults. These are described below and are illustrated in the models (Figs. 7–10).

En Echelon Folds

En échelon folds are the most attractive prospective structures in wrench zones because they form early and thus provide traps during early hydrocarbon migration, and because they commonly afford the largest closures that are genetically related to wrenching (Harding, 1973). As the amount of displacement on the wrench zone increases, the initial folds are broken first by fractures and then by faults. In later stages of wrenching the folds may become shattered (Fig. 9C), and parts of the folds on either side of the wrench fault may be offset (Fig. 10C). As movement of crustal blocks continues over long periods of geologic time, the half-folds on one block can be removed completely away from the area, and the wrench fault itself may provide updip closure.

The term *"en échelon"* refers to the arrangement of structures along a linear zone so that individual folds or faults of the same kind are parallel with each other and are inclined equally to the strike of the zone. The nomenclature for describing *en échelon* fold sets is similar to that for wrench displacements. Right-lateral wrenches produce right-handed fold sets (Fig. 11A), where a traverse along the axis of any fold to its terminus would turn right to reach the next fold in the *en échelon* set (Campbell, 1958). A left-handed

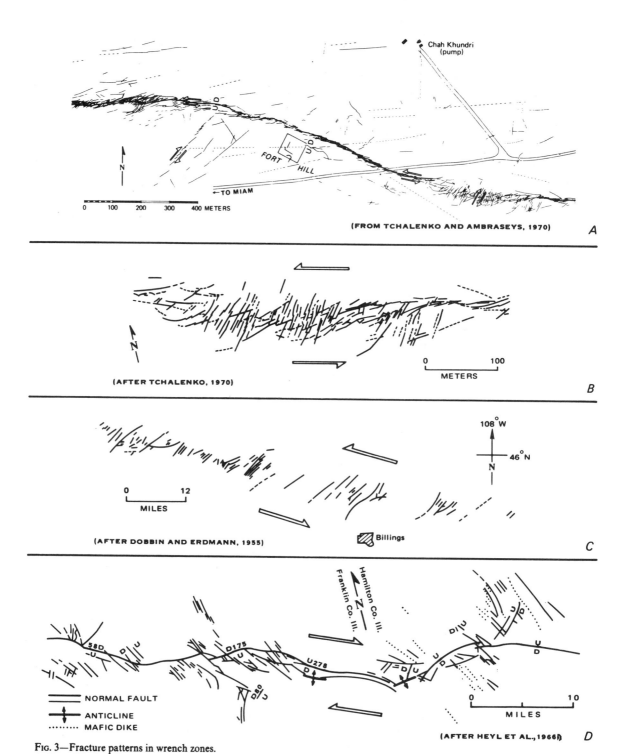

Fig. 3—Fracture patterns in wrench zones.
A. Part of Dasht-e Baÿaz (Iran) earthquake fracture zone along left-lateral wrench. Western part shows development of synthetic *en échelon* faults; eastern end shows both antithetic and synthetic *en échelon* faults. **B.** Synthetic and antithetic *en échelon* fractures (enlargement of east end of wrench zone in **A** above). **C.** Lake Basin fault zone, Montana, showing *en échelon* normal faults along indicated wrench zone. **D.** Cottage Grove fault zone, Illinois; note *en échelon* normal faults, parallel mafic dikes, and reversal of vertical separation sense on throughgoing strike-slip fault. Vertical components indicated in feet.

set of *en échelon* folds in eastern Panama (Fig. 11B) is probably related to a left-lateral wrench.

All *en échelon* folds in one zone are usually of similar shape and extent. The folds in Figure 9 are more distinct and more uniform than is usual for clay-wrench models, because a thin sheet of plastic film (0.0005-in. thick) was interlayered in the clay 0.25 in. below the surface. Several larger *en échelon* folds developed in the other two models (Figs. 7, 8, 10), which are homogeneous clay cakes without plastic film. The folds in Figures 7 and 8 are low and only faintly visible, whereas those in Figure 10 are larger. This difference probably is explained by the rates of deformation; the model with distinct folds (Fig. 10) was deformed 2.5 times faster than the other model (Figs. 7-8).

A close examination of Figure 9A reveals a small difference between the average fold trend and the trend of the longer axes of the ellipses. This difference probably ,is accentuated by the presence of the thin plastic sheet, which has influenced strongly the folding. Other similar experiments have shown that the fold size and fold spacing in the wrench zone are related to the depth of burial of the plastic film below the clay surface. Shallower plastic sheets produce smaller, more closely spaced folds. Another characteristic unique to models with plastic film layers is the rapidity of folding after slow deformation begins. In the extreme case of the plastic sheet directly on the clay surface, a very slight distortion by wrenching immediately causes folding in the plastic sheet and in the clay just below.

In models without the plastic sheet (*e.g.*, Fig. 10B), the longer ellipse axes are nearly parallel with the axes of the clay folds. This is similar to the ellipse diagram (Fig. 6), but the model ellipses are not so elongate as the ellipse in Figure 6.

For a true simple shear the angle between the fold axis (long axis of the ellipse) and the strike of the wrench zone is always less than 45°. For most wrench-fault experiments with clay, the angle between *en échelon* fold axes and the wrench fault approximates 30°. Folds that form later during the deformation have lower angles.

Fortunately, in the early stages of exploration in an area where wrenching is suspected, the recognition of several typical wrench-zone structures will serve to define the trend of the zone itself and probably also the sense of wrench displacement. By extrapolation from models, the axes of *en échelon* folds, which may be subtle, low-relief closures, should lie at an angle of 30° ± 15° to the wrench trend, either in a clockwise direction (left-handed folds) or in a counterclockwise direction (right-handed folds). If the

wrench-zone trend is known or suspected, and the displacement sense is unknown, folds still could be anticipated along the wrench trend with their axes inclined about 30° to that trend.

In nature (Figs. 1-5), fold orientations in a wrench zone can be different for several folds along the same fault trend. Some folds, or parts of folds with irregular axial trends, may parallel the wrench fault or cross the wrench zone at a low angle. Several factors that can influence the shape and trend of *en échelon* folds include convergence of blocks during wrenching, changes in strike of the wrench fault, large components of vertical displacement, differences in kind and thickness of sediments, and mobility of basement near the folds.

Conjugate Strike-Slip Faults

Wrenching causes two sets of intersecting, vertical fractures to form in a predictable orientation along the wrench zone. One set, the low-angle fractures (*C-C'*, Fig. 6), makes an angle between 10 and 30° with the wrench strike (*X-X'*), whereas the high-angle set (*D-D'*) intersects the wrench at an angle between 70 and 90°. These conjugate fractures can be either joints or faults, or both, depending on the magnitude of wrenching.

The acute angle of intersection of the two fracture sets is dependent on the nature of the rocks and the deformation; it is usually in the range of 60-70°. This angle is bisected by the direction of maximum compression (*B-B'*, Fig. 6). On the clay model in Figure 7C, one fracture of each set forms an "X" cutting the center small ellipse. The wedge in the acute angle of the intersection is displaced (Fig. 8D) toward the center of the ellipse as deformation continues. Two important aspects of the deformation are illustrated by this wedging: (1) the opposite senses of lateral displacement on the two intersecting strike-slip faults; and (2) contemporaneous plastic deformation and faulting.

The low-angle faults (Fig. 7C) intersect the wrench strike (line of ellipse centers) at 12° and have the *same* sense of displacement (left) as that of the main wrench zone (Figs. 7B-C, 8D-E) and the final wrench fault (Fig. 8F). These low-angle faults are called synthetic strike-slip faults, or simply synthetic faults. In contrast, the high-angle set of conjugate strike-slip faults has a displacement sense *opposite* that of the wrench; these are known as antithetic strike-slip faults, and they are right-lateral in this left-lateral wrench model. They form angles of 78° with the wrench and 66° with the synthetic fault in the center ellipse (Fig. 7C). The low- and high-angle

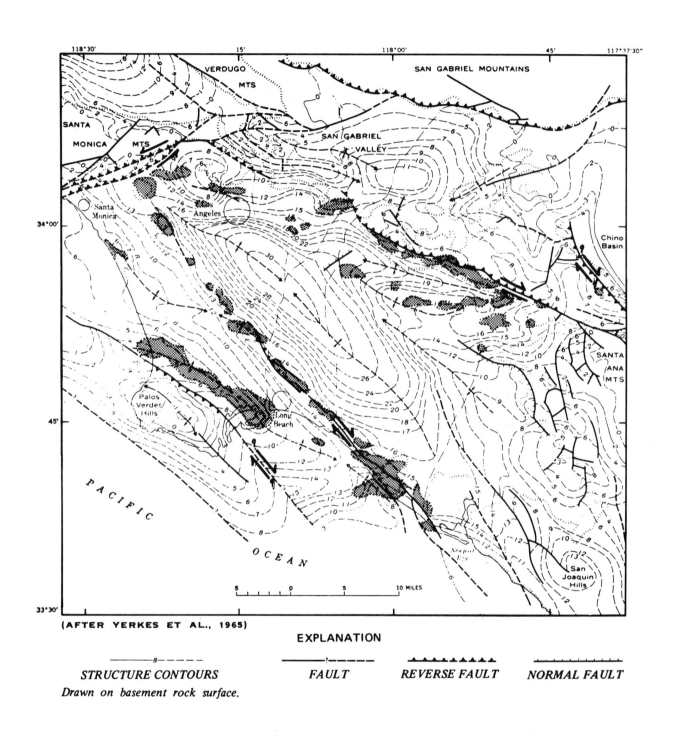

(AFTER YERKES ET AL., 1965)

EXPLANATION

————8————   ————?————   ▲▲▲▲▲▲▲▲   —┴——┴——┴—

STRUCTURE CONTOURS   *FAULT*   *REVERSE FAULT*   *NORMAL FAULT*

Drawn on basement rock surface.

ANTICLINE   *SYNCLINE*   *OIL FIELD*

Showing direction of plunge   *Showing direction of plunge*

Fig. 4—Major wrench structures and oil fields, Los Angeles basin, California.

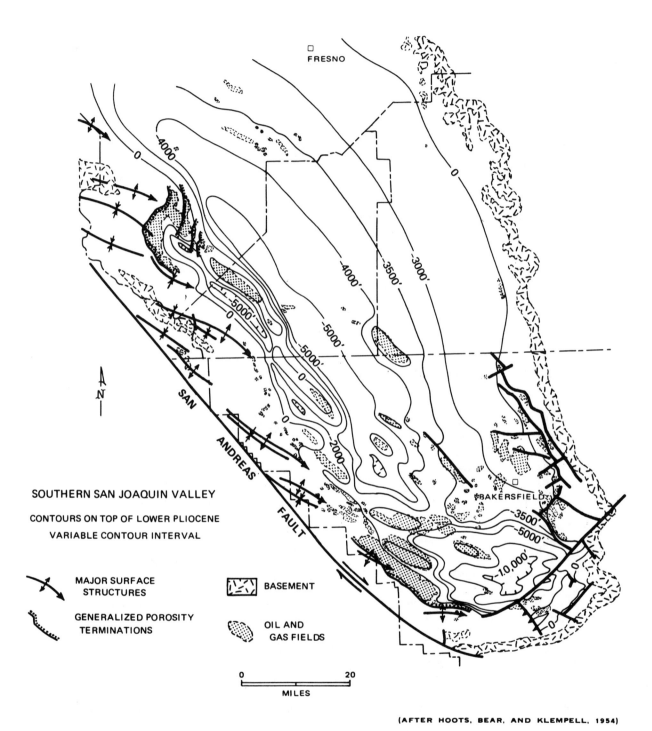

SOUTHERN SAN JOAQUIN VALLEY

CONTOURS ON TOP OF LOWER PLIOCENE
VARIABLE CONTOUR INTERVAL

MAJOR SURFACE
STRUCTURES

GENERALIZED POROSITY
TERMINATIONS

BASEMENT

OIL AND
GAS FIELDS

0   20

MILES

(AFTER HOOTS, BEAR, AND KLEMPELL, 1954)

Fig. 5—Major wrench structures and oil fields, San Joaquin Valley, California.

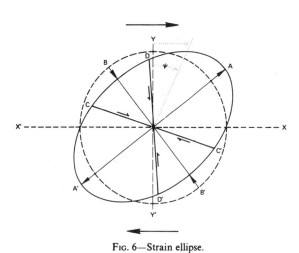

Fɪɢ. 6—Strain ellipse.

conjugate fractures have been termed Riedel shears and conjugate Riedel shears, respectively, by Tchalenko and Ambraseys (1970).

Continuing deformation after the conjugate fractures have developed proceeds as a combination of strike-slip faulting and plastic distortion. The acute angle between the two faults enlarges as the two faults rotate away from each other. The supplementary obtuse angles decrease as the larger wedges bounded by them move outward along the long-ellipse axis (A-A', Fig. 6), which marks the direction of extension (or minimum compression).

The rotation of the conjugate faults is an internal (local) rotation caused by compressive deformations and is not related uniquely to wrenching. The same conjugate fault pattern, wedging, and internal fault rotation are possible when rocks (or clay) are subjected to straight external compression, that is, when the compressive forces are opposed on a straight line (Ramsay, 1967, p. 60).

Wrenching, however, also produces external (regional) rotational deformation. The wrenching forces, which result from regional simple shear, act in opposite directions as if on separate, parallel lines so as to form a couple. The resulting deformation generally is restricted to a linear wrench zone parallel with the couple and to the edges of the moving crustal blocks. A left-lateral wrench has an external sense of rotation that is counterclockwise (Figs. 7, 8), whereas right-lateral wrenches have clockwise external rotation (Figs. 9, 10). This can be seen in the models by noting the rotation of the ellipse axes as wrenching proceeds.

The effects of both the internal rotation due to wedging and the external rotation due to wrenching further distinguish synthetic and antithetic faults. For a left-lateral wrench (Figs. 7, 8), external rotation tends to move the synthetic fault counterclockwise away from the wrench trend as the internal rotation tends to move the fault clockwise toward the main wrench. The result is little rotation of the synthetic fault in either direction. It originally formed nearly parallel with the strike of the main wrench zone and, therefore, remains in this favorable orientation to accommodate additional wrench displacements.

The antithetic faults, however, formed at a high angle to the wrench, and the continuing deformation cause both the external and the internal rotations to be counterclockwise (Figs. 7, 8). This tends to increase further the original high angle to around 90° to the wrench zone. As a consequence, lateral displacements on antithetic faults are generally small compared with those on either their synthetic counterparts or the main wrench fault. In some cases, the high-angle position of the antithetics is so poorly favored for displacements as to preclude their formation. In all the clay models (Figs. 7-10) synthethics are much better developed and account for much more wrench displacement than the antithetics.

The combined effects of external and internal rotation on the fault sets are compared in Table 1 for the left-lateral wrench model (Figs. 7, 8). Note a second set of conjugate shears, nearly parallel with the first set, cutting the center ellipse (Fig. 8D-F).

A useful clue to interpretation is provided by the antithetic faults that have been rotated. Their original planar attitude becomes bent by the combined internal and external rotations acting in opposite directions on either side of the wrench zone. The map view of the twisted faults is a flat S with the arcs of the S pointing toward the direction of displacement, i.e., S for left wrenches (Fig. 8D-F), and a reverse S for right wrenches (Fig. 10C).

Wrench Faults

The development of the main, throughgoing wrench fault is the last stage in the early phase of wrench-zone deformation. The entire early phase of wrenching usually constitutes a brief and tran-

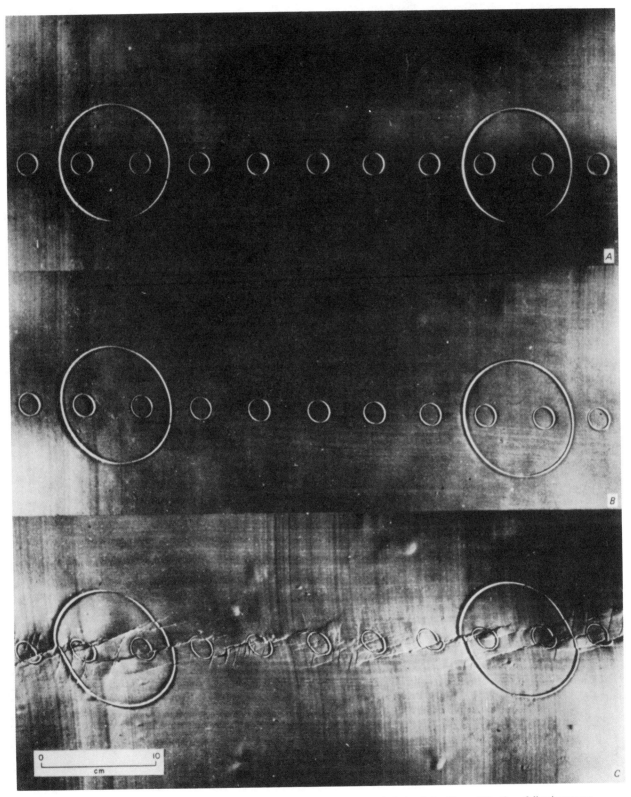

FIG. 7—Clay model of parallel left-lateral wrench fault (**A–C** = three stages, vertical views). See Figure 8 for three following stages.

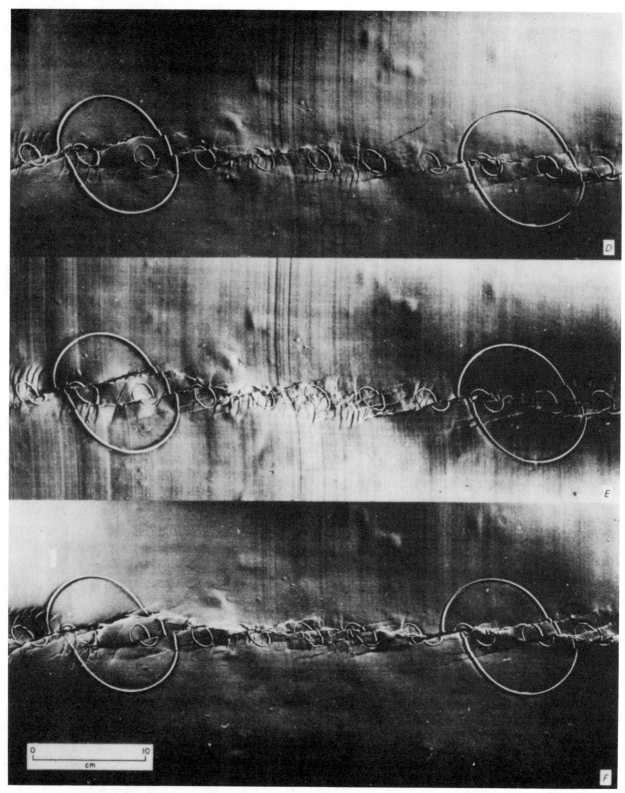

FIG. 8—Clay model of parallel left-lateral wrench fault (**D-F** = three stages, vertical views). See Figure 7 for first three stages.

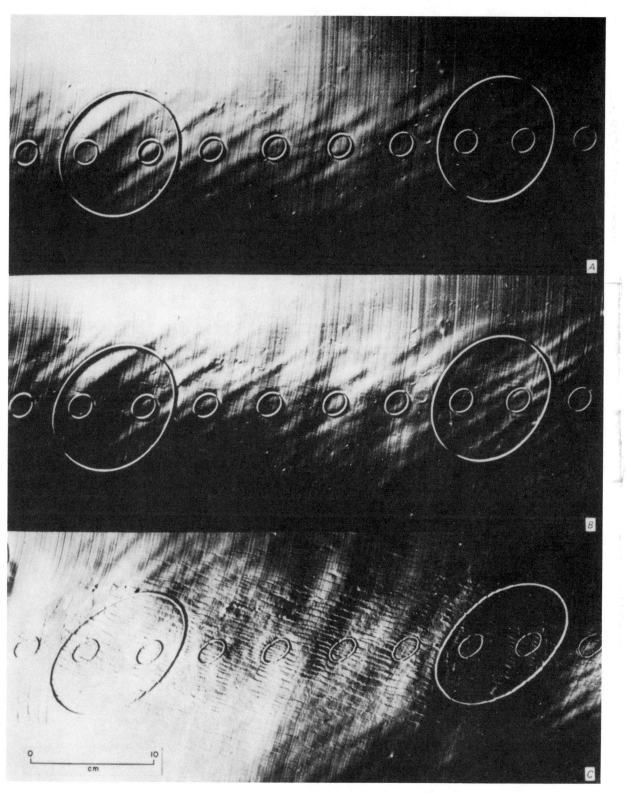

FIG. 9—Clay model of parallel right-lateral wrench fault with layer of thin plastic film embedded 0.25 in. below surface to enhance *en échelon* folds (**A-C** = three stages, vertical views).

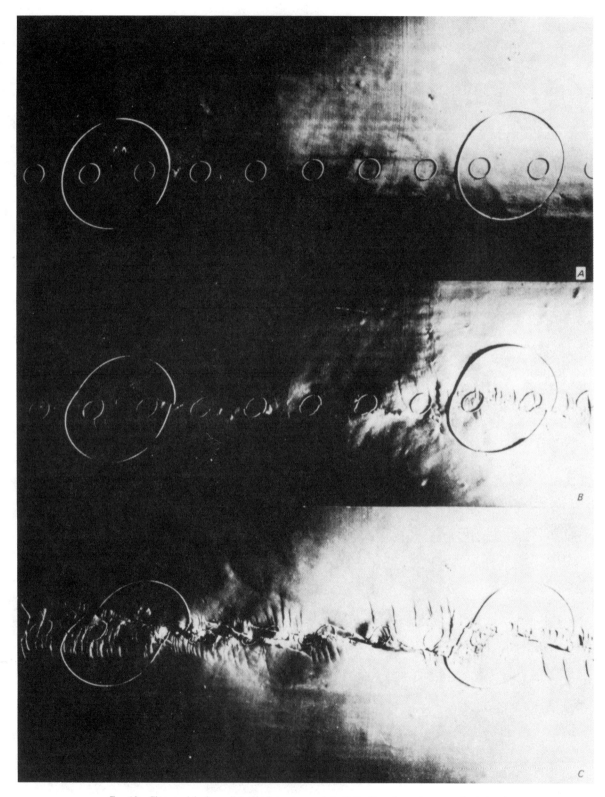

FIG. 10—Clay model of parallel right-lateral wrench fault (**A–C** = three stages, vertical views).

sitory period in the long history of a major wrench fault, but this early phase is of great importance in the process of hydrocarbon-trap formation.

After a short interval of concurrent folding and conjugate faulting, the rocks (or clay) fracture in a relatively narrow zone within the overall deformational swath, and the master wrench fault is created. This process of rock failure begins at several points along the wrench zone (e.g., see Fig. 8E, between small circles 4 and 5, 7 and 8, 9 and 10). At some locations a synthetic fault deviates into the incipient wrench-fault trend, and at others a new fracture forms more nearly parallel with the strike of the wrench zone and at a small angle to the nearby synthetic faults. As this process continues, the main wrench fault gradually emerges as an interconnected series of these earlier fractures. (The plastic film prevented the formation of the single wrench fault in the model in Fig. 9.)

A great variety of fault blocks is produced within the wrench zone. Some large blocks are caught between early formed branches of the main wrench (Fig. 8F, near large ellipse at left), and many smaller blocks are sliced and deformed into horsts and grabens between the main wrench fault and the conjugate faults. Once individual fault blocks are separated by faulting, they tend to deform somewhat independently; some rise, some sink, some are folded, and some are faulted again.

As displacement on the main wrench fault increases, slip diminishes on the other faults in the zone. The active fault "plane," or a relatively thin, crush zone along the active part of the fault, commonly shifts from side to side of the wrench zone. Distortion and faulting of the whole zone become complex, and this results in a braided fault pattern that is typical of major wrench zones (Fig. 12C).

Changes in the strike of the active fault lead to additional deformation of the wall rocks as strike slip continues. The parallel wrench becomes a convergent or a divergent wrench, at least locally. The size and extent of the resulting compressional or extensional structures depend on the amount of change in fault strike and the amount of displacement along the curved fault surface within the braided system (see Fig. 12 and accompanying text discussion of convergent and divergent wrenching).

Tension Fractures

The orientation of tension joints or normal faults parallels the short axis of the strain ellipse (Fig. 6, B-B'), crosses the en échelon fold axes at right angles, and bisects the acute angle between the conjugate shears. En échelon tension fractures may form along a wrench zone in the initial stage of deformation, but they easily are destroyed as wrench displacement increases and compressive structures (folds and conjugate faults) become more prominent. In clay models of wrenching, tension fractures are uncommon because of the strong cohesion within the clay. Water placed on the clay surface eliminates this cohesion, and large, open, en échelon tension cracks form to the exclusion of other fractures and folds.

Two examples of en échelon normal faults that are presumed to lie above buried wrench faults are the Lake Basin fault zone, Montana (Fig. 3C), and the Cottage Grove fault zone, Illinois (Fig. 3D). In both these fault zones, the amount of wrench movement of the basement blocks after sedimentation has been small—just enough to fracture the overlying sedimentary rocks without causing significant lateral offset. Additional linear zones which may represent wrenching have been recognized near the Lake Basin zone (Smith, 1965).

The Cottage Grove zone displays two other features of wrenching. The northern block of the main east-west fault is downthrown in the western part of the zone and upthrown in the eastern part. This kind of change in the vertical displacement sense along strike is typical of wrenches. The tensional component of wrenching is marked in the eastern area around the fault zone by mafic dikes. Such intrusions and vein fillings in tension fractures are well known in mineral deposits and plutonic terranes, and they fit the fracture pattern for wrenching along this zone.

Antithetic fractures inherit some of the tensional component of a wrench deformation and commonly become nearly vertical normal faults with negligible lateral displacements. A downward displacement on either of the conjugate strike-slip faults tends to be toward the acute wedge. This is well shown on one of the models (Fig. 10C), where there are many closely spaced antithetic faults at both ends of the wrench zone. Such concentrations of "antithetic-normal" faults impart a pseudoplasticity to the clay (or rocks) that permits these zones to deform more or less uniformly without being cut by one main wrench fault. Thus, a wrench fault with measurable strike slip can pass into one of these fracture zones along its strike where there is the same regional shift across the zone but no single fault of large lateral displacement.

CONVERGENT AND DIVERGENT WRENCHING

Opposed crustal blocks that do not move par-

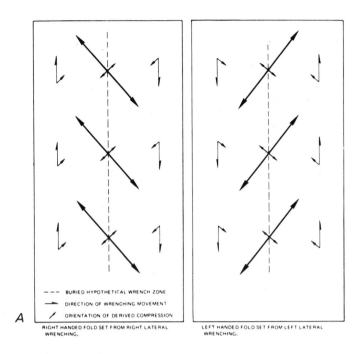

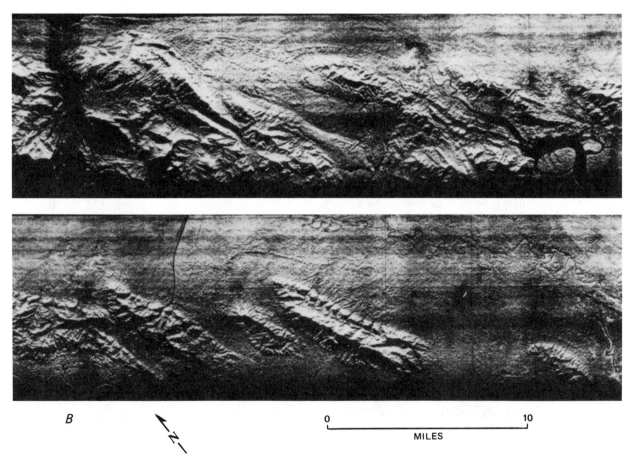

FIG. 11—*En échelon* folds. **A.** Diagrammatic right- and left-handed fold sets (caused by right- and left-lateral wrenching, respectively). **B.** Radar image of surface folds, Darien basin, eastern Panama, caused by left-lateral wrenching. (Imagery acquired by Westinghouse Electric Corporation, under contract from U. S. Army TOPOCOM, Ft. Belvoir, Virginia.)

Table 1. Orientation of Conjugate Fractures
Cutting Center of Small Ellipse

Figure	Approximate Shear Angle*	Angle Between Wrench Strike, and Synthetic Fault	Angle Between Wrench Strike, and Antithetic Fault	Angle Between Synthetic and Antithetic Faults
7c	20°	12°	78°	66°
8d	28°	12°	82°	70°
8e	36°	14°	87°	73°
8f	48°	15°	93°	78°

*See figure 6.

allel with a wrench fault either converge or diverge as wrenching proceeds. These oblique movements may be related to nonparallel displacements of crustal blocks on a regional scale, or they may be due to local changes in strike of a generally parallel wrench. It is common for both convergence and divergence to develop locally along a wrench. Convergent wrenching, on whatever scale, tends to enhance compressive wrench-zone structures, namely, folds and conjugate strike-slip faults, and strong convergence can cause reverse faulting and thrusting. The formation of tensional structures, mainly normal faults, is typical of divergent wrenching.

A particularly good example of both convergence and divergence is seen north of Los Angeles, California, in the San Andreas wrench system (Figs. 12, 15). A pod-shaped block, which is about 100 mi long and 20 mi wide, lies southwest of the San Andreas fault and northeast of the curving San Gabriel fault (Fig. 12A). Both faults are well-documented, right-lateral wrenches.

The pod-shaped block has moved southeast along the curved San Gabriel fault and has caused convergence on its southern and southeastern margin. Reverse faults with strike-slip components characterize this margin and attest to the lateral wrenching combined with compression and high-angle thrusting.

Concurrently, the northwestern part of the pod was under tension as it diverged from the curving northern end of the San Gabriel fault, and the Ridge basin was formed (Fig. 12A). Sediments filled this basin as faulting continued, and they record the fault movements by preserving several unique rock types whose source areas were displaced alongside the basin (Fig. 12B).

One such suite of gneissic rocks is preserved as coarse blocks in the Violin Breccia (Fig. 12A, B), which accumulated along the northeast side of the San Gabriel fault scarp as wrenching continued from the late Miocene to the late Pliocene (Crowell, 1954a, b). The Ridge basin illustrates

how major wrench faults can influence basin development and sedimentation as well as the tectonic history and structural style of a region.

En échelon folds in a clay model are enhanced by even a slight convergence of only 2° (Fig. 13). In the early stage of movement, the folds are well developed throughout most of the central part of the model (Fig. 13A), and a few synthetic fractures have formed. At a later stage (Fig. 13B), the folds have been offset along the synthetic faults and the incipient throughgoing wrench. A few antithetic faults also formed, but their importance in this deformation was minimal.

A more intensive *en échelon* zone of compression develops along a model wrench with a convergence of 15° (Fig. 14A). Good *en échelon* folds form in the narrow zone that later is uplifted, and both sets of conjugate shears are well developed. Nearly all wrench displacement is concentrated on the synthetic faults, along which the fold axes are offset. A side view of the same model (Fig. 14B) reveals the complex thrusting of the wedges squeezed up and out of the wrench zone by the strong convergence. As these blocks rose, they were bounded by vertical or high-angle reverse synthetic faults, and they resemble upthrust blocks.

Just south of the San Gabriel fault in the Little Tujunga Canyon area, upthrusts out of the San Gabriel fault zone are exposed (Fig. 15). (Reverse faulting in this area accompanied the San Fernando earthquake of February 9, 1971; see Palmer and Henyey, 1971.)

Layered-sand models (Emmons, 1969) are also instructive in studying the cross-sectional characteristics of wrench faults. The fault zone widens as the wrench fault splays upward, and individual faults have normal or reverse dip-slip separation, depending on how adjacent fault blocks are displaced within the wrench zone (Fig. 16).

An important result of divergent wrenching is an overlay of extensional block faulting on the

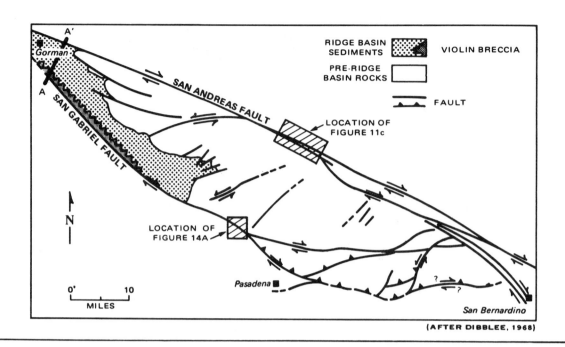

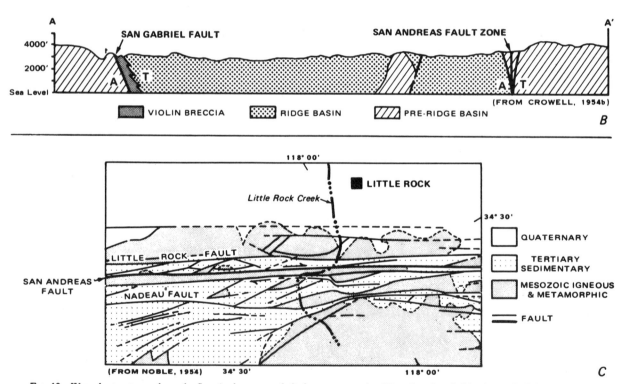

Fig. 12—Wrench structures along the San Andreas wrench-fault system, north of Los Angeles, California. **A.** Pod-shaped major slice between San Andreas and San Gabriel wrench faults. **B.** Cross section of Ridge basin, formed and filled with sediments in the northern part of "pod" during wrenching. **C.** Braiding of faults along San Andreas wrench-fault zone on northeastern side of "pod"; note right-lateral shift of Little Rock Creek and tilted fault blocks, evidenced by varied outcrop pattern.

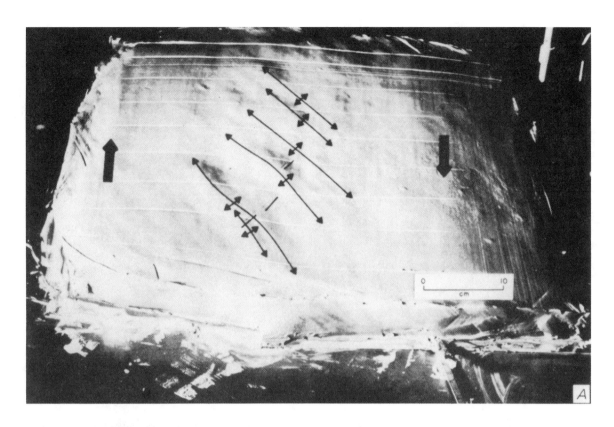

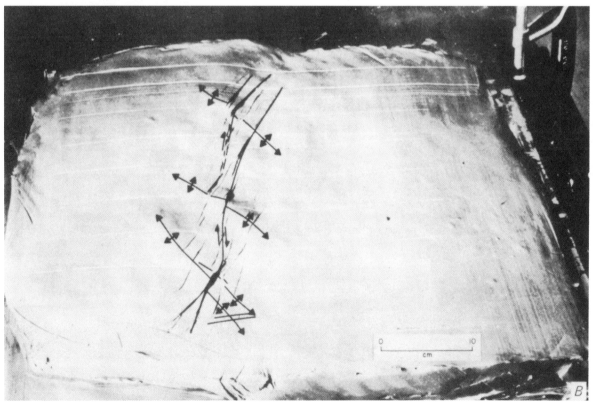

Fig. 13—Clay model of 2°-convergent right-lateral wrench fault (**A, B** = two stages, oblique views). (From unpublished work by P. G. Temple.)

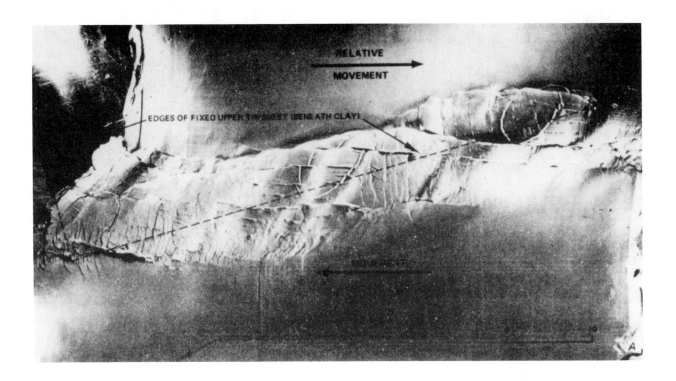

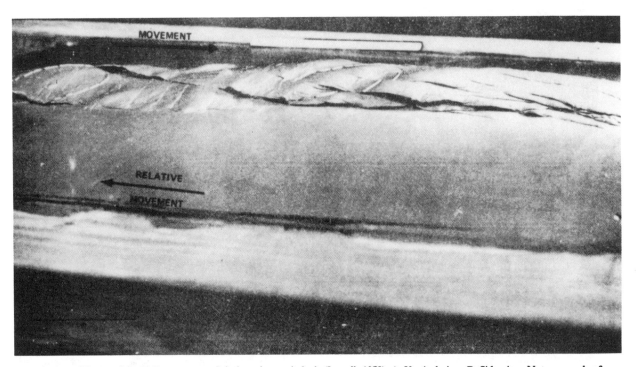

Fig. 14—Clay model of 15°-convergent right-lateral wrench fault (Lowell, 1972). A. Vertical view. B. Side view. Note reversals of vertical separation on synthetic faults in foreground and dominant strike-slip offset of fold axes.

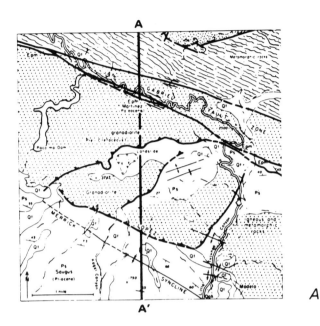

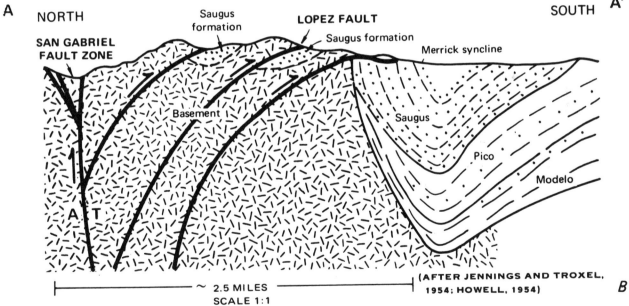

Fig. 15—Upthrust structures caused by wrenching. **A.** Map of upthrusts (high-angle reverse faults) along San Gabriel fault zone, Little Tujunga Canyon area, north of Los Angeles, California. **B.** Cross section of upthrusts, Little Tujunga Canyon area.

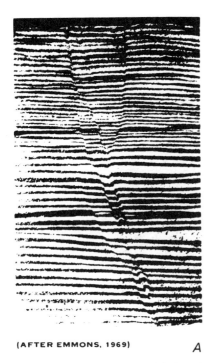
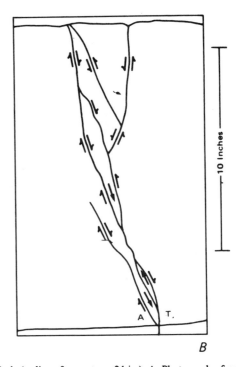

A   *B*

FIG. 16—Layered-sand model of a curved, right-lateral wrench fault (radius of curvature, 24 in.). **A.** Photograph of cross section through center of model (15 in. high and 11 in. wide). **B.** Line drawing of faults in model; right-lateral wrench movement shown by *A* (away) and *T* (toward).

simple wrench pattern (Fig. 17A). Grabens form in preference to horsts, and nearly all fractures have a tendency to develop into high-angle normal faults with oblique slip. *En échelon* folds are poorly developed and have low relief along divergent wrenches, but warping of fault blocks to produce closures between the faults is possible.

The Fitzroy trough in northwestern Australia (Fig. 17B) is probably a divergent wrench graben. It appears that wrenching formed the trough, which filled with sediments, and a final episode of minor wrenching deformed the basin fill. *En échelon* folds in the trough and a zone of *en échelon* normal faults in the adjoining but shallower Northeast Canning basin are properly oriented for the inferred right-lateral wrench zone along the trend of the trough (Rattigan, 1967; Smith, 1968).

CONCLUSIONS

Large quantities of oil and gas are trapped in structures caused by wrenching or influenced by some aspect of wrench tectonics. Knowledge of the wrenching structural style is especially useful in exploration because the basic structural patterns of wrenching are simple and consistent and are well documented from many areas. The struc-

tures and structural traps to be expected in a wrench terrane generally can be predicted with a high degree of confidence.

The principal elements of the basic wrench pattern are (1) *en échelon* folds inclined at a low angle to the wrench zone; (2) conjugate strike-slip faults, including synthetic faults inclined at a low angle to the wrench zone but in the opposite direction from the folds, and antithetic faults nearly perpendicular to the wrench zone; (3) the main wrench fault, parallel or subparallel with the wrench zone; and (4) normal faults or tension joints oriented perpendicular to the fold axes. Any combination of these structures may form within a given wrench zone, and the recognition of any one or a combination of them usually will serve to define the trend and displacement sense of the wrench zone.

Three general styles of wrenching are recognized: (1) simple parallel wrenching, in which crustal blocks move parallel with the wrench fault; (2) convergent wrenching, caused by blocks moving obliquely toward the wrench; and (3) divergent wrenching, resulting from oblique movements of the blocks away from the wrench. All three styles develop on both local and regional scales.

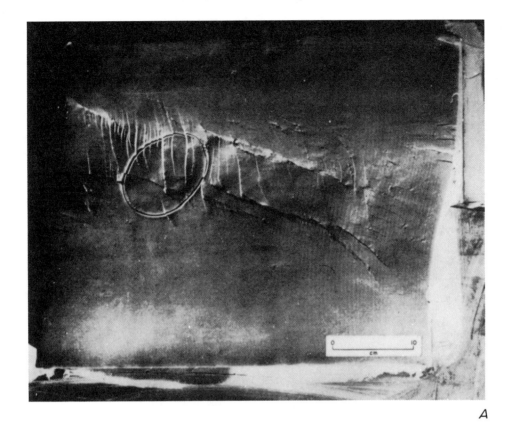

A

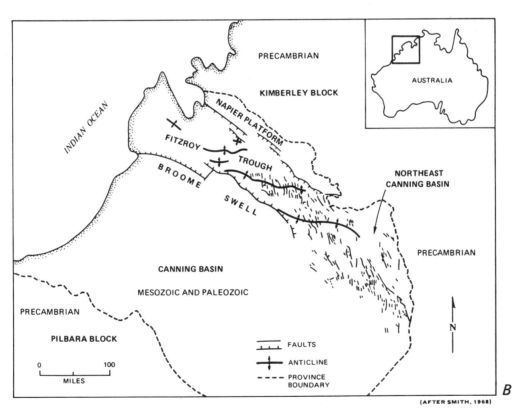

B

(AFTER SMITH, 1968)

FIG. 17—Divergent wrenching. **A.** Clay model of 15°-divergent right-lateral wrench fault. **B.** *En échelon* folds and faults in the Fitzroy trough, western Australia.

REFERENCES CITED

Aharoni, E., 1966, Oil and gas prospects of Kurnub Group (Lower Cretaceous) in southern Israel: Am. Assoc. Petroleum Geologists Bull., v. 50, no. 11, p. 2388-2403.

Anderson, E. M., 1951, The dynamics of faulting and dyke formation, with applications to Britain, 2d ed.: Edinburgh, Oliver and Boyd, 206 p.

Bishop, D. G., 1968, The geometric relationships of structural features associated with major strike-slip faults in New Zealand: New Zealand Jour. Geology and Geophysics, v. 11, no. 2, p. 405-417.

Campbell, J. D., 1958, En échelon folding: Econ. Geology, v. 53, no. 4, p. 448-472.

Cloos, E., 1955, Experimental analysis of fracture patterns: Geol. Soc. America Bull., v. 66, no. 3, p. 241-256.

Crowell, J. C., 1954a, Strike-slip displacement of the San Gabriel fault, southern California, pt. 6, Chap. 4, in R. H. Jahns, ed., Geology of southern California: California Div. Mines Bull. 170, p. 49-52.

——— 1954b, Geology of the Ridge basin area, Los Angeles and Ventura Counties, in R. H. Jahns, ed., Geology of southern California: California Div. Mines Bull. 170, Map Sheet 7.

Dibblee, T. W., Jr., 1968, Displacements on the San Andreas fault system in the San Gabriel, San Bernardino, and San Jacinto Mountains, southern California, in W. R. Dickinson and A. Grantz, eds., Proceedings of conference on geologic problems, San Andreas fault system: Stanford Univ. Pubs. Geol. Sci., v. 11, p. 260-278.

Dobbin, C. E., and C. E. Erdmann, 1955, Structure contour map of the Montana plains: U.S. Geol. Survey Oil and Gas Inv. Map OM 178A, scale 1:500,000.

Emmons, R. C., 1969, Strike-slip rupture patterns in sand models: Tectonophysics, v. 7, no. 1, p. 71-87.

Hamilton, W., 1972, Preliminary tectonic map of the Indonesian region, scale 1:500,000: U.S. Geol. Survey Open File Rept.

Harding, T. P., 1973, The Newport-Inglewood trend, California—an example of wrenching style of deformation: Am. Assoc. Petroleum Geologists Bull., v. 57, no. 1 (in press).

Heyl, A. V., M. R. Brock, J. L. Jolly, and C. E. Wells, 1966, Regional structure of the southeast Missouri and Illinois-Kentucky mineral districts: U.S. Geol. Survey Bull. 1202-B, p. 1-20.

Hoots, H. W., T. L. Bear, and W. D. Kleinpell, 1954, Geological summary of the San Joaquin Valley, California, pt. 8, Chap. 2, in R. H. Jahns, ed., Geology of southern California: California Div. Mines Bull. 170, p. 113-129.

Howell, B. F., Jr., 1954, Geology of the Little Tujunga area, Los Angeles County, in R. H. Jahns, ed., Geology of southern California: California Div. Mines Bull. 170, Map Sheet 10.

Jaeger, J. C., and N. G. W. Cook, 1969, Fundamentals of rock mechanics: London, Methuen and Co. Ltd., 513 p.

Jennings, C. W., and B. W. Troxel, 1954, Geologic guide through the Ventura basin and adjacent areas, southern California, in R. H. Jahns, ed., Geology of southern California: California Div. Mines Bull. 170, Geologic Guide No. 2, 63 p. (San Gabriel Mountains Section, p. 15-19).

Kennedy, W. Q., 1946, The Great Glen fault: Geol. Soc. London Quart. Jour., v. 102, pt. 1, p. 41-76.

Lowell, J. D., 1972, Spitsbergen Tertiary orogenic belt and the Spitsbergen fracture zone: Geol. Soc. America Bull., v. 83, (in press).

Moody, J. D., and M. J. Hill, 1956, Wrench-fault tectonics: Geol. Soc. America Bull., v. 67, no. 9, p. 1207-1246.

Noble, L. F., 1954, The San Andreas fault zone from Soledad Pass to Cajon Pass, California, pt. 5, Chap. 4, in R. H. Jahns, ed., Geology of southern California: California Div. Mines Bull. 170, p. 37-48.

Palmer, D. F., and T. L. Henyey, 1971, San Fernando earthquake of 9 February 1971: pattern of faulting: Science, v. 172, no. 3984, p. 712-715.

Quennell, A. M., 1959, Tectonics of the Dead Sea rift: Asociación de Servicios Geológicos Africanos 20th Internat. Geol. Cong., México, D.F., 1956, Actas y Tr., p. 385-405.

Ramsay, J. G., 1967, Folding and fracturing of rocks: New York, McGraw-Hill, 568 p.

Rattigan, J. H., 1967, Fold and fracture patterns resulting from basement wrenching in the Fitzroy depression, Western Australia: Australasian Inst. Mining and Metallurgy Proc., no. 223, p. 17-22.

Salvador, A., and R. M. Stainforth, 1968, Clues in Venezuela to the geology of Trinidad, and vice versa: 4th Caribbean Geol. Conf. Trans., 1965, p. 31-40.

Sigit, Soetarjo, 1962, Geologic map of Indonesia, scale 1:2,000,000: U.S. Geol. Survey, Misc. Geol. Inv. Map I-414.

Smith, J. G., 1965, Fundamental transcurrent faulting in northern Rocky Mountains: Am. Assoc. Petroleum Geologists Bull., v. 49, no. 9, p. 1398-1409.

——— 1968, Tectonics of the Fitzroy wrench trough, Western Australia: Am. Jour. Sci., v. 266, no. 9, p. 766-776.

Tchalenko, J. S., 1970, Similarities between shear zones of different magnitudes: Geol. Soc. America Bull., v. 81, no. 6, p. 1625-1640.

——— and N. N. Ambraseys, 1970, Structural analysis of the Dasht-e Baȳaz (Iran) earthquake fractures: Geol. Soc. America Bull., v. 81, no. 1, p. 41-60.

Yerkes, R. F., T. H. McCulloch, J. E. Schoellhamer, and J. G. Vedder, 1965, Geology of the Los Angeles basin, California, an introduction: U.S. Geol. Survey Prof. Paper 420-A, 57 p.

Reprinted by permission of the International Association of
Sedimentologists from P. F. Ballance and H. G. Reading,
eds., *Sedimentation in Oblique-Slip Mobile Zones,* IAS Special Publication 4 (1980), p. 27-41.

Spec. Publ. int. Ass. Sediment. (1980) **4,** 27–41

Analysis of pull-apart basin development produced by *en echelon* strike-slip faults

DONALD A. RODGERS

*Cities Service Company, Energy Resources Group,
Exploration and Production Research Laboratory, Box* 3908,
Tulsa, Oklahoma 74102, *U.S.A.*

ABSTRACT

Mathematical models are used to study the fault patterns and shapes of pull-
apart basins produced by right-stepping *en echelon* right-lateral strike-slip faults.
The basin shapes and fault patterns are controlled by: (1) the amount of overlap
between the faults; (2) the amount of separation between the faults; and (3)
whether or not the tops of the faults intersect the ground surface. Secondary
strike-slip faults will develop in some part of any pull-apart basin, but secondary
normal faults develop only when the tops of the master faults are near the ground
surface. The models suggest that, as pull-apart basins evolve, faults in the basement
may change their sense of movement. Fault patterns in the basement may be quite
different from those in the sedimentary fill of the basin, but as the basin evolves
the faults in the centre of the basin will tend to become strike-slip both in the basement and in the sedimentary fill.

INTRODUCTION

Basins associated with major strike-slip fault zones such as the San Andreas fault
zone in California are generally elongate parallel to the strike of the fault zone and
can be filled with substantial accumulations of sediment. The size of these basins
ranges from that of a sag pond a few hundred metres long and filled with a few metres
of sediments to the Salton Trough in southern California which is about 200 km long,
80 km wide, and filled with up to 10 km of sediments.

Wilcox, Harding & Seely (1973), Harding (1976), Norris, Carter & Turnbull
(1978) and Blake *et al.* (1978) discuss the structural development associated with a
single strike-slip fault but do not consider in detail the problem of *en echelon* faults.
Burchfiel & Stewart (1966) and Crowell (1974) discuss qualitatively the effects of
en echelon faults. As Fig. 1 suggests, left-stepping right-lateral strike-slip faults produce
a zone of compression and uplift between the two faults, and right-stepping right-
lateral strike-slip faults produce a zone of tension and depression between the two

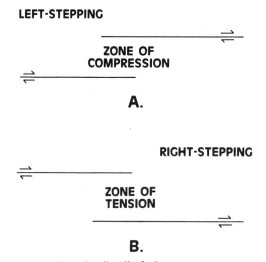

Fig. 1. Terms for *en echelon* right-lateral strike-slip faults.

A. Left-stepping (looking along one *en echelon* fault towards the adjacent fault, the offset is to the left); produces compression between faults.

B. Right-stepping; produces tension between faults.

faults. The zone of tension and depression has been called a pull-apart basin by Burchfiel & Stewart (1966) and Crowell (1974).

Pull-apart basins have been postulated along the Hope fault in New Zealand (Clayton, 1966; Freund, 1971), along the Levant fault zone (Dead Sea Rift) in Israel (Quennell, 1959; Freund, 1965), in Death Valley in California (Hill & Troxel, 1966), and in Norton Sound offshore of western Alaska (Fisher *et al.*, 1979). The crustal thinning, high heat flow, and potentially thick sedimentary accumulations within pull-apart basins make the basins prospective targets for geothermal and hydrocarbon exploration. Thus, it would be useful to study in some detail the structural and sedimentological development of pull-apart basins.

Crowell (1974) presents a qualitative model of the structures within a pull-apart basin (Fig. 2). The model is generalized from field studies of pull-apart basins in California.

In this paper mathematical models of pull-apart basins of various configurations are used to suggest the structures which might develop within the basins. The suggested models will be compared to Crowell's model and field examples, and the effects of the predictions on sedimentation will be discussed.

MODELS

The mathematical models are based on elastic dislocation theory (Chinnery, 1961, 1963). It has been shown that fault models based on this theory produce surface deformations which are similar to those deformations observed after earthquakes (Savage & Hastie, 1969; Alewine, 1974). The fault model is produced by making a cut within a body and moving one side of the cut relative to the other, then welding the sides of the cut together. This results in a discontinuity in the displacement field

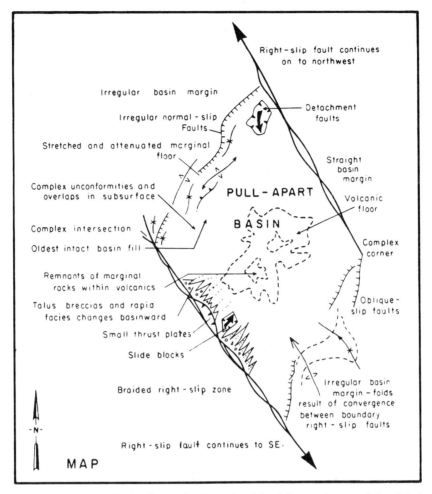

Fig. 2. Sketch map of idealized pull-apart basin produced by right-stepping *en echelon* right-lateral strike-slip faults (from Crowell, 1974, fig. 3).

across the cut. The body containing the deformed cut will be strained, and it is relatively easy mathematically to obtain the displacements, strains and stresses produced by the cut.

The equations of Chinnery (1961, 1963) can be used to calculate the elastic displacement, strain, and stress fields in an isotropic, homogeneous, linearly elastic half-space which are produced by a rectangular displacement discontinuity (or fault) in the half-space. The earth, of course, is not an isotropic, homogeneous, linearly elastic half-space. However, the equations of Chinnery (1961, 1963) are algebraically fairly simple and are fully integrated. Thus, many models can be generated in a small amount of time on any computer. By using a model that is physically simple, we are able to investigate a large number of geologically interesting situations. However, we must always keep in mind the limitations of the model and not attempt to make too detailed predictions from a first-order theory.

In Chinnery's model the offset across the fault is constant over the whole fault

30 *Donald A. Rodgers*

and falls instantaneously to zero at the edges of the fault. This, of course, produces a singularity around the edges of the fault. Chinnery & Petrak (1968) have shown that eliminating the singularity with an offset that goes to zero in a finite distance will smooth out the elastic fields produced by the constant offset model but will not affect in any important way the conclusions drawn from the constant offset model. Using a non-constant fault offset introduces complexities into the solutions which require numerical integration of the solutions. Thus, the constant offset model will be used in this paper in order to run a large number of models.

Finally, the equations of Chinnery (1961, 1963) are based on infinitesimal strain theory and thus are strictly valid only when the offsets on the faults are small compared to the fault dimensions. Infinitesimal strain models can describe reasonably well finite strain field problems (Johnson, 1970; Couples & Stearns, 1978; Reches & Johnson, 1978) and laboratory models (Sanford, 1959; Reches & Johnson, 1978). Rodgers (1979) has shown that a dislocation model of the San Andreas fault in southern California can explain the Holocene and Quaternary fault pattern in the Transverse Ranges. Thus it seems reasonable to suggest that the infinitesimal strain theory in this paper can be used to draw valid conclusions about large-scale deformations.

CALCULATED RESULTS

The postulated examples of pull-apart basins are not well documented in terms of the structures within the basin. This is not unexpected, for the thick sedimentary fills of the basins obscure the structural pattern. However, the shapes of the basins are known, and it is possible to compare the calculated and observed basin shapes. In this section the vertical displacement field of the ground surface will first be presented, as calculated from several models, and then the predicted fault patterns determined from the stress fields calculated from the same models. The models are chosen to show the effects of varying the amount of overlap and separation (Fig. 3).

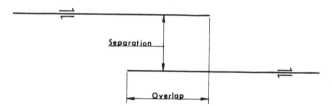

Fig. 3. Definition of separation and overlap for *en echelon* faults.

In order to look at the simplest case, the faults extend to infinity at depth and along strike away from the zone of overlap. This is a theoretical convenience which will not appreciably affect the results in the upper 10–15 km of the crust. In all the models, the master *en echelon* strike-slip faults have equal right-lateral offsets of 1 m. Changing the value of the offsets on the faults affects the magnitude of the calculated displacement and stress fields but does not affect the shapes of the calculated fields. It is possible to change the shapes of the calculated fields by making the offsets on the master faults unequal or by allowing the offsets across the master faults to decay slowly to zero over a finite length of the fault. Unless stated explicitly, the tops of the master faults intersect the ground surface.

Vertical displacements

When the separation of the master faults is twice the overlap, there is a basin between the faults (Fig. 4a). The deepest part of the basin is along a line joining the ends of the two faults with small uplifts at the ends of the master faults. Since the offset

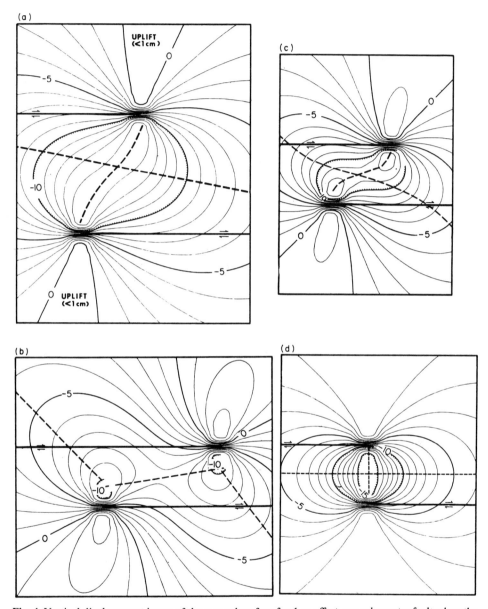

Fig. 4. Vertical displacement in cm of the ground surface for 1 m offset on each master fault when the tops of the master faults intersect the ground surface. Heavy lines show master faults. Negative values are down, and positive values are up. Contour interval 1 cm. Dashed lines show the approximate locations of the axes of the basin.

(a) Separation (20 km) equals twice the overlap; (b) separation (10 km) equals 1/2 the overlap; (c) separation (10 km) equals the overlap; (d) overlap is zero, separation equals 10 km.

on each of the master faults is 1 m, the plotted displacements are percentages of the offset on the master faults. Thus for this model, the depth of the basin is about 15% of the offset on the master faults.

When the overlap is twice the separation of the *en echelon* faults (Fig. 4b), two basins develop between the ends of the two faults, and the uplifts at the ends of the faults are larger than in Fig. 4a. Basin depths are only about 10% of the offset on the master faults.

When the overlap equals the separation (Fig. 4c), a basin develops with the deeper part of the basin along a line joining the ends of the two faults. However, as in Fig. 4b, smaller, less distinct basins develop near the ends of the faults. The greatest depth of the basin is about 14% of the offset on the two faults. The uplift at the ends of the master faults is more than in Fig. 4a but less than Fig. 4b.

When the overlap is zero (Fig. 4d), the deepest part of the basin is exactly perpendicular to the faults, and the greatest depth is about 15% of the offset on the master faults. There is no uplift at the ends of the master faults.

From the vertical displacements, it can be concluded that: (i) a basin will form between *en echelon* right-stepping right-lateral strike-slip faults; (ii) the axis of the deepest part of the basin is on the line joining the ends of the master faults; (iii) the depth and shape of the basin are related to the amount of overlap and the amount of separation between the *en echelon* faults; (iv) there can be small amounts of uplift near the ends of the faults; (v) the amount and shape of the uplift was also a function of the amount of overlap and the separation between the *en echelon* faults; and (vi) the depth of the basin is less than 20% of the offset on the master faults, although this latter conclusion is probably strongly dependent on the elastic properties of the body.

Recalling that the model is for an elastic half-space with infinitesimal displacements on the master faults, the question arises as to the applicability of the models to the earth where the basins are kilometres deep and the master faults have tens of kilometres of offset. Strictly speaking, the models presented above are only valid when describing the initiation of a pull-apart basin. However, once the initial depression forms, it will start to fill with sediments. As movement continues on the master faults, the basin will continue to deepen and fill with sediments. Thus, it seems reasonable to expect that a basin which begins due to an elastic (and hence recoverable) effect will become a permanent feature due to geological processes.

Predicted fault patterns

It is possible to predict fault patterns in the basin from the stresses produced by the *en echelon* faults (Jaeger & Cook, 1969, ch. 4). I will not present a complete analysis of the potential fault pattern as I feel that a simple approach will illustrate the important points. Also, the simplicity of the model probably makes a more detailed analysis unnecessary.

I assume that shear failure will occur at an angle of $\pm 45°$ to the most compressive principal stress axis. Experimental evidence indicates that failure in rock specimens tends to occur at angles between 25–40° to the most compressive principal stress axis (Jaeger & Cook, 1969), so the fault orientations predicted by my assumption could be off by at most 20°. Considering the lack of complexity of the model, this seems to be an acceptable error.

Once again, the tops of the master faults intersect the ground surface unless explicitly stated otherwise.

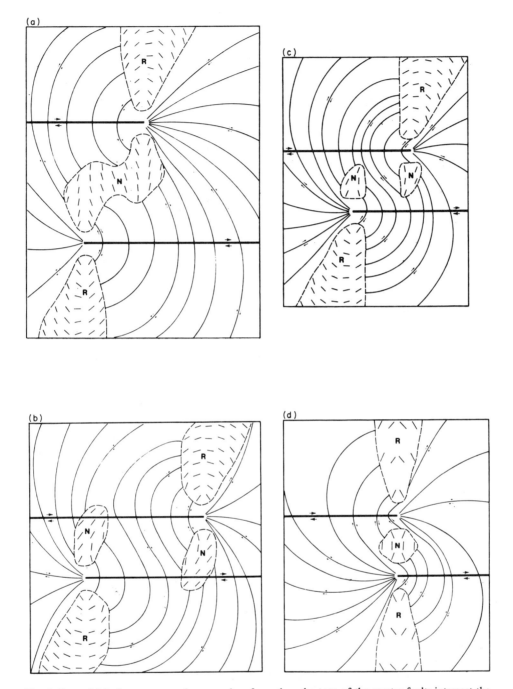

Fig. 5. Potential fault pattern on the ground surface when the tops of the master faults intersect the ground surface. Offset on each master fault is 1 m. Heavy lines show master faults. Possible right-lateral strike-slip fault pattern is shown; the left-lateral fault pattern would be perpendicular to the right-lateral fault pattern. Possible strikes of normal faults shown in area denoted 'N', and possible strikes of reverse faults shown in areas denoted 'R'. Normal and reverse faults would dip 45°.

(a) Separation (20 km) equals twice the overlap; (b) separation (10 km) equals 1/2 the overlap; (c) separation (10 km) equals the overlap; (d) overlap is zero, separation equals 10 km.

When the separation is twice the overlap (Fig. 5a), there is a zone of normal faulting in the central part of the basin and zones of reverse faulting near the ends of the master faults outside of the basin. The model predicts strike-slip faulting in part of the basin. For clarity, only the predicted right-lateral strike-slip faults are shown, but the left-lateral faults would be perpendicular to the right-lateral faults shown. The fault lines drawn are not intended to represent individual faults but rather show schematically the fault patterns which could develop in the model. Thus as in Crowell's model (Fig. 2), there is an extensive zone of normal faulting on the basin sides of the ends of the master faults, and the normal faulting probably dies out in a zone of oblique-slip faulting as one moves along a line perpendicular to one master fault from the end of the fault into the basin. However, there is clearly strike-slip faulting between the zones of normal faulting and the basin sides of the ends of the master faults.

Note that the right-lateral fault trends are not parallel to the master faults. The dislocation model predicts the stress field which is produced by movement on the master faults. In the absence of any other stress field, the resulting stresses would try to remove the initial displacement on the master faults, and this would require left-lateral movement on the master faults. Presumably, the initial right-lateral displacement on the master faults is caused by some regional stress field, but this is not the problem considered in this paper.

When the overlap is twice the separation between the master faults (Fig. 5b), the zone of normal faulting extends from near the end of one master fault across the other master fault, and there is no normal faulting in the centre of the basin. There is strike-slip faulting just at the basin side of the ends of the master faults, and the centre of the basin is dominated by strike-slip faulting. This is in contrast to the fault pattern predicted by Crowell's model (Fig. 2). There are still zones of reverse faulting at the ends of the master faults outside of the basin.

When the overlap equals the separation between the two faults (Fig. 5c), there are zones of normal faulting at the ends of the basin, but the faulting in the centre of the basin is strike-slip. Once again, there is strike-slip faulting between the ends of the master faults and the normal faults toward the basin, and there is a zone of reverse faulting at the ends of the master faults outside of the basin.

When the overlap is zero (Fig. 5d), there is a well developed zone of normal faulting in the centre of the basin which is again separated from the ends of the master faults by zones of strike-slip faulting. There is strike-slip faulting on the flanks of the basin, and zones of reverse faulting are seen at the ends of the master faults outside of the basin.

These models show that the predicted secondary fault patterns in pull-apart basins are dependent on the amount of overlap and the separation between the master faults. In two cases (Fig. 5a and d), normal faulting develops in the centre of the basin and strike-slip develops on the flanks, while in the other two cases (Fig. 5b and c), normal fault zones develop on the flanks of the basin near the ends of the master faults and strike-slip faulting develops in the central part of the basin. In all cases, the faulting on the basin side of the ends of the master faults starts out as strike-slip faulting and changes to normal faulting as one moves into the basin. This suggests that the faulting on the basin side of the ends of *en echelon* faults is generally going to have oblique-slip. The zones of reverse faulting near the ends of the master faults outside of the basin occur in each model. However, it would be unexpected to find the reverse faulting to be well developed in nature because the stresses in these zones will probably not be high enough to cause the rocks to fail.

Once again, the question of applicability of the models can be raised. The fault patterns predicted by the models only apply to the initiation of faulting in the basin. However, once a secondary fault pattern is established in the basin, subsequent movements on the master faults should produce movement on the previously formed secondary fault pattern rather than produce a new fault pattern. This is because a lower stress level is generally required to cause movement on a pre-existing fault than to fracture new rock. Thus the initial secondary fault pattern would be structurally important for a significant part of the basin's history.

Buried faults

The basin shapes discussed in the previous section and the secondary fault patterns discussed here strictly apply only to the basement rocks. As movement continues on the master faults, the basins will deepen and movement will continue on the secondary fault system. The deepening basin will fill with sediments, and the secondary faults will propagate up into the sedimentary fill. If the sedimentation rate is slow enough, the initial secondary fault pattern may propagate through the sediments and be seen at the surface of the basin. If the basin is filled faster than the secondary faults can propagate upward through the sediments, then a tertiary structural pattern may develop in the sediments within the basin. The structural development within the basin would then be very complex and rather difficult to model. In this section a model which is a first approximation to the above problem will be presented.

Consider the case where the overlap is zero (Figs 4d and 5d) and the top edges of the master faults are at a distance below the ground surface equal to the separation. This model is equivalent to assuming that the *en echelon* faults are overlain by a uniformly thick sedimentary layer with the same elastic properties as the semi-infinite

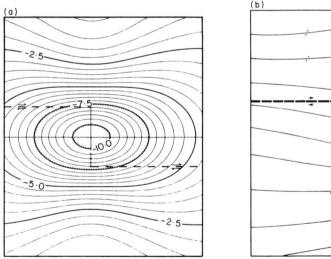
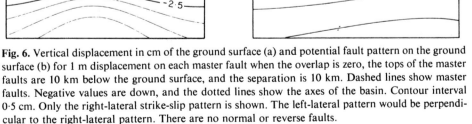

Fig. 6. Vertical displacement in cm of the ground surface (a) and potential fault pattern on the ground surface (b) for 1 m displacement on each master fault when the overlap is zero, the tops of the master faults are 10 km below the ground surface, and the separation is 10 km. Dashed lines show master faults. Negative values are down, and the dotted lines show the axes of the basin. Contour interval 0·5 cm. Only the right-lateral strike-slip pattern is shown. The left-lateral pattern would be perpendicular to the right-lateral pattern. There are no normal or reverse faults.

body. Figure 6a shows contours of the displacement of the ground surface for this model and should be compared with Fig. 4d. Note that the uplifted zones at the ends of the master faults in Fig. 4d are not seen in Fig. 6a. The greatest depth of the basin is about 10% of the offset on the master faults.

Figure 6b shows the predicted fault pattern for this model and should be compared to Fig. 5d. The faulting throughout the basin in Fig. 6b is strike-slip, and there is none of the normal or reverse faulting seen in Fig. 5d. This is important, for it suggests that strike-slip faults should occur in the sediments of a pull-apart basin in the early stages of development before the master faults have propagated to the surface.

The conclusions drawn from this model are also valid for the other models of Figs 4 and 5 with buried master faults. That is, the zones of uplift are not seen and the basins are characterized by strike-slip faulting. Models with varying depths to the tops of the master faults suggest that normal faulting only appears at the ground surface when the fault tops are very near to the ground surface.

It is worth discussing the effects of layering briefly at this time. Previous work on a straight strike-slip fault in a semi-infinite body overlain by a layer which is less rigid than the semi-infinite body (Rodgers, 1976) shows that the layer tends to concentrate the deformation closer to the fault. However, the layer does not affect the predicted fault pattern at the surface. I would expect the same thing to happen for the *en echelon* faults. That is, if the material above the master faults is less rigid than that containing the master faults (e.g. sediments over a crystalline basement), then the basin will be smaller than that shown in Fig. 6 but will still be characterized by strike-slip faulting.

DEVELOPMENT OF PULL-APART BASINS

As the offset across the master faults increases, we might expect the master faults to extend themselves in some direction. If we assume that the master faults extend themselves parallel to their strike, this would have the effect of increasing the overlap while keeping the separation more or less constant. If we start with the configuration of no overlap, then presumably the basin would evolve toward increasing overlap of the master faults. At the same time, a secondary fault system should form in the basin in response to the initial configuration of the master faults. At any later time we would expect the secondary fault system to have the orientation predicted from the current stress field *if* the basin were unfaulted. However, the secondary fault system produced by the initial configuration of the master faults would presumably be present at the later stages of deformation and would thus represent a pre-existing fabric for the basin. Thus the stress fields produced by successive configurations of the master faults would be acting on a pre-existing secondary fault pattern, and it might be expected for movement to continue on the original secondary fault system even though the stress field might be quite different at some later time. We should then be able to use Figs 4 and 5 to suggest the structural evolution of a pull-apart basin.

Figure 7 shows schematically the changes which might be expected if we start with *en echelon* faults with no overlap. The basin shape and secondary fault pattern would be similar to those shown in Figs 4d and 5d. There would be normal faulting in the centre and strike-slip faulting on the flanks of the basin. If the master faults extend themselves along strike, then when the offset on the master faults is about one-half

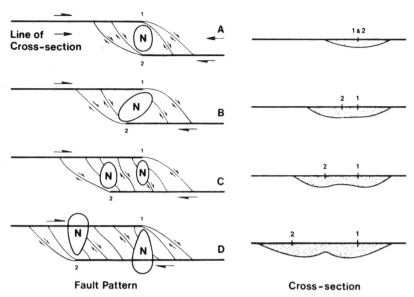

Fault Pattern   **Cross-section**

Fig. 7. Sketch of possible evolution of a pull-apart basin based on Figs 4 and 5. Left side is a map view of basement fault development. Numbers refer to locations of ends of master faults. Zones labelled 'N' are zones of normal faulting. Left-lateral strike-slip faults not shown. Right side is a vertical cross-section parallel to master faults through the centre of the basin. Vertical and horizontal scales are arbitrary. Stippled areas are sedimentary fill of basins. Thus cross-sections show suggested evolution of basin shape.

the separation of the faults we have nearly the same situation (Fig. 7B). The zone of normal faulting in the centre of the basin is somewhat larger, and the basin has deepened and broadened. If the basin has been filled by sediments, the secondary fault pattern near the ground surface will probably be dominated by strike-slip faulting such as that shown in Fig. 6.

When the offset across the master faults is about equal to the separation between the faults, some interesting changes occur. Two distinct depocentres begin to develop near the ends of the master faults, and the zones of normal faulting are separated by a zone of strike-slip faults. If the orientations of the secondary faults remain similar to the initial orientations, this means that the sense of offset on some of the secondary faults must change from strike-slip to normal or vice versa as the basin develops. The fault patterns in Fig. 7 still represent the faulting in or near the basement, and, as Fig. 6 suggests, the faulting in the basin fill may still be dominated by strike-slip. Figure 7C suggests that the faulting in the centre of the basin may become strike-slip throughout the section as the basin grows. Figure 7D demonstrates the expected situation when the offset on the master faults is twice their separation. Figure 7C and 7D are similar; and two depocentres are more separated and the zone of strike-slip faulting occupies more of the basin centre in Fig. 7D.

The models suggest that the basement secondary fault patterns will be very complex throughout the history of a pull-apart basin and may include changes in the sense of movement on many of the faults. The fault patterns in the sedimentary fill of the basin should be fairly simple near the ground surface, but there should be a zone within the sediments where the basement and near-surface fault patterns merge. This zone would be structurally complex.

FIELD EXAMPLES

Sharp (1975) suggests that several areas along the San Jacinto fault in California may be pull-apart basins. San Jacinto Valley (Fig. 8, inset) is a deep, narrow valley

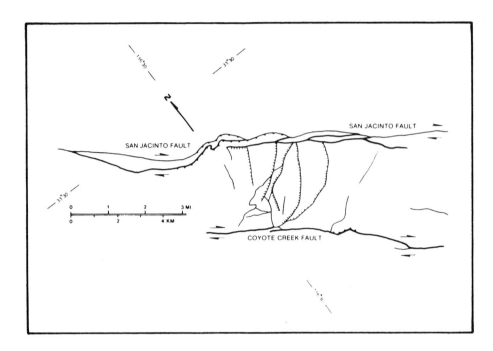

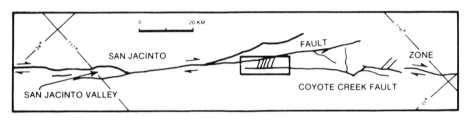

Fig. 8. Generalized fault maps showing pull-apart basins on the San Jacinto fault. Hachures show down-thrown sides of normal faults. Barbs are in the upper plates of low-angle reverse faults. Arrows show sense of movement on strike-slip faults. Box in inset shows location of larger part of figure. Inset generalized from Sharp (1975, Fig. 1). Larger figure generalized from a portion of Sharp (1967, plate 1).

between right-stepping right-lateral faults (Lofgren & Rubin, 1975; Lofgren, 1976), and, according to Lofgren & Rubin (1975), in the last 15 000 years the valley has subsidided at a rate of about 6 mm/year. The fault pattern in the basement is not exposed in the San Jacinto Valley. However, the basement fault pattern is exposed between the San Jacinto and Coyote Creek faults (Fig. 8). Note that there are several normal faults exposed between the San Jacinto and Coyote Creek faults with the northwestern sides generally downthrown. There appears to be reasonable agreement between the fault patterns seen in Figs 5c and 8.

Other areas that have been suggested as pull-apart basins include the Dead Sea in the Levant fault zone (or Dead Sea Rift; Quennell, 1959) and Death Valley in California (Burchfiel & Stewart, 1966). Certainly both basins are properly located between *en echelon* faults, and both basins are substantially lower than their surroundings (about 1300 m for the Dead Sea and about 3400 m for Death Valley). The structures mapped in Death Valley are not indicative of extension (Hill & Troxel, 1966) as might be expected if Death Valley were a typical Basin and Range structure.

The Salton Trough in southern California lies between NW–SE trending dextral faults, the San Jacinto and Imperial faults on the southwest and the Banning–Mission Creek members of the San Andreas fault zone on the northeast. The trough is filled with about 6100 m of sediment and is characterized by very high temperatures (300°F below about 1300 m) and high seismicity (Hileman, Allen & Nordquist, 1973; Hill, Mowinckel & Peake, 1975). Elders *et al.* (1972) have suggested that there are several small spreading centres in the Salton Trough orientated NE–SW, perpendicular to the bounding faults. However, the fault map of California (Jennings, 1975) does not show any faults in the trough with NE–SW orientation. In addition, the trend of epicentres in the recent earthquake swarms within the trough is nearly N–S, along the line of the Brawley fault. Fault plane solutions from these swarms (Hill *et al.*, 1975) suggest that faulting within the trough is not normal, but is strike-slip and that opening of the spreading centre takes place in a diffuse zone of *en echelon* strike-slip faults which trend parallel to the main bounding faults.

The orientation of the current seismicity together with the lack of mapped faults with the trend of the proposed spreading centres suggests that the Salton Trough may be a complex pull-apart basin. The currently active N–S trending strike-slip faulting is consistent with the fault patterns suggested in the discussion of Fig. 7D. Also, high precision level surveys in the area (Lofgren, 1978) suggest that the northwestern end of the Salton Sea is deepening faster than the southeastern end. Once again, this is consistent with the predictions of Figs 7D and 4b. A spreading centre in the vicinity of Brawley might cause the southeastern end of the Salton Sea to deepen faster than the northwestern end. Of course, none of these data conclusively prove that the Salton Trough is a pull-apart basin, and, if the Trough is a pull-apart basin, it is certainly more complex than the models in this paper. However, the models together with the observational data do clearly allow a pull-apart origin for the Salton Trough.

CONCLUSIONS

Mathematical models have been used to study surface deformation and fault patterns associated with pull-apart basins produced by *en echelon* strike-slip faults. The model studies suggest that the basin shape and the fault patterns are closely related to: (1) the amount of overlap between the faults; (ii) the amount of separation between the faults; (iii) whether or not the faults intersect the ground surface. In particular, strike-slip faults should be well developed in any pull-apart basin, and normal faulting will occur only when the tops of the master strike-slip faults are near the ground surface. In general, the models suggest that pull-apart basins may have forms different from that proposed by Crowell (1974).

The models also suggest that the long-term evolution of a pull-apart basin may be structurally complex, and some of the secondary faults may show different types of movement through time. The faults in the sedimentary fill of the basins may show somewhat simpler patterns than those in the basement.

Examples of pull-apart basins from California show fault patterns which are in reasonable agreement with those predicted by the models. General patterns of basin shape and topographic relief for pull-apart basins in California and the Middle East also agree reasonably well with predictions of the models.

ACKNOWLEDGMENTS

This paper is based on a Cities Service Co. Research Report, and I am grateful to the company for permission to publish the paper. The work benefited from discussions with Rick Groshong. Jack Etter, Rick Groshong, Bill Rizer and Martha Withjack reviewed the manuscript of the company report and made many helpful suggestions. Anonymous reviewers made suggestions which helped to improve the final manuscript. I am grateful to Dr J. C. Crowell and the Society of Economic Paleontologists and Mineralogists for permission to reproduce Fig. 2.

REFERENCES

ALEWINE, R.W. III (1974) *Application of Linear Inversion Theory toward the Estimation of Seismic Source Parameters.* Unpublished Ph.D. Thesis, California Institute of Technology.

BLAKE, M.C., JR, CAMPBELL, R.H., DIBBLEE, T.W., JR, HOWELL, D.G., NILSEN, T.H., NORMARK, W.R., VEDDER, J.C. & SILVER, E.A. (1978) Neogene basin formation in relation to plate-tectonic evolution of San Andreas fault system, California. *Bull. Am. Assoc. Petrol. Geol.* **62**, 344–372.

BURCHFIEL, B.C. & STEWART, J.H. (1966) 'Pull-apart' origin of the central segment of Death Valley, California. *Bull. geol. Soc. Am.* **77**, 439–442.

CHINNERY, M.A. (1961) The deformation of the ground around surface faults. *Bull. seismol. Soc. Am.* **51**, 355–372.

CHINNERY, M.A. (1963) The stress changes that accompany strike-slip faulting. *Bull. seismol. Soc. Am.* **53**, 921–932.

CHINNERY, M.A. & PETRAK, J.A. (1968) The dislocation fault model with a variable discontinuity. *Tectonophysics*, **5**, 513–529.

CLAYTON, L. (1966) Tectonic depressions along the Hope fault, a transcurrent fault in North Canterbury, New Zealand. *N.Z.J. Geol. Geophys.* **9**, 95–104.

COUPLES, G. & STEARNS, D.W. (1978) Analytical solutions applied to structures of the Rocky Mountains foreland on local and regional scales. In: *Laramide Folding Associated with Basement Block Faulting in the Western United States* (Ed. by V. Matthews III). *Mem. geol. Soc. Am.* **151**, 313–335.

CROWELL, J.C. (1974) Origin of late Cenozoic basins in southern California. In: *Tectonics and Sedimentation* (Ed. by W. R. Dickinson). *Spec. Publ. Soc. econ. Paleont. Miner. Tulsa*, **22**, 190–204.

ELDERS, W.A., REX, R.W., MEIDAV, T., ROBINSON, R.T. & BIEHLER, S. (1972) Crustal spreading in southern California. *Science*, **178**, 15–24.

FISHER, M.A., PATTON, W.W., JR, THOR, D.R., HOLMES, M.L., SCOTT, E.W., NELSON, C.H. & WILSON, C.L. (1979) The Norton basin of Alaska. *Oil and Gas Journal*, **77**, 96–98.

FREUND, R. (1965) A model of the structural development of Israel and adjacent areas since Upper Cretaceous times. *Geol. Mag.* **102**, 189–205.

FREUND, R. (1971) The Hope fault, a strike-slip fault in New Zealand. *Bull. N.Z. geol. Surv. (new series)* **86**, 49 pp.

HARDING, T.P. (1976) Predicting productive trends related to wrench faults. *World Oil*, **182**, 64–69.

HILEMAN, J.A., ALLEN, C.R. & NORDQUIST, J.M. (1973) *Seismicity of the southern California region, January 1, 1932, to December 1, 1972.* Seismological Laboratory, California Institute of Technology.

HILL, M.L. & TROXEL, B.W. (1966) Tectonics of Death Valley region, California. *Bull. geol. Soc. Am.* **77**, 435–438.

HILL, D.P., MOWINCKEL, P. & PEAKE, L.G. (1975) Earthquakes, active faults and geothermal areas in the Imperial Valley, California. *Science*, **188**, 1306–1308.

JAEGER, J.C. & COOK, N.G.W. (1969) *Fundamentals of Rock Mechanics*. Methuen, London.

JENNINGS, C.W. (1975) Fault map of California, *California Division of Mines and Geology Geologic Data Map Number* 1, Scale 1:750,000.

JOHNSON, A.M. (1970) *Physical Processes in Geology*. Freeman, Cooper & Co., San Francisco. 577 pp.

LOFGREN, B.E. (1976) Land subsidence and aquifer-system compaction in the San Jacinto Valley, Riverside County, California—A progress report. *J. Res. U.S. geol. Surv.* **4**, 9–18.

LOFGREN, B.E. (1978) Salton Trough continues to deepen in Imperial Valley, California (Abstract). *EOS, Trans. Am. geophys. Union.* **59**, 1051.

LOFGREN, B.E. & RUBIN, M. (1975) Radiocarbon dates indicate rates of graben downfaulting, San Jacinto Valley, California. *J. Res. U.S. geol. Surv.* **3**, 45–46.

NORRIS, R.J., CARTER, R.M. & TURNBULL, I.M. (1978) Cainozoic sedimentation in basins adjacent to a major continental transform boundary in southern New Zealand. *J. geol. Soc. Lond.* **135**, 191–205.

QUENNELL, A.M. (1959) Tectonics of the Dead Sea Rift. *Int. geol. Congr.* 1956. **20**, 385–405.

RECHES, Z. & JOHNSON, A.M. (1978) Development of monoclines: Part 11. Theoretical analysis of monoclines. In: *Laramide Folding Associated with Basement Block Faulting in the Western United States* (Ed. by V. Matthews III). *Mem. geol. Soc. Am.* **151**, 273–311.

RODGERS, D.A. (1976) Mechanical analysis of strike slip faults. 11. Dislocation model studies (Abstract). *EOS, Trans. Am. geophys. Union.* **57**, 327.

RODGERS, D.A. (1979) Vertical deformation, stress accumulation, and secondary faulting in the vicinity of the Transverse Ranges of southern California. *Bull. California Div. Mines Geol.* **203**, 74 pp.

SANFORD, A.R. (1959) Analytical and experimental study of simple geological structures. *Bull. geol. Soc. Am.* **70**, 19–51.

SAVAGE, J.C. & HASTIE, L.M. (1969) A dislocation model for the Fairview Peak, Nevada, earthquake. *Bull. seismol. Soc. Am.* **59**, 1937–1948.

SHARP, R.V. (1967) San Jacinto fault zone in the Peninsular Ranges of southern California. *Bull. geol. Soc. Am.* **78**, 705–730.

SHARP, R.V. (1975) En echelon fault patterns of the San Jacinto fault zone. In: *San Andreas fault in southern California* (Ed. by J. C. Crowell) *Spec. Rep. California Div. Mines Geol.* **118**, 147–152.

WILCOX, R.E., HARDING, T.P. & SEELY, D.R. (1973) Basic wrench tectonics. *Bull. Am. Ass. Petrol. Geol.* **57**, 74–96.

TECTONICS, VOL. 1, NO. 1, PAGES 91-105, FEBRUARY 1982

EVOLUTION OF PULL-APART BASINS AND THEIR SCALE INDEPENDENCE

Atilla Aydin[1] and Amos Nur

Department of Geophysics, Stanford University, Stanford, California
94305

Abstract. Pull-apart basins or rhomb grabens and horsts along major
strike-slip fault systems in the world are generally associated with
horizontal slip along faults. A simple model suggests that the width
of the rhombs is controlled by the initial fault geometry, whereas the
length increases with increasing fault displacement. We have tested
this model by analyzing the shapes of 70 well-defined rhomb-like pull-
apart basins and pressure ridges, ranging from tens of meters to tens
of kilometers in length, associated with several major strike-slip
faults in the western United States, Israel, Turkey, Iran, Guatemala,
Venezuela, and New Zealand. In conflict with the model, we find that
the length to width ratio of these basins is a constant value of
approximately 3; these basins become wider as they grow longer with
increasing fault offset. Two possible mechanisms responsible for the
increase in width are suggested: (1) coalescence of neighboring
rhomb grabens as each graben increases its length and (2) formation
of fault strands parallel to the existing ones when large displacements need to be accommodated. The processes of formation and growth
of new fault strands promote interaction among the new faults and
between the new and preexisting faults on a larger scale. Increased
displacement causes the width of the fault zone to increase resulting
in wider pull-apart basins.

INTRODUCTION

Many rhomb grabens and rhomb horsts have been recognized along major
strike-slip faults throughout the world (see Table 1). Pull-apart
basins or rhomb grabens are depressional basins, while pressure ridges
or rhomb horsts are uplifted terranes. Basins associated with active
strike-slip faults can be readily identified because of their morphological expressions as elongated lakes and sag ponds, which often
contain young sedimentary deposits and sometimes involve volcanic and
geothermal activities [Clayton, 1966; Freund, 1971; Elders et al.,
1972; Clark, 1973; Crowell, 1974; Hill, 1977]. The horst-like ridges
usually form conspicuous rectilinear hills along strike-slip faults
and are characterized by en echelon folds [Sharp and Clark, 1972].
The geometry of some pull-apart basins and pressure ridges has been
inferred from associated seismicity and focal mechanism solutions
[Johnson and Hedley, 1976; Johnson, 1979] and from surface faulting
associated with major earthquakes on strike-slip faults [Clark, 1972;
Sharp, 1976, 1977; Arpat et al., 1977; Tchalenko and Ambrasyes, 1970].
Mechanical aspects of pull-apart basins and pressure ridges have been
recently investigated by Segall and Pollard [1980] and Rodgers [1980].

---

[1]Now at Department of Geosciences, Purdue University, West Lafayette
Indiana 47907

TABLE 1. Strike-Slip Faults Associated Grabens (G) or Horsts (H) and Their Dimensions

Fault and/or Location	Basin or Mountain Range	Graben (G) or Horst (H)	Dimension (M) Length	Dimension (M) Width	Reference
Motagua, Guatemala	Motagua Valley	G	50,000	20,000	Schwartz et al. [1979]
	Rio El Tambor	G	25	8	
Polochic	Lago de Izabal	G	80,000	30,000	Bonis et al. [1970]; Plafker [1976]; this study
Dead Sea Rift, Israel	Hula	G	20,000	7,000	Freund et al. [1968]
	Lake Kineret	G	17,000	5,000	
	Ayun	G	6,600	1,600	
	East of Timna	G	1,000	250	
	North of Ayun	G	1,200	400	Garfunkel et al. [1982]
		G	1,200	400	
		G	1,600	450	
		G	5,000	1,200	
		G	2,000	500	
	South of Timna	G	8,800	3,000	
		G	20,000	6,000	
	West of the Dead Sea	G	3,500	750	Garfunkel [1982]
		G	3,000	750	
		G	3,000	800	
		G	6,000	1,500	
		G	7,500	1,800	
		G	3,000	750	
	East of the Dead Sea	G	4,500	1,500	
Paran	Karkom	G	18,000	6,000	Bartov [1979]
		G	6,000	1,500	
Bir Zrir, Sinai		G	5,000	2,000	Eyal et al. [1980]
Gulf of Elat	Elat	G	45,000	10,000	Ben-Avraham et al. [1979]
	Aragonese	G	40,000	9,000	
	Tiran-Dakor	G	65,000	8,000	
Dasht-e Bayaz, Iran		G	1,200	500	Freund [1974]
Hope, New Zealand	Medway-Karaka	G	700	230	Freund [1971]
	Glynnwye	G	980	210	
	Glynnwye Lake	G	1,800	550	
	Poplars Station	G	2,300	900	
	Hanmer Plains	G	13,000	3,500	Freund [1974]
Hope, New Zealand	Medway-Karaka	H	90	30	Freund [1971]
	Glynnwye Lake	H	300	90	
	Poplars Station	H	300	150	
	Hanmer Plains	H	4,500	2,700	
North Anatolian, Turkey	Niksar	G	25,000	10,000	Seymen [1975]; this study
	Erzincan	G	40,000	12,000	Ketin [1969]
	Susehri	G	23,000	6,000	
San Andreas, Calif., USA	Cholame Valley	G	17,000	3,000	Jennings [1959]; Brown [1970]
	San Bernardino Mountains	H	32,000	14,000	Dibblee [1975]
Imperial	Brawley	G	10,000	7,000	Johnson and Hadley [1976]; Sharp [1976, 1977]
Elsinore	Elsinore Lake	G	12,000	3,000	Rogers [1965]
Garlock	Koehn Lake	G	40,000	11,000	Jennings et al. [1969]; Smith [1964]; Clark [1973]; this study
		G	300	150	
		G	600	110	
		G	600	100	
	West of Quail Mountain	G	240	90	Clark [1973]
		G	900	220	
	Searleys Valley	G	1,600	380	
	East of Christmas Canyon	G	1,250	250	
San Jacinto,	Hog Lake	G	680	170	Sharp [1972]
	Hemet	G	22,000	5,000	Sharp [1975]
Buck Ridge	Santa Rosa Mountain	G	6,000	1,700	Sharp [1972]
Coyote Creek	Ocotillo Badlands	H	5,500	1,800	Sharp and Clark [1972]
	Borrega Mountain	H	4,000	1,600	
	Bailey's Well	G	500	200	Clark [1972]
		G	190	80	
Olinghouse, Nevada	Tracy-Clark Station	G	70	40	Sanders and Slemmons [1979]; this study
		G	160	90	
		G	450	175	
		G	980	250	
Bocono, Venezuela	La Gonzales	G	23,000	6,200	Schubert [1980a]
	Merida-Mucuchies	G	6,200	1,700	Schubert [1980b]
		G	700	200	
		G	280	70	
		G	1,000	280	
Valencia	Lake Valencia	G	30,000	11,500	Schubert and Laredo [1979]
El Pilar	Casanay	H	3,000	1,200	Schubert [1979]

    In this paper, we first review the kinematics of strike-slip
faulting from the viewpoint of basin and ridge formation in strike-slip
environments. We then examine the scale dependence of the geometry
of many pull-apart basins and pressure ridges. Finally, we suggest
tectonic models for the growth and development of these basins and
ridges.

KINEMATICS OF STRIKE-SLIP FAULTING
AND BASIN AND RIDGE FORMATION

    It is generally thought that both pull-apart basins and pressure
horsts near strike-slip faults are associated with geometrical and
possibly mechanical irregularities of these faults. This concept
implies that motion on discrete fault strands within a strike-slip
fault system is responsible for the creation of pull-aparts and
horsts.

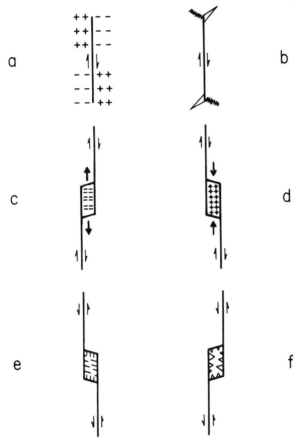

Fig. 1. (a) Extensional (minus) and compressional (plus) quadrants
around a right lateral strike-slip fault; (b) tail cracks (open) in
the extensional quadrant and pressure solutions or folds (zig-zag
line) in the compressional quadrants; (c) rhomb graben on a right
stepover; (d) rhomb horst on a left stepover for right lateral strike-
slip faults; (e) normal faults (barbs on downthrown side) and major
strike-slip fault segments with normal slip component bounding a
rhomb graben at a left stepover; and (f) reverse faults (teeth on upthrown side) and major strike-slip fault segments with reverse slip
component bounding a rhomb horst at a right stepover for left lateral
strike-slip faults.

Horizontal slip on a single strike-slip fault will induce extension in two quadrants and compression in the other two quadrants (Figure 1a). Structures reflecting the extension (cracks) and compression (pressure solution and folds) are sometimes observed in the proper quadrants (Figure 1b) in the field [Rispoli, 1981; P. Segall and D. D. Pollard, manuscript in preparation, 1982]. When strike-slip faults are arranged in en echelon pattern, the extensional or compressional quadrants of the neighboring faults partially overlap, thereby enhancing either extensional or compressional deformation. For example, right and left lateral strike-slip faults with right (Figure 1c) and left (Figure 1e) stepovers, respectively, produce depressions at the stepover regions. While two sides of such depressions are bounded by the segments of the strike-slip faults that have significant normal slip components, the other two sides are defined predominantly by normal faults trending diagonally to the strike-slip faults [Clayton, 1966; Freund, 1971; Sharp, 1976, 1977].

Pressure ridges or rhomb horsts are associated with right and left lateral strike-slip faults with left stepover (Figure 1d) and right stepover (Figure 1f), respectively. Like pull-apart basins, pressure ridges are bounded on two sides by segments of strike-slip faults and by reverse or thrust faults on the remaining two sides (Figure 1f).

DIMENSIONS OF PULL-APART BASINS
AND PRESSURE RIDGES

Pull-apart basins and pressure ridges of various sizes have been reported by several authors (reference list, see Table 1). Figure 2 illustrates some examples of basin and ridge structures ranging from 0.6 m to 80,000 m in length and from 0.17 m to 30,000 m in width. Figure 2a shows a small pull-apart structure in Sierra Nevada granite, which is similar to those studied by P. Segall and D. D. Pollard

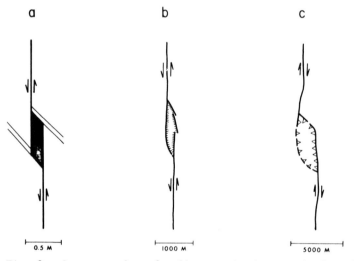

Fig. 2. Some examples of pull-apart basins or rhomb grabens in various scales and one rhomb horst. (a) Courtesy of Paul Segall, (b) from Fruend et al. [1968], (c) from Sharp and Clark [1972], (d) from Ketin [1969]; NAF (inset): North Anatolian Fault; F: faults (arrows indicating sense of displacement on strike-slip faults and plus and minus indicating upthrown and downthrown blocks, respectively, on normal fault; HS: hot springs; VC: volcanic cones; Q: Quaternary; T: Tertiary; pM&M: pre-Mesozoic and Mesozoic, (e) slightly modified from Bonis et al. [1970] and Plafker [1976]; Q & T; Quaternary and Tertiary; and pT: pre-Tertiary.

d

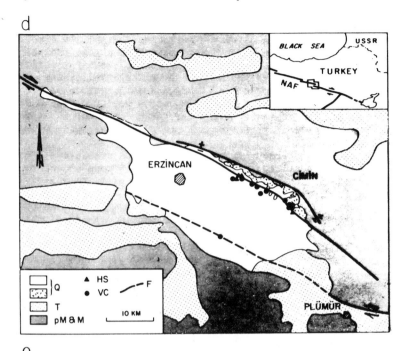

e

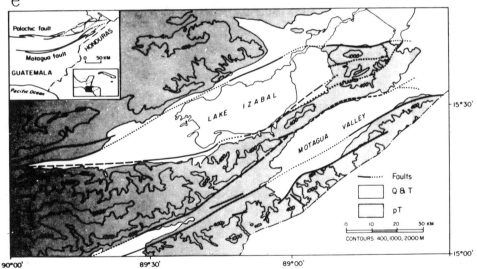

Fig. 2.   (continued)

(manuscript in preparation, 1982). The pull-apart, which is well
defined by quartz filling, is comparable in length to the horizontal
offset of a preexisting vein. Figure 2b shows a graben in recent
alluvium along the Dead Sea Fault near Timna, Isreal [Freund et al.,
1968]. Here the normal faults and normal component of the left
lateral strike-slip faults bounding the graben are illustrated.
Figure 2c is one of few examples of horst structure; the thrust nature
of the bounding fault is supported by the observed displacement on the
active breaks associated with the Borrego Valley earthquake of 1968
in southern California [Sharp and Clark, 1972]. Figure 2d is a large
basin along the North Anatolian Fault at the Erzincan region, Turkey
[Ketin, 1969]. The Erzincan Basin is filled with young detrital

deposits and volcanic rocks. Several volcanic cones and hot springs conspicuously aligned along the major faults indicate high heat flow and thinning of the crust at the pull-apart region. The last example shown in Figure 2e includes two huge pull-apart basins, the Polochic Valley, mostly occupied by Lake Izabal, and the Motague Valley in Guatemala [Bonis et al., 1970; Plafker, 1976]. Both appear to be composite pull-apart basins, the former being about 80,000 m long and 30,000 m wide. Still larger features interpreted as structures somewhat analogous to pull-apart basins and pressure ridges were reported by Carey [1958].

One implication of the concept that pull-aparts and pressure ridges are formed by predominantly strike-slip motion along en echelon faults is that their length should increase proportionally to the amount of offset (Figure 5). In contrast, the widths of pull-aparts and ridges should remain roughly fixed at the initial value of fault strand spacing, which must be due to an earlier tectonic process, the nature of which is not well understood.

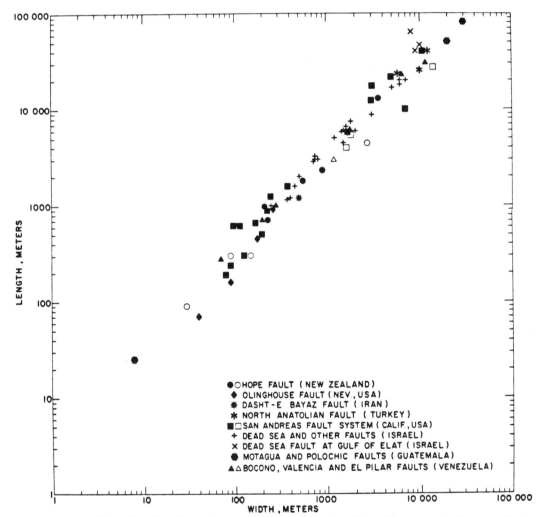

Fig. 3. Log length versus log width for 70 pull-apart basins or rhomb grabens and horsts associated with major strike-slip faults of the world. Full symbols grabens and empty symbols horsts.

The most cursory survey of known pull-apart basins indicates that their widths vary extensively from tens of meters in small sag ponds to tens of kilometers for large basins.   The range suggests that the process responsible for the initial spacing between fault strands is spatially variable.   Furthermore, because pull-apart basins of significantly different widths and lengths are often found along the same fault system (e.g., the San Andreas system), we conclude that new strands must be forming while slip occurs and that all of the basins were not formed during the onset of strike-slip motion.

These conclusions lead to a contradiction.   On the one hand, we must invoke an independent process for the spacing between strands prior to the process responsible for slip on these strands.   On the other hand, this independent process must be creating new strands while slip occurs on existing ones.

To resolve this apparent contradiction concerning the nature and origin of pull-aparts, horsts, and strike-slip fault systems in general, we examine the widths and lengths of 70 pull-apart basins and pressure ridges associated with the San Andreas Fault system and the Olinghouse Fault in the western United States, the Dead Sea Fault and other faults in Israel, the North Anatolian Fault in Turkey, the Dasht-e Bayaz Fault in Iran, the Polochic and Motagua faults in Guatemala, the Bocono, El Pilar, and Victoria faults in Venezuela, and the Hope Fault in New Zealand.

We have studied well-documented pull-apart basins or ridges along these fault systems, using data from written reports or detailed published geological maps (see Table 1) to determine the position of the bounding faults.   Some of the data have been obtained from aerial photos and from direct field observations.   For the 70 best documented

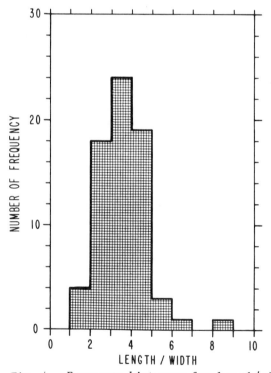

Fig. 4.   Frequency histogram for length/width ratio showing that most of the length-width ratios calculated directly from Table 1 fall between 2 and 5 and that the maximum frequency value is somewhere between the ratios 3 and 4.

cases, we have measured the length, ℓ, and width, w, of the basins or
ridges and plotted them against each other, as shown in Figure 3.
These data are analyzed by using a least square fit of the function

$$\log \ell = c_1 \log w + \log c_2 \tag{1}$$

The best-fitting constants in equations (1) were found to be $c_1 = 1.0$
and $c_2 = 3.2$. The variance of c_1 at 95% confidence level is .04, and
95% confidence interval for c_2 is defined by an upper limit of 4.3
and a lower limit of 2.4.

By taking antilogs, equation (1) can be rewritten as

$$\ell \simeq 3.2 \, w \tag{2}$$

Equation (2) shows that a well-defined linear correlation exists
between length and width, with a ratio of approximately 3. A 95%
confidence interval for the ratio (2.4-4.3) obtained from the linear
regression is in agreement with the most common range of ratios
(3-4) illustrated graphically in Figure 4, a relative frequency
histogram constructed by using ratios calculated directly from Table 1.

The persistence of this correlation ranging over scales from meters
to 100 kilometers, not only between length and width within a given
fault system, but also among systems, is particularly remarkable. The
strong correlation leaves little doubt as to the reality and gener-
ality of this observation: basins and ridges associated with strike-
slip faults become wider as they grow longer due to increasing fault
offsets with time.

This observation is inconsistent with the simplest model for the
formation of basins and ridges, which dictates that their widths are
initially fixed, whereas their lengths increase with offset (Figure 5).
If this were the case, we would expect no correlation between width and
length. We would expect to observe many very long and narrow basins
and ridges. Furthermore, the results do not easily favor a model
involving the development of strands independent of the slip along
them. In that case, we would anticipate a random width distribution
uncorrelated with the lengths of basins and ridges. The strong
correlation implies that as slip along the strike-slip fault increases,
so does the width as well as the length of the basin or ridge; therefore, the process which controls the width is not independent of the
slip but is an integral part of it.

MODELS

We propose two models for the evolution of pull-apart basins and
ridges. A simple process with a constant ratio of length to width,
a ratio that is independent of the magnitude of slip, is illustrated
in Figure 6. We imagine, for example, a right lateral fault system

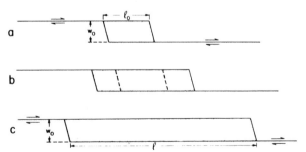

Fig. 5. A simple model illustrating increasing length (from ℓ_0 to ℓ)
with increasing fault offset. The width, w_0, is constant.

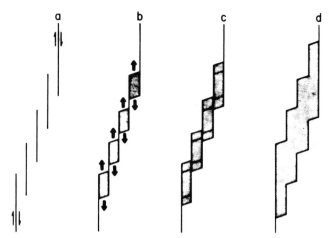

Fig. 6.   Model 1 showing coalescence of rhomb grabens associated with
en echelon strike-slip faults.   The end product is a composite
pull-apart basin.

consisting of numerous right-stepping echelon strands (Figure 6a).
Many small grabens appear initially (Figure 6b).   As slip increases,
the grabens begin to coalesce into composite ones (Figure 6c), finally
leading to a large basin (Figure 6d) having a length comparable to the
offset and the width that is the sum of the spacing between the
fault strands involved in the process.   Examples of basins that show
the elements of this process are common and can be recognized easily
based on the elbow-shaped geometry of the diagonal faults.   The Koehn
Lake Basin along the Garlock Fault in southern California (Figure 7)

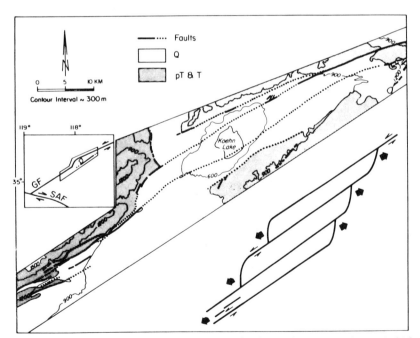

Fig. 7.   Koehn Lake basin on the Garlock Fault in southern California,
envisioned as a composite pull-apart basin formed by coalescence of
en echelon rhombs (insert).   GF: Garlock Fault; SAF: San Andreas
Fault; pT&T: pre-Tertiary and Tertiary; and Q: Quaternary.

can be thought of as a composite of at least three major strands, as
is shown in the inset. Additional examples can be seen along the
Elsinore Fault around Elsinore Lake in southern California [Rogers,
1965] and in the Gulf of Elat (Aqaba) in the Northern Red Sea [Ben-
Avraham et al., 1979].

The second model which is based on more random coalescense and
interaction processes is illustrated in Figure 8. The initial fault
configuration (Figure 8a) develops gradually, or perhaps the sites
of strike-slip faults are controlled by preexisting tensile fractures
[Segall, 1981]. At the initial stage grabens and horsts are produced
by interaction among closer and longer fault strands (Figure 8b).
Faults that are further away grow longer as more slip is accommodated,
and new strands form to promote further interaction and coalescense
resulting in the formation of longer and wider complex basins and
ridges (Figure 8c). Figure 9 shows a spectacular example of the
development of a composite basin, as envisioned in this model, along
the Olinghouse Fault in western Nevada (see Sanders and Slemmons
[1979] for more information about the fault). Here exceptionally
exposed smaller basins occur within larger basins and each basin
has similar length/width ratio (Figure 3).

The two processes described above may operate separately, or they
may operate more or less simultaneously at the same site complementing
each other. Large structures, such as the Imperial Valley, California
[Crowell and Sylvester, 1980; Fuis et al., 1981], exhibit not only
coalescense of neighboring basins, but also a more intricate composite
of basins within basins, horsts within basins, and perhaps basins
within horsts. Figure 10 is a cartoon illustrating the main features
of these tectonics, which, we believe, characterize a typical strike-
slip environment. Strike-slip faults together with connecting normal
and reverse or thrust faults divide the region into blocks or domains
or terranes, which, while moving in the general direction of horizontal shear, also rise or subside depending on the nature of the interaction between the discrete fault segments that make up the system.

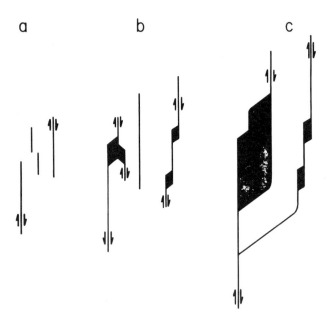

Fig. 8. Model 2 illustrates formation of composite pull-apart basin,
which includes rhomb grabens and horsts of various size.

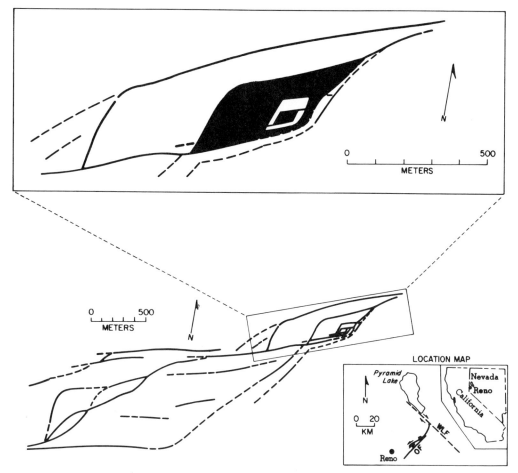

Fig. 9. Small basins along the Olinghouse Fault, which is a left
lateral strike-slip fault in the western Nevada (lower insert). Small
basins within basins (upper insert) on the top right of the figure
appear to be formed by the second mechanism shown in Figure 8. OF:
Olinghouse Fault; WLF Walker Lane Fault.

CONCLUSIONS AND IMPLICATIONS

    The clear, global correlation between the width and length of pull-
apart basins and ridges associated with strike-slip systems suggests
that smaller basins coalesce into bigger ones as slip continues to
take place. This conclusion has important implications for our
understanding of (1) fault systems and (2) formation of basins. The
two mechanisms suggested for the growth of basins and ridges provide
an understanding of the nature of strike-slip faulting as an evolutionary process. Basins and ridges of various sizes in the same
strike-slip fault system should be expected if the interaction and
coalescense processes leading to the formation of the basins and ridges
occur in a long time span.
    Faults that are traditionally classified into different groups such
as strike-slip, dip-slip normal, and reverse or thrust, and which are
believed to have distinct environments, can occur next to each other
in the same tectonic environment under the same remote stress condition. Normal and thrust faults associated with active strike-slip
faults should be recognized as potential active faults.

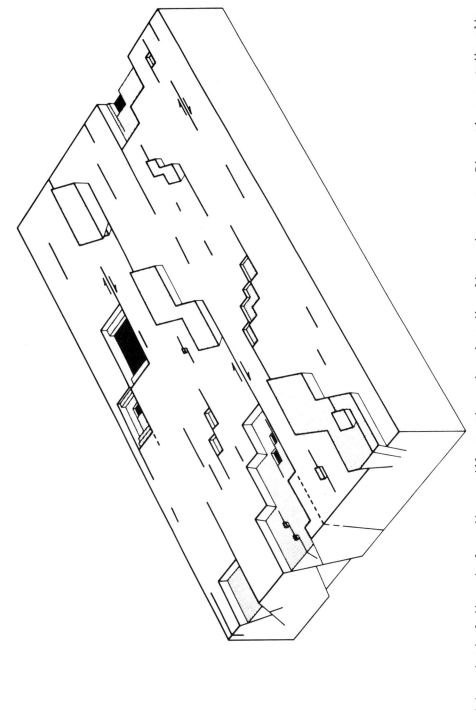

Fig. 10. A hypothetical dimensionless diagram illustrating a broad strike-slip environment. Discontinuous strike-slip faults with varying length and displacement interact and form basins and ridges of various size by an evolutionary process. Darker areas: grabens or basins; lighter areas: horsts or ridges.

The processes of coalescense and interaction imply that the width of
the fault system itself must also tend to grow with time, incorporating
old and new fault strands as well as a complex arrangement of basins
and ridges. These broad zones, which are broken by faults, are
likely to be mechanically weaker than normal crust. The presence of
a weak, brittle upper crust around major faults limits the shear
stress level that can be supported by such faults. This limitation
may account for the low stresses inferred, for example, from in situ
stress measurement around the San Andreas Fault system [Zoback and
Roller, 1979].

The dimensional and geometric features of the basins and ridges described in this study, together with the nature of deformation in these
tectonic domains, can be used to interpret ancient basins and ridges
in terms of strike-slip tectonics. The fact that pull-apart basins
become wider as they grow longer may provide a mechanism for the
initiation and the enlargement of sedimentary basins. Sedimentary
basins and back arc basins probably develop as a result of crustal
stretching [Sclater and Christie, 1980; Dewey, 1980] followed by the
rise of hot and light mantle material. As this material cools, the
surface above it subsides, creating a basin that is usually filled
with sediments. The most viable process for crustal stretching is a
pull-apart basin, which must be large enough (tens of kilometers in
width) to interact with the upper mantle. Our observations suggest
that a large pull-apart basin can develop from small ones if the
associated fault displacements are large enough and the fault strands
are numerous enough.

Acknowledgments. We thank D. D. Pollard, P. Segall, R. V. Sharp,
M. M. Clark, R. E. Wallace, G. Mavko, G. Plafker, Z. Ben-Avraham, and
G. Aral for many fruitful discussions and for their encouragement,
C. Sanders, who loaned us the aerial photographs of the Olinghouse
Fault, and Z. Garfunkel and G. Fuis, who have made available the
preprints of their recent papers. The manuscript was reviewed by
D. D. Pollard, R. V. Sharp, and C. Sanders. This study was supported
by research grants from U. S. Geological Survey and NASA's geodynamics
program.

REFERENCES

Arpat, E., F. Şaroğlu, and H. B. Iz, 1976 Çaldıran depremi Yeryuvarı
    ve İnsan, Şubat 1977, 29-41, 1977.
Bartov, Y. (Compiler), Israel-geological map, scale 1:500,000, Surv.
    of Israel, Tel Aviv, 1979.
Ben-Avraham, Z., G. Almagor, and Z. Garfunkel, Sediments and structure
    of the Gulf of Elat (Aqaba) - Northern Red Sea, Sediment. Geol., 23,
    239-267, 1979.
Bonis, S., O. H. Bohnenberger, and G. Dengo (Compilers), Mapa geologico
    de la Republica de Guatemala, escale 1:500,000, Inst. Geograf. Nac.
    Guatemala, Guatemala, 1970.
Brown, R. D., Jr., Map showing recently active breaks along the San
    Andreas and related faults between the northern Gabilan Range and
    Cholame Valley, California, U. S. Geol. Surv. Misc. Geol. Invest.
    Map, I-579, 1970.
Carey, S. W., The Tectonic approach to continental drift, in
    Continental Drift: A Symposium Held in the Geology Department,
    University of Tasmania, March 1956, edited by S. W. Carey, pp. 177-
    355, Univ. of Tasmania, Hobart, 1958.
Clark, M. M. Surface rupture along the Coyote Creek Fault, The
    Borrego Mountain Earthquake of April 9, 1968, U. S. Geol. Surv.
    Prof. Pap., 787, 55-86, 1972.
Clark, M. M., Map showing recently active breaks along the Garlock

and associated faults, California, U. S. Geol. Surv. Misc. Geol.
    Invest. Map, I-741, 1973.
Clayton, L., Tectonic depressions along the Hope Fault, a transcurrent
    fault in North Canterbury, New Zealand, N. Z. J. Geol. Geophys., 9,
    95-104, 1966.
Crowell, J. C., Origin of Late Cenozoic basins in southern California,
    Spec. Publ. Soc. of Econ. Paleontol. and Mineral., 22, 190-204,
    1974.
Crowell, J. C., and A. G. Sylvester, Introduction to the San Andreas
    -Salton Trough juncture, Tectonics of the Juncture Between the
    San Andreas Fault System and the Salton Trough, Southern California
    - A Guidebook, edited by J. C. Crowell and A. G. Sylvester, Publ.
    1-13, Univ. of Calif. Dept. of Geol. Sci., Santa Barbara, 1980.
Dewey, J. F., Episodicity, sequence, and style at convergent plate
    boundaries, Geol. Assoc. Can. Spec. Pap., 20, 553-573, 1980.
Dibblee, T. W., Jr., Late Quaternary Uplift of the San Bernardino
    Mountains on the San Andreas and related faults, Spec. Rep. Calif.
    Div. Mines. Geol., 118, 127-135, 1975.
Elders, W. A., R. W. Rex, T. Meidav, P. T. Robinson, and S. Biehler,
    Crustal spreading in southern California, Science, 178, 15-24,
    1972.
Eyal, M., Y. Eyal, Y. Bartov, and G. Steinitz, Sinistral faulting
    in eastern Sinai, in Programs, Abstracts, Annual Reports, Explanatory
    Notes on the Excursions, Israel Geological Society, Jerusalem, 1980.
Freund, R., The Hope Fault, a strike-slip fault in New Zealand, N. Z.
    Geol. Surv. Bull., 86, 1-48, 1971.
Freund, R., Kinematics of transform and transcurrent faults,
    Tectonophysics, 21, 93-134, 1974.
Freund, R., I. Zak, and Z. Garfunkel, Age and rate of the sinistral
    movement along the Dead Sea Rift, Nature, 220 (5164), 253-255, 1968.
Fuis, G. S., W. D. Mooney, J. H. Healy, G. A. McMechan, and W. J.
    Lutter, Crustal structures of the Imperial Valley region, in the
    Imperial Valley earthquake of October 15, 1979, U. S. Geol. Surv.
    Prof. Pap., in press, 1982.
Garfunkel, Z., Internal structure of the Dead Sea leaky transform
    (rift) in relation to plate kinematics, Tectonophysics, in press,
    1982.
Garfunkel, Z., I. Zak, and R. Freund, Active faulting in the Dead Sea
    rift, Tectonophysics, in press, 1982.
Hill, D. P., A model for earthquake swarms, J. Geophys. Res., 82,
    1347-1352, 1977.
Jennings, C. W. (Compiler), Geologic Map of California, San Luis
    Obispo sheet, scale 1:250,000, Calif. Div. of Mines and Geol.,
    San Francisco, 1959.
Jennings, C. W., J. L. Burnett, and B. W. Troxel (Compilers),
    Geologic Map of California, Trona sheet, scale 1:250,000, Calif. Div.
    of Mines and Geol., San Francisco, 1969.
Johnson, C. E., Seismotectonics of the Imperial Valley of southern
    California, part 2, Ph.D. thesis, 208 pp., Calif. Inst. of Technol.,
    Pasadena, 1979.
Johnson, C. E., and D. M. Hadley, Tectonic implication of the Brawley
    earthquake swarm, Imperial Valley, California, January 1975, Seismol.
    Soc. Am. Bull., 66, 1133-1144, 1976.
Ketin, I., Kuzey Anadolu fayı hakkında (in Turkish and German), Maden
    Tetkik Arama Enst. Derg., n 72, 1-27, 1969.
Plafker, G., Tectonic aspects of the Guatemala earthquake of February
    4, Science, 193, 1201-1208, 1976.
Rispoli, R., Stress fields about strike-slip faults inferred from
    stylolites and tension gashes, Tectonophysics, 75, T29-36, 1981.
Rodgers, D. A., Analysis of pull-apart basin development produced by
    en echelon strike-slip faults, Spec. Publ. Int. Assoc. Sedimentol.,
    4, 27-41, 1980.

Rogers, T. H. (Compiler), Geologic map of California, Santa Ana sheet, scale 1:250,000, Calif. Div. of Mines and Geol., San Francisco, 1965.

Sanders, C. O. and D. B. Slemmons, Recent crustal movements in the central Sierra Nevada - Walker Lane region of California-Nevada, 3, The Olinghouse Fault zone, Tectonophysics, 52, 585-597, 1979.

Schubert, C., El Pilar Fault zone, northeastern Venezuela: Brief review, Tectonophysics, 52, 447-455, 1979.

Schubert, C., Late-Cenozoic pull-apart basins, Bocono fault zone Venezuelan Andes, J. Struct. Geol., 2, 463-468, 1980a.

Schubert, C., Morfologia neotectonica de una fallon rumbodeslizante e informe preliminar sobre Ra fallon de Bocono (with English abstract), Acta Cient. Venez., 31, 98-111, 1980b.

Schubert, C., and M. Laredo, Late Pleistocene and Holocene faulting in Lake Valencia basin, north-central Venezuela, Geology, 7, 289-292, 1979.

Schwartz, D. P., L. S. Cluff, and T. W. Donnelly, Quaternary faulting along the Caribbean-North American Plate boundary in Central America, Tectonophysics, 52, 431-445, 1979.

Sclater, J. G. and P. A. F. Christie, Continental stretching: An explanation of the post-mid-Cretaceous subsidence of the central North Sea Basin, J. Geophys. Res., 85, 3711-3739, 1980.

Segall, P., The development of joints and faults in granitic rocks, Ph.D. thesis, 233 pp., Stanford University, Stanford, Calif., 1981.

Segall, P., and D. D. Pollard, Mechanics of discontinuous faults, J. Geophys. Res., 85, 4337-4350, 1980.

Seymen, I., Kelkit vadisi kesiminde Kuzey Anadolu fay zonunun tektonik özelliği (with English abstract), Ph.D. thesis, 192 pp., Istanbul Teknik Universitesi, Istanbul, Turkey, 1975.

Sharp, R. V., Map showing recent active breaks along the San Jacinto Fault zone between the San Bernardino area and Borrego Valley, California, U. S. Geol. Surv. Misc. Geol. Invest. Map, I-675, 1972.

Sharp, R. V., En echelon fault patterns of the San Jacinto fault zone, San Andreas Fault in southern California, Spec. Rep. Calif. Div. Mines Geol., 118, 147-152, 1975.

Sharp, R. V., Surface faulting in Imperial Valley during the earthquake swarm of January-February, 1975, Seismol. Soc. Am. Bull., 66 1145-1154, 1976.

Sharp, R. V., Map showing the Holocene surface expression of the Brawley Fault, Imperial County, California, U. S. Geol. Surv. Misc. Field Stud. Map, MF-838, 1977.

Sharp, R. V., and M. M. Clark, Geological evidence of previous faulting near the 1968 rupture of the Coyote Creek Fault, in the Borrego Mountain earthquake of April 9, 1968, U. S. Geol. Surv. Prof. Pap., 787, 131-140, 1972.

Smith, A. R. (Compiler), Geological map of California, Bakersfield sheet, scale 1:250,000, Calif. Div. of Mines and Geol., San Francisco, 1964.

Tchalenko, J. S., and N. N. Ambraseys, Structural analysis of the Dasht-e Bayaz (Iran) earthquake fractures, Geol. Soc. Am. Bull., 81, 41-60, 1970.

Zoback, M. D., and J. C. Roller, Magnitude of shear stress on the San Andreas Fault, implications from a stress measurement profile at shallow depth, Science, 206, 445-447, 1979.

(Received August 25, 1981;
revised November 16, 1981;
accepted November 16, 1981.)

THE AMERICAN ASSOCIATION OF PETROLEUM GEOLOGISTS BULLETIN

DECEMBER 1976   VOLUME 60, NUMBER 12

Tectonic Transpression and Basement-Controlled Deformation in San Andreas Fault Zone, Salton Trough, California[1]

ARTHUR G. SYLVESTER[2] and ROBERT R. SMITH[3]
Santa Barbara, California 93106, and Houston, Texas 77001

Abstract The San Andreas fault zone in the Salton trough is characterized by subparallel, high-angle faults, en-echelon folds, and systematically arranged arrays of reverse and normal faults which are typical of major wrench faults elsewhere in the world. Detailed field mapping in the Mecca Hills shows that the fault zone is delineated by two northeast-striking high-angle faults which subdivide pre-Cenozoic crystalline and schistose basement and overlying late Cenozoic and Quaternary nonmarine sedimentary rocks into three structural domains: (1) a high-standing, relatively undeformed block northeast of the fault zone; (2) a 1.5 km-wide zone of folded sedimentary rocks between the two faults; and (3) a sparsely exposed, deeply down-faulted block on the southwest. Locally the two faults branch upward and flatten abruptly outward into low-angle thrusts which carry thrust slices and gravity slides of sedimentary rocks from the central block short distances on to the adjacent blocks. Between the two faults the sedimentary succession is folded into broad west-northwest-trending open folds which tighten and are overturned and truncated against the faults. Where exposed in the fault zone the basement–sedimentary rock contact is a buttress unconformity which also is folded and overturned locally. Field observations reveal that the basement in the core of the fold is fractured and sheared pervasively, indicating that the mechanism of basement deformation was one of cataclastic flow by piecemeal slip on the closely spaced fracture and shear planes, whereas the folding of the overlying sedimentary rocks was largely a passive consequence of deformation at the basement level. Although horizontal shear strain predominated, local vertical uplift of the basement occurred as a result of compression between two convergently and laterally slipping rigid crustal blocks.

INTRODUCTION

En-echelon anticlines associated with major wrench faults constitute the structural style of some of the most important and productive oil and gas fields in California, as well as other parts of the world (Harding, 1973, 1974, 1976). Together with related subsidiary reverse faults and low-

angle thrusts, these folds are evidences of a shortening component of strain associated with the more dominant shear component. Crowell (1974) has shown that these structures form typically in the vicinity of bent or braided parts of wrench faults where the slip of the two crustal blocks is simultaneously lateral and convergent. Harland (1971) proposed the term "transpression" for the concept of lateral and convergent slip along wrench faults.

Recent clay-model studies of the geometry and kinematic history of the assemblages of structures formed in transpression (Wilcox et al, 1973), show that en-echelon folds typically form at low angles to the main trace of a wrench fault, and that the geometry of the fold axial traces depends on the sense of horizontal shear strain (Fig. 1).

Although these model studies, together with those of Hafner (1951) and Sanford (1959) on dip-slip faults, give bases for understanding the origin of upthrusts and associated folds in thick

[1]Manuscript received, December 18, 1975; accepted, June 1, 1976.

[2]Department of Geological Sciences, University of California.

[3]Shell Oil Company.
The writers are indebted to the Shell Development Company, for which initial phases of the study were conducted, for permission to publish this paper. Earlier versions of the manuscript were reviewed critically by J. K. Arbenz, A. W. Bally, J. H. Howard, N. L. McIver, and S. Snelson. Several of the tectonic models were developed subsequently by the writers while the senior author was at the University of California, Santa Barbara. We also acknowledge with pleasure the critical comments upon the present manuscript by T. P. Harding who kindly made available manuscripts of some of his papers prior to their publication.

2081

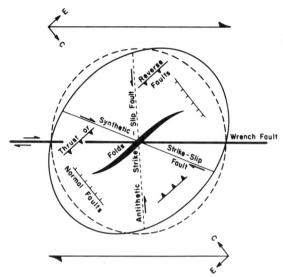

FIG. 1—Geometric relation of folds and faults to right-slip wrench fault combined schematically with strain ellipse and principal strain directions, contraction, *C*, and extension, *E*. Axial traces of folds have flattened S-shapes, corresponding to field and laboratory observations. Geometry for left-slip wrench fault is mirror image of this diagram. Adapted from Harding (1974, p. 1291).

sequences of layered sedimentary rocks, little attention has been focused on the nature and mechanism of deformation at the basement level in wrench-fault zones, or on the effect of basement deformation on the overlying sedimentary succession. This lack of attention has led to erroneous assumptions and questionable interpretations. For example, it is common to find in the literature structure sections across wrench faults which imply a moderately to tightly folded crystalline basement–sedimentary rock interface that is concentric to the observed folds in the overlying sedimentary rocks, as if the basement deformed plastically. However, palinspastic reconstructions of the depth to basement in most studies show that the basement was relatively shallow during deformation—too shallow to have deformed ductily by plastic flow and recrystallization. As Prucha et al (1965) showed in upthrust areas and from experimental studies in rock deformation, it is clear that basement composed of rock types such as gneiss and massive plutonic rocks cannot be folded simply, but must deform by pervasive brittle fracturing leading to cataclastic flow at the relatively shallow depths inferred from the palinspastic reconstructions.

In this paper we document the geometry and tectonics of structures in an unusually well-exposed part of the southern San Andreas fault zone where the mechanism of basement deformation is displayed clearly and can be related to

folding of the overlying sedimentary cover. Whereas Crowell (1974) provided models for wrench-fault tectonics on a regional scale, this investigation provides models on the scale of an oil field for the geometry and origin of subsidiary folds and faults which form as a result of transpression and which, in other areas, have been found to be important traps for hydrocarbons.

TECTONIC FRAMEWORK

The Coachella and Imperial Valleys (Fig. 2) comprise an elongated structural depression known as the Salton trough which is considered to be the landward extension of the Gulf of California. The trough is believed to be a complex rift formed by crustal spreading at the northern extension of the East Pacific Rise; the plate motion is thought to be transformed to the San Andreas fault near the south end of the Salton Sea, south of which the fault has not been traced with confidence (Carey, 1958; Hamilton, 1961; Allison, 1964; Biehler et al, 1964; Hamilton and Myers, 1966; Elders et al, 1972; Sharp, 1972).

Salton trough is bounded by northwest-trending mountains which are underlain on the northeast by plutonic, metamorphic, and volcanic rocks ranging in age from Precambrian to late Tertiary (Table 1). Massive plutonic rocks of the southern California batholith of middle Cretaceous age and prebatholithic metamorphic rocks are exposed in the Santa Rosa Mountains and other ranges southwest of the trough.

Gravity studies in Coachella Valley indicate that this part of the trough is a half-graben tilted northeastward with the deepest part (4,000 to 5,000 m deep) against the San Andreas fault zone adjacent to the Mecca Hills (Biehler, 1964).

The broad outlines of the stratigraphy of the Cenozoic rocks filling the trough have been summarized by Tarbet and Holman (1944), Tarbet (1951), and Dibblee (1954), but the details are not well known because obscure well records cannot be correlated confidently with sparse exposures along the trough margins (Dibblee, 1954; Muffler and Doe, 1968). In general, the distribution of sedimentary facies is asymmetric as shown in Figure 3. Preceding and following a marine incursion in the Pliocene, represented by shale and mudstone of the Imperial Formation, deposition was characterized by intermittent and interfingering alluvial-fan, flood-plain, and lacustrine sediments derived from the surrounding mountains and the Colorado River (Merriam and Bandy, 1965). Volcanic rocks of late Cenozoic age crop out sparingly around the edges of the trough and are intercalated with the sedimentary succession (Fig. 3).

The San Andreas fault zone is the northeasternmost of a number of major northwest-

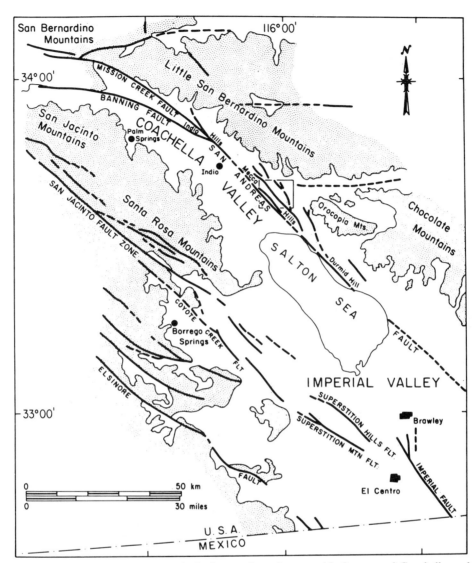

FIG. 2—Index map showing principal tectonic and geographic features of Coachella and Imperial Valleys. Together, these two valleys and their tectonic elements comprise Salton trough. Stippled areas are underlain by pre-Cenozoic crystalline and schistose basement rocks.

trending faults of the San Andreas fault system (Fig. 2) which transect the Salton trough and trend more westerly than the axis and borders of the trough (Sharp, 1972). The San Andreas fault zone is exposed discontinuously as far south as the Durmid area in the Indio Hills and the Mecca Hills; between these exposed areas, however, the fault can be traced by discontinuous fault scarps in Quaternary alluvium.

Detailed mapping in the Indio Hills (Dibblee, 1954; Popenoe, 1959; Stotts, 1965), the Mecca Hills (Dibblee, 1954; Hays, 1957; Ware, 1958; Sylvester and Smith, 1975; and this paper), and the Durmid area (Dibblee, 1954; Babcock, 1974) shows that the San Andreas fault zone is a relatively narrow, northwest-trending zone of anastomosing high-angle faults and associated west-

northwest-trending, locally overturned en-echelon folds. The axial traces of the folds are either straight or curvilinear in flattened S patterns, indicative of right slip on the wrench fault. Subsidiary high-angle faults have arcuate traces trending from nearly north a few kilometers northeast of the zone, to northwest within the zone itself. Their geometric relation to the main trace of the San Andreas fault corresponds to the orientation of the synthetic set of Riedel shears (in the terminology of Tchalenko and Ambraseys, 1970) characteristic of a wrench-fault system (Fig. 1).

The Mecca Hills, on the northeast margin of the Salton trough (Fig. 2), consist of pre-Cenozoic crystalline basement rocks overlain by late Cenozoic nonmarine sedimentary rocks. The San Andreas–Skeleton Canyon fault and the Painted

TABLE 1. ROCK TYPES AND STRUCTURES OF BASEMENT ROCKS

Name	Lithology	Structures
Cottonwood Granite	Porphyritic quartz monzonite	Plutons intruded into batholitic basement complex
Orocopia Schist	Schist, light greenish-gray to black, composed chiefly of chlorite-albite-quartz, quartz-biotite-muscovite-albite, muscovite-actinolite; some quartzite; cut by numerous veins of massive white quartz, cut locally by diabase and rhyolite	Broad east-west trending anticlinal arch with lesser folds with amplitudes up to 20 m
Prebatholithic Complex	Hornblende-biotite gneiss, chiefly, dark greenish-gray; intruded locally by diorite, anorthosite, gabbro, syenite, and dikes of granite, pegmatite, diabase, and felsite (Crowell and Walker, 1962)	Disharmonic flow folds; shear fractures; low- and high-angle faults which do not extend into the overlying sedimentary rocks

Canyon fault subdivide the Mecca Hills into two structurally distinct domains or blocks which, for the purposes of this study, are termed the platform block and the central block (Fig. 6). A third domain, the basin block, is inferred from gravity data (Biehler, 1964) beneath a thick cover of alluvium southwest of the San Andreas fault. Each block is distinguished by marked differences in deformation style, lithostratigraphy, and stratigraphic thicknesses of sedimentary cover (Table 2). Detailed field investigations were limited to the central part of the area (Fig. 4) where the structural relief is greatest. The structures of each block are discussed in the following sections. Structures related to the major faults are considered in the discussion of the central block.

The only documented historic fault movement in the Mecca Hills was associated with the Borrego Mountain earthquake of April 8, 1968, which had a Richter magnitude of 6.4 (Allen et al, 1968; Allen and Nordquist, 1972). Whereas the primary displacements took place along the Coyote Creek fault on the southwest side of Salton trough (Fig. 2), minor right-slip (from 1 to 2 cm) or creep took place at about the same time on other faults in the Salton trough, including the San Andreas fault in the Mecca Hills. Seismologists concluded that these minor displacements were triggered dynamically by the Borrego Mountain earthquake (Allen et al, 1968, 1972). This was the first time such a triggering phenomenon was demonstrated convincingly.

LITHOLOGY

Basement Rocks

Banded migmatitic gneiss and anorthosite and related rocks of Precambrian age, Mesozoic granitic rocks, and Orocopia Schist (Table 2) are ex-

TABLE 2. LITHOSTRATIGRAPHIC AND STRUCTURAL CONTRASTS AMONG THREE STRUCTURAL BLOCKS OF MECCA HILLS

Basin Block	Central Block	Platform Block
	Pre-Cenozoic Basement Rocks	
Not exposed	Gneiss and granite, highly sheared, of Chuckawalla Complex	Gneissic and plutonic rocks of Chuckawalla Complex, moderately sheared to unsheared; Orocopia Schist
	Basement-sedimentary rock surface steeply tilted to southwest	Basement-sedimentary rock surface gently inclined to southwest
	Cenozoic Sedimentary Rocks	
Alluvium	Arkose and conglomeratic arkose	Arkose, conglomeratic, and conglomerate
Thickness: 3,000-5,000 m (12,000-15,000 ft)	Thicker stratigraphic sequence than in eastern block (approximately 1,750 m or 5,000 ft)	Relatively thin stratigraphic sequence (<750 m; <2,000 ft)
Structure of sedimentary rocks beneath alluvial cover is not known	Broad open folds, locally appressed, and overturned, with axes oblique to traces of major faults	Virtually unfolded except for minor drag folds with axes slightly oblique to fault trends
	Steep northwest-trending normal cross faults	Steep to gently inclined northwest-trending normal faults

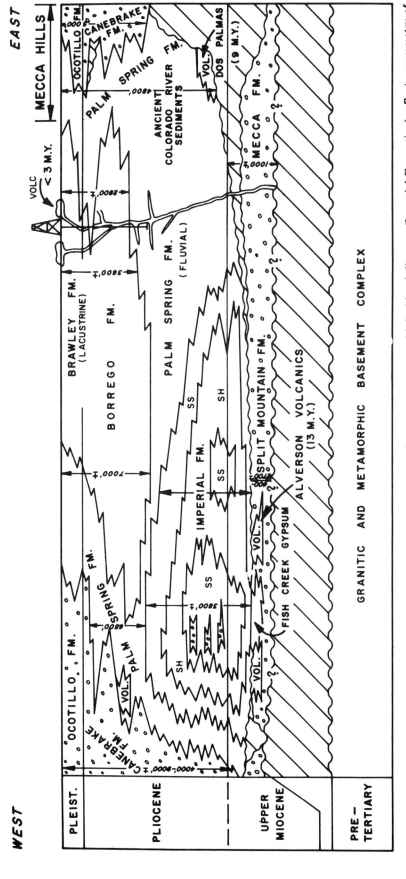

FIG. 3—Time-stratigraphic diagram for Salton trough. Presence of marine sedimentary sequence of left side of diagram (Imperial Formation) reflects asymmetry of sedimentation in Imperial Valley. All formations except Imperial Formation are nonmarine. Adapted from Abbott (1968).

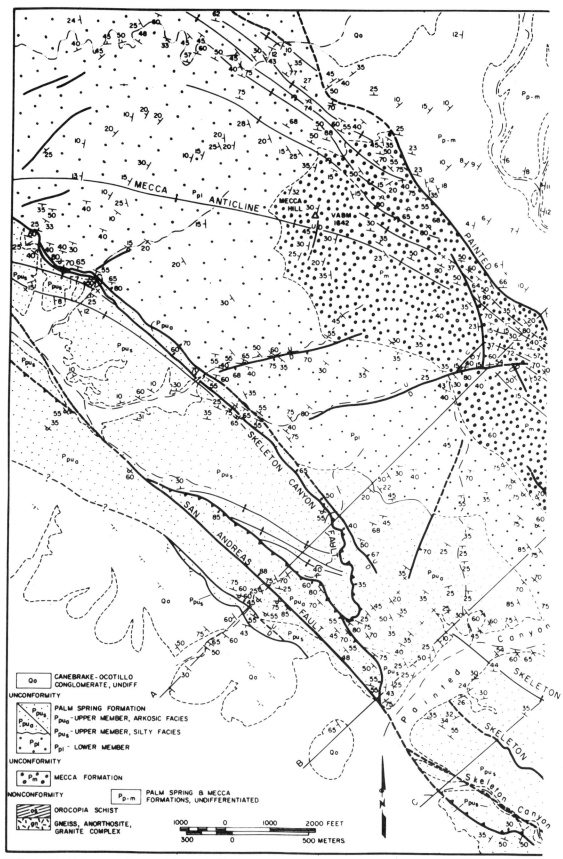

FIG. 4—Geologic map of central Mecca Hills. *AA'*, *BB'*, *CC'* are locations of cross sections shown in Figure 5.

362

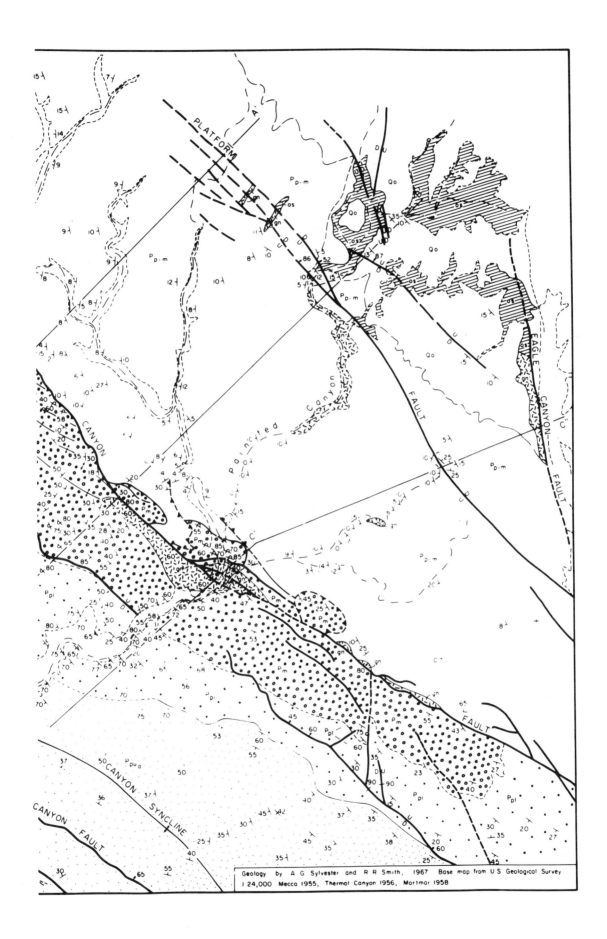

Geology by A G Sylvester and R R Smith, 1967   Base map from U S Geological Survey
1 24,000 Mecca 1955, Thermal Canyon 1956, Mortmar 1958

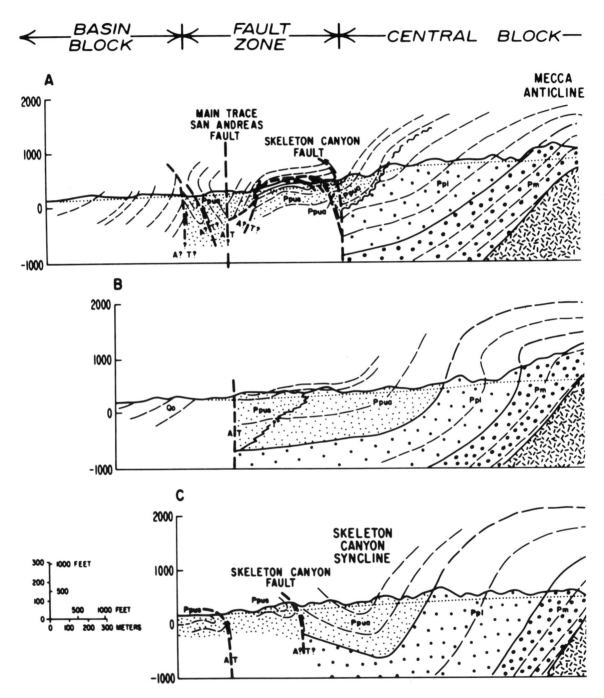

FIG. 5—Structural cross sections of central Mecca Hills. For locations see Figure 4.

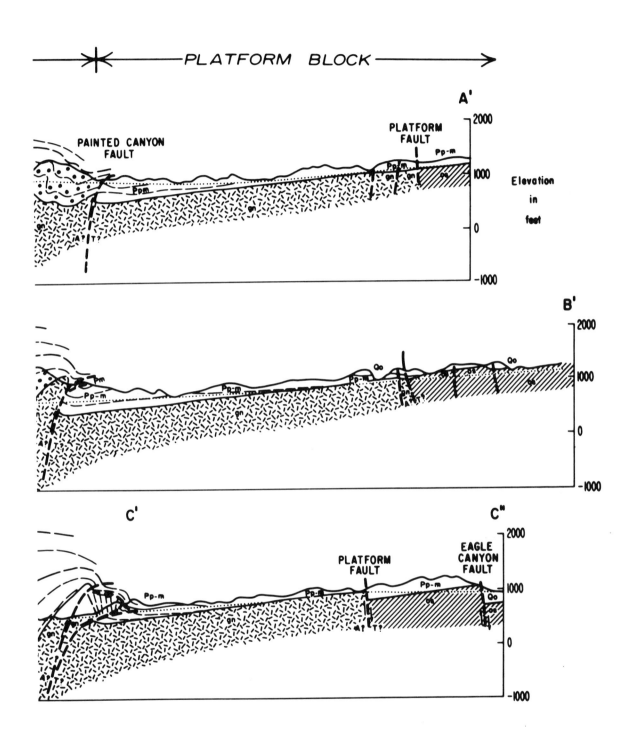

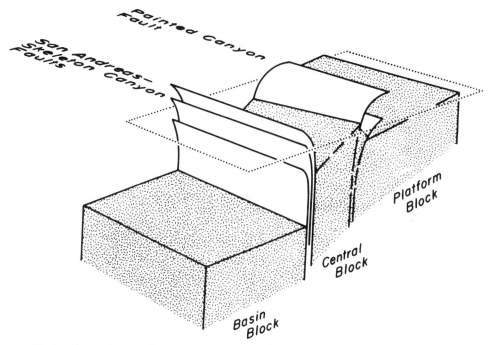

FIG. 6—Idealized block diagram of basement and principal faults in Mecca Hills. Dotted parallelogram represents surface.

posed sparingly in the Mecca Hills where deep canyons have cut through the overlying sedimentary sequence (Figs. 4, 5). Felsite dikes cut the gneiss in Painted Canyon at the Painted Canyon fault; they have been dated at 24 m.y. by K-Ar methods (G. Edwards, personal commun., 1967) and represent an early Miocene plutonic-volcanic belt that is widespread in southern California and southwestern Arizona (Olmsted, 1966; Miller and Morton, 1974; Dillon, 1975a). The age and relation of the Orocopia Schist to other basement rock types are not well known, although the schist is considered to have been metamorphosed in late Mesozoic time (Ehlig, 1968). In the nearby Orocopia and Chocolate Mountains, the Orocopia Schist is overthrust by the Precambrian gneissic complex (Crowell, 1962, 1975; Dillon, 1975b; Dillon and Hazel, 1975); in the Mecca Hills these two rock units are separated by the high-angle Platform fault.

Sedimentary Rocks

The basement rocks are overlain by nonmarine late Cenozoic and Quaternary sedimentary rocks representing alluvial-fan, braided-stream, and lacustrine environments of deposition (Table 3). Stratigraphic thicknesses, age relations, and correlation of various lithologic units along strike and across faults are tenuous because of numerous diastems, abrupt facies changes, and the lack of fossils and distinctive marker beds. However, the gross nature of the sequence records a history of nonmarine deposition near an active basin margin. This is illustrated especially in Painted Canyon by the lowermost unit of the sequence, the Mecca Formation, which thickens abruptly and coarsens markedly across the Painted Canyon fault (Fig. 5). The strata are comprised chiefly of coarse, locally derived, torrentially deposited detritus of gneiss, Orocopia Schist, and granite. The Mecca Formation lies nonconformably on basement northeast of the fault and in buttress unconformity southwest of it. The contact is not exposed for any great distance laterally from Painted Canyon, precluding a more confident reconstruction of the depositional-basin framework.

The Palm Spring Formation (Table 3) appears to mark an abrupt change in provenance from that of the older Mecca Formation, in that strata in the central Mecca Hills are comprised almost entirely of granitic detritus; Orocopia Schist detritus is predominant in the eastern Mecca Hills near the Orocopia Mountains. The Palm Spring Formation appears to record spreading of the central and distal parts of alluvial fans from the Little San Bernardino Mountains on the northeast which are underlain almost entirely by Mesozoic granitic plutons, and from the Orocopia Mountains on the east. Like the Mecca Formation, the Palm Spring Formation thickens abruptly southwest of the Painted Canyon fault (Fig. 5).

TABLE 3. THICKNESS, AGES, AND LITHOLOGY OF CENOZOIC FORMATIONS IN MECCA HILLS*

Formation	Lithology
Canebrake-Ocotillo Conglomerate (Pleistocene) 0-750 m (0-5,000 ft)	Conglomerate, gray, of granitic debris in central Mecca Hills, conglomerate, reddish, of schist in eastern Mecca Hills.
Palm Spring Formation (Pliocene? and Pleistocene) 0-1,200 m (0-4,800 ft)	Upper memmber: sandstone, thin-bedded buff arkosic, grading basinward into light greenish sandy siltstone. Lower member: conglomerate, thick-bedded buff arkosic and arkose with thin interbeds of gray-green siltstone.
Mecca Formation (Pliocene) 0-225 m (0-800 ft)	Arkose, reddish, conglomerate, claystone; chiefly metamorphic debris in basal strata.

*After Dibblee, 1954

and is progressively finer grained southwestward toward the axis of the Salton trough. Numerous diastems within the formation southwest of the Painted Canyon fault probably reflect Pliocene-Pleistocene episodes of folding and faulting along the trough margin.

The Canebrake-Ocotillo conglomerates range in age from Pliocene to recent (Table 3) and are the coarse coalescent alluvial-fan facies of the Palm Spring Formation (Dibblee, 1954).

STRUCTURE

Platform Block

The platform block is a discrete structural domain between the Painted Canyon fault and the front of the Little San Bernardino Mountains. It is a pediment with a moderately incised planate erosion surface cut upon gneiss, granite, and Orocopia Schist; the erosion surface dips about 10° to the southwest and is thinly veneered with fluvial strata of the Mecca, Palm Spring, and Canebrake-Ocotillo Formations.

The basement rocks in the platform block are relatively massive and unfractured in contrast to the pervasively fractured and sheared exposures in the central block that are described below. The gneiss is characterized by disharmonic flow folds. Low to high-angle faults lacking gouge locally cut the basement, but they do not extend upward into the overlying sedimentary strata.

The sedimentary strata are undeformed except where they are dragfolded adjacent to several northwest-trending faults which break the pediment into small horsts and grabens (Figs. 4, 5). The faults are generally planar and dip from 50 to 90° with gouge zones ranging from a few centimeters to as much as a meter in width. The drag folds in relatively competent beds of arkose and conglomerate are gentle open structures, whereas in claystone and siltstone, they are smaller, tighter, and moderately overturned (Fig. 7). The synclines in the downfaulted blocks are more open than the corresponding anticlines. Axes of the drag folds are inclined gently and, together with

gently inclined slickensides on fault surfaces, indicate a complex history of oblique displacement of the fault blocks.

The Platform fault is a major feature of the platform block, because it separates gneiss and related Precambrian basement rocks from the Orocopia Schist (Fig. 8). The overlying sedimentary strata also are dissimilar across the fault, but their relative position in the stratigraphic succession could not be determined with confidence, so that the sense and amount of displacement could not

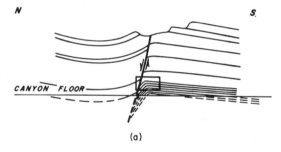

(a)

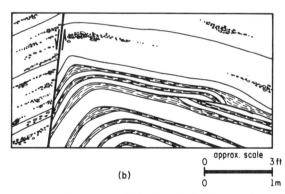

(b)

FIG. 7—Geometry of Platform fault 1 km northwest of intersection with Eagle Canyon fault. a, Diagrammatic sketch of east wall of canyon. Drag folding is tighter in footwall than in hanging wall. Detail of rectangular area is shown in b, thin-bedded sequence of siltstone and claystone (dashed) is dragfolded adjacent to fault and bedding-plane fault in footwall block superposes some beds so that section is dilated locally.

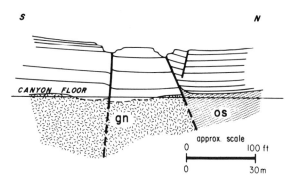

FIG. 8—Diagrammatic sketch of Platform fault in the east wall of upper Painted Canyon. Platform fault separates gneiss, *gn*, from Orocopia Schist, *os*.

be determined. Nearly horizontal slickensides in the fault plane show that strike-slip displacement occurred most recently.

Central Block

The central block is a structural domain from 1 to 3 km wide between the Painted Canyon fault and the Skeleton Canyon–San Andreas fault zone (Figs. 3-5). It is characterized by broad folds, most of whose axial traces trend west-northwest, oblique to the bounding faults. Adjacent to the faults, the folds are appressed, locally overturned, and in some places truncated by faults. The geometry and structural style of this domain is similar to those documented in proximity to other wrench faults in the Salton trough (Dibblee, 1954; Popenoe, 1959; Stotts, 1965; Sharp and Clark, 1972; Babcock, 1974) and elsewhere (Wilcox et al, 1973).

Basement of gneiss and granite is exposed only along the Painted Canyon fault in Painted Canyon (Figs. 4, 5). The contact between the basement and the Mecca Formation is a buttress unconformity. The angle between the strata and the erosion surface on the basement is about 15°. Locally the contact between the Mecca and Palm Spring Formations is a minor dip-slip fault.

Folds—The Mecca and Palm Spring Formations are folded into several broad major westnorthwest-trending folds. Mecca anticline, with a core of highly fractured and sheared basement, is flanked on the south by the Skeleton Canyon syncline and on the north by the Mecca syncline (Figs. 4, 5). Mecca anticline plunges about 15° west-northwestward, so that its structurally deepest exposures are at its northeastern end adjacent to the Painted Canyon fault in Painted Canyon where the anticline is breached to the basement. The north flank of Mecca anticline is folded into a series of smaller folds which are overturned locally to the north-northwest and are truncated by the fault (Figs. 4, 5, 9). Skeleton Canyon syncline

is a broad open fold whose axial trace is approximately parallel with the Skeleton Canyon–San Andreas faults. Between the two faults, about 1 km northwest of the mouth of Painted Canyon, two asymmetric anticlines with an intervening syncline are exposed with axial traces subparallel with that of Mecca anticline. They are overturned to the south-southwest.

The tectonic relation of the folded sedimentary sequence and the basement is visible northwest of Painted Canyon adjacent to the Painted Canyon fault where Mecca anticline plunges 30° west-northwest. There the gneissic and granitic basement core is fractured and sheared so closely that it has been reduced to a granulated mass of rock fragments ranging typically from 0.5 to 5 cm in diameter. The degree of fracturing increases toward the fault plane until, at the fault itself, the basement is pulverized and gouged; but even in these highly deformed exposures, vestiges of migmatitic banding still are preserved in the gneiss, showing that the internal disruption is slight but pervasive. However, the overlying sedimentary strata are sheared only adjacent to the fault plant itself, and the basement–sedimentary rock contact is not disrupted, that is, it has not functioned as a plane of detachment. The relations clearly show that the basement rocks deformed cataclastically by piecemeal slip along fractures and shear planes, whereas the apparently more pliable sedimentary strata merely deformed passively as a consequence of deformation at the basement level, analogous to the draping of a pliable material over a constrained and deformed mass of buckshot (Fig. 10). From these observations we extrapolate and postulate that the large-scale folding of strata elsewhere within the central block reflects cataclastic deformation of basement within the entire San Andreas fault zone, an interpretation which is discussed more fully in a following section.

On a smaller scale, folding is accomplished within the sedimentary strata by buckling and

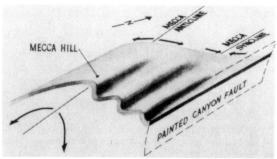

FIG. 9—Idealized sketch of buckled northern limb of Mecca anticline between Mecca Hill and Painted Canyon fault.

bedding-plane slip, particularly in oversteepened flanks of the major anticlines and in the cores of synclines. Such relatively minor structures die out vertically and laterally over short distances. The mechanism of shortening, or crowding, of beds by small folds and minor faults is illustrated in Figure 11. Small disharmonic folds tighten gradually upsection so that strata on the oversteepened flanks of the folds are overturned, but beds in the cores of folds dip steeply or vertically. The faulted strata are superposed by bedding-plane faults so that the hinge is thickened; stratigraphically higher beds are pushed upward and outward toward the syncline. Superposition of faulted strata by bedding-plane faults and consequent thickening of the stratigraphic section are common in folds in the Mecca Hills, especially where siltstone and claystone are intercalated within more competent beds of arkose and conglomeratic arkose.

Painted Canyon fault—The Painted Canyon fault is a major structural discontinuity which separates the platform and central blocks (Figs. 4-6). The fault is at least 20 km long and is defined by a zone of crushed rock and fault gouge from a few centimeters to several meters wide. The crush zone is wider in basement rocks (approx. 25 to 40 m) than in the juxtaposed sedimentary rocks (approx. 5 to 10 m). The fault surface is

FIG. 11—Folding and faulting in core of overturned part of south limb of Mecca anticline, approximately 0.5 km northwest of Painted Canyon. Vertical bedding-plane fault truncates and superposes beds in hinge of fold, resulting in thickening of hinge.

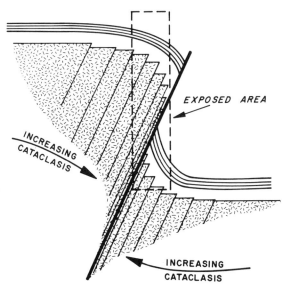

FIG. 10—Idealized sketch of mechanism whereby sedimentary cover folds passively over fractured and sheared basement adjacent to fault. For simplicity only one set of possible shear planes is shown. In reality, many sets pervade basement so that basement-sedimentary rock interface is smooth, rather than stepped as shown here. Dashed rectangle indicates approximate part of sketch which is documented by field exposures adjacent to Painted Canyon fault.

curviplanar, dipping more steeply in canyon bottoms than on adjacent ridges (Fig. 12), showing that it is convex-upward in cross section. Beneath the low-angle segments, footwall strata (platform block) are dragged abruptly to vertical and overturned attitudes (Fig. 13). Locally, strata of the central block are carried short distances northeast over the platform block, either as discrete nappes or as gravity slides, and some of these thrust slices are isolated klippe too small to show on Figure 4.

The central block has been uplifted more than 150 m relative to the platform block as determined from the offset of the basement–sedimentary rock interface in Painted Canyon; the sense and total magnitude of slip could not be determined, however, because of the low angle of intersection made by the fault and the truncated strata, and because marker beds could not be matched confidently across the fault.

The geometry of the Painted Canyon fault and its associated structures is displayed best in the Painted Canyon area (Figs. 4, 5) where the plunge of the structures is steep enough (approx. 25° west-northwest) to permit study of the geometry at several structural levels. Of particular interest is a sequence of strata in the overturned footwall syncline which appears to have been buckled between older and younger strata like pages of a flat-lying book between its covers (Fig. 14). The buckled beds are bounded by a triangular arrangement of low-angle detachment faults which have the following geometries (Fig. 15): (a) those which are convex-upward, flatten gradually away from the main fault, and die out in the bedding (fault *A*, Fig. 15c); (b) those which are concave-upward, steepen gradually with distance from the main fault, and are truncated at higher structural levels by other low-angle faults (fault *B*, Fig. 15c); and (c) those which are mainly convex-upward and flatten toward the main fault (fault *C*, Fig. 15a). The opposite end of the fault either flattens asymptotically into a low-angle fault of the first kind and dies out in the bedding as indicated in Figure 15a, or is truncated by a low-angle fault (Fig. 15c).

Each fault, except fault *C*, carries older rocks upon younger ones, causing thickening of the stratigraphic · section beneath the main fault. Faults *A, B,* and *C* bound a triangular domain of tight folds whose axes parallel the main fault

plane and plunge 25° west-northwest. Field observations of the folds, faults, and unfolded beds show that faults *A* and *C* are detachment surfaces along which the folded sequence in the triangular domain separated from overlying and underlying beds; the observations support the interpretation that the overlying beds merely were pushed up by the buckled wedge of folded strata (Fig. 14). The three interpretive cross sections of this structure (Fig. 15) differ from one another only in the way in which the faults are projected above and below the levels of exposure. We favor interpretation *B* (Fig. 15) in which the Painted Canyon fault is depicted as a high-angle reverse fault rather than a low-angle thrust fault for two reasons: (1) the Painted Canyon fault dips steeply where observed except where it flattens locally upward into thin nappe-like structures; and (2) the linearity of the trace of the Painted Canyon fault throughout the Mecca Hills strongly suggests that the fault dips steeply or vertically, at least within the realm of its intersection with the surface. The evolution of the structure is given in the following and is illustrated in Figure 16. A basic assumption is that the

FIG. 12—View toward west-northwest of Painted Canyon fault, *PCF*, 0.5 km northwest of Painted Canyon. Fault, which dips 70° southwest in center of picture, flattens abruptly upward to 20° on skyline ridge. Dotted lines indicate bedding in Mecca, *Pm*, and Palm Spring, *Pp-m*, Formations.

vertical component of oblique slip predominated during development of the structure:

1. The nonconformity and overlying strata (Fig. 16a) were folded into a sigmoidal configuration (Fig. 16b). The strata defined an asymmetric anticline in the central block and an overturned asymmetric syncline in the platform block.

2. With continued flexuring, a master fault (the Painted Canyon fault) developed; simultaneously or following even more uplift, a secondary fault split off the master fault and extended into the strata of the overturned syncline (Fig. 16c).

3. As the fold continued to grow and tighten, the wedge of basement in the footwall, together with part of the overlying sequence of vertical and overturned strata, detached on secondary low-angle thrust faults and moved northeastward like a "piston" into the gentle, southwest-dipping limb of the footwall syncline (Fig. 16d).

4. The "piston" shoved relatively northeastward and head-on into the southwest-dipping beds, causing them to buckle, uplift, and dilate the entire section; higher in the structure, overturned beds from the central block were thrust short distances on fault D (Fig. 15) over the uplifted, flat-lying beds (Fig. 16e).

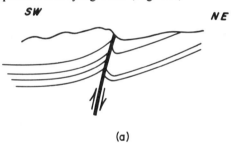

(a)

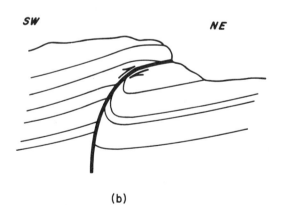

(b)

FIG. 13—Idealized cross-sectional geometries of Painted Canyon fault within sedimentary strata. **a**, Southeast of Painted Canyon where footwall syncline is tight, hanging wall syncline is open; **b** northeast of Painted Canyon where hanging wall anticline is broad, footwall syncline is overturned.

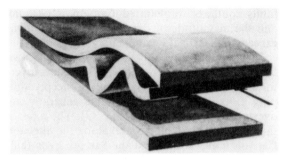

FIG. 14—Interpretive model showing buckling of beds within stratified sequence in footwall (Platform) block of Painted Canyon fault.

Intuitively, it seems clear that such a structure must have formed at shallow depths under low overburden pressure so that the strata could be uplifted bodily above the dilated section. This is supported by the youthfulness of the relatively thin sequence of strata on the platform block where the main effect of erosion has been to cut deep canyons rather than to strip off great thicknesses of strata for which there is no evidence of previous existence.

San Andreas–Skeleton Canyon fault zone—A zone of complexly deformed strata lies between Skeleton Canyon and San Andreas faults (Figs. 4, 5). As shown in Figure 6, the central block has been uplifted relative to the basin block along these two prominent faults. The relatively low structural and topographic relief and the lack of distinctive marker beds permit only the following generalizations:

1. The most recently active trace of the San Andreas fault is clearly marked northwest of Painted Canyon by aligned gulches and ridge notches, offset stream courses, crushed rock and phacoid-bearing gouge, nearly vertical shear surfaces with horizontal slickensides, and en-echelon cracks and fault scarps in alluvium. All of these features are complementary and consistent in indicating right-slip displacement, and it was on this trace about 3 km northwest of Painted Canyon where en-echelon cracks and 1.3 cm of right slip on the fault were triggered dynamically by the Borrego Mountain earthquake in 1968 (Allen et al, 1972).

2. Several distinct, low-angle, west-northwest-trending zones of brick-red, phacoid-bearing fault gouge are parallel with low-angle reverse faults along the San Andreas fault. The faults are convex-upward in cross section and steepen with depth into the main trace of the San Andreas fault. They appear to be similar in geometry and origin to the low-angle segments of the Painted Canyon fault described previously. According to Hays (1957), the displacement on the low-angle

faults southeast of Painted Canyon is oblique, and the magnitude of the horizontal component exceeds the vertical component.

3. The footwall strata southwest of the two faults are folded into asymmetric synclines overturned southwestward. Adjacent to steeply dipping parts of the faults, the folds generally are overturned and appressed (Fig. 17). The axes of the folds are nearly horizontal along the Skeleton Canyon fault, but against the San Andreas fault they plunge from horizontal to as steep as 70°. Locally in such steeply plunging folds, thin beds of mudstone and arkose are deformed into tight, disharmonic folds where they are crowded in the cores of the main syncline. Other investigators have cited the steeply plunging folds, which are about 400 m southeast of the mouth of Painted Canyon, as evidence of strike-slip on the San Andreas fault (Hamilton and Meyers, 1966, p. 520).

4. The exposures of Canebrake-Ocotillo conglomerates 1 km northwest and southeast of the mouth of Painted Canyon show lithostratigraphic evidence of having been displaced about 15 km right-laterally from their source since deposition in Quaternary time (Ware, 1958). The basal part of the stratified sequence consists almost entirely of Orocopia Schist clasts; granitic clasts increase in abundance upsection so that, in the stratigraphically highest exposed parts of the sequence, the strata are comprised completely of granitic clasts. Pebble imbrications show that the sources of the clasts were northeast, across the San Andreas fault. Ware (1958) pointed out that the only monolithologic source for Orocopia Schist would have been the Orocopia Mountains, the nearest drainage from which is now 15 km southeast. Thus, as the depositional center was transported northwestward away from the Orocopia Mountains by right slip on the San Andreas fault, the strata received decreasing amounts of Orocopia Schist and increasing amounts of granitic detritus from fans shedding from the Little San Bernardino Mountains.

5. The Skeleton Canyon fault closely parallels and may have been related genetically to the San Andreas fault (Figs. 4, 5). It is a nearly vertical fault which flattens locally upward into low-angle thrust segments carrying the arkosic facies of the upper Palm Spring Formation upon the silty facies for distances up to 500 m (Fig. 18). Evidence for significant horizontal displacement was not found, owing largely to the nearly parallel strikes of the fault and the truncated beds.

6. About 1 km northwest of Painted Canyon (Figs. 4, 5) the Skeleton Canyon fault is folded into an anticline which plunges gently southeast.

Beneath the thrust, the strata are deformed into three west-northwest-trending asymmetric folds which are overturned southwestward. Not only are they oblique to and truncated by high-angle segments of the fault, but the tops of the anticlines are "decapitated" by low-angle segments of the fault (Fig. 19).

Cross faults—Steeply dipping cross faults, trending from N 70° E to nearly north, cut the sedimentary rocks of the central block southwest

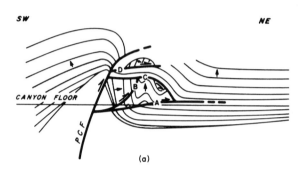

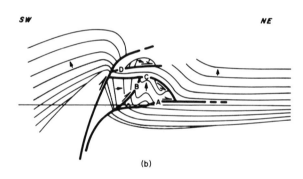

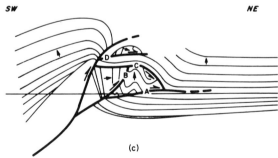

FIG. 15—Possible fault geometry below level of observation in Painted Canyon area. *PCF*, Painted Canyon fault; *A, B, C, D*, secondary faults. Arrows indicate tops of beds. Shaded bed is lithologically distinct stratum. **a,** Painted Canyon fault steepens with depth, faults *A* and *B* project directly into Painted Canyon fault. **b,** Painted Canyon fault steepens with depth, fault *B* bends asymptotically parallel with Painted Canyon fault. **c,** Painted Canyon fault flattens with depth, faults *A* and *B* are asymptotic to it.

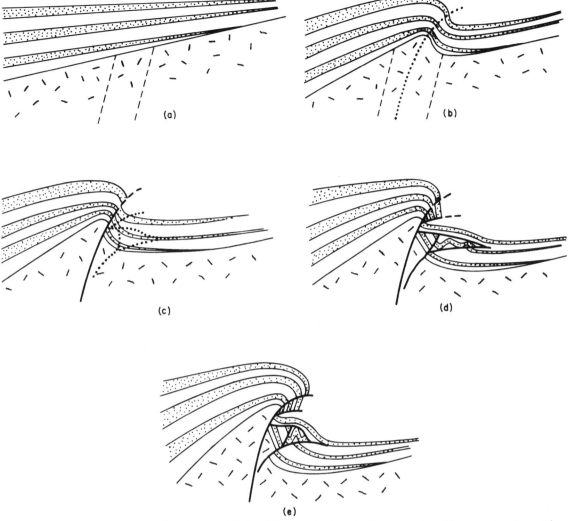

FIG. 16—Postulated structural evolution of Painted Canyon fault. **a**, Initial geometry of basement and overlying sedimentary strata. **b**, Flexure of basement by movement along closely spaced fractures and shear planes within zone indicated by dashed lines. **c**, Rupture of basement along Painted Canyon fault. **d**, Secondary faulting in syncline, incipient buckling of beds in "bookcover structure." **e**, Continued vertical displacement on Painted Canyon fault, rotation of secondary faults, and continued buckling of strata in "bookcover structure."

FIG. 17—View southeast, parallel with Skeleton Canyon fault, of folded upper Palm Spring Formation, silty facies, in footwall of Skeleton Canyon fault, *SCF*. Geologist stands in core of overturned syncline.

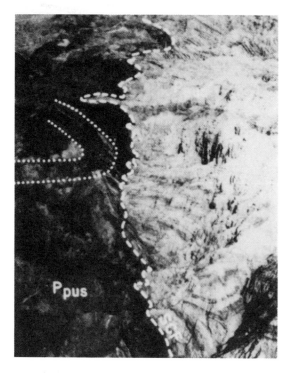

FIG. 18—Oblique aerial view toward northwest of Skeleton Canyon fault, *SCF*, southwest of Painted Canyon. Low-angle segment of fault carries upper Palm Spring arkosic facies, *Ppua*, upon overturned syncline in silty facies, *Ppus*.

of Mecca Hills (Figs. 4, 5). They dip from 70 to 80° southeast and are expressed as discrete fractures or narrow zones of sheared rock. Dip separation across most of them amounts to less than 30 m and diminishes along strike northeastward away from the Skeleton Canyon fault. Associated minor drag folds consistently indicate a relative southeast-side-down dip separation on all of the faults.

The geometry of the cross faults with respect to the total structure of the central block resembles a similar structural geometry in the Alpine fault zone of New Zealand (Kingma, 1958). There they are considered to be extension faults formed as a result of finite horizontal separation within a strike-slip fault zone, an interpretation which is confirmed in clay-model laboratory studies (Wilcox et al, 1973) in which extension fractures and normal faults develop parallel with the short axis of the strain ellipse, perpendicular to en-echelon folds. Most of the cross faults shown in Figure 4 are not oriented ideally according to the typical wrench-fault geometry, suggesting that they may have been rotated.

Basin Block

The structure of the area southwest of the San Andreas fault zone adjacent to the Mecca Hills was determined partly from detailed gravity studies (Biehler, 1964). The data show a steep gravity gradient of 4,000 m or more in the basement but, because of the complete lack of density-depth control in Coachella Valley (Biehler, 1964, p. 78) and the paucity of surface exposures and well data, a more detailed interpretation of the structure of this domain is not warranted here.

DISCUSSION AND INTERPRETATIONS

This study of structural geometries in the Mecca Hills documents three principal observations: (1) a narrow zone of crustal shortening (central block), coinciding with the basin margin, is separated from adjacent, relatively undeformed areas by major strike-slip faults which dip steeply toward the shortened zone, which are convex-upward in cross section, and which have lesser components of reverse-slip; (2) the major strike-slip faults have low-angle thrust segments which extend outward short distances from the shortened

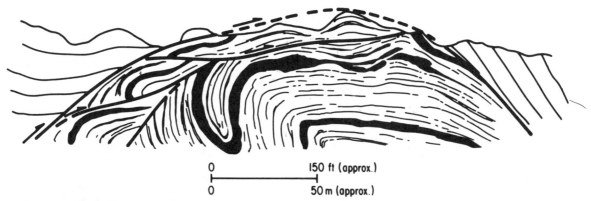

| 0 | 150 ft (approx.) |
| 0 | 50 m (approx.) |

FIG. 19—Field sketch of folded arkosic facies of upper Palm Spring Formation thrust over strongly folded silty facies by folded trace of Skeleton Canyon fault. Shaded beds are sandstone. East wall of canyon 1 km northwest of Painted Canyon.

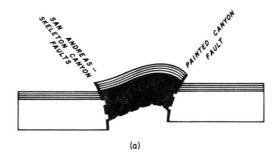

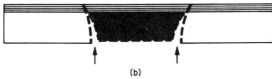

FIG. 20—Schematic illustration showing apparent geometric impossibility of two-dimensional palinspastic restoration of Mecca anticline. **a,** Idealized present-day cross section. **b,** Restoration of strata to prefolding configuration and resultant space problem at basement level. Area of shaded parts are equal in both diagrams.

zone; and (3) the basement behaved in a nonrigid, cataclastic manner in the shortened zone. The deformation of the overlying sedimentary rocks was largely a passive consequence of basement-level deformation at shallow depths under low overburden pressure.

Convex-Upward Faults

The faults bounding the central block, where observed, steepen with depth and dip consistently toward one another and toward the axis of the central block. This relation is illustrated in Figures 6 and 20a. In both of these illustrations it is assumed that the faults are nearly vertical at depth. As shown in Figure 20b, two-dimensional palinspastic restoration by "unrolling" the folds and restoring the strata back to their probable prefolding, prefaulting configurations is geometrically impossible if only dip slip is assumed, if the basement is rigid, and if the sedimentary rocks are not capable of tectonic thickening and thinning. In other words, if the folded strata and the underlying "keystone" of basement are restored to the assumed predeformation positions between the bounding faults, and if the sense of displacement on the faults is purely dip slip and nonrotational, then basement must be "created" to satisfy areal considerations of the restoration. But if oblique slip is invoked, together with nonrigid cataclastic flow of the basement, then the problem is three dimensional, and complex structural reconstructions are possible.

The low-angle segments of the San Andreas, Skeleton Canyon, and Painted Canyon faults are similar to structures which are common along many major wrench faults elsewhere, especially where the faults delimit the base of a steep mountain front. That these low-angle faults steepen abruptly with depth has been documented for other wrench faults having a component of dip slip, such as the Alpine fault in New Zealand (Wellman, 1955), and in California on the San Jacinto fault (Sharp, 1967) and along part of the San Gabriel Mountain front (Whitcomb et al, 1973). Wellman (1955) considered that the surficial thrusting results from downslope creep under gravity, but Allen (1965, p. 84) suggested:

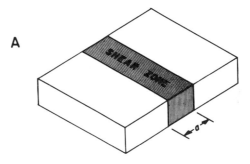

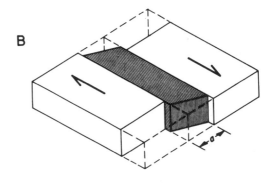

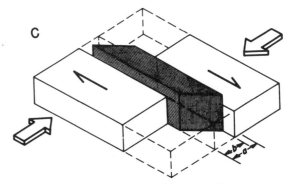

FIG. 21—Conceptual model of transpression basement deformation. See text for explanation.

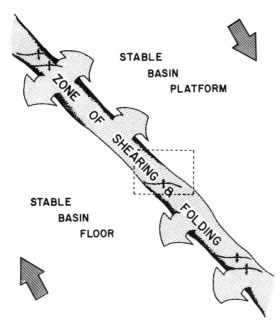

FIG. 22—Conceptual model of transpressive deformation, northeast margin of Salton trough. Rectangle shows study area (Fig. 4) relative to model.

Another important causal factor here and elsewhere may be related to the origin of the steep mountain front itself; if, as seems likely, the presence of the mountain front is caused by a local vertical component of displacement along the predominantly transcurrent fault, then vertical motion constrained at depth to a vertical plane must necessarily result in localized low-angle thrusting at the surface, as has been demonstrated analytically and in models by Sanford (1959).

A representative oil producing analog for the "basin block–central block–platform block" of the Mecca Hills is in the central Puente Hills in the Los Angeles basin, California (Harding, 1974, Figs. 9, 10). There oil has been trapped beneath low-angle, outward-spreading segments of the high-angle Whittier fault on the west and the Chino fault on the east. Such a style of trapping also is common, with variations, along wrench faults and oblique-slip faults elsewhere in California.

Crustal Shortening

One of the main structural features of the Mecca Hills is the juxtaposition along the Painted Canyon fault of two crustal blocks with markedly different lithostratigraphic and structural characteristics (Table 1). The relatively simple structure and stratigraphy of the platform block evidently are reflections of the tectonic stability of the basement and contrast sharply with the central block. Clearly, the high-angle reverse faults, local thrusts, and en-echelon folds which characterize the central block in the Mecca Hills, the Indio

Hills, and the Durmid area are evidences of crustal shortening concentrated within a remarkably narrow zone which, from geologic and geophysical data, appears to be less than 3 km wide.

Nonrigid Basement

The limited exposures of basement rocks in the central block are more highly fractured and sheared than their counterparts exposed in the platform block. Moreover, the basement–sedimentary rock contact is not a plane of detachment, even where the sedimentary rocks are folded. These relations are interpreted above as evidence that the basement and overlying sedimentary rocks behaved mechanically as a unit, and that the basement deformed by cataclastic flow by piecemeal slip on the pervasive shear fractures.

We maintain that these observations and interpretations support extrapolation of the deformation mechanism to at least those parts of the central block which are overlain by folded sedimentary rocks. Thus, we postulate that the basement beneath much of the central block has been so pervasively fractured and sheared during a prolonged period of movement in the San Andreas fault zone during Cenozoic time (Crowell, 1962, 1975) that it is no longer mechanically rigid, but as the exposures show in Painted Canyon, it has flowed cataclastically on a large scale in response to regional transpression. By this hypothesis, folding, uplift, and local outward-directed thrusting of the sedimentary succession are consequences of transpressive distortion and resultant uplift of wedges of basement. The mechanism of basement deformation is illustrated in Figure 21. A narrow volume of nonrigid material is constrained between two rigid blocks (Fig. 21A); if the blocks merely were displaced laterally by strike slip (Fig. 21B) the material would deform by simple shear; in transpression (Fig. 21C), however, the material would be sheared and compressed between the two laterally and convergently slipping blocks, so that it would be uplifted out of the shear zone.

A summary diagram of the observations and interpretations of this study (Fig. 22) resembles that of Lowell (1972, Fig. 9) and shows, in a conceptual way, the tectonic relations of the various geometric elements observed in the field and in clay-model laboratory studies of wrench faulting in which transpression is the dominant mechanism of strain.

CONCLUSIONS

Field observations in the Mecca Hills support clay-model laboratory studies in showing clearly that finite zones of oblique, inward-directed crustal shortening form along wrench faults which, themselves, are driven by deep crustal decoupling in transform margin setting. En-echelon folds and high-angle reverse faults are typical in those parts of the wrench fault zone which are dominated by a transpressive mode of deformation at shallow crustal levels, and they are produced and controlled by cataclastic flow of the underlying crystalline basement. Although the overall resultant geometry of the various folds and faults conforms closely to theoretical and laboratory models, the folds and faults combine locally to form discontinuous complex structures which form small but favorable structural traps for hydrocarbons.

REFERENCES CITED

Abbott, W. O., 1968, Salton basin, a model for Pacific rim diastrophism (abs.): Geol. Soc. America Abs. with Programs, v. 1, p. 1.

Allen, C. R., 1965, Transcurrent faults in continental areas, *in* A symposium on continental drift: Royal Soc. London Phil. Trans., v. 258, p. 82-89.

———— and J. M. Nordquist, 1972, Foreshock, main shock, and larger aftershocks of the Borrego Mountain earthquake, *in* The Borrego Mountain earthquake of April 9, 1968: U.S. Geol. Survey Prof. Paper 787, p. 16-23.

———— Max Wyss, J. N. Brune, et al, 1972, Displacements on the Imperial, Superstition Hills, and San Andreas faults triggered by the Borrego Mountain earthquake, *in* The Borrego Mountain earthquake of April 9, 1968: U.S. Geol. Survey Prof. Paper 787, p. 87-104.

———— A. Grantz, J. N. Brune, et al, 1968, The Borrego Mountain, California, earthquake, 9 April 1968: a preliminary report: Seis. Soc. America Bull., v. 58, p. 1183-1186.

Allison, E. C., 1964, Geology of areas bordering Gulf of California, *in* Marine geology of the Gulf of California: AAPG Mem. 3, p. 3-29.

Babcock, E. A., 1974, Geology of the northeast margin of the Salton trough, Salton Sea, California: Geol. Soc. America Bull., v. 85, p. 321-332.

Biehler, S., 1964, Geophysical study of the Salton trough of southern California: PhD thesis, California Inst. Technology, 139 p.

———— R. L. Kovach, and C. R. Allen, 1964, Geophysical framework of the northern end of the Gulf of California structural province, *in* Marine geology of the Gulf of California: AAPG Mem. 3, p. 126-143.

Carey, S. W., 1958, The tectonic approach to continental drift, *in* S. W. Carey, ed., Continental drift, a symposium: Univ. Tasmania, p. 177-355.

Crowell, J. C., 1962, Displacement along the San Andreas fault, California: Geol. Soc. America Spec. Paper 71, 61 p.

———— 1974, Origin of late Cenozoic basins in southern California, *in* Tectonics and sedimentation: Soc. Econ. Paleontologists and Mineralogists Spec. Pub. 22, p. 190-204.

———— 1975, The San Andreas fault in southern California, *in* San Andreas fault in southern California— a guide to the San Andreas fault from Mexico to Carrizo Plain: California Div. Mines and Geology Spec. Rept. 18, p. 7-27.

———— and J. W. R. Walker, 1962, Anorthosite and related rocks along the San Andreas fault, southern California: Univ. California Pubs. Geol. Sci., v. 40, p. 219-288.

Dibblee, T. W., Jr., 1954, Geology of the Imperial Valley region, California: California Div. Mines Bull. 170, p. 21-28.

Dillon, J., 1975a, Remnants of a late Oligocene–early Miocene plutonic-volcanic belt in southeastern California and southwestern Arizona (abs.): Geol. Soc. America Abs. with Programs, v. 7, p. 605.

———— 1975b, Geology of the Chocolate and Cargo Muchacho Mountains, southeasternmost California (abs.): Geol. Soc. America Abs. with Programs, v. 7, p. 311.

———— and G. Hazel, 1975, The Chocolate Mountain-Orocopia-Vincent thrust system as a tectonic element of late Mesozoic California (abs.): Geol. Soc. America Abs. with Programs, v. 7, p. 311-312.

Ehlig, P. L., 1968, Causes of distribution of Pelona, Rand, and Orocopia Schists along the San Andreas and Garlock faults, *in* Conference on geologic problems of San Andreas fault system, Stanford, Calif., 1967, Proc.: Stanford Univ. Pubs. Geol. Sci., v. 11, p. 294-305.

Elders, W. A., R. W. Rex, Tsvi Meidav, et al, 1972, Crustal spreading in southern California: Science, v. 178, p. 15-24.

Hafner, W., 1951, Stress distribution and faulting: Geol. Soc. America Bull., v. 62, p. 373-398.

Hamilton, W., 1961, Origin of the Gulf of California: Geol. Soc. America Bull., v. 72, p. 1307-1318.

———— and W. B. Myers, 1966, Cenozoic tectonics of the western United States: Rev. Geophysics, v. 4, p. 509-549.

Harding, T. P., 1973, Newport-Inglewood trend, California—an example of wrenching style of deformation: AAPG Bull., v. 57, p. 97-116.

———— 1974, Petroleum traps associated with wrench faults: AAPG Bull., v. 58, p. 1290-1304.

———— 1976, Tectonic significance and hydrocarbon trapping consequences of sequential folding synchronous with San Andreas faulting, San Joaquin Valley, California: AAPG Bull., v. 60, p. 356-378.

Harland, W. B., 1971, Tectonic transpression in Caledonian Spitsbergen: Geol. Mag., v. 108, p. 27-42.

Hays, W. H., 1957, Geology of the central Mecca Hills, Riverside County, California: PhD thesis, Yale Univ., 324 p.

Kingma, J. T., 1958, Possible origin of piercement structures, local unconformities, and secondary basins in the eastern geosyncline, New Zealand: New Zealand Jour. Geology and Geophysics, v. 1, p. 269-274.

Lowell, J. D., 1972, Spitsbergen Tertiary orogenic belt and the Spitsbergen fracture zone: Geol. Soc. America Bull., v. 83, p. 3091-3102.

Merriam, R., and O. L. Bandy, 1965, Source of upper Cenozoic sediments in Colorado delta region: Jour. Sed. Petrology, v. 35, p. 911-916.

Miller, F. K., and D. M. Morton, 1974, Comparison of granitic intrusions in the Orocopia and Pelona Schists, southern California (abs.): Geol. Soc. America Abs. with Programs, v. 6, p. 220-221.

Muffler, L. J. P., and B. R. Doe, 1968, Composition and mean age of detritus of the Colorado River delta in the Salton trough, southeastern California: Jour. Sed. Petrology, v. 38, p. 384-399.

Olmsted, F. H., 1966, Tertiary rocks near Yuma, Arizona (abs.): Geol. Soc. America Spec. Paper 101, p. 153-154.

Popenoe, F. W., 1959, Geology of the southeastern portion of the Indio Hills, Riverside County, California: Master's thesis, Univ. California, Los Angeles, 153 p.

Prucha, J. J., J. A. Graham, and R. P. Nickelsen, 1965, Basement-controlled deformation in Wyoming province of Rocky Mountains foreland: AAPG Bull., v. 49, p. 966-992.

Sanford, A. R., 1959, Analytical and experimental study of simple geologic structures: Geol. Soc. America Bull., v. 70, p. 19-52.

Sharp, R. V., 1967, San Jacinto fault zone in the Peninsular Ranges of southern California: Geol. Soc. America Bull., v. 78, p. 705-730.

——— 1972, Tectonic setting of the Salton trough, in The Borrego Mountain earthquake of April 9, 1968: U.S. Geol. Survey Prof. Paper 787, p. 3-15.

——— and M. M. Clark, 1972, Geologic evidence of previous faulting near the 1968 rupture on the Coyote Creek fault, in The Borrego Mountain earthquake of April 9, 1968: U.S. Geol. Survey Prof. Paper 787, 131-140.

Stotts, J. L., 1965, Stratigraphy and structure of northwest Indio Hills, California: Master's thesis, Univ. California, Riverside, 208 p.

Sylvester, A. G., and R. R. Smith, 1975, Structure section across the San Andreas fault zone, Mecca Hills, in San Andreas fault in southern California—a guide to the San Andreas fault from Mexico to Carrizo Plain: California Div. Mines and Geology Spec. Rept. 118, p. 111-118.

Tarbet, L. A., 1951, Imperial Valley, in Possible future petroleum provinces of North America: AAPG Bull., v. 35, p. 260-263.

——— and W. H. Holman, 1944, Stratigraphy and micropaleontology of the west side of Imperial Valley (abs.): AAPG Bull., v. 28, p. 1781-1782.

Tchalenko, J. S., and N. N. Ambraseys, 1970, Structural analysis of the Dasht-e Baȳaz (Iran) earthquake fractures: Geol. Soc. America Bull., v. 81, p. 41-60.

Ware, G. C., 1958, The geology of a portion of the Mecca Hills, Riverside County, California: Master's thesis, Univ. California, Los Angeles, 60 p.

Wellman, H. W., 1955, The geology between Bruce Bay and Haast River, South Westland, 2d ed.: New Zealand Geol. Survey Bull., n.s. 48, 46 p.

Whitcomb, J. H., C. R. Allen, J. D. Garmany, and J. A. Hileman, 1973, San Fernando earthquake series, 1971: focal mechanisms and tectonics: Rev. Geophysics and Space Sci., v. 11, p. 593-730.

Wilcox, R. E., T. P. Harding, and D. R. Seely, 1973, Basic wrench tectonics: AAPG Bull., v. 57, p. 74-96.

THE AMERICAN ASSOCIATION OF PETROLEUM GEOLOGISTS BULLETIN
V. 52, NO. 10 (OCTOBER, 1968), P. 2016-2044, 52 FIGS.

CRUSTAL MECHANICS OF CORDILLERAN FORELAND DEFORMATION:
A REGIONAL AND SCALE-MODEL APPROACH[1]

JOHN K. SALES[2]
Oneonta, New York 13820

ABSTRACT

Foreland Rockies deformation was caused by compressive stresses transmitted tangentially through the continental basement and the overlying geosynclinal prism from the Pacific continental margin. The stresses probably originated from relative overriding of that ocean block by the continental block as a result of drag on the bottoms of the blocks by convection cells in the mantle.

Dissipation of these stresses by deformation has altered through time from the continental margin to the foreland, depending on the ability of the geosyncline to transmit the stress. Through Mississippian time, Pacific-margin deformation continually thickened the geosynclinal prism. Permo-Pennsylvanian stress transmission through the thickened prism caused deformation of the Ancestral Rocky Mountain foreland. A change in Mesozoic time to hydraulic stresses while batholiths developed in the geosyncline halted foreland deformation by relieving tangential stresses. Stress transmission resulting from Late Cretaceous solidification of the batholiths caused Laramide breakup of the foreland. Tertiary development of northwest oblique yield at the continental margin (San Andreas and Basin-Range structure) again dissipated tangential stress, halting Laramide breakup.

A west-northwest left-lateral couple superimposed on this compression, and caused by greater resistance of the Canadian shield to compression, caused weakening and deformation of the foreland. Eastward movement of the Colorado Plateau block caused the couple to be accentuated in Wyoming and diminished in Colorado. This caused crust-thick, drag fold-slabs in Wyoming. North-south compression between the "Wyoming couple" and the Colorado Plateau caused the Uinta uplift. Eastward jamming of the Colorado Plateau against the Mid-Continent caused the Front Range uplift. Release along the south margin was by left-lateral yield on the Wichita lineament. The deformation has been duplicated in models using similar stresses.

The Wyoming crust was flexed into combinations of three basic configurations: (1) a slab compressed tangentially into a sine curve with negative parts subsiding as positive parts rose; (2) overlapping slabs caused by failure of the mutual limb of this sine curve; and (3) isostatically collapsed slabs. Basin-margin depression by surrounding uplifts caused the Rock Springs uplift. Isostatic collapse of overlapping slabs during and after compression caused the high south rim of the Sweetwater uplift. Post-compression "rheid" crustal thinning, failure of the mutual limb (fold-thrusting), late removal of basin sediments, and the effect of the left-lateral couple have caused collapse. Bounding structures are aggregates of monoclines, upthrusts, sheared-out and overturned limbs, basin-block swells, second-order faults and folds, isostatically reversed thrusts, and several types of collapse structure. The mechanical functions of these diverse structures can be isolated.

INTRODUCTION

GEOGRAPHIC AND STRUCTURAL SETTING

The Wyoming structural province of the Cordilleran foreland includes much of Montana, Colorado, eastern Utah, Arizona, New Mexico, and western Texas (Fig. 1). The Wyoming foreland crust is similar to, and an extension of, the Mid-Continent craton except that it has been deformed much more severely. It is thus intermedi-

[1] Manuscript received, April 1, 1968; accepted, June 6, 1968.

[2] Department of Earth Science, State University College. Formerly geologist, Mobil Oil Corporation, Casper, Wyoming. The writer thanks Mobil Oil Corporation for permission to publish this paper. Acknowledgment is made of the guidance of James Keenan in the early stages of the study and of the continuing help of Jack Peters. O. B. Shelburne and H. A. Sellin read the manuscript critically and suggested many improvements. Help with photography was given by Wallace Lumb, John Lumb, and Don Pearson.

ate in position between the geosyncline and the craton.

The Wyoming foreland is not an isolated mechanical entity. It is part of an integrated mechanical system that includes at least the eastern Pacific, the entire Cordillera from the Arctic to the Antarctic, and possibly much of the continental crust at least 1,000 mi east of the Rocky Mountains. The Wyoming foreland differs from the Cordilleran geosyncline on the west by its characteristically thinner, more persistent, more unconformity-bounded Paleozoic and Mesozoic units. The foreland is distinguished structurally from the geosyncline by large broad-backed uplifts in which the deep basement is exposed. These uplifts are separated from adjacent basins by bounding reverse faults which steepen and project into the deep basement, giving the appearance of cutting the entire crust. It is of prime significance to interpretation of the basic mechanism that during deformation the basin blocks went *down* at least as much as the adjacent lands were uplifted. This is substantiated by the fact that the top of the basement is 25,000–30,000 ft below sea level under the deeper basins despite the fact that, in the later Tertiary, postorogenic uplift of several thousand feet occurred in the entire region. Such movement cannot be attributed to basin loading by sediments eroded from adjacent areas of uplift created by purely vertical forces without problems of isostasy. Structural relief between basins and adjacent uplifts is as much as 40,000 ft (Prucha *et al.*, 1965, p. 970), very near the maximum known within a continental framework. Recent drilling, and seismic and gravimetric studies suggest that in places the Precambrian cores of the uplifts overhang the sedimentary section in the basins many miles, about 12 mi in the case of the Wind River uplift (Berg, 1961). These uplifts and basins appear to be huge crust-thick blocks or slabs that have been "jostled" severely enough to have been tilted, structurally overlapped, bowed, and partly collapsed.

Tectonics of the Cordilleran geosyncline are of an entirely different aspect. With the exception of

the fused crust in the batholithic blocks, the thick geosynclinal sediments have been deformed much more plastically than the foreland blocks. The basement is almost nowhere brought to the surface by the geosynclinal thrusts, a fact which suggests that the sediments have been peeled from it and forced eastward on a regional glide surface or surfaces above the deeply buried basement (Misch, 1960; Armstrong and Oriel, 1965; Bally *et al.*, 1966).

STUDY PURPOSE AND METHOD

Many descriptive papers have been written on the geologic history, depositional geometry, and deformational geometry of the United States Cordilleran foreland, and many important advances have been made in recent years in the fields of oceanography, crustal geophysics, and rock mechanics. These have provided the necessary data from which the writer has derived an integrated working hypothesis for evolution of the foreland in terms of crustal mechanics.

The writer has attempted to show that some of the most basic and long-standing tectonic arguments about the origin of foreland structural relief (*i.e.*, gravity sliding *versus* a primary push for geosynclinal thrusts and vertical *versus* horizontal primary stresses) are moot. Some evidence that has been used for years as "proof" of an assumption is found to be inconclusive. For example, the fact that many foreland thrusts are high angle or seem to steepen downward is not proof that foreland tectonics is the result of wholly differential vertical stresses. Plaster models used by the writer contain many nearly vertical reverse faults and reverse faults that steepen downward, even though they were generated by tangential stresses.

Models have been very valuable both in development and substantiation of the hypothesis presented. Of the several types used, the plaster models have been most noteworthy. The technique used consists of laying down one or more layers of soupy, pigmented plaster in a hinged box or other stressing apparatus. In a short time the layers begin to "set," and there is a period during which the strength relations of the layers are "in scale" with the part of the crust that they simulate. They are deformed at this time and thus the major factor in scaling the model is timing.

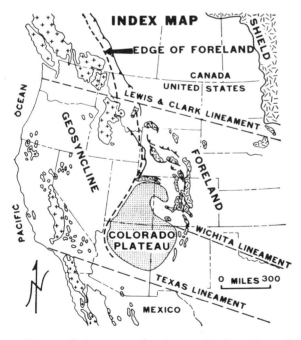

Fig. 1.—Index map showing regional setting of United States Cordilleran foreland and distribution of Colorado Plateau block, shield, exposed batholiths, west-northwest-trending megashears, and foreland uplifts.

SOUTH AMERICAN STRUCTURE
EARTHQUAKE SEQUENCES

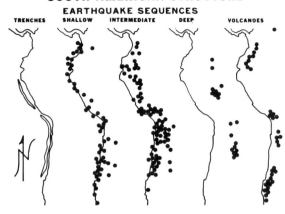

FIG. 2.—South American structure and relation of tectonic features to Pacific continental margin. Each deeper earthquake sequence (shallow, intermediate, deep) is farther inland, and trenches and volcanoes parallel continental margin. Modified from Benioff (1954).

A basic structural axiom used throughout this study is that of spatial compatibility; *i.e.,* any change in shape of the crust must trigger and be compensated for by other changes in shape. One of the advantages of the model approach is that the models emphasize the space relations and space requirements of the area of study.

The organization of the part of the paper concerning the source of the stresses that deformed the Wyoming province was dictated by the mechanics involved. A deformation must be caused by a force; the force in question enters the continental prism largely at the continental margin, which is explained first, and is transmitted through the Cordilleran geosyncline, discussed second, into the Wyoming province. A force will not cause deformation until it is met by resistance and, therefore, the resistance is treated last.

The organization of the part of the paper concerning the foreland itself was dictated mostly by the fact that part of the material is visualized best in map view and part is visualized best in cross section.

STRESS HYPOTHESIS

DRIVING FORCE

One of the most significant tectonic features of the Western Hemisphere is the Cordillera, which hugs and seems to be related mechanically to the Pacific margin of the continent. In contrast, the stable shield areas of the continents are

far east, near the eastern margins of North and South America.

South American model.—To understand the tectonics of the Wyoming foreland, it is instructive to examine first the South American Cordillera because it is in a developmental stage that is more primitive than that of the western United States. The mountain ranges of the South American Cordillera are narrow, high, concentrated, and close to and parallel with the Pacific margin of the continent. Deep oceanic trenches lie at the base of a steep continental slope. Each successively deeper sequence of earthquake foci is farther beneath the continent, along a plane which dips eastward and intersects the surface at the trenches (Benioff, 1954). The distribution of volcanoes also is strongly controlled by tectonic elements along the continental margin (Fig. 2).

In cross section (Fig. 3) the shallow earthquake sequence is near the trench, the intermediate sequence is farther inland, and the deep-focus earthquakes are still farther inland east of the Cordilleran crest. To explain this, Benioff (1954) postulated that the earthquakes are aligned along a huge shear on which the continental block is being thrust relatively westward over the Pacific block. Along this "megathrust," earthquakes occur to a depth of 700 km. This suggests that this type of large structure may represent the most severe deformation on the surface of the earth. A major postulate of this paper is that the interplay between the continental and oceanic

EARTHQUAKE SEQUENCES
WESTERN SOUTH AMERICA

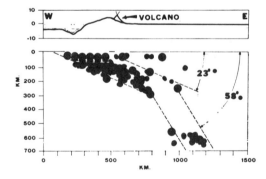

FIG. 3.—Cross section of Pacific continental margin of South America showing concentration of earthquake foci along "Benioff" megashear along which continent is being thrust over Pacific block. Modified from Benioff (1954).

blocks is the driving mechanism for the entire Cordilleran orogenic belt. Recent oceanographic studies suggest that the energy for this mechanism is derived from the spreading of the sea floor away from ascending parts of convection cells in the mantle, forcing the sea-floor crust under the continent as it spreads toward the descending part of the cell (Wilson, 1963). Features in the sprawling United States Cordillera can be explained in terms of this South American model and two deviant tectonic styles (crustal spreading in the Basin-Range province and breakup of the foreland) superimposed on it.

TRANSMISSION OF STRESS THROUGH GEOSYNCLINE

A mechanical basis for geosynclinal versus foreland timing.—The sequential development of the more complicated United States Cordillera is shown in Figure 4. Figure 4D, shows an early stage in the formation of the megashear along the edge of the continent. A deformed prism of debris eroded from the craton gradually accumulated against the wedge edge of the continental block as the oceanic crust was forced under it.

There are very few records of both the early stages of this orogenic process and the west side of the deformed geosynclinal prism, because the vulnerable western edge of the orogenic belt was remobilized by the "cement-mixer" type of tectonic activity along the outer margin. By Missis-

CAUSE AND EFFECT

GEOSYNCLINE VS. FORELAND

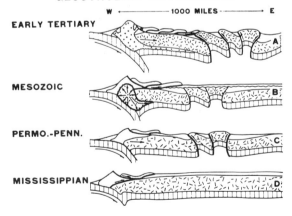

FIG. 4.—Time cross sections of western United States suggesting mechanical linkage between foreland and geosynclinal deformation. Deformation alternates between continental margin and foreland.

sippian time in central Nevada the prism had grown sufficiently large and had been pushed inland far enough that its eastern part was preserved as the Antler orogenic belt (Roberts *et al.*, 1958). The extremely uniform and widespread Madison-Redwall-Leadville lithology indicates that the foreland in the Wyoming area and on the south remained stable during this time (Mallory, 1967).

Shortly after the Antler orogeny, or after the deformed geosynclinal prism had grown to a certain size and the crust had been thickened orogenically far enough inland, sufficient force was transmitted eastward to break up the cratonic crust in Colorado and form the Ancestral Rockies (Fig. 4C). In Permian time deformation of the Ancestral Rockies ceased and stable foreland conditions resumed. Significantly, at about this time batholiths began to form on the western margin of the Cordillera (Gilluly, 1963). Previously, a direct transfer of compressive stress from the oceanic block to the foreland, through the deformed prism at the outer edge of the continental block, had caused the foreland to break up. With the development of molten batholiths, stresses were distributed in a hydraulic fashion (Moody, 1962) and relieved some of the tangential stress (Fig. 4B). When the batholiths solidified during the Late Cretaceous and early Tertiary, stress again was transmitted directly to the craton (Fig. 4A) and caused the Laramide breakup of the foreland. This transmission of tangential stress was aided by the eastward shift of thrusting on the east side of the deformed prism as it grew. The resulting thrusts not only directed a component of stress against the foreland, but also thickened the crust and allowed more tangential stress to be transmitted through it from the oceanic block.

In the late Tertiary, right-lateral strike-slip movements along the western continental margin, as evidenced by the San Andreas fault (Crowell, 1962) and Basin-Range system (Sales, 1966), seem to have relieved some of the interior stresses and again allowed the foreland to become stable. This situation has continued to the present as evidenced by the relative lack of seismic activity in the Wyoming-Colorado area of the foreland in contrast to the high seismic activity just west in the Basin-Range province and along the continental margin.

EAST SIDE OF THE DEFORMED GEOSYNCLINAL PRISM

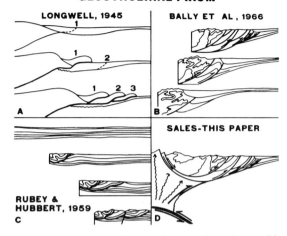

FIG. 5.—Interpretations of deformation of east side of geosyncline by several authors. They have in common eastward migration of thrust belt with time and resultant crustal thickening.

Mechanics of east side of geosynclinal prism. —Interpretations by several authors suggest eastward shift of the belts of geosynclinal thrusting and thickening of the geosynclinal prism by this thrusting (Fig. 5). Figure 6 is a composite cross section of the prism. The development of a hydraulic character in the batholithic zone caused uplift and unroofing of the batholiths at the same time that it caused a decrease in tangential stress against the foreland. Because of the strength of the material being uplifted, there is a definite height or upper limit above which the prism cannot grow without collapsing. This can be termed the "upper morphologic limit" of th eprism. As the material is plastered against the western edge of the cratonic crust, part of it is pushed upward across the top of that crust toward the more interior foreland on a basal thrust. This basal thrust is accompanied by megalandsliding or gravity sliding from the top of the prism as it collapses after reaching the morphologic limit. Hence these thrusts clearly originated by two important mechanisms. Figure 6 shows that neither the basal thrust nor the gravity thrusts significantly involve the Precambrian basement. It is unlikely that thrusts of the magnitude of the Roberts thrust of Nevada, the Lewis thrust, and the overthrust belts of Wyoming and Canada, were created by either mechanism separately.

Specific-gravity and strength factors limit the distance to which the sialic crustal material can be depressed into the more basic mantle material. This distance is referred to as the "lower morphologic limit." At depths below this limit, lighter material is forced to move horizontally. Material that is being dragged under the continental prism along the Benioff megashear, therefore, is forced inland at some level below the continental crust (Fig. 6). This morphologic line may be the cause of the mantle-crust mix or the intermediate velocity layer which has been observed in the western part of the United States. Late stage epeirogenic uplift of the western United States may have been caused by isostatic imbalance resulting from the addition of lighter material under the continental prism.

The bottom schematic cross section of Figure 6 shows a plunger effect with the oceanic block moving "X" distance toward the continental block. Because the upper and lower morphologic limits control the vertical extent of displaced material, the excess is forced horizontally a greater distance. Because the area of squeezing at the interface of the two blocks is crust-thick, there is greater volume displaced per unit of shortening than is displaced either above or below the cratonic crust. The displaced material is forced horizontally between the upper and lower morphologic limits and the continental crust. Thus, for an amount of total shortening of the

COMPOSITE CONTINENTAL MARGIN

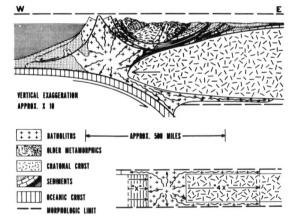

FIG. 6.—Composite cross section of deformed geosynclinal prism showing its internal structures and relation to oceanic and cratonic crust. Wavy arrows suggest hydraulic nature of stresses in batholith zone. Lower diagram suggests that apparent shortening above and below cratonic crust can be greater than total crust-thick shortening. Partly after Bally *et al.* (1966).

two big blocks there can be a greater distance of supracrustal thrusting and subcrustal eastward flow of mantle-crust mix.

RESISTANCE TO STRESS

In order to explain the deformation of the Cordillera, the driving force for Cordilleran orogeny must be resisted by material on the east. The nature of this resistance strongly controls the geometry of the deformation.

If the driving force had been uniform along the entire length of the North and South American Pacific margin and the resistance of the continents also had been uniform (*i.e.*, if the North and South American cratons had been continuous the entire length of the hemisphere), a very uniform Cordillera would be expected (Fig. 7). There should have been a uniform batholithic zone east of the Benioff shear zone, a uniform set of thrusts interior to that, and an undeformed foreland farther east. A situation such as that in the Canadian Cordillera (Bally *et al.*, 1966) should have been generated throughout the length of the hemisphere.

However, the resistance is nonuniform. It consists of the North and South American continents with a large Central American area between, which according to geophysical data has a "more oceanic" crust. It is postulated that there is much less resistance to the compressive stress from the

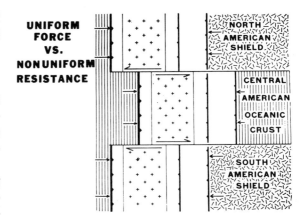

FIG. 8.—Modification of Figure 7 to suggest that effect of decreased resistance provided by Central American oceanic crust to uniform Pacific force should be greater eastward yield in that area and consequent development of transitional horizontal couples.

UNIFORM FORCE VS. UNIFORM RESISTANCE

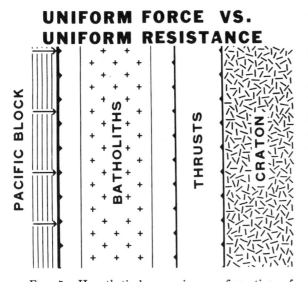

FIG. 7.—Hypothetical map-view configuration of Cordillera if both force from Pacific block and resistance from cratons had been uniform throughout length of hemisphere. Under this assumption there is no reason for horizontal couples.

west in the Central American area because of the easier yield of the thinner crust. Therefore, the push from the Pacific block can cause yield farther east in the Caribbean-Central America-Gulf of Mexico area than it can on the north or south where it is buttressed against the shields of the two continents (Fig. 8).

Although the driving force also may have been nonuniform, this is difficult to prove. In contrast, the concept of a nonuniform resistance seems mechanically probable. It is assumed, therefore, that variable resistance rather than variable driving force is the dominant reason for the abnormal United States segment of the Cordillera.

The hemisphere map (Fig. 9) shows the geographic distribution of the tectonic features that were portrayed schematically in Figure 8. The dashed lines represent the crest of compressive "Benioff" overthrust blocks along the continental margin or similar island-arc-type structures in the oceanic crust. The solid lines represent some of the major strike-slip faults of North and South America which distribute the compressive shortening from one zone to another. In Central America there is a zone of eastward yield around the south side of the North American craton. A similar accentuated zone of eastward yield at the south end of South America separates it from Antarctica. At the east end of each zone of yield, the arcuate dashed lines represent island arcs or overthrusts in oceanic crust where the push from the Pacific crust has been transmitted far east of the normal Pacific margin. Thus, there is a defi-

WESTERN HEMISPHERE MEGA SHEARS

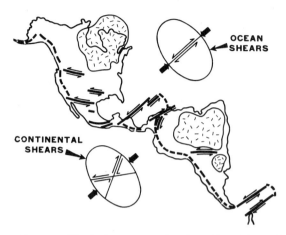

Fig. 9.—Hemisphere map showing disposition of megashears (solid lines with strike-slip arrows), combined compressive Cordilleran crest-island arc type structure (dashed lines), and shield areas. Shears transfer compressive stress from one compressive zone to another. Strain ellipsoids show appropriate shear angles in two types of crust. Partly after Moore (1963) and Rod (1960).

nite zone of greater eastward yield between the cratonic areas. A significant part of this yield is distributed throughout the outer regions of the cratonic block in North America.

In Figure 9 the long axes of the strain ellipsoids approximately parallel the gross Pacific margin of the two continents. Assuming an average 30° angle of shearing between the conjugate directions of maximum shearing stress and the direction of primary compression (normal to the Pacific margin), the shearing directions seen in the strain ellipsoid on the left should be developed. The left-lateral direction coincides very well with the large shears south of the shields on the continental blocks in North and South America. However, this direction does not correspond well to the direction of the large shears in the oceanic blocks in the Caribbean and off the tip of South America because there is a considerably different angle of shearing in the oceanic blocks. A strain ellipsoid with a 0° angle between the conjugate shears (Fig. 9, upper right) is more appropriate, probably because the continents are semi-uncoupled floating masses free to expand laterally, whereas the oceanic crust is a more continuous confined sheet both in the ocean and below the continents. Lack of room for expansion within the oceanic crust forces a smaller conju-

gate angle of shearing. Thus, deviant trends occur between oceanic shears and continental shears even though they result from the same basic stresses.

FORELAND DEFORMATION

Stresses in the Canadian and Wyoming forelands differ significantly (Fig. 10). In the Canadian foreland, the force directed eastward through the geosyncline is buttressed directly by the resistance of the Canadian shield, with the result that all the compressive stress is "in-line." There are no horizontal couples and no discernible reason for the existence of horizontal couples. The crust should be much stronger under this condition of direct compression than under the condition of direct compression and a lateral couple. The result in Canada was the formation of a high, narrow, linear Cordillera with an undeformed basement sloping gently westward beneath it.

On the south, there has been distributive eastward yield around the south side of the Canadian shield because of the greater eastward movement in Central America. As a result, a huge west-northwest-oriented left-lateral couple was established in the United States, especially in Wyoming. The Lewis and Clark, Wichita, and possibly the Texas lineaments (Figs. 1, 9) are

**STRESSING — CANADIAN VS.
UNITED STATES FORELAND**

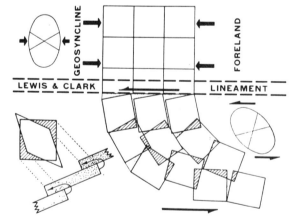

Fig. 10.—Difference in effect on segmented crustal blocks between wholly foreland compression in Canada and left-lateral couple that is chiefly responsible for breakup of United States foreland. Lower left shows space change caused by couple. Compare with shape of Wind River Range (Figs. 19, 20).

left-lateral shears associated with this couple, the effects of which were most severe during the Laramide orogeny. This couple created differential stress conditions on the boundaries of the large foreland crustal blocks, leaving them susceptible to being uplifted, structurally overlapped, and otherwise contorted by the compression. The diagram on the lower left of Figure 10 shows the space change from a predeformational rectangle to a postdeformational parallelogram which results from a left-lateral couple. The space change cannot be accommodated unless the two corners of the rectangle are thrust alternately upward over and downward under the adjacent blocks represented by the parallelogram. This is the cause of the overlap of crustal slabs in the Wyoming foreland province. The configurations of the Green River basin, Wind River Range–Wind River basin, and Owl Creek Mountain blocks are classic examples of this relation.

A factor in addition to coupling, which may have aided in breakup of the Wyoming foreland rather than the Canadian foreland, is related to the fact that in Canada the late Precambrian Belt Supergroup has been brought to the surface on large thrusts (Fig. 11). This sequence of Precambrian strata has ridden far upward out of its geosynclinal depositional position on such structures as the Lewis thrust and is now structurally high.

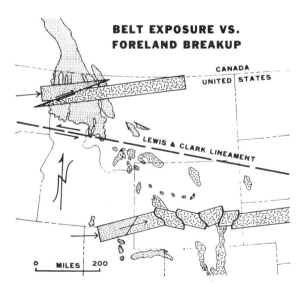

BELT EXPOSURE VS. FORELAND BREAKUP

CANADA
UNITED STATES

LEWIS & CLARK LINEAMENT

0   MILES   200

Fig. 11.—Thrusting of Belt series may have helped decrease tangential crustal compression and consequent breakup of foreland in Canada. In United States stress has been transmitted through Belt Supergroup into cratonic crust.

In contrast, south of the Lewis and Clark lineament (Smith, 1965) there is no comparable exposure of the Belt Supergroup. The Belt is present in Utah and Idaho and can be seen in the Uinta Range and in isolated outcrops in the geosyncline. It is buried very deeply except for these outcrops. The thrusting in western Wyoming has taken the form of a *décollement* over the top of the Belt Supergroup, and has not brought these rocks to the surface.

The thrusting of the Belt Supergroup in Canada and Montana has been a stress-dissipating mechanism. In contrast, in the United States south of the Lewis and Clark lineament the stress was directed through the Belt Supergroup as well as the deeper basement into the foreland crust, and caused a greater total tangential stress and greater deformation of the foreland.

STRESS DISTRIBUTION CAUSING SPECIFIC FORELAND FEATURES

Front Range–Denver basin.—The distribution of stresses that caused the configuration of the United States foreland is shown in Figure 12. One of the most noticeable features in the tectonics of the area, clearly shown on the tectonic map of the United States (Cohee, 1961), is the Colorado Plateau block which has remained intact and is less deformed than surrounding areas. Stress transmitted through the geosynclinal prism has pushed the Colorado Plateau block bodily eastward as shown in the stress-strain diagram of Figure 12. The wet-paper-towel structural model shown in this illustration was formed by moving the part representing the Colorado Plateau block eastward with hand pressure. This simplest of structural models duplicates most of the gross structural features of the Wyoming province, including the Front Range-Laramie, Wind River, Big Horn, and Uinta uplifts. It also shows analogs for the northeast-trending tears at the south end of the Wind River uplift.

The eastward movement of the Colorado Plateau block against the Mid-Continent craton caused a direct compression at its front which resulted in the Front Range uplift. On the south, the Front Range uplift becomes less pronounced at about the place where it would be intersected by a westward projection of the Wichita lineament. This may result from left slip along the Wichita lineament (Tanner, 1967) that relieved

FORELAND STRESS AND YIELD

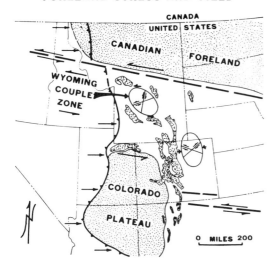

Fig. 12.—Foreland stresses and yield. Southern strain ellipsoid represents direct compression between eastward-moved Colorado Plateau block and Mid-Continent craton. Northern strain ellipsoid results from left-lateral "Wyoming" couple between that same block and stationary Canadian foreland. Photograph shows damp paper towel subjected to similar stresses by placing hand on lower left quarter and moving it toward right to simulate eastward movement of Colorado Plateau block. Compare "range" distribution in paper towel with that of Wyoming province.

some of the compressive stresses between the Colorado Plateau block and the "Texas craton" south of this lineament. Thus, stresses were transmitted through the Colorado Plateau and the Texas craton as a unit with relatively less shortening between them.

The block diagram (Fig. 13) viewed southward depicts the interplay between the Front Range, Denver basin, and Wichita lineament. The fold-slab configuration of the Denver basin and the Front Range uplift, and the termination of this deformation by left slip along the Wichita lineament, are shown. The diagram applies specifically to the Laramide, but stresses during formation of the Ancestral Rockies were similar.

Wyoming couple.—North of the Colorado Plateau block, and south of the unyielding Canadian foreland, a concentrated left-lateral "Wyoming couple" was created which has resulted in the ranges of central Wyoming.

Figure 14 is a model of the central Wyoming couple consisting of a powder-dried barite crust over a more fluid barite subcrust. The hinged containing box was started as a rectangle and was deformed to a parallelogram.

A map of the structurally analogous area in central Wyoming shows many similarities (Fig. 15). On the left side of the model the "Wind River Range" can be visualized with the southwestward overthrust, and northeast-trending tear faults near its south end. The "Rawlins uplift" is in approximately the right place and is convex westward, as is the "Wind River Range." The "Sweetwater uplift" is structurally high in the model, but has collapsed in nature. It has a large

YIELD EAST OF THE

COLORADO PLATEAU BLOCK

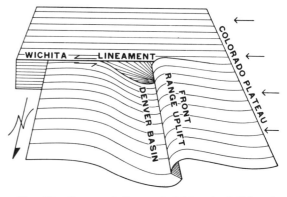

Fig. 13.—Space relation and yield east of Colorado Plateau block. View is south. Note fold-slab relation of Front Range uplift–Denver basin and southward termination of these structures by eastward yield along Wichita lineament. Lines on faults simulate slickensides.

 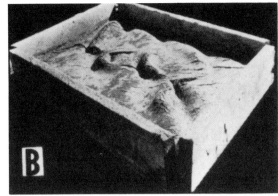

FIG. 14.—Vertical (A) and oblique (B) photograph of structural model which duplicated structure of central Wyoming by application of left-lateral couple. Longest dimension of model is about 2 ft. Model was made of viscous barite drilling mud with upper crust made more rigid by sprinkling on powdered barite.

thrust on the south flank. On the lower right is the "Laramie Range." The "Green River," "Hanna," and "Wind River" basins are in their correct positions. The model contains many northeast-trending tear faults which also are found in many places in the Wyoming area. It is proposed that the Wyoming couple is the generating force for the ranges of central Wyoming. By analogy with the model, the ranges of central Wyoming are a type of crust-thick drag fold. Because of the equally well-developed tendency of the crust to act as structurally overlapping tilted

slabs, the most accurate descriptive term is "drag fold-slab."

Three basic hypotheses for the configuration of major foreland thrusts have been proposed (Fig. 16): (1) vertical uplift with sagging over the basin, (2) thrusting, and (3) fold-thrusting. The writer's model studies support Berg's (1961) concept that the range-bounding structures are a cross between a true fold and a true thrust. The barite model (Fig. 14) showed an excellent correspondence in map view with the ranges of central Wyoming. Models also show very realistic analogs

CENTRAL WYOMING GENERALIZED STRUCTURE

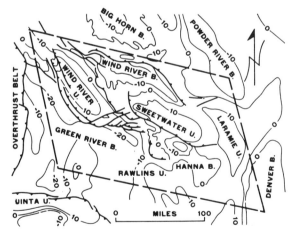

FIG. 15.—Index map outlining area of central Wyoming duplicated in Figure 14 model. All major structural features within outline are present in structural model, including tear faults at south end of Wind River Range. Generalized structure contour datum is approximately basement. Contour interval in thousands of feet.

THREE CONCEPTS FORELAND THRUSTS

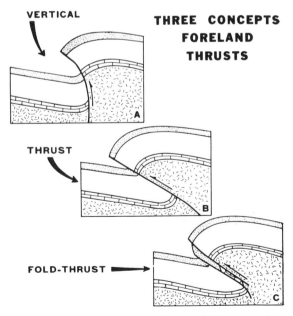

FIG. 16.—Three concepts of foreland thrusts. Writer supports fold-thrust interpretation of Berg (1961). Note overturned, sheared-out limb in fold-thrust interpretation. Modified from Berg (1961).

FORELAND MODEL

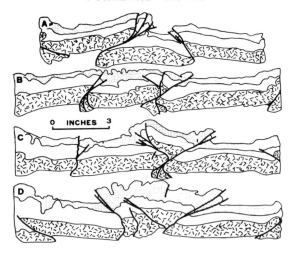

FIG. 17.—Cross sections of plaster model deformed in same manner and degree as model shown in Figure 14. These suggest that stresses which produced good map-view similarity in Figure 14 also produce very good structural similarities in cross section.

for cross sections of the boundary structures in the Wyoming province. Figure 17 shows cross sections of a model of layered plaster lying on "soupy" barite mud that was created in the hinged box used for the barite model in Figure 14. Particularly well shown are the steepening of some thrusts into the lower parts of the crust, and the general change from thrusting to folding upward with decreasing competence of the section.

The bottom (plaster-barite interface) of these models (Fig. 18) shows the relation between tear faulting (A) and thrusting (B), as it would appear in a "worm's eye view" of the base of the crust.

Wind River Range.—The Wind River Range is one of the best examples of structure caused by the Wyoming couple. A map of the Wind River Range (Fig. 19) shows a dominant zone of thrusting on the southwest side, and a well-developed group of tear faults at the south end of the range. The bounding structures blend tangentially with structures that pass eastward and westward from the ends of the range. The range has the overall cross-sectional configuration of a huge tilted fold-slab that has been thrust over the Green River basin, and has a dip slope northeast under the Wind River basin.

A fold-slab generated in one of the plaster models by a "Wyoming couple" experiment is shown for comparison in Figure 20. The trends are the same, the overthrust side is the same, the tear faults are present, and the model structure also blends tangentially into structures which pass east and west of it. If a less competent layer had been present above the "basement" layer shown, the fold-thrust would have "ridden up" through these "sediments" and would have been less recumbent.

Figure 21 is a comparison of model cross sections with the structural interpretation of the

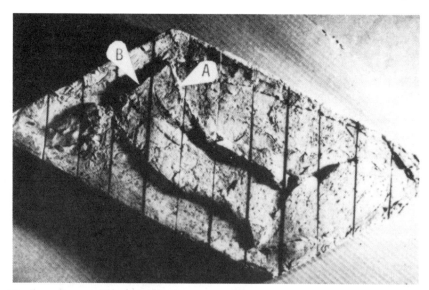

FIG. 18.—Photograph of bottom of plaster model shown in Figure 17 showing "root-zone" relations of tear faults (A) and thrusts (B). This surface overlay "soupy" barite drilling mud. Note how tear faults distribute stress from one compressive zone to another.

Wind River Range by Berg (1961). They have in common the overturned sheared-out limb, the overall slab configuration, and the tendency for upper layers of the crust to be more folded and less thrust than the lower layers.

Uinta Range.—The generalized view of the east end of the Uinta Range (Fig. 22) shows the thrust development on the north flank, the domal shape, and the blending of this thrust zone eastward into a lineament passing east of the range. If the figure is rotated 180° its geometry is similar to that of the Wind River Range. In the case

COMPARISON OF WIND RIVER THRUST AND THRUST IN MODEL

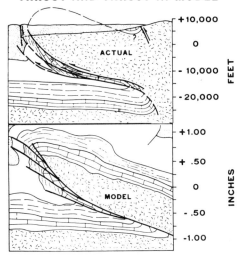

WIND RIVER RANGE

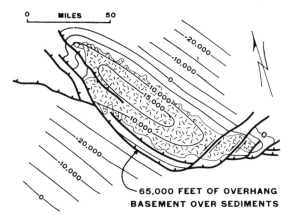

FIG. 19.—Wind River uplift and adjacent basins modified from Berg (1961). Note gross tilted, structurally overlapping slab configuration, postulated 12 mi of overhang, and well-developed tear faults at south end of range. Structural contour datum is top of Precambrian. C.I. = 5,000 ft.

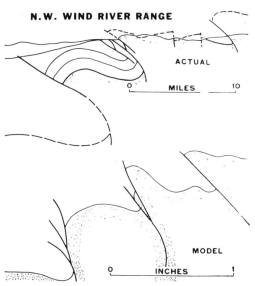

FIG. 21.—Comparison of two cross sections from Berg (1961) with analogs from plaster models deformed by left-lateral "Wyoming" couple. Note overturned sheared-out limbs, splayed fault patterns, and steepening of some faults downward. Model cross sections are from same relative position in model as their counterpart in nature.

FIG. 20.—Vertical photograph of plaster model deformed by left-lateral couple. Central fold is analog of Wind River uplift. Compare with Figure 19. Note trend, asymmetry, "southwestward" overthrusting, and tear faults at south end of fold-slab. If there had been an additional "sedimentary" layer above it, "basement" fold-slab would have been less recumbent and would have been bounded by a fold-thrust.

of a horizontal couple, this does not violate principles of symmetry. Therefore, by analogy, it can be seen that the eastern end of the Uinta Range, like the Wind River Range, is essentially a crust-thick, drag fold-slab related to the Wyoming couple. The Uinta Range, however, has been modified by special features of position and rock composition. The unique 24,000-ft-thick pod of late Precambrian Uinta Quartzite (Belt Supergroup

EAST DOME
UINTA RANGE

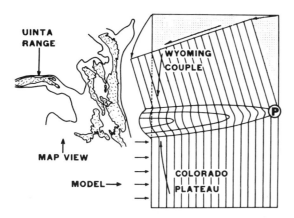

Fig. 22.—Simplified configuration of east dome of Uinta Range modified from Hansen (1965). Rotate 180° and compare with Wind River Range map (Fig. 19). Configuration and direction of overthrusting are similar.

equivalent) has exerted much control on the configuration of this range.

Eastward movement of the Colorado Plateau block and development of the Wyoming couple have resulted in an abnormally large component of north-south compression between the two, forming the Uinta Range. The space relation and the resulting stresses are shown diagrammatically in Figure 23. It can be seen in the damp-

paper-towel structural model (Fig. 12) that the Uinta uplift is a space requirement of the regional strain pattern.

In total, several factors have contributed to the configuration of the Uinta Range. The abnormal geometry of the late Precambrian Uinta Quartzite, the lateral component on the northern bounding structure (Hansen, 1965), and similarities in trend suggest that the Uinta Range may lie along a very ancient and recurrent zone of left-lateral yield similar to the Wichita and Lewis and Clark lineaments (Fig. 24A). East-west shortening of the basins north and south of the uplift has been differentially greater than the cross-folded east-west-trending uplift (Fig. 24C). North-south-trending compressive features are present north of the Uinta uplift. In contrast, the cross-folded Uinta uplift crust should have provided greater resistance to east-west shortening. This differential east-west shortening also contributed to the lateral component along some of the frontal structures. Figure 24B is a repeat of Figure 23 and suggests the pivotal action between the Wyoming couple and the Colorado Plateau as the main north-south compression-generating mechanism. The draping of thrusts around the west end of the Uinta uplift may have added a

ORIGIN OF UINTA RANGE
NORTH-SOUTH COMPRESSION

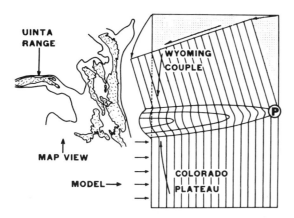

Fig. 23.—Diagram illustrating source of north-south compression responsible for Uinta Range which was developed between eastward-moved Colorado Plateau crust and left-laterally coupled crust of Wyoming. Note fine geographic-structural analog created in damp paper-towel experiment (Fig. 12). Uinta Range can be seen to be space requirement of regional stress pattern and not an "enigma."

FACTORS—UINTA RANGE FORMATION

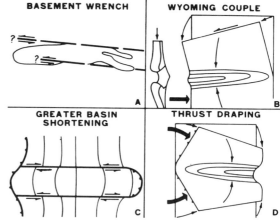

Fig. 24.—Composite diagram of factors affecting development of Uinta Range. A.—Ancient and recurrent west-northwest "Lewis and Clark"-type lineament. B.—North-south compression generated between Wyoming couple zone and Colorado Plateau. C.—Greater east-west shortening of adjacent basins relative to cross-folded Uinta arch. D.—Possible north-south component caused by thrust-draping around west end of incipient arch.

component of north-south compression inside the area of draping, with a pivotal zone at the west end of the range (Fig. 24D). All of these factors and the abnormal mechanical properties of the thick east-west-trending pod of late Precambrian Uinta Quartzite have contributed to the unique structure and orientation of the Uinta Range.

Beartooth, Big Horn, and Black Hills uplifts. —Bounding the north side of the province are the Beartooth, Big Horn, and Black Hills uplifts (Fig. 25). They are aligned with their north edges abutting the Lewis and Clark lineament zone of Montana. On the opposite side of that lineament the smaller uplifts and basins of northern Montana represent a transition between the severely deformed foreland crust in Wyoming and the much less deformed foreland in Canada.

Figure 26 is a plaster model of a left-lateral wrench fault showing drag folds and the second-order faults that are developed along such a wrench. The folds meet the zone obliquely and are aligned in the same manner with approximately the same angle as the three uplifts in nature. This model suggests that the actual uplifts are crust-thick drag folds or fold-slabs formed along the south side of the Lewis and Clark linea-

FORELAND STRUCTURE MAP

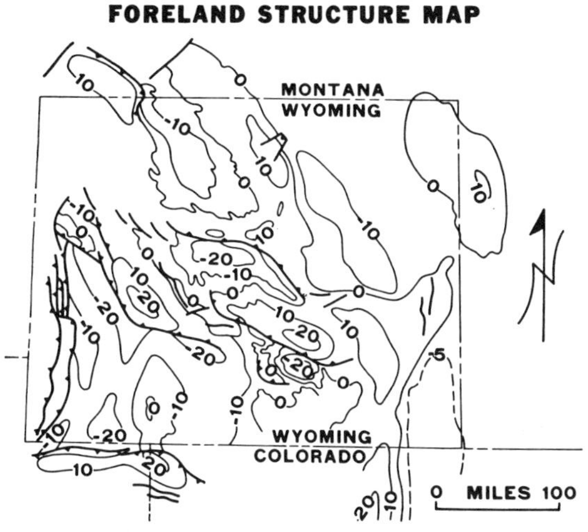

FIG. 25.—Index map of Wyoming area showing major uplifts and basins and major boundary thrusts. Names shown on Figure 15. Contour interval in thousands of feet.

ment as a result of left-lateral coupling. The facts that the angle of contact with the lineament, structural relief, exposure of the basement, and severity of the bounding structures become more pronounced westward also suggest that the source of stress was from the west. Each range is the re-

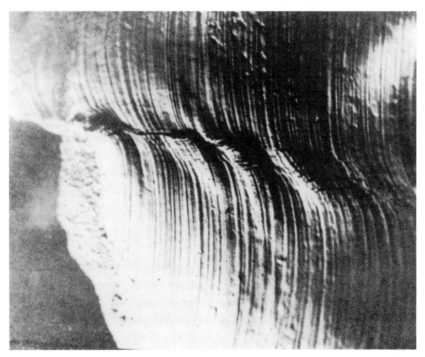

FIG. 26.—Vertical photograph of plaster model deformed by left-slip "Lewis and Clark" basement wrench. Note northeast-trending second-order tear faults, and three drag-fold analogs of the Beartooth, Big Horn, and Black Hills uplifts. Flexed reference lines were "swept on" with a brush and were straight and normal to wrench zone prior to movement.

**FORELAND SPACE RELATIONS
(ACTUAL)**

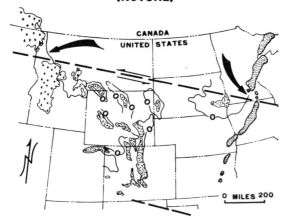

FIG. 27.—Index map showing features associated with Lewis and Clark lineament (heavy dashed line). Note batholith offset (left arrow), Mid-Continent gravity-high offset (right arrow), and alignment of uplifts along south side of lineament.

**FORELAND SPACE RELATIONS
(RESTORED)**

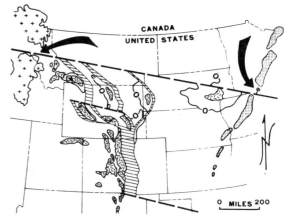

FIG. 28.—Figure 27 map, cut and restored by eliminating batholith and gravity offsets and shortening that occurred during foreland deformation (horizontally ruled areas). This duplicates (in reverse) movements that took place during deformation of foreland.

sult of the total tangential shortening minus the amount expended in formation of each of the ranges west of it.

Space changes during deformation.—Important features of the Lewis and Clark lineament, which have been described by Smith (1965), are shown in Figure 27. On the west the lineament appears to displace the Cretaceous batholiths by 100-150 mi. In Iowa, the Mid-Continent gravity high, which is the strongest gravity feature on the continent, seems to be displaced about 80 mi left laterally along a projected extension of the Lewis and Clark lineament zone. South of the lineament, the foreland is broken up into a series of huge uplifts, whereas north of the zone the foreland is relatively undeformed.

Figure 28 is a duplicate of Figure 27 in which the map has been cut along the lineaments and basin-uplift bounding structures in such a way as to restore the terrane to its predeformational position. Parts of the map have been pushed back along these zones to align the offset segments of the Mid-Continent gravity high and the batholiths. The ruled areas adjacent to the uplifts represent the crustal space that remains after palinspastic restoration, and show the shortening that has taken place in the formation of the uplifts. If the effects of the Wyoming couple and the Front Range compression are added as in the illustration, there is a reasonable distribution of the shortening which has accounted for the ranges of

the Wyoming province. This shortening can be carried as far south as the Wichita lineament. This lineament again is torn left laterally and the eastward-directed stresses have been transmitted along it all the way to the Gulf Coast.

MECHANICS AND EXAMPLES OF DRAG FOLDING

Surface reference lines.—The degree to which flexure of the sedimentary section occurs at the side of the basement wrench zone may differ greatly from place to place. "Surface" reference lines that are draped across such a wrench zone by strike-slip movement show this variation. Before deformation the lines are straight and normal to the wrench (Fig. 29). These reference lines are comparable with those that have been "swept" onto the surface of a plaster model before deformation (Fig. 26).

In the basement such a wrench may be a distinct shear, but where it has been propagated upward to the surface it is a wide, distributively flexed zone. Total horizontal offset at both basement and surface levels must be equal. The patterned areas of Figure 29 represent "glued-down" or coupled zones in which the upper crust moves rigidly and in unison with the basement, resulting in no flexure. The white area represents zones in which the sediments are uncoupled from the basement at one or many horizons and have been flexed.

394

UNCOUPLING OVER BASEMENT WRENCH SURFACE REFERENCE LINES

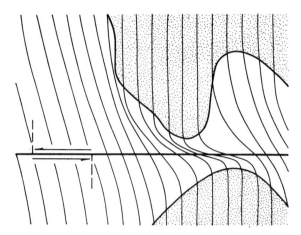

FIG. 29.—Surface reference lines deformed over basement wrench. Sediments are uncoupled from basement (clear area) and coupled with basement (stippled areas). Lines are equally spaced and normal to wrench prior to movement and duplicate "swept-on" lines on models. Note zone of concentrated flexure between two coupled salients.

Reference lines in the couple zones have a definite spacing. Wherever these lines exhibit a closer spacing in the uncoupled zones there must have been compression normal to them. The figure shows that the closest spacing and greatest compression are in areas on the "upstream side" of coupled salients which converge on the wrench zone. Here more material is being crowded into a shorter space than in the areas with the gently sweeping reference lines.

Residuals of surface reference lines.—If the swept reference lines are considered to be contour lines, it is possible to derive a residual map from this "trend surface." Essentially these residuals measure the degree of curvature of the lines or deviations from the regional trend of the reference lines. The residual map looks like a structure contour map of a drag fold generated by the basement wrench (Fig. 30). It provides a means of cross-comparison between the deformation of the basement wrench and the configuration of the overlying drag folds. It also may provide a basis for quantifying drag-fold mechanics.

Controlled drag-fold model.—On the basis of this predeveloped hypothesis, a series of model experiments exemplified by Figures 26 and 31 was performed. A "blob" of plaster was placed on top of two adjacent boards and one board was pushed past the other. By putting fluid lubricating mud

RESIDUALS OF SURFACE REFERENCE LINES

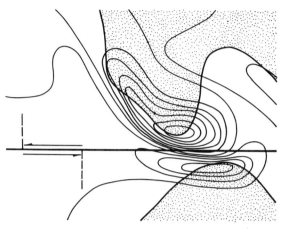

FIG. 30.—"Residual" map formed by treating reference lines of Figure 29 as trend surface contours. Note that it looks like drag fold which is centered over area of most severe flexure in Figure 29.

in position to simulate the uncoupled zones on Figures 29 and 30, it was possible to control the position of formation and configuration of the drag fold (Fig. 31).

A "worm's eye" photograph of the bottom of this drag-fold model (Fig. 32) shows the darker areas where soupy barite drilling mud was placed between the board and the plaster to cause uncoupling. The lighter areas are places where the plaster was in contact with the boards and no uncoupling took place. When the boards were moved, the two opposite coupled salients came closer together causing compression of the overlying plaster and formation of the drag fold.

Good fold structures also can be generated by a wholly distributive lateral couple. Figure 33 shows a model consisting of a layer of plaster that was "sandwiched" between two barite layers and deformed in the type of box that was successful in duplicating the major range structures of central Wyoming (Figs. 14, 17, 20, 21). One of the important aspects of this experiment is that good folds and good shear patterns were formed simultaneously.

Drag-fold examples.—A base map of the Wyoming area with an overlay simulating the swept-reference-line concept (Fig. 34) generalizes this hypothesis to regional scope. A reasonable picture results of the shortening which has occurred during deformation. The map shows that the deformation is concentrated at the boundaries

FIG. 31.—Plaster model in which drag-fold position and configuration were predetermined by using "soupy" drilling mud to force uncoupling by hypothesis presented in Figures 29 and 30. Note similarity of this fold to Big Horn uplift.

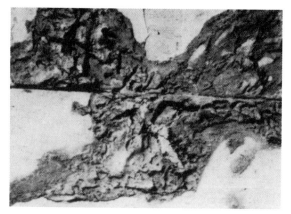

FIG. 32.—Photograph of bottom of Figure 31 model inverted to correspond with that figure and showing distribution of drilling mud (darker areas) and axis of fold. Compare with stippled and clear areas of Figures 29 and 30.

between the large uplifted and downdropped blocks.

A comparison of the Big Horn uplift with the smaller Elk Basin anticline on its west flank (Fig. 35) shows that they are very similar in configuration, though one is about one order of magnitude larger than the other. They are also alike in terms of mechanics except that the smaller fold involves only the sedimentary layers and possibly some of the upper basement and the larger fold

involves the entire crust. In the formation of the smaller fold the required uncoupling occurs in or on the upper basement or, more likely, along one or many incompetent zones throughout the sedimentary section. In formation of the larger fold the same uncoupling occurs in the plastic layer at the base of the crust. The "one order of magnitude" factor also seems to apply to the northeast-trending second-order tear faults which characteristically terminate drag folds, whether in

FIG. 33.—Vertical photograph of folds and shear patterns created in plaster slab "sandwiched" between layers of "soupy" barite drilling mud and deformed by left-lateral couple. Longest dimension of model is about 2 ft; amplitude of folds is about 2 in.; thickness of slab is about ½ in.

models (Fig. 26), large anticlines (Fig. 35), or the huge crust-thick uplifts shown on the tectonic map of the United States (Cohee, 1961). For example the Fromberg fault, which partly terminates the Big Horn uplift on the northwest, is about 50 mi long. This is about 10 times as long as the *en échelon* tear faults just north and south in the Lake basin and Nye-Bowler zones, respectively, and as those associated with the Elk Basin fold. This relation might provide a basis for quantification of related crustal properties.

Many of the anticlinal structures of the foreland can be explained in terms of this drag-fold hypothesis. Figure 36 shows an area in the southwest Piceance basin of northwest Colorado containing *en échelon* folds and northeast-trending *en échelon* faults. At the upper right is an oriented insert of Moody and Hill's (1956) diagram of wrench-fault tectonics which shows the directions of the folds and faults that they believed should result from a left-lateral basement wrench of similar orientation. There is good correspondence of the basement wrench direction, the fold direction, and the second-order fault direction with their diagram. Figure 37 is another plaster basement-wrench model which shows a good correspondence with the geometry shown in the Piceance basin in Figure 36.

A generalized structural map of the Piceance basin with some of the larger oil fields superim-

ELK BASIN ANTICLINE AND
BIG HORN UPLIFT COMPARED

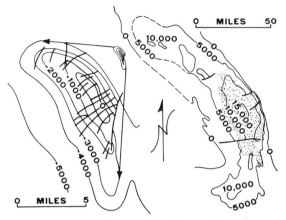

FIG. 35.—Comparison of Big Horn uplift with Elk Basin anticline on its flank. They differ in size and layers of crust involved, but are both drag folds. Big Horn uplift bears same space relation to Lewis and Clark lineament as Elk basin fold bears to Nye-Bowler zone of Lewis and Clark lineament. Compare with Figures 30 and 31.

posed is shown in Figure 38. Of particular interest is the distribution of oil fields and the shape of the bounding structure on the east side of the basin. If a similar map of the northern Big Horn basin (Fig. 39) is rotated 180° and compared with the Piceance basin map, there is good geometric similarity. The bounding structures have the same orientation and angularity. The shape and distribution of oil-field structures in the Piceance basin coincide with those in the Big Horn basin. Northeast-trending tear faults are abundant in both areas. The two structural situations more than 300 mi apart are similar in symmetry and probably mechanics. It is significant that these areas are near the north and south boundaries of the concentrated Wyoming couple zone. The northern area includes a long-known transcurrent zone, the Nye-Bowler zone. The southern area contains the zone documented in Figures 36 and 37 in about the same relative geometric position. In both areas the geometry can be explained by the combination drag fold-tilted slab hypothesis presented by the writer.

It has been suggested that most of the ranges of the Wyoming foreland are essentially crust-thick drag folds that result from the left-lateral Wyoming couple. This concept might be expanded to describe a crustal tendency toward even bigger drag folds (Fig. 40). All of the deformed ranges

FORELAND STRUCTURE MAP

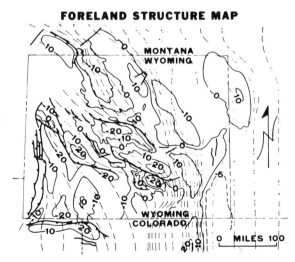

FIG. 34.—Portrayal of deformation of Wyoming area by expansion of reference-line theory of Figure 29 to regional scale. Deformation is concentrated at boundaries of crustal blocks. Compare with Figures 28 and 29. Contour datum is basement. Contour interval in thousands of feet.

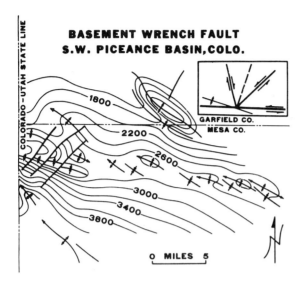

FIG. 36.—Map of left-lateral basement wrench in southwest Piceance basin, Colorado. Note agreement with wrench-fault tectonic diagram in upper right corner. Datum is top of "F" zone, Dakota Sandstone, Lower Cretaceous. CI = 200 ft. Modified from Krey (1962) and Moody and Hill (1956).

of the Wyoming foreland, including the Front Range, in total, have the shape of a huge drag fold between the Colorado Plateau block and the Lewis and Clark lineament. This drag fold, because of strength properties of the crust, was much too big to hold together during deformation and it broke into the series of smaller Laramide uplifts. The remains of the bigger uplift, however, are still discernible.

The Coast Range batholiths in British Columbia are draped eastward around the undeformed Canadian foreland in the correct manner for a drag fold associated with the Lewis and Clark lineament. South of the Colorado Plateau block, the Precambrian outcrops in the Basin-Range area of Arizona have the shape of a drag fold between the Colorado Plateau block and the Texas lineament. In the southern example, the fold has been broken up more severely by late-stage Basin-Range structure.

EXPLANATION OF COLORADO PLATEAU BLOCK

One reason why the Colorado Plateau block remained intact during Laramide deformation may be the unique position of this block on the "exposed" outer corner of the craton relative to the deforming forces from the Pacific block. Because it was less confined by surrounding cratonic crust, it was more free to be pushed bodily without undergoing severe internal deformation. Another reason may be the position of the Ancestral Rockies. Experience with structural models shows that compressively deformed material resists further deformation and, therefore, is self limiting. The more the model is squeezed, the more difficult it is to squeeze it further or to make it shear. Therefore, it might be expected that, after part of the crust is deformed by compressive deformation, it should resist further compressive deformation and horizontal shearing. A large-scale sub-

FIG. 37.—Plaster-model analog of basement wrench shown in Figure 36.

GENERALIZED STRUCTURE
PICEANCE BASIN

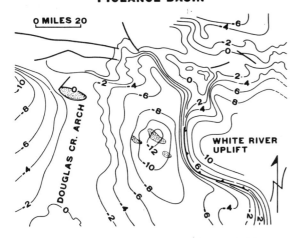

FIG. 38.—Generalized structure map of Piceance basin, Colorado, showing oil structures (stippled) and nearly right-angle curvature of thrust 'on east flank of basin. Datum approximate top of Cretaceous. CI = 2,000 ft.

stantiation of this is the fact that maturely deformed geosynclines eventually assume cratonic crustal properties. If the Colorado Plateau block is examined in this light, it is seen to be "cradled" between the force from the geosyncline and the previously deformed Ancestral Rockies (Fig. 41). This cradling effect should have tectonically

GENERALIZED STRUCTURE
BIG HORN BASIN

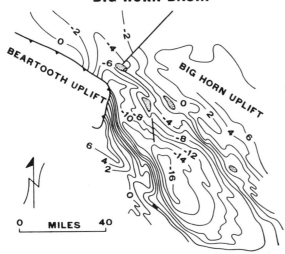

FIG. 39.—Generalized structure map of Big Horn basin. Note geometric similarity with Piceance basin if it is rotated 180°. Bounding structures of White River and Beartooth uplifts and major oil structures occupy similar positions in both basins.

REALLY BIG DRAG FOLDS

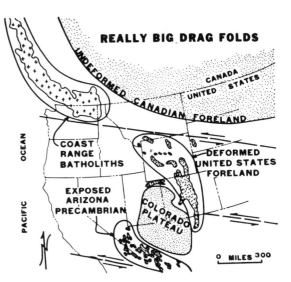

FIG. 40.—Big drag folds of western North America. Two southern folds collapsed during formation because of excessive size relative to crustal strength. Note relative positions of lineaments, stable crustal blocks, and drag folds. Compare with Figure 29 and 30, and compare foreland drag fold with paper-towel experiment (Fig. 12).

"sheltered" the Colorado Plateau area during the later Laramide deformation.

ANCESTRAL ROCKIES AND LARAMIDE
DEFORMATION COMPARED

The Ancestral Uncompahgre uplift and the Laramide Big Horn uplift are compared in Figure 42 which shows that the shapes, trends, and sizes are nearly identical. There is probably no basic mechanical difference between the ancestral Permo-Pennsylvanian deformation and the Laramide

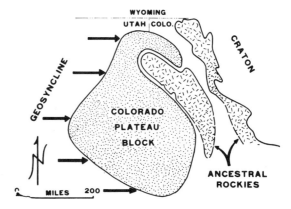

FIG. 41.—"Cradling" of Colorado Plateau block between force from geosyncline and predeformed Ancestral Rockies crust is one reason for its lesser Laramide deformation.

ANCESTRAL ROCKIES AND LARAMIDE UPLIFTS COMPARED

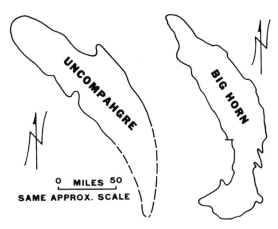

FIG. 42.—Diagram illustrating great similarity between Ancestral Uncompahgre uplift and Laramide Big Horn uplift which suggests similar mechanics of formation. Partly after Mallory (1958) and Bucher (1964).

deformation except for a difference in location of the concentrated couple because of the Laramide interplay of the Colorado Plateau block and the undeformed foreland in Canada. The geometry suggests that there may have been a northward migration of foreland deformation between the episodes of Ancestral Rockies and Laramide deformation.

ANALYSIS OF FORELAND DEFORMATION IN CROSS SECTION

Basic deformational configurations.—One of the most instructive ways to examine foreland deformation is in terms of tangential pressures applied to semi-elastic slabs that are floating on a viscous fluid subcrust (Fig. 43). Such a slab (Fig. 43A) deformed by tangential compression into a sine curve (Fig. 43B) shows elastic behavior; half of the curve is positive and half is negative. Similarly, on the foreland, the basins went down at least as much as the uplifts went up. This vertical movement is well explained in terms of a slab flexed into a sine curve by tangential compression. Wholly vertical uplift does not explain the abnormally great basin subsidence without producing problems of isostasy.

At some stage, the flexed slab breaks and the uparched part of the curve generally will ride over the negatively flexed part and approximate the overlapping-slab configuration (Fig. 43C).

This configuration can be considered in terms of isostasy. It is assumed, for easy visualization only, that the slabs have half the specific gravity of the subcrustal material on which they are floating. At the ends of the cross section the slabs must be in isostatic equilibrium because they are floating half out of the subcrustal material. In the middle, there is a double thickness of crust because the slabs have been overlapped by thrusting. The upper slab is riding entirely out of the subcrustal material but the lower slab is completely immersed; the central section, as a whole, is in isostatic equilibrium. At the ends of the overlap zone, however, the crust is severely out of isostatic equilibrium, and stress in that area generally will break or sag the slabs. Figure 43D represents the same sequence of slabs after these isostatic stresses have been relieved and isostatic equilibrium has been attained by collapse of the overthrust slab and uplift of the basin parts of the subthrust slab at the ends of the overlap zone. In the Wyoming foreland, the end result has been a combination of all the basic cross-section configurations shown.

The crustal slabs have been flexed into the sine-curve configuration, probably dominantly before and during the early stages of development

DEFORMATION CONFIGURATIONS

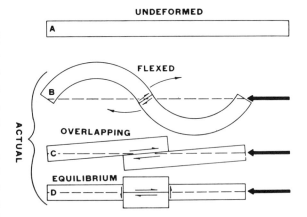

FIG. 43.—Basic configurations of foreland deformation. *A.*—Undeformed crustal slab; *B.*—Slab deformed into unbroken sine curve; *C.*—Uncollapsed structurally overlapped and tilted slabs; and *D.*—Structurally overlapped slabs that have collapsed to complete isostatic-equilibrium state. It is assumed for easy visualization that crustal slabs have half density of tectonically fluid subcrust, upper surface of which is represented by dashed line. Foreland deformation is combination of *B, C,* and *D.*

of the slab-bounding fold-thrusts. The sine-curve configuration is, in a structural sense, a series of linked alternating negative and positive arches; the buttressing or strut action of the mutual limb is required to create and maintain the arches. If the mutual limb fails, by development of fold-thrust flank structures, the required compression no longer can be transmitted to accentuate further the curvature of the adjacent positive and negative arches. The remnant arching that is seen in most of the large uplift cores even after development of fold-thrust flank structures is a function (1) of the imperfect elasticity of the upper crust and (2) of the fact that even a fully developed flank structure only partly dissipates the tangential compression required to maintain the arch. The Uinta uplift is the best-documented example of such arching because of the unusual condition of having well-stratified Uinta Quartzite exposed in its core (Hansen, 1965; Fig. 44).

Although the crustal slabs do not "unflex" geometrically from the arched sine-curve configuration to attain the overlapping-slab configuration, the stress changes are those suggested in the basic cross sections of Figure 43. With failure of the mutual limb the segments of the sine curve lose the structural properties of an arch and take on the structural properties of a beam (in terms of the cross-sectional view), even though they still may be flexed. This beam must span from one support point where the crust is in isostatic equilibrium to the next such point. Disregarding major sagging of these "beams," the support points have to be in the overlap zones of the overlapping-slab configuration or, in terms of the uplifts, under the flank structures. With regard to an uplift such as the central Uinta Range with a large flank structure on both sides, this concept of support suggests a partial beam effect across the width of the range.

The Green River basin, Wind River uplift, Wind River basin, and Owl Creek uplift in central Wyoming represent a series of three crustal slabs that are tilted north and east and are overlapped structurally by fold-thrusting. This results in severe overthrusting of the south and west flanks of the uplifts, with long smooth dip slopes on the north and east sides of the uplifts plunging under the adjacent basin. The isostatic-structural conditions should be those represented in Figure 43C. The midpoints of the structurally overlapped

ISOSTATIC THRUST REVERSAL

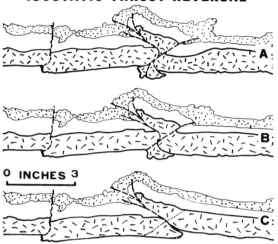

Fig. 44.—Stresses caused by isostatic imbalance can cause direction of thrusting to reverse during compressive phase. This model evolved from pure structurally overlapping slab configuration (C). Isostatic stresses at ends of the overlap zone caused thrust in opposite direction which dragged and offset older thrust, and created "two-sided" uplift and "mirror-image" mass at bottom of crust. This has analogies with "two-sided" uplifts of foreland, such as Uinta Range.

(fold-thrust) zones, and the midpoints of the long dip slope between them, should be in isostatic equilibrium. Between these neutral points the slabs should be out of isostatic equilibrium, with collapse being prevented by the "beam strength" of the slabs.

In materials of low tensile strength such as the crust, a beam should be inherently weaker than an arch. It follows that transition from arch to beam characteristics with the development of bounding fold-thrusts should lead to collapse. In at least two examples (the sunken top of the Sweetwater uplift and the Brown's Park graben of the eastern Uinta Range) collapse occurred. In both cases the collapse is adjacent to a large frontal structure which may have allowed the collapse. However, the reverse argument also may apply; "keystone" extension of a collapsing uplift top should force extension of the bounding structure.

One puzzling feature is that the Wind River Range, which has at least as much structural relief as the two collapsed areas and a huge fold-thrust on its southwest flank, did not collapse. This competence may be related to trend, because the two greatest examples of collapse are asso-

ROCK SPRINGS UPLIFT ORIGIN

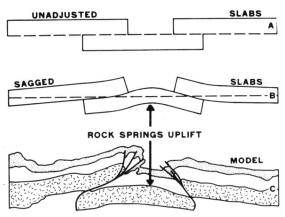

FIG. 45.—Rock Springs uplift can be explained in terms of structurally overlapped slabs partially sagged to obtain isostatic equilibrium (*B*). *C,* model analog with (from left to right) "Wind River uplift," "Green River basin," and "Uinta uplift." Bowed basin block simulates Rock Springs uplift.

ciated with the two most nearly east-west-trending uplifts. The mechanical reason may be a "catalyst effect" from a greater horizontal shear component along these east-west zones. Hansen (1965) documented a significant horizontal component along the north-flank structure of the eastern Uinta Range. Analysis of the Wyoming couple scale models and of regional stresses suggests that there should be a greater horizontal component along the south Sweetwater uplift than along the southwest Wind River Range uplift. This horizontal component may have the effect of producing an "environment of sliding friction" which allows the frontal structure to develop more fully and the uplift to assume greater beam *versus* arch characteristics, leading to collapse. It should be emphasized that the mechanics presented for the uplifts apply in reverse to the basins, which are subjected to approximately opposite isostatic stresses.

Isostatic thrust-reversal hypothesis.—The deformation seen in some model cross sections suggests that isostasy exerts a strong controlling influence on the shape of bounding structures between foreland slabs. Of special interest is the "isostatic thrust reversal history" shown in the model in Figure 45. This structure is a result of slab stresses related to both the overlapping-slab (Fig. 43C) and the isostatic-equilibrium (Fig. 43D) configurations. As the slabs attained the overlapping configuration (Fig. 44C) isostatic stresses generally collapsed the upper slab downward at

the right end of the overlap zone and tilted the lower slab upward at the left end of the overlap zone. If the slabs had remained intact until the compressive phase was completed, this adjustment would have been accomplished by normal faulting on the right and left of the overlap zone. If, however, gravity collapse of the slabs occurred while the slabs were being compressed, the result would be that shown in Figure 44B and C.

The stress fields in the areas of maximum shearing stress at the ends of the overlap zone are in such a position that the most effective mechanism for dissipating these stresses during the compressive phase would be a thrust moving opposite to the thrust that created the original overlapping-slab configuration. The initiation of this thrust is shown in Figure 44B and the model end product in Figure 44A. Of interest are the offset and drag of the earlier thrust by the later thrust. This model cross section is a nearly perfect example of an isostatic-equilibrium configuration. The formerly uplifted slab has collapsed to the same elevation as the original basin slab. The "broken off" ends of the slabs are superimposed one over the other, with the lower providing a "nearly mirror-image root" which compensates for the elevation of the uplift.

Isostatic thrust-reversal application.—This concept applies directly to the evolution of foreland uplifts because most, with the possible exception of the Wind River Range, show some degree of mirror-image compressive structures on both sides of the uplift. As an example, Figure 44A and the mechanics which it suggests could represent a north-south cross section through the Green River basin–Uinta uplift–Uinta basin or an east-west cross section through the Denver basin–Front Range uplift–West Slope basins.

The isostatic thrust reversal hypothesis has specific application to the collapse history of the south-bounding structure of the Sweetwater uplift. The fact that the upturned flanking sedimentary rocks and a capping veneer of granitic debris are now higher than the crest of the uplift has been interpreted in two ways: (1) the core of the uplift collapsed along normal faults after the compressive phase (Carpenter and Cooper, 1951); or (2) the uplift collapsed by upthrusting of the flank structure toward the core of the uplift during the compressive phase (Bell, 1956). The analysis of the model (Fig. 45) suggests that

(1) both hypotheses are mechanically sound, (2) both are collapse-mechanism deviations from the pure overlapping-slab configuration, (3) they are mechanically identical except for their relation to the time of compression, and (4) there is good reason to expect that both mechanisms have contributed to collapse of the Sweetwater uplift. It is mechanically significant that the location of the cross section of Figure 44 in the model is exactly analogous to the Ferris Mountain area along the high south flank of the Sweetwater uplift where the most severe collapse has occurred in nature.

Four other areas that are characterized by downfaulting of the core of the uplift relative to the higher flank are the eastern Uinta Range, northern Medicine Bow Range, northern Laramie Range, and the Owl Creek Mountains. These may be smaller examples of the overlapping-slab configuration partly collapsed toward the isostatic-equilibrium configuration.

Rock Springs uplift origin.—The Rock Springs uplift can be explained in terms of flexed-slab mechanics (Fig. 45). It is assumed that the slabs have a specific gravity half as great as that of the subcrustal material on which they are floating. In Figure 45A the slabs are completely out of isostatic equilibrium. The slabs on the side are floating entirely above the subcrustal material, and the center slab is entirely immersed. The only areas in isostatic equilibrium are the overlapped parts of the slabs. The slabs will sag as shown in Figure 45B. At the ends of this cross section the slabs are in isostatic balance because they are half immersed. The middle of the central slab is also in isostatic equilibrium, floating half above the line representing the top of the subcrust. Similarly, in both of the overlap zones, the slabs are in isostatic equilibrium because there is a full slab above and a full slab below the reference line. The result demands a bowing in the center of the basin block. The peripheral areas of the basin block have been depressed by the weight of the Wind River and Uinta overthrusts on the north and south, respectively.

Figure 45C is a cross section of a plaster-model analog of the Wind River Range, the Rock Springs uplift, and the Uinta Range. Similarity to cross section B configuration as well as to the Green River basin is apparent. Approximately the same relation obtains with regard to an east-west cross section through the Rock Springs uplift, where the Wyoming overthrust belt and the Raw-

RHEID CONCEPT

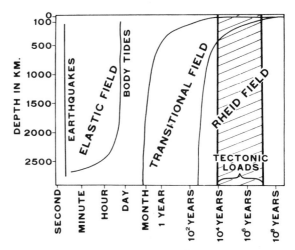

FIG. 46.—Rheid concept, modified from Carey (1953), suggests that thickness of crust which is acting as a solid above tectonically fluid subcrust is directly proportional to rate of application of deforming stresses.

lins and Sierra Madre uplifts at the eastern edge of the Green River basin act as the depressing weights. This adjustment may account for the "turtle back" shape and the distinct north-south Rock Springs uplift axis superimposed on a distinct east-west Wamsutter arch. Apparently none of the other basins were large enough or surrounded by sufficiently large flank structures to allow such a medial swell to form.

Application of rheid concept.—In order to understand this type of crustal mechanics it is necessary to comprehend the relation between the rigidity of the outer crust and the tectonic fluidity of the lower crust. This relation was presented by Carey (1953) as the rheid concept (Fig. 46). The rheid concept suggests that, for long-term stresses in the time span of tectonic application of stresses, the subcrust will act essentially as a viscous fluid and the upper crust will act as a rigid solid. Glacial rebound in particular and isostasy in general substantiate this hypothesis. In contrast, with short-term stresses such as earthquake shear waves the crust acts as a solid to the depth of the outer core.

The thickness of crust that acts as a solid above the fluid subcrust is a function of the rate of application of the tectonic force. The generalization can be made that the faster the tectonic force is applied the thicker will be the part of outer crust that acts as a solid. This factor has

RHEID CONCEPT AS IT APPLIES TO CONFIGURATION OF UPLIFTS

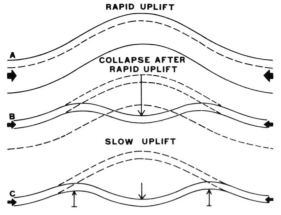

FIG. 47.—Application of rheid concept to foreland suggests that rapidly applied stress should result in thick solid crust and uplifts of large wave length and amplitude (*A*) which later should collapse to an amplitude and wave length that can be sustained in a time-independent manner.

direct application to the size of the uplifts in Wyoming and to the later collapse of uplifts.

It is assumed in Figure 47A that the tangential compression was applied rapidly enough that a fairly thick outer crust acted as a solid, causing uplifts and basins of large amplitude and large wave length. After the compressive stresses cease, the rheid interface between the tectonically fluid subcrust and the solid upper crust should migrate upward until the crust attains a thickness representing a time-independent stage that can be maintained indefinitely. This thinner crust cannot sustain uplifts and basins of the amplitude and wave length of those in the previous thicker crust and the thinner crust will collapse to an amplitude and wave length which can be sustained in a time-independent manner (Fig. 47B). Likewise, if the deformation has been much slower, uplifts of somewhat smaller wave length and amplitude may have resulted (Fig. 47C). It is very possible that collapse of such elements as the Sweetwater uplift and the Brown's Park graben in the eastern Uinta uplift is partly a function of this crustal property. Symmetry and isostasy demand that similar but opposite considerations be valid for the basins.

Isostasy, tangential compression, and abnormal vertical movements.—A composite series of isopach maps from Hansen (1965) suggests that the Uinta Range area has been vertically hyperactive through a great part of geologic time (Fig. 48).

There was abnormally thick sedimentation during the late Precambrian, Mississippian, Early Pennsylvanian, Middle Pennsylvanian, Permian, Triassic, and Cretaceous times. There were also intervening times of abnormally thin sedimentation or greater-than-normal erosion, indicating uplift. This positive tendency was most severe during the time of Laramide orogeny.

The hyperactive vertical histories of the Uinta Range (Fig. 48) and the analogous Central Montana uplift along the Lewis and Clark lineament zone (Norwood, 1965) can be explained in terms of tangential compression and isostasy (Fig. 49). The left side of Figure 49 shows a series of schematic cross sections which show the isostatic characteristics of a crust which has been buckled by tangential compression rather than being bent by vertical tectonics. Simple whole numbers are used to suggest the relative weights of the crustal segments. They are not specific-gravity figures. The number sequence in Figure 49 suggests that, because a second-generation sediment is lighter than its parent material and because much material is removed completely by trunk rivers, a significant upward movement will cause an excess mass in the uplift relative to the adjacent basins. If the basic deforming stresses are tangential, relatively small isostatic stresses superimposed normal to the tangential stress should be required to trigger flexure in the opposite vertical direction. The two bottom cross sections on the left in Fig-

UINTA VERTICAL HISTORY

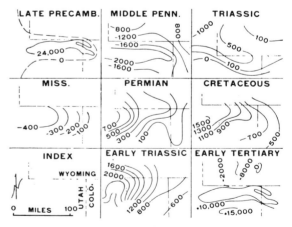

FIG. 48.—Uinta Range area vertical history summarized. After Hansen (1965). Note that area of Uinta Range (heavy outline in first and last subdivision) has had abnormal sediment thickness through much of geologic time. Isopachs are of selected units of age indicated and not of total sequence.

ure 49 suggest that as this process is repeated it becomes accentuated and results in even greater positive and negative anomalies in the next cycle.

This vertical reversal makes sense in terms of tangential compression; it makes no sense in terms of vertical uplift. Under tangential compression a relatively small isostatic imbalance could trigger a major reversal of vertical flexure. This movement in the reverse direction would then continue until the opposite isostatic imbalance caused flexure to reverse again. With only vertical forces as the deforming mechanism, vertical reversal would necessitate complete reversal of the deforming mechanism.

This type of mechanics may explain why areas such as central Montana and the Uinta Range have had such a hyperactive vertical history. It is probable that a basement wrench zone would be especially vulnerable to these vertical oscillations because of the predominance of nearly vertical faults and the triggering effect of the horizontal component.

The right side of Figure 49 may help explain why 24,000 ft of Uinta Quartzite is exposed in the Uinta Range, but no similar rocks are known on the north and south (Haun and Kent, 1965; Ritzma, 1956). If a normally shaped basin is

squeezed by tangential compression causing uplift, as explained, the flanks of the basin are shortened. When the reverse occurs, downdropping of the central segment causes further shortening. Finally, the result might be the formation of a thick, elevated pod with the flanks of the basin essentially eliminated—the configuration seen today in the Uinta Range. Such a phenomenon is much more easily explained by tangential compression than by vertical uplift for at least two reasons. As explained, numerous vertical reversals are likely to occur under conditions of tangential compression in combination with the isostatic imbalance caused by uplift and erosion. Moreover, each flexure would cause significant tangential shortening, eliminating some of the basin margin.

Effect of morphologic limits.—Another mechanical factor which may bear on the configuration of the cross sections of the uplifts is related to the concept of the upper and lower morphologic limits. Strength of crustal materials is such that they cannot be pushed beyond a certain height into the air or depth into the subcrust.

If limited tangential compression is applied to this crustal slab, so that it can flex freely within these limits, an undisturbed sine curve can result (Fig. 50A). If, however, additional tangential shortening is applied, so that the upper and lower morphologic limits begin to interfere with the ideal sine curve, two situations can result. The center of the uplift can act as a straight strut, forcing development of overturning by fold-thrusting on the edge of the uplift (Fig. 50B), or the strut can collapse and absorb the shortening as the slab tends to flex into sine curves with half the wave length of the previous curve (Fig. 50C) The latter situation reduces the necessity for overturning on the periphery of the system and would represent collapse of the top of the uplift rather than overthrusting of the edge of the uplift. The latter mechanism is related to the isostatic thrust reversal previously described.

Summary of cross-sectional evolution.—The sequential development of a range-bounding structure from inception to a full 40,000 ft of throw, as on the Wind River and Uinta thrusts, is shown in Figure 51. In cross section A the crust is undeformed. In cross section B the crust is flexed into a monoclinal fold such as those that are well displayed on the Colorado Plateau. Several thousand feet of structural relief can be at-

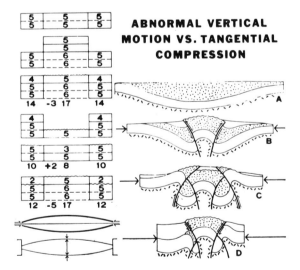

FIG. 49.—Abnormal vertical motion *vs.* tangential compression. Left side shows alternating positive and negative gravity anomalies that result from uplift and subsidence caused by tangential compression in conjunction with erosion and deposition. Simple whole numbers represent relative density of crustal segments and not specific gravity. Right side illustrates possible elimination of flank of sedimentary basin by alternating uplift and subsidence of central basin block subjected to tangential compression.

MORPHOLOGICAL LIMITS VS. UPLIFT SHAPE

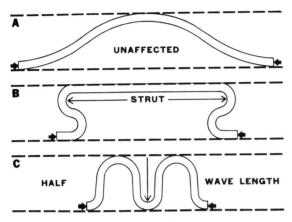

FIG. 50.—Upper and lower morphologic limits affect uplift configuration. Tangential shortening in excess of that which can be absorbed by perfect sine curve within morphologic limits must be absorbed by overturning on flanks (*B*) or by development of a half-wave-length configuration (*C*).

tained during this stage. At some point the fold starts to break as a rather steep thrust separating the lower and upper limbs of the monocline. As this thrust develops, the hanging wall begins to sag over the basin. The amount of sag varies directly with structural relief and inversely with the competence of the basin sedimentary section. Finally, the thrust may sag so far out over the basin and become so far out of the line of direct shear from the root zone that a new, steeper thrust forms. This shift of movement from one thrust to another may have occurred along the flanks of the Uinta Range.

In the later stages of development of the structure, collapse of the uplift occurs down the root zone of the fault. Examples are Brown's Park graben in the eastern Uinta Range (Hansen, 1965) and the Sweetwater uplift of central Wyoming (Bell, 1956; Carpenter and Cooper, 1951). A fault set that is the basin analog of the down-to-the-uplift faults seen in the uplifts should be present in the basin basement under the lip of the overhang of the thrust. It should also be down-to-the-uplift and probably would dip predominantly toward the uplift. This fault type, however, is mechanically independent of near-surface faults that are down-to-the-uplift and rim the lip of the overthrust block. These shallow faults die out downward and are a direct result of sagging of the heavy overthrust block into the incompetent basin sediments. A prime example is

the Continental fault which parallels the basinward limit of the Wind River thrust (Berg, 1961, p. 71).

It is possible that a significant part of this late normal faulting, much of which is Miocene-Pliocene or even Pleistocene, is at least partly a result of isostatic stresses caused by the exhuming of younger sediments during the late Tertiary and the recreation of topographic relief. The reasoning is as follows. Remnants of well-graded high-level erosion surfaces that project nearly to the tops of the ranges show that (1) the area was very stable for a long period after the compressive deformation and (2) the ranges were nearly buried by their debris. These interpretations suggest that at this time both primary compressive and secondary isostatic stresses had dissipated to below (probably just below) the threshold required for deformation. With exhuming of the basins caused by late Tertiary regional uplift, air was substituted for sediment and isostatic stresses again exceeded the threshold required for deformation, causing the late normal faulting. This concept may apply on a whole-crust basis to down-to-the-uplift basement-penetrating normal faults, and on a shallow basis to such shallow features as the Continental fault.

Figure 52 is a representative schematic view of the evolution of Wyoming foreland uplifts, taking into consideration all of the principles enumerated by the writer.

CONCLUSIONS

1. Laramide and pre-Laramide deformation of the entire western United States and much of the

UPTHRUST EVOLUTION

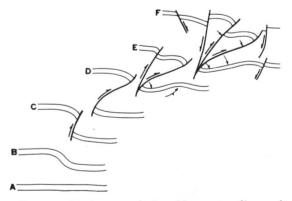

FIG. 51.—Upthrust evolution. Note monocline and torn-monocline stages, sagging of uplifts over basin, and development of late-stage gravity-adjustment faults.

SUMMARY EVOLUTION OF FORELAND UPLIFTS

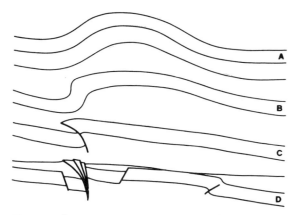

FIG. 52.—Summary evolution of foreland uplifts based on concepts of writer.

Mid-Continent is a mechanically integrated tectonic system which resulted from the 'Pacific block being relatively driven under the North American craton.

2. The first-order parent structure of the Cordilleran orogenic belt is the "Benioff" megashear along which the continent relatively overrides the Pacific block.

3. Stress generated along this megashear has been transmitted through the geosyncline into the adjacent foreland cratonic crust, and into the Mid-Continent. This tangential stress transmission is confined to the "tectonically solid" crust above the rheid interface or the asthenosphere. Such a boundary may be gradational and may not coincide with any mineralogic phase boundary or seismic-velocity boundary.

4. The basic differences between the Canadian, United States, and the Central American segments of the Cordillera are the manner and degree in which stress from the Pacific block is resisted by the crust east of the Cordillera. In Canada there is a direct "in-line" abutment of stresses between the Pacific block and the Canadian shield, resulting in a narrow, highly compressed tectonic prism which is devoid of significant oblique lateral shears. In Central America, "in-line" forces meet the lesser resistance of the thinner, more oceanic Caribbean crust, resulting in greater eastward yield. Because the position of the United States is intermediate between those of Canada and the Caribbean, a huge distributive west-northwest-oriented left-lateral couple is created. This couple is a major cause of the ab-

normally wide Cordillera and of the abnormally broken foreland crust. The Lewis and Clark, Wichita, and Texas lineaments are concentrated basement tear zones associated with this regional couple.

5. The regional left-lateral couple is subdued in the latitude of the Colorado Plateau block because that block has held together and has been translated bodily eastward. The Front Range uplift and the Denver basin downwarp are direct results of eastward movement of the Colorado Plateau block.

6. The regional left-lateral couple is accentuated in Wyoming by the eastward translation of the Colorado Plateau block relative to the unyielding Canadian foreland. Model studies affirm both in map view and in cross section that this "Wyoming couple" caused the uplifts and basins of the Wyoming foreland, including the enigmatic Uinta Range. There is no significant mechanical difference between the Ancestral Rockies deformation and the later Laramide deformation.

7. Most of the uplifts of the Wyoming province are crust-thick drag folds. Because of crustal rigidity, however, there is also a tendency to break into overlapping tilted slabs. Therefore, the best descriptive term for these uplifts is "drag fold-slabs."

8. Transfer of shortening from the geosyncline in British Columbia to the foreland in Wyoming has been accomplished by left-lateral yield on the Lewis and Clark lineament. Similarly, the Wichita lineament transfers Front Range-Denver basin shortening to the eastern edge of the Texas craton.

9. Configuration of the foreland deformation, model studies, and the apparent transmission of tangential stress great distances through the cratonal crust suggest that the crust is acting as a rigid semi-elastic slab floating on a heavier, very viscous, but tectonically fluid subcrust. In terms of these gross mechanical properties, foreland deformation in cross section can be seen to be a mixture of several discrete basic configurations: undeformed slabs, slabs flexed into a sine curve, and tilted overlapping slabs.

The configuration of the deformation is controlled strongly by isostasy. Geometry of collapse as a result of excessive structural relief can be divided into several discrete basic configurations: structurally overlapped slabs completely collapsed to isostatic equilibrium, partly sagged structurally

overlapping slabs, strut folds, and half-wave-
length folds. Collapse was both concomitant
with compression, causing isostatic thrust reversal, and subsequent to compressive deformation,
causing normal faulting.

10. Model experiments suggest that Berg's
fold-thrust configuration of major block-bounding
structures is correct and that the folding tendency increases upward through the crust. The attitude of the fault plane is a function of structural relief, competence of the section, depth of
erosion, relative amount of horizontal component,
and timing and geometry of collapse mechanisms.

SELECTED REFERENCES

Armstrong, F. C., and S. S. Oriel, 1965, Tectonic development of Idaho-Wyoming thrust belt: Am.
Assoc. Petroleum Geologists Bull., v. 49, no. 11, p.
1847–1866.
Bally, A. W., P. L. Grody, and G. A. Stewart, 1966,
Structure, seismic data, and orogenic evolution of
the southern Canadian Rocky Mountains: Bull.
Canadian Petroleum Geology, v. 14, no. 3, p.
337–381.
Bell, W. G., 1956, Tectonic setting of Happy Springs
and nearby structures in the Sweetwater uplift
area, central Wyoming: Am. Assoc. Petroleum Geologists, Rocky Mountain Sec., Geol. Record, p.
81–86.
Benioff, Hugo, 1954, Orogenesis and deep crustal
structure: additional evidence from seismology:
Geol. Soc. America Bull., v. 65, p. 385–400.
Berg, R. R., 1961, Laramide tectonics of the Wind
River Mountains, in Symposium on Late Cretaceous rocks, Wyoming and adjacent areas: Wyoming Geol. Assoc. 16th Ann. Field Conf. Guidebook,
p. 70–80.
Bucher, W. H., 1964, Deformation of the earth's
crust: New York, Hafner Pub. Co., 518 p.
Carey, S. W., 1953, The rheid concept in geotecton-
ics: Geol. Soc. Australia Jour., v. 1, p. 67–117.
Carpenter, L. C., and H. T. Cooper, 1951, Geology of
the Ferris-Seminoe Mountain area: Wyoming Geol.
Assoc. Guidebook 6th Ann. Field Conf., p. 77–79.
Cohee, G. V., 1961, Tectonic map of the United
States exclusive of Alaska and Hawaii: U.S. Geol.
Survey and Am. Assoc. Petroleum Geologists, scale
1:2,500,000.
Crowell, J. C., 1962, Displacement along the San An-
dreas fault, California: Geol. Soc. America Spec.
Paper 71, 61 p.
Gilluly, James, 1963, The tectonic evolution of the
western United States: London Geol. Soc. Quart.
Jour., v. 119, p. 133–174.
Hansen, W. R., 1965, Geology of the Flaming Gorge
area, Utah, Colorado, and Wyoming: U.S. Geol.
Survey Prof. Paper 490, 196 p.
Haun, J. D., and H. C. Kent, 1965, Geologic history
of Rocky Mountain region: Am. Assoc. Petroleum
Geologists Bull., v. 49, no. 11, p. 1781–1800.

Krey, Max, 1962, North flank Uncompahgre arch,
Mesa and Garfield Counties, Colorado, in Exploration for oil and gas in northwestern Colorado:
Rocky Mt. Assoc. Geologists Guidebook, p.
111–113.
Longwell, C. R., 1945, The mechanics of orogeny:
Am. Jour. Sci., v. 243-A, Daly volume, p. 417–447.
Mallory, W. W., 1958, Pennsylvanian coarse arkosic
redbeds and associated mountains in Colorado, in
Symposium on Pennsylvanian rocks of Colorado
and adjacent areas: Rocky Mt. Assoc. Geologists,
p. 17–20.
——— 1967, Mississippian and Pennsylvanian stratigraphy in middle and southern Rocky Mountains
(abs.): Am. Assoc. Petroleum Geologists Bull., v.
51, no. 9, p. 1903.
Misch, Peter, 1960, Regional structural reconnaissance
in central northeast Nevada and some adjacent
areas, observations and interpretations, in Geology
of east central Nevada: Intermountain Assoc. Petroleum Geologists Guidebook, p. 17–42.
Moody, J. D., 1962, Comments on "The origin of
folding in the earth's crust" by V. V. Beloussov,
Jour. Geophys. Res., July, 1961: Houston Geol.
Soc. Bull., v. 4, no. 4, n. p.
——— and M. J. Hill, 1956, Wrench-fault tectonics:
Geol. Soc. America Bull., v. 67, p. 1207–1248.
Moore, R. C., 1963, Tectonic summary of the backbone of the Americas, in Backbone of the Americas: Am. Assoc. Petroleum Geologists Mem. 2, p.
297–311.
Norwood, E. E., 1965, Geological history of central
and south-central Montana: Am. Assoc. Petroleum
Geologists Bull., v. 49, no. 11, p. 1824–1832.
Prucha, J. J., J. A. Graham, and R. P. Nickelsen,
1965, Basement-controlled deformation in Wyoming
province of Rocky Mountains foreland: Am. Assoc.
Petroleum Geologists Bull., v. 49, no. 7, p.
966–992.
Ritzma, H. R., 1956, Structural development of the
eastern Uinta Mountains and vicinity, Colorado,
Utah, and Wyoming: Am. Assoc. Petroleum Geolo-
gists, Rocky Mt. Sec., Geol. Record, p. 119–128.
Roberts, R. J., et al., 1958, Paleozoic rocks of north-
central Nevada: Am. Assoc. Petroleum Geologists
Bull., v. 42, no. 12, p. 2813–2857.
Rod, Emile, 1960, Strike-slip fault of continental importance in Bolivia: Am. Assoc. Petroleum Geolo-
gists Bull., v. 44, no. 1, p. 107–108.
Rubey, W. W., and M. K. Hubbert, 1959, Role of
fluid pressure in mechanics of overthrust faulting:
Geol. Soc. America Bull., v. 70, p. 167–205.
Sales, John K., 1966, Structural analysis of the Basin
Range province in terms of wrench faulting:
unpub. Ph. D. thesis, Nevada Univ., 178 p.
Smith, J. G., 1965, Fundamental transcurrent faulting
in northern Rocky Mountains: Am. Assoc. Petroleum Geologists Bull., v. 49, no. 9, p. 1398–1409.
Tanner, J. H., III, 1967, Wrench fault movements
along Washita Valley fault, Arbuckle Mountain
area, Oklahoma: Am. Assoc. Petroleum Geologists
Bull., v. 51, no. 1, p. 126–134.
Wilson, J. T., 1963, Continental drift: Sci. American,
v. 208, no. 4, p. 86–100.
Woollard, G. P., and H. R. Joesting, 1964, Bouguer
gravity map of the United States: U. S. Geol. Survey, scale 1:2,500,000.

Reprinted by permission of the Rocky Mountain Association of Geologists from *The Mountain Geologist*, v. 6, no. 2 (1986), p. 67-79.

WRENCH FAULTING AND ROCKY MOUNTAIN TECTONICS

DONALD S. STONE: Baumgartner Oil Company,

Denver, Colorado

ABSTRACT: Wrench faults are near-vertical fractures characterized by important lateral slip components. They form under regional compression when the greatest and least principal pressures (axes) lie in the horizontal plane. With the aid of conceptual models (block diagrams) and realizing that faults are finite and confined within an essentially incompressible crust, it can be shown that both vertical separation and an important component of vertical slip must accompany lateral movement along wrench faults. Thus, separation and slip along wrench faults may vary importantly, a geometric condition considered singularly characteristic. The reversal of fault dip along a fault line ("propeller faulting"), the development of belts of en echelon parafolds ("drag folds") or pinnate tension fractures, and abrupt stratigraphic changes across the fault zone are considered indicative of wrench faulting.

In the central Rocky Mountain foreland area, many of these wrench fault features can be recognized along major west-northwest (left wrench) and northeast (right wrench) trending fault zones. Considering the nearly north-south orientation of primary folds and of the low-angle thrust faults of the disturbed belt, it appears that a single north-northeast — south-southwest direction of principal horizontal stress (PHS) can account for essentially all of the Laramide structural features of the central Rocky Mountain foreland. An idealized tectonic diagram and a regional tectonic map of the central Rocky Mountain area are presented to show graphically the writer's interpretation of the various faults and folds of this area according to the wrench fault concept. These diagrams portray the genetic relationships between first and second order wrench fault zones and other structural features.

A check list of characteristic features also is presented to aid in identification of other wrench fault zones in the Rocky Mountain region. Such identifications are considered economically important inasmuch as the structures related to zones of wrench faulting commonly are the habitat of petroleum and mineral deposits.

INTRODUCTION

E. M. Anderson (1951, p. 2) has defined *wrench fault* as "a nearly vertical fracture, along which the separated segments have slid in a horizontal or nearly horizontal manner." The term *wrench* fault is applied in a genetic sense following Anderson's (1942, 1951) ternary classification of faults (Fig. 1): that is, according to the theoretical orientation of the three principal stress axes. By definition, these faults form under compression when the maximum principal pressure, P, is horizontal; the intermediate principal pressure, Q, is vertical; and the third, or least, principal pressure, R, is horizontal.

It can be seen from Figure 1 that the resulting wrench fractures theoretically are

vertical and form at an acute angle β (Moody and Hill, 1956) to the maximum pressure (or principal horizontal stress, PHS of Lensen, 1958b, 1960). This angle β may vary considerably from as low as 15 degrees to more than 45 degrees, depending upon the coefficient of internal friction ("Navier's principle"). However, in accordance with the Mohr-Coulomb theory of fracture, angle β has been shown to be about 30 degrees for average rocks at ordinary temperatures, pressures, and time-rate application of stress (for example, see Hubbert, 1937; Handin and Hager, 1957).

Once the initial fractures are formed, continued compression, as well as reorientation of stress along the resulting faults, can fold or

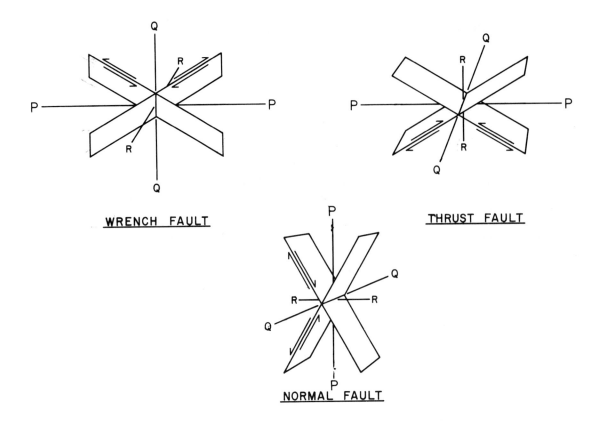

WRENCH FAULT

THRUST FAULT

NORMAL FAULT

Fig. 1.   Anderson's Genetic Classification of Faults

distort the original fault surfaces in both the vertical and the horizontal plane, and can widen the angle enclosing the primary stress axes. This angle β also may become obtuse at great depths, where the melting point of rocks is approached, allowing them to behave as ductile material. However, this is not the regime in which petroleum geologists normally work.

Wrench faults develop in response to compressive stress and nearly always are associated closely with other compressional features, such as folds and thrust faults. But because the particular orientation of the three stress axes prerequisite to the formation of wrench faults can exist only at some depth below the surface (the minimum stress axis must become vertical at the surface), wrench fractures probably are initiated at depth and propagated upward. In areas characterized by thick sedimentary sections, the normally straight traces of near-vertical wrench faults are not always

seen at the surface because shallower en echelon fold-thrust complexes may cover the deeper wrench plane, or the wrench fault may die out upward or "uncouple" before reaching the surface.

Before the two separated segments cut by a wrench fault can move horizontally past each other any appreciable distance, their bases must be detached from the underlying material. With very large wrench faults, this level of detachment or uncoupling, is relatively deep within the crust. On smaller wrenches, the level of detachment must be much shallower — for example, at the basal overthrust plane, or plane of décollement, in the case of "tear" faults (cf. Perry, 1935). Detachment also may occur across a larger vertical section by slippage along bedding planes or numerous small thrust surfaces.

Wrench faults may be divided into two classes: right, or dextral, and left, or sini-

68

410

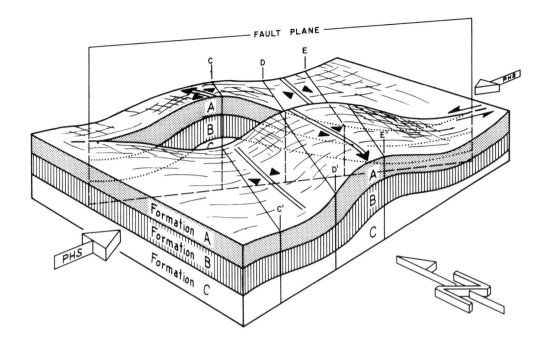

CONCEPTUAL MODEL I : SIMPLE LEFT WRENCH FAULT

Figure 2a

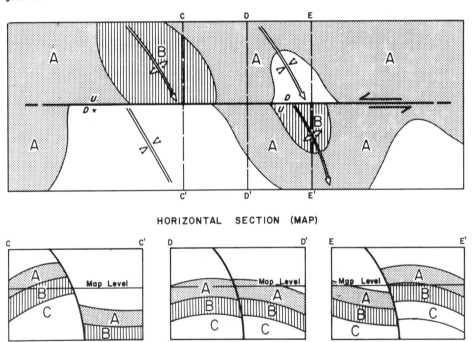

HORIZONTAL SECTION (MAP)

VERTICAL SECTIONS

Figure 2b

Fig. 2. Conceptual model of Simple Left Wrench Fault: top - block diagram; middle-plan view; bottom - cross sections.

stral. When a *right wrench* fault is viewed in plan, the block opposite the observer has been displaced relatively to the right. If the movement of the opposite block has been to the left, then the fault is termed a *left wrench.* These two classes of wrench faults can form syngenetically as a complementary set in response to the same principal horizontal stress.

Such complementary right and left wrenches are oriented so that their traces form an acute angle twice the angle β (i. e., about 60°), opening toward the direction of principal horizontal stress, and bisected by this stress (Fig. 35). In the Rocky Mountain area, as in most regions, only one of these "sets" of wrench faults is well-developed locally; but the dominance of one set over the other sometimes reverses from one area to another. This predominance of one set or class of wrench faults over the other in a given area suggests that a regional couple may be the cause of wrench faulting more commonly than direct compression (DeSitter, 1956, p. 120; Maxwell and Wise, 1958, p. 928).

In the following discussion, the terms *separation* and *slip* will be used frequently and need to be clearly defined. Separation is the "distance between the *surfaces* of the two parts of a disrupted bed, vein, or of any recognizable surface measured in any indicated direction" (Reid et al., 1913, p. 169). Slip is "the relative displacement of formerly adjacent *points* on opposite sides of the fault, measured in the fault surface" (Reid et al., 1913, p. 168). The meaning of the terms *net slip, lateral slip* (component), and *vertical slip* (component) are illustrated in Fig. 3.

CONCEPTUAL MODELS

Perhaps the best way to illustrate the characteristics of wrench faulting is to consider a typical wrench fault in a simple, three-dimentional block diagram. Figure 2a is a *conceptual model* of a typical, finite wrench fault. By definition, and for simplicity in construction, the fault surface is shown to be vertical, and the orientation of the fault trace is set at 30 degrees to the primary stress direction. In this model, the three formations cut by the fault do not exhibit any detectable changes in facies or thickness, and no homologous elements are present within the model from which lateral slip component can be established definitely. This is a usual condition in nature. Now, if it is assumed that, instead of being

vertical as shown, the fault dips steeply to the southwest, then the three vertical cross-sections, C-C', D-D', and E-E', will appear as illustrated in Figure 2b.

More often than not, if this hypothetical wrench fault were several miles in length and the observer were to study only that segment of the fault in the vicinity of cross-section C-C', he would label the fault "normal" and probably would infer dip-slip movement. If the observer worked only along the segment in the vicinity of cross-section E-E', he probably would term it a "reverse", or "thrust", fault (compare discussion by Ransome, 1906, p. 785). And along the line of section D-D', it is unlikely that any faulting at all would be recognized, because neither vertical nor lateral separation, or slip, can be measured in the model, although lateral slip is present by design.

Further complexities could be added to the model. For example, where more intense, contemporaneous deformation has taken place along a wrench fault, it is not uncommon to find the direction of dip of the fault surface reversed along the fault trace. Where this has happened, geologists frequently have failed to recognize the lateral continuity of the fault.

The second conceptual model (Fig. 3) illustrates the syngenetic development of a wrench fault and a high-angle thrust fault, which combine to form an arcuate uplift, or so-called "trapdoor" structure (Nelson and Church, 1943). This model illustrates one way in which a high-angle wrench fault can appear to change its strike from the theoretical orientation imposed by regional stress direction. It should be noted that the angle between the wrench and thrust portions of this fault zone is obtuse and opens away from the direction of lateral movement.

In gross aspect, an analogy can be made between this second conceptual model and the structure of many of the mountain uplifts of the central Rocky Mountains province. Wrench fault tectonics provides a very attractive genetic explanation for these seemingly anomalous "trapdoors", as well as the many other, perhaps less spectacular, compressional structures of diverse trend so commonly found in the central Rocky Mountain area.

FUNDAMENTAL PRINCIPLES

As illustrated in, or inferred from, the two conceptual models discussed above, the following fundamental principles can be stated.

70

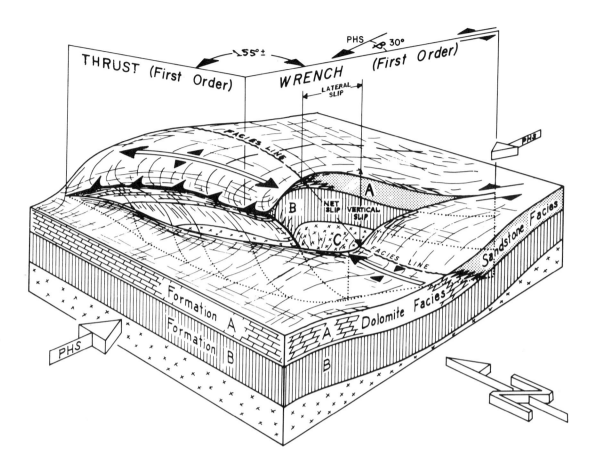

Fig. 3. Conceptual model of "trapdoor" structure.

1. *Faults are finite.* Thus, both separation and slip along faults theoretically *must* range from zero to a maximum value and back to zero.

2. Since the lithified rocks confined within the earth's outer crust essentially are incompressible and cannot change their density or volume greatly (according to the elastic theory), vertical separation and an important vertical component of slip *must* accompany lateral movement along wrench faults. Also, a "level of detachment", or uncoupling, at some depth must exist.

3. Along wrench faults, separation and slip can vary considerably both in amount and direction. (Therefore, the nature of the separation, or net slip, along a *segment* of *any* fault is not a valid

criterion for classification of the whole fault.)

4. Because of the requirement that the intermediate stress axis (Q) be essentially vertical before wrench faulting can occur, wrench faulting probably is initiated at some depth below the surface. But being of compressional origin, wrench faults are associated with, or give way vertially or laterally to, other compressional features such as folds and thrust faults, the development of which requires a nearly horizontal intermediate stress axis (Q). Many factors may determine which type of compressional structure is seen at the surface. Possibly the most important factor is the level of erosion.

71

413

CHARACTERISTIC FEATURES OF WRENCH FAULTS

It seems only natural to expect documentation of lateral slip in the identification of wrench faulting. By definition, there must be some component of lateral slip on all wrench faults. When an important lateral slip component can be identified, a wrench fault designation generally is acceptable to most geologists. But too often geologists have not insisted on the same documentation when labeling faults "normal" or "thrust", implying predominantly dip-slip movement. Moreover, it has been pointed out by many authors that factual data on slip along faults is extremely hard to find. Crowell (1959, p. 2654) has cautioned that "since many faults have significant components of strike-slip, no longer can we assume logically that a fault has only dip-slip until proved otherwise". Haites (1960, p. 37) has gone even further by stating that "field evidence suggesting the presence of a normal or reverse fault is not conclusive unless absence of strike-slip movements is proved." Thus, while lateral slip is an important and necessary feature of wrench faulting, it is *not* the most practical criterion in the identification of this type of faulting, for it seldom can be recognized definitely and measured.

In addition to lateral slip evidence, twelve characteristic features of wrench faults are listed below. Each of these features has been described and illustrated with field examples from the Rocky Mountains and other areas of the western United States in oral presentations during 1968 before the Rocky Mountain Association of Geologists and the Wyoming Geological Association. These examples will be dealt with in detail in a forthcoming paper. Positive identification of several of the following features usually will be sufficient to indicate wrench faulting:

1. Lateral separation with slip significance.
2. Near-vertical fault dip (and relatively straight fault trace).
3. Reversal of fault dip along the fault line (propeller faulting).
4. Associated "drag" structures (parafolds and first order thrusts).
5. Reversal of vertical separation along the fault line.
6. Variable "reverse drag" (in cross-section).
7. Abrupt stratigraphic change or mismatch across the fault zone (in cross-

section).
8. Displacement involving the basement; association with igneous activity.
9. "Trapdoor" structures.
10. Certain surface and near-surface features such as "shutter-ridges"; horizontal slickensides; pinnate tension, or shear fractures; and unusually complex or wide fault zones.
11. Certain geophysical evidence, particularly first-motion earthquake solutions, and geodetic measurements (recent wrench faults).
12. Orientation of the fault trace in conformity with predictable wrench fault directions.

SECOND-ORDER WRENCH FAULTS

Second-order, or secondary, wrench faults (including "splay" faults) are faults which "have arisen as a direct result of the redistribution of stress that accompanied the movement on the master fault" (Chinnery, 1966, p. 175). In 1956, Moody and Hill proposed a multi-order world-wide regmatic wrench fault system resulting from meridianal compression based upon an extension of ideas presented earlier by Anderson (1951) and McKinstry (1953). Although both Billings (1960) and Chinnery (1966, p. 1979) have stated that "there is no mechanical basis for the arguments of Moody and Hill (1956), "the relatively persistent angular relationships between master wrench, associated parafolds, and second-order wrench faults reported by Moody and Hill can be recognized in many parts of the world. However, these areas may have PHS directions of orientations other than meridianal (Moody and Hill, 1964, p. 115, have more recently admitted to at least *two!*). The angle between the master wrench and associated parafolds also may vary widely. Moreover, through the application of the Mohr-Coulomb theory of fracture, it seems to this writer that the development of this type of second-order wrench fault is related to the locally reoriented stress associated directly with the formation of these parafolds and thus only secondarily to the master wrench. If this is the case, then the angle that these second-order wrench faults make with the master fault must be a function of the variable angle γ; and only in those singular cases where this angle bears a consistent relationship to the master wrench fault, or PHS direction, will the second-order wrench

72

414

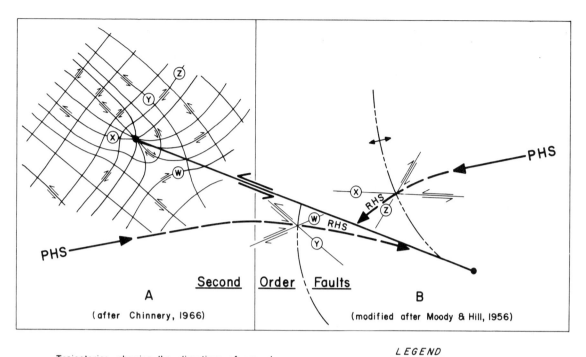

Second Order Faults

A
(after Chinnery, 1966)

B
(modified after Moody & Hill, 1956)

Trajectories showing the directions of second
order faulting (maximum shearing stress)
resulting from movement on master left wrench
fault under uniaxial compression.

Fig. 4. Trajectories showing the directions of second-order faulting (maximum shearing
stress) resulting from movement on master left wrench fault under uniaxial compression.

fault orientations to be predictable.

The Moody and Hill type of second-order
faults generally cut parafold axes at the typical
60° angle (i.e., 90°-β), as shown in Figure
4B. These faults clearly are of much smaller
magnitude (i.e., of second-order) than the
master fault. Often they are referred to as
"tear faults" or "transverse faults". Under
this concept, the formation of third-order
faults would seem to be unlikely.

Chinnery (1966a, 1966b) has taken an
entirely different approach to the problem of
second-order or secondary wrench faulting.
Through the use of dislocation theory developed
by Steketee (1958a, 1958b), and a IBM 7040
computer at the University of British Columbia,
he has calculated the pattern of final maximum
shearing stresses around a hypothetical
wrench fault which would theoretically result

from movement along it (Fig. 4A). In these
calculations he has made a number of polemic
assumptions, most important of which is that
"the effect of movement on a fault is to reduce
the initial shear stress everywhere except in
the vicinity of the ends of the fault, where it
causes complex additional stress" (Chinnery,
1966b, p. 175). Thus, his mathematical model
"has a displacement which is constant, which
is equivalent to assuming that the displacement
falls off quite rapidly near the end of the fault"
(p. 180). On the other hand, if these assumptions are invalid — that is, if displacement
decays slowly along a fault, a more likely
situation — then second-order faulting near the
end of a master wrench fault of the type
described by Chinnery "is found to be much
less likely" (Chinnery, 1966b, p. 180).

Even though some of the fundamental

73

415

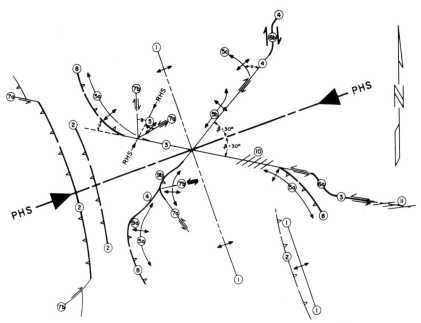

TECTONIC DIAGRAM FOR CENTRAL ROCKY MOUNTAIN AREA

TECTONIC ELEMENTS	EXAMPLES FROM CENTRAL ROCKY MOUNTAINS
1. Primary folds	1. Front Range; Cedar Creek anticline
2. Primary thrusts	2. Wyoming thrust belt; Rock Springs fault
3. First-order left wrench faults	3. Lake Basin; Tensleep; Seminole
4. First-order right wrench faults	4. Weldon-Brocton; Nash Fork-Hartville
5. Parafolds a) acute ($\angle\gamma \cong 90°$) b) "Compression ridge", or parallel ($\angle\gamma \sim 0°$) c) Perpendicular ($\angle\gamma \cong 90°\pm$)	5. a) Grass Creek dome, Miller Hill anticline b) Neiber anticline; Grenville dome c) Quealy dome
6. Second-order faults (after Chinnery) a) left wrench b) right wrench	6. a) Fish Creek zone b) Nash Fork-Quealy zone
7. Second-order faults (after Moody and Hill) a) left wrench b) right wrench	7. a) Cellers Ranch b) North Rawlins
8. First-order thrusts	8. Wind River; Moffat; Horse Creek
9. Trapdoor structures, formed by the joining of: a) Type I: First-order wrench fault with first-order thrust fault b) Type II: First-order wrench fault with second-order wrench fault having opposite sense of lateral slip	9. a) Wind River uplift; Sage Creek anticline b) Rawlins uplift
10. Pinnate tension fractures	10. Lake Basin and Sussex left wrench fault zones
11. Pinnate shear fractures	11. Duchesne left wrench fault zone

SYMBOLS		IN CENTRAL ROCKY MTS.:
PHS	Principle horizontal stress direction	(ENE-WSW)
RHS	Reoriented horizontal stress direction	(variable)
β	Angle between PHS or RHS and wrench fault traces	($30°\pm$)
γ	Angle between wrench fault and associated parafolds	(variable; 0-90°)

Fig. 5.   Tectonic Diagram for Central Rocky Mountain Area

74

assumptions made by both Moody and Hill (1956) and Chinnery (1966a and b) are probably not entirely legitimate, the orientations and character of second-order wrench faults predicted under either hypothesis (after incorporating the limitations noted) can be recognized in nature in the Rocky Mountain area and elsewhere. In fact, reference to Figure 4 suggests that either concept could explain the second-order faults shown (note the similar orientations of the faults marked w, x, y and z, under both systems). Moody and Hill were impressed by and have emphasized, the dynamic geometry of folded rocks in the vicinity of wrench faults, drawing upon their experience in petroleum exploration. Chinnery, on the other hand, has utilized a theoretical mathematical model, which necessarily is simplified. Apparently, he was concerned only with the consequences of movement along the fault surface and not with associated parafolding in the surrounding terrain (personal communication, 1967). Both concepts appear to have sufficient validity to be very useful in regional tectonic analysis and structural classification.

APPLICATION OF THE WRENCH FAULT CONCEPT TO TECTONICS OF CENTRAL ROCKY MOUNTAINS

It has been pointed out that wrench faults develop in response to a compressive stress which has acted at an angle (β) of about 30° to the resulting wrench fault planes. In a geological province occupied by rocks whose physical characteristics and elastic response do not vary too greatly, a relatively uniform direction of Principal Horizontal Stress (PHS) application produces relatively consistent regional strain patterns. Thus, in most structural provinces, the angular relationships that folds and the three types of faults bear to one another should fit certain set patterns. When analysed in terms of the principles of fault and fold genesis, these patterns usually will reveal a consistent PHS direction for the province. This type of regional study has been referred to as regional "tectonic analysis" (Dennis, 1967).

Until the development of the wrench fault concept in North America, regional tectonic analysis consisted largely of surmising a perpendicular PHS direction for linear fold and/or thrust belts such as those of the Appalachians or Canadian Rockies. In areas exhibiting multiple structural trends, such as California or the central Rocky Mountain foreland, the

regional tectonics often were considered too complex to be worked out satisfactorily in terms of a uniform PHS. There still are many geologists who have not fully recognized the value of wrench fault tectonics in seeking a possible unique solution to multi-trend regional tectonic problems.

It is the author's belief that the diverse structural patterns of the Rocky Mountain foreland can be explained neatly and perhaps uniquely in terms of fundamental wrench fault tectonics. Figure 5, "Tectonic Diagram for the Central Rocky Mountain Area," graphically illustrates the writer's interpretation of the various faults and folds of this area; it portrays the genetic relationships between first-order wrench faulting and other structural features.

Analysis of the various structural features depicted indicates a regional northeast-southwest PHS direction. Only a slight rotation of this tectonic diagram is required locally to bring it into satisfactory fit with all of the major structural features of the Central Rocky Mountain province. (A clear film of the diagram moved over the Tectonic Map, Plate I, should convince the reader that this agreement is real.) Most of the first-order wrench faults of the Central Rocky Mountains can be identified by matching them, within 10 to 15 degrees, to one of the two directions shown in the diagram.

This basic concept of wrench fault segmentation of the Central Rocky Mountains (viz. Cordilleran Foreland of Osterwald, 1961) is not new. R. T. Chamberlin, in two early classic papers on Central Rocky Mountain tectonics, fully developed the idea that in this area "accommodation was chiefly by horizontal shearing along a succession of nearly vertical east-west shear planes" (1945, p. 116). He suggested that on these shear planes "differential movement might perhaps have developed the succession of short, scattered mountain uplifts, which display in some places an echelon arrangement of minor folds on their flanks" (1940, p. 698). "The relief upward being, in general, easier than lateral relief, the dominant type of yielding was the rise of ranges and the sinking of basins more or less normal to the shortening" (p. 709). He recognized that northeast-southwest compression (the PHS direction) was expressed as "a strain strongly rotational in horizontal plan" (p. 708), resulting in "east-west shifting along steep faults or shear zones," (p. 701) ". . . toward the west on the north, toward the east on the south" (1940, p. 708), i.e., left-lateral; "short (para)folds, more or less en echelon, have developed near the

75

417

CHARACTERISTIC FEATURE	1. Lateral slip (predominant)	2. Lateral separation (w/slip significance)	3. Near-vertical fault dip and straight trace	4. Reversal of fault dip along fault line	5. Associated "drag" structures Parafolds = P First-Order Thrusts = T	6. Reversal of vertical separation ("hinge")	7. Variable "reverse drag"	8. Abrupt stratigraphic change or mismatch	9. (a) Displacement on basement (b) Assoc. igneous activity	10. Trapdoor structure(s) Type I; Type II	11. Surface features P; S; TF; SF; CFZ*	12. Geophysical evidence Seis. (S), Mag. (M), Gravity (G), Geodetic (Ge)	13. Predictable orientation	REFERENCES
FAULT NAME														
First-Order Left Wrench														
Example: Tensleep fault	X	X	X		P	X	X	X	(a)	--	SF, CFZ	G	X	1
First-Order Right Wrench													,	
Example: Nash Fork-Quealy-Hartville fault zone	X	X	X		P, T	X	X	X	(a)	(I)	SF, CFZ	S, G	X	2

*Symbols: P=Physiographic; S=Slickensides; TF=Tension fractures; SF=Shear fractures; CFZ=Complex fault zone

1. Chamberlin (1940), Palmquist (1967), Wilson (1938).

2. Droullard (1963), Houston (1967), Houston and Parker (1963), Houston, et al (1965), McCallum (1964).

TABLE 1

CHECKLIST OF CHARACTERISTIC FEATURES OF WRENCH FAULTS

rifts in consequence" (p. 698). A general analogy to structural conditions associated with California's San Andreas fault system also was proposed by Chamberlin (1940, p. 698). Moreover, many other authors, both before and after Chamberlin (e. g., Badgely, 1960, 1965; Bell, 1956; Sales, 1968; Thom, 1923; Warner, 1956), have recognized the probable importance of wrench faulting in the Central Rocky Mountains.

A Regional Tectonic Sketch Map, Plate I, on which most of the probable major wrench fault zones of the Central Rocky Mountain area are shown and identified, and on which many of the more important folds and subsidiary faults also appear, accompanies this paper. The reader is encouraged to investigate further those faults which which he is familiar, or in which he has a special interest, and decide for himself whether or not the available data support a wrench fault classification.

A checklist of characteristic features, such as that shown in Table 1, may be helpful in this analysis. Generally, a positive identification of three or four of these criteria should substantiate a wrench fault assignment; however, this will vary, as the process generally involves personal judgment. Whereas proof of predominant lateral slip is all that is needed for a wrench fault identification in one case; the combination of near-vertical fault dip, abrupt stratigraphic change, displacement on basement, and aeromagnetic expression, might not suffice in another case, because each criterion carries different weight under different circumstances. It is fully expected that

76

418

sedulous efforts to identify these wrench fault criteria will reward the investigator with a more accurate and useful understanding of fault geometry and will result in many new wrench fault assignments.

CONCLUSIONS

The writer must agree with Moody and Hill (1956, p. 1241) that "wrench faulting is much more prevalent in the earth's crust than ordinarily supposed." Historically, recognition of this type of faulting has been seriously impeded by two-dimensional thinking, and by the current emphasis being placed on descriptive fault classification. Wrench faulting fundamentally is a genetic concept, because lateral fault slip must be accompanied by vertical separation on finite faults in an earth-bound environment. This fact, coupled with the usual absence of correlatable linear elements across faults, which could prove the true nature of net slip, explain why most wrench faults, covered, buried, or not recently active, generally have been labeled "normal" or "thrust" by geologists, before the total geometry of faulting has been fully delineated. It is true that before a fault can be classified genetically, knowledge of the geometry along the entire length of the fault usually must be acquired. But this conclusion is considered to be an argument in favor of genetic fault classifications. It is the writer's belief that the recognition of wrench faulting, with or without specific slip information, requires analysis of the other characteristic features listed in this paper. Such an analysis should promote more complete and accurate structural interpretations, thereby resulting in more successful economic applications.

In the Central Rocky Mountain foreland area, many wrench fault characteristics can be recognized along major west-northwest (left wrench) and northeast (right wrench) trending fault zones. The north-northeast, south-southwest direction of principal horizontal stress (PHS) required by these wrench fault orientations also fits well with the nearly north-south orientation of primary folds and low angle thrust faults of the disturbed belt, and can account for essentially all of the Laramide structural features of the Central Rocky Mountain foreland. The idealized Tectonic Diagram (Fig. 5) and the Tectonic Sketch Map of the Central Rocky Mountains (Plate I) show graph-ically the writer's interpretation of the various faults and folds of this region. These exhibits portray the genetic relationships between first-order wrench fault zones and associated structures.

In addition to the above conclusions, the following generalizations have particular economic significance in exploration:

1. Within sedimentary basins, wrench faulting produces parafold anticlines, "trapdoors", or half-domal structural features that serve to localize oil and gas accumulations. These parafolds develop along either side of, or vertically above, wrench faults as a consequence of lateral movement. Many of the oil field structures in California, the Rocky Mountain states, Oklahoma, and other areas of the western United States have such an origin. Once the approximate location and orientation of a wrench fault zone is indicated, geophysical surveys may be more efficiently programmed if designed to investigate structural conditions along either side of the fault line. The presence of parafold structures hidden beneath unconformities or covered by shallow thrusts may be detected by such surveys.

2. Major wrench fault zones in igneous and metamorphic terrains provide avenues for intrusions and magnetic solutions that may localize economic mineral deposits (e.g., the Colorado Mineral Belt). Where sedimentary rocks are cut by important wrench faults, the fault zones also may provide channels for vertical migration and distribution of oil and gas into multiple reservoirs cut by faulting.

3. The fundamental nature and relative antiquity of many wrench fault zones commonly are reflected by important stratigraphic changes which occur along them. These localized stratigraphic changes, together with variable structural deformation, provide traps for petroleum. Knowledge of the amount of lateral offset of facies lines or pre-fault structures (associated with oil accumulations) may suggest new areas of potential accumulation in the opposite fault block.

77

REFERENCES CITED

Anderson, E. M., 1942, The dynamics of faulting and dyke formation, with application to Britain: London, Oliver and Boyd, 191 p.; 2d ed., 1951, 206 p.

Badgely, P. C., 1960, Tectonic relationships in central Colorado; in Guide to the Geology of Colorado: Geol. Soc. America, Colo. Sci. Soc., Rocky Mtn. Assoc. Geologists, p. 165-169.

————, 1965, Structural and Tectonic Principles: New York, Harper and Row, 521 p.

Bell, W. G., 1956, Tectonic setting of Happy Springs and nearby structures in the Sweetwater uplift area, Central Wyoming; in Geological Record: Rocky Mtn. Sec. Am. Assoc. Petroleum Geologists, Denver, Colo., Petroleum Information, p. 81-86.

Billings, M. P., 1960, Diastrophism and mountain building: Geol. Soc. America Bull., v. 71, p. 363-397.

Chamberlin, R. T., 1940, Diostrophic behavior around the Bighorn Basin: Jour. Geology, v. 48, p. 673-716.

————, 1945, Basement control in Rocky Mountain deformation: Am. Jour. Sci., v. 243-A (Daly Volume), p. 98-116.

Chinnery, M. A., 1965a, The vertical displacements associated with transcurrent faulting: Jour. Geophys. Res., v. 70, p. 4627-4631.

————, 1965b, Secondary faulting; Part I, Theoretical aspects; Part II, Geological aspects: Canadian Jour. Earth Sci., v. 3, p. 163-190.

Crowell, J. C., 1959, Problems of fault nomenclature: Am. Assoc. Petroleum Geologists Bull., v. 43, p. 2653-2674.

Dennis, J. G. (ed.), 1967, International Tectonic Dictionary: Am. Assoc. Petroleum Geologists Memoir 7, 196 p.

DeSitter, L. U., 1956, Structural Geology: New York, McGraw-Hill Book Co., 552 p.

Droullard, E. K., 1963, Tectonics of the southeast flank of the Hartville uplift, Wyoming, in Guidebook to the Geology of the Northern Denver Basin and Adjacent Uplifts: Rocky Mtn. Assoc. Geologists, p. 176-178.

Haites, T. B., 1960, Transcurrent faults in western Canada: Jour. Alberta Soc. Petroleum Geologists, v. 8, p. 33-78.

Handin, J., and R. V. Hager, Jr., 1957, Experimental deformation of rocks under confining pressure: tests at room temperature on dry samples: Am. Assoc. Petroleum Geologists Bull., v. 41, p. 1-50.

Hoppin, R. A., J. C. Palmquist, and L. O. Williams, 1965, Control by Precambrian basement structure on the location of the Tensleep-Beaver Creek fault, Bighorn Mountains, Wyoming: Jour. Geology, v. 60, p. 183-215.

Houston, R. S., 1967, Geologic map of the Medicine Bow Mountains, Albany and Carbon Counties, Wyoming: Geol. Surv. Wyo. Memoir 1.

————, and R. B. Parker, 1963, Structural analysis of a folded quartzite, Medicine Bow Mountains, Wyoming: Geol. Soc. America Bull., v. 74, p. 197-202.

————, A. Hills, and P. W. Gast, 1965, Regional aspects of structure and age of rocks of the Medicine Bow Mountains, Wyoming (Abstract): Geol. Soc. America Program, Ann. Meeting Rocky Mtn. Sec., p. 35-36.

Hubbert, M. K., 1937, Theory of scale models as applied to the study of geologic structures: Geol. Soc. America Bull., v. 48, p. 1459-1519.

Lensen, G. J., 1958, Rationalized fault interpretation: New Zealand Jour. Geol. and Geophys., v. 1, p. 307-317.

————, 1960, Principal horizontal stress directions as an aid to the study of crustal deformation: in A Symposium on Earthquake Mechanisms: J. H. Hodgson, ed., Dominion Obs. Ottawa, v. 24, p. 389-397.

Maxwell, J. C., and D. U. Wise, 1958, Wrench-fault tectonics: a discussion: Geol. Soc. America Bull., v. 69, p. 927-928.

McCallum, M. E., 1964, Cataclastic migmatites of the Medicine Bow Mountains, Wyoming: Contributions to Geology (Univ. Wyo.), v. 3, p. 78-88.

McKinstry, H. E., 1953, Shears of the second order: Am. Jour. Sci., v. 251, p. 401-414.

Moody, J. D., and M. J. Hill, 1956, Wrench fault tectonics: Geol. Soc. America Bull., v. 67, p. 1207-1246.

————, 1964, Moody and Hill system of

78

420

wrench fault tectonics: reply: Am.
Assoc. Petroleum Geologists Bull., v. 48,
p. 112-122.

Nelson, V. E., and V. Church, 1943, Critical
structures of the Gros Ventre and northern
Hoback Ranges, Wyoming: Jour. Geology,
v. 51, p. 143-166.

Osterwald, F. W., 1961, Critical review of
some tectonic problems in Cordilleran
Foreland: Am. Assoc. Petroleum
Geologists Bull., v. 45, p. 219-237.

Palmquist, J. C., 1967, Structural analysis of
the Horn area, Bighorn Mountains,
Wyoming: Geol. Soc. America Bull.,
v. 78, p. 283-298.

Perry, E. L., 1935, Flaws and tear faults:
Am. Jour. Sci., v. 29, p. 112-124.

Ransome, F. L., 1906, The directions of
movement and the nomenclature of faults:
Econ. Geol., v. 1, p. 777-787.

Reid, H. F., W. M. Davis, and A. C. Lawson,
1913, Report of the Committee on the
Nomenclature of Faults: Geol. Soc.
America Bull., v. 24, p. 163-186.

Sales, J. K., 1968, Crustal mechanics of
Cordilleran foreland deformation: a
regional and scale-model approach:
Amer. Assoc. Petroleum Geologists
Bull., v. 52, p. 2016-2044.

Steketee, J. A., 1958a, On Volterra's dislocations in a semi-infinite medium:
Can. Jour. Phys., v. 36, p. 192-205.

_____, 1958b, Some geophysical applications
of the elasticity theory of dislocations:
Can. Jour. Phys., v. 36, p. 1168-1197.

Thom, W. T., Jr., 1923, The relation of
deep-seated faults to the surface structural
feature of central Montana: Am. Assoc.
Petroleum Geologists Bull., v. 7, p. 1-13.

Warner, L. A., 1956, Tectonics of the
Colorado Front Range, in Geol. Record:
Rocky Mtn. Sec. Am. Assoc. Petroleum
Geologists, Petroleum Information,
Denver, Colorado, p. 129-144.

Wilson, C. W., 1938, The Tensleep fault,
Johnson and Washakie Counties, Wyoming:
Jour. Geology, v. 46, p. 868-881.

79

The American Association of Petroleum Geologists Bulletin
V. 58, No. 7 (July 1974), P. 1305-1322, 13 Figs.

Lineament-Block Tectonics: Williston-Blood Creek Basin[1]

GILBERT E. THOMAS[2]
Denver, Colorado 80210

Abstract A series of basement-weakness zones, represented at the surface as lineaments, trend northeasterly and northwesterly, and define a framework of possible basement blocks in the Williston–Blood Creek basin of North Dakota and Montana. These basement blocks and bounding weakness zones appear to have influenced the development of structural features in that the basement-weakness zones apparently have adjusted laterally in response to regional compressive forces, thereby coupling the blocks and the overlying sediments. This simple-shear block coupling during the Laramide orogeny has resulted in such drag folds as Nesson anticline, Cedar Creek anticline, and the Big Snowy Mountains uplift, among others. In influencing the stratigraphic and structural conditions the basement-block framework has affected directly the localization of oil and gas in the area.

INTRODUCTION

A detailed photogeologic-photogeomorphic mapping project carried out by Geophoto Services from 1964 to 1970 in the Williston–Blood Creek basin and adjacent areas (Fig. 1) provided the necessary detailed information to document this paper. The project was carried out on 1:20,000-scale air photographs and the mapped results, which include surface structure, stratigraphic contacts, geomorphic anomalies, and lineaments, were compiled on 1:96,000-scale final maps. At the completion of the project, the mapped results were reduced photographically to a scale of 1:250,000, allowing a regional view of the approximately 125,000-sq mi (323,749 sq km) area.

These regional photogeologic-photogeomorphic maps show surface lineaments that trend discontinuously across the entire area in northeasterly and northwesterly directions (Fig. 2). These features are for the most part not evident on the conventional geologic map (Fig. 1).

The two directions of surface lineaments define a block pattern (Fig. 2) presumably within the basement rocks.

It is the purpose of this paper to describe this block framework and to explain its very pronounced effect on the stratigraphic and structural conditions of the Williston–Blood Creek basin. In addition, the localizing influence of the framework on oil and gas deposits will be pointed out.

LINEAMENTS

Before going into detail concerning the block framework, it is useful to establish the structural validity of the surface lineaments that define the 'framework. Hobbs used the term "lineament" in 1912 (p. 227) to describe significant landscape lines "which reveal the hidden architecture of the rock basement." For many years after the work of Hobbs, lineaments were considered essentially as physiographic or geomorphic in nature. Lineaments being somewhat vague geomorphic features, there was a natural reluctance on the part of geologists to accept them as valid indicators of the "hidden architecture" of the basement rocks.

In the 1940s and 1950s, however, several articles appeared emphasizing the structural significance of geomorphic lineaments and extending the term to include narrow linear zones of structural features. Vening Meinesz (1947) and Sonder (1947) discussed lineaments as part of a worldwide system of basement fractures referred to by Sonder as the regmatic fracture system.

In 1956, Hills described the major lineaments of Australia, relating them to the tectonic style of that continent. Moody and Hill (1956), perhaps more than any other authors, popularized lineaments as possible wrench faults. Maxwell and Wise (1958) raised the question of whether pure-shear or simple-shear mechanics are the means by which wrench faults or lineaments are activated, whereas

[1] Manuscript received, June 30, 1972; revised manuscript received, April 14, 1973; accepted, November 23, 1973. Read before a regional meeting of the Rocky Mountain Section of the Association at Billings, Montana, May 5, 1971. Published by permission of Geophoto Services of Texas Instruments Incorporated.

[2] Thomas and Associates, P.O. Box 10384.

The writer thanks Geophoto Services for the use of proprietary maps in preparing this study and Don B. Gould, who reviewed the manuscript. John D. Harper of Shell Oil Company provided valuable criticism of the manuscript and suggestions for rewriting. The editors of the *Bulletin* contributed greatly to the clarification of the material.

1305

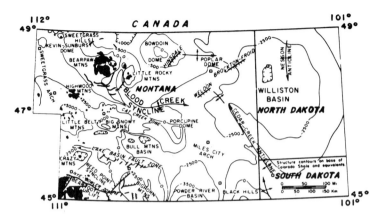

Fig. 1—Index map, Williston–Blood Creek basin (modified from Tectonic Map of U.S., 1961). Tertiary igneous rocks indicated by black areas.

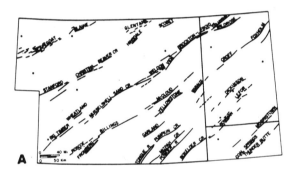

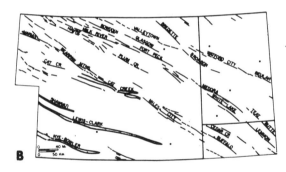

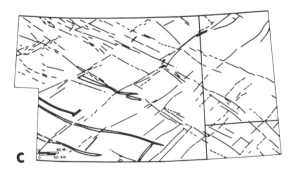

Fig. 2—Lineament pattern of Williston–Blood Creek basin. Thick lines are surface faults; stippled zones are prominent structural lineaments. **A,** Northeast-trending lineaments; **B,** Northwest-trending lineaments; **C,** Combined lineaments.

Carey (1958) utilized lineaments as megashears in his worldwide evaluation of continental drift in response to earth expansion.

More recently, Sales (1968) has shown the structural and petroleum significance of lineaments, primarily in Wyoming, and Stone (1969) has used lineaments in delineating wrench faults in the Rocky Mountains. O'Driscoll (1971) has gone a step further and developed a comprehensive system of deformational mechanics incorporating intersecting lineaments in western Australia. The term "lineament," therefore, has evolved to include both geomorphic and structural connotations.

In the Williston–Blood Creek basin area, surface lineaments are expressed both geomorphically and structurally. Geomorphically they are present as colinear stream courses, topographic alignments, and/or soil tonal lines. This expression is believed to be caused by a zonal arrangement of smaller fracture traces that cause differential erosion phenomena, thereby producing a lineament (see Lattman and Matzke, 1961). The very well-expressed Brockton-Froid lineament (Fig. 3), for example, is a surface geomorphic lineament in northeastern Montana. Subsurface faults that caused this lineament are believed to have controlled Mississippian Ratcliffe production localities (Hansen, 1966).

Lineaments in the studied area also are expressed structurally as zonal fault trends (Lake Basin fault zone defining the Lewis-Clark lineament), as monoclines (Sand Creek monocline), or as en echelon fold trends (Cat Creek lineament).

However lineaments are expressed at the surface, it should be stressed that the term "lineament" is a term used to define a *surface* linear feature. The exact subsurface cause of

Fig. 3—Brockton–Froid lineament (between large dark arrows) in northeast Montana.

lineaments is not known. Major lineaments, however, such as the Lewis-Clark of Montana have been interpreted as being over basement-weakness zones (Badgley, 1965). These weakness zones may be Precambrian fault or fracture zones or even lithologic boundaries between Precambrian igneous and metamorphic rocks.

Whatever the exact cause of surface lineaments, the regularity in pattern of the lineaments in the study area is quite pronounced (Fig. 2). Further, the many lineaments suggest an essential weakness of the basement, an observation which has many ramifications in the structural mechanics of the area as well as in the stratigraphic conditions.

STRATIGRAPHIC SIGNIFICANCE

For convenience in the discussion to follow, the many individual lineament names (Fig. 2) will be replaced by a general name for each lineament zone. For example, the northeast-trending Scapegoat-Blaine lineament (Fig. 2) will be referred to as the Scapegoat lineament zone. These names have been used to designate

the northeast and northwest block sets (Figs. 4, 5). The northern-bounding lineament zone of each block has been chosen as the name of the block. Thus in Figures 4 and 5, the lineament zone defining the northern boundary of the Scapegoat block is the Scapegoat lineament zone; the Cat Creek lineament zone is the northern boundary of the Cat Creek block, *etc.*

To show the stratigraphic significance of the block framework, several isopach maps and one lithofacies distribution map have been selected from the literature. These maps were selected for stratigraphic variety and also because they demonstrate a nonbias toward the block framework, having been published prior to its introduction.

Red River Formation

A formation of important petroleum interest in the area is the widely dolomitized Ordovician Red River limestone (Fig. 4A). Its isopachs show a change in trend at the Weldon lineament zone from a general northerly direction to a northeast-southwest direction parallel with the Weldon block. The northwest-trending thick

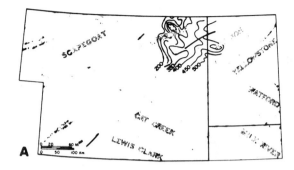

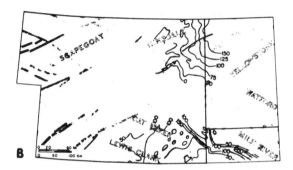

FIG. 4—Comparison of isopach maps with block frame-work.
A. Ordovician Red River Formation (Ballard, 1969).
B. Devonian Winnipegosis Formation (Kinard and Cronoble, 1969) in northeast Montana; Cretaceous Muddy Formation (Bolyard, 1969) in southeast Montana.

and thin lobes of the unit appear to be associated with the northwest-trending Watford lineament zone (the northern-bounding lineament of the Watford block). Ballard (1969, p. 16, 17) speculated that the "thicks" and "thins" are related genetically to tectonic movement on the northwest-southeast and northeast-southwest basement alignments common to the area. Lateral adjustment on the Watford lineament zone could have produced a series of drag folds, thereby affecting Ordovician sea-floor topography. Such mechanics will be explained in the structural discussion to follow.

Winnipegosis and Muddy Formations

Isopachs of the Devonian Winnipegosis Formation (a limestone, locally dolomitized) in northeastern Montana and the Cretaceous Muddy Formation (a littoral sandstone) in southeastern Montana (Fig. 4B) demonstrate more clearly that a relation may exist between stratigraphic trends and lineament zones. The Winnipegosis isopachs trend essentially north-south as they enter Montana from Canada but

change to a northwest-southeast trend on intersecting the Weldon lineament zone (the northern bounding feature of the Weldon block). Note, also, the offset of the isopach lines along the Weldon lineament zone and the coincidence of the zero isopach with the Milk River lineament zone, as well as the general parallelism of the isopach lines with the Watford block.

An even more pronounced influence of the lineament zone distribution on stratigraphy is indicated by the Cretaceous Muddy Formation isopachs (Fig. 4B). Isopach contours enter Montana from Wyoming more or less parallel

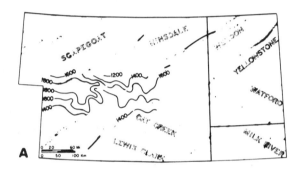

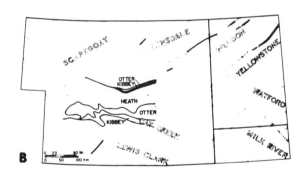

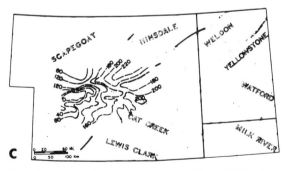

FIG. 5—Comparison of isopach (**A, C**) and lithofacies (**B**) maps with block framework; A, Lower Mississippian Madison Group (Michelson, 1956); B, Upper Mississippian Big Snowy Group (Sonnenberg, 1956); C, Jurassic Rierdon Formation (Rayl, 1956).

with the northeast-trending Tongue River and Powder River lineament zones (within the Yellowstone block) but change trend abruptly to a northwest-southeast direction at the Cat Creek lineament zone. Although this northwest-southeast trend of the unit is considered a channel-delta sand buildup (Bolyard, 1969), it would appear that the Cat Creek lineament zone may have controlled the position of the major channel and subsequent delta during deposition.

Madison Group, Big Snowy Group, and Rierdon Formation

The stratigraphic significance of lineament-zone location and the tectonic-paleotopographic evolution of a block through time are illustrated in Figure 5. The thicker part of the Lower Mississippian Madison Group (Fig. 5A; Mission Canyon and Lodgepole limestones) is confined essentially to a block formed by the intersection of the Cat Creek and Hinsdale blocks. This condition suggests that the Cat Creek–Hinsdale block may have been structurally and topographically negative at this time. The Lower Mississippian "trough" continues east-southeastward, paralleling the Cat Creek lineament zone, and changes trend to a northeasterly direction at the Weldon block.

The Cat Creek–Hinsdale block and the Weldon block still were structurally and topographically negative in Late Mississippian time, as indicated by the distribution of the Big Snowy Group (the Heath and Otter Shales and the Kibbey Sandstone; Fig. 5B). Again, a sharp directional change takes place at the intersection of the Weldon and Cat Creek blocks.

By Late Jurassic time, however, during the deposition of shales in the Rierdon Formation, the Cat Creek–Hinsdale block became a positive feature, as indicated by the isopachs in Figure 5C. The large negative basin then was confined to the Milk River–Hinsdale block. One might say that the basin environment had migrated through time from one block locality to the adjacent block. The cause of the migration can be interpreted as Jurassic reactivation of the basement-weakness zones which, in the Cat Creek–Hinsdale block, began to form an uplift (indicated by the zero isopach) that may have acted as an island affecting Rierdon deposition or may later have affected Rierdon erosion. Figure 5C illustrates quite clearly that isopach trends of the Rierdon Formation conform strongly to the trends and location of the Cat Creek and Weldon lineament zones.

Stratigraphic Conclusions

From the preceding figures comparing the lineaments of the area to various isopach maps, it is evident that the block framework has had some type of effect on depositional conditions throughout the rock section. Depositional sub-basins appear to be controlled in extent and configuration by the many blocks defined in the framework. Presumably this stratigraphic control is the result of the basement blocks and weakness zones controlling paleotopographic conditions. This relation could occur if the paleotopographic conditions were the result of paleostructures generated within the blocks and along the lineament zones by tectonic reactivation of the block framework. Thus, the block system of Precambrian age could to some degree control depositional conditions throughout the stratigraphic section.

STRUCTURAL SIGNIFICANCE

A comparison of the major structural features of the Williston–Blood Creek basin with the block framework as defined by the lineaments of the area (Fig. 6) suggests that the framework also has structural significance. There is a pronounced confinement of major structure features within the subblocks of the framework. For example, both the Bearpaw Mountains uplift and Cedar Creek anticline are within rectangular subblocks, as is the Blood Creek basin. Nesson anticline also appears to start and stop against lineaments defining the northeast-trending Weldon block. The extreme northern part of the Black Hills uplift "disappears" against the northwest-trending Cat Creek lineament zone, which also appears to influence the trend of the Miles City "arch."

Other major features also terminate against major lineaments. Kevin-Sunburst dome, Poplar dome, Wibaux nose, Rough Rider ridge, Porcupine dome, Big Coulee dome, the Bighorn Mountains, and the Reed Point syncline all start or stop against the lineaments of the block framework.

Furthermore, the termination of these structures against lineaments seems to involve a remarkably consistent angle of termination. The Kevin-Sunburst dome, for example, terminates against the northwest-trending Cat Creek lineament zone at an angle very similar to angles produced by the Bearpaw Mountains uplift, the Cleveland and Little Rocky Mountains uplifts, Poplar dome, Cedar Creek anticline, Big Coulee dome, Reed Point syncline, and the Bighorn Mountains uplift against other lineaments.

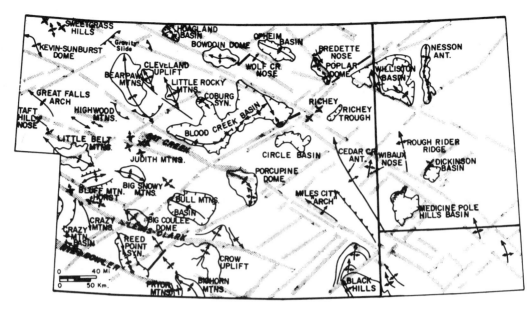

Fig. 6—Comparison of block framework with structural features of Williston–Blood Creek basin area.

There also appears to be a consistency of termination angles for the more northerly trending structures such as Nesson anticline, Wibaux nose, and the northern part of the Black Hills.

From these observed comparisons, the question has to be asked whether the observed relations are merely coincidental or were, in fact, produced in some fashion by the lineament-block framework during the last orogeny affecting the area.

To answer this question, cardboard models were constructed (Fig. 7) to simulate, in a general sense, the block framework of the studied area. The lineaments reflecting possible basement-weakness zones were cut into the cardboard (dashed lines) and solid lines were added to indicate any horizontal or vertical offset along the weakness zones during compression experiments. The models were constructed with left and right edges perpendicular to three general force directions: (1) a bisectrix direction to the intersection angle between the weakness zones (Fig. 7B); (2) north of the bisectrix direction (Fig. 7C); and (3) south of the bisectrix direction (Fig. 7D). These three directions represent the three unique compression directions possible, given the directional trends of the lineament-block framework. Compressive forces affecting the framework can be only from the bisectrix direction or from various degrees north or south of the bisectrix. Compressive-force experiments were carried out rather than vertical-force experiments because of the structural evidence in the area (Nye-

Bowler lineament, for example) suggestive of shearing.

Although a cardboard model bears no resemblance whatsoever to actual basement lithologic conditions, it was hoped that the model's spatial-adjustment behavior under compression might reflect a certain degree of basement-adjustment as a result of continental compressive forces.

The only structural features that could be used to judge the validity of the model experiments are the three prominent structural lineaments in the southwest part of the area: the Nye-Bowler, Lewis-Clark, and Cat Creek lineaments, all of which display evidence for left-lateral adjustment during the Laramide orogeny. In addition, actual lineament conditions in the Williston–Blood Creek basin are characterized by sparse surface evidence for northeast-trending lineaments. In other words, if the spatial adjustment behavior of the cardboard models represents some degree of actual basement adjustment, then there should be some direction of experimental force which would produce dominant left-lateral adjustment on northwest lineaments and subordinate adjustment on northeast lineaments.

Of the three model studies only the "direction south of the bisectrix" (Fig. 7D; correlative to a southwest regional compressive force for the actual region) appeared to produce tectonic effects similar to the Laramide effects. The "bisectrix direction" (Moody and Hill, 1956; pure-shear premise) repeatedly produced

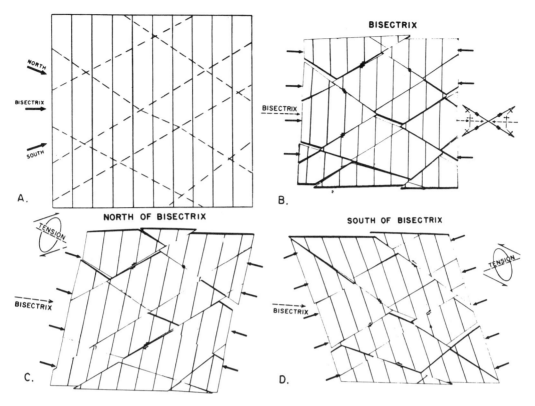

FIG. 7—Cardboard model studies: **A,** initial model; dashed lines are precut weakness zones; solid lines are displacement indicators; arrows represent three compression experiments; **B,** bisectrix direction compression; **C,** north of bisectrix compression; **D,** south of bisectrix compression.

equal lateral adjustment on the two weakness zones. Repeated experiments with the "direction north of the bisectrix" (Fig. 7C) produced lateral adjustment effects exactly opposite to those found in the area; *i.e.,* dominant right-lateral adjustment on the northeast-trending weakness zones and subordinate left-lateral adjustment on the northwest weakness zones.

The conclusion drawn from the model studies is that weakness zones in a medium subjected to compressive force adjust laterally in a simple-shear fashion; the relative sense of the simple-shear adjustment is determined by the direction of the compressive force in relation to the direction of the weakness zones. The term "simple shear" is used here in the sense of differential horizontal shear stress along a basement-weakness zone, horizontal shear stress along two weakness zones giving rise to coupling of the block between the two zones.

Further, at least with the cardboard models, the individual subblocks appear not to rotate as the lateral-adjustment arrows might tend to indicate, but rather to become part of larger northeast-trending (Fig. 7C) or northwest-trending-blocks (Fig. 7D), adjusting laterally

to the new spatial conditions required by the compressive forces.

As Sales (1968) has shown in his study of the Wyoming couple during the Laramide orogeny, lateral adjustment on basement-weakness zones produces· coupling in the blocks defined by the weakness zones. It would follow that, if the lineament-basement weakness zones of the Williston–Blood Creek basin adjusted laterally when subjected to orogenic forces, coupling might have resulted in the blocks and subblocks defined by the lineaments. A comparison of the results of simple-shear block coupling with actual structural features of the basin suggests that block coupling during the Laramide orogeny did take place in the basin and was, indeed, the major type of deformation in the area.

SIMPLE-SHEAR BLOCK-COUPLE MECHANICS

Although simple-shear block coupling long has been recognized as a valid deformational process, it often has been relegated to a secondary role to pure-shear compressional deformation during major orogenies. Sales (1968) has demonstrated, however, that block coupling

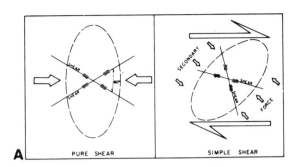

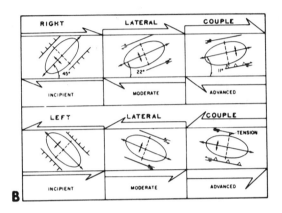

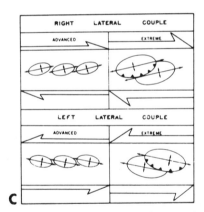

FIG. 8—Comparison between pure-shear and simple-shear mechanics (**A**). Various phases of simple shear mechanics (**B, C**).

A. Pure shear characterized by inline compression; simple shear by differential horizontal movement.

B. Lineament simple shear produces block coupling which results in drag-fold uplifts with associated flank faults and crossfold tension. Phases of coupling classified as to degree of drag-fold rotation.

C. En echelon drag folds, characteristic of advanced phase, overthrust one another in extreme phase.

could be the deformational mechanics responsible for most of the Rocky Mountain structure in Wyoming and parts of Montana and Colorado. The two types of deformation are illustrated in Figure 8A.

As related to this study, compressional pure-shear mechanics has a prerequisite assumption that is not always recognized; *i.e.*, that the basement is a relatively strong, unbroken medium through which orogenic forces can be propagated directly. If the basement, however, is a segmented medium with many zones of weakness as this study indicates, compressional pure-shear mechanics would not come about in a primary sense because the orogenic forces constantly would be deflected along the weakness zones. This deflection of orogenic forces would set up a system of structure mechanics herein labeled lineament simple-shear, block coupling, or lineament-block tectonics.

Atwater (1972) stated much the same premise concerning conventional (pure-shear) mechanics and a faulted area:

The conventional methods for relating stress and strain apply to a homogeneous medium, so that they certainly are not applicable to regions containing weak faults bounding strong plates.

O'Driscoll (1971) has recognized that basement-weakness zones as shear zones are an integral part of deformational mechanics as illustrated by his statement (p. 351):

The essentially meridional fold trends in the "greenstone" belts of southwestern Australia are represented as resulting from right-lateral movements along a fundamental system of basement shears trending northeasterly, with subsequent modification by left-lateral movements along the complementary northwesterly cross-shear system.

If simple-shear block-couple mechanics are the structural mechanics responsible for the Laramide deformation in the study area, it should be possible to examine the structural features and relate them to the block framework via coupling. The guide for relating the features should be the theoretical fold and fault results for simple-shear block-coupling mechanics (Figs. 8B, C). These structural results are deceivingly similar to those produced by pure-shear mechanics except for the following factors.

1. The couple-generated drag fold, flank fault, and/or crossfold tension fault or fracture are confined *within* the coupled block in which they were formed.

2. As simple shear on the bounding weakness zones of the block increases in intensity, the drag fold and associated faults tend to rotate in the direction of the shear couple. (The various stages of rotation herein are labeled incipient, moderate, advanced, and extreme for discussion purposes.) This rotation phenomenon

is especially true in the stratigraphic section if it becomes detached from the basement (Sales, 1968). With the increasing rotation, the flank faults and normal faults as subsequent weakness zones can become shear zones and eventually high-angle reverse faults in the later phases of orogeny. New crossfold tensional directions can be produced in these later phases.

3. If basement rocks are involved in the coupled-generated drag fold, rotation within the block appears seldom to proceed beyond a 40–35°-gamma angle (intersection angle of the drag fold with the bounding weakness zone) during the incipient stage. The "toe" of such basement drag folds, however, in proximity to the simple-shear zone commonly displays considerable "drag" so that the local gamma angle is suggestive of advanced coupling (Wind River Mountains, Wyoming; Thomas, 1971). In some blocks, such as the Cat Creek block, the simple shear is apparently of such great intensity that even drag folds incorporating basement rocks appear to have rotated to an advanced stage of coupling (the Little Belt Mountains, for example, Fig. 11B). Such advanced to extreme rotation necessitates a theoretical secondary compressive reduction in the width of the block at right angles to the drag fold.

4. Block coupling produces a diagnostic en echelon arrangement of drag folds either within the block proper or along the bounding simple-shear zones. For right-lateral coupling a clockwise en echelon pattern is produced, whereas left-lateral coupling produces a counterclockwise pattern (Fig. 8C). Rotation of the drag folds within the clockwise or counterclockwise patterns can produce an extreme stage of coupling whereby one drag fold overrides the next. Such conditions are common in the left-lateral Nye-Bowler simple-shear zone.

5. Crossfold tension faults or fractures occur at 45° to the bounding weakness zones in the incipient stage of block coupling (Pease, 1969). This relation is of use in defining ancient tension zones in the basement, as well as in recognizing tensional reactivation of ancient shear zones.

Northeast Blocks and Associated Structures

From the theoretical characteristics for block coupling, the structural features of the area can be interpreted as the result of simple shear rather than conventional pure-shear mechanics. The northeast-trending lineament zones, for example (Fig. 9A), should have adjusted right laterally to the regional Laramide compressive

stresses from the southwest, as hypothesized from the cardboard model studies (Fig. 7D). This right-lateral adjustment would have been subordinate to the left-lateral adjustment on the northwest lineament zones. Consequently, resultant structural features from the right-lateral adjustment should not be so numerous or extensive as those from the left-lateral adjustment. This seems to be valid in the area (Fig. 9A).

Right-lateral adjustment on the northeast lineament zones should have produced right-lateral simple shear along the zones and right-lateral coupling within the blocks. The sense of block coupling is indicated by the *outside* lateral adjustment arrows; the Hinsdale block, for example, would have been subjected to the coupling sense of the "squared" arrows (Fig. 9A).

Right-lateral coupling of the northeast blocks during the Laramide orogeny is suggested by two lines of evidence. First, the confinement of large uplifts such as Nesson anticline and the Crow uplift to specific northeast blocks is characteristic of coupling; and, secondly, the north-northeasterly trend of the block drag folds is nearly parallel with the incipient stage of block coupling as shown in Figure 9A. Right-lateral simple shear along the northeast zones appears also to have formed advanced drag folding in the stratigraphic section adjacent to the zones. The Richey and Dickinson anticlines, both northeast-trending features, are interpreted as examples of this local drag folding.

Furthermore, the existence of at least the Weldon and Powder River basement blocks is suggested by the confinement of the subbasins in the area to these blocks. It probably is no coincidence that the Reed Point syncline, the Bull Mountains basin, the Circle basin, Richey trough, and the central low of the large Williston basin are all located in the Weldon block (Fig. 9A). The Paleozoic negative topographic effect of this northeast block (suggested by the isopach maps) may have had a strong enough effect during the Laramide orogeny to localize basin lows, even though these basins were being formed by regional left-lateral coupling on northwest-trending blocks (Fig. 9B).

Northwest Blocks and Associated Structures

According to the results of the cardboard model experiments incorporating a southwest compressive force (south of the bisectrix, Fig. 7D), the northwest-trending weakness zones of the area (Fig. 9B) should have adjusted left laterally to the Laramide southwest stresses. This adjustment in fact should have been the

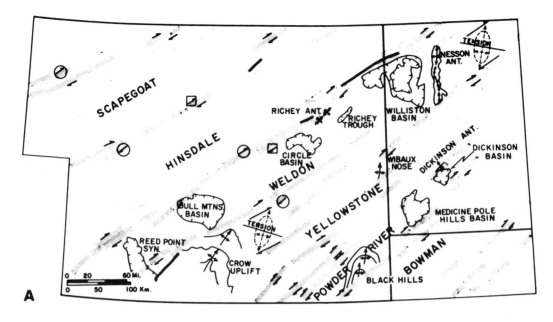

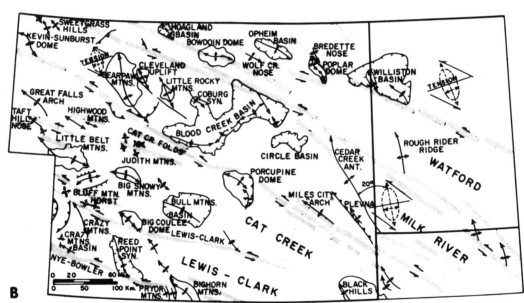

Fig. 9—Lineament simple-shear block-couple mechanics, Williston–Blood Creek basin area.
A. Structures related to northeast lineaments. Sense of coupling on blocks is determined by *outside* arrows. "Squared" arrows indicate right-lateral couple on Hinsdale block; "circled" arrows on Scapegoat and Weldon blocks, *etc.*
B. Structures related to northwest lineaments.

dominant structural mechanics in the area.

The strong structural expression of three northwest weakness zones (the Cat Creek, Lewis-Clark, and Nye-Bowler) attests to these dominant mechanics, as do the multitude of northwesterly-trending structural features (Fig. 9B). These features show various degrees of drag folding confined to specific blocks or lineament zones from incipient (Rough Rider ridge) to moderate (Cedar Creek anticline), to advanced (Big Snowy–Little Belt Mountains), to the extreme stage along the Nye-Bowler lineament zone where anticlines are thrust southwestward over the next en echelon drag-fold anticline. In this sequence of drag fold stages the more severe stages occur progressively toward the west-southwest, where Laramide forces were most intense.

Other local simple shears also occurred in proximity to the northwest weakness zones. This is indicated by the shear-zone drag-fold complexes forming the Miles City "arch," Wolf Creek "nose," and the Cat Creek drag folds.

Both the Pryor Mountains and Porcupine dome appear to be primary folds (as defined by Moody and Hill, 1956) formed within block wedges, but made more complex by "dragging" along the lineaments of the wedges. The Pryor Mountains terminate and are "dragged" along the Nye-Bowler shear zone, whereas Porcupine dome terminates and is "dragged" along the Cat Creek shear zone. The two folds are unique in the area because of their steeper eastern flanks and their eastward convexity of trend. Most other uplifts have steeper south-southwestern flanks and a south-southwestern convexity of trend. This observation indicates also that the Moody and Hill concept of pure-shear wrench faulting was not the dominant structure mechanics active in the area, inasmuch as many more folds with eastward convexity would be expected.

Alternation of drag folds within a block can be seen in the Milk River northwest block (Fig. 9B). The Cedar Creek anticline, the Bearpaw–Little Rocky Mountains complex, and the Kevin-Sunburst dome are three major anticlinal drag folds generated by Laramide left-lateral coupling on the Milk River block (igneous rock activity in the Bearpaw–Little Rocky Mountains is interpreted here to be post-Laramide. See Bearpaw Mountains discussion to follow). Cedar Creek anticline was probably in existence in the Paleozoic as a Milk River block drag fold, as the Devonian and upper part of the Silurian Interlake Formation are known to be absent over the buried Paleozoic crest of the fold (Davis and Hunt, 1956). This fact, however, also can be interpreted as the result of Paleozoic vertical adjustment rather than compressive block coupling.

The apparent 20° rotation of the present Cedar Creek anticline from the incipient drag-fold position (Fig. 9B) may have occurred during the Laramide orogeny or may be the cumulated result of earlier Paleozoic deformational events plus the Laramide forces. This apparent 20° rotation is hypothesized on the basis that any block-drag fold begins to form at the incipient 45° gamma angle to the bounding weakness zones (de Sitter, 1956). Whatever the correct interpretation, counterclockwise southwestward rotation of the fold in the stratigraphic rock section during its develop-

ment may explain the westward high-angle reverse faulting found at depth on the west side of the fold (Davis and Hunt, 1956) in that such rotation theoretically will lead to this type of faulting. It also may explain possible thrusting farther west of the fold that may be the cause of such anticlines as Plevna, Westmore, and Medicine Rocks (not all shown on Fig. 9B).

On the basis of the similarity of actual structural conditions with theoretical structural conditions predicted by the mechanics of simple-shear block coupling, it is hypothesized that a block framework, presumably in the basement, has controlled the direction and extent of structural features in the Williston–Blood Creek basin area.

Block Framework and Combined Structure

One of the advantages of recognizing the block framework and the simple-shear coupling within the framework is an explanation for the diverse structural trends of the area (Fig. 10). Regional compression acting on a basement without weakness zones would not be expected to produce such diverse trends.

The block-framework premise also can explain structurally puzzling apparent relations. For example, in the Williston–Blood Creek basin, the structural relation between Nesson anticline and Cedar Creek anticline has been a puzzle, as has the northeast-trending Richey anticline interrupting a possible connection of the Cedar Creek anticline with Poplar dome.

In both cases it is clear from Figure 10 that Nesson and Richey anticlines are related genetically to the northeast-trending Weldon block, whereas Cedar Creek anticline and Poplar dome were formed in different northwest-trending blocks. In other words, there is no direct relation between Cedar Creek anticline and Nesson anticline or Poplar dome. Each fold was formed independently of the others in distinct separate blocks.

Another structural problem has been the regional "connecting" of separate structural features into broad "arches" or basins as, for example, the relation between the Black Hills uplift and the Miles City "arch" (Fig. 1). Regional structure maps show a broad saddle between the Black Hills structure and Porcupine dome on the northwest. This saddle is called the Miles City "arch." As Figure 10 shows, however, detailed photogeologic maps indicate the Black Hills uplift actually ceases as an entity at the Cat Creek lineament zone,

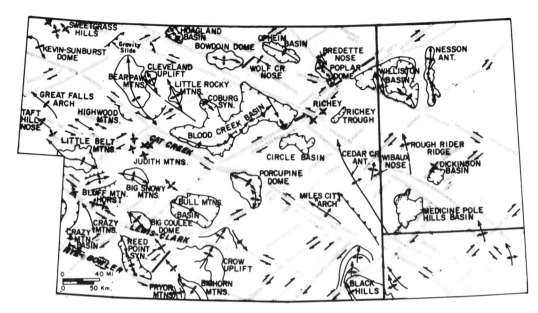

FIG. 10—Combined structures and block-adjustment pattern during Laramide orogeny.

whereas the Miles City "arch" is a distinct shear-zone complex of small drag folds generated by left-lateral movement on the Cat Creek megalineament.

Another example, the well-known Sweetgrass "arch," is a "connection" of the Taft Hill nose, the Great Falls arch, and the Kevin-Sunburst dome, whereas the "connection" of the Bearpaw Mountains uplift and the subtle, adjacent Cleveland uplift (defined by photogeologic information) produces an apparent northeasterly axis for the uplift. The following discussion of the Bearpaw Mountains within the block framework illustrates the importance of recognizing the existence of the blocks in interpreting any structure and provides an alternate interpretation.

Block Framework and Bearpaw Mountains

Figure 11A shows the main structural-lithologic features of the Bearpaw Mountains uplift and adjacent features including the block framework. The uplift has been interpreted to have a northeast-trending axis by Maher (1969; shown by the dashed-line axis). The uplift also has been interpreted to be a post-Laramide feature caused primarily by Tertiary intrusive and extrusive activity. A glance, however, at Figure 11A indicates that the uplift has a much larger extent, as shown by the gravity-slide faults, and that the Bearpaw Mountains proper are across the crest of the broad uplift.

The broad uplift can be interpreted to have

a Laramide northwesterly axis if the shallow, gravity-slide faults surrounding the structure (Reeves, 1953) define the true shape of the structure, for a northwest feature is compatible with the other Laramide features in the region. Such a configuration allows an axis nearly parallel with the left-lateral incipient drag-fold stage (Fig. 11A) and also suggests some "dragging" on the "toe" of the fold adjacent to the Cat Creek zone. Although the drag fold proper is confined to the Milk River block, the gravity-slide faults extend beyond the block, a condition apparently compatible with the lithologic strength of the stratigraphic rock units.

The crossfold northeasterly-axis interpretation can be understood by the "connecting" of the Cleveland uplift to the Bearpaws (Fig. 11A), as well as by the Eocene-Miocene igneous activity in the Bearpaws. The post-Laramide igneous activity is interpreted here to have been controlled by the Missouri River lineament zone which extends across the Bearpaw Mountains uplift coincident with the extrusive-rock outcrops. These crossfold lineaments should have been subjected to tension during the Eocene-Miocene as the coupling intensity increased on the Milk River block, inasmuch as the lineaments are nearly parallel with the incipient 45°, cross-block tensional direction (illustrated by the strain ellipsoid diagram in Fig. 11A). Upwelling of volcanic rocks along these tension zones probably contributed to the northeasterly crossfold elongation. Any eventual subsidence in the volcanic-rock belt presumably

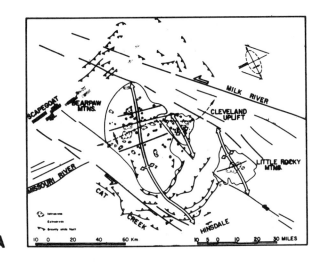

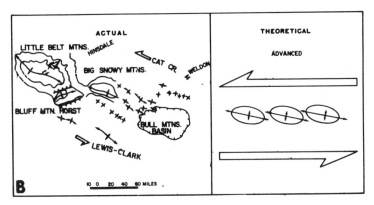

Fig. 11—Examples of Laramide couple-generated drag folds.
A. Bearpaw Mountains uplift: Laramide left-lateral drag fold. Northeast-trending dashed axis is currently recognized trend of uplift.
B. Little Belt-Big Snowy Mountains: Laramide left-lateral, advanced drag folds.

was localized by these same northeast weakness zones.

In short then, the Bearpaw uplift is interpreted as a northwest-trending Laramide drag fold across which post-Laramide igneous activity took place along northeast weakness zones subjected to tension.

This northeasterly crossfold elongation produced in the northwest blocks over northeast zones of weakness is not unique to the Bearpaws. The Little Rocky Mountains (Fig. 11A) also contain a northeast crossfold "bulge," as do the Judith and Moccasin Mountains directly south of the area in Figure 11A.

Other igneous-rock mountain complexes in the region also may be related to the mechanics of lineament-block coupling. Both the Highwood and Crazy Mountains display an overall northwest (Highwoods) or north-northwest (Crazy Mountains) trend apparent on detailed photogeologic maps of the area. Earlier, Lara-

mide drag folds may have helped to localize the post-Laramide igneous activity. More study is needed in these areas to demonstrate the relation.

Block Framework and Big Snowy Mountains

Another example of block-framework structural influence is the Big Snowy and Little Belt Mountains in the highly active Cat Creek northwest block (Fig. 11B). The en echelon spatial arrangement of both mountain ranges when compared to theoretical left-lateral advanced coupling suggests that the mountain ranges were formed as couple-generated drag folds. The drag-fold rotation implied in their formation is attested by the steep southern flank of the Big Snowy Mountains and by the extreme structural complexity involving Precambrian Belt Supergroup rock present on the southwest-flank of the Little Belts.

This en echelon drag-fold arrangement is extremely common in smaller folds within the Cat Creek block also. The Big Wall and Devils Basin anticlines, for example, as well as the Gage anticline, are all asymmetrically steeper on their southern flanks and form a series of drag folds en echelon southwest. The evidence suggests that the entire block was subjected to advanced coupling during the Laramide orogeny.

One complexity in this interpretation is the snoutlike southeast nose of the Little Belt Mountains. This feature is believed to have been caused by the tensional activation of northeast-trending basement-weakness zones, producing a horst. It should be remembered from the Bearpaw Mountains discussion that left-lateral coupling on the northwest blocks subjected intrablock northeast weakness zones to tension during the *latter* phases of the Laramide orogeny and during post-Laramide events.

An interesting point concerning right-lateral movement on the northeast weakness zones during the *early* phases of the Laramide orogeny is indicated in Figure 11B, where two drag folds generated along the Hinsdale lineament zone are present on the northern flank of the Little Belts. These drag folds are at the correct intersection angle (Moody and Hill's gamma angle) with the lineament zone, indicating that they were formed as simple-shear drag folds superimposed on the Little Belts. The interesting aspect, however, is that these drag folds and others in the locality contain intrusive-rock cores. It generally has been considered that intrusive rocks form the fold or dome in which they are found, but the fact that the folds on the north flank of the Little Belts are present as drag folds with a consistent intersection angle indicates that the drag folds localized the intrusive bodies and not the other way around. This hypothesis is supported by the fact that the Cretaceous-Paleocene folds are intruded by igneous rocks of Eocene age.

Regional Block Tectonics

The block framework and local tectonics described previously are not unique to the Williston–Blood Creek basinal area (Fig. 12). Indeed, lineament-block tectonics appears to be the rule rather than the exception in Rocky Mountain and High Plains Laramide tectonics. Sales (1968) was the first to demonstrate this with the introduction of his Wyoming couple defined by the Lewis-Clark and Uinta lineament zones. The present writer recognizes a Colorado couple and the Montana couple (the

northern bounding lineament zone is unknown).

Structural Conclusions

The basement-weakness zones in the Williston–Blood Creek basinal area, judging from the trend and spatial arrangement of the larger structural features, have deflected Laramide regional orogenic forces into a block framework defined by the weakness zones. The pattern regularity of this block framework and the persistent direction of incoming orogenic forces combined to produce tectonics best described as simple-shear on the weakness zones translated into coupling on the basement blocks.

Simple shear on the basement-weakness zones has produced in the stratigraphic section narrow zones of en echelon drag folds (Cat Creek zone), extreme rotation and overriding of drag folds (Nye-Bowler zone), and tensional faulting (Lake Basin fault zone). Simple shear on the weakness zones translated into coupling on the blocks between weakness zones has produced intrablock drag folds such as Nesson anticline, Cedar Creek anticline, and the Bearpaw Mountains uplift, as well as cross-block northeast normal faults such as the Weldon and Hinsdale faults. Advanced coupling on the northwest-trending Cat Creek block is interpreted to have produced the drag-fold development of the Big Snowy and Little Belt Mountains. In addition, simple shear on the northwest-trending weakness zones, translated into block coupling, subjected intrablock northeast-trending weakness zones to tension during the Eocene-Miocene, thereby providing conduits for intrusive and extrusive igneous activity.

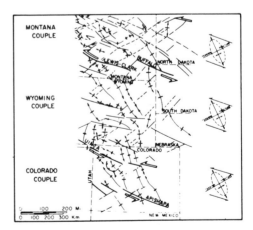

Fig. 12—Laramide, lineament simple-shear block coupling in Rocky Mountain region.

BLOCK FRAMEWORK AND OIL AND GAS LOCALIZATION

Geologic features or groups of features, such as basement-weakness zones which control structural development of any area during every orogeny and influence the stratigraphy between tectonic pulses, must directly affect oil and gas occurrences. This is thought to be true because, as pointed out in the structural discussion, the subblocks defined by the weakness zones can be coupled by orogenic forces to form intrablock folds. These folds and lineaments in turn, as noted in the stratigraphic discussion, can influence the development of paleotopographic highlands and subbasins within the regional basinal area, thereby affecting shelf development and trend. Paleostrand lines, facies changes, and thickness variations may, in turn, be "controlled" by this shelf development on the flanks of the highlands and subbasins and along lineaments. Because these weakness zones are the same tectonically mobile zones producing drag folds in the stratigraphic section, it is logical to find many oil and/or gas localizations in proximity to the weakness zones in addition to those fields surrounding or within drag folds.

Figure 13 illustrates the importance of the block framework in relation to oil and gas in the studied area. The large drag-fold features such as Nesson and Cedar Creek anticlines, Poplar, Bowdoin, and Kevin-Sunburst domes contain, of course, the most localizations, but it is interesting to note the many stratigraphic, structural, and combination traps in proximity to the Weldon block and its bounding lineament zones. The Lake Basin field, Big Coulee Dome field, the Bull Mountains Basin and Sumatra fields, Weldon, Richey and Duck Creek fields, the Brorson and Fairview fields, the Grenora and Dwyer fields, and the many Mississippian Mission Canyon stratigraphic fields of north-central North Dakota are all confined to the block or its bounding features.

A smaller northeast-trending block that also may control oil and gas localizations is the Powder River block. Several recent oil discoveries have been made in or adjacent to this block, including the Bell Creek field.

It is also interesting to speculate on the block framework's control of oil and gas migration. For example, the large nonproducing area east of Nesson anticline and southwest of the updip Mission Canyon fields in the Weldon block

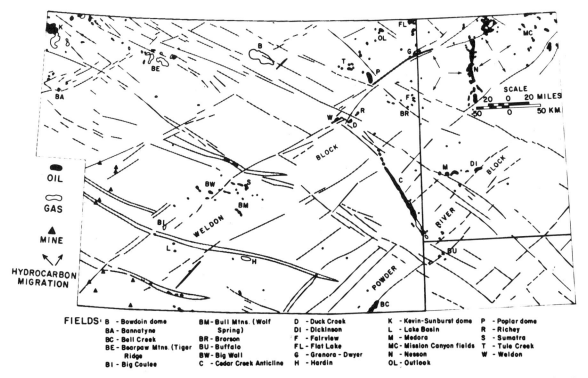

FIELDS:
B - Bowdoin dome
BA - Bannatyne
BC - Bell Creek
BE - Bearpaw Mtns. (Tiger Ridge
BI - Big Coulee
BM - Bull Mtns. (Wolf Spring)
BR - Brorson
BU - Buffalo
BW - Big Wall
C - Cedar Creek Anticline
D - Duck Creek
DI - Dickinson
F - Fairview
FL - Flat Lake
G - Grenora - Dwyer
H - Hardin
K - Kevin-Sunburst dome
L - Lake Basin
M - Medora
MC - Mission Canyon fields
N - Nesson
OL - Outlook
P - Poplar dome
R - Richey
S - Sumatra
T - Tule Creek
W - Weldon

FIG. 13—Block framework compared with oil, gas, and some mineral localizations. Oil and gas closely associated with drag folds (B, BE, BI, BM, BR, BW, C, D, DI, F, K, L, N, OL, P, R, and S) or lineaments (BA, BC, BU, G, H, M, and W). Ore localities suggest similar associations.

might have lost its oil and gas, as migration proceeded updip in all directions away from this area. Thus, Nesson anticline, the north- and south-bounding lineament zones, and the updip Mission Canyon facies-change zones would be expected to receive and trap the migrating oil and gas.

The same situation may be present in the central low of the Williston basin west of Nesson anticline. There the migration would be expected to transport oil to Nesson anticline, the bounding Weldon lineament zone (Grenora and Dwyer fields), and the bounding Watford lineament zone (Brorson and Fairview fields).

If this migration theory is valid, the importance of the basement block framework to the occurrence of oil and gas in any area becomes obvious.

Mineral localizations also appear to be controlled in part by the block framework because of the northwest-block coupling that opens up intrablock northeast zones of weakness as tensional conduits to migrating mineralized solutions. A proper host rock for mineral emplacement is, of course, as important to mineral localization as is the proper reservoir rock for oil and gas. But here, too, mining localities as reported from topographic maps in the area are found more commonly in proximity to lineaments (Fig. 13) than not. Admittedly the number of mines that show this relation is sparse but intriguing, nevertheless.

SUMMARY AND CONCLUSIONS

On the basis of detailed, photogeologic surface mapping, isopach subsurface map comparisons, and structural analyses, it has been demonstrated that a system of basement blocks is present in the region of the Williston–Blood Creek basin. This system is indicated at the surface by northeast- and northwest-trending lineaments. When subjected to Laramide orogenic forces, the basement weakness zones that define the block framework apparently adjusted laterally by simple shear to the regional compressive forces. Differential simple shear on opposite boundaries of a block produced coupling across the block which, in turn, produced dragfold uplifts and downwarps, faulting, and fracturing. These end-result Laramide structural features control the present topography of the region.

Although this study is based to a large extent on surface-lineament evidence to define the basement-weakness zones, the fact that the weakness zones generally are believed to be of early Precambrian age also suggests that earlier Paleozoic deformational events, probably of an epeirogenic nature in the region, probably were controlled to some extent by the block framework. This means in turn that paleotopographic conditions were block-framework controlled. As paleotopographic conditions to a large extent control paleodeposition, the block framework can be used to investigate subsurface stratigraphic relations.

From the many Laramide structural features in the region and their demonstrated relation to the block framework, it is concluded that during the Laramide orogeny the Williston–Blood Creek basin region should be referred to as a moderately deformed foreland contiguous with the more intensely deformed Rocky Mountains in the west rather than as a craton with its connotation of merely epeirogenic deformation.

A Laramide foreland designation seems more apt in view of the fact that the Williston–Blood Creek basin region is adjacent to the intensely deformed Wyoming couple foreland of Sales (1968). The block framework appears to be present in both regions (Fig. 12) and both share the Lewis-Clark lineament as a major boundary element. The only deformational difference between the two regions is in the intensity of the coupling. Indeed, because the Williston–Blood Creek basin foreland is not so deformed as the Wyoming couple foreland of the Rocky Mountains, it is the ideal area to work out simple-shear block-coupling mechanics that can be applied to Rocky Mountain tectonics.

Although this study concerns the Williston–Blood Creek basin area, lineament-block tectonics as defined herein for that area should not be considered as unique to the area or to the Laramide orogeny. Figure 12 illustrates a similar lineament-block framework for much of the Rocky Mountains region and, as the basement-weakness zones are considered to be early Precambrian in age, they probably had a part in all deformational events in the Rocky Mountains region. (Thomas, 1972, stated that lineaments are common in all of North America and figured prominently in Paleozoic, Mesozoic, and Cenozoic deformation).

For these reasons it is concluded that lineament simple-shear block-couple mechanics offer a plausible alternative to compressional, pure-shear mechanics for orogenic events. The following phases of a simple-shear, block-couple orogeny are presented for use in tectonic analyses wherever lineaments are recognized as being an integral part of the deformational process.

Simple-Shear, Block-Couple Orogeny

1. Regional compressive stress intersects basement block framework, resulting in vertical (epeirogenic) adjustment of blocks to new spatial conditions, and initial channeling of stress into lateral-adjustment zones in basement.

2. Increased channeling of stress into lateral-adjustment zones brings local drag folding of sedimentary rocks in proximity to zones, build up of dominant coupling forces on blocks or block sets, and perhaps increased epeirogenic adjustment of blocks in alignment with increasing dominant couple.

3. In the incipient drag-fold phase, flank ("step") faults and crossfold tensional faults develop as basement rocks are involved in uplifts; igneous plutons are intruded along drag-fold uplifts and associated faults in response to decreased heat-pressure gradients along uplifts. If uplifts develop in previous deposition sites (shelf, trough, or even geosyncline), gravity sliding of sedimentary rocks is initiated.

4. In the moderate drag-fold phase, coupling forces increase—drag folds rotate; flank faults become local shear zones; secondary crossfold tension fractures develop; large-scale igneous activity probably declines in response to more shearing than tensional activity; gravity sliding in depositional sites is more extensive as structural relief increases; low-grade metamorphism may develop in narrow depositional sites subjected to greater shearing.

5. Advanced to extreme drag-folding phases are dependent on intensity of stress and local competence of basement and sedimentary rocks. Flank faults may become high-angle reverse faults and drag folds thrust on one another may merge as a continuous mountain chain in the extreme phase; only local igneous activity occurs unless cross-block tension increases (see phase 6); higher grades of metamorphism may occur in response to greater shearing forces.

6. In the cross-block tensional phase, increased dominant coupling on one block set produces increased tensional forces on cross-block weakness zones parallel or nearly parallel with tensional direction of dominant couple; outpouring of igneous extrusive rocks is accompanied locally by igneous plutons along tensional conduits. The tensional phase may be a concluding phase of the entire orogeny or may occur at the end of incipient, moderate, or

advanced phases if basement resistance to further lateral adjustment is high.

In the Williston–Blood Creek basin, lineament-block deformation during the Mesozoic and Cenozoic proceeded through the following general phases: (1) initial compressive stress leading to epeirogenic uplift (possibly a Jurassic event); (2) increased channeling of stress; epeirogenic uplift more pronounced; local drag folding in sedimentary rocks (Jurassic-middle Cretaceous); (3) incipient drag-fold phase with local, moderate drag folding (Laramide orogeny; Late Cretaceous–Paleocene); and (4) cross-block tensional phase (Eocene-Miocene).

In a lineament-block orogenic system, the Laramide orogeny proper is interpreted here as the intense phase of a larger, longer time-extensive orogenic event which probably began in the Jurassic–Early Cretaceous as essentially a period of epeirogenic uplift and local drag folding, and culminated in the Eocene-Miocene as a maximum tensional period. Actually, the region is probably still on the "downward" slope of tectonic activity related to this larger orogenic event as evidenced by tensional faults in Quaternary deposits and Quaternary volcanic rocks in the Rocky Mountains.

Thus all aspects of an orogeny can be related via lineament simple-shear block coupling without the necessity of separating initial epeirogenic and culmination-tensional events from the Laramide event that appears compressional in nature.

REFERENCES CITED

Atwater, T., 1972, Test of new global tectonics: discussion: Am. Assoc. Petroleum Geologists Bull., v. 56, no. 2, p. 385–388.

Badgley, P. C., 1965, Structural and tectonic principles: New York, Harper and Row, p. 521.

Ballard, W. W., 1969, Red River of northeast Montana and northwest North Dakota, *in* Eastern Montana symposium: Montana Geol. Soc. 20th Ann. Conf. p. 15–24.

Bolyard, D. W., 1969, Muddy Sand oil potential in South Dakota, *in* Eastern Montana symposium: Montana Geol. Soc. 20th Ann. Conf. p. 85–94.

Carey, S. W., 1958, Continental drift: a symposium: Univ. Tasmania Geology Dept., p. 177–355.

Davis, W. E. and R. E. Hunt, 1956, Geology and oil production on the northern portion of the Cedar Creek anticline, Dawson County, Montana, *in* 1st Internat. Williston basin symposium: North Dakota and Saskatchewan Geol. Socs., p. 121–129.

De Sitter, L. U., 1956, Structural geology: New York, McGraw-Hill, 552 p.

Hansen, A. R., 1966, Reef trends of Mississippian Ratcliff zone, northeast Montana and northwest North Dakota: Am. Assoc. Petroleum Geologists Bull., v. 50, no. 10, p. 2260–2268.

Hills, E. S., 1956, The tectonic style of Australia, *in*

Geotektonisches Symposium zu ehren von Hans Stille: Stuttgart, Kommissions-Verlag Ferdinand Enke, p. 336–346.

Hobbs, W. H., 1912, Earth features and their meaning, an introduction to geology for the student and general reader: New York, 506 p.

Kinard, J. C., and W. R. Cronoble, 1969, The Winnipegosis Formation in northeast Montana, *in* Eastern Montana symposium: Montana Geol. Soc. 20th Ann. Conf., p. 33–44.

Lattman, L. H., and R. H. Matzke, 1961, Geological significance of fracture traces: Photogramm. Eng., v. 27, no. 3, p. 435–438.

Maher, P. D., 1969, Eagle gas accumulations of the Bearpaw uplift area, Montana, *in* Eastern Montana symposium: Montana Geol. Soc. 20th Ann. Conf., p. 121–127.

Maxwell, J. C., and D. U. Wise, 1958, Wrench-fault tectonics: a discussion: Geol. Soc. America Bull., v. 69, no. 7, p. 927–928.

Michelson, J. C., 1956, Madison Group in central Montana, *in* Judith Mountains, central Montana: Billings Geol. Soc. 7th Ann. Field Conf., p. 68–72.

Moody, J. D., and M. J. Hill, 1956, Wrench-fault tectonics: Geol. Soc. America Bull., v. 67, no. 9, p. 1207–1246.

O'Driscoll, E. S. T., 1971, Deformational concepts in relation to some ultramafic rocks in western Australia: Geol. Soc. Australia Spec. Pub. 3, p. 351–366.

Pease, R. W., 1969, Normal faulting and lateral shear in northeastern California: Geol. Soc. America Bull., v. 80, no. 4, p. 715–720.

Rayl, R. L., 1956, Stratigraphy of the Nesson, Piper, and Rierdon Formations of central Montana, *in* Judith Mountains, central Montana: Billings Geol. Soc. 7th Ann. Field Conf., p. 35–45.

Reeves, F., 1953, Bearpaw thrust-faulted area, *in* The Little Rocky Mountains and southwestern Saskatchewan: Billings Geol. Soc. 4th Ann. Field Conf., p. 114–117.

Sales, J. K., 1968, Crustal mechanics of Cordilleran foreland deformation: a regional and scale-model approach: Am. Assoc. Petroleum Geologists Bull., v. 52, no. 10, p. 2016–2044.

Sonder, R. A., 1947, Shear patterns of the earth's crust (by F. A. Vening Meinesz): Am. Geophys. Union Trans., v. 28, no. 6, p. 939–945.

Sonnenberg, F. P., 1956, Tectonic patterns of central Montana, *in* Judith Mountains, central Montana: Billings Geol. Soc. 7th Ann. Field Conf., p. 73–81.

Stone, D. S., 1969, Wrench-faulting and central Rocky Mountain tectonics: Mtn. Geologist, v. 6, p. 67–79.

Thomas, G. E., 1971, Continental plate tectonics; southwest Wyoming, *in* Symposium on Wyoming tectonics and their economic significance: Wyoming Geol. Assoc. Guidebook 23, p. 103–123.

———— 1972, Continental plate tectonics: North America (abs.); Am. Assoc. Petroleum Geologists Bull., v. 56, p. 658.

Vening Meinesz, F. A., 1947, Shear patterns of the earth's crust: Am. Geophys. Union Trans., v. 28, no. 1, p. 1–61.

U.S. Geological Survey and American Association Petroleum Geologists, 1961, Tectonic map of the United States, exclusive of Alaska and Hawaii, scale 1:2,500,000, 2 sheets.

Reprinted by permission of the Geological Society of America
from V. S. Mount and J. Suppe, *Geology*, v. 15 (1987), p.
1143-1146.

State of stress near the San Andreas fault: Implications for wrench tectonics

Van S. Mount, John Suppe
Department of Geological and Geophysical Sciences, Princeton University
Princeton, New Jersey 08544

ABSTRACT

Borehole elongations or breakouts in central California show that the direction of regional maximum horizontal stress is nearly perpendicular to the San Andreas fault and to the axes of young thrust-related anticlines. This observation resolves much of the controversy over shear-stress magnitude in the crust and around the San Andreas fault specifically. A low shear stress of 10–20 MPa (100–200 bar) or less on the San Andreas fault, suggested by heat-flow and seismic observations, is compatible with a high regional deviatoric stress (100 MPa, 1 kbar) when the observed principal stress directions are considered. Therefore, the San Andreas fault is a nearly frictionless interface, which causes the transpressive plate motion to be decoupled into a low-stress strike-slip component and a high-stress compressive component. These observations suggest that standard concepts of transpressive wrench tectonics—which envisage drag on a high-friction fault—are wrong. The thrust structures are largely decoupled from the strike-slip fault.

INTRODUCTION

Among the important unknown quantities in tectonophysics and structural geology are the magnitudes of deviatoric stresses in the crust and the magnitudes of shear stresses on major active faults, such as the San Andreas. Estimates of stress magnitude have ranged over an order of magnitude, roughly from 10 to 100 MPa (100 bar to 1 kbar). For example, the lack of a heat-flow anomaly over the San Andreas fault and seismic radiation estimates suggest that shear stress on the fault is extremely low, 10 to 20 MPa (100 to 200 bar) or less averaged over the top 10–20 km of the crust (Brune et al., 1969; Lachenbruch and Sass, 1980). In contrast, Brace and Kohlstedt (1980) showed that laboratory measurements of rock strength suggest relatively high deviatoric stresses (100 MPa or 1 kbar) to be widely present in the upper crust. Hanks (1977) proposed shear stresses on active crustal fault zones of 100 MPa (1 kbar) on the basis of laboratory measurements of frictional rock strength at mid-crustal conditions. Hanks (1977) contended that if frictional stresses on active faults were low, then the crustal deviatoric stress would also be low and of comparable magnitude.

In this paper we show that a low shear stress of 10–20 MPa (100–200 bar) on the San Andreas fault, which is the upper limit indicated by heat-flow and seismic observations, is compatible with a high regional deviatoric stress (100 MPa, 1 kbar) when the observed principal stress directions are considered. Furthermore, the observed orientation of principal stresses accounts for the regional compressional deformation in a 50–100-km-wide zone spanning the San Andreas fault in central California, deformation that previously had been attributed to the fault-drag concepts of wrench tectonics (Harding, 1976).

STATE OF STRESS NEAR THE SAN ANDREAS FAULT
Principal Stress Orientation

Previous studies (e.g., Zoback and Zoback, 1980) reported a north-south direction of maximum horizontal stress in central California largely

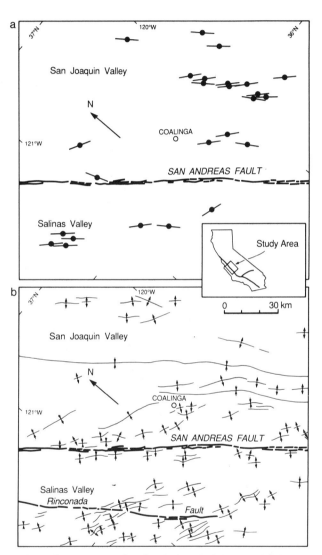

Figure 1. Region surrounding San Andreas fault in central California. This segment of fault includes northernmost section of 1857 rupture, southern part of creeping section, and intermediate stick-slip and creeping Parkfield section. a: Bars with circles indicate direction of borehole elongation (minimum horizontal stress). Direction of regional maximum horizontal stress is at high angle (nearly perpendicular) to San Andreas fault. b: Young fold axes and faults in study area (from Jennings et al., 1977). Most folds are Pliocene and younger. Northwest of Coalinga, two Miocene folds strike at higher angle ($\theta = 25°–30°$) to fault than active folds such as those southeast of Coalinga ($\theta = 6°$).

on the basis of interpretation of strike-slip earthquake focal mechanisms. In contrast, new data on borehole elongations or breakouts summarized in Figure 1 indicate that the direction of maximum horizontal stress is actually northeast-southwest, approximately normal to the strike of the San Andreas fault. The fault-plane solutions along the San Andreas are apparently controlled by the orientation of the fault and do not simply reveal the principal stress directions.

Breakouts in nearly vertical boreholes are thought to form by brittle failure or spalling in response to a concentration of compressive stress at the borehole wall (Bell and Gough, 1979; Zoback et al., 1985). The wall is insufficiently supported by the pressure of the drilling mud and fails at the points of greatest stress concentration, which occur in the minimum horizontal compressive stress direction. Breakout orientations are consistent over large regions (Cox, 1970; Suppe et al., 1985; M. Suter, in prep.; Plumb and Cox, 1987) and show excellent agreement with other in situ stress measurements (Plumb and Hickman, 1985).

Breakout orientations were determined for 108 wells in a study of regional stress in California by V. S. Mount (in prep.) using four-arm dipmeter logs and following the methods of Plumb and Hickman (1985). The azimuth of individual elongations is determined to a precision of $\pm 5°$. The results from 25 of the wells in central California are shown in Figure 1. The study area encompasses sections of the San Andreas fault that are locked, creeping, and displaying intermediate stick-slip and creep behavior. The wells range in depth from 1.3 to 4.6 km; the mean depth is 2.9 km. Breakouts were observed from depths of 0.2 to 4.6 km. The number of borehole elongations per well ranged from 4 to 33, the mean being 15. The mean elongation azimuth for each well is plotted in Figure 1. Figure 2 shows a rose diagram for 394 borehole elongations from the 25 wells studied for which the sample mean breakout azimuth is $133.9° \pm 1.9°$ (95% confidence interval), which is taken as the regional direction of minimum horizontal stress. Therefore, the regional direction of maximum horizontal stress is $43.9° \pm 1.9°$, at a high angle (84°) to the $139.8° \pm 1.0°$ strike of the San Andreas fault in the study area (Brown, 1970; Vedder and Wallace, 1970; Fig. 2). Breakout data north of the study area indicate a similar direction of maximum horizontal stress. South of the study area, in the "Big Bend" region, breakout data indicate that the direction of maximum horizontal stress is approximately north-south, again at a high angle to the approximately east-west San Andreas fault (V. S. Mount, in prep.).

The direction of maximum horizontal stress determined from borehole breakouts is in agreement with other geologic and geophysical evidence: (1) The trends of Pliocene to Holocene fold axes are approximately perpendicular to the direction of maximum horizontal stress (Figs. 1 and 2). (2) The fault-plane solutions for the 1983 Coalinga and 1985 Avenal earthquakes in the study area indicate northeast-southwest compression on thrust or high-angle reverse faults (Stein, 1983; Eaton, 1985). (3) Geologic cross sections based on well and seismic data through the active northwest-southeast–trending Coalinga and nearby Lost Hills anticlines (Namson and Davis, 1988; Medwedeff and Suppe, 1986) show that they are formed by slip on northeast- and southwest-vergent thrust faults.

In light of the predominant thrust deformation away from the San Andreas proper, it is concluded that the vertical principal stress is less than the horizontal, and is regionally

$$\sigma_1(\text{NE-SW}) > \sigma_2(\text{NW-SE}) > \sigma_3(\text{vertical}). \qquad (1)$$

Magnitude of the Deviatoric Stress

The regional horizontal deviatoric stress $\frac{1}{2}(\sigma_1 - \sigma_2)$ can be estimated if we assume an approximately homogeneous state of stress over the region of Figure 1, equivalent to shrinking the region to a point. The consistency of breakout and fold-axis data on both sides and immediately adjacent to the San Andreas fault within the 160×135-km study area suggests that the assumption is valid. The stresses may refract through a weak zone of finite width (1–5 km) along the San Andreas fault zone, but for the purposes of

this regional analysis, the San Andreas is treated as a plane of zero width. Consideration of possible stress refraction is important mainly for understanding the mechanical properties of the fault zone, but it is not our concern here. Using the conditions for equilibrium, we can obtain an equation for shear stress σ_τ on a plane—the San Andreas fault in our example—in terms of the principal stresses σ_1 and σ_2:

$$\sigma_\tau = \frac{1}{2}(\sigma_1 - \sigma_2)\sin 2\theta, \qquad (2)$$

where θ is the angle between the σ_1 direction and the normal to the fault. We determine $\theta = 5.9°$ from the mean breakout direction and the mean strike of the San Andreas fault (Fig. 2). If we assume the shear stress σ_τ on the fault is 20 MPa (200 bar), the upper limit on the basis of heat-flow and seismic observations (Brune et al., 1969; Lachenbruch and Sass, 1980), a maximum horizontal deviatoric stress $\frac{1}{2}(\sigma_1 - \sigma_2)$ of approximately 100 MPa (1 kbar) is computed (Fig. 3). The magnitude of the full deviatoric stress $\frac{1}{2}(\sigma_1 - \sigma_3)$ is not known because the relative magnitudes of σ_2 and σ_3 are not well constrained.

It is apparent from equation 2 and Figure 3 that a small change in the shear stress on the fault will cause a large change in the estimated deviatoric stress magnitude. However, a change in the orientation of the stress field, θ, of only a couple of degrees will also cause a large change in the computed deviatoric stress magnitude. Therefore, variations in shear stress along the fault could be compensated by a slight (1°–2°) adjustment in the orientation of the stress field, thus avoiding large fluctuations in the regional deviatoric stress magnitude.

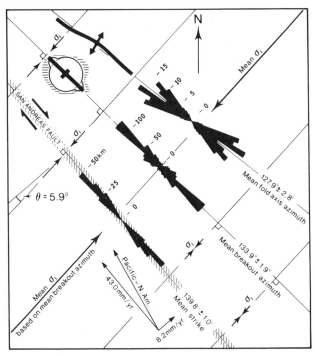

Figure 2. Schematic map showing angular relation between mean strike of San Andreas fault, mean breakout azimuth, and mean fold-axis azimuth for area in Figure 1. Mean strike of San Andreas fault was determined from strikes of recently active fault breaks (Brown, 1970; Vedder and Wallace, 1970) weighted by length. Mean breakout azimuth was determined from 394 breakouts from 25 wells in study area (Fig. 1a). Mean fold-axis azimuth was determined from 100 fold axes shown in Figure 1b. Means were computed by using methods of Mardia (1972) and are reported with 95% confidence intervals for true mean. Pacific-North America relative plate motion is plotted (F. Pollitz, in prep.) with components parallel and normal to San Andreas fault.

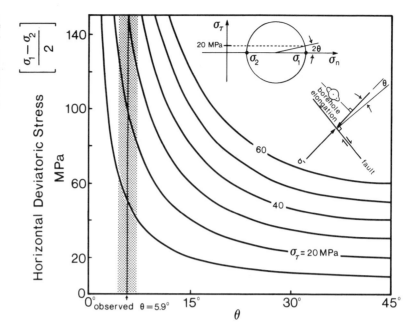

Figure 3. Plot of σ_1 stress orientation θ vs. horizontal component of deviatoric stress $\frac{1}{2}(\sigma_1 - \sigma_2)$ for lines of constant shear-stress magnitude σ_τ. For $\theta = 5.9°$ (angle between normal to San Andreas fault and σ_1 based on breakout orientation) and shear stress on San Andreas fault of 20 MPa (upper bound imposed by heat-flow and seismic constraints of Lachenbruch and Sass, 1980), horizontal deviatoric stress of approximately 100 MPa (1 kbar) is predicted.

A regional deviatoric stress on the order of 100 MPa seems reasonable; it is consistent with fracture strengths at moderate fluid pressures (Brace and Kohlstedt, 1980) and agrees with the interpretation of Lachenbruch and Sass (1980) that the broad Coast Ranges heat-flow anomaly is due to mechanical heat generation, which requires that the plate interaction along western California be a high-stress system, regardless of the shear-stress magnitude on the San Andreas fault.

We have used the simple analysis above because we believe that it contains the essence of the problem. It incorporates two key conditions that we believe dominate the state of stress in western California: (1) the strength of the San Andreas fault, which is at failure in strike slip and extends through the lithosphere and the length of California, and (2) the strength of the surrounding Coast and Transverse ranges, which are at failure by thrust faulting and also extend the length of California. The strengths of both are heterogeneous, of course, but we assume that it is useful to consider mean values. It is the contrast in strength between the San Andreas and the Coastal Range that is assumed to control the observed stress orientation, the simplest model of which is equation 2. The conclusion of this analysis is that the San Andreas is very weak relative to the Coast Ranges and acts like an "almost free surface" causing the principal stresses to be almost perpendicular and parallel to the fault, which agrees with the observation that the stress field bends as the San Andreas enters the "Big Bend" of the Transverse Ranges, maintaining its high angle to the fault (V. S. Mount, in prep.).

IMPLICATIONS FOR TRANSPRESSIVE WRENCH TECTONICS

The San Andreas fault system in central California has been considered a classic example of active wrench tectonics (Harding, 1976; Wilcox et al., 1973). However, the state of stress documented in the previous section on the basis of heat-flow and breakout data differs from conceptual models of transpressive tectonics, which envisage substantial frictional coupling and drag.

Wrench tectonics features a zone of distributed shear characterized by en echelon folds surrounding a throughgoing wrench fault (Wilcox et al., 1973; Harding, 1976), as shown schematically in Figure 4a; σ_1 is expected to be oriented perpendicular to primary fold axes and thrust fronts in the

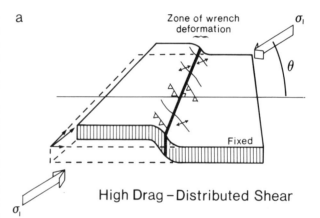

High Drag – Distributed Shear

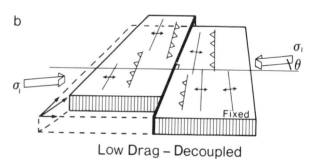

Low Drag – Decoupled

Figure 4. a: Deformation style predicted by distributed-shear, high-drag concepts of wrench tectonics. Note that zone of wrench deformation is confined to zone of distributed shear adjacent to fault. Angle between σ_1 and normal to fault is 30°–45°. **b:** Deformation style predicted for decoupled transcurrent and thrust deformation along low-drag wrench fault having little distributed shear. Angle between σ_1 and normal to fault is small. This style of deformation is similar to Pleistocene and Holocene deformation adjacent to San Andreas fault in central California (cf. Fig. 1).

early stages of wrenching ($\theta = 30°–45°$) (Fig. 4a). The fold axes may subsequently rotate into parallelism with the high-drag wrench fault through distributed shear. By this concept younger folds would be close to the ideal wrench orientation ($\theta = 30°–45°$), whereas progressively older folds would more closely parallel the fault ($\theta = 5°–30°$). In fact, just the opposite is observed near the San Andreas; Miocene folds have an orientation close to the ideal wrench orientation ($\theta = 25°–30°$; northwest of Coalinga, Fig. 1), whereas the Pleistocene folds are nearly parallel (Figs. 1 and 2). Perhaps with increase in age and gouge development, the San Andreas has undergone progressive strain softening, accompanied by a drop in shear stress and rotation of principal-stress directions.

In contrast with these concepts of distributed-shear, high-drag wrench tectonics, the heat-flow constraints on the shear stress (Lachenbruch and Sass, 1980) combined with the breakout data (Fig. 3) indicate that present-day relative plate motion in central California (Fig. 2) is factored into two nearly independent components: (1) a low-shear-stress strike-slip component (43 mm/yr parallel to the San Andreas fault) and (2) a high-deviatoric-stress thrust component (8 mm/yr normal to the San Andreas fault) (rates from F. Pollitz, in prep.). The strike-slip component is accommodated along a narrow (<3–10 km) zone of nearly pure strike-slip faulting. The compressive component is accommodated in a wide (50–100 km) zone of thrust faults, which verge approximately perpendicular to the strike of the San Andreas fault; fault-bend folding above these faults (Namson and Davis, 1988; Medwedeff and Suppe, 1986) produces folds with axes trending approximately parallel to the San Andreas fault (Fig. 1). We conclude that present-day compressional deformation outside the narrow strike-slip zone is not controlled by distributed shear associated with a high-drag San Andreas fault. Transpressive tectonics in central California can be better described as decoupled transcurrent and compressive deformation, operating simultaneously and largely independently (Fig. 4b).

How typical the present-day situation in central California is of wrench tectonics in general remains to be seen. Preliminary breakout data near the Philippine fault suggest compression nearly perpendicular to the fault (R. Dorsey and J. Suppe, unpub. data), and young fold axes are nearly parallel to the fault (Barcelona, 1986).

REFERENCES CITED

Barcelona, B.M., 1986, The Philippine fault and its tectonic significance: Geological Society of China Memoir 7, p. 31–44.
Bell, J.S., and Gough, D.I., 1979, Northeast-southwest compressive stress in Alberta: Evidence from oil wells: Earth and Planetary Science Letters, v. 45, p. 475–482.
Brace, W.F., and Kohlstedt, D.L., 1980, Limits on lithospheric stress imposed by laboratory experiments: Journal of Geophysical Research, v. 85, p. 6248–6258.
Brown, R.D., Jr., 1970, Map showing recently active breaks along the San Andreas and related faults between the northern Gabilan Range and Cholame Valley, California: U.S. Geological Survey Miscellaneous Geologic Investigations Map I-575, scale 1:62,500.
Brune, J.N., Henyey, T.L., and Roy, R.F., 1969, Heat flow, stress, and rate of slip along the San Andreas fault, California: Journal of Geophysical Research, v. 74, p. 3821–3827.
Cox, J.W., 1970, The high resolution dipmeter reveals dip-related borehole and formation characteristics: Society of Professional Well Log Analysts, Eleventh Annual Logging Symposium, Paper D, 25 p.

Eaton, J.P., 1985, The North Kettleman Hills earthquake of August 4, 1985, and its first week of aftershocks—A preliminary report: U.S. Geological Survey Monthly Data Review.
Hanks, T.C., 1977, Earthquake stress drops, ambient tectonic stresses and stresses that drive plate motions: Pure and Applied Geophysics, v. 115, p. 441–458.
Harding, T.P., 1976, Tectonic significance and hydrocarbon consequences of sequential folding synchronous with San Andreas faulting, San Joaquin Valley, California: American Association of Petroleum Geologists Bulletin, v. 60, p. 356–378.
Jennings, C.W., Strand, R.G., and Rogers, T.H., 1977, Geologic map of California: California Division of Mines and Geology Geologic Map Series, no. 2.
Lachenbruch, A.H., and Sass, J.H., 1980, Heat flow and energetics of the San Andreas fault zone: Journal of Geophysical Research, v. 85, p. 6185–6222.
Mardia, K.V., 1972, Statistics of directional data: London, Academic Press Inc., 357 p.
Medwedeff, D.A., and Suppe, J., 1986, Growth fault-bend folding—Precise determination of kinematics, timing, and rates of folding and faulting from syntectonic sediments: Geological Society of America Abstracts with Programs, v. 18, p. 692.
Namson, J.S., and Davis, T.L., 1988, Seismically active fold and thrust belt in the San Joaquin Valley, central California: Geological Society of America Bulletin, v. 100 (in press).
Plumb, R.A., and Cox, J.W., 1987, Stress directions in eastern North America determined to 4.5 km from borehole elongation measurements: Journal of Geophysical Research (in press).
Plumb, R.A., and Hickman, S.H., 1985, Stress-induced borehole elongation: A comparison between the four-arm dipmeter and the borehole televiewer in the Auburn geothermal well: Journal of Geophysical Research, v. 90, p. 5513–5521.
Stein, R.S., 1983, Reverse slip on a buried fault during the 2 May 1983 Coalinga earthquake: Evidence from geodetic elevation changes, in Bennett, J.H., and Sherburne, R.W., eds., The 1983 Coalinga, California, earthquake: California Division of Mines and Geology Special Publication 66, p. 151–163.
Suppe, J., Hu, C., and Chen, Y., 1985, Present-day stress directions in western Taiwan inferred from borehole elongation: Petroleum Geology of Taiwan, no. 21, p. 1–12.
Vedder, J.G., and Wallace, R.E., 1970, Map showing recently active breaks along the San Andreas fault between Cholame Valley and Tejon Pass, California: U.S. Geological Survey Miscellaneous Geologic Investigations Map I-574, scale 1:24,000.
Wilcox, R.E., Harding, T.P., and Seely, D.R., 1973, Basic wrench tectonics: American Association of Petroleum Geologists Bulletin, v. 57, p. 74–96.
Zoback, M.D., and Zoback, M., 1980, State of stress in the conterminous United States: Journal of Geophysical Research, v. 85, p. 6113–6156.
Zoback, M.D., Moos, D., Mastin, L.G., and Anderson, R.N., 1985, Well bore breakouts and in situ stress: Journal of Geophysical Research, v. 90, p. 5523–5530.

ACKNOWLEDGMENTS

Supported by ARCO Oil and Gas Company and the Princeton University Department of Geological and Geophysical Sciences. Van Mount thanks the California Division of Oil and Gas District Offices for assistance and generous hospitality during data acquisition. We also thank Tony Dahlen, Kathy Hansen, Don Medwedeff, Richard Plumb, and Max Suter for contributing ideas and insight.

Manuscript received May 26, 1987
Revised manuscript received September 11, 1987
Manuscript accepted September 23, 1987

Reprinted by permission of the Geological Society of America
from J. S. Tchalenko, *Geological Society of America Bulletin*,
v. 81 (1970), p. 1625-1640.

J. S. TCHALENKO   *Imperial College of Science and Technology, London S.W.7., England*

Similarities between Shear Zones of Different Magnitudes

ABSTRACT

An examination is made of the formation and development of shear zone structures on (1) the microscopic scale in the shear box test, (2) an intermediate scale in the Riedel experiment, and (3) the regional scale in the earthquake fault. On the basis of the resistance to shear, three structural stages are chosen for detailed study: the peak structure occurring at peak shearing resistance, the post-peak structure occurring after peak shearing resistance, and the residual structure occurring at residual shearing resistance. Most of the similarities in structure between the different scales at each of these stages are interpreted in terms of the mechanical properties of the material, the Coulomb failure criterion, and the kinematic restraints inherent in the type of deformation. Other similarities which are not as yet understood are described and suggested as topics for future research.

INTRODUCTION

Most soils and rocks, when deformed in direct shear, develop narrow shear zones within which the major displacements take place. Depending on the scale of the deformation and on the amount of material involved, these shear zones may vary in size by many orders of magnitude, ranging from the microscopic deformation bands in crystal aggregates to the regional tectonic faults in the earth's crust. In general, a shear zone at a particular scale is formed by a system of "shears" on a lower scale. The pattern of these shears is referred to here as the "shear zone structure."

Similarities between shear zones of different scales have been recognized since the earliest days of structural geology and used in model studies of tectonic processes. One of the models most frequently cited in this connection is the Riedel experiment. This consists of a slab of "plastic" material, placed horizontally over two adjoining boards, one board being then slowly displaced in a horizontal sliding motion past the other. When, as is often the case, the material used is a clay paste, the slab usually measures a few millimeters to a few centimeters in thickness, and from one to several decimeters in length. The characteristic en echelon shears formed at an angle to the board's interface during the early stages of the deformation were first investigated in detail by Cloos (1928) and

Riedel (1929)[1]. The en echelon shears and their conjugates are commonly referred to as "Riedel shears" (or "Riedels") and "conjugate Riedel shears" (or "conjugate Riedels") (for example, Hills, 1963). They are denoted by R and R', respectively, in Figure 1. The width of the shear zone and the length of each individual shear in the experiment may vary from a few millimeters to a few centimeters, the exact dimensions depending on the properties of the material used and on the thickness of the slab.

Riedel shears have also been observed on a smaller scale in clay samples subjected to direct shear in the shear box. The inner dimensions of the conventional apparatus are 6 x 6 x 4 cm, and the width of the shear zone and the length of individual shears vary in this case from a few microns to a few millimeters. Tests by Hvorslev (1937) were the first to reveal the en echelon structure, and Morgenstern and Tchalenko (1967a) observed that this structure prevailed on all scales down to the magnification limit of the optical microscope. Microscopic Riedel shears were also found in the failure zones of landslides in clays (Morgenstern and

---

[1] Fujiwhara (1924, 1925) performed similar experiments, but does not describe the details of the shear zone structure. Fath's model (Fath, 1920) is also of the same type, even though probably without boards, but the structures illustrated are mainly tension gashes. See also Brown (1928) for experiments on materials other than clay.

Geological Society of America Bulletin, v. 81, p. 1625–1640, 12 figs., June 1970

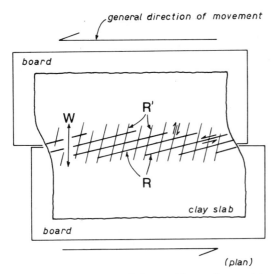

Figure 1. Diagram of the Riedel experiment. (R) Riedel shear. (R′) conjugate Riedel shear. (W) width of shear zone.

Tchalenko, 1967b), and their similarity with the shear box structures was shown to reflect a similarity in the deformation mechanism (Tchalenko, 1968).

On a much larger scale, the en echelon arrangement of fractures in shear zones produced by strike-slip ground movements associated with earthquakes is fairly well documented in the literature on earthquake effects. In this case, movements along faults situated in the bedrock deform the sedimentary overburden or weathered rock and cause fracturing at the ground surface. The shear features are of the order of meters and hundreds of meters in length. The movements are known with much less precision than on the smaller scales, but the similarity of the structures with the Riedel and shear box cases suggests, at least for the simpler faults, a similar deformation mechanism (Tchalenko and Ambraseys, 1970).

This paper examines the extent of the structural similarity between shear zones of different magnitudes. Some aspects of this similarity are explained in terms of the mechanical properties of the materials involved. Other aspects, however, remain as yet unexplained, and they are presented here as observations. As particular structural features are more pronounced at some magnifications than at others, the simultaneous study of shear zones on more than one scale leads to a better understanding of the general deformation mechanism.

FORMATION AND DEVELOPMENT OF SHEAR ZONES

The method of study and the most important results obtained at each scale are summarized in this section. For further details of the experimental procedure, the reader is referred to the previous work mentioned in the text.

Riedel Experiment

In the Riedel experiment, briefly described in the Introduction, the clay adheres sufficiently to the boards to insure that the movement is transmitted to the slab (Fig. 2). The sample is unconfined at its upper surface, but the capillary forces acting on the free surface give rise to an equivalent ambient pressure throughout the specimen. The adjunction of a proving ring to one of the boards enables the variation of the lateral force T to be measured (Tchalenko, 1967). The force-displacement curve (Fig. 3A) shows a rapid increase of shearing resistance to a maximum value, at which point the first shears appear, then a decrease until a stable

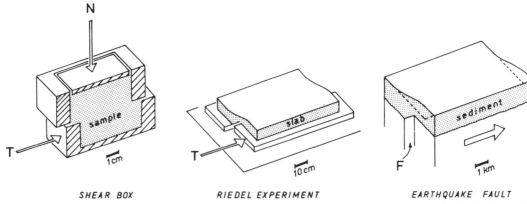

Figure 2. Deformation in the shear box, Riedel experiment and earthquake fault. (N) normal effective force. (T) horizontal shear force. (F) tectonic fault in bedrock.

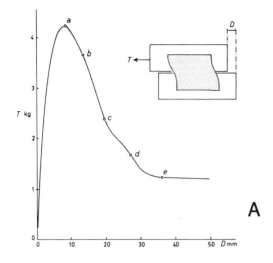

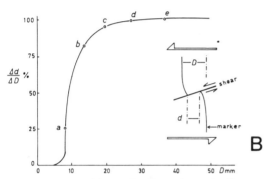

Figure 3. Riedel experiment results on kaolin. (A) force-displacement curve. (B) proportion of total movement taken up by shears. Water content = 56 percent (D) total (board) movement. (T) horizontal shear force. (a,b,c,d,e,) stages in the deformation, also shown in Figure 4.

value is reached at large displacements when the shear zone is fully formed. By analogy with the stress-displacement behavior which will be demonstrated for the shear box, the maximum shearing resistance is referred to as the "peak shear strength," and the final stable resistance as the "residual shear strength."

The appearance and evolution of the shear zone structure may be followed visually or photographically. By observing the development of individual shears and the distortion of markers inscribed on the clay surface, the approximate proportion of the board movement (or total displacement, D) taken up by movement on individual shears (d) can be measured. Figure 3B shows the rapid increase of movement on shears after peak shear strength, a characteristic feature of all clays.

Figure 4 shows the sequence of structures

observed in the experiment, with kaolin mentioned in connection with Figure 3. The structures, drawn on the basis of photographs taken during the experiment, are shown at five stages of deformation. Displacements measured on the shears are plotted in the form of histograms indicating the cumulative amounts of displacement d_i having taken place along shears of different inclinations i. The histograms were constructed by adding the offsets incurred by each marker, and by noting the average inclination of the shears responsible for each offset. Stages a to e in the outline which follows, are common to Figures 3 and 4.

Pre-Peak Strength Deformation. (Not shown in Fig. 4). The initial movement of the boards causes a homogeneous straining in the region of the future shear zone. Circles inscribed on the clay slab are transformed into ellipses, indicating that the deformation is of the simple shear type. No shears are discernible during this stage.

Stage a, Peak Structure. The first shears, the Riedels, appear just before peak shear strength is reached, at an average inclination of $12° \pm 1°$. At peak strength, the Riedels have been bodily rotated to a maximum inclination of about $16°$. During this stage, the proportion of total board displacement accommodated by individual shears increases rapidly from 0 to around 50 percent.

Stage b, Post-Peak Structure. Some Riedels are extended into a more horizontal direction, and a few new shears appear at angles of about $8°$. The proportion of total displacement accommodated by shears attains about 75 percent.

Stage c, Post-Peak Structure. New shears, referred to as the "P shears," are formed at an average inclination of $-10°$, that is, approximately symmetrical to the Riedels. They interconnect pairs of Riedels, at times forming characteristic "bull nose" structures (Skempton, 1966). More than half of the shears are now inclined at $4°$, and nearly all the total movement at this stage is taken up by displacements along shears. This stage is illustrated in Figure 7B.

Stage d, Pre-Residual Structure. The first continuous horizontal shears, the "principal displacement shears," are formed, isolating elongated lenses of essentially passive material between them. Most of the shears are inclined at about 0 to $4°$.

Stage e, Residual Structure. Nearly all displacements take place along a single principal

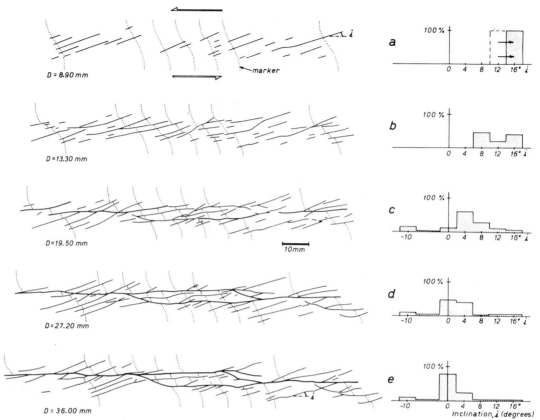

Figure 4. Sequence of structures in the Riedel experiment. (*D*) total (board) movement. (*i*) inclination of shear in degrees with respect to general direction of movement. (Σd_i) cumulative amount of displacements on shears inclined at *i* at each stage of movement. (*a,b,c,d,e,*) stages in the deformation, also shown in Figure 3. For stage *c*, see Figure 7B.

displacement shear superimposed on the interface between the two boards. The shearing resistance is stable, and at its residual value.

When kaolin or other clays are used at lower water contents than in the experiment described above, a second family of shears is observed during stage *a*. These are the conjugate Riedels, inclined at 78° ± 1° for kaolin, and appearing simultaneously with, or just before, the Riedels. Due to the large angle they make with the general direction of movement, the conjugate Riedels soon become passive and distorted into an S shape (Fig. 5). The general characteristics of stages *a* to *e* are however, not modified by the presence of the conjugate Riedels.

Shear Box Test

In the shear box apparatus, the sample is confined between rigid sides and it can thus be tested under larger ambient stresses (Fig. 2). The test is performed by immobilizing the upper half of the box and displacing the lower

half horizontally at a constant rate. The tests described here are performed under drained conditions at a rate of 5 x 10^{-6} cm/sec. The normal effective force *N*, is maintained constant throughout each test, and the horizontal shear force *T* is measured; the normal effective stress σ_n' and shear stress τ are then computed for each stage of the test. The stress-displacement curves and the Mohr failure envelopes for two tests of the same kaolin as used in the Riedel experiments are given in Figure 6. For each normal effective stress, the shear stress increases rapidly at first to the peak shear strength, then decreases gradually to the residual shear strength (Fig. 6A). The peak and the residual angles of shearing resistance, ϕ' and ϕ_r, are obtained from the slopes of the corresponding Mohr envelopes (Fig. 6B).

The structures formed in the shear box may be studied by interrupting tests on initially identical samples at different points of the stress-displacement curve, and by preparing thin sections of the entire sample (Morgenstern

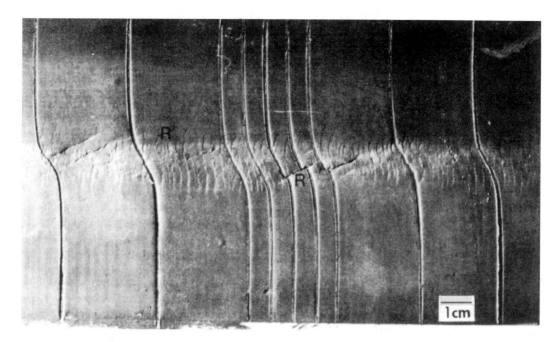

Figure 5. Detail of Riedel experiment structure. Note inhomogeneous strain around a Riedel shear (R), and distortion of conjugate Riedel shears (R'). Thick vertical lines are markers inscribed on the slab surface prior to deformation.

and Tchalenko, 1967a). The increase in particle parallelism occurring in the shears enables them to be distinguished, in polarized light, from the surrounding material. Structural observations are carried out on two different scales: the entire thin section is viewed directly or at a small magnification, and a petrographic microscope is used to examine details at magnifications of up to ×500.

The sequence of structures observed in the shear box test, both on the scale of the entire sample and on the microscopic scale, is essentially the same as the one outlined for the Riedel experiment (*see* Morgenstern and Tchalenko, 1967a, Figs. 5 and 8). Figure 8A shows a longitudinal thin section of a sample which has nearly attained residual shear strength. Both Riedels and conjugate Riedels which appeared at peak shear strength can be seen, as well as some interconnecting *P* shears oriented approximately symmetrically to the Riedels with respect to the general direction of movement. The sample shown is at a stage corresponding to stage *d* of the Riedel experiment when the horizontal principal displacement shears appeared across the whole clay slab.

On a smaller scale, some of the individual shears described above undergo a similar development which is illustrated in Figure 8B for one of the Riedel shears. Each substructure is

typically less than 10 microns wide and formed of strongly parallel particles lying on an average in the general direction of the substructure. An example of such a microscopic shear zone will be described in greater detail in a later section.

Earthquake Fault

Little is known about how ground surface deformations relate to bedrock movements at depth. For the rectilinear segments of the major strike-slip faults, the type of deformation shown in Figure 2 is generally postulated. In these segments it is often found that fracturing associated with earthquakes occurs along traces of pre-existing faults, suggesting that surface fractures are caused by the slipping of pre-sheared blocks of bedrock material.

With earthquake faults it is not possible to follow the structural evolution continuously, as in the Riedel experiment, or even at intermittent stages, as in the shear box test. With the exception of a few rare cases where displacement measurements were carried out immediately after an earthquake (Wallace and Roth, 1967; Smith and Wyss, 1968), the usual data is pertinent to a structure as observed some considerable time after the main shock, that is, when most of the induced movement has already taken place. Even in these cases, however, detailed mapping may reveal segments having

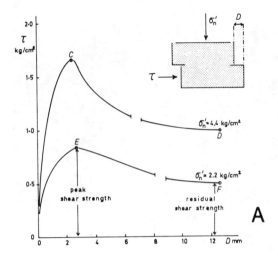

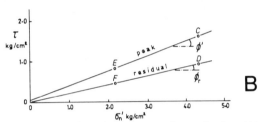

Figure 6. Shear box tests results on kaolin. (A) stress-displacement curves. (B) Mohr envelopes. σ_n' normal effective stress. (τ) shear stress. (D) total displacement. (ϕ') peak angle of shearing resistance (ϕ_r) residual angle of shearing resistance. (The stress-displacement curves are interrupted to show that the sense of movement has been reversed; *see* footnote 6).

undergone different amounts of relative displacements, and under favorable conditions, a time sequence and stages of a structural evolution become evident. Such is the case for the fractures associated with the Dasht-e Bayāz earthquake of 1968, from which most of the examples quoted in this paper are taken. Surface ground movements associated with this earthquake were essentially strike-slip, and over-all vertical displacements were negligible (Ambraseys and Tchalenko, 1969). The ground features, in the form of ridges and cracks, when seen from the air, were found to be concentrated in shear zones which are the equivalent of the shears described for the smaller scales (Fig. 7A). These zones varied in length from several meters to several hundreds of meters, and in width from about a meter to some tens of meters. They combined to form a "principal displacement zone" in which Riedel directions

and some P structures could be recognized. Conjugate Riedels were also sometimes found. The reconstructed structural evolution followed a pattern basically similar to the one observed in the Riedel experiment and shear box test (Tchalenko and Ambraseys, 1970).

STRUCTURAL SIMILARITIES

The relative arrangement of the Riedel, conjugate Riedel and P shears, which constitute the shear zone structure, is seen to be similar in the three deformations described in the preceding section. The extent of this similarity is best illustrated by comparing the Riedel experiment, shear box test, and earthquake fault at three distinctive stages of their structural evolution, the peak structure (stage a), the post-peak structure (stage c) and the residual structure (stage e). The basic shear zone dimensions and relative displacements corresponding to the examples quoted here are given in Table 1.

Stage a, which in the shear box and Riedel experiment occurs just before or at peak shear strength, is shown in Figure 9. The over-all displacements measured at the boundaries are 6 mm in the shear box[2], 8.5 mm in the Riedel experiment, and about 150 cm in the earthquake fault. The conjugate Riedels are the first shears to appear; their larger angle to the general movement direction causes them to be subsequently distorted and rotated by a few degrees. During this stage the en echelon Riedels are still at an early phase of their development. The examples illustrated are taken from cases where the conjugates were particularly well developed.

At increased displacements, P shears interconnect the Riedels at inclinations approximately symmetrical to the Riedels (Figure 10). This is stage c, which in the shear box and Riedel experiment, occurs about half-way between peak and residual shear strengths. A continuous line can now be drawn through the different shears along most of the length of the shear zone. Over-all displacements in the examples illustrated are 5 mm in the shear box, 20 mm in the Riedel experiment, and about 250 cm in the earthquake fault. "Bull nose" structures and wedges are common features at all scales.

---

[2] The example chosen in Figure 9C is, in fact, of a later stage of the stress-displacement curve (hence, the presence of some P shears), but the peak structure formed at about 3 mm displacement is still perfectly retained in the clay.

TABLE 1. BASIC SHEAR ZONE DIMENSIONS AND RELATIVE DISPLACEMENTS

| Type of Shear Zone | Material | Shear Zone | | Shears | | |
		Width (W)	Relative Displacement* (D)	Width	Length	Relative Displacement (d)
Earthquake fault	clayey silt and colluvial material	2–100 m	max 4.5 m	1–10 m	meters to hundreds meters	≤ 4 m
Riedel experiment	kaolin	5–20 mm	~30 mm	≤ 1 mm	≤ 50 mm	≤ 30 mm
Shear box test (entire sample)	kaolin	20–500 μ	~10 mm	20–200 μ	≤ 30 mm	≤ 10 mm
Shear box test (substructure)	kaolin	20–200 μ	≤ 10 mm	≤ 10 μ	~1 mm	

*for shear zone structure to be fully formed (stage d and e)

At stage e, corresponding to the residual shear strength in the laboratory cases, the shears formed in the previous stages have undergone small modifications to accommodate larger movements (Fig. 11). The over-all displacement, 8 mm in the shear box, 30 mm in the Riedel experiment, and about 300 cm in the earthquake fault, is now primarily concentrated in one or sometimes in two parallel principal displacement shears. Elongated wedges are formed between them in the general direction of movement.

The strong similarities between the structures of different magnitudes are evident from Figures 9 to 11. The examples were, of course, chosen to emphasize these similarities, but segments of the shear zone chosen at random invariably show the trends described above.

INTERPRETATION

The shear zones described in the previous sections occurred in soils undergoing active deformations which could be followed visually or reconstructed with some degree of confidence. Many fossil shear zones, that is, shear zones formed in the past but now inactive, also display substructures of the same type (see, for example, Norris and Barron, 1968; Skempton, 1966; Bishop, 1968), but their interpretation depends essentially on the accurate knowledge of the forces having generated these features. Such a knowledge is rarely available from sources independent of the structural analysis. Indeed, substructures and small scale structures often constitute the only available clue to the deformation mechanism. In contrast, some of the stress and strain characteristics of the simpler active shear zones are known, and the study of these shear zones may thus provide

a basis for the structural analysis of fossil shear zones.

Our knowledge of the deformations considered here varies with each case. In the Riedel experiment and shear box test, the over-all strain rate is controlled and measured; in the case of earthquake faulting, rates of movement are usually unknown, but there is growing evidence that ground displacements take place over a period of time well in excess of the duration of the main bedrock rupture (Scholz and others, 1969). Over-all normal and tangential stresses in the shear box can be measured, whereas the horizontal force in the Riedel experiment can only be used to indicate relative values of the resistance to shear. For the earthquake fault, the stresses responsible for the surface fractures are largely unknown and not directly deducible from microseismic data. The structural interpretation which follows is therefore mainly related to the shear box and Riedel experiment cases, with occasional reference to earthquake fractures.

Peak Structure

The distortion of the markers in the Riedel experiment and of the initial fabric in the shear box tests indicates that the homogeneous strain which characterizes the ascending part of the stress-displacement curve can be considered as a simple shear deformation. This stage is followed by the Riedel and conjugate Riedel shears, which appear just prior to peak strength, regardless of the type of clay or its water content (Tchalenko, 1967).

The failure condition for the majority of soils can be adequately expressed by the Coulomb failure criterion

$$\tau = c' + \sigma_n' \tan \phi' \qquad (1)$$

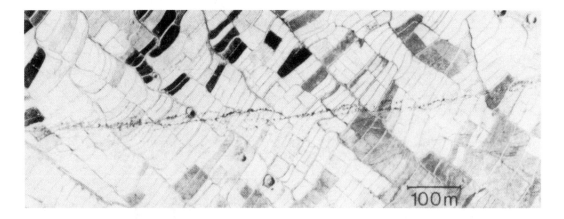

A

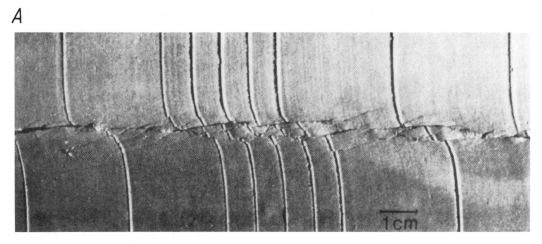

B

Figure 7.   Shear zones of different magnitudes. (A) Dasht-e Baȳaz earthquake fault: total displacement = 250 cm (*see also* Fig. 10A). (B) Riedel experiment on kaolin: total displacement = 19.5 mm (*see also* Fig. 4C).

where c' is the cohesion intercept, ϕ' the peak angle of shearing resistance, and σ_n' and τ are, respectively, the effective normal stress and shear stress acting across the failure surface. The Coulomb criterion also predicts that the directions of the failure surfaces with respect to the major principal stress σ_1 are given by

$$\beta = 45° - \phi/2 . \qquad (2)$$

Hvorslev (1937), Peynircioglu (1939) and Gibson (1953) showed that the inclination of failure surfaces in the triaxial test are in closer agreement with the Coulomb prediction if the "true angle of friction" ϕ_e (Hvorslev, 1936, 1937) is used instead of ϕ' in equation (2). The difference $\phi' - \phi_e$ varies from a maximum of about 10° for highly active clays such as bentonites, to about 0° for kaolins and coarser minerals. The deformation being of the simple shear type, σ_1 is taken to be at 45° to the general direction of movement,[3] and the failure surfaces should be inclined at $\phi_e/2$ and $90° - \phi_e/2$ to the movement.

The value of ϕ_e for kaolin is about 23°. Average inclinations of the Riedels and conjugate Riedels in the shear box were observed to be, respectively, 12° and 80° − 85°, hence in reasonable agreement with the directions predicted by the Coulomb criterion interpreted in terms of the true angle. Riedel experiments on kaolin as well as on clays of different ϕ_e, such as London Clay ($\phi_e = 13°$) and Wyoming Bentonite ($\phi_e = 2.5°$), showed the same relationship between peak structure inclinations and the true angle of friction (Table 2). Thus,

[3] This can be demonstrated in the Riedel experiment by flooding the slab with a thin film of water. The surface tension is in this way eliminated, and the clay fails in tension with the formation of open gashes at 45° to the movement.

A

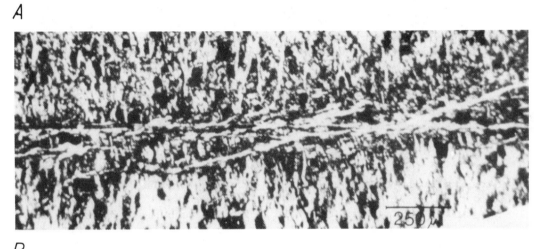

B

Figure 8. Shear zones of different magnitudes. (A) shear box test on kaolin, entire sample: total displacement = 8 mm (*see also* Fig. 11C). The NW-SE hairline cracks are due to slight shrinkage associated with thin sectioning procedure. (B) shear box test on kaolin, detail: estimated total displacement = 3 mm.

the relationship (2), already known to hold for triaxial stress conditions, is also verified for the peak stage in the shear box test and in the Riedel experiment. Detailed mapping of the Dasht-e Bayaz earthquake fractures has shown that this relationship holds reasonably well for the larger scale features, providing that an over-all angle of shearing resistance between $35°$ and $40°$ was adopted for the material. This value of ϕ' was considered to be a good approximation for the type of overburden sediment deformed by faulting (Tchalenko and Ambraseys, 1970).

Post-Peak Structure

The post-peak structure is associated with the decrease in shearing resistance from the peak to the residual strength value. The Riedels and conjugate Riedels created at the peak stage are both unfavorably oriented to sustain large

relative displacements, and further straining and shearing must take place to accommodate increased over-all movements. The conjugate Riedels in particular, being nearly at right angles to the general movement, respond passively to post-peak deformations by rotation and distortion (Fig. 5). Likewise, the Riedels extend into undeformed material on either side of the shear zone where their relative displacements, from a maximum in their central portion, are rapidly reduced to zero. Markers engraved on the clay slab of the Riedel experiment and truncated by the Riedels illustrate this inhomogeneity in strain (Fig. 5). The same process is also characteristic of this stage on the microscopic and earthquake fault scales (Morgenstern and Tchalenko, 1967a, Fig. 13; Tchalenko and Ambraseys, 1970, Fig. 10).

The new structures formed as a consequence of this kinematic restraint are the *P* shears

TABLE 2. OBSERVED AND PREDICTED INCLINATIONS OF SHEARS AT PEAK STRENGTH IN THE RIEDEL EXPERIMENT

Type of Clay	ϕ_e	Water Content %	Type of Shear	Observed Inclination	Predicted Inclination
Kaolin	23°	45, 50, 54, 56	R	12° ± 1°	11.5°
(LL = 60, PL = 24)			R'	78° ± 1°	79.5°
London Clay	13°	40 and 62	R	7.5° ± 1°	6.5°
(LL = 79, PL = 34)					
Wyoming B Bentonite	2.5°	150 and 167	R	2° ± 1°	1°
(LL = 579, PL = 45)					

LL—Atterberg liquid limit.　　PL—Atterberg plastic limit.

oriented approximately opposite to the Riedels with respect to the general direction of movement.[4] For the shear box, it was postulated that the formation of the P shears involved the following two processes: (1) a reduction of the shearing resistance along the Riedel shears toward the residual strength value;[5] and (2) a local increase and rotation of the principal stresses in the sense opposite to that of the general movement. Further work is needed to show whether a similar mechanism can also be postulated for the larger scale features.

Residual Structure

The combination of displacement along the Riedel and P shears leads to the formation of the principal displacement shears oriented in the general direction of movement. Simultaneously, the active portion of the shear zone decreases in width until all the movement is concentrated in a very thin zone or on a single slip surface.[6] At this stage, the deformation approximates direct shear conditions, and the shear stress τ_r necessary to cause sliding is given by:

$$\tau_r = c_r' + \sigma_n' \tan \phi_r \qquad (3)$$

where c_r' is the residual cohesion intercept, ϕ_r the residual angle of shearing resistance, and σ_n' the effective normal stress acting on the slip

surface. For clays and soft rocks, c_r' is generally small or nil. The major principal stress is now predicted at $45° - \phi_r/2$ to the direction of the slip surface, which is also the general direction of movement, and the microscopic substructure of the shear zone seems to confirm this prediction (Tchalenko, 1968).

"Riedel Within Riedel" Structure

It was seen that some of the individual structures in the shear box, when viewed under the microscope, presented, on a smaller scale, a further structural arrangement basically similar to the one governing the main structure. An example of this structure is given in Figure 12B. ZZ is a Riedel shear formed at peak and crossing a good portion of the entire sample. Under the microscope it is observed that this shear consists of a system of shears ss oriented in the Riedel direction with respect to ZZ. Preliminary results on the electron microscope indicate that these shears are formed by parallel clay particles lying approximately in the ss direction. At this ultimate scale, the deformation mechanism is primarily an interparticle basal plane slip in the direction of the structure.

Similar cases of this structural arrangement were found on the larger scales. Figure 12A shows a segment ZZ of the principal displacement zone of the Dasht-e Bayaz earthquake fault. The segment is in the Riedel attitude with respect to the whole fault trace on the regional scale. At greater magnifications, it is observed that the segment is formed of individual Riedel shears such as ss oriented in the Riedel direction, this time with respect to ZZ. At an even greater magnification, for example as seen by an observer on the ground, these Riedel shears are composed of ridges and cracks, the latter (denoted c) being short narrow openings arranged in an en echelon manner and indicating a tensional formation mechanism.

In both these examples, a Riedel structure

---

[4] These structures have been called "thrust shears" by Skempton (1966) and Tchalenko (1968), and "restraint shears" by Morgenstern and Tchalenko (1967a). Both terms invoke specific mechanisms which, although reasonable for the scale considered in each case, have not yet been proved to be general to all scales. The notation P was adopted by Skempton (1966) to indicate the probability that the material affected by these shears was in the "passive" Rankine state.

[5] This is known to occur along structural discontinuities in stiff clays (Skempton and Petley, 1967).

[6] For many soils, the displacements permissible in the conventional shear box are not sufficient to attain in one traverse the residual stage. Special testing techniques are then required (Skempton, 1964).

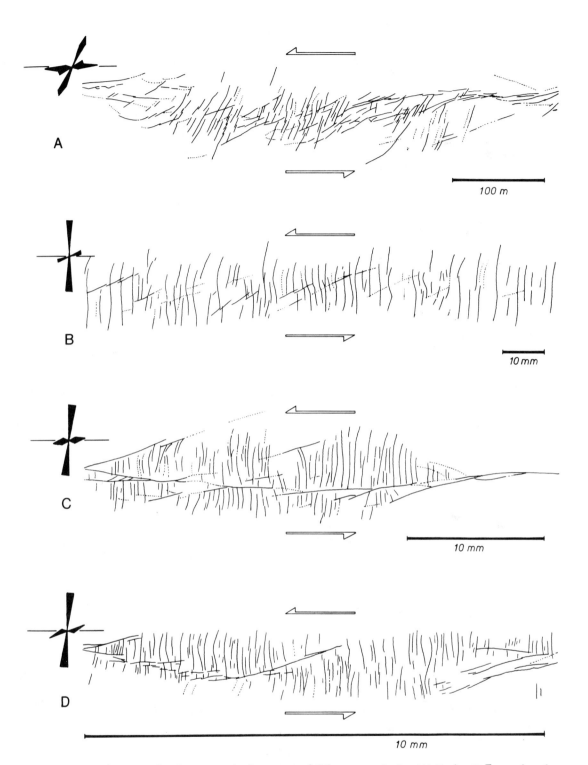

Figure 9. Comparison of peak structure in shear zones of different magnitudes. (A) Dasht-e Bayaz earthquake fault (*after* Tchalenko and Ambraseys, 1970, Fig. 8). (B) Riedel experiment (*after* Tchalenko, 1967, Fig. 132). (C) entire shear box (*after* Morgenstern and Tchalenko, 1967a, Fig. 5). (D) detail of shear box sample (*after* Morgenstern and Tchalenko, 1967a, Fig. 12). Total displacements are given in the text. Dotted lines indicate less prominent shears. The structures plotted in the form of rose diagrams show Riedel and conjugate Riedel directions.

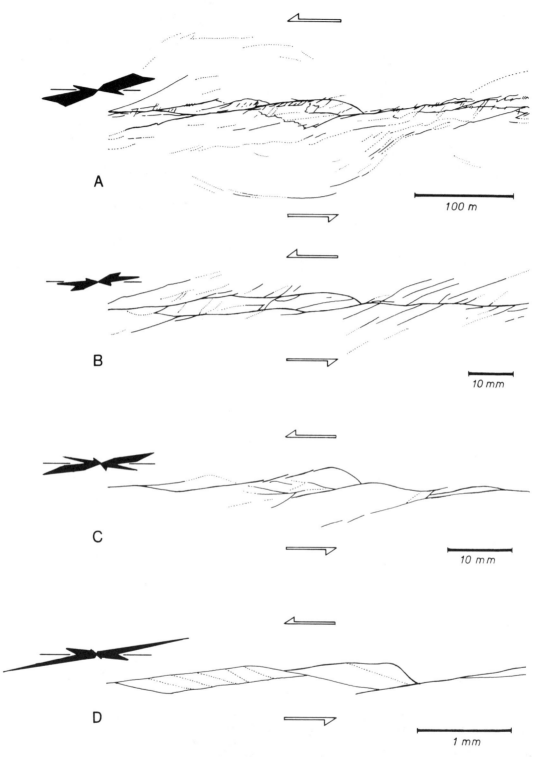

Figure 10. Comparison of post-peak structure in shear zones of different magnitudes. (A) Dasht-e Baȳaz earthquake fault (*see also* Fig. 7A). (B) Riedel experiment. (C) entire shear box sample. (D) detail of shear box sample (*after* Tchalenko, 1968, Fig. 5). Total displacements are given in the text. Dotted lines indicate less prominent shears. The structures plotted in the form of rose diagrams show Riedel and *P* shear directions.

contained within it similar Riedels on a smaller scale. This resolution of a structure into substructures (also known for some cases of kinkbands in foliated rocks: Ramsay, 1962; Anderson, 1968; Dewey, 1969), is a process about which very little is known. In the cases considered here, it seems to be arrested, and an ultimate structure seems to be produced by the emergence of an altogether different mechanism, basal plane slip in the first example and tensional in the second, to the mechanism operating on the larger scale. No interpretation is offered here for these "Riedel within Riedel" structures, the elucidation of which, both in

terms of the movement picture and the stress history, may provide an important element in the understanding of shear zones.

CONCLUSIONS

Three characteristic stages in the evolution of a shear zone were defined on the basis of the stress-displacement behavior and the structures in the Riedel experiment. They are: (1) the peak stage, during which the resistance to shear is maximum and the structures formed are the Riedels and conjugate Riedels; (2) the post-peak stage, during which the resistance to shear decreases, and the structures formed are

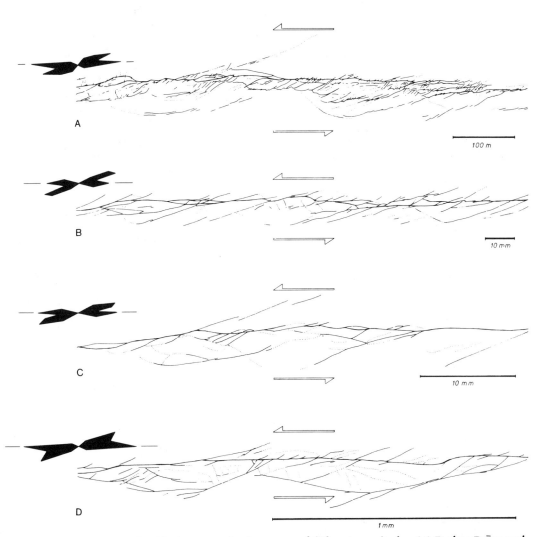

Figure 11. Comparison of residual structure in shear zones of different magnitudes. (A) Dasht-e Bayaz earthquake fault (*after* Tchalenko and Ambraseys, 1970, Fig. 5). (B) Riedel experiment. (C) entire shear box sample (*see also* Fig. 8A). (D) detail of shear box sample. Total displacements are given in the text. Dotted lines indicate less prominent shears. The structures plotted in the form of rose diagrams show Riedel, *P* shear and principal displacement shear directions.

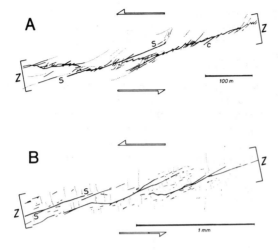

Figure 12. "Riedel within Riedel" structures. (A) Dasht-e Bayaz earthquake fault. (B) detail of shear box sample (*after* Morgenstern and Tchalenko, 1967a, Fig. 9). Over-all trace and general direction of movement in the shear box are in the direction of the sense arrows.

the P shears lying approximately symmetrically to the Riedels with respect to the general direction of movement; and (3) the residual stage, during which the resistance to shear attains a stable though smaller value than that of the peak stage, and the characteristic structure consists of one or several parallel principal displacement shears in the general direction of movement.

These three stages were shown to be common to deformations taking place on the microscopic (shear box), intermediate (Riedel experiment), and regional (earthquake fault) scales. The range in linear scales of the shear zones described is of the order of 10^5. The similarities in structure were interpreted as indicating similarities in the deformation mechanism. At the peak stage, the mechanism is essentially of the simple shear type, at the post-peak stage it is governed by the kinematic restraints inherent in the strain field, and at the residual stage it is of the direct shear type. The Coulomb criterion predicts adequately the peak stage structures.

The relationships between shearing resistance and structural evolution, and between direction of shears and mechanical properties of the material, explain the most perceptible similarities between the shear zones of different magnitudes. There are, however, further similarities which have not been explained, for example, in the relative dimensions of structures and substructures, in the spacing of shears, and in the development of "Riedel within Riedel" structures. The study of these points will require new parameters to be taken into account, such as the degree of homogeneity and the volume of the material participating in the deformation, the variations in local ambient pressures and pore-pressures, and so forth. The fact that these and other parameters undoubtedly vary greatly from one scale to another makes the similarity in structure all the more remarkable, and shows the necessity of continued research into the mechanism of shear zones.

ACKNOWLEDGMENTS

The author acknowledges the support of the Natural Environment Research Council. He is also grateful to Professor A. W. Skempton, F. R. S., and to Doctors N. N. Ambraseys and R. J. Chandler for helpful discussions.

REFERENCES CITED

Anderson, T. B., 1968, The geometry of a natural orthorhombic system of kink-bands, p. 200–220, *in* Baer, A. J., and Norris, D. K., *Editors*, Conference on Research in Tectonics Proc., (Kink-bands and brittle deformation): Ottawa, Geol. Soc. Canada Paper 68–52.

Ambraseys, N. N., and Tchalenko, J. S., 1969, The Dasht-e Bayaz (Iran) earthquake of August 31st, 1968: A field report: Seismol. Soc. America Bull., v. 59, p. 1751–1792.

Bishop, D. G., 1968, The geometric relationships of structural features associated with major strike-slip faults in New Zealand: New Zealand Jour. Geology and Geophysics, v. II, p. 405–417.

Brown, R. W., 1928, Experiments relating to the results of horizontal shearings: Am. Assoc. Petroleum Geologists Bull., v. 12, p. 715–720.

Cloos, H., 1928, Experimente zur inneren Tektonik: Centralbl. f. Mineral. u. Pal., v. 1928 B, p. 609–621.

Dewey, J. F., 1969, The origin and development of kink-bands in a foliated body: Geol. Jour., v. 6, p. 193–216.

Fath, E. A., 1920, The origin of faults, anticlines, and buried "granite ridges" of the northern part of the mid-continent oil and gas field: U.S. Geol. Survey Prof. Paper 128 C, p. 75–89.

Fujiwhara, S., 1924, Torsional form of the earth surface and the great earthquake of Sagami Bay: Jour. Meteorological Soc. Japan, Ser. 2, v. 2, p. 32–36.

——1925, Torsional form on the face of the earth: Japanese Jour. Astronomy and Geophysics Trans., v. 3, p. 103–114.

Gibson, R. E., 1953, Experimental determination of the true cohesion and true angle of internal friction in clays: 3rd Internat. Conf. Soil Mech. and Foundation Engineering Proc., Zurich, v. I, p. 126–130.

Hills, E. S., 1963, Elements of Structural Geology: London, Methuen and Co. Ltd., 483 p.

Hvorslev, M. J., 1936, Conditions of failure for remolded cohesive soils: 1st Internat. Conf. Soil Mech. and Foundation Engineering Proc., Cambridge, v. 3, p. 51–53.

——1937, Über die Festigkeitseigenschaften gestörter bindiger Böden: Ingeniorvidenkabelige Skrifter, Ser. A., No. 35.

Morgenstern, N. R., and Tchalenko, J. S., 1967a, Microscopic structures in koalin subjected to direct shear: Geotechnique, v. 17, p. 309–328.

——1967b, Microstructural observations on shear zones from slips in natural clays: Geotechnical Conf. Oslo Proc., v. I, p. 147–152.

Norris, D. K., and Barron, K., 1968, Structural analysis of features on natural and artificial faults, p. 136–167, in Bear, A. J., and Norris, D. K., Editors, Conf. on Research in Tectonics Proc. (Kink-bands and brittle deformation), Ottawa, Geol. Soc. Canada Paper 68–52.

Peynircioglu, H., 1939, Über die Scherfestigkeit bindiger Böden: Berlin, Springer-Verlag, Degebo no. 7.

Ramsay, J. G., 1962, The geometry of conjugate fold systems: Geol. Mag., v. 99, p. 516–526.

Riedel, W., 1929, Zur Mechanik geologischer Brucherscheinungen: Centralbl. f. Mineral. Geol. u. Pal., v. 1929 B, p. 354–368.

Scholz, C. H., Wyss, M., and Smith, S. W., 1969, Seismic and aseismic slip on the San Andreas fault: Jour. Geophys. Research, v. 74, p. 2049–2069.

Skempton, A. W., 1964, Long-term stability of clay slopes: Geotechnique, v. 14, p. 77–101.

——1966, Some observations on tectonic shear zones: 1st Cong. Internat. Soc. Rock Mech. Proc., Lisbon, v. 1, p. 329–335.

Skempton, A. W., and Petley, D. J., 1967, The strength along structural discontinuities in stiff clays: Geotechnical Conf. Proc., Oslo, v. 2, p. 3–20.

Smith, S. W., and Wyss, M., 1968, Displacement on the San Andreas fault subsequent to the 1966 Parkfield earthquake: Seismol. Soc. America Bull., v. 58, p. 1955–1973.

Tchalenko, J. S., 1967, The influence of shear and consolidation on the microstructure of some clays: Ph.D. thesis, London Univ., 395 p.

——1968, The evolution of kink-bands and the development of compression textures in sheared clays: Tectonophysics, v. 6, p. 159–174.

Tchalenko, J. S., and Ambraseys, N. N., 1970, Structural analysis of the Dasht-e Bayaz (Iran) earthquake fractures: Geol. Soc. America Bull., v. 81, p. 41–60.

Wallace, R. E., and Roth, E. F., 1967, Rates and patterns of progressive deformation, p. 23–40, in the Parkfield-Cholame, California, earthquakes of June-August, 1966: U.S. Geol. Survey Prof. Paper 579.

MANUSCRIPT RECEIVED BY THE SOCIETY AUGUST 18, 1969
REVISED MANUSCRIPT RECEIVED JANUARY 12, 1970

The American Association of Petroleum Geologists Bulletin
V. 72, No. 6 (June 1988), P. 738-757, 16 Figs.

Interpretation of Footwall (Lowside) Fault Traps Sealed by Reverse Faults and Convergent Wrench Faults[1]

T. P. HARDING and A. C. TUMINAS[2]

ABSTRACT

Lowside (footwall) closures sealed by reverse-slip faults and convergent strike-slip faults offer opportunities for significant field extension and new field prospects in basins deformed by contraction. The faults have reverse separation in cross section and transverse closure (in the direction of reservoir dip) is often provided by dip of beds away from the fault at structural upturns. The upturns are common and form at the edge of the footwall block as a consequence of block-edge folding, fault drag, and shortening transverse to fault strike. Effective fault seal and longitudinal closure (parallel to reservoir strike) are the most uncertain trap controls. Fault seal may be provided by the juxtaposition of older, less permeable rocks against the down-dropped reservoir or by impermeable material within the fault zone. Fault-zone barriers to fluid flow include shaly smear gouge, cataclastic gouge, mineral deposits, or asphalt or tar impregnation. Longitudinal closure is most commonly formed by a broad positive warp or bowing at the edge of the footwall block or by stratigraphic reservoir terminations. Secondary faults, intersections of primary block faults, and en echelon folds may also provide longitudinal closure.

Prospects can range in importance from secondary extensions of existing highside closures to large traps unrelated to hanging-wall structure. The variety of geometries, relationships that provide transverse and longitudinal closure, and important geologic parameters that determine fault seal are illustrated with examples from oil fields in Sumatra and southern California. These fields can be used as models for the recognition and delineation of prospects in other basins.

INTRODUCTION

Early exploration of a basin typically concentrates on the delineation of anticlinal culminations. Later stages of exploration must consider more subtle secondary structural or stratigraphic plays around the flanks of these folds and in other areas. A synthesis of producing fields in several basins deformed by shortening reveals that upturned beds on the lowside (footwall) of faults with reverse profile separation may offer opportunities for large undiscovered reserves. The Ventura and Los Angeles basins of southern California and the Central Sumatra basin are used to illustrate our conclusions because the investigation was helped by our past knowledge of these basins (e.g., Harding, 1973, 1974, 1983) and because of the wealth of published field studies available for the California areas. Examples from the latter are also chosen because they demonstrate a wide variety of hydrocarbon trapping geometries. The models of prospects drawn from the synthesis include potential new fields in concealed structural settings (Figure 1a, b), and either lateral extensions (pools at left margin of Figure 1c, d) or deeper pools at existing fields (Figure 1a; hanging-wall pool omitted). The upturn of footwall strata provides the basic closure geometry, and in the basins studied these upturns are common. The upturns can be anticipated elsewhere as natural and predictable results of several types of contractional deformation (Figure 2).

Exploration experience demonstrates that the fault closures are commonly secondary in number and total reserves to highside fold culminations, but sometimes the fault traps can have very significant reserve volumes. Our study also suggests that the risk of finding an effective lowside fault trap must be higher than that for traps associated with highside fold culminations.

The structural styles in which most of the productive footwall fault traps have been recognized are contractional fault blocks (Figure 2a, b) and convergent wrench zones (Figure 2c). Geometrically similar traps may be present in fold-thrust belts (structurally similar to Figure 1a), but productive examples are much less common here possibly because footwall upturns are less likely to be an integral part of the deformation (Suppe, 1983; Suppe and Medwedeff, 1984), exploration for footwall traps in these settings has been limited, and subthrust closure is often more difficult to predict in fold-thrust belts.

Harding and Lowell (1979) have presented overviews of these three styles, and their examples can be used as general guides for interpreting the structural frameworks within which the footwall traps can occur. The distinctive map patterns of the contractional fault-block style in the Wyoming foreland have been presented by Prucha et al (1965). Lowell (1985) has also discussed various interpretational and genetic models for the fault-block style, mostly from a cross-sectional viewpoint. Seismic examples and the implications of well control were described

[1]Manuscript received, April 13, 1987; accepted, February 24, 1988.
[2]Exxon Production Research Company, P.O. Box 2189, Houston, Texas 77252-2189.

This investigation was conducted for Exxon Production Research Company, and grateful appreciation is extended to this company for its permission to publish. K. T. Biddle, T. R. Bultman, L. H. Fairchild, S. H. Lingrey, P. S. Koch, D. W. Phelps, and R. C. Vierbuchen reviewed the manuscript. L. H. Fairchild made helpful suggestions concerning its organization. Appreciation is also extended to Exxon Company, U.S.A., for their release of the seismic profile across the Palos Verdes fault.

by Bally (1983) and Gries (1983), respectively. Gries (1983) discussed the results of subthrust exploration at major foreland block uplifts in the Rocky Mountains. The structural style of convergent wrench zones has also been presented by Wilcox et al (1973) and by Sylvester and Smith (1976). Examples of footwall traps within one convergent wrench system were discussed in general terms by Harding (1974). The geometries of fold-thrust belt structures have been documented by numerous authors (e.g., Boyer and Elliott, 1982). Other relevant discussions of style are referenced in these works.

In the synthesis that follows, we do not attempt to present a comprehensive review of the three styles nor a precise description of the great variety of footwall structures observed. Instead, we document the most common relationships that relate directly to the recognition and delineation of fault-sealed hydrocarbon traps. Examples of producing traps illustrate our conclusions and can be used as models for the development of drillable prospects in other areas. We first discuss the deformational geometries that form the footwall fault traps, the common features that provide dip and strike closure, and the controls of effective fault seal. Producing fields from the Ventura basin demonstrate these trap controls in settings dominated by reverse-slip and overthrust faults. An oil field from the Central Sumatra basin illustrates a trap sealed by a reverse-separation fault thought to have had convergent strike-slip. Two producing trends in the Los Angeles basin, California, document lowside traps at reverse-slip faults formed on the flanks of en echelon folds and additional traps sealed by major wrench faults.

The cross sections and maps of producing fields used to illustrate trap geometries have been accepted without modification from the sources credited with each figure. The examples of California oil fields are taken from authors of the California Division of Oil and Gas and their work is a significant contribution to our understanding of hydrocarbon traps. References to more detailed maps and cross sections by these authors are given in the California Division of Oil and Gas (1974) compilation of California field data. The structural interpretations in all three basins are mostly derived from well control, which is relatively dense. Well-location maps are also available in separate field studies referenced in California Division of Oil and Gas (1974). Seismic reflection profiles are another source of structural control and are particularly comprehensive at the Viking graben examples.

One of the strike-slip systems, the Whittier fault trend (Figure 1c), is notable for the abundance of oil seeps along faults. This occurrence raises the possibility that the faults may act both as the seal and as a migration pathway. Large hydrocarbon columns, nevertheless, are trapped along the edge of the footwall block and the Whittier trend fields are included because they illustrate additional types of closure geometries.

Fault seal aspects of the producing reservoirs were evaluated using the static geometric criteria outlined by Smith (1966, 1980). These criteria state that a fault is considered to be laterally sealing (i.e., prevents leakage across the fault zone) where a hydrocarbon-bearing res-

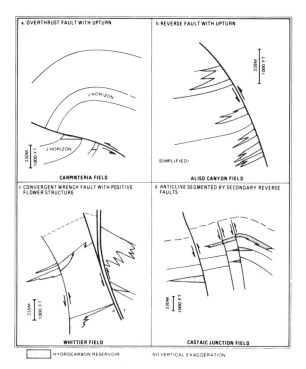

Figure 1—Profile characteristics of typical footwall (lowside) traps at faults with reverse separation. Updip closure is provided by fault seal in (a), (b), and left margin of (d), and mostly by reservoir pinch-out in footwall blocks at (c). (d) is secondary fault trap on flank of larger fold closure. Displacements may be either reverse or convergent strike-slip. Cross sections generalized from productive oil fields in California basins (after California Division of Oil and Gas, 1974). Hanging-wall accumulation in (a) is omitted. T = displacement toward viewer, A = displacement away from viewer.

ervoir lies against a water-bearing reservoir, or where the hydrocarbon-water contacts within juxtaposed reservoirs are different across the fault. The fault zone is considered to be laterally sealing because some type of lower permeability rocks must be present along the fault to inhibit lateral fluid communication. Where a hydrocarbon reservoir is juxtaposed against nonporous or porous but nonpermeable rock, the fault may be a barrier to hydrocarbon leakage simply because of the lithology of the juxtaposed rocks. For the purposes of our closure descriptions, however, the fault plane still is described as laterally sealing.

The vertical sealing capability of a fault (i.e., its ability to prevent hydrocarbon leakage upward along the fault surface) is determined by the permeability of the top seal bed at the fault and any fault zone material. The presence of a finite buoyant column of hydrocarbon beneath a faulted top seal is taken as evidence that some capacity for vertical seal is present along the faults.

The static geometric appraisal of fault seal is limited because dynamic phenomena such as the relative rates of migration into a trap vs. rates of leakage out of the trap are not considered. Faults and other possible seals also

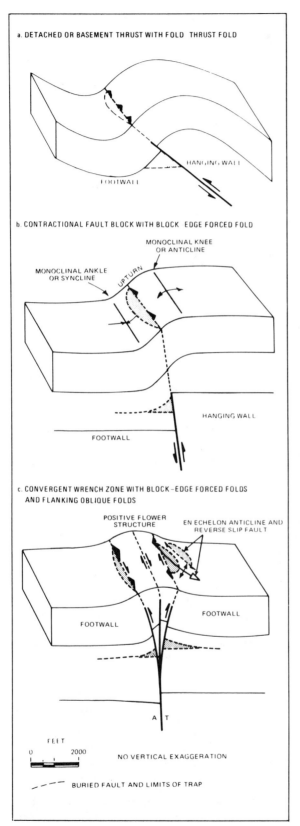

a. DETACHED OR BASEMENT THRUST WITH FOLD THRUST FOLD

HANGING WALL

FOOTWALL

b. CONTRACTIONAL FAULT BLOCK WITH BLOCK EDGE FORCED FOLD

MONOCLINAL KNEE
OR ANTICLINE

MONOCLINAL ANKLE
OR SYNCLINE

UPTURN

HANGING WALL

FOOTWALL

c. CONVERGENT WRENCH ZONE WITH BLOCK-EDGE FORCED FOLDS
AND FLANKING OBLIQUE FOLDS

POSITIVE FLOWER
STRUCTURE   EN ECHELON ANTICLINE AND
                   REVERSE SLIP FAULT

FOOTWALL   FOOTWALL

A T

FEET

0   2000

NO VERTICAL EXAGGERATION

- - - - BURIED FAULT AND LIMITS OF TRAP

Figure 2—Deformational geometries that commonly form structural upturns along footwall of faults with reverse profile separation. Cross-sectional characteristics of traps produced by geometries (a), (b), and (c) are approximated in Figure 1a-c; T = displacement toward viewer, A = displacement away from viewer.

are not absolute barriers to fluid flow and, with sufficient time, all seals would eventually leak. Nevertheless, where traps with similar geologic settings are evaluated, we believe this geometric approach is acceptable for comparing the varying potentials for fault seal.

TRAPPING CONTROLS

Formation of Structure

Closures along the footwall block of faults with reverse separation are termed subthrust closures. The most frequently observed subthrust closures lie within the down-dropped limb of synclines or within structural upturns (i.e., one synclinal limb) that commonly are present at the edge of the footwall block of contractional fault blocks (Figures 1b; 2a, b) and convergent wrench zones (Figures 1c; 2a, c). Similar upturns can be components of certain structures associated with detached overthrust faults but they may be relatively minor in breadth (Figure 1a) (Suppe, 1983; Suppe and Medwedeff, 1984).

Improved seismic data (Bally, 1983) and deeper drilling at the frontal edges of basement uplifts (Gries, 1983) in the Wyoming foreland have demonstrated that the contractional block faults here have moderate-to-low dips at depth and that basement shortening is the most important deformation component. These relationships, particularly at major basement uplifts, are best illustrated by the fold-thrust model of Berg (1962) (Figure 2a) and have been discussed recently by Lowell (1985). According to Berg's and Lowell's analyses, the structures in the sedimentary cover began as a contractional fold pair and with increased shortening are offset by a thrust fault that breaks through the fold limb joining the anticline and its flanking syncline. The resultant geometry is a footwall upturn that is potentially closed updip by the thrust.

The mechanism of block-edge forced folding (Stearns, 1978) is thought to be a more graphic explanation of several other important characteristics of some block-shaped uplifts. This geometry is illustrated in Figure 2b but with a shallower dipping boundary than originally shown by Stearns (1978, his Figure 1). In this deformation, the fold geometry within the sedimentary cover generally conforms to the shape and orientation of an underlying forcing member, which in Figure 2b is a contractional fault block. The blocklike multidirectional aspect of the fault boundaries, the multidirectional trend of the overlying block-edge paralleling folds, and the distinctive slaplike planar profile of the long back limb (hanging-wall limb in Figure 2b) are all inherited from the underlying fault block. A lowside syncline or monoclinal ankle ("half syncline" or upturn) is present at the base of the block, and a highside rollover or monoclinal knee ("half anticline") develops above the upthrown

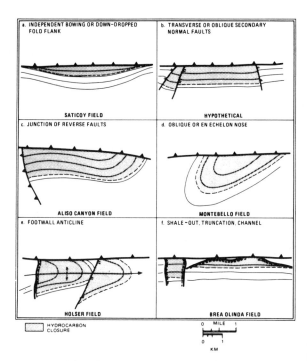

Figure 3—Map geometries of common footwall fault traps at faults with reverse profile separation. Critical controls for longitudinal (strike) closure parallel to sealing fault noted by headlines. Traps similar to (a) and (f) were most frequently observed in our study; (c) and (e) were rare. (e) is fold trap modified by faulting. Displacements may be either reverse or convergent strike-slip. Illustrations derived from productive fields in Sumatra (Crostella, 1981; Villarroel, 1985) and California (after California Division of Oil and Gas, 1974). Limits of two different reservoirs facies are shown in (f).

edge of the block. The fault may ultimately offset the limb connecting the fold pair, and because of its structural position between the folds, is optimally located to provide updip closure within the footwall upturn. This deformational geometry may apply best to trap-door blocks and dogleg structures similar to those in the Pryor Mountains and northern Bighorn Mountains of Montana and Wyoming (Harding and Lowell, 1979, their Figure 8; Lowell, 1985).

A more complex deformation geometry is present at convergent wrench faults. Bending and braiding in the map trace of the wrench fault can cause convergence between fault slices during strike slip (Crowell, 1974). On a larger scale, regional contraction usually accompanies convergent strike slip and also contributes to shortening across the fault slices. In either case, the shortening may be partially accommodated by an upturn within beds at the edge of the footwall block and possibly by the development of a positive flower structure (Figure 2c) (Harding, 1985). The main fault strands in this wrench style commonly have reverse separation in cross section and the resulting footwall structure can have geometric profile characteristics that resemble reverse-slip faults. Ideally, however, the dip of the strike-slip faults is

expected to steepen at depth (Crowell, 1974), whereas the reverse-slip faults described previously are mostly planar or flatten with depth (Bally, 1983; Gries, 1983; Lowell, 1985).

At convergent wrench systems, contraction generates en echelon or parallel folds along the side of some fault zones (Wilcox et al, 1973; Harding, 1976; Aydin and Page, 1984). Individually, these structures have the same geometry as folds formed by other contractional deformations and include basement involved (Schoellhamer and Woodford, 1951; California Division of Oil and Gas, 1974) and detached (Morse and Purnell, 1987) fold types. The distinguishing characteristics of the fold set are the clustering of folds along the wrench fault and their en echelon arrangement. The folds generally plunge away from the wrench (Harding, 1974, 1976), offering opportunities for structural noses sealed obliquely up plunge against the footwall of the wrench fault (crest of en echelon anticline, Figure 2c). Where intense shortening has occurred across the fold set, the flank of individual anticlines may be displaced by a reverse-slip fault. The latter faults trend generally parallel to the anticlines and, thus, repeat the en echelon or parallel pattern of the fold set (flank of en echelon anticline, Figure 2c). The down-dropped flank of the anticlines can contain closures geometrically similar to footwall traps in other reverse-slip systems (similar to Figure 2a).

In some settings, an anticlinal axis may form away from the edge of the footwall block and a fold flank will abut the sealing fault in a geometry resembling interference folds (similar to Figure 3e). Alternatively, development of the sealing fault may occur after folding, and the fault may offset the axial plane of an anticline. Additional types of footwall fault traps can result in either case, but these may be only secondarily dependent on fault seal for closure (Figure 3e).

Short, reverse-slip, or reverse strike-slip faults can develop at individual folds in response to continued shortening of the fold profile. The faults are restricted to a single fold and can seal extensions of major fold closures (pool at left margin, Figure 1d). The strike of the faults can be parallel, oblique, or transverse to the fold axis.

Transverse (Dip) Closure

Transverse closure (i.e., closure in a direction perpendicular to the strike of the sealing fault and in the general direction of reservoir dip) in most of the fault traps to be discussed is provided in one direction by dip of the reservoir strata away from the fault and in the other direction by the updip fault surface (e.g., footwall blocks in Figure 1). (Factors controlling the fault's sealing capacity are discussed in a separate section.) At several examples, syndepositional displacements restricted deposition of reservoir rocks mostly to the footwall block. In these instances, a combined structural-stratigraphic trap was produced in which upturned reservoirs pinch out just downdip from the dislocation surface (central block, Figure 1c). In a third type, a footwall rollover provides the major part of the transverse closure (Figure 3e).

Longitudinal (Strike) Closure

Longitudinal closure (i.e., closure in the direction of the trend of the sealing fault and in the general direction of reservoir strike) can result from a greater number of factors but is the most difficult trap dimension to identify. Longitudinal closure may result from the following (Figure 3).

- Broad bowing or warping of the reservoir strata parallel to fault strike.
- Secondary transverse or oblique normal faults.
- Intersection of two or more block faults.
- Dip reversal at oblique or transversely oriented noses.
- Plunge of block paralleling footwall anticline.
- Stratigraphic termination of reservoir.
- Combinations of the preceding or additional factors.

The most common type of longitudinal closure is a broad positive basement warp at the edge of the footwall block (Figure 3a). The warp may be the down-dropped flank of a highside anticlinal culmination or may form independently of the geometry of the hanging-wall structure.

In some instances, longitudinal extension appears to have occurred parallel to the fault zone, perhaps in compensation of the shortening across the footwall upturn. The secondary transverse or oblique normal faults that result are restricted to the down-dropped block and can be another important source of strike closure (Figure 3b).

The reverse faults that bound contractional fault blocks usually have several orientations. These include longitudinal (parallel to bed strike) and two main oblique directions oriented to either side of the longitudinal faults (Harding and Lowell, 1979). The faults abut at junctions that have a variety of patterns that can provide both updip and longitudinal closure (Figure 3c).

The dip reversal at anticlines that trend oblique or transverse to the sealing fault will provide closure parallel to the trace of that fault (Figure 3d). Obliquely oriented folds are common in the structural assemblages that accompany faults with reverse separation. In the contractional fault-block style, a fold associated with one fault direction can obliquely intersect a block fault that has a second orientation (similar to Figure 3d, fault associated with the oblique fold would parallel the fold axis and is buried below level of this figure) (Prucha et al, 1975; Harding and Lowell, 1979). The en echelon folds and strike-slip faults of the convergent wrench-fault style have a similar oblique intersection. Figure 3d in this context would illustrate part of a right-lateral wrench system (Wilcox et al, 1973).

The plunge of displaced anticlines or other folds lying within and parallel to the edge of the footwall block will also provide strike closure parallel to the trace of the sealing fault (Figure 3e). Folds or faults superimposed by several tectonic events can form more complex interference patterns in which structures of different ages cross.

Shale-out, erosional truncation, and other permeability barriers within the reservoir interval are additional effective sources of strike closure (Figure 3f). In the examples studied, the stratigraphic closure most com-

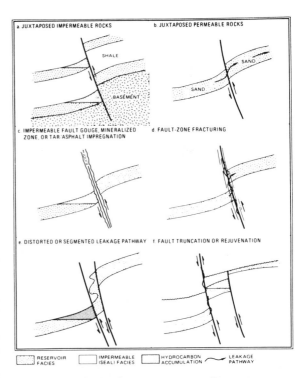

Figure 4—Schematic cross sections that illustrate important geologic controls of fault seal at footwall traps. Juxtaposed lithology may either (a) enhance seal or (b) provide a pathway for leakage. Fault-zone materials or deformation can also either enhance fault zone impermeability (c, e) or permeability (d), and different displacement histories can have similar effects (f). See discussion in text.

monly occurs in combination with footwall bowing and transverse normal faulting. The stratigraphic changes (and secondary faults) can also provide the up-plunge closure at footwall anticlines that trend parallel to the block fault (Figure 3e). More complicated trap configurations are possible from other combinations of the structural and stratigraphic controls.

Fault Seal

For an effective fault trap, the boundary fault zone or the juxtaposed lithologies must inhibit leakage both up and across the fault. Several geological variables are recognized that control fault seal (Figure 4a, c, e, f) and fault leakage (Figure 4b, d, f). The capacity for sealing provided by these factors has finite limits (Weber et al, 1978; Schowalter, 1979). If these limits are exceeded by the volume of the closure, additions of hydrocarbons can escape along the sealing fault until equilibrium is reached (Schowalter, 1979). Our general conclusions are summarized below from the published studies of other authors and from our own studies.

The nature of the rock type faulted against the reservoir is the single most important seal parameter (Figure 4a, b) (Smith, 1966, 1980). In lowside closures, stratigraphically older rocks of the hanging wall are typically juxtaposed against younger reservoirs of the footwall block (Figure 2). In very general terms, the hanging wall rocks may be less permeable because of their longer compaction and cementation histories. If shale on the highside is juxtaposed against the lowside reservoir, lateral seal is probable, but seal up the fault surface needs to be evaluated independently (upper reservoir, Figure 4a) (Smith, 1966, 1980; Weber and Daukoru, 1975; Seeburger, 1981). If displacements juxtapose basement rock against down-dropped reservoirs, the possibility of lateral fault seal is also increased where the basement is not densely fractured (lower reservoir, Figure 4a). Where reservoir facies are juxtaposed across the fault, the chance of lateral leakage is high (Figure 4b) (Weber and Daukoru, 1975; Smith, 1980). Impermeable gouge or mineral deposits can be present within the fault zone, however, increasing the potential for lateral seal (Figure 4c) (Smith, 1980; Seeburger, 1981). In all cases, the effects of the juxtaposed lithology are best appraised by constructing a cross section along the trace of the potentially sealing fault (Allan, 1980).

The presence of cataclastic gouge has been documented along many reverse-slip and strike-slip faults (Higgins, 1971; Engelder, 1974; Logan, 1981; Robertson, 1982). This type of gouge normally has very low permeability and, therefore, can provide an effective fault seal (Friedman, 1969; Smith, 1980; Seeburger, 1981; Pittman, 1981). Factors that favor formation of cataclastic gouge include large displacements (Robertson, 1982), relatively brittle rock types (Engelder, 1974), and nonplanar fault surfaces (Higgins, 1971). Also, rock cement (Smith, 1980; Seeburger, 1981) or asphalt/tar deposited along the fault surface can form a seal after the initial flow of material into the fault (Figure 4c). Asphalt is believed to seal several lowside accumulations in the Los Angeles and Ventura basins (A. C. Tuminas and C. A. Dengo, 1984, unpublished field work).

Some fault zones are composed of multiple fault strands. Potential pathways for leakage can be greatly distorted or cut off entirely by drag folding and by the complex mixing of permeable and impermeable rock types between the strands (Figure 4e) (Seeburger, 1981). Multiple fault slices are probably most common in wrench zones (Dibblee, 1967; others).

The factors most likely to increase the incidence of leakage up the fault are permeable gouge, fractures, and fault rejuvenation. Gouge formed by smearing the displaced lithologies along the fault surface often results from displacements having the characteristics of a ductile deformation. Smear gouge having a large sand fraction is the most prone to promote leakage up the fault (Weber et al, 1978), and shaly smear gouge is the least likely (Weber and Daukoru, 1975; Weber et al, 1978; Smith, 1980).

Fracture systems can parallel or nearly parallel the fault surface and are most prevalent at faults within consolidated rock sequences (Figure 4d) (Seeburger, 1981). The fracture type, whether shear or extension, appears to be critical (Friedman, 1969). Extension fractures are thought to enhance fault zone permeability much more readily than do shear fractures. We infer that in very general terms the possibility of leakage may decrease with older faults because, with time, fractures may gradually

fill with cements and cease to be conduits for fluid migration.

Fault displacements are often renewed during episodic or continuous tectonic activity. Commonly, however, the displacement renewal is selective and is not distributed uniformly throughout the faulted terrain. The variations in displacement histories have influenced fault seal on normal, reverse, and strike-slip faults in the Los Angeles (present study) and Central Sumatra basins (Villarroel, 1985). If faults are old and do not cut high in the section, the likelihood for recent vertical leakage is decreased (fault to right, Figure 4f) (T. P. Harding and A. C. Tuminas, 1987, personal communication). If faults are young or rejuvenated and, particularly, if they extend across unconformities or upward to shallow reservoirs, the chance of leakage up the fault is increased considerably (fault to left, Figure 4f).

VENTURA BASIN EXAMPLES

Basin Overview

The complex tectonic history of the Ventura basin (Figure 5) began during the Cretaceous and early Tertiary when the region lay within or near the forearc of a subduction zone sited at the margin of the North American plate (Crowell, 1976, his Figure 6). Extension in the middle or late Miocene led to the inception of a graben that contains the present basin deep (Blake et al, 1978). Clockwise rotation of the region occurred during the middle to late Miocene (Hornafius et al, 1986). In approximately the latest Miocene, major right slip started on northwest-trending faults northeast of the basin (Crowell, 1981) and the region experienced intensive north-south shortening starting in the Pleistocene (Crowell, 1976, 1981). Hydrocarbon traps were formed during the last event and are mostly a result of the contractional deformation (California Division of Oil and Gas, 1974); the reservoirs and fault-juxtaposed rock types, however, include horizons deposited in the forearc and subsequent settings.

Productive footwall traps in the Ventura basin are associated with the Oak Ridge, Santa Susana, and San Cayetano thrusts, and the Venture Avenue–Dos Cuadras trend (fields with footwall traps are named in Figure 5) (California Division of Oil and Gas, 1974). Lowside pools along the Oak Ridge and San Cayetano thrusts have an essentially uninterrupted pathway for hydrocarbon migration from the Neogene basin deep, which presumably is the site of hydrocarbon maturation (Figures 5, 6) (N. Greene, 1985, personal communication).

Saticoy Oil Field

The Saticoy oil field lies within the footwall of the Oak Ridge fault and has a structural setting similar to the Bridge pool illustrated in Figure 6. The Saticoy trap has the following characteristics.

• Transverse (dip) closure (Figure 7): northwest dip of structural upturn on the northwest; updip pinch-out of reservoir and footwall of reverse fault on southwest.

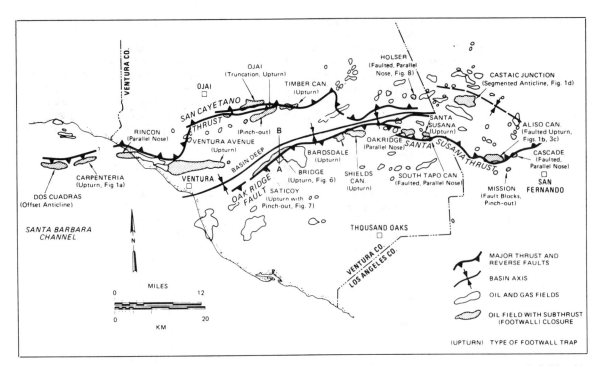

Figure 5—Producing fields and major thrust faults, Ventura basin and adjacent portion of the Santa Barbara channel, California. Main trends with productive subthrust (footwall) closures are the Oak Ridge (reverse) fault, Santa Susana thrust, San Cayetano thrust, and the Ventura Avenue–Dos Cuadras trend. Types of footwall traps are denoted by terminology illustrated in Figures 1-3. Offshore fields farther west in Santa Barbara channel were not included in the present study. Structures after Bailey and Jahns (1954).

- Longitudinal (strike) closure (Figure 7a): broad positive warp.
- Fault displacement: approximately 20,000 ft (6,000 m) reverse separation at top of Sespe Formation (Figure 6).
- Fault seal type: cataclastic gouge (Friedman, 1969).

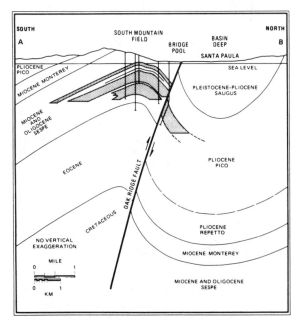

- Estimated ultimate recovery as of December 31, 1983: 22.8 MMbbl of oil, 42.2 bcf of gas (California Division of Oil and Gas, 1984).

The footwall trap at the Saticoy field lies down the flank of the southwest plunge of the South Mountain anticline (Figure 5). The positive bulge or warp that provides longitudinal closure may be the down-dropped flank of a culmination on this anticlinal trend. Alternatively, the bowing of reservoir strike may have formed independently of the hanging-wall geometry. The dip of reservoirs away from the Oak Ridge fault appears to have been greatly accentuated by subsidence of the basin deep judging from the size of the upturn shown on the regional cross section (Figure 6). In this same area, Yeats (1975) interpreted that the southwest-trending segment of the Oak Ridge fault near Santa Paula has both reverse and left-lateral components.

The updip pinch-out of Pliocene turbidite sandstones is the major transverse seal for the hydrocarbon accumu-

Figure 6—Regional cross section across axis of Ventura basin and Oak Ridge fault trend near Bridge pool (see Figure 5 for location). Four fields along the Oak Ridge trend (indicated by field names in Figure 5) have similar association of productive hanging-wall anticline and footwall fault traps. Lowside traps have unobstructed migration pathway from present basin deep, thought to be maturation site for hydrocarbons in Pliocene reservoirs (C. N. Green, 1985, personal communication).

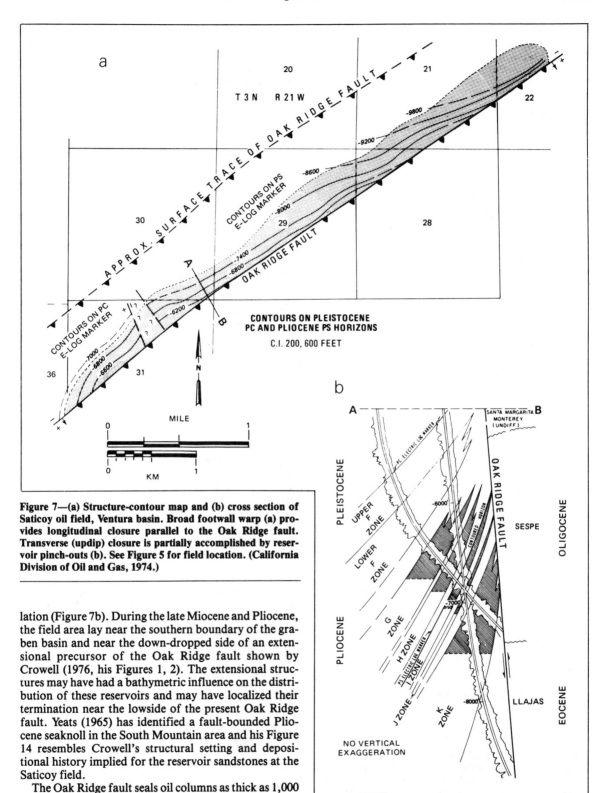

Figure 7—(a) Structure-contour map and (b) cross section of Saticoy oil field, Ventura basin. Broad footwall warp (a) provides longitudinal closure parallel to the Oak Ridge fault. Transverse (updip) closure is partially accomplished by reservoir pinch-outs (b). See Figure 5 for field location. (California Division of Oil and Gas, 1974.)

lation (Figure 7b). During the late Miocene and Pliocene, the field area lay near the southern boundary of the graben basin and near the down-dropped side of an extensional precursor of the Oak Ridge fault shown by Crowell (1976, his Figures 1, 2). The extensional structures may have had a bathymetric influence on the distribution of these reservoirs and may have localized their termination near the lowside of the present Oak Ridge fault. Yeats (1965) has identified a fault-bounded Pliocene seaknoll in the South Mountain area and his Figure 14 resembles Crowell's structural setting and depositional history implied for the reservoir sandstones at the Saticoy field.

The Oak Ridge fault seals oil columns as thick as 1,000 ft (300 m) in some reservoirs (Figure 7b). Friedman (1969) has documented the presence of cataclastic gouge along this fault and has shown that the percentage of

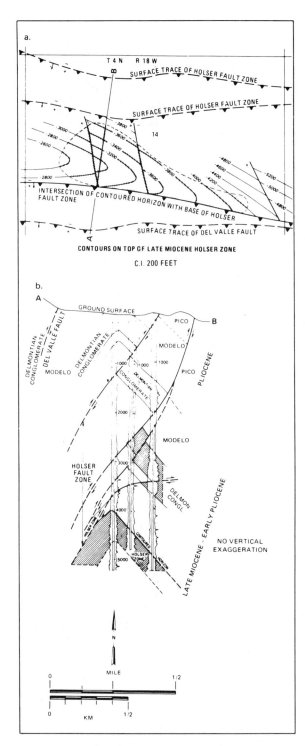

Figure 8—(a) structure-contour map and (b) cross section of Holser oil field, Ventura basin. Transverse (dip) closure within Delmontian conglomerate is provided by a footwall upturn sealed against the Holser fault zone (b). Rollover (a, b) and up-plunge pinch-out (a) forms footwall trap in Holser reservoir. See Figure 5 for field location. (California Division of Oil and Gas, 1974.)

gouge increases progressively toward the main fault strand. Even though the cataclastic gouge is spatially associated with shear fractures, the permeability of the juxtaposed strata is markedly decreased by the presence of the gouge (Friedman, 1969).

Holser Oil Field

The Holser oil field is located at the eastern end of the Ventura basin and near the eastern Ventura subbasin (Figure 5). The field has two productive subthrust traps; characteristics of the deeper pool include the following:

• Transverse closure (Figure 8): footwall of Holser fault zone on south; rollover of parallel anticline to north.
• Longitudinal closure (Figure 8a): plunge of footwall anticline on east; westward pinch-out of reservoir facies on west.
• Fault displacement: approximately 800-1,000 ft (250-300 m) reverse separation at level of Delmontian conglomerate (computed across lower strands of Holser fault zone in Figure 8b).
• Fault-seal type: juxtaposed impermeable lithology (thick Modelo shale against the down-dropped pebbly sandstone reservoir; Figure 8b) and possible cataclastic gouge.
• Estimated ultimate recovery as of December 31, 1983: 1.2 MMbbl of oil, 0.8 bcf of gas (includes Holser and Delmontian conglomerate reservoirs) (California Division of Oil and Gas, 1984).

Although the hydrocarbon reserves are quite small, the Holser field is important because the Holser reservoir illustrates a distinct type of footwall closure. As shown in the cross section of Figure 8, the shallow and deep traps are faulted anticlinal fold segments. We infer that the anticline formed first and was subsequently offset by strands of the Holser fault. At the level of the Delmontian conglomerate, the closure is within the down-dropped flank of the highside rollover segment. This shallower pool is sealed directly updip to the south by the Holser thrust. The lowside fold crest forms the trap within the Holser reservoir and its structure is dominantly a faulted nose. The Holser fault, according to the cross section (Figure 8b), provides the top seal for the south flank of this anticline and prevents oil migration into the strata of the overlying plate.

The Holser fault seals at several oil and gas fields in the vicinity of the Holser trap and does so even where reservoir sands are juxtaposed across the fault. Relatively impermeable cataclastic gouge has been observed within the fault zone at surface exposures (A. C. Tuminas and C. A. Dengo, unpublished field studies).

EXAMPLE FROM CENTRAL SUMATRA BASIN

Basin Overview

The Central Sumatra basin lies within the back arc of the Java trench subduction system. Development of the Tertiary basin was initiated in the Eocene when north-south grabens formed oblique to the subduction zone

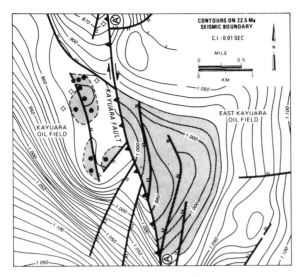

Figure 9—Time-structure map of East Kayuara oil field and adjacent area, Central Sumatra basin. Structure crests of Kayuara and East Kayuara pools are apparently right-laterally offset 1 mi (1.6 km) (Villarroel, 1985). A-A ' is location of cross section along Kayuara fault shown in Figure 10. Well locations in East Kayuara area are not shown.

(Eubank and Makki, 1981; Crostella, 1981). Passive regional downwarping, which included areas between the grabens, followed during the Miocene to about the early Pliocene (de Coster, 1974). Back-arc contraction began in the early Pliocene and lead to the development of most of the hydrocarbon traps (de Coster, 1974). Some of the Paleogene grabens in Central and South Sumatra basins were inverted by this contractional deformation (Harding, 1985), and several grabens have experienced superimposed strike-slip. The Bengkalis trough, a deep within the northeast shelf of the Central Sumatra basin (Eubank and Makki, 1981), is thought by Crostella (1981) to be an example of the latter history. This trough is also thought to be the site of hydrocarbon maturation for fault-sealed traps that lie updip on either side of its axis (Villarroel, 1985). The East Kayuara oil field produces from one of these accumulations. Eubank and Makki (1981) presented a detailed overview of the region's tectonics.

East Kayuara Oil Field

The East Kayuara oil field is sealed against the lowside of the Kayuara fault, a convergent strike-slip fault that trends northward within the Bengkalis trough (Villarroel, 1985). The hydrocarbon trap at East Kayuara has the following characteristics:

• Transverse closure (Figure 9): east dip of displaced flank of anticline on east, footwall of Kayuara fault to west.
• Longitudinal closure (Figure 9): north and south plunge components of faulted anticline.
• Fault displacement: 1-1.5 mi (1.6-2.5 km) apparent

right slip of fold segments (Figure 9) and approximately 450 ft (137 m) (0.1 sec two-way time) downthrow of fold crest on 22.5 m.y. seismic marker (Figure 10).
• Fault-seal type: probably impermeable fault gouge.

The East Kayuara trap is formed by the down-dropped eastern flank of a north-south trending anticlinal culmination. The fold crest (Kayuara oil field, Figure 9) and the faulted flank (East Kayuara field, Figure 9) trend parallel to the Kayuara fault and appear to be offset 1 mi (1.6 km) right-laterally by this fault (Villarroel, 1985). A second fold culmination farther north has a similar apparent offset and additional structural characteristics that suggest a wrench-fault interpretation have been reported by Villarroel (1985): a 30-mi (50-km) long broadly curvate fault trace, flower structures observable on seismic profiles, and a right-stepping en echelon arrangement of folds near the zone's northern end. Another trap-sealing wrench fault with similar structural features at the northern portion of the Bengkalis trough has been described by Crostella (1981).

The reservoir sandstones of the upper Tualang Formation have been juxtaposed against lower Tualang sandstones across the Kayuara fault (Figure 10). The juxtaposition of water-bearing sandstones against the oil column indicates that a seal must be provided by impermeable material within the fault zone. On the basis of the anticipated mechanical behavior of the clastic rock sequence, we believe that the material is cataclastic gouge. The strike-slip displacement apparently created sufficient fault gouge for an effective seal, even though the dip-slip component was not sufficient to completely separate the reservoir formation across the fault.

LOS ANGELES BASIN EXAMPLES

Basin Overview

The early and middle Tertiary history of the Los Angeles basin has several similarities to the Ventura basin—a general forearc setting during early Tertiary subduction and initiation of subsidence of the present basin during an early middle Miocene extensional event (Crowell, 1974; Blake et al, 1978). Cross sections of basement anticlines on the southwest flank of the basin and their en echelon map pattern (Schoellhamer and Woodford, 1951; California Division of Oil and Gas, 1974) suggest that strike-slip motion started in the beginning of the late Miocene or possibly earlier. The strike-slip deformation culminated in the Pleistocene with a marked increase in contraction (Crowell, 1974). The presently preserved portion of the basin is bordered on three sides by convergent wrench zones, each of which has had a large component of reverse slip. All three zones contain footwall fault traps that are richly productive (e.g., Figure 11). Wright (1987, personal communication) presented a comprehensive review of the basin's development and structural geology.

Santa Monica Fault Trend

The Santa Monica–Raymond Hill fault forms the present northern boundary of the Los Angeles basin (Fig-

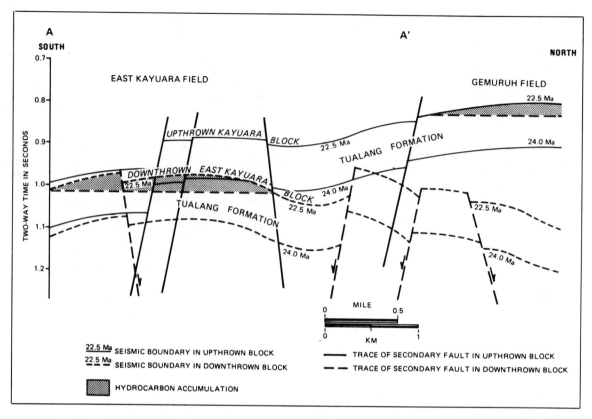

Figure 10—Fault-plane cross section along the Kayuara fault, Central Sumatra basin. Downthrown hydrocarbon containing sands of Tualang Formation (dashed lines) are juxtaposed against water-wet older Tualang sands of upthrown block (solid lines). Upthrown block apparently has been displaced 1 mi (1.6 km) right relative to downthrown pool (see discussion in text) (Villarroel, 1985). Location of section shown on Figure 9.

ure 12). This fault was active in the latest Miocene and early Pliocene (Lang and Dressen, 1975; Wright, 1987) and has had an estimated left slip of 8 mi (13 km) (Wright, 1987) to 14 mi (22 km) (Dibblee, 1987). The fault has a convergent wrench style, and left-stepping en echelon folds extend obliquely away from the zone's footwall (Figure 12) (Dibblee, 1987). Several folds are down-dropped on their south flanks by reverse-separation faults that trend parallel to the fold axes (similar to reverse-slip fault in Figure 2a, c). The most important hydrocarbon traps are hanging-wall anticlinal culminations, but footwall fault traps are productive at both the down-dropped flank of the en echelon folds and directly against the Santa Monica fault. Most structures within the trend lie directly updip from the apparent site of hydrocarbon maturation along the basin axis (basin deep, Figure 12).

San Vicente Oil Field

The San Vicente oil field is an example of the en echelon folds here and at adjacent fields that are displaced by reverse faults that subparallel the fold axes (see west-northwest–trending faults, Figure 12). The field has the following trap characteristics.

•Transverse closure (Figure 13): rollover at east-west-trending anticlines on the south and footwall of northern east-west reverse fault on north.

•Longitudinal closure (Figure 13a): mostly eastern plunge of folds to east and termination of reservoir or possibly footwall of Santa Monica fault to west.

•Fault displacement: about 600 ft (200 m) reverse separation on northern reverse fault (Figure 13b); 8 mi (13 km) (Wright, 1987) to 14 mi (22 km) (Dibblee, 1987) of left-slip and approximately 3,000 ft (1,000 m) reverse separation (Wright, 1987) on Santa Monica fault.

•Fault-seal type: seals at reverse faults probably due to cataclastic fault gouge.

•Estimated ultimate recovery as of December 31, 1983: 21.0 MMbbl of oil, 19.5 bcf of gas (total for all segments of field) (California Division of Oil and Gas, 1984).

The productive anticlines are part of an en echelon fold set that extends across the northern end of the present Los Angeles basin (Figure 12). The northern reverse fault within the field has sealed and trapped columns of oil over 1,000 ft (300 m) thick. The fault seals laterally even where water-bearing Puente Formation sands are juxtaposed against the oil-bearing intervals of this formation

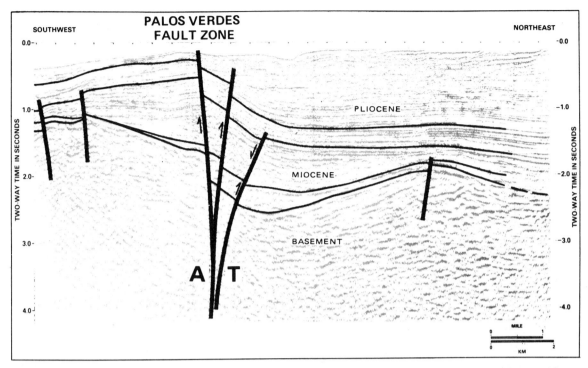

Figure 11—Seismic profile across Palos Verdes fault, Los Angeles basin, California. Fault strikes northwest-southeast and forms southwest border of productive portion of basin. We interpret this to be a right-lateral wrench zone because of adjacent right-stepping en echelon folds (Jennings, 1977), parallelism of fault trace to nearby known right-slip faults, and inconsistent (both normal and reverse at northeast strand) and mixed (normal at southwest strand and reverse at central strand) profile separations (Fischer et al, 1987). Along this trend, stratal upturn at edge of footwall block contains a giant hydrocarbon accumulation, the Beta oil field, sealed updip by convergent wrench fault. Basement-involved flexure is present northeast of fault and some folds trend parallel to fault. T = displacement toward viewer, A = displacement away from viewer. (P. R. Bowden, 1985, personal communication.)

(Figure 13b). Because of this relationship, we infer that fault seal is at least partly controlled by impermeable cataclastic gouge. Seal up the fault is believed to be protected by the truncation of the fault at the lower Pliocene unconformity. The truncation has blocked direct upward leakage along the fault into the unfolded Repetto reservoirs (Figure 13b).

Sawtelle Oil Field

The Sawtelle oil field lies west of the San Vicente field (Figure 12) and illustrates sealing directly against basement at the Santa Monica wrench zone. The trap has the following characteristics.

•Transverse (dip) closure (Figure 14): footwall of Santa Monica fault on northeast and southwest dip of structural upturn on southwest within northern segment of field; rollover of en echelon anticline in southwest areas of field.
•Longitudinal (strike) closure (Figure 14a): positive footwall warp or bowing (analogous to Figure 3a) within northern area and double plunge of fold culminations at southwest pools.

•Fault displacement: 8 mi (13 km) (Wright, 1987) to 14 mi (22 km) (Dibblee, 1987) of left-slip and approximately 8,000 ft (2,500 m) reverse separation (computed from Figure 14b).
•Fault-seal type: impermeable basement juxtaposed against down-dropped sandstone reservoir.
•Estimated ultimate recovery as of December 31, 1983: 15.3 MMbbl of oil, 13.0 bcf of gas (total for all segments of field) (California Division of Oil and Gas, 1984).

The northern segment of the Sawtelle field is distinguished from previous examples by the similar dip direction of the reservoir and sealing fault (Figure 14b). The reservoir beds are overturned and are truncated updip and above by the shallower dipping Santa Monica fault. The upper plate of this thrust profile has placed impermeable slate of the basement across the upturned reservoir. The effectiveness of the fault seal may have been increased by the disruption of potential drainage pathways within the tightly folded upturn above the reservoir. We speculate that the shaly facies of the Modelo Formation that overlie the reservoir could have been contorted into small secondary folds along the Santa Monica fault analogous to those illustrated in Figure 4e.

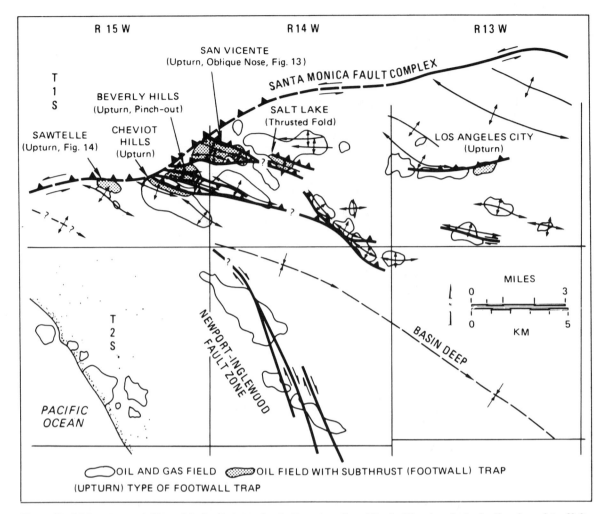

Figure 12—Main structures of Santa Monica fault trend and adjacent northwest flank of Los Angeles basin. Trend consists of left-stepping en echelon anticlines and en echelon reverse faults that parallel anticlines. Types of footwall traps are denoted by terminology illustrated in Figures 1-3. Footwall traps are also productive at Newport-Inglewood zone (Harding, 1973) but were not included in the present study. (California Division of Oil and Gas, 1974.)

Whittier Fault Trend

The Whittier fault forms the present northeast flank of the Los Angeles basin and has been active in at least the middle Miocene to the Holocene (Figure 15) (Ledingham, 1973; Wright, 1987). The fault has had approximately 3 mi (5 km) right-lateral displacement since the late Miocene (Troxel, 1954; Yerkes et al, 1965) and may displace lower and middle Tertiary rocks as much as 25 mi (40 km) (Lamar, 1961). The zone has a consistent down-to-the-south reverse separation along its trace opposite the eastern shelf of the basin, and this separation reaches a maximum of about 14,000 ft (4,300 m) near the shelf's mid-point (near the Sansinena field, Figure 15) (Yerkes et al, 1965). Gourley (1975) interpreted upper Miocene sediments in Whittier Hills as restricting Neogene displacement there to 5,500 ft (1,700 m) reverse slip and 2,500 ft (800 m) right slip. An overview of the structural style of the eastern portion of the Los Angeles basin and descriptions of the several different types of hydrocarbon traps were presented by Harding (1974). Several examples are repeated here to emphasize the role of fault seal.

Two distinctly different groups of productive subthrust fault traps are present and both lie updip from the nearby basin deep (Figure 15). One group is present on the downfaulted south flank of en echelon anticlines that extend obliquely westward from the Whittier fault (fold trends extending 8-12 mi or 13-19 km westward from fault in Figure 15). The footwall fault traps at the East Coyote and Richfield oil fields lie opposite larger anticlinal culminations within the hanging-wall block and are secondary to the latter in size (right margin, Figure 16a) (see California Division of Oil and Gas [1974] for cross section of Richfield oil field). A similar pattern of lowside fault traps and highside fold culmination traps was

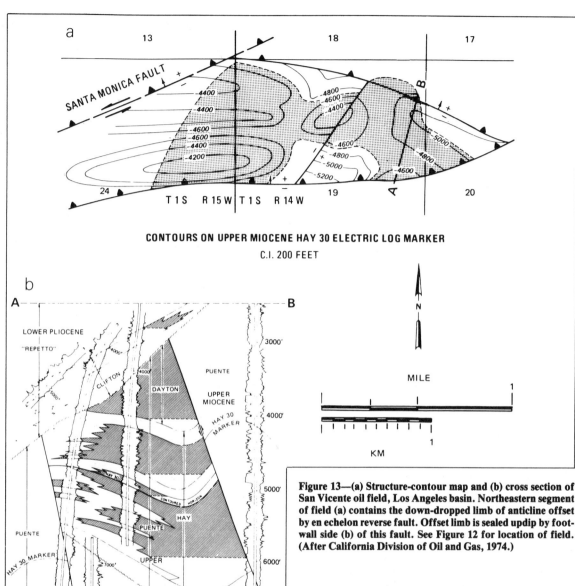

CONTOURS ON UPPER MIOCENE HAY 30 ELECTRIC LOG MARKER
C.I. 200 FEET

Figure 13—(a) Structure-contour map and (b) cross section of San Vicente oil field, Los Angeles basin. Northeastern segment of field (a) contains the down-dropped limb of anticline offset by en echelon reverse fault. Offset limb is sealed updip by foot-wall side (b) of this fault. See Figure 12 for location of field. (After California Division of Oil and Gas, 1974.)

described at the Santa Monica fault trend (compare Figures 12, 15).

The second group of subthrust traps lie directly against the down-dropped side of the Whittier fault and are larger and more varied in their geometry (fields with productive limits that abut Whittier fault in Figure 15). In this trend, an en echelon anticline is sealed obliquely up plunge by the Whittier fault at the Montebello oil field

(Figures 3d, 15, 16b) (California Division of Oil and Gas, 1974). At the Whittier oil field (Figures 1c, 15), a positive flower structure plunges southeastward, parallel to the zone of strike slip. The northeast limb of the flower structure is juxtaposed against the Whittier fault and tar impregnations at outcrops of the reservoir sands provide the up-plunge closure (Gaede, 1964). The seal is incomplete; early wells were located at oil seeps in the crushed fault zone (Norris, 1930) and the small accumulation of oil in the hanging-wall block may have migrated across the fault from the main pools of the footwall block (right margin, Figure 1c).

At the main area of the Sansinena oil field, the anticlinal culmination of a small en echelon drag fold is closed partially against the Whittier wrench zone (Figures 15, 16c) (California Division of Oil and Gas, 1974). Seeps have been active within the general field area but are poorly located with respect to the Whittier fault

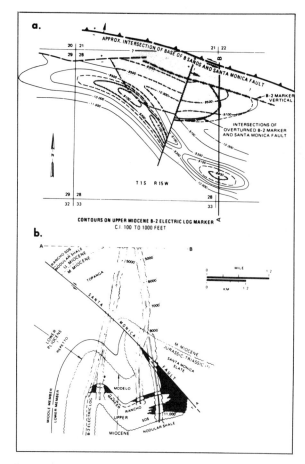

Figure 14—(a) Structure-contour map and (b) cross section of Sawtelle oil field, Los Angeles basin. Northeast segment of field contains footwall upturn closed updip against Santa Monica wrench fault. Beds within upturn (a) are steeply overturned in north-northeast direction within area lying northeast of line labeled "B-2 Marker Vertical." Reservoir beds and sealing fault dip in same direction at this locality and, in this way, trapping geometry differs from previous examples. T = displacement toward viewer, A = displacement away from viewer. See Figure 12 for location of field. (California Division of Oil and Gas, 1974.)

(Hodgson, 1980). Farther southeast, updip closure at the giant Brea Olinda oil field is mostly due to shale-out of sandstone reservoirs (e.g., first Pliocene zone in Figure 16d) (Gaede et al, 1967). The shale-outs may have been localized by Neogene bathymetric expression of the Whittier fault (Kundert, 1952). Longitudinal closure at the field is also provided by the pinch-outs and by gentle changes in reservoir strike. Oil sands outcrop within the limits of the Brea Olinda field and contain seeps (Gaede et al, 1967). Small accumulations of oil are present within the hanging-wall block, suggesting leakage across the Whittier fault, but structural data for the hanging-wall block are not available for analysis (Gaede et al, 1967).

Major reverse and strike-slip strands associated with the Whittier fault have generally been effective for

retaining large hydrocarbon accumulations. Along most of the basin flank, the overall fault zone juxtaposes older less permeable Miocene strata on the northeast against down-dropped younger Miocene and lower Pliocene reservoirs on the southwest (Woodford et al, 1954). Individual faults commonly segment the fields into numerous pressure-isolated production blocks and trap columns of oil as thick as 1,500 ft (450 m) (California Division of Oil and Gas, 1974). In many places, the individual faults juxtapose reservoir facies against reservoir facies (either sandstones of the same formation or sandstones of different formations) and still retain oil laterally. Tight fault zones are implied and their impermeability may be due to the structural disruption of potential drainage pathways (Figure 16d; compare with Figure 4e), to cataclastic gouge (Yerkes et al, 1965), or to the presence of immovable asphalt within the fault zones.

The sealing capacity of strike-slip fault zones in the Los Angeles basin is demonstrated further by groundwater movement (Poland et al, 1959). For instance, the Newport-Inglewood wrench zone directly west of the Whittier trend is the boundary between saline ground water on the west and fresh ground water to the east (Poland, 1959).

EXPLORATION APPLICATIONS

The examples of fault-sealed traps at reverse and convergent strike-slip faults provide analogs for exploration in basins that have been deformed by contraction. The most critical factors in developing and risking such prospects are the delineation of longitudinal (strike) closure and the prediction of fault seal.

Because seismic energy is generally dispersed by the complex structure of the hanging-wall block, footwall prospects are often located in a seismic shadow zone. In addition, the steep dips may be beyond seismic resolution. At the Bridge pool, for example, very steeply upturned Pliocene reservoirs lie directly below the steep flank of an anticline (Figure 6). Both structural levels make it difficult to delineate closure. The third component of closure, typically dip away from the fault, is most generally provided or accentuated by the structural upturn commonly found at the edge of the footwall block. This dip may be predicted, with a reasonable degree of confidence, to extend to the sealing fault. Furthermore, in a number of examples (Oak Ridge fault trend, Figures 5 and 6, north end of Los Angeles basin in Figure 12, and Whittier fault trend in Figure 15) the required footwall dip corresponds to a continuation of the regional into-the-basin dip. This correspondence increases the assurance of transverse (dip) closure, and provides a broader area for drainage of matured hydrocarbons and for development of structural closure.

In some instances, the structure at the edge of the footwall block may be inferred from the positions and geometry of more distant structures. Hanging-wall folds formed before or during faulting (as in Figure 2a, b) may extend across the sealing fault. At the Oak Ridge fault trend, several lowside pools lie directly opposite productive highside anticlinal culminations (Bridge Pool and

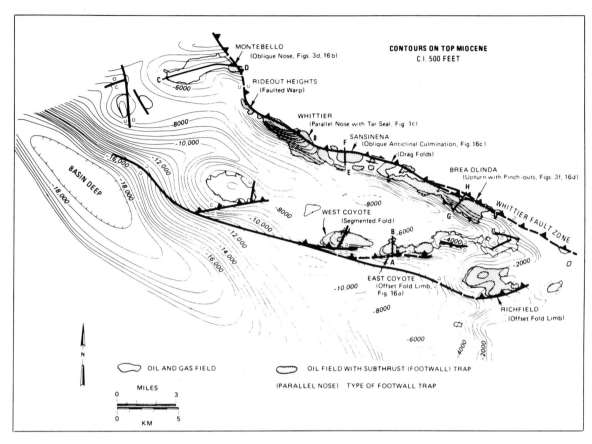

Figure 15—Structure-contour map of Whittier fault trend and eastern shelf, Los Angeles basin. At Whittier fault, a steep structural upturn is disrupted locally by right-stepping en echelon folds. Folds plunge away from fault and both folds and upturn provide effective footwall traps. Southwest of Whittier fault, right-stepping en echelon fold culminations are segmented by associated en echelon reverse faults, which also seal footwall closures. Types of footwall traps are denoted by terminology illustrated in Figures 1-3 (also see cross sections in Figure 16). Map adapted from unpublished compilation by J. H. McDonald (1957).

parts of Bardsdale and Shields Canyon fields, Figure 5). This association suggests that the outward bowing of beds around the apex of a hanging-wall fold culmination may extend across the fault and result in a positive low-side warp similar to that in Figure 3a.

Structures within the down-dropped plate that are more distant from the fault can be identified with conventional exploration programs. Oblique folds are commonly part of the contractional fault block and convergent wrench-fault styles (Harding and Lowell, 1979; Wilcox et al, 1973, respectively). The tectonics of either style suggest that these folds may reasonably be expected to continue through the seismic shadow zone to the sealing fault where they can form a faulted-nose prospect similar to Figure 3d. Gentle bowing of regional strike, oblique or transverse faults, or reservoir terminations may also be identified within the down-dropped region and may be projected to the footwall of the sealing fault. These projections can result in prospects resembling those in Figures 3a-c and the left side of f. For example, changes in trend of the adjacent basin axis or reversals of basin plunge can identify discrete basin deeps

that would commonly be separated by structurally higher saddles, and these saddles may extend to the fault to form a positive-warp trap similar to Figure 3a. However, changes in strain at the fault may alter the geometry or alignment of the projected structures resulting in a loss of closure.

Structural style, sense, magnitude, and timing of fault displacement and fault-zone deformation are all important factors in assessing the probability of fault seal at the trap-bounding fault (Seeburger, 1981; Villarroel, 1985; Harding and Tuminas, 1987, personal communication). Our studies suggest that strike-slip faults are more likely to seal than are reverse faults; reverse faults appear to seal somewhat more effectively than normal faults. This apparent correlation of fault type and seal efficiency is not well understood, but we can make several speculations. Seal capacity is believed to be increased by the greater development of shear fractures and cataclastic gouge along strike-slip and reverse faults (Figure 4c). Large strike-slip faults commonly are braided and contain internal drag folds and slices that can severely disrupt the continuity of a leakage pathway (Figures 4e,

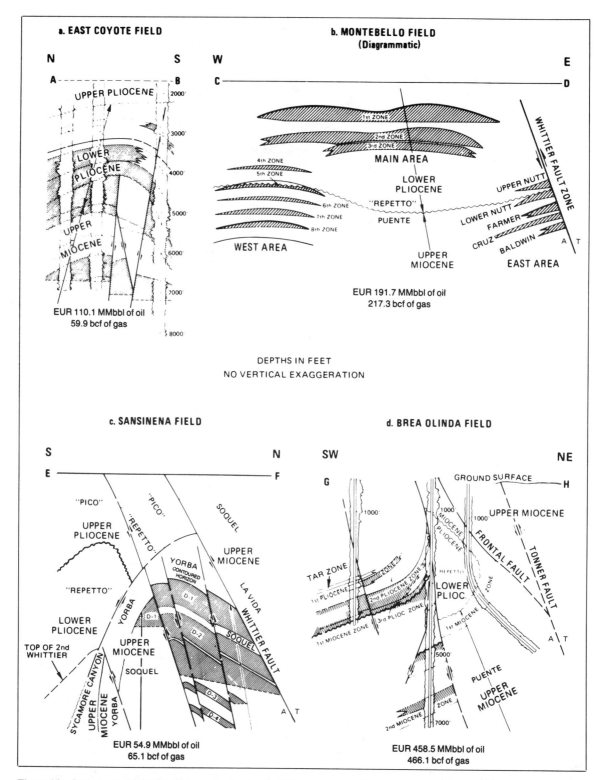

Figure 16—Cross sections of fields with productive footwall closures within (a) eastern shelf and (b-d) Whittier fault trend, Los Angeles basin. Traps along Whittier fault trend have great structural variation, but all footwall pools within eastern shelf closely resemble (a). See Figure 15 for location of cross sections. EUR is estimated ultimate recovery for field, and in (a) and (b) includes significant reserves outside of footwall fault trap. T = displacement toward viewer, A = displacement away from viewer.

16d) (Seeburger, 1981). The stratigraphic juxtapositions that are a natural consequence of reverse separation faults help lateral seal at the footwall block of these types of faults. The younger lowside hydrocarbon reservoirs are juxtaposed against upthrown, older, and, therefore, generally less permeable strata (Figure 4a). The potential for juxtaposition of impermeable rock in the hanging wall increases with increased separation.

In many examples, impermeable fault-zone material must be present to explain the limited escape of hydrocarbons along the fault (Figure 4c) (Smith, 1980; Seeburger, 1981). If strata were relatively unlithified during faulting, deformation along the fault may have been ductile and would have enhanced the development of smear gouge (Weber et al, 1978). Brittle deformation and, hence, cataclastic gouge are more likely to be present if the strata were well lithified at the time of faulting. In addition, faults with large amounts of displacement have wider and more continuous cataclastic gouge zones (Robertson, 1982) and, therefore, would be expected to be better seals. In regions with very viscous oils or where water washing is anticipated, tar or asphalt deposits may form in the fault zone and also increase the potential of subsequent fault seal.

Where faults of varying age occur, those most likely to leak, if all other factors are equal, are faults that cut the youngest strata (Figure 4f) (Harding and Tuminas, 1987, personal communication). Conversely, the possibility of updip closure may be enhanced if displacements were syndepositional and contributed to a decrease in reservoir permeability in a direction up the stratal dip (Figure 7b; first Pliocene zone, Figure 16d).

A wide range of closure sizes and hydrocarbon reserves were observed in the basins studied. In areas with the contractional fault-block style, the footwall traps were decidedly smaller (mostly with ultimate recoveries of 25 MMbbl of oil or less) than the highside fold-culmination traps (e.g., Figure 6). Furthermore, fault-sealed footwall accumulations are notably rare in some regions deformed with contractional block faults such as the Permian basin (T. P. Harding, unpublished study) and the Rocky Mountains foreland (Gries, 1983). Subthrust exploration at major Rocky Mountain foreland uplifts, however, has been mostly for traps independent of footwall upturn or fault seal (Gries, 1983). Regional dip at a number of the drilled locations is toward the thrust fault and not away from it.

In the convergent wrench settings, pools range from a modest size to the giant category (over 100 MMbbl). The largest reserves were contained within a very long closure formed by a broad regional upturn of the basin flank that is faulted against a regional strike-slip zone (Brea Olinda field, Figure 15). Reverse faults are typically neither as long nor as straight as regional wrench zones and either characteristic could cause a loss of longitudinal closure or segmentation into small closures at the reverse faults. Significant closures and reserves were also present where individual en echelon anticlines intercepted the wrench zone (e.g., Montebello field, Figure 15). The smallest closures in the wrench fault settings were formed by reverse faults that down-dropped the flank of en echelon fold culminations (e.g., East Coyote field, Figure 15).

The Whittier fault trend is noteworthy for the abundance of oil seeps associated with large accumulations. The fault has had recent seismic activity (Lamar, 1973) and offsets modern stream courses right-laterally (Troxel, 1954). Upturn of beds along the footwall block has brought reservoirs to the outcrop and accounts for many of the seeps (Norris, 1930; Gaede et al, 1967), but other seeps have been reported from the fault zone itself (Norris, 1930). We speculate as to several possible implications of seepage from the fault. First, rupture of fault seal and the loss of hydrocarbons may be a consequence of the active nature of the fault (Sibson et al, 1975). Second, the highly efficient maturation-migration systems present in the California basins may refill the accumulations as rapidly as the hydrocarbons leak along the trap-bounding faults and leak at the reservoir outcrops. If true, this dynamic hydrocarbon movement implies that the accumulations could be quite transient in geologic time and questions the applicability of some California fields as models for fault seal, particularly as applicable to older hydrocarbon systems. Alternatively, the faults may be long-term seals and may have reached their full capacity to retain hydrocarbons (Schowalter, 1979; Smith, 1980). Excess input would then currently be leaked from the reservoirs while the storage capacity would ultimately remain intact (Schowalter, 1979). We are unable to evaluate these important differences at present, but the value of the Whittier fault trend for structural examples of different closure geometries remains in any case.

REFERENCES CITED

Allan, U.S., 1980, A model for migration and entrapment of hydrocarbons (abs.): AAPG Research Conference on Seals for Hydrocarbons (unpublished conference abstracts), 16 p.

Aydin, A., and B. M. Page, 1984, Diverse Pliocene-Quaternary tectonics in a transform environment, San Francisco Bay region, California: GSA Bulletin, v. 95, p. 1303-1317.

Bailey, T. L., and R. H. Jahns, 1954, Geology of the Transverse Range province, southern California, *in* R. H. Jahns, ed., Geology of southern California, chapter II; geology of the natural provinces: California Division of Mines Bulletin 170, p. 83-106.

Bally, A. W., ed., 1983, Seismic expression of structural styles—a picture and work atlas: AAPG Studies in Geology 15, v. 3, p. 3.1-1 to 4.2-29.

Berg, R. R., 1962, Mountain flank thrusting in Rocky Mountain foreland, Wyoming and Colorado: AAPG Bulletin, v. 46, p. 2019-2032.

Blake, M. C., Jr., R. H. Campbell, T. W. Dibblee, Jr., D. G. Howell, T. H. Nilsen, W. R. Normark, J. C. Vedder, and E. A. Silver, 1978, Neogene basin formation in relation to plate-tectonic evolution of San Andreas fault system, California: AAPG Bulletin, v. 62, p. 344-372.

Boyer, S. E., and D. Elliott, 1982, Thrust systems: AAPG Bulletin, v. 66, p. 1196-1230.

California Division Oil and Gas, 1974, California oil and gas fields, volume II—south, central coastal and offshore California: California Division of Oil and Gas, Sacramento, no page numbers available.

——— 1984, 69th annual report of the state oil and gas supervisor: California Department of Conservation, Division of Oil and Gas Publication PR 06, 147 p.

Crostella, A., 1981, Malacca Strait wrench fault controlled Lalang and Mengkapan oil fields: Offshore Southeast Asia Conference, Exploration III—Geology Session, p. 1-12.

Crowell, J. C., 1974, Origin of late Cenozoic basins in southern California, *in* W. R. Dickinson ed., Tectonics and sedimentation: SEPM Special Publication 22, p. 190-204.

—— 1976, Implications of crustal stretching and shortening of coastal Ventura basin, California, in D. G. Howell, ed., Aspects of the geologic history of the California continental borderland: AAPG Pacific Section Miscellaneous Publication 24, p. 365-382.

—— 1981, An outline of the tectonic history of southeastern California, in W. G. Ernst, ed., The geotectonic development of California, Rubey Symposium, v. 1: New Jersey, Prentice-Hall, p. 583-600.

de Coster, G. L., 1974, The geology of the Central and South Sumatra basins: Indonesian Petroleum Association Third Annual Convention Proceedings, p. 77-110.

Dibblee, T. W., Jr., 1967, Areal geology of the western Mojave Desert, California: USGS Professional Paper 522, 153 p.

—— 1987, Geology of Santa Monica Mountains and Simi Hills, California (abs.): AAPG Bulletin, v. 71, p. 548.

Engelder, J. T., 1974, Cataclasis and the generation of fault gouge: GSA Bulletin, v. 85, p. 1515-1522.

Eubank, R. T., and A. C. Makki, 1981, Structural geology of the Central Sumatra back-arc basin: P. T. Caltex Pacific Indonesia, Indonesian Petroleum Association Tenth Annual Convention (proceedings), 53 p.

Fischer, P. J., G. Simila, R. H. Patterson, A. C. Darrow, and J. H. Rudat, 1987, The Palos Verdes fault zone—onshore to offshore: SEPM Geology of the Palos Verdes Peninsula and San Pedro Bay Field Trip Guidebook 55, p. 91-134.

Friedman, M., 1969, Structural analysis of fractures in cores from Saticoy field, Ventura County, California: AAPG Bulletin, v. 53, p. 367-389.

Gaede, V. F., 1964, Central area of Whittier oil field: California Division of Oil and Gas Summary of Operations—California Oil Fields, v. 50, no. 1, p. 59-67.

—— R. V. Rothermel, and L. H. Axtell, 1967, Brea-Olinda oil field: California Division of Oil and Gas Summary of Operations—California Oil Fields Part 2, v. 53, p. 5-24.

Gourley, J. W., 1975, Upper Miocene Puente Formation in the Whittier Hills and surrounding region, in J. N. Truex, ed., A tour of the oil fields of the Whittier fault zone, Los Angeles basin, California: Pacific Section AAPG, SEPM, SEG Field Trip Guidebook, p. 13-24.

Gries, R., 1983, Oil and gas prospecting beneath Precambrian of foreland thrust plates in Rocky Mountains: AAPG Bulletin, v. 67, p. 1-28.

Harding, T. P., 1973, Newport-Inglewood trend, California—an example of wrenching style of deformation: AAPG Bulletin, v. 57, p. 97-116.

—— 1974, Petroleum traps associated with wrench faults: AAPG Bulletin, v. 58, p. 1290-1304.

—— 1976, Tectonic significance and hydrocarbon trapping consequences of sequential folding synchronous with San Andreas faulting, San Joaquin Valley, California: AAPG Bulletin, v. 60, p. 356-378.

—— 1983, Structural inversion at Rambutan oil field, South Sumatra basin, in A. W. Bally, ed., Seismic expression of structural styles—a picture and work atlas: AAPG Studies in Geology 15, v. 3, p. 3.3-13 to 3.3-18.

—— 1985, Seismic characteristics and identification of negative flower structures, positive flower structures, and positive structural inversions: AAPG Bulletin, v. 69, p. 582-600.

—— and J. D. Lowell, 1979, Structural styles, their plate-tectonic habitats, and hydrocarbon traps in petroleum provinces: AAPG Bulletin, v. 63, p. 1016-1058.

Hardoin J. L., 1960, Holser oil field: California Department of Natural Resources, California Division of Oil and Gas Summary of Operations, California Oil Fields, v. 46, no. 2, p. 53-60.

Higgins, M. W., 1971, Cataclastic rocks: USGS Professional Paper 687, 97 p.

Hodgson, S. F., 1980, Onshore oil and gas seeps in California: California Division of Oil and Gas Publication TR26, 95 p.

Hornafius, J. C., B. P. Luyendyk, R. R. Terres, and M. J. Kamerling, 1986, Timing and extent of Neogene tectonic rotation in the western Transverse Ranges, California: GSA Bulletin, v. 97, p. 1476-1487.

Jennings, C. J., 1977, Geologic map of California: California Division of Mines and Geology California Geologic Data Map Series, 1:750,000.

Kundert, C. J., 1952, Geology of the Whittier-La Habra area, Los Angeles County, California: California Division of Mines Special Report 18, 22 p.

Lamar, D. L., 1961, Structural evaluation of the northern margin of the Los Angeles basin: unpublished PhD dissertation, University of California, Los Angeles, California, 142 p.

—— 1973, Microseismicity and recent tectonic activity, Whittier fault area, California (abs.): AAPG Bulletin, v. 57, p. 789-790.

Lang, H. R., and R. S. Dressen, Jr., 1975, Subsurface structures of the northwestern Los Angeles basin: California Division of Oil and Gas Technical Papers TP01, p. 15-21.

Ledingham, G. W., Jr., 1973, The east area of Sansinena oil field: California Division of Oil and Gas Summary of Operations, v. 59, no. 1, p. 5-20.

Logan, J. M., 1981, Laboratory and field investigations of fault gouge: USGS Open-File Report 81-883, 80 p.

Lowell, J. D., 1985, Structural styles in petroleum exploration: Tulsa, Oklahoma, Oil and Gas Consultants International, 477 p.

McCulloh, T. H., 1969, Geologic characteristics of the Dos Cuadras offshore oil field: USGS Professional Paper 679-C, p. 29-46.

Morse, P. F., and G. W. Purnell, 1987, An atlas of compressional structural styles (abs.), in Interpretation of reflection seismic data workshop: Fourth Annual SEG Summer Research Workshop, p. 10.

Norris, B. B., 1930, Report on the oil fields adjacent to the Whittier fault: California Division of Oil and Gas Summary of Operations, v. 15, no. 4, p. 5-27.

Pittman, E. D., 1981, Effect of fault-related granulation on porosity and permeability of quartz sandstones Simpson Group (Ordovician), Oklahoma: AAPG Bulletin, v. 65, p. 2381-2387.

Poland, J. F., 1959, Hydrology of Long Beach–Santa Ana area, California: USGS Water-Supply Paper 1471, 257 p.

—— A. A. Garrett, and A. Sinnott, 1959, Geology, hydrology and chemical character of ground water in the Torrance–Santa Monica area, California: USGS Water-Supply Paper 1461, 425 p.

Prucha, J. J., J. A., Graham, and R. P. Nickelsen, 1965, Basement-controlled deformation in Wyoming province of Rocky Mountains foreland: AAPG Bulletin, v. 49, p. 966-992.

Robertson, E. C., 1982, Continuous formation of gouge and breccia during fault displacement: 23rd Symposium of Rock Mechanics Association of America, p. 397-404.

Schoellhamer, J. E., and A. O. Woodford, 1951, The floor of the Los Angeles basin, Los Angeles, Orange and Riverside Counties, California: USGS Oil and Gas Investigations Map OM-154, 1:24,000.

Schowalter, T. T., 1979, Mechanics of secondary hydrocarbon migration and entrapment: AAPG Bulletin, v. 63, p. 723-760.

Seeburger, D. A., 1981, Studies of natural flexures, fault zone permeability, and a pore space–permeability model: part II permeability and fault zone structure: PhD dissertation, Stanford University, Stanford, California, p. 76-128.

Sibson, R. H., M. J. Moore, and A. H. Rankin, 1975, Seismic pumping, a hydrothermal fluid transport mechanism: Geological Society of London Journal, v. 131, p. 653-659.

Smith, D. A., 1966, Theoretical considerations of sealing and nonsealing faults: AAPG Bulletin, v. 50, p. 363-379.

—— 1980, Sealing and nonsealing faults in Louisiana Gulf Coast basin: AAPG Bulletin, v. 64, p. 145-172.

Stearns, D. W., 1978, Faulting and forced folding in the Rocky Mountains foreland, in V. Matthews, ed., Laramide folding associated with block faulting in western United States: GSA Memoir 151, p. 1-37.

Suppe, J., 1983, Geometry and kinematics of fault-bend folding: American Journal of Science, v. 283, p. 684-721.

—— and D. A. Medwedeff, 1984, Fault-propagation folding (abs.): GSA Abstracts with Programs, v. 16, p. 670.

Sylvester, A. G., and R. R. Smith, 1976, Tectonics transpression and basement-controlled deformation in San Andreas fault zone, Salton trough, California: AAPG Bulletin, v. 60, p. 2081-2101.

Troxel, B. W., 1954, Geologic guide for the Los Angeles basin, southern California, in R. H. Jahns, ed., Geology of southern California: California Division of Mines Bulletin 170, Geologic Guide no. 3, 46 p.

Villarroel, T., 1985, Observations on the sealing properties of faults in Sumatra: Indonesian Petroleum Association 14th Annual Convention Proceedings, p. 105-116.

Ware, G. G., Jr., and R. D. Stewart, 1958, Bridge area, South Mountain oil field, in J. W. Higgins, ed., A guide to the geology and oil fields of the Los Angeles and Ventura regions: AAPG Pacific Section, p. 180-181.

Weber, K. J., and E. Daukoru, 1975, Petroleum geology of the Niger delta: Ninth World Petroleum Congress Transactions, v. 2, p. 209-221.

—— G. Mandl, W. F. Pilaar, F. Lehner, and R. G. Precious, 1978, The role of faults in hydrocarbon migration and trapping in Nigerian growth fault structures: Offshore Technology Conference Paper 3356, p. 2643-2652.

Wilcox, R. E., T. R. Harding, and D. R. Seely, 1973, Basic wrench tectonics: AAPG Bulletin, v. 57, p. 74-96.

Woodford, A. D., J. E. Schoellhamer, J. G. Veddes, and R. F. Yerkes, 1954, Geology of the Los Angeles basin, in R. H. Johns, ed., Geology of southern California: California Division of Mines Bulletin 170, ch. II, pt. 5, p. 65-81.

Wright, T. L., 1987, Structural geology and tectonic evolution of the Los Angeles basin (abs.): AAPG Bulletin, v. 71, p. 629.

Yeats, R. S., 1965, Pliocene seaknoll at South Mountain, Ventura basin, California: AAPG Bulletin, v. 49, p. 526-546.

——— 1975, High-angle reverse faults of the Ventura basin, Transverse Ranges, California (abs.): GSA Abstracts with Programs, v. 7, p. 391.

Yerkes, R. F., T. H. McCulloh, J. E. Schoellhamer, and J. G. Vedder, 1965, Geology of the Los Angeles basin, California—an introduction: USGS Professional Paper 420-A, 57 p.